Lecture Notes in Computer Science 945
Edited by G. Goos, J. Hartmanis and J. van Leeuwen

Springer

Berlin
Heidelberg
New York
Barcelona
Budapest
Hong Kong
London
Milan
Paris
Tokyo

Bernadette Bouchon-Meunier
Ronald R. Yager Lotfi A. Zadeh (Eds.)

Advances in Intelligent Computing - IPMU '94

5th International Conference
on Information Processing and Management
of Uncertainty in Knowledge-Based Systems
Paris, France, July 4-8, 1994
Selected Papers

Series Editors

Gerhard Goos, Universität Karlsruhe, Germany

Juris Hartmanis, Cornell University, NY, USA

Jan van Leeuwen, Utrecht University, The Netherlands

Volume Editors

Bernadette Bouchon-Meunier
CNRS-LAFORIA-IBP, University of Paris 6
Boîte 169, 4 place Jussieu, F-75252 Paris Cedex 05, France

Ronald R. Yager
Machine Intelligent Institute, Iona College
New Rochelle, NY 10801, USA

Lotfi A. Zadeh
Computer Science Division, University of California
Berkeley, CA 94720, USA

CIP data applied for

Die Deutsche Bibliothek - CIP-Einheitsaufnahme

Advances in intelligent computing : selected papers / IPMU '94, 5th International Conference on Information Processing and Management of Uncertainty in Knowledge Based Systems, Paris, France, July 1994. Bernadette Bouchon-Meunier ... (ed.). - Berlin ; Heidelberg ; New York : Springer, 1995
(Lecture notes in computer science ; Vol. 945)
ISBN 3-540-60116-3
NE: Bouchon-Meunier, Bernadette [Hrsg.]; IPMU <5, 1994, Paris>; GT

CR Subject Classification (1991): I.2, F.2.1, H.3.4, G.3, H.1.1, H.4, F.4.1

ISBN 3-540-60116-3 Springer-Verlag Berlin Heidelberg New York

Typesetting: Camera-ready by author
SPIN 10486397 06/3142 - 5 4 3 2 1 0 Printed on acid-free paper

Foreword

Starting in the late 1950s with the work of Marvin Minsky, John Mc Carthy, Herbert Simon and Allan Newell, serious attempts have been made to construct intelligent computer-based systems. The pioneering efforts of these researchers lead to the development of the field of Artificial Intelligence. This approach is most noted for its interesting paradigms and structures such as knowledge-based systems. These efforts have been most noted for there reliance on symbolic information processing techniques and the use of logic as a representation and inference mechanism. Furthermore, the approach of Artificial Intelligence can be seen to be at a very high level of human cognition.

At about the same time, another approach to the modeling of human intelligence, neural networks, was also being initiated. As opposed to the Artificial Intelligence approach, this effort is noted for its reliance on numerical representations and was based on modeling low level cognitive processes. A strong emphasis on learning is very characteristic of this methodology.

At a slightly later time, the early 1960s, another tool for the modeling of human cognition was being initiated. This approach, the development of fuzzy logic, can be more generally seen as the beginning of the concern for the inclusion of the modeling of uncertainty in the construction of intelligent systems. Furthermore, this approach provides a linkage between the two previous paradigms in that it provides a machinery for the numeric representation of the types of constructs developed in Artificial Intelligence.

As we now approach the middle of the 1990s, we see the coming together of these three directions. The purpose of this book is to look at the current state of this unification in the construction of intelligent computing. This very diverse and fast moving endeavour cannot be covered in one volume and we concentrate on the role of uncertainty in the construction of intelligent systems, which is the main topic of the IPMU Conference.

Imperfect knowledge is at the center of all the computing methods for the construction of intelligent systems and its representation is a key point. The management of uncertainty has been one of the main domains of research in this direction since the 1970s and different streams have been developed. Two of them, theory of evidence and causal networks, are extensively presented in part 1 and parts 2 and 3 are respectively devoted to contributions related to these topics. More classical are the probabilistic approaches, developed in part 4, but their involvement in inference strategies as well as knowledge representation is really a question of the day. Another approach for the management of uncertainty, discussed in part 5 is possibility theory, which was introduced in the framework of fuzzy set theory, but has found an autonomy in possibilistic logic. Part 6 presents contributions in the domain of classical or non standard logics. Finally, the management of imperfect knowledge can also find solutions in chaotic modeling, presented in part 7. The last part of the book is devoted to more applied problems : software reusability in part 8, management of uncertainty in image and speech processing, scheduling, decision-making and scientific discovery in part 9.

In part 1, G. Shafer presents precise conditions under which Bayes nets can be said to have a causal interpretation. P. Smets gives a survey of the mathematical models proposed to represent quantified beliefs and their comparison.

Part 2 contains new elements on the foundations of evidence theory, in a reasoning setting (Kohlas and Brachinger) or as algebraic structures (Daniel). The problems of coherence (Wang and Wang) and aggregation of belief measures (Ramer et al.) are addressed, as well as their utilization in decision-making (Mellouli).

Part 3 presents developments on causal networks, proposing several methodologies to perform approximate computations in complex graphs (David et al., Xu, Acid and Campos), and heuristic algorithms for the propagation of uncertainty (Cano and Moral). A link is established between Bayesian networks and relational database models (Wong et al.). Decision influence diagrams are considered with values of evidence (Ezawa) or fuzzy utilities (Lopez). Finally, causal networks are introduced in a knowledge system environment (Liang et al.).

Part 4 deals with concepts derived from probability theory: qualitative methods to represent uncertainty (Wellman, Parsons and Saffiotti), probability intervals (Campos) or bounds (Gilio). Probabilities are regarded in a fuzzy framework from various points of view (Ralescu, Georgescu et al., Bertoluzza et al.). Informational methods are used for statistical problems (Morales et al., Pardo et al.) and for the management of dynamical systems.

In part 5, a possibilistic representation of uncertainty is applied to the updating problem for the management of constraints (Dubois et al.), to the definition of reliability when statistical information is not relevant to the situation (Cappelle and Kerre) and to semantic nets (Sandri and Bittencourt). Possibilistic logic is related to modal interpretability logic (Hajek) and to the use of default rules (Benferhat). Abductive reasoning is presented (Gebhart and Kruse), as well as approaches to the notion of dependence (Farinas del Cerro and Herzig) in a possibilistic environment.

Part 6 presents various works in classical or non-classical logic. Recent non-standard logics are first presented or developed (Trillas et al., Abar, Guiasu, Gasquet and Herzig, Besnard and Moinard). In artificial intelligence, the problem of consistency (Studeny), the introduction of assertions (da Silva and Costa Pereira) and the generation of explanations (Bigham) are addressed. Approximate reasoning is a framework for researches regarding the handling of partially sound rules

(Gottwald) and theorem-proving techniques (Brüning and Schaub). A multiple-valued propositional logic (Escalada-Imaz and Manya) and a modal many-valued logic (Godo et al.) are finally studied.

Chaos is the central problem studied by authors in part 7. Image processing (Gan) and medical applications underlie most of the work on modeling (Sakai et al., Cohen et al.) and control (Bernard-Weil) of dynamical systems subject to chaotic behavior.

Part 8 is devoted to software reusability. Reliability (Park), uncertainty management (Klösch, Simos), data representation properties (Bakhouch), fuzzy measures of understandability (Balentine et al.), and natural representation of programs (Mittermeir) are key points in these papers.

Finally, part 9 presents applications of uncertainty management methods: in image processing, a pixel-based labeling method (Minoh and Maeda) and fuzzy summarization or classification of data (Zhang and Ralescu, Bothe, Botticher) are proposed. Fuzzy methods are also presented for multisensor information processing (Nimier). Associative memories are used for the control of disturbed processes (Ferreiro Garcia). Temporal reasoning based on intervals is used for real-time scheduling problems (Anger and Rodriguez). Orderings are constructed for choicc-making when the available information is partial (Guénoche). Discovery of scientific laws is approached with a management of inaccuracies (Moulet). Practical applications of stochastic fractals are presented in image analysis (Höfer, Pandit).

Thanks must be expresses to the contributors to this volume, as well as the organizers of invited sessions at IPMU in which several of these papers have been presented: C. Bertoluzza, M. Cohen, D. Hudson, S. Ovchinnikov, A. Ralescu, M. Samadzadeh, K. Zand.

March 1995 B. Bouchon-Meunier, R.R. Yager, L.A. Zadeh

Contents

1. FUNDAMENTAL ISSUES

Philosophical Foundations for Causal Networks

Glenn Shafer

Department of Accounting and Information Systems
Faculty of Management, Rutgers University
180 University Avenue, Newark, New Jersey 07102 USA
gshafer@andromeda.rutgers.edu.

Abstract. Bayes nets are seeing increasing use in expert systems [2, 6], and structural equations models continue to be popular in many branches of the social sciences [1]. Both types of models involve directed acyclic graphs with variables as nodes, and in both cases there is much mysterious talk about causal interpretation. This paper uses probability trees to give precise conditions under which Bayes nets can be said to have a causal interpretation. Proofs and elaborations are provided by the author in [4].

1 Introduction

In spite of the impression of action conveyed by its arrows, a Bayes net is not necessarily a dynamic model. It merely expresses conditional independence relations for a single probability distribution, and this is static information. So what do we mean when we claim causal status for a Bayes net?

When we say that a Bayes net is causal, we are apt to elaborate by saying that its arrows point in the direction of time and causality. When there is an arrow from variable X to variable Y, X precedes and causes Y. But this remains a bit mysterious. Perhaps it means that (1) there is a point in time where X is determined and Y is not yet determined, and (2) if the value of X at that point is x, then the probabilities for Y at that point are the conditional probabilities for Y given X=x. But functional dependence, which is generally accepted as an important example of causality and is often used in Bayes nets, violates condition (1). If Y is a function of X, say $Y=X^2$, then Y is determined at least as soon as X and perhaps sooner.

This article uses an overtly dynamic framework for probability—the probability tree—to clarify these issues. A Bayes net can be thought of as an incomplete description of a probability tree, and we may say that a Bayes net is causal if it is a incomplete description of nature's probability tree.

As it turns out, there is more than one way a Bayes net can usefully describe nature's probability tree. We may choose to draw a net with the property that whenever there is an arrow from X to Y, X is always determined at least as soon as Y as nature moves through the tree. Or we may instead draw a net such that the arrows to Y indicate that Y is a function of the variables from which the arrows come, in which case these variables—some or even all of them—may be determined after Y.

In the probability-tree framework, temporal precedence and functional dependence are unified by the concept of *tracking*. We say that X tracks Y if the probabilities for Y are the same at any two points in the tree where X is just settled to have a particular value, say x. This definition allows for the possibility that Y itself is already settled at all the points where X has just been settled, in which case the probabilities at those points are zeros and ones, and Y is indeed a function of X. It also allows for the possibility that Y is not yet settled at any of these points, in which case X may be said to precede Y. It even allows for the possibility that X may be settled sometimes before Y and sometimes after Y.

The word "cause" can be used in several ways in a probability tree, but it when we are speaking of nature's tree, it is most natural to say that steps in the tree are causes of how things come out in the end. The causes of a variable Y are the steps that change Y's probability. When we use "cause" in this way, we may say that tracking is a causal concept. As it turns out, X tracking Y means that until X is settled Y has no causes other than the causes of X.

Some readers will want to say "X causes Y" instead of "X tracks Y," but this is a recipe for confusion and even mischief. Truth be told, variables are not causes. Things that happen and things one does are causes. The total number of cigarettes smoked is not a cause of lung cancer. It is the smoking—the steps in nature's tree where one smokes or where things happen that lead to more smoking—that raise the probability of lung cancer and hence can be said to be among its causes. The total number of cigarettes measures the effect of these steps, but this might also be measured by the stain on one's fingers. Moreover, tracking is only one of many concepts that we can use to relate causes of one variable to causes of another.

Another very important causal concept is *independence*. Two variables X and Y are independent in the probability tree sense if they have no common causes—i.e., if there are no steps in the tree that change both the probabilities for X and the probabilities for Y. This definition of independence is stronger than the usual one; if X and Y are independent in this probability-tree sense, then their probabilities multiply at every point in the tree.

The thesis of this article is that one way to interpret a Bayes net causally is to assert two things about nature's probability tree: (1) each variable is tracked by its parents, and (2) after its parents are settled, it is independent of its nondescendants.

The ideas outlined here are explained more fully the author's book, *The Art of Causal Conjecture* [4] which also develops the probability-tree understanding of causality in several other directions. It uses concepts analogous to tracking (tracking in mean, linear tracking, and linear sign) and generalizations of independence (uncorrelatedness and unpredictability in mean) to account for structural equation models and path diagrams, probability structures analogous to Bayes nets that have been used widely in the social sciences. It also shows how the ideas of tracking, sign, and uncorrelatedness generalize from probability trees and Bayes nets to decision trees and influence diagrams.

2 Situations, Events, and Variables in a Probability Tree

A probability tree represents a plan for performing a sequence of experiments. The plan specifies the possible results of each experiment. At each step, it also specifies what experiment is to be performed next, depending, perhaps, on how earlier experiments have come out. As we perform the sequence of experiments, we follow some path through the tree. Figure 1 shows an example. Here we begin by spinning a fair coin. If it comes out heads, we spin another fair coin. If it comes out tails, we spin a coin biased to heads, and so on.

The sequence of experiments can be thought of as a composite experiment. The stop signs constitute the sample space of this composite experiment. Any subset of the sample space is an event. The circles and stop signs are *situations*. The circles are situations in which an experiment is performed, and the stop signs are terminal situations—situations in which experimentation has stopped. The initial situation is designated by Ω.

Every situation corresponds to an event—the set of stop signs below it. In Figure 2, for example, the situation W corresponds to the event {a,b,c}.) We often find it convenient to identify the situation with this event. Most events, however, are not situations.

When an event E contains a situation S, we say that E is *certain* in S. In Figure 2, {a,d,e} is certain in U. When $E \cap S = \emptyset$, we say that E is *impossible* in S. In Figure 2, {k,m} is impossible in W. When $E \cap F \cap S = \emptyset$, we say that E and F are *incompatible* in S. When E is impossible or certain in S, it is *determinate* in S. Otherwise, it is *indeterminate* in S.

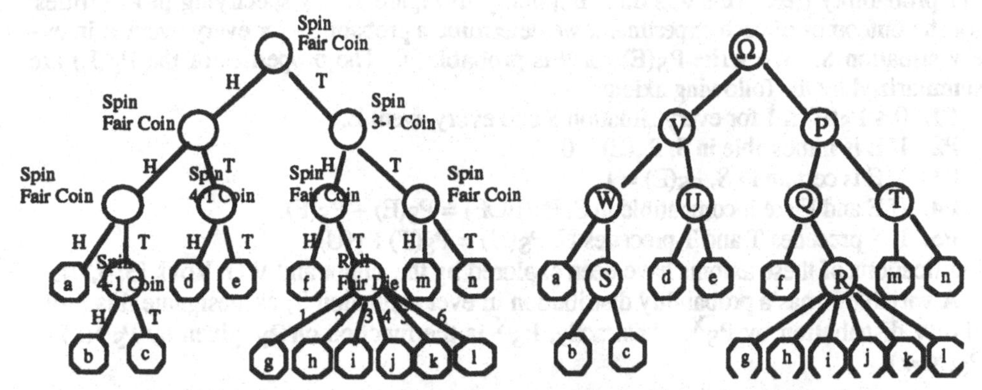

Fig. 1. A probability tree.

Fig. 2. Names for the situations.

Except for the initial situation Ω, every situation has a situation immediately above it, which we call its *mother*. We say that E *happens at* S if E is certain in S but not in S's mother. We say that E *fails at* S if E is impossible in S but not in S's mother. In Figure 2, the event {a,d,e} happens at {a} and at U and fails at S and at P. We say that S *resolves* E if E happens or fails at S. There are two events, the impossible event ∅ and the sure event Ω, that are determinate in the initial situation Ω and hence never happen or fail.

We call an event not equal to ∅ or Ω a *proper* event. Any proper event E is the disjoint union of the situations at which it happens; its complement is the disjoint union of the situations at which it fails. (In Figure 2, {a,d,e} is equal to $\{a\} \cup U$ and its complement is equal to $S \cup P$.) Thus the situations that resolve E constitute a partition of the entire sample space Ω. Another way to put this is to say that no matter what path we take down the tree—no matter what stop sign we end up in—E is resolved exactly once. We call the set consisting of the situations that resolve E the *resolving partition* for E, and we designate this partition by Ξ_E.

We say that an event E *precedes* an event F if E is determinate in any situation in which F is determinate. If E and F are both proper, then E precedes F if and only if E is always resolved at least as soon as F. In other words, for every path down the tree, the situation that resolves E comes before (or equals) the situation that resolves F. If E is improper (equal to ∅ or Ω), then E precedes any event F.

Variables are defined in the event-tree framework just as they are in the sample-space framework; a variable X is a mapping from the sample space onto another set, say Θ_X. Variables generalize events, for an event can be thought of as a variable with only two values.

A variable X is *determinate* in S if it assigns the same value to all elements of S; otherwise it is *indeterminate* in S. If X is determinate in every situation in which Y is determinate, then we say that X *precedes* Y. We say that S *resolves* X if X is determinate in S but not in S's mother. If S resolves X and X's value for elements of S is x, then we say that S *resolves* X to x.

Any variable X determines two partitions of the sample space Ω, both of which are important: its *value partition* Π_X and its *resolving partition* Ξ_X. The *value partition* consists of events of the form {X=x}. The resolving partition consists of the situations that resolve X. If X is constant, it never resolves, but we call {Ω} its resolving partition.

3 Probability

We assume that a probability has been specified for the outcome of each experiment in our probability tree. This was done implicitly in Figure 1. By specifying probabilities for the outcomes of each experiment, we determine a probability for every event E in every situation S. We write $P_S(E)$ for this probability. The properties of the $P_S(E)$ are summarized by the following axioms.

P1. $0 \le P_S(E) \le 1$ for every situation S and every event E.

P2. If E is impossible in S, $P_S(E) = 0$.

P3. If E is certain in S, $P_S(E) = 1$.

P4. If E and F are incompatible in S, $P_S(E \cup F) = P_S(E) + P_S(F)$.

P5. If S precedes T and T precedes U, $P_S(U) = P_S(T)\, P_T(U)$.

Implications of these axioms have been explored by the author and V.G. Vovk [3, 4, 7].

A variable X has a probability distribution in every situation S; we designate this probability distribution by $P_S{}^X$. Abstractly, $P_S{}^X$ is the function on Θ_X given by $P_S{}^X(x) = P_S(X{=}x)$.

4 Independence

We say that two events E and F are *independent* if the probability of only one of them can be changed by a given experiment. In other words, for each non-terminal situation S, either $P_T(E) = P_S(E)$ for all daughters T of S, or else $P_T(F) = P_S(F)$ for all daughters T of S. Independence is a causal concept; it says that E and F have no common causes. The experiments that influence one of the events do not influence the other. This is a strong but qualitative condition. It constrains only when probabilities change, not the values they take or the amount by which they change.

Figure 3, in which a fair coin is spun three times, provides the setting for a familiar example of independence. If we let E be event we get heads on the first spin (E={a,b,c,d}) and F the event we get heads on the second spin (F={a,b,e,f}), then E and F are independent. Other examples are provided by the sure event Ω and the impossible event $\varnothing$. Since their probabilities never change, each is independent of any other event.

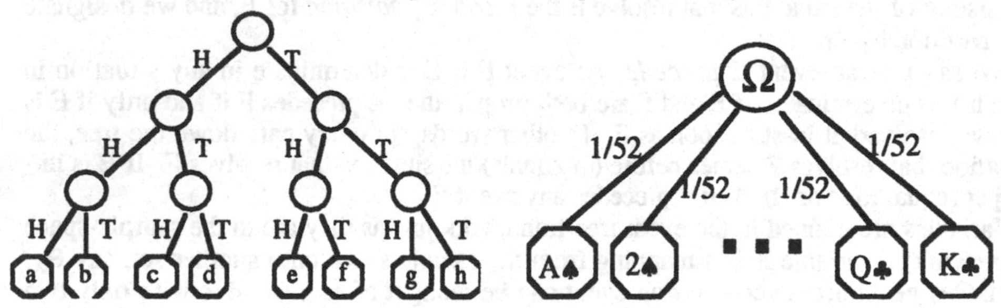

Fig. 3. Three spins of a fair coin.

Fig. 4. Choosing one card at random.

We say that E and F are *formally independent* if $P_S(E \cap F) = P_S(E\ P_S(F)$ for every situation S. In other words, E and F are independent in the usual sense with respect to each probability measure P_S. The events E and F in Figure 3 are formally independent as well as independent. Figure 4 gives an example of events E and F that are formally independent but not independent; E is the event we get a spade, and F is the event we get an ace.

The definitions of independence and formal independence extend immediately from events to variables. Two variables X and Y are *independent* if only one of their probability distributions can change in a given experiment. They are *formally independent* if $P_S(X{=}x \,\&\, Y{=}y) = P_S(X{=}x)P_S(Y{=}y)$ for all x and y and every situation S.

The following proposition reveals how independence and formal independence are related: independence is stronger.

Proposition 1 If two variables (or events) are independent, then they are formally independent.

In general, we do not stand beside nature and watch how her probabilities change as events unfold in her probability tree, and hence we may be unable to tell whether given variables are independent in her tree. Proposition 1 sometimes gives us grounds, however, for conjecture. If we observe that two variables are independent in the conventional sense in many different situations, then we may conjecture that they are independent in the probability tree sense in nature's probability tree.

In the event-tree framework, as in the sample-space framework, independence can be generalized to conditional independence. But there are several distinct concepts of conditional independence in probability trees. We say that events X and Y are *independent in* S if the probability distribution of only one of them can change in any given experiment at S or below. We say that Y and Z are *independent posterior to* X if they are independent in every situation in which X is determinate—i.e., if the probability distribution of only one of them can change in any experiment after X is resolved. We say that X and Y are *formally independent posterior to* X if $P_S(Y=y\&Z=z) = P_S(Y=y)P_S(Z=z)$ holds in every situation S in which X is determinate. Finally, we say that Y and Z are *formally independent given* X if

$$P_S(Y=y\&Z=z|X=x) = P_S(Y=y|X=x)\, P_S(Z=z|X=x) \tag{1}$$

holds for every situation S and all values x, y, and z such that $P_S(X=x)>0$. This is stronger than formal independence posterior to X, which only requires that Equation (1) hold in situations in which X is determinate.

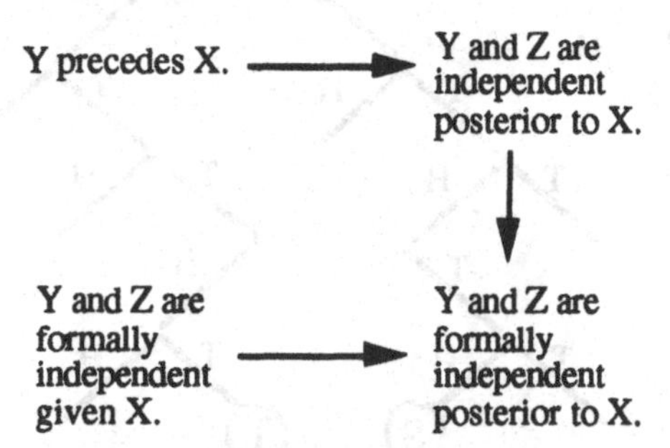
Y precedes X.
Y and Z are independent posterior to X.
Y and Z are formally independent given X.
Y and Z are formally independent posterior to X.

Fig. 5. Logical implications.

Figure 5 summarizes the logical relations among the conditional independence concepts. For the purposes of causal conjecture, Figure 5 is disappointing because there is no arrow from the probability-tree concepts, precedence and independence, to formal independence. Proposition 1 says that when we observe conventional sample-space independence of X and Y (their probabilities multiply), we can explain this observation by the conjecture that X and Y are independent in the probability-tree sense (they have no common causes). Figure 5 does not give us the means to extend this kind of explanation to the conditional case. Since there is no arrow from independence posterior to X to independence given X, independence posterior to X (Y and Z have no common causes after X) will not be an adequate explanation of an observed conditional independence given X.

5 Tracking

Consider two variables X and Z. Suppose that for any possible value x of X, the probability distribution of Z is the same in any two situations that resolve X to x; if S and T are both situations that resolve X to x, then $P_S^Z = P_T^Z$. Then we say that X *tracks* Z.

When X tracks Z, we write $P_{X=x}{}^Z$ for the probability distribution of Z in the situations that resolve X to x. A constant variable tracks every variable and is tracked by every variable.

Coin spinning provides another familiar example of tracking. Suppose we spin a fair coin repeatedly, and let X_i designate the number of heads in the first i spins. Then X_i tracks X_j whenever i<j. Figure 6 illustrates this for the case where i=6 and j=8.

It turns out that when X tracks Z, we can write

$$P_S{}^Z = \sum_{x \in \Theta_X} P_S(X=x)\, P_{X=x}{}^Z \qquad (2)$$

in every situation S that resolves X and in every situation S in which X is indeterminate. Since the situation S appears on the right-hand side of Equation (2) only to tell us its probabilities for X, we may say that the probabilities for Z are initially governed, in a certain sense, by the probabilities of X. Of course, after X is resolved, thc equation no longer applies, and the probabilities of Z are left to evolve on their own. In Figure 6, for example, the probabilities for X_8 are tied to those for X_6 only through the first six spins; after that X_6 is settled and X_8 is on its own.

The message of Equation (2) can also be expressed using the language of causality. It says that until is X settled, Z has no causes distinct, in identity or magnitude, from X's causes.

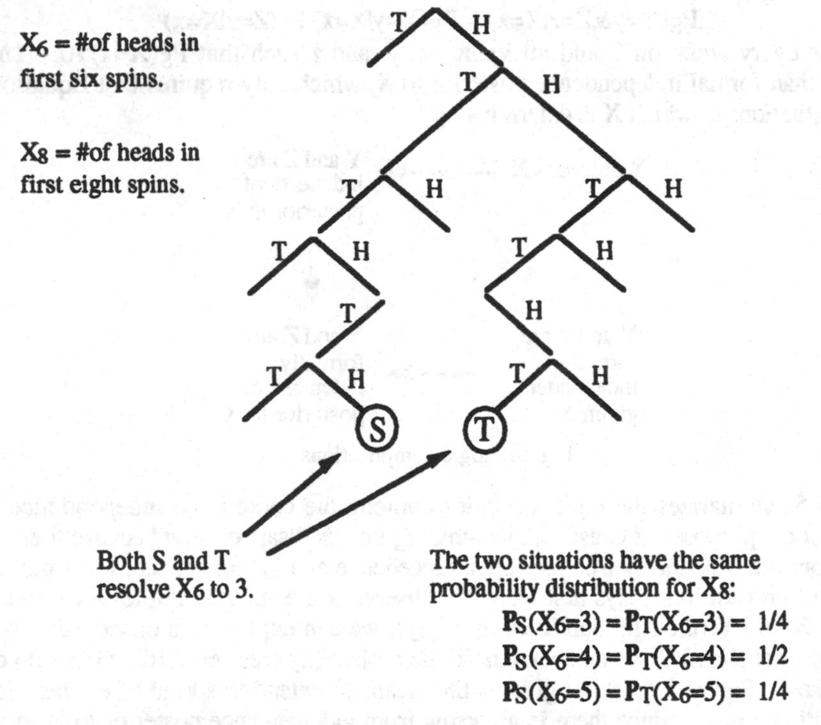

X_6 = #of heads in first six spins.
X_8 = #of heads in first eight spins.
T
H
S
T
Both S and T resolve X_6 to 3.
The two situations have the same probability distribution for X_8:
$P_S(X_6=3) = P_T(X_6=3) = 1/4$
$P_S(X_6=4) = P_T(X_6=4) = 1/2$
$P_S(X_6=5) = P_T(X_6=5) = 1/4$

Fig. 6. Fragment of a probability tree for coin spinning.

Although Figure 6 is a natural example of tracking, the definition of "X tracks Z" does not require that X precede Z. It is possible that Z might instead actually precede X. An example of this is provided by Figure 7. Here we have three piles of cards; the first contains the deuce and three of spaces, the second contains the four and five of hearts, and the third contains the six and seven of hearts. We choose a pile at random and then choose a card at random from the pile. The variable X is the number on the card chosen, while Z is the suit. Since every situation that resolves X resolves it to a different value, X strongly tracks every other variable, including Z. But Z obviously precedes X.

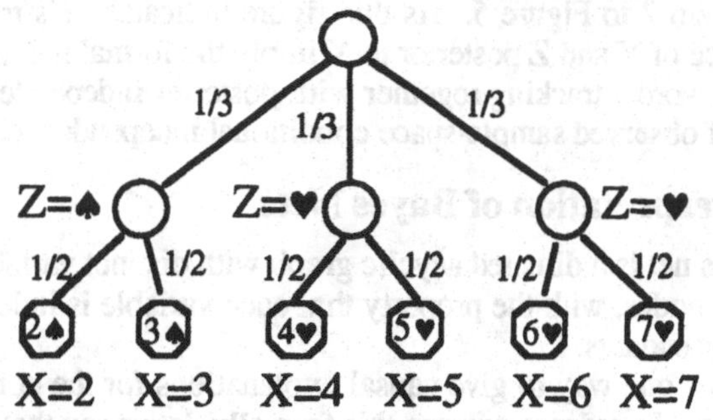

Fig. 7. An example where X tracks Z but Z precedes X.

The example in Figure 7 has another interesting feature: Z is a function of X (if X is 2 or 3, then Z is spades; if X is 4, 5, 6, or 7, then Z is hearts). This feature is not accidental, as the following proposition makes clear.

Proposition 2. The variable Z is a function of the variable X if and only if Z precedes X and X tracks Z.

A certain clash of intuitions is evoked by the statement that a function of X must precede X. When Z is a function of X, we often think of using X to find Z, and in this picture, X comes first. In order to avoid confusion, we must bear in mind the distinction between the observer in the probability tree, who goes down the tree finding out events as they happen, and a person outside the tree, who uses values of certain variables to compute values of others. The second person may also be an observer in a probability tree, but it will be a different probability tree.

In general, we cannot tell for certain whether a given variable X tracks another variable Z in nature's probability tree. But the following theorem allows us to sometimes conjecture that it does.

Theorem 1. If X tracks Z and S is a situation in or before X's resolving cut such that $P_S(X=x) > 0$, then $P_S(Z=z|X=x) = P_{X=x}{}^{Z}(z)$.

This theorem tells us that X tracking Z can serve as a causal explanation of the observation that the probability distributions of Z given X are the same in many different situations.

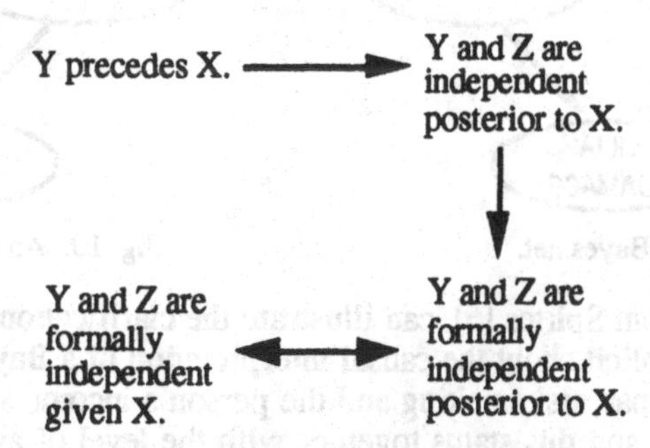

Fig. 8. Tracking and conditional independence.

The following theorem indicates another way in which tracking can serve as part of a causal explanation.

Theorem 2. If X tracks Z and Y and Z are formally independent posterior to X, then Y and Z are formally independent given X.

Figure 8 adds Theorem 2 to Figure 5. As this figure indicates, X's tracking Z together with the independence of Y and Z posterior to X imply the formal independence of Y and Z given X. In other words, tracking together with posterior independence can provide a causal explanation of observed sample-space conditional independence.

6 A Causal Interpretation of Bayes Nets

Recall that a Bayes net is a directed acyclic graph with distinct variables from a probability space (P,Ω) as nodes, with the property that each variable is independent, given its parents, of its nondescendants.

Theorem 2 provides one way to give causal explanations for the conditional independencies in a Bayes net. In order to express this formally, let us say that a directed acyclic graph with variables as nodes is a *causal net in the tracking sense* if in nature's probability tree,

- each variable is tracked by its parents, and
- each variable is independent, posterior to its parents, of its nondescendants.

Theorem 2 tells us that a causal net in the tracking sense is a Bayes net in the conventional sense with respect to the probability distribution for every situation.

This result provides a framework within which users of Bayes nets can express, elaborate, examine, and defend their causal assumptions. Three points should be kept in mind in this process.

- Saying that a Bayes net is causal net in the tracking sense does not necessarily mean that X is settled before Y when there is an arrow or a directed path from X to Y. If this is intended, it should be made explicit.
- In many cases, the claim that a Bayes net is causal net in the tracking sense is not plausible.
- Causality in the tracking sense is not the only possible causal interpretation for Bayes nets. But whatever interpretation is used, it should be explicitly stated and examined.

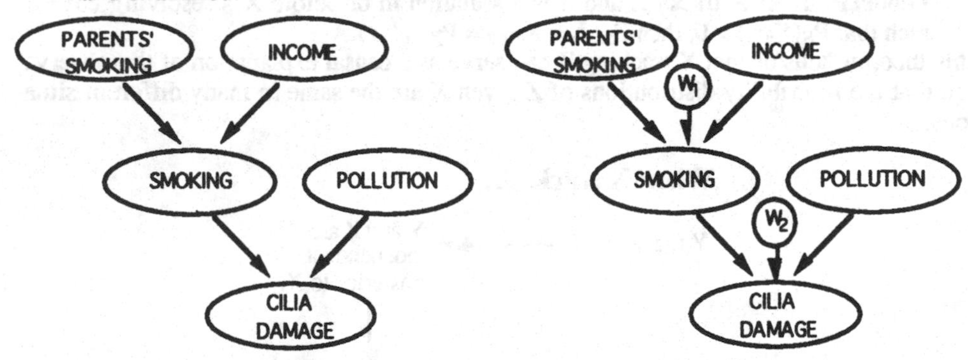

Fig. 9. A hypothetical Bayes net.

Fig. 10. An elaboration.

Figure 9, adapted from Spirtes [5], can illustrate the clarification that is possible when we undertake to be explicit about the causal interpretation of a Bayes net. In this purely hypothetical example, parents' smoking and the person's income are said to be "causes" of her smoking status, and this status together with the level of air pollution where she lives are said to be "causes" of the extent of damage to the cilia in her lungs. A little thought shows, however, that these assertions cannot be plausibly interpreted in terms of tracking and posterior independence. At the point in time where a person's income and her parent's smoking habits have been determined, there are surely other factors that have also already come into play in influencing her smoking status.

One way to get a causal interpretation of Figure 9 might be to enlarge it, as in Figure 10. Here W_1 represents the other factors influencing smoking status by the time its parents in Figure 9 are resolved, and W_2 similarly represents the other factors influencing cilia damage. If we are willing to believe the additional independencies indicated by Figure 10, then it will be a causal net in the tracking sense, and since Figure 10 being a Bayes net implies Figure 9 is also a Bayes net, the causal explanation of Figure 10 can be taken as a causal explanation of Figure 9.

This example also illustrates the point that tracking provides only one causal explanation of Bayes's nets, and that others are often needed. Since the damage that smoking does to the lungs is progressive, it may be more useful to think of a causal interpretation in which each additional act of smoking merely proportionally raises the expected level of lung damage. This leads to the concept of linear sign, which is weaker than tracking. There are results for linear sign analogous to Theorems 1 and 2:

- if X is a linear sign of Z, then the linear regression coefficient of Z on X is the same in every situation, and
- if X is a linear sign of Z, and Y and Z do not change in mean when X does not change in mean, then Y and Z have zero partial correlation linearly accounting for X in every situation.

7 Conclusion

This paper has demonstrated one way in which a Bayes net can arise from qualitative and overtly causal conditions in a probability tree. The demonstration should be taken only as illustrative, for many Bayes nets are not causal nets in the tracking sense, and there are other conditions on a probability tree that can lead to the conditional independence relations that define Bayes nets.

The existence of precise causal interpretations for Bayes nets raises a challenge to anyone who wishes to attribute causality to any Bayes net—whether the person be a social scientist working with a Bayes net inferred from data or an AI engineer constructing a Bayes net for use in decision making. It is no longer enough to say vaguely that the Bayes net is causal; instead, one must provide a precise conjecture about the causal conditions that might give rise to the Bayes net.

The ideas developed here go beyond networks encoding conditional independence relations to similar directed acyclic graphs that encode partial uncorrelatedness or unpredictability relations. This includes the structural equation models used in the social sciences. In these cases, too, we can give causal conditions in terms of an underlying probability tree that will explain the observed or conjectured conditional independence, conditional unpredictability, or partial uncorrelatedness relations, and before we accept a causal claim, we will want to ask that it be made precise in probability-tree terms.

Acknowledgment

Research for this note was partially supported by National Science Foundation grant SBE9213674.

References

1. Kenneth A. Bollen: Structural Equations with Latent Variables. New York: Wiley. 1988.
2. Pearl, J. Probabilistic Reasoning in Intelligent Systems. San Mateo, California: Morgan Kaufmann 1988.

3. Shafer, Glenn: Can the various meanings of probability be reconciled? In: G. Keren and C. Lewis (eds.): A Handbook for Data Analysis in the Behavioral Sciences: Methodological Issues. Hillsdale, New Jersey: Lawrence Erlbaum 1993, pp. 165-196.
4. Shafer, Glenn: The Art of Causal Conjecture. Cambridge: MIT Press 1995.
5. Spirtes, Peter, Clark Glymour, and Richard Scheines. Causation, Prediction, and Search. Lecture Notes in Statistics 81. New York: Springer 1993.
6. D.J. Spiegelhalter, A.P. Dawid, S. L. Laurtizen, and R. G. Cowell: Bayesian analysis in expert systems (with discussion). Statistical Science, 8 219-283 (1993).
7. Vovk, V.G. : The logic of probability. Journal of the Royal Statistical Society, Series B, 55 317-351 (1993).

Non Standard Probabilistic and Non Probabilistic Representations of Uncertainty.

Philippe SMETS[1]
I.R.I.D.I.A. Université Libre de Bruxelles
50 av. Roosevelt, CP 194-6. B-1050, Brussels, Belgium.

Summary: Survey of the mathematical models proposed to represent quantified beliefs, and their comparison. The models considered are separated into non standard probability and non probability models, according to the fact they are based on probability theory or not. The first group concerns the upper and lower probability models, the second the possibility theory and the transferable belief model.

Keywords: uncertainty, belief representation, probability functions, upper and lower probability functions, possibility functions, belief functions.

1. Introduction.

This paper surveys the mathematical models that have been proposed to represent quantified beliefs. The mathematical representation of the 'real world' has been a permanent subject of research as it provides an objective and unambiguous formalization of the 'reality'. That they represent idealized reality does not reduce their value. The 19th. century has seen the development of mathematical models to represent physical realities. In the 20th. century, randomness became omnipresent. More recently, the modelization has been extended toward the representation of subjectivity[2], and in particular of uncertainty.

Uncertainty can take numerous forms (Smets, 1991a, 1993a, Smithson, 1989) and usually induces 'beliefs', i.e. the graded dispositions that guide 'our' behavior. Proposed models of belief can be split into symbolic models as in modal logic (Hintikka, 1962, Chellas, 1980) or numerical models as those studied here. The model we study are normative, the

---

[1] Research work has been partly supported by the Action de Recherches Concertées BELON funded by a grant from the Communauté Française de Belgique and the ESPRIT III, Basic Research Action 6156 (DRUMS II) funded by a grant from the Commission of the European Communities.
[2] Nguyen H., 1994, personal communication.

agent that holds the beliefs, called You hereafter, is an 'ideal rational subject'. That humans hardly behave as predicted by the normative model reflects essentially their lack of 'rationality', a well known fact that does not bear negative connotation. The 'rationality' that underlies the models and the so called human rationality do not cover the same concepts.

The Bayesian model based on probability functions was the first model proposed to represent quantified beliefs. In spite of its success, alternative models have been proposed. They can be split into two families: the non-standard probability and the non-probability models (Kohlas, 1994b). The models based on non-standard probabilities are extensions of models based on probability functions. They include the upper and lower probabilities models (Good, 1950, Smith, 1961, Kyburg, 1987, Walley, 1991, Voorbraak, 1993), Dempster-Shafer's models (Dempster, 1967, Shafer, 1976a, Smets, 1994b), the Hints models (Kohlas and Monney, 1994a), the probability of provability models (Ruspini, 1986, Pearl, 1988, Smets, 1991c),. The non-probabilistic models are not based on probability functions, but on alternative functions like the possibility functions and the belief functions. They include the transferable belief model (Smets, 1988, 1990a, Smets and Kennes, 1994), the possibility theory model (Zadeh, 1978, Dubois and Prade, 1985). The difference between these models and their domain of application is discussed hereafter.

We first define what concept of belief is covered. Belief represent the opinion of the agent, You, at time t given what You know. We can say "You believe at level .7 that the name of Philippe Smets' daughter is Karine". It means that "You believe at level .7 that the proposition 'the name of Philippe Smets' daughter is Karine' is true". So beliefs weight the strength given by the agent to the fact that a given proposition is true. The domain of beliefs is the truth status of a proposition. When the truth domain is restricted to True and False, shortcuts can be used without danger, and we can just say "You believe p at .7" where p is a proposition. When the truth domain becomes more exotic, like in multi-valued logics and in fuzzy logics, these shortcuts can be misleading, and the use of the full expressions will avoided ambiguities, if not plain stupidities.

Coming back to classical logic, the meaning of the sentence 'You believe at level .7 that it will rain tomorrow' can be either: 'the measure of the belief held by You that the proposition "it will rain tomorrow" is true is .7' or: 'You believe at level .7 that "tomorrow" belongs to the set of rainy days'. So beliefs given to 'propositions' can equivalently be given to the subset of worlds that denote the propositions. Defining beliefs on propositions or on sets of possible worlds is equivalent. We will restrict ourselves to propositions and possible worlds, and left aside more elaborate structures.

What are the needed ingredients for a model for the representation of quantified beliefs? We must be able to represent static states of belief. Such static states representation should be able to cover every states of belief, from the state of total ignorance up to the state of absolute knowledge. We must describe the dynamic of beliefs, i.e., how a new piece of information changes the belief states. Decisions are the only observable outcomes of a belief state and the most important practical components of any model for uncertainty. So we must also explain how decisions will be made. We will see how the various models answer to these questions.

In Section 2, we present and criticize the justifications that lead to the traditional Bayesian model based on probability functions. In section 3, we compare various non standard probability models. In section 4, we present the major non probability models: the possibility theory and the transferable belief model. In section 5, we just conclude with small remarks.

2. The Probability Models.

The representation of quantified beliefs really started in the eightieth century with the work of Bernoulli, De Moivre and Lambert (Shafer, 1978) culminating in the famous essay of Bayes (1763) where the author defends that beliefs should be mathematically represented by probability functions. Today three types of justifications are usually proposed in defense for such a representation: the decision-oriented, the preference-ordering and Cox algebraic justifications.

2.1. Decision-oriented justifications.

Ramsey (1931) defines a probability as the price one would be ready to pay to bet on an event which reward is $1 if the event occurs and nothing if the event does not occur. Such a definition is usually based on a betting scheme in which there is a player and a banker. The player must pay to the banker a certain amount of money (say $p) to enter a game where the banker will pay $1 to the player if event A occurs and $0 if A does not occur. The probability assessor (You) must decide the price $p of the game and I (Your opponent) decide which position will be held by You, banker or player. Besides, the bet is "forced" in that You must play. The prices You give to each game are arbitrary except that they must be coherent, i.e. they must satisfy the axioms of probability theory. If they did not, then I could always build a set of games where You would loose for sure.

For instance, if You decide to pay $.6 on a game where the winning event is A (Game 1) and $.3 on a game where the winning event is $\overline{A}$ (Game 2), then I would force You to be the banker for both games. If A occurs You loose \$.4 on Game 1 (You receive \$.6 and pay \$1) and You win \$.3 on Game 2 (You receive \$.3 and pay \$0), hence Your net loss is \$.1. If $\bar{A}$ occurs You win \$.6 on Game 1 and You loose \$.7 on Game 2, hence Your net loss is also \$.1. Therefore in both cases, You suffer a loss.

This strategy that leads to a sure loss is called a Dutch Book strategy. In order to avoid such a Dutch Book, it has been shown that the prices given to the various games must satisfy the axioms of probability theory, in particular, the additivity and the conditioning rules. In order to derive the conditioning rule, the set of potential events are partitioned in 3 subsets: A, B, C. Bets are posted such that You win if A occurs, You loose if B occurs and the game is canceled if C occurs in which case the banker returns the ticket price to the player. Such bet is called a conditional bet and leads to the definition of the conditional probability of A given A∪B: $P(A|A\cup B) = P(A)/P(A\cup B)$

The decision-oriented approach seems quite convincing, except for the fact that it is based on "forced" bets. Some critics were raised against it as people feel they are not forced to bet. It leads to a generalization of the probability models where the concepts of upper and lower probabilities are introduced (Walley, 1991). The lower probability of an event is defined as the maximum amount You would be ready to pay to enter the game where You win \$1 if the event occurs and nothing otherwise. It does not require that You would accept to be the banker. The upper probability is defined as the minimum amount You would require that the player pays to enter the game if you were the banker. So when the probability assessor can assign prices without committing himself to accept to be the player or the banker according to his opponent's orders, the Dutch Book argument collapses (Smith, 1961, Williams, 1976, Jaffray, 1989, Voorbraak, 1993).

Another critic is oriented towards the idea that what Dutch Book argument leads to are not a quantified representation of our belief but a quantified representation of our belief qua basis of decision (Ramsey, 1931). Beliefs held outside any decision context are not considered in the Dutch Book argument and therefore the argument does not justify that quantified belief in general must be represented by a probability function. It only says that when decision must be made, beliefs will induce a probability measure that will be used for the decision-maker. Therefore, the question is to decide if beliefs can exist outside any decision process.

2.2. Credal versus pignistic beliefs.

When developing the transferable belief model (see section 4.2), we introduced a two-level mental model for representing belief. At one level beliefs are entertained without any regard to decision-making and at the other level beliefs are used to make decision. We call these two mental levels, the credal level and the pignistic level (respectively from credo = I believe and pignus = a bet, both in Latin, (Smith, 1961)). That beliefs are necessary ingredients for our decisions does not mean that beliefs cannot be entertained without any revealing behavior manifestations (Smith and Jones, 1986, p.147). A belief is a disposition to feel that things are thus-and-so. It must be contrasted with the concept of acceptance (Cohen, 1993). A probability measure is a tool for action, not for assessing strength of evidence (Sahlin, 1993).

Some authors will just discard the existence of a credal level arguing that only the pignistic level exists and indeed, at that level, beliefs should better be represented by a probability measure if one wants to avoid Dutch Books or similar sub-optimal behaviors. We will defend the distinction between the two levels and the whole paper is devoted to the development of a quantified representation of belief at the credal level. We will illustrate the impact of considering the two-level model in the transferable belief model through the 'Peter, Paul and Mary Saga'.

2.3. The preference-ordering justification.

Arguments based on preference-ordering have been proposed in order to support the use of probability measures for quantified beliefs. They are based on Koopman's initial justification of probability measure from qualitative requirements (Koopman, 1940, Fine, 1973). He introduces an order $\geq$ on the events where $A \geq B$ means that the probability of A is larger or equal to the probability of B. His major axiom is the Disjoint Union Axiom:

$$A \cup (B \cup C) = \emptyset \Rightarrow (B \geq C \Leftrightarrow A \cup B \geq A \cup C)$$

This axiom is central to the proof that the ordering relation is represented by a probability measure (Fine, 1973). Wong et al. (1990) propose to replace the Disjoint Union Axiom by a more general axiom.

$$C \subsetneq B,\ A \cap B = \emptyset \Rightarrow (B > C \Rightarrow A \cup B \geq A \cup C)$$

In that case, they show that the ordering cannot always be represented by a probability measure, that it can always be represented by a belief function, a family of functions more general than the family of probability functions, but it can also be represented by other functions. Which of Koopman axiom or Wong et al. less demanding axiom must be satisfied is hardly obvious. They both seem reasonable but this cannot provide a definitive justification for their acceptation. Many other variants of these axioms have been suggested, but they always encountered the same problems. So whatever its beauty, the ordering justification is not compelling.

2.4. Cox algebraic justification.

Finally, Cox (1946) proposes an algebraic justification for the use of probability measure that has become quite popular in artificial intelligence domain (Cheeseman, 1985). He assumes that any measure of beliefs should essentially satisfy:

$$P(\overline{A}) = f(P(A)) \tag{2.1}$$

$$P(A \cap B) = F(P(A|B), P(B)) \tag{2.2}$$

for some f and F functions where f is decreasing in its argument. These requirements lead easily to the use of probability measure. But even the first requirement is criticized (Dubois et al., 1991). Suppose a medical context where a symptom is very frequent for a patient in disease class A and quite frequent for patient not in disease class A. Its observation should support the fact that the patient could belong to group A, this being translated by an increase in the probability that the patient belongs to group A. By (2.1) the fact that the patient belongs to group $\overline{A}$ becomes automatically less believable (in the sense of probable) even though the symptom is quite frequent in that group (but not as frequent as in group A). This unsatisfactory conclusion led some to reject (2.1) and (2.2). This was the origin of the certainty factor in artificial intelligence and developed in the 70's by Shortliffe and Buchanan (1975). This was also rejected in possibility theory (Zadeh, 1978, Dubois and Prade, 1985) and in Shafer's theory of evidence (Shafer, 1976a, Smets, 1978). Cox's axioms provide a nice characterization of probability measures. Nevertheless they are not really necessary.

2.5. Objective and subjective probabilities.

We deal with subjective probabilities but their link to objective probabilities must be considered. Hacking's Frequency Principle states that (Hacking, 1965):

> If the objective probability of A is p then the subjective probability of A is p.

It provides a very reasonable scale for Your subjective probability but it is based on the assumption that the objective probability exists. When You toss a coin and say that the probability of heads is 0.5, what is the 0.5 about? Hacking suggests it is a property of the chance set-up, i.e. the coin and the procedure that tosses the coin.... In that case, the .5 would preexist to the belief holder and a rational belief holder should use the .5 to quantify his belief about observing heads on the next toss. So in Hacking's perspective, there exists something called the objective probability and the subjective belief induced by a chance set-up and its related objective probability should share the same scale. Again, this does not justify that beliefs in general are quantified by probability measure. It only justifies the representation of beliefs by probability functions when they are induced by a chance set-up. Life is not just a bunch of chance set-ups, and the domain of our beliefs is much larger then those relate to some chance set-ups.

De Finetti takes a much more extreme position by rejecting any form of probability that is not subjective. For De Finetti, the .5 is not even a property of the chance set-up but it is a property of the belief holder. De Finetti goes as far as claiming that You are free to assign whatever probability You would like to heads under the provision that You allocate the remaining probability to tails. A .7 to heads is perfectly valid (if tails receive .3).

We will not further discuss of the existence of objective probability. It seems that this is an acknowledged property in quantum mechanics (d'Espagnat, 1976). We feel that Hacking's ideas of objective probability, chance set-up and his Frequency Principle are ideas that could be accepted, what we will do in our development. Nevertheless, it does not lead to the use of probability function to quantify all forms of beliefs.

2.6. Conclusion about the probabilistic models.

None of the justifications proposed for the use of probability functions to represent quantified beliefs leads to the necessity of their use. We present in this paper alternative models that will represent belief at the credal level, i.e. at the mental level where beliefs are only "entertained" outside of any consideration for some underlying decision. I feel I can have some belief about the weather in Hawaii...... even though I do not take any decision based on that weather. Degrees of belief can be assimilated to degree of conviction, degree of support, degree of justified belief.... (Voorbraak, 1993).

3. Non-Standard Probability Models.

We successively examined models based on upper and lower probabilities, sets of probability functions, Dempster-Shafer theory, second order probabilities and probabilities of provability. All these models have in common the existence of some additive measure that underlies somehow the agent's beliefs.

3.1. Static components.

Smith (1961, 1965), Good (1950, 1983) and Walley (1991) suggested that personal degrees of belief cannot be expressed by a single number but that one can only assess intervals that bound them. The interval is described by its boundaries called the upper and lower probabilities. Such interval can easily be obtained in a two-person situation when one person, Y_1, communicates the probability of some events in Ω to a second person, Y_2, by only saying, for all $A \subseteq \Omega$, that the probability P(A) belong to an interval. Suppose Y_2 has no other information about the probability on Ω. In that case, Y_2 can only build a set $\mathcal{P}$ of probability measures on Ω compatible with the boundaries provided by Y_1. All that is known to Y_2 is that there exists a probability measure P and that $P \in \mathcal{P}$. Should Y_2 learn then that an event $A \subseteq \Omega$ has occurred, $\mathcal{P}$ should be updated to $\mathcal{P}_A$ where $\mathcal{P}_A$ is this set of conditional probability measures obtained by conditioning the probability measures $P \in \mathcal{P}$ on A. (Smets, 1987, Fagin and Halpern, 1990, Jaffray, 1992).

One obtains a similar model by assuming that one's belief is not described by a single probability measure as do the Bayesians but by a **family of probability measures** (usually the family is assumed to be convex). Conditioning on some event $A \subseteq \Omega$ is obtained as in the previous case.

A special case of upper and lower probabilities has been described by Dempster (1967, 1968) and underlies most interpretations of **Dempster-Shafer theory**. He assumes the existence of a probability measure on a space X and a one to many mapping M from X to another space Y. Then the lower probability of A in Y is equal to the probability of the largest subset of X such that its image under M is included in A. The upper probability of A in Y is the probability of the largest subset of X such that the images under M of all its elements have a non empty intersection with A.

Example 1: Dempster model.
Suppose a space $X = \{x_1, x_2\}$ and $P_X(\{x_1\}) = .3$, $P_X(\{x_2\}) = .7$. Suppose a mapping M form X to Y where $Y = \{y_1, y_2\}$ and $M(x_1) = \{y_1\}$, $M(x_2) = \{y_1, y_2\}$. What can we say about the probability P_Y over the space Y. We know $P_Y(\{y_1\})$ is at least .3, and might be 1.0 if x_2 was indeed mapped onto y_1. We know $P_Y(\{y_2\})$ is at most .7 as the probability .3 given to x_1 cannot be given to y_2. Finally $P_Y(\{y_1,y_2\}) = 1$. Our knowledge about P_Y can be represented by the lower probability function P_{Y*}, with $P_{Y*}(\{y_1\}) = .3$, $P_{Y*}(\{y_1\}) = .7$, $P_{Y*}(\{y_1,y_2\}) = 1$. It happens that the lower probability function obtained through the M mapping is a belief function, what does not necessary mean that the lower probability function quantifies the agent's beliefs.

But what is a belief function? It is a Choquet capacity monotone of order infinite (Choquet, 1953) that Shafer (1976a) introduces in order to represent beliefs. Let Ω be a set and let 2^Ω be the Boolean algebra of subsets of Ω. A belief function is a function bel from 2^Ω to [0, 1] such that :

1) $bel(\emptyset) = 0$
2) $\forall A_1, A_2, \ldots A_n \in \Re$,

$$bel(A_1 \cup A_2 \cup ... A_n) \geq \sum_i bel(A_i) - \sum_{i>j} bel(A_i \cap A_j) .... -(-1)^n bel(A_1 \cap A_2 \cap ... A_n) \quad (3.1)$$

Notice that we do not require that $bel(\Omega) = 1$ as is usually accepted. We can have $bel(\Omega)<1$, the difference $1-bel(\Omega)$ quantifies the amount of internal conflict in the information that leads to the construction of bel (Smets, 1992a). This unnormalization of bel is not very important at this point and will not be further discussed.

Associated to every belief function bel, the is another function m from 2^Ω to [0, 1], called the basic belief assignment, which values m(A) for $A \subseteq \Omega$ are called the basic belief masses.

$$m(A) = \sum_{B: B \subseteq \Omega, \emptyset \neq B \subseteq A} (-1)^{|A|-|B|} . bel(B) \qquad \forall A \subseteq \Omega, A \neq \emptyset$$

$$m(\emptyset) = 1 - bel(\Omega)$$

and, $$bel(A) = \sum_{B: B \subseteq \Omega, \emptyset \neq B \subseteq A} m(B) \qquad \forall A \in \Re, A \neq \emptyset$$

Coming back to the upper and lower probability models, their generalization into **second-order probability** models is quite straightforward. Instead of just acknowledging that $P \in \mathcal{P}$, one can accept a probability measure P* on $\mathbb{P}_\Omega$, the set of probability measures on Ω. So for all $\mathcal{A} \subseteq \mathbb{P}_\Omega$, one can define the probability $P^*(\mathcal{A})$ that the actual probability P on Ω belongs to the subset $\mathcal{A}$ of probability measures on Ω. In that case, the information $P \in \mathcal{P}$ induces a conditioning of P* into $P^*(\mathcal{A} | \mathcal{P}) = P^*(\mathcal{A} \cap \mathcal{P}) / P^*(\mathcal{P})$.

Second-order probabilities, i.e. probabilities over probabilities, do not enjoy the same support as subjective probabilities. Indeed, there seems to be no compelling reason to conceive a second-order probability in terms of betting and avoiding Dutch books. So the major justification for the subjective probability modeling is lost. Further introducing second-order probabilities directly leads to a proposal for third-order probabilities that quantifies our uncertainty about the value of the second-order probabilities.... Such iteration leads to an infinite regress of meta-probabilities that cannot be easily avoided.

Ruspini (1986) and Pearl (1988) have suggested that the degree of belief given to a proposition that is quantified by a belief function could be understood as the probability that the proposition in know, is provable. Indeed, if one applies probability theory to modal propositions, one discovers easily that the 'probability of proving' is a belief function. This idea is also what Kohlas models in his hints models (Kohlas and Monney, 1994a). He assumes assumptions and hypothesis, knowing under which assumptions each hypothesis can be proved. These models are in fact part of the Dempster models, as each assumes a space endowed with a probability function and the provability relation plays the role of the one-to-many mapping M. The problems with these problems appear once conditioning is introduced. This will be further discussed in section 3.2, where the dynamic of beliefs is analyzed.

3.2. Dynamic components.

Suppose the agent had a given beliefs at time t_0 belief. At time $t_1>t_0$, the agent learns for sure that 'the proposition A is true', or equivalently that 'the actual world is in A' for some A. We denote this information by Ev_A. We also suppose that, between t_0 and t_1, the agent has not learned anything relevant to his beliefs under consideration. So Ev_A is the first information relevant to the agent' beliefs since t_0. This situation corresponds to the one described by the conditioning process in probability theory.

How does the agent change his beliefs once he learns Ev_A? This is the first question to be answered when the dynamic of beliefs is studied.

In probability theory, conditioning is achieved by Bayes rule of conditioning. In the upper and lower probability models, the second-order probability models and the family of probability functions models, the conditioning is performed for each probability function compatible with the constraints imposed by the static representation. A new set of probability functions is computed from these conditional probability functions. This method is quite immediate. In strict upper and lower probability models, one must be cautious as conditioning cannot be simply iterated. In fact, the knowledge of the upper and lower probability functions is not sufficient to represent a state of belief, one must replace them by upper and lower expectations in which case conditioning can be iterated (Walley, 1991).

With Dempster models as well as with the probability of provability models, conditioning becomes more delicate. The so-called Dempster's rule of conditioning proposed by Shafer is not blindly applicable. Suppose the agent learns that the real value of Y belongs to some $B \subseteq Y$. Does it means that the real value of X belongs to the set of $x \in X$ such $M(x) \subseteq B$? If it is the case, then the appropriate rule of conditioning is the so-called geometric rule (Shafer, 1976b). If the information means that the agent has learned a new information that states that the true value of Y belongs to B, and if this information means that the mapping M between X and Y must be revised into M_B such that for every $x \in X$, $M_B(x) = M(X) \cap B$, then Dempster's rule of conditioning is the appropriate rule (except for the problem of normalization).

In the probability of provability model, these two forms of conditioning correspond to:

case 1: the probability that the assumption selected randomly according to P_X will prove the hypothesis C given it proves the hypothesis B.
case 2: the probability that the assumption selected randomly according to P_X will prove the hypothesis C given You know that the hypothesis B has been proved by another pieces of evidence.

Once probabilities are generalized into non-standard probability models, conditioning can takes many forms, and the selection of the appropriate rules is delicate (Smets, 1991b). Most criticisms against Dempster-Shafer theories are essentially due to an inappropriate use of Dempster's rule of conditioning.

3.3. Decisions Making.

The most important quality of the Bayesian model is that is provides an excellent, if not the best tool, for optimal decision making under risk. But the decisions considered by Bayesians, an in particular in the Dutch Books arguments they favor so often, are all based on forced bets. The players must bet, and must be ready to be in either of the two positions, player or banker. In that case, betting without respecting the probability axioms would lead the agent to a sure loss, turning him into a money pump, a situation nobody could accept (see section 2.2). But the argument collapses once bets are neither forced nor symmetrical. Models based on upper and lower probabilities claim that the lower probability is the maximal prize the agent is ready to pay if he is the player, and the upper probability is the minimum prize the agent would ask from the player if the agent happens to be the banker. Such a model presents the weakness that alternatives are no more ordered as in the probability approach. Some pairs of decisions cannot be compared any more in order to decide which is best. Is it a good or a bad property is a matter of personal opinions.

3.4. Conclusions about the non standard probability models.

The models answer to the questions that must be answered by any model for the representation of quantified beliefs (see section 1). They can even solve some of the weaknesses of the probability model. Total ignorance can be represented by just accepting that for any $A\subseteq\Omega$, $A\neq\Omega$, its lower probability is null. Such a state cannot be represented in pure probability model and the use of the Principle of Insufficient Reason is the best way to get trapped into paradoxes, inconsistencies, if not plain non-sense. As a matter of fact, Bayesians usually claim that a state of total ignorance does not exist, a preposterous attitude. If I ask You what are Your beliefs about the location of Pa'acal burial on Earth, I hardly think You could give any reasonable and justified answer, as far as You don't even know who Pa'acal is[3]. Claiming total ignorance does not exist is nothing more than sweeping it under the carpet.

4. Non probability Models.

We analyze the two major non probability models based respectively on possibility functions and on belief functions.

4.1. Possibility Theory.

4.1.1. Possibility and Necessity Measures.

Incomplete information such as "John's height is above 170" implies that any height h above 170 is possible and any height equal or below 170 is impossible. This can be represented by a 'possibility' measure defined on the height domain whose value is 0 if h < 170 and 1 if h is ≥ 170 (with 0 = impossible and 1 = possible). Ignorance results from the lack of precision, of specificity of the information "above 170".

When the predicate is vague like in 'John is tall', possibility can admit degrees, the largest the degree, the largest the possibility. But even though possibility is often associated with fuzziness, the fact that non fuzzy (crisp) events can admit different degrees of possibility is shown in the following example. Suppose there is a box in which you try to squeeze soft balls. You can say: it is possible to put 20 balls in it, impossible to put 30 balls, quite possible to put 24 balls, but not so possible to put 26 balls...These degrees of possibility are degrees of realizability and they are totally unrelated to any supposedly underlying random process.

Identically ask a salesman about his forecast about next year sales. He could answer: it is possible to sell about 50K, impossible to sell more than 100K, quite possible to sell 70K, hardly possible to sell more than 90K... His statements express what are the possible values for next year sales. What the values express are essentially the sale capacity. Beside, he could also express his belief about what he will actually sell next year, but this concerns another problem for which the theories of probability and belief functions are more adequate.

Let $\Pi:2^{\Omega}\rightarrow[0, 1]$ be the **possibility measure** defined on a space Ω with $\Pi(A)$ for $A\subseteq\Omega$ being the degree of possibility that A (is true, occurs...). The fundamental axiom is

[3]Pa'acal is the Mayan king burried at Palenque.

that the possibility $\Pi(A \vee B)$ of the disjunction of two propositions A and B is the maximum of the possibility of the individual propositions $\Pi(A)$ and $\Pi(B)$. (Zadeh, 1978, Dubois and Prade, 1985):

$$\Pi(A \vee B) = \max (\Pi(A) , \Pi(B)). \qquad (4.1)$$

Usually one also requires $\Pi(\Omega) = 1$.

As in modal logic, where the **necessity** of a proposition is the negation of the possibility of its negation, one defines the necessity measure N(A) given to a proposition A by:

$$N(A) = 1 - \Pi(\neg A)$$

In that case, one has the following:

$$N(A \wedge B) = \min (N(A) , N(B))$$

Beware that one has only:

$$\Pi(A \wedge B) \leq \min (\Pi(A) , \Pi(B))$$

$$N(A \vee B) \geq \max (N(A) , N(B)).$$

Let Ω be the universe of discourse on which a possibility measure Π is defined. Related to the possibility measure $\Pi: 2^{\Omega} \rightarrow [0, 1]$, one can define a **possibility distribution** $\pi: \Omega \rightarrow [0, 1]$,

$$\pi(x) = \Pi(\{x\}) \qquad \text{for all } x \in \Omega.$$

Thanks to (4.1), one has

$$\Pi(A) = \max_{x \in A} \pi(x) \qquad \text{for all A in } \Omega.$$

A very important point in possibility theory (and in fuzzy set theory) when only the max and min operators are used is the fact that the values given to the possibility measure (or to the grade of membership) are not intrinsically essential. The only important element of the measure is the order they create among the elements of the domain. Indeed the orders are invariant under any strictly monotonous transformation. Therefore a change of scale will not affect conclusions. This property explains why authors insist on the fact that possibility theory is essentially an ordinal theory, a nice property in general. This robustness property does not apply once addition and multiplication are introduced as is the case with probability and belief functions, or when operators different from the min-max operators are used.

Example 2: Probability versus possibility. Hans eggs.

As an example of the use of possibility measure versus probability measure, consider the number of eggs X that Hans is going to order tomorrow morning (Zadeh, 1978). Let $\pi(u)$ be the degree of ease with which Hans can eat u eggs. Let p(u) be the probability that Hans will eat u eggs at breakfast tomorrow. Given our knowledge, assume the values of $\pi(u)$ and p(u) are those of table 1.

Table 1 : The possibility and probability distributions associated with X.

u	1	2	3	4	5	6	7	8
$\pi(u)$	1	1	1	1	.8	.6	.4	.2
p(u)	.1	.8	.1	0	0	0	0	0

We observe that, whereas the possibility that Hans may eat 3 eggs for breakfast is 1, the probability that he may do so might be quite small, e.g., 0.1. Thus, a high degree of possibility does not imply a high degree of probability, nor does a low degree of probability imply a low degree of possibility. However, if an event is impossible, it is bound to be improbable. This heuristic connection between possibilities and probabilities may be stated in the form of what might be called the **possibility/probability consistency principle** (Zadeh, 1978).

4.1.2. Physical and Epistemic Possibility.

Two forms of (continuous valued) possibility have been described: the physical and the epistemic. These 2 forms of possibility can be recognized by their different linguistic uses: it is **possible that** and it is **possible for** (Hacking, 1975). When I say it is possible that Paul's height is 170, it means that for all I know, Paul's height may be 170. When I say it is possible for Paul's height to be 170, it means that physically, Paul's height may be 170. The first form, 'possible that', is related to our state of knowledge and is called epistemic. The second form, 'possible for', deals with actual abilities independently of our knowledge about them. It is a degree of realizability. The distinction is not unrelated to the one between the epistemic concept of probability (called here the credibility) and the aleatory one (called here chance). These forms of possibilities are evidently not independent concepts, but the exact structure of their interrelations is not yet clearly established.

4.1.3. Relation between fuzziness and possibility.

Zadeh has introduced both the concept of fuzzy set (1965) and the concept of possibility measure (1978). The first allows us to describe the grade of membership of a well-known individual to an ill-defined set. The second allows us to describe what are the individuals that satisfy some ill-defined constraints or that belong to some ill-defined sets.

For instance $\mu_{Tall}(h)$ quantifies the membership of a person with height h to the set of *Tall* men and $\pi_{Tall}(h)$ quantifies the possibility that the height of a person is h given the person belongs to the set of *Tall* men. **Zadeh's possibilistic principle** postulates the following equality :

$$\pi_{Tall}(h) = \mu_{Tall}(h) \qquad \text{for all } h \in H$$

where H is the set of height = $[0, \infty)$

This writing of Zadeh's possibilistic principle is the one most usually encountered but its meaning should be interpreted with care. It states that the possibility that a tall man has a height h is equal numerically to the grade of membership of a man with height h to the set of tall men. The writing is often confusing and would have been better written as

$$\pi(h|Tall) = \mu(Tall|h) \qquad \text{for all } h \in H$$

or still better

$$\text{If } \mu(Tall|h) = x \text{ then } \pi(h|Tall) = x \quad \text{for all } h \in H$$

The last expression avoids the confusion between the two concepts. Just as with Hacking Frequency Principle, it shows that they share the same scale without implying that a possibility is a membership and *vice versa*. The previous expression clearly indicates the domain of the measure (sets for the grade of membership μ and height for the possibility distribution π) and the background knowledge (the height h for μ and the set *Tall* for π). The difference between μ and π is analogous to the difference between a probability distribution $p(x|\theta)$ (the probability of the observation x given the hypothesis θ) and a likelihood function $l(\theta|x)$ (the likelihood of the hypothesis θ given the observation x) in which case Zadeh's possibilistic principle becomes the likelihood principle:

$$l(\theta|x) = p(x|\theta)$$

The likelihood of an hypothesis θ given an observation x is numerically equal to the probability of the observation x given the hypothesis θ.

4.2. The Transferable Belief Model.

4.2.1. Static and Dynamic Representations.

We are now exploring a model for representing quantified beliefs based on belief functions, the transferable belief model. No concept of randomness, or probability, is involved. We want to study the appropriate model that should be used to represent beliefs at the credal level. When randomness is not involved, there is no necessity for beliefs at the credal states (the psychological level where beliefs are entertained) to be quantified by probability measures (Levi, 1984). The coherence principle advanced by the Bayesians to justify probability measures is adequate in a context of decision (Degroot, 1970), but it cannot be used when all we want to describe is a cognitive process. Beliefs can be entertained outside any decision context. In the transferable belief model (Smets, 1988) we assume that beliefs at the credal level are quantified by belief functions (Shafer, 1976a). When decisions must be made, our belief held at the credal level induces a probability measure held at the so-called 'pignistic' level (the level at which decisions are made). This probability measure will be used in order to make decisions using expected utilities theory. But it is important to stress that this probability measure is not a representation of our belief, but is only induced from it when decision is involved.

The next example illustrate the concept we want to cover with the transferable belief model.

Example 3. Let us consider a somehow reliable witness in a murder case who testifies to You that the killer is a male. Let $\alpha = .7$ be the reliability You give to the testimony. Suppose that *a priori* You have an equal belief that the killer is a male or a female. A classical probability analysis would compute the probability P(M) of M = 'the killer is a male'. $P(M) = .85 = .7 + .5 \times .3$ (the probability that the witness is reliable (.7) plus the probability of M given the witness is not reliable (.5) weighted by the probability that the witness is not reliable (.3)). The transferable belief model analysis will give a belief .7 to M: bel(M) = .7. In $P(M) = .7 + .15$, the .7 value can be viewed as the *justified* component of the probability given to M (called the belief or the support) whereas the .15 value can be viewed as the *aleatory* component of that probability. The transferable belief model deals only with the justified components. (Note: the Evidentiary Value Model (Ekelof, 1982, Gärdenfors et al., 1983) describes the same belief component, but within a strict probability framework, and differs thus from the transferable belief model once conditioning is introduced.)

Further suppose there are only two potential male suspects: Phil and Tom. Then You learn that Phil is not the killer. The testimony now supports that the killer is Tom. The reliability .7 You gave to the testimony initially supported 'the killer is Phil or Tom'. The new information about Phil implies that .7 now supports 'the killer is Tom'. □

Hence the major component of the transferable belief model are the parts of beliefs that supports some proposition without supporting any strictly more specific propositions. These parts of beliefs are called the basic belief masses. A basic belief mass given to a set A supports also that the actual world is in every subsets that contains A. The degree of belief bel(A) for $A \in \Re$ quantifies the total amount of *justified specific support* given to A. It is obtained by summing all basic belief masses given to subsets $X \in \Re$ with $X \subseteq A$ (and $X \neq \emptyset$)

$$\text{bel}(A) = \sum_{\emptyset \neq X \subseteq A, X \in \Re} m(X)$$

We say *justified* because we include in bel(A) *only* the basic belief masses given to subsets of A. For instance, consider two distinct atoms x and y of $\Re$. The basic belief mass m({x,y}) given to {x,y} could support x if further information indicates this. However given the available information the basic belief mass can only be given to {x,y}. In example 3, the .7 was given to the set {Phil, Tom} and not split among its elements. We say *specific* because the basic belief mass $m(\emptyset)$ is not included in bel(A) as it is given to the subset $\emptyset$ that supports not only A but also $\bar{A}$.

Given a belief function bel, we can define a dual function that formalizes the concept of plausibility. The degree of plausibility pl(A) for $A \in \Re$ quantifies the maximum amount of *potential specific support* that could be given to A. It is obtained by adding all those basic belief masses given to subsets X compatible with A, i.e., such that $X \cap A \neq \emptyset$:

$$\mathrm{pl}(A) = \sum_{X \cap A \neq \emptyset, X \in \Re} m(X) = \mathrm{bel}(\Omega) - \mathrm{bel}(\bar{A})$$

We say *potential* because the basic belief masses included in pl(A) could be transferred to non-empty subsets of A if new information could justify such a transfer. It would be the case if we learn that $\bar{A}$ is impossible.

The function pl is called a plausibility function. It is in one-to-one correspondence with belief functions. It is just another way of presenting the same information and could be forgotten, except inasmuch as it often provides a mathematically convenient alternate representation of our beliefs.

If some further evidence becomes available to You and implies that B is true, then the mass m(A) initially allocated to A is transferred to $A \cap B$. This transfer of the basic belief masses characterizes the conditioning process described in the transferable belief model. It is called the Dempster's rule of conditioning, and provides the major element to describe the dynamic of beliefs.

Total ignorance is represented by a vacuous belief function, i.e. a belief function such that $m(\Omega) = 1$, hence $\mathrm{bel}(A) = 0\ \forall A \in \Re$, $A \neq \Omega$, and $\mathrm{bel}(\Omega) = 1$. The origin of this particular quantification for representing a state of total ignorance can be justified. Suppose that there are three propositions labeled A, B and C, and You are in a state of total ignorance about which is true. You only know that one and only one of them is true but even their content is unknown to You. You only know their number and their label. Then You have no reason to believe any one more than any other; hence, Your beliefs about their truth are equal: $\mathrm{bel}(A) = \mathrm{bel}(B) = \mathrm{bel}(C) = \alpha$ for some $\alpha \in [0,1]$. Furthermore, You have no reason to put more or less belief in $A \cup B$ than in C: $\mathrm{bel}(A \cup B) = \mathrm{bel}(C) = \alpha$

(and similarly $bel(A \cup C) = bel(B \cup C) = \alpha$). The vacuous belief function is the only belief function that satisfies equalities like: $bel(A \cup B) = bel(A) = bel(B) = \alpha$. Indeed the inequalities (3.1) are such that $bel(A \cup B) \geq bel(A) + bel(B) - bel(A \cap B)$. As $A \cap B = \emptyset$, $bel(A \cap B) = 0$. The inequality becomes $\alpha \geq 2\alpha$ where $\alpha \in [0,1]$, hence $\alpha = 0$. The basic belief assignment related by the vacuous belief function is called a vacuous basic belief assignment.

In general, the basic belief assignment looks similar to a probability distribution function defined on the power set 2^{Ω} of the frame of discernment Ω. This analogy led several authors to claim that the transferable belief model is nothing but a probabilistic model on 2^{Ω}. Such an interpretation does not resists once conditioning is introduced, as far as it does not lead to Dempster's rule of conditioning we derive in section 4.4 (Smets, 1992b).

Even though the conditioning process is by far the most important form of belief dynamic, other forms of belief dynamic have been developed within the transferable belief model. Most famous is the Dempster's rule of combination rules that allows us to combine conjunctively the belief functions induced by two distinct pieces of evidence. The disjunctive combination rule has been studied in Smets (1993b). Cautious rules applicable when the two pieces of evidence are not distinct are being developed. Generalization of the Bayesian theorem, so important in any inferential procedure within the Bayesian approach, has also been developed and justified in Smets (1993b).

4.2.2. Decision Making.

When a decision must be made that depends on the actual world, the agent constructs a probability function in order to make the optimal decision, i.e., the one that maximizes the expected utility (Savage, 1954, DeGroot, 1970). As far as beliefs guide our actions, the probability function is a function of the belief function bel that describes the agent's belief at the credal level. Hence one must transform bel into a probability function, denoted BetP. This transformation is called the pignistic transformation. We call BetP a pignistic probability to insist on the fact that it is a probability measure used to make decisions (Bet is for betting). Of course BetP is a classical probability measure.

The structure of the pignistic transformation is derived from and justified by the following scenario.

Example 4: Buying Your friend's drink. Suppose You have two friends, G and J. You know they will toss a fair coin and the winner will visit You tonight. You want to buy the drink Your friend would like to have tonight: coke, wine or beer. You can only buy one drink. Let D = {coke, wine, beer}.

Let $bel_G(d)$, for all $d \subseteq D$, quantifies Your belief about the drink G is liable to ask for. Given bel_G, You build the pignistic probability $BetP_G$ about the drink G will ask by applying the (still to be defined) pignistic transformation. You build in identically the same way the pignistic probability $BetP_J$ based on bel_J, Your belief about the drink J is liable to ask for. The two pignistic probability distributions $BetP_G$ and $BetP_J$ are the conditional probability distributions about the drink that will be asked for given G or J comes. The pignistic probability distributions $BetP_{GJ}$ about the drink that Your visitor will ask for is then:

$$BetP_{GJ}(d) = .5\, BetP_G(d) + .5\, BetP_J(d) \qquad \text{for all } d \in D$$

You will use these pignistic probabilities $BetP_{GJ}(d)$ to decide which drink to buy.

But You might as well reconsider the whole problem and first compute Your belief about the drink Your visitor (V) would like to have. It can be proved that:

$$bel_V(d) = .5\, bel_G(d) + .5\, bel_J(d) \qquad \text{for all } d \subseteq D$$

Given bel_V, You could then build the pignistic probability $BetP_V$ You should use to decide which drink to buy. It seems reasonable to assume that $BetP_V$ and $BetP_{GJ}$ must be equal. In such a case, the pignistic transformation is uniquely defined. □

Given a belief function defined on Ω, its pignistic transformation BetP is:

$$BetP(\omega) = \sum_{A:\omega \in A \subseteq \Omega} \frac{m(A)}{|A|(1-m(\emptyset))} \qquad \text{for } \omega \in \Omega. \tag{4.2}$$

and $$BetP(A) = \sum_{\omega \in A} BetP(\omega)$$

where $|A|$ is the number of atoms of $\Re$ in A (Smets, 1990b, Smets and Kennes, 1994).

It is easy to show that the function BetP is a probability function and the pignistic transformation of a probability function is the probability function itself.

Historical note. In a context close to ours, Shapley (1953) derived relation (4.2). Amazingly, the model he derived was called the 'transferable utility model' whereas, independently, we called our model the 'transferable belief model'.

4.2.3. The impact of the two-level model.

In order to show that the introduction of the two-level mental model based on the credal and the pignistic levels, is not innocuous, we present an example where the results will be different if one takes the two-level approach as advocated in the transferable belief model or a one-level model like in probability theory.

Example 5: The Peter, Paul and Mary Saga.
Big Boss has decided that Mr. Jones must be murdered by one of the three people present in his waiting room and whose names are Peter, Paul and Mary. Big Boss has decided that the killer on duty will be selected by a throw of a dice: if it is an even number, the killer will be female; if it is an odd number, the killer will be male. You, the judge, know that Mr. Jones has been murdered and who was in the waiting room. You know about the dice throwing, but You do not know what the outcome was and who was actually selected. You are also ignorant as to how Big Boss would have decided between Peter and Paul in the case of an odd number being observed. Given the available information at time t_0, Your odds for betting on the sex of the killer would be 1 to 1 for male versus female.

At time $t_1 > t_0$, You learn that if Big Boss had not selected Peter, then Peter would necessarily have gone to the police station at the time of the killing in order to have a perfect alibi. Peter indeed went to the police station, so he is not the killer. The question is how You would bet now on male versus female: should Your odds be 1 to 1 (as in the transferable belief model) or 1 to 2 (as in the most natural Bayesian model).

Note that the alibi evidence makes 'Peter is not the killer' and 'Peter has a perfect alibi' equivalent. The more classical evidence 'Peter has a perfect alibi' would only imply P('Peter is not the killer' | 'Peter has a perfect alibi') = 1. But P('Peter has a perfect alibi' | 'Peter is not the killer') would stay undefined and would then give rise to further discussion, which would be useless for our purpose. In this presentation, the latter probability is also 1.

The transferable belief model solution.
Let k be the killer. The information about the waiting room and the dice throwing pattern induces the following basic belief assignment m_0:

$k \in \Omega = \{\text{Peter, Paul, Mary}\}$

$m_0(\{\text{Mary}\}) = .5 \qquad m_0(\{\text{Peter, Paul}\}) = .5$

The .5 belief mass given to {Peter, Paul} corresponds to that part of belief that supports "Peter or Paul", could possibly support each of them, but given the lack of further information, cannot be divided more specifically between Peter and Paul.

Let $BetP_0$ be the pignistic probability obtained by applying the pignistic transformation to m_0 on the betting frame which set of atoms is {{Peter], {Paul}, {Mary}}. By relation (4.2), we get:

$BetP_0(\{\text{Peter}\}) = .25 \qquad BetP_0(\{\text{Paul}\}) = .25 \qquad BetP_0(\{\text{Mary}\}) = .50$

Given the information available at time t_0, the bet on the killer's sex (male versus female) is held at odds 1 to 1.

Peter's alibi induces an updating of m_0 into m_2 be Dempster's rule of conditioning:

$$m_2(\{Mary\}) = m_2(\{Paul\}) = .5$$

The basic belief mass that was given to "Peter or Paul" is transferred to Paul.

Let $BetP_2$ be the pignistic probability obtained by applying the pignistic transformation to m_2 on the betting frame which set of atoms is {{Paul}, {Mary}}.

$$BetP_2(\{Paul\}) = .50 \qquad BetP_2(\{Mary\}) = .50$$

Your odds for betting on Male versus Female would still be 1 to 1.

The probabilistic solution:

The probabilistic solution is not obvious as one data is missing: the value α for the probability that Big Boss selects Peter if he must select a male killer. Any value could be accepted for α, but given the total ignorance in which we are about this value, let us assume that α =.5, the most natural solution. Then the odds on male versus female before learning about Peter's alibi is 1 to 1, and after learning about Peter's alibi, it becomes 2 to 1. The probabilities are then:

$$P_2(\{Paul\}) = .33 \qquad P_2(\{Mary\}) = .66$$

The 1 to 1 odds of the transferable belief model solution can only be obtained in a probabilistic approach if $\alpha = 0$. Some critics would claim that the transferable belief model solution is valid as it fits with $\alpha = 0$. The only trouble with this answer is that if the alibi story had applied to Paul, than we would still bet at 1 to 1 odds. Instead the probabilistic solution with $\alpha = 0$ would lead to a 0 to 1 bet, as the probabilities are:

$$P_2(\{Peter\}) = .0 \quad P_2(\{Mary\}) = 1.$$

So the classical probabilistic analysis does not lead to the transferable belief model solution.

We are facing two solutions for the bet on male versus female after learning about Peter's alibi: the 1 to 1 or at 1 to 2 odds? Which solution is 'good' is not decidable, as it would require the definition of 'good'. Computer simulations have been suggested for solving the dilemma, but they are impossible. Indeed for every simulation where the killer must be a male, one must select Peter or Paul, and as far as simulation are always finite, the proportion of cases when Peter was selected in the simulations when a male has to be selected will be well defined. The value of the two solutions under comparison will only reflect the difference between that proportion and the missing probability α. Such a comparison is irrelevant for what we look for. We are only left over with a subjective comparison of the two solution... or an in depth comparison of the theoretical foundations that led to these solutions, an alternative that explains why we try to develop a full axiomatization of the transferable belief model (Smets, 1993c)

Example 6: The Five Breakable Sensors.
To show that the choice between a Bayesian and a transferable belief model approach is important in practice, we analyze the story of the five breakable sensors where the two approaches completely disagree in their conclusions, and would led to acting differently. These examples might help the reader in deciding which approach is better.

Suppose I must check the temperature of a process. To do it I have five sensors, and I have the same information for each of them.

Each sensor can check the temperature of the process. The temperature can only be hot or cold. If the temperature is hot (TH), the sensor light is red (R) and if the temperature is cold (TC), the sensor light is blue (B). Each sensor is made of a thermometer and a device that turns the blue or the red light on according to the temperature reading. Unfortunately the thermometer can be broken.

The only known information is what is written on the box containing each sensor. "Warning: the thermometer included in this sensor can be broken. The probability that it is broken is ...%. (a number is written there that depends on the box). When the thermometer is not broken, the sensor is a perfectly reliable detector of the temperature. When the thermometer is not broken, red light means the temperature is hot, blue light means that the temperature is cold. When the thermometer is broken, the sensor answer is unrelated to the temperature".

I am a new technician and I never saw the five sensors before. On box 1, the probability is 1% that the thermometer is broken, on the boxes 2 to 5, the probability is 35%. I know nothing about these sensors except the warnings written on the boxes. I use them and the red light gets on for the box 1 sensor, and the blue light gets on for boxes 2 to 5 sensors. How do I make up my mind about the temperature status? What is my opinion (belief) that the temperature status is hot or cold? Let us assume that the consequences (utilities) of the good and bad decisions are symmetrical: I am either right or wrong. We can thus avoid discussions about possible consequences. The problem is: does the agent belief that the temperature is hot increases or decreases given the five observations. With the transferable belief model, the probability increases, with the Bayesian analysis it decreases. Bayesians could conclude that the temperature is cold. The transferable belief model could concludes that the temperature is hot. To decide which is more natural is left to the reader!

The lesson of the examples 5 and 6 is that the choice of the model is important and interferes with real decisions.

5. Conclusions.

No definitive conclusions can be taken from our presentation, except that uncertainty can take several forms, and that the choice of the appropriate model is a necessity. Any claim like 'my model is the best and the only valid model for representing quantified belief' is just nonsensical. It characterized a dogmatic approach without any relation with a scientific attitude.

We did not tackle the problem of assessing the values of the belief. This problem is well solved in probability theory, thanks to the exchangeable bets schema. The same solution is used within the transferable belief model. For upper and lower probability models, the assessment is not so well defined as forced symmetrical bets are not available. The assessment problem is not acute in ordinal possibility theory, i.e., the one where only the ordering is relevant, not the value themselves. In cardinal possibility theory, the problem is not yet resolved.

The problem of finding which is the appropriate model for which model is hardly solved, and justifies further researches. We hope we have provided some tools in that direction by presenting the various models that have been so far proposed for the quantified representation of beliefs.

Bibliography.

CHEESEMAN P. (1985) In defense of probability. IJCAI-85, Morgan Kaufman, San Mateo, Ca, 1002-1009.

CHELLAS B.F. (1980) Modal logic. Cambridge Univ. Press, G.B.

CHOQUET G. (1953) Theory of capacities. Annales de l'Institut Fourier, Université de Grenoble, 5 131-296.

COHEN L. J., (1993) What has probability to do with strengh of belief. in DUBUCS J.P. (ed.) Philosophy of Probability, Kluwer, Dordrecht, pg. 129-143.

COX R.T. (1946) Probability, frequency and reasonable expectation. Amer.J.Phys. 14:1-13.

DEGROOT M.H. (1970) Optimal statistical decisions. McGraw-Hill, New York.

DEMPSTER A.P. (1967) Upper and lower probabilities induced by a multplevalued mapping. Ann. Math. Statistics 38: 325-339.

DEMPSTER A.P. A generalization of Bayesian inference. J. Roy. Statist. Soc. B.30 (1968) 205-247.

D'ESPAGNAT B. (1976) Conceptual foundations of quantum mechanics. W.A. Benjamin, Reading, Mass.

DUBOIS D., GARBOLINO P., KYBURG H.E., PRADE H. and SMETS P. (1991) Quantified Uncertainty. J. Applied Non-Classical Logics 1:105-197.

DUBOIS D. and PRADE H. (1985) *Theorie des possibilités.* Masson, Paris.

EKELOF P.O. (1982) Rättegång IV. Fifth edition, Stockholm.

FAGIN R. and HALPERN J. (1990) A new approach to updating beliefs. 6th Conf. on Uncertainty in AI.

FINE T. (1973) Theories of probability. Academic Press, New York.

GÄRDENFORS P., HANSSON B. and SAHLIN N.E. (1983) Evidentiary value: philosophical, judicial and psychological aspects of a theory. C.W.K. Gleerups, Lund.

GÄRDENFORS P. (1988) Knowledge in flux. Modelling the dynamics of epistemic states. MIT Press, Cambridge, Mass.

GOOD I.J. (1950) Probability and the weighting of evidence. Hafner.

GOOD I.J. (1983) Good thinking: the foundations of probability and its applications. Univ. Minnesota Press, Minneapolis.

HACKING I. (1965) Logic of statistical inference. Cambridge University Press, Cambridge, U.K.

HACKING I.(1975) The emergence of probability. Cambridge University Press, Cambridge.

HINTIKKA J. (1962) Knowledge and belief. Cornell Univ. Press, Ithaca, NY.

JAFFRAY J.Y. (1989) Coherent bets under partially resolving uncertainty and belief functions. Theory and Decision 26:99-105.

JAFFRAY J.Y. (1992) Bayesian updating and belief functions. IEEE Trans. SMC, 22: 1144-1152.

KOHLAS J. and MONNEY P.A (1994a) Representation of Evidence by Hints. in Advances in the Dempster-Shafer Theory of Evidence. Yager R.R., Kacprzyk J. and Fedrizzi M., eds, Wiley, New York, pg. 473-492.

KOHLAS J. and MONNEY P.A. (1994b) Theory of Evidence. A Survey of its Mathematical Foundations, Applications and Computations. ZOR-Mathematical Methods of Operatioanl Research 39:35-68.

KOOPMAN B.O. (1940) The bases of probability. Bull. Amer. Math. Soc. 46:763-774.

KYBURG H. (1987) Objective probabilities. IJCAI-87, 902-904.

LEVI I. (1984) Decisions and revisions: philosophical essays on knowledge and value. Cambridge University Press, Cambridge.

PEARL J. (1988) Probabilistic reasoning in intelligent systems: networks of plausible inference. Morgan Kaufmann Pub. San Mateo, Ca, USA.

RAMSEY F.P. (1931) Truth and probability. in Studies in subjective probability, eds. KYBURG H.E. and SMOKLER H.E., p. 61-92. Wiley, New York.

RUSPINI E.H. (1986) The logical foundations of evidential reasoning. Technical note 408, SRI International, Menlo Park, Ca.

SAHLIN N. E. (1993) On higher order beliefs. in DUBUCS J.P. (ed.) Philosophy of Probability, Kluwer, Dordrecht, pg. 13-34.

SAVAGE L.J. (1954) Foundations of Statistics. Wiley, New York.

SHAFER G. (1976a) A mathematical theory of evidence. Princeton Univ. Press. Princeton, NJ.

SHAFER G. (1976b) A theory of statistical evidence. in Foundations of probability theory, statistical inference, and statistical theories of science. Harper and Hooker ed. Reidel, Doordrecht-Holland.

SHAFER G. (1978) Nonadditive probabilities in the work of Bernoulli and Lambert. Arch. History Exact Sci. 19:309-370.

SHAPLEY L.S. (1953) A value for n-person games. In Contributions to the Theory of Games, vol. 2, eds. H. Kuhn and A.W. Tucker. Princeton University Press, pp. 307-317.

SHORTLIFFE E.H. and BUCHANAN B.G. (1975) A model of inexact reasoning in medicine. Math. Biosci. 23:351-379.

SMETS P. (1978) Un modèle mathématico-statistique simulant le processus du diagnostic médical. Doctoral dissertation, Université Libre de Bruxelles, Bruxelles, (Available through University Microfilm International, 30-32 Mortimer Street, London W1N 7RA, thesis 80-70,003)

SMETS P. (1987) Upper and lower probability functions versus belief functions. Proc. International Symposium on Fuzzy Systems and Knowledge Engineering, Guangzhou, China, July 10-16, pg 17-21.

SMETS P. (1988) Belief functions. in SMETS Ph, MAMDANI A., DUBOIS D. and PRADE H. ed. Non standard logics for automated reasoning. Academic Press, London pg 253-286.

SMETS P. (1990a) The combination of evidence in the transferable belief model. IEEE-Pattern analysis and Machine Intelligence, 12:447-458.

SMETS P. (1990b) Constructing the pignistic probability function in a context of uncertainty. Uncertainty in Artificial Intelligence 5, Henrion M., Shachter R.D., Kanal L.N. and Lemmer J.F. eds, North Holland, Amsterdam, , 29-40.

SMETS P. (1991a) Varieties of ignorance. Information Sciences. 57-58:135-144.

SMETS P. (1991b) About updating. in D'Ambrosio B., Smets P., and Bonissone P.P. eds, Uncertainty in AI 91, Morgan Kaufmann, San Mateo, Ca, USA, 1991, 378-385.

SMETS P. (1991c) Probability of provability and belief functions. Logique et Analyse, 133-134:177-195.

SMETS P. (1992a) The nature of the unnormalized beliefs encountered in the transferable belief model. in Dubois D., Wellman M.P., d'Ambrosio B. and Smets P. Uncertainty in AI 92. Morgan Kaufmann, San Mateo, Ca, USA, 1992, pg.292-297.

SMETS P. (1992b) The transferable belief model and random sets. Int. J. Intell. Systems 7:37-46.

SMETS P. (1993a) Imperfect information: imprecision - uncertainty. in Smets P. and Motro A. eds. Uncertainty Management in information systems: from needs to solutions. UMIS Workshop, Puerto Andraix.Volume 2. pg. 1-28.

SMETS P. (1993b) Belief functions: the disjunctive rule of combination and the generalized Bayesian theorem. Int. J. Approximate Reasoning 9:1-35.

SMETS P. (1993c) An axiomatic justifiaction for the use of belief function to quantify beliefs. IJCAI'93 (Inter. Joint Conf. on AI), San Mateo, Ca, pg. 598-603.

SMETS P. (1994a) The representation of quantified belief by belief functions: an axiomatic justificiation. TR/IRIDIA/94.3. Submitted for publication.

SMETS P. (1994b) What is Dempster-Shafer's model? in Advances in the Dempster-Shafer Theory of Evidence. Yager R.R., Kacprzyk J. and Fedrizzi M., eds, Wiley, New York, pg. 5-34.

SMETS P. and KENNES R. (1994) The transferable belief model. Artificial Intelligence 66:191-234.

SMITH C.A.B. (1961) Consistency in statistical inference and decision. J. Roy. Statist. Soc. B23:1-37.

SMITH C.A.B. (1965) Personal probability and statistical analysis. J. Roy. Statist. Soc. A128, 469-499.

SMITH P. and JONES O.R. (1986) The philosophy of mind, an introduction. Cambridge University Press, Cambridge.

SMITHSON M. (1989) Ignorance and Uncertainty: Emerging Paradigms. Springer-Verlag, New York.

VOORBRAAK F. (1993) As Far as I Know: Epistemic Logic and Uncerttaiunty. Dissertation, Utrecht University.

WALLEY P. (1991) Statistical reasoning with imprecise probabilities. Chapman and Hall, London.

WILLIAMS P.M. (1976) Indetermiinate probabilities. in Formal methods and the methodology of empirical sciences, Przlecki M., Szaniawski K. and Wojciki R. (eds.) Reidel, Dordrecht.

WONG S.K.M., YAO Y.Y., BOLLMANN P. and BÜRGER H.C. (1990) Axiomatization of qualitative belief structure. IEEE Trans. SMC, 21:726-734.

ZADEH L.A. (1965) Fuzzy sets. Inform.Control. 8:338-353.

ZADEH L. (1978) Fuzzy sets as a basis for a theory of possibility. Fuzzy Sets and Systems 1: 3-28.

2. THEORY OF EVIDENCE

Argumentation Systems and Evidence Theory

Jürg Kohlas[1] and Hans Wolfgang Brachinger[2]

[1] Institute of Informatics, University of Fribourg, Switzerland. Research supported by the Swiss National Foundation for Research, grant nr. 21-32660.91 and the Swiss Federal Office for Science and Education, Esprit BRA DRUMS2, Defeasible Reasoning and Uncertainty Management

[2] Seminar of Statistics, University of Fribourg, Switzerland

Abstract. The Dempster-Shafer theory of evidence is developed in a very general setting. Its algebraic part is discussed as a body of arguments which contains an allocation of support and an allowment of possibility for each hypothesis. A rule of combination of bodies of arguments is defined which constitutes the symbolic counterpart of Dempster's rule. Bodies of evidence are introduced by assigning probabilities to arguments. This leads to support and plausibility functions on hypotheses, which constitute the numerical part of evidence theory. Combination of evidence based on the combination of bodies of arguments is discussed and Dempster's rule is derived.

1 Introduction

Evidence theory has been introduced by G. Shafer (1976) essentially as a theory of belief functions. Another axiomatic approach to belief functions has been proposed by Smets (1988). Evidence theory and belief functions have also been equivalently defined on the base of probabilistic models, originally by Dempster (1967) using multivalued mappings, by Nguyen (1978) based on random sets and by Kohlas (1993a; Kohlas, Monney 1994) as a theory of hints.

On the other hand, Laskey, Lehner (1988), Pearl (1988), Provan (1990) and Kohlas (1991) showed that evidence theory can also be seen as a theory of the probability of provability of hypotheses by uncertain assumptions or arguments. Kohlas (1993b) finally pointed out, that there is a qualitative version of evidence theory based on the latter approach, which uses no probabilities nor other numerical degrees of uncertainty. This last point of view is the starting point of this paper. First, a general *qualitative* theory of evidence will be developed. Only in a second step probability measures will be introduced, leading to a *numerical* theory of evidence. Thereby general frames will be covered, not only finite ones as in most papers so far. In this respect, this paper is a continuation of the first attempt of a general theory of evidence by Shafer (1979) and of the subsequent work on evidence theory on general frames (Nguyen, 1978; Hestir et al. 1991; Kohlas 1993a; Kohlas, Monney 1994). In this paper proofs are omitted. For reference see Kohlas, Brachinger (1993).

2 Bodies of Arguments

2.1 Allocation of Support

The theory of evidence is concerned with the representation of both incomplete and uncertain information relative to certain questions. Hypotheses can be formulated about the possible answers to these questions. These hypotheses are the first elements one is interested in. The knowledge available may permit to construct arguments in favour of and against those hypotheses. But in general there is not enough information to assure these arguments. They are only to be considered as possible and the possibility that certain arguments may prove not to be valid is not excluded. That is why arguments in favour of and arguments against a hypothesis may be considered simultaneously. These arguments are the second elements of the theory to be developed.

In talking about hypotheses it is desirable to be able to talk about the negation of a hypothesis. The negation of a hypothesis should also be a valuable hypothesis. In the same vein, if several hypotheses are considered, it should be possible to consider the question if all or at least one of these hypotheses be true. Therefore, it is sensible to assume that these hypotheses form a Boolean algebra, denoted by $\mathcal{B}$. By the same consideration, if a is an argument, then its negation should also be an argument (if the argument does not hold, then its negation holds necessarily), also the conjunction and disjunction of two or several arguments should be an argument. So, arguments should also form a Boolean algebra, denoted by $\mathcal{A}$. Conjunction (meets), disjunction (joins) and negation (complement) will be denoted by $\vee, \wedge$ and ^{c} respectively, except in cases of Boolean fields of subsets, where the usual symbols $\cap$ and $\cup$ for intersection and union are used. The zero element is denoted by Λ (the empty set by $\emptyset$ in the case of fields) and the one element (the full set) by V. The partial order relation in the Boolean algebra will be denoted by $\leq$ or $\geq$ ($\subseteq$ and $\supseteq$ in the case of fields). The reader is refered to the books of Sikorski (1960) and Halmos (1963) for accounts on the theory of Boolean algebras.

It is assumed that knowledge is available which allows to allocate to every hypothesis some arguments in the sense that these arguments imply the hypothesis they are allocated to. Thereby implication means that, given some knowledge, the hypothesis is true if one of the arguments allocated is true. More precisely assume that, given some knowledge, for any hypothesis $b \in \mathcal{B}$ there is a least specific, largest argument $\rho(b) \in \mathcal{A}$ which implies b, such that all $a \leq \rho(b)$ also imply b, but no other $a \in \mathcal{A}$ implies b. We impose the following conditions on ρ:

$$\rho(V) = V \tag{1}$$

$$\rho(\Lambda) = \Lambda \tag{1'}$$

$$\rho(b_1 \wedge b_2) = \rho(b_1) \wedge \rho(b_2) \tag{2}$$

A mapping $\rho : \mathcal{B} \rightarrow \mathcal{A}$ which satisfies conditions (1) and (2) is called an ***allocation of support*** and the triple $(\mathcal{A}, \mathcal{B}, \rho)$ a ***body of arguments***. An allocation of support which additionally satisfies condition (1') and the corresponding body of arguments will be called ***normalized***. The intuitive justification of conditions (1) through (2) is as follows:

(1) The sure hypothesis V should be implied by every argument because this hypothesis is true anyhow.

(1') The impossible hypothesis Λ should be implied only by the impossible argument Λ, because this argument, according to the conventions of logic, implies everything. However, condition (1') will not always be required. That is, we may well admit that there are arguments different from Λ which imply Λ, $\rho(\Lambda) > \Lambda$. In this case ρ will be called *nonnormalized*. Arguments implying the impossible will be called *contradictions*. A nonnormalized body of arguments $(\mathcal{A}, \mathcal{B}, \rho)$ can always be *normalized* by replacing $\mathcal{A}$ by the algebra $\mathcal{A}' = \mathcal{A} \wedge \rho^c(\Lambda)$ which is the algebra of all elements $a' = a \wedge \rho^c(\Lambda)$ with $a \in \mathcal{A}$. The mapping $\rho'(b) = \rho(b) \wedge \rho^c(\Lambda)$ from $\mathcal{B}$ into $\mathcal{A}'$ is then a normalized allocation of support, as is easily seen.

(2) Any argument for $b_1 \wedge b_2$ should also be an argument for both, b_1 and b_2. Inversely, any argument which implies both b_1 and b_2 should also imply $b_1 \wedge b_2$.

Axiom (2) implies monotonicity

$$b_1 \leq b_2 \Longrightarrow \rho(b_1) \leq \rho(b_2) \tag{3}$$

and superadditivity

$$\rho(b_1 \vee b_2) \geq \rho(b_1) \vee \rho(b_2) \qquad (4),$$

properties which are in fact equivalent.

These properties of ρ as well as their equivalence are easy to prove. Monotonicity means that if a hypothesis b_1 is contained in another one b_2 then all arguments implying b_1 also imply b_2. Superadditivity signifies that any argument for b_1 or for b_2 is also an argument for $b_1 \vee b_2$. However there may be arguments for $b_1 \vee b_2$ which imply neither b_1 nor b_2.

An allocation of support describes which arguments imply, hence support a hypothesis. If some argument a proves to be true, then clearly all hypotheses b with $a \leq \rho(b)$ are necessarily proved to be true. However, in general nothing is known about which arguments are true or not. So, a body of arguments permits only hypothetical reasoning, that is the determination of arguments in favour of or against a hypothesis.

2.2 Allowment of Possibility

The arguments against a hypothesis b are the arguments which support the complement of b and are given by $\rho(b^c)$. The complement of this argument represents the arguments under which the hypothesis b is not excluded, remains possible, even if not necessarily true. So the mapping

$$\xi : \mathcal{B} \rightarrow \mathcal{A},\ \xi(b) := \rho^c(b^c)$$

is called an *allowment of possibility*. Arguments a such that $a \wedge \xi(b) \neq \Lambda$, or, equivalently, $a \not\leq \rho(b^c)$ may be called *nondominated* by $\rho(b^c)$. Whenever an argument a which is nondominated by $\rho(b^c)$ is true then b is possible. By DeMorgan's laws, the following properties of ξ are easily derived form conditions (1) and (2) for an allocation of support:

$$\xi(\Lambda) = \Lambda \quad (1)$$
$$\xi(V) = V, \quad (1')$$

$$\xi(b_1 \vee b_2) = \xi(b_1) \vee \xi(b_2) \quad (2).$$

It is easy to show that property (2) implies

$$\xi(b_1 \wedge b_2) \leq \xi(b_1) \wedge \xi(b_2) \quad (3)$$

$$b_1 \leq b_2 \Longrightarrow \xi(b_1) \leq \xi(b_2) \quad (4).$$

A body of arguments may also equivalently be defined by means of an allowment of possibility ξ instead of an allocation of support, as $\rho(b) := \xi^c(b^c)$.

There are a number of important and interesting special cases of bodies of arguments, some of which shall be mentioned here:

(i) If, in addition to conditions (1) and (2) of an allocation of support, one imposes

$$\rho(b_1 \vee b_2) = \rho(b_1) \vee \rho(b_2)$$

then the allocation is called *additive*. In this case it follows easily that $\rho(b) = \xi(b)$ for all b.

(ii) If for all $b \in B$ $\rho(b) > \Lambda$ implies $\xi(b) = V$ or, equivalently, $\xi(b) < V$ implies $\rho(b) = \Lambda$, then the body of arguments is called *consonant*. The arguments are in this case consistent in the sense that if a hypothesis is supported by some argument, then its complement has no support. There are no arguments pointing into contradictory directions.

(iii) Select a $b' \in B$, an a in $\mathcal{A}$ and define $\rho(b) = a$ if $b' \leq b \leq V$, $\rho(V) = V$ and $\rho(b) = \Lambda$ in all other cases. This is clearly an allocation of support, it is called a *simple support* pointing to b'. If $a = V$, then it is a *deterministic* support of b'. In this case the body of arguments states that b' is necessarily true. If $b' = V$, then it is a *vacuous* body of arguments, which supports in fact no hypothesis (except the sure one).

A number of ways how bodies of arguments may arise in applications is sketched in Kohlas, Brachinger (1993).

2.3 Combination of Bodies of Arguments

Suppose that two bodies of arguments $(\mathcal{A}, \mathcal{B}, \rho_1)$ and $(\mathcal{A}, \mathcal{B}, \rho_2)$ are available. How can these two bodies be integrated and synthesized into a new, single body, enclosing the information of the two original bodies. This is the question discussed in this section.

If a_1 is an argument in the first body of arguments which implies b_1, and a_2 an argument in the second body which implies b_2, then $a_1 \wedge a_2$ surely implies $b_1 \wedge b_2$. Therefore, if $b_1 \wedge b_2 \leq b$, the combined argument $a_1 \wedge a_2$ from the two bodies should imply b. Thus, if $\rho(b)$ is the allocation of support of the combined bodies of arguments, then

$$\rho(b) \geq \rho_1(b_1) \wedge \rho_2(b_2)$$

should hold. It seems reasonable to define $\rho(b)$ as the smallest element of $\mathcal{A}$ dominating $\rho_1(b_1) \wedge \rho_2(b_2)$ whenever $b_1 \wedge b_2 \leq b$. This leads to the following tentative definition

$$\rho(b) = \vee \{\rho_1(b_1) \wedge \rho_2(b_2) : b_1 \wedge b_2 \leq b\} \qquad (2.1).$$

This supremum does not necessarily exist. Yet this definition should not be dismissed. Rather $\mathcal{A}$ should be extended to a Boolean algebra containing these suprema. Fortunately it turns out that there is a unique minimal completion $\mathcal{A}^*$ of $\mathcal{A}$, that is a complete Boolean algebra $\mathcal{A}^*$ containing $\mathcal{A}$ and contained in every other complete Boolean algebra containing $\mathcal{A}$ (loosely speaking, see Halmos, 1963 for a more precise formulation). A complete Boolean algebra is one which contains all the suprema over arbitrary families of its arguments. Thus (2.1) is an element of $\mathcal{A}^*$. Therefore we define the combined body of arguments to be $(\mathcal{A}^*, \mathcal{B}, \rho)$, with ρ defined by (2.1). In order that this definition makes sense it must be verified that ρ is indeed an allocation of support.

There is the possibility, that $\rho_1(b_1) \wedge \rho_2(b_2) = \Lambda$ for all b_1 and b_2. If this is the case, the two bodies of arguments are said to be not combinable. They are in fact completely contradictory in this case.

Theorem 2.1. If the two bodies of arguments are combinable, then (2.1) defines an allocation of support $\rho : \mathcal{A}^* \rightarrow \mathcal{B}$ (in general nonnormalized).

There is a similar formula for the allowment of possibility:

Theorem 2.2 The allowment of possibility ξ associated with the combined allocation of support is given by

$$\xi(b) = \wedge \{\xi_1(b_1) \vee \xi_2(b_2) : b_1 \vee b_2 \geq b\} \qquad (2.2)$$

where ξ_1 and ξ_2 are the allowments of possibility associated with ρ_1 and ρ_2 respectively.

Depending on how bodies of arguments are obtained specific ways to define their combination may be considered. It is shown in Kohlas/Brachinger (1993) that at least in the most interesting cases these ways lead to formula (2.1). This constitutes a further justification of this definition of combination.

3 Bodies of Evidence

3.1 Support and Plausibility Functions

A body of arguments $(\mathcal{A}, B, \rho)$ in itself permits not yet to establish which hypotheses are true, because it is unknown which arguments in $\mathcal{A}$ hold. If it would become known, that a certain argument $a \in \mathcal{A}$ is true, then clearly all $b \in \mathcal{B}$ with $a \leq \rho(b)$ must necessarily also be true. As an intermediate stage between knowing nothing about which arguments are true and knowing that a certain specified argument is true, one might assume that a probability measure indicating the likelihood at least of some arguments to be true, is given. The combination of a *body of arguments* together with a *probability measure* in $\mathcal{A}$ will be called a *body of evidence.*

More precisely let $\mathcal{A}_0$ be a subalgebra of $\mathcal{A}$ which is assumed to be a σ-algebra, that is suprema and infima of countable families of elements of $\mathcal{A}_0$ are also in $\mathcal{A}_0$. Let μ be a probability measure on $\mathcal{A}_0$ (a normalized measure in the terminology of Halmos, (1963). With these elements one can measure how likely it is that b can be derived (proved) using the body of arguments. At least this is the case for some $b \in \mathcal{B}$. Define

$$sp(b) = \mu(\rho(b)) \tag{3.1}$$

if $\rho(b) \in \mathcal{A}_0$. $sp(b)$ is called the *degree of support* of b (relative to a body of evidence). It measures the degree to which b is supported by the body of evidence. Similarly, one can measure how likely it is that b is not excluded by the body of arguments. Define

$$pl(b) = \mu(\xi(b)) \tag{3.2}$$

if $\xi(b) \in \mathcal{A}_0$. $pl(b)$ is called the *degree of plausibility* of b (relative to a body of evidence). It measures to what degree b is plausible in the light of the body of evidence.

It seems that the degrees of support and of plausibility are not defined for all hypotheses $b \in \mathcal{B}$. In the next section it will however be seen how this deficiency can be removed. For the time being let $\mathcal{E}_s = \{b \in \mathcal{B} : \rho(b) \in \mathcal{A}_0\}$ and $\mathcal{E}_p = \{b \in \mathcal{B} : \xi(b) \in \mathcal{A}_0\}$. $\mathcal{E}_s$ and $\mathcal{E}_p$ are called the sets of *s- and p-measurable hypotheses.* It follows from property (2) of ρ or ξ respectively that $\mathcal{E}_s$ is a multiplicative class (a class closed under intersection), $\mathcal{E}_p$ an additive class (a class closed under unions). V belongs always to $\mathcal{E}_s$ and so does Λ, if ρ is normalized; on the other hand Λ belongs always to $\mathcal{E}_p$, but V only if ρ is normalized. sp and pl can be considered as functions from $\mathcal{E}_s$ and $\mathcal{E}_p$ to $[0, 1]$ and as such are called *support* and *plausibility* functions.

If ρ is not normalized, then $\rho(\Lambda)$ represents *contradictory* arguments. It can be argued, that those arguments cannot be true, because Λ is known not to be true. This implies that these arguments can be removed from $\mathcal{A}$, i.e. the *normalized* body of arguments $(\mathcal{A} \wedge \rho^c(\Lambda), \mathcal{B}, \rho \wedge \rho^c(\Lambda))$ can be considered (see section 2.1). At the same time, the probability measure μ can also be conditioned on the information $\rho^c(\Lambda)$: $\mathcal{A}_0 \wedge \rho^c(\Lambda)$ is still a σ-algebra and $\mu'(a) = \mu(a)/\mu^*(\rho^c(\Lambda))$ is a probability measure on $\mathcal{A}_0 \wedge \rho^c(\Lambda)$, if μ^* is the outer measure of μ (Neveu, 1964). In this way any body of evidence can be transformed into a *normalized* one by conditioning on the possible arguments. The corresponding support and plausibility functions are sometimes called *proper support and plausibility functions.*

It is however convenient to continue to consider general, possibly nonnormalized bodies of evidence. The following theorem summarizes the main properties of support and plausibility functions.

Theorem 3.1 Support functions $\mathcal{E}_s \to [0,1]$ *and plausibility functions* $\mathcal{E}_p \to [0,1]$ *have the following properties*

$$sp(V) = 1, pl(\Lambda) = 0 \quad (1)$$

$$sp(\Lambda) = 0, pl(V) = 1 \quad \textit{in the case of normalized bodies,} \quad (1')$$

$$sp(b) = 1 - pl(b^c), \; pl(b) = 1 - sp(b^c) \quad (2)$$

$$sp(b) \leq pl(b) \quad \textit{in the case of normalized bodies,} \quad (2')$$

$$\textit{monotonicity: if } \; b_1, b_2, \ldots, b_n \leq b, \; \textit{all in } \; \mathcal{E}_s, n \geq 1, \; \textit{then} \quad (3)$$

$$sp(b) \geq \sum \{(-1)^{|I|+1} sp(\wedge_{i \in I} b_i) : \emptyset \neq I \subseteq \{1, \ldots, n\}\} \quad (3.3),$$

and if $b_1, b_2, \ldots, b_n \geq b$, *all in* $\mathcal{E}_p$, $n \geq 1$, *then*

$$pl(b) \leq \sum \{(-1)^{|I|+1} pl(\vee_{i \in I} b_i) : \emptyset \neq I \subseteq \{1, \ldots, n\}\} \quad (3.4)$$

The question arises whether any function sp from a multiplicative class $\mathcal{E}_s$ in a Boolean algebra to $[0,1]$, which satisfies properties (1) and (3) is generated by a body of evidence. In case $\mathcal{E}_s$ is a multiplicative class of subsets of some set, it has been shown that sp is generated by a generalized hint (Kohlas, 1993a) and b), see also Shafer, 1979).

3.2 Allocation and Allowment of Probability

It is unsatisfactory that only certain hypotheses in $\mathcal{B}$ have a degree of support or of plausibility, depending on ρ and especially on the σ-algebra $\mathcal{A}_0$. It will be shown in this section that this is only superficially so. In fact, it is possible to associate with every b in $\mathcal{B}$ very naturally a degree of support and of plausibility.

Let $\mathcal{I}_0$ be the σ-ideal of the μ-null sets in $\mathcal{A}_0$ and $\mathcal{M}_0 = \mathcal{A}_0/\mathcal{I}_0$ the quotient algebra of all equivalence classes ($a' \sim a''$ iff $a' - a'' \in \mathcal{I}_0$ and $a'' - a' \in \mathcal{I}_0$). If $[a]$ denotes the equivalence class of the element a, then $[a]^c = [a^c]$ and

$$\vee \{[a_i] : i = 1, \ldots\} = [\vee \{a_i : i = 1, \ldots\}],$$

$$\wedge \{[a_i] : i = 1, \ldots\} = [\wedge \{a_i : i = 1, \ldots\}].$$

$h(a) = [a]$ is a σ-homomorphism from $\mathcal{A}_0$ to $\mathcal{M}_0$, called the projection. Furthermore by $\nu([a]) = \mu(a)$ a *normalized, positive* measure is defined on $\mathcal{M}_0$ (that is $\nu([a]) > 0$ if $[a] \neq \Lambda$). Thus, $(\mathcal{M}_0, \nu)$ is a *measure algebra* (Halmos, 1963) and therefore $\mathcal{M}_0$ is a *complete* algebra, which satisfies the so-called countable chain condition. That is, every family of disjoint elements of $\mathcal{M}_0$ is countable.

Define for every $a \in \mathcal{A}$

$$\rho_0(a) = \vee \{[a'] : a' \leq a, a' \in \mathcal{A}_0\}.$$

This is a well defined element of $\mathcal{M}_0$ because of the completeness of the latter.
Theorem 3.2 $\rho_0 : \mathcal{A} \to \mathcal{M}_0$ *is a normalized allocation of support.*

Now, the allocations of support $\rho : \mathcal{B} \to \mathcal{A}$ and $\rho_0 : \mathcal{A} \to \mathcal{M}_0$ can be chained to $\pi = \rho_0 \circ \rho : \mathcal{B} \to \mathcal{M}_0$. π is clearly still an allocation of support and it allocates to every hypothesis b in $\mathcal{B}$ an element of the *probability algebra* $\mathcal{M}_0$. Such an allocation of support in a measure algebra will be called an *allocation of probability* (Shafer, 1979). The allocation of probability π is normalized iff the allocation of support ρ is normalized.

There is of course a dual definition for the allowment of possibility. $\eta = \xi_0 \circ \xi : \mathcal{B} \to \mathcal{M}_0$ is the dual allowment of possibility associated with π, if ξ_0 is the allowment of possibility corresponding to ρ_0. An allowment of possibility in a measure algebra will be called *allowment of probability* (Shafer, 1979).

It becomes now possible, to define the degree of support and the degree of plausibility for *every* hypothesis b in $\mathcal{B}$:

$$sp_e(b) = \nu(\pi(b)), \; pl_e(b) = \nu(\eta(b))(3.5).$$

For every $b \in \mathcal{E}_s$ there is equality between $sp(b)$ and $sp_e(b)$, $sp(b) = sp_e(b)$, just as $pl(b) = pl_e(b)$ for every $b \in \mathcal{E}_p$. This is so because, for $b \in \mathcal{E}_s$, $\nu(\pi(b)) = \nu(\rho_0(\rho(b))) = \nu([\rho(b)]) = \mu(\rho(b))$, as $\rho_0(a) = [a]$ whenever $a \in \mathcal{A}_0$. sp_e and pl_e are support and plausibility functions enjoying the properties of theorem 1.1. They are *extensions* of the support and plausibility functions sp and pl from $\mathcal{E}_s$ or $\mathcal{E}_p$ respectively to $\mathcal{B}$. Note that there may be several extensions of support and plausibility functions (Shafer, 1979) and only one of them is determined by the body of evidence. Different bodies of evidence may induce the same support and plausibility functions sp and pl, but different extensions sp_e and pl_e (Kohlas, 1993 b).

3.3 Combining Bodies of Evidence

Consider two bodies of arguments $(\mathcal{A}, \mathcal{B}, \rho_1)$ and $(\mathcal{A}, \mathcal{B}, \rho_2)$. Furthermore let $\mathcal{A}_0$ be a σ-subalgebra of $\mathcal{A}$ and μ a probability measure on $\mathcal{A}_0$ such that we have two bodies of evidence. The problem addressed in this section is how to derive the combined body of evidence from these two bodies.

The problem as it is posed here may seem very special at first sight. But note that it encloses for example the following very important case: $\mathcal{A}$ is the product algebra of two algebras $\mathcal{A}_1$ and $\mathcal{A}_2$, $\rho_1(\mathcal{B}) \subseteq \mathcal{A}_1, \rho_2(\mathcal{B}) \subseteq \mathcal{A}_2$, and $\mathcal{A}_0$ is the product of two σ-subalgebras $\mathcal{A}_{01}$ and $\mathcal{A}_{02}$ of $\mathcal{A}_1$ and $\mathcal{A}_2$ respectively, each with its own probability measure μ_1 and μ_2 and μ is finally the product measure of μ_1 and μ_2. In this case, the two bodies of evidence are called *independent*. This is the case which is almost exclusively considered in the literature. Although independence and related notions are undoubtedly important, only the general case will be treated here.

The way to combine two (or more) bodies of evidence is clear: First the bodies of arguments are combined and then support and plausibility functions are derived relative to the combined body of evidence. However a moment's notice shows that there are two presumably equivalent ways to do this: Either the two bodies $(\mathcal{A}, \mathcal{B}, \rho_1)$ and $(\mathcal{A}, \mathcal{B}, \rho_2)$ are first combined and then the allocation of probability relative to the combined body is derived. Or else, the two allocations of probability derived individually for each body of evidence are combined.

Let's first pursue the second proposal, which is the more direct one. It has the undeniable advantage that the completion of the algebra $\mathcal{A}$ which is needed in the combination of the two bodies of arguments $(\mathcal{A}, \mathcal{B}, \rho_1)$ and $(\mathcal{A}, \mathcal{B}, \rho_2)$ can be avoided. In fact, let $(\mathcal{M}_0, \mu)$ as before be the measure algebra derived from $\mathcal{A}_0$ and μ. $\mathcal{M}_0$ is complete, such that the two allocations $\pi_1 = \rho_0 \circ \rho_1$ and $\pi_2 = \rho_0 \circ \rho_2$ can be combined without extending $\mathcal{M}_0$.

According to (2.1) of section 2.3, the combined allocation π of π_1 and π_2 is defined by

$$\pi(b) = \vee \{\pi_1(b_1) \wedge \pi_2(b_2) : b_1 \wedge b_2 \leq b\}.$$

Similarly, the combined allowment of probability η is according to theorem 2.2 given by

$$\eta(b) = \wedge \{\eta_1(b_1) \vee \eta_2(b_2) : b_1 \vee b_2 \leq b\}.$$

This permits then, to determine the degrees of support and plausibility relative to the combined body of evidence. There is the following nice result in the special case of independent bodies of evidence.

Theorem 3.3 If sp_{e1}, sp_{e2} and pl_{e1}, pl_{e2} are the support and plausibility functions of two independent bodies of evidence then the support and plausibility functions of the combined body of evidence are given by

$$sp_e(b) = sup \; \{ \sum_{\emptyset \neq I \subseteq \{1,\ldots,n\}} (-1)^{|I|+1} sp_{e1}(\wedge_{i \in I} b_{1i}) sp_{e2}(\wedge_{i \in I} b_{2i})\}$$

where the supremum is to be taken over all finite sets of $b_{1i}, b_{2i} \in \mathcal{B}, i = 1, \ldots, n, n \geq 1$, such that $b_{1i} \wedge b_{2i} \leq b$, and

$$pl_e(b) = inf \; \{ \sum_{\emptyset \neq I \subseteq \{1,\ldots,n\}} (-1)^{|I|+1} pl_{e1}(\vee_{i \in I} b_{1i}) pl_{e2}(\vee_{i \in I} b_{2i})\}$$

where the infimum is to be taken over all finite sets of $b_{1i}, b_{2i} \in \mathcal{B}, i = 1, \ldots, n, n \geq 1$ such that $b_{1i} \vee b_{2i} \geq b$.

Independent bodies of evidence arise especially in the case of hints, where the interpretations of two hints may often be assumed to be stochastically independent. This leads then to the usual rule of Dempster (Kohlas, 1993 a) and b)).

4 Conclusion

Evidence theory is here seen as a theory of reasoning with uncertain arguments. Allocations of support define arguments which imply hypotheses. If the likelihood of these arguments can be measured by probabilities, then degrees of support and of plausibility can be assigned to hypotheses. These support and plausibility functions turn out to be Choquet capacities, monotone or alternating of infinite order. This qualifies this evidence theory as a Dempster-Shafer theory.

Argument-based reasoning is proposed as a very powerful paradigm which very naturally leads to a new synthesis and integration of probability and logic. It offers a unifying approach which is able to cover such apparently divers fields as statistical inference and logic based reasoning.

References

1. Dempster, A.: Upper and lower probabilities induced by a multivalued mapping. Ann. Math. Stat. **38** (1967) 325-339
2. Halmos, P.R.: Lectures on Boolean Algebras. Van Nostrand, Princeton, N.J. (1963)
3. Hestir, K., Nguyen, H.T., Rogers, G.S.: A Random Set Formalism for Evidential Reasoning. In: Goodman, I.R. et al. (eds.) (1991) 309-344
4. Kohlas, J.: The Reliability of Reasoning with Unreliable Arguments. Ann. of Op. Res. **32** (1991) 67-113
5. Kohlas, J.: Support- and Plausibility Functions Induced by Filter - Valued Mappings. Int. J. Gen. Syst. **21**, (1993a) 343-363
6. Kohlas, J. (1993b): Symbolic Evidence, Arguments, Supports, and Valuation Networks. In: Clarke, M., Kruse, R., Moral, S. (eds.): Symbolic and Quantitative Approaches to Reasoning and Uncertainty Springer Berlin (1993b) 186-198
7. Kohlas, J., Brachinger, H.W.: Mathematical Foundations of Evidence Theory. Institute of Informatics University of Fribourg Internal Working Paper**93-06**
8. Kohlas, J.; Monney, P.A.: Probabilistic Assumption-Based Reasoning. In: Heckermann, Mamdani, (eds.): Proc. 9th Conf. on Uncertainty in Artificial Intelligence Kaufmann Morgan Publ. (1993)
9. Kohlas, J.; Monney P.A.: Representation of Evidence by Hints. In: Yager, R. et al. (eds.): Recent Advances in Dempster-Shafer Theory Wiley New York 473-492
10. Laskey, K.B.; Lehner, P.E.: Belief Maintenance: An Integrated Approach to Uncertainty Management. Proc. Amer. Ass. for AI (1988) 210-214
11. Nguyen, H.T.: On Random Sets and Belief Functions. J. Math. Anal. and Appl. **65** (1978) 531-542
12. Pearl, J.: Probabilistic Reasoning in Intelligent Systems: Networks of Plausible Inference. Morgan Kaufman San Mateo Cal. (1988)
13. Provan. G.M.: A Logic-Based Analysis of Dempster-Shafer Theory. Int. J. Approx. Res. **4** (1990) 451-495
14. Shafer. G.: A Mathematical Theory of Evidence. Princeton University Press Princeton N.J. (1976)
15. Shafer, G.: Allocations of Probability. Ann. of Prob. **7** (1979) 827-839
15. Sikorski, R.: Boolean Algebras. Springer Berlin (1960)
16. Smets, P.: Belief functions. In: Smets, P. et al. (eds.): Non-Standard Logics for Automated Reasoning. Academic Press London (1988) 253-286

Algebraic Structures Related to Dempster-Shafer Theory *

Milan Daniel

Institute of Computer Science, Academy of Sciences of the Czech Republic
Pod vodárenskou věží 2, 182 07 Prague 8, Czech Republic

Abstract. There are described some algebraic structures on a space of belief functions on a two-element frame, namely so called Dempster's semigroup (with Dempster's operation ⊕), dempsteroids, and their basic properties. The present paper is devoted to the investigation of automorphisms of Dempster's semigroup. Full characterization of general and ordered automorphisms is obtained, their parametric description is stated both in intuitive and explicit forms. There is also full characterization of ordered endomorphisms, and other related results.

1 Introduction

Dempster-Shafer (D-S) theory of belief functions (see [14]) is well recognized as one of the main directions in uncertainty management in AI. The first attempts at applying D-S theory in AI were related to MYCIN-like systems (see e.g. [6]), but now D-S theory is applied to uncertainty in general (see e.g. [15]).

Belief functions are numeric (quantitative) characteristics of belief. Investigation of their general properties is of great importance; the study of comparative properties (non-numeric comparison of belief) may serve as an example (cf. e.g. [3]). A related example is the study of algebraic properties of systems of belief functions, showing in particular the behaviour of important properties of belief functions under isomorphisms of the systems in question. Results of this kind are relevant to the attempts to make the user of a concrete AI-system (dealing with uncertainty) free from relying heavily on concrete numbers (weights, numerical beliefs, etc.), and free to make his/her decision dependent on more general and more stable things (e.g., things not dependent on transition from one belief structure to an isomorphic one - rather loosely worded).

Algebraic aspects were investigated by Hájek and Valdés [11] (for basic belief assignments on two-element frames), see also [9] chapter IX. We can also mention later papers axiomatizing, in an algebraic manner, Dempster's rule on general (finite) frames: [7, 12, 16].

The present paper is devoted to the investigation of automorphisms of the semigroup of (non-extremal) belief functions on a two-element frame (with Dempster's operation ⊕ - we refer to Dempster's semigroup). Full characterization is obtained and there are various related results. In particular, the question

* Partial support by the COPERNICUS grant 10053 (MUM) is acknowledged.

of existence of non-trivial automorphisms of Dempster's semigroup is answered positively: there are lots of them.

This paper has the following structure:

In the second section are briefly introduced used basic algebraic notions. The third section recalls main definitions and results from [11], in particular, the definition of Dempster's semigroup and of the class of algebras called dempsteroids.

Sections 4 and 5 deal with the standard dempsteroid $\mathbf{D}_0$ (Dempster's semigroup with an additional structure, the usual algebraic structure underlying belief functions on a two-element frame of discernment); the main result is full characteratization of all automorphisms of $\mathbf{D}_0$ (sect. 4) and also full characteratization of all of its o-endomorphisms (sect. 5). The latter result gives us a characteratization of all dempsteroids constructed by o-isomorphic embedding of $\mathbf{D}_0$ to itself.

The sixth section summarizes some conclusions for comparative (non-numeric) theories of uncertainty. Some tasks for future research in the theory of Dempster's semigroup are also mentioned.

2 Preliminaries

Let us recall some basic algebraic notions before we begin algebraic description of D-S theory.

A *commutative semigroup* (called also an *Abelian semigroup*) is a structure $X = (X, \oplus)$ formed by set X and a binary operation $\oplus$ on X which is commutative and associative ($x \oplus y = y \oplus x$ and $x \oplus (y \oplus z) = (x \oplus y) \oplus z$ holds for all $x, y, z \in X$). A *commutative group* is a structure $Y = (Y, \oplus, -, o)$ such that $(Y, \oplus)$ is a commutative semigroup, o is a neutral element ($x \oplus o = x$) and $-$ is a unary operation of inverse ($x \oplus -x = o$). An *ordered commutative semigroup* consists of a commutative semigroup Y as above and a linear ordering $\leq$ of its elements satisfying monotonicity ($x \leq y$ implies $x \oplus z \leq y \oplus z$ for all $x, y, z \in X$). A subset of Y which is a (semi)group itself is called a *sub(semi)group*.

A *homomorphism* $p : (X, \oplus_1) \longrightarrow (Y, \oplus_2)$ is a mapping which preserves structure, i.e. $p(x \oplus_1 y) = p(x) \oplus_2 p(y)$ for each $x, y \in X$. The special cases are *automorphisms*, which are bijective morphisms from a structure onto itself; *endomorphisms* are morphisms to a substructure of the original one. Morphisms which also preserve ordering of elements are called *ordered morphisms* (*o-automorphisms, o-endomorphisms*, ...).

Ordered structures and ordered morphisms are very important for a comparative approach to uncertainty management.

3 On Dempster's semigroup

Now we introduce some principal notions according to [9]. For a two-valued frame $\Theta = \{0, 1\}$ each basic belief assignment determines a d-pair $(m(1), m(0))$ and conversely, each d-pair determines a basic belief assignment:

Definition 1. A *Dempster pair* (or *d-pair*) is a pair of reals such that $a, b \geq 0$ and $a + b \leq 1$. A d-pair (a, b) is *Bayesian* if $a + b = 1$, (a, b) is *simple* if $a = 0$ or $b = 0$, in particular, *extremal* d-pairs are pairs (1,0) and (0,1). (Definitions of Bayesian and simple d-pairs correspond evidently to the usual definitions of Bayesian and simple belief assignments [9, 14]).

Definition 2. Dempster's semigroup $D_0 = (D_0, \oplus)$ is the set of all non extremal Dempster pairs, endowed with the operation $\oplus$ and two distinguished elements $0 = (0, 0)$ and $0' = (\frac{1}{2}, \frac{1}{2})$, where the operation $\oplus$ is defined by

$$(a, b) \oplus (c, d) = \left(1 - \frac{(1-a)(1-c)}{1-(ad+bc)}, 1 - \frac{(1-b)(1-d)}{1-(ad+bc)}\right).$$

Remark. It is well known that this operation corresponds to Dempster's rule of combination of basic belief assignments.

Definition 3. For $(a, b) \in D_0$ let

$$\begin{aligned} -(a, b) &= (b, a), \\ h(a, b) &= (a, b) + 0' = \left(\tfrac{1-b}{2-a-b}, \tfrac{1-a}{2-a-b}\right), \\ h_1(a, b) &= \tfrac{1-b}{2-a-b}, \\ f(a, b) &= (a, b) \oplus (b, a) = \left(\tfrac{a+b-a^2-b^2-ab}{1-a^2-b^2}, \tfrac{a+b-a^2-b^2-ab}{1-a^2-b^2}\right). \end{aligned}$$

For $(a, b), (c, d) \in D_0$ we define
$(a, b) \leq (c, d)$ iff $h_1(a, b) < h_1(c, d)$ or if $h_1(a, b) = h_1(c, d)$ and $a \leq c$.

Let G denote the set of all Bayesian nonextremal d-pairs. Let us denote the set of all simple d-pairs such that $b = 0$ ($a = 0$) as S_1 (S_2). Furthermore, put

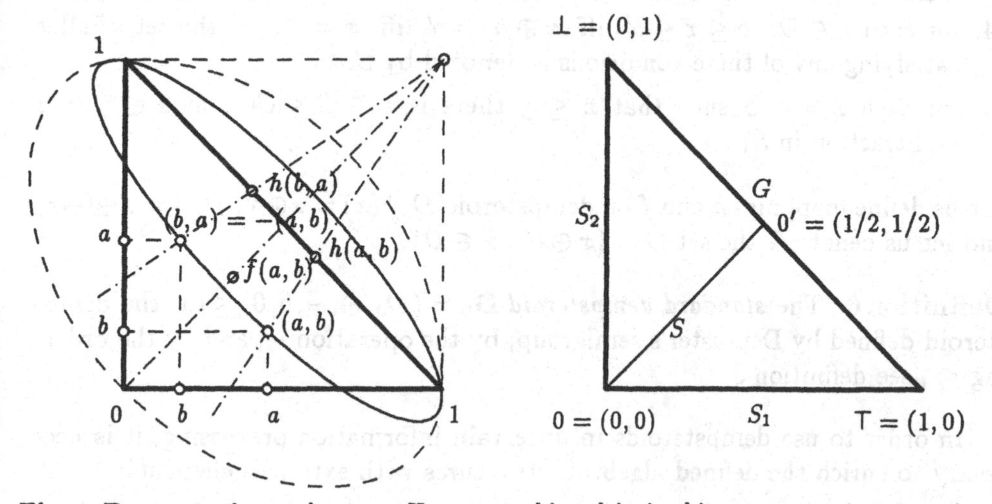

Fig. 1. Dempster's semigroup. Homomorphism h is, in this representation, a projection to group G along the straight lines running through the point $(1, 1)$. All Dempster pairs laying on the same ellipse are, by homomorphism f, mapped to the same d-pair in semigroup S.

$S = \{(a, a) : \leq a \leq 0.5\}$.
(Note: $h(a, b)$ is an abbreviation for $h((a, b))$, etc.)

Theorem 4. *(i) Dempster's semigroup with the relation $\leq$ is an ordered commutative semigroup with neutral element 0; $0'$ is the only nonzero idempotent of it.*

(ii) The set G with ordering $\leq$ is an ordered Abelian group $(G, \oplus, -, 0', \leq)$ which is isomorphic to the PROSPECTOR group **PP** *(cf. [9]) and consequently isomorphic to the additive group of reals with usual ordering.*

(iii) The sets S, S_1 and S_2 with the operation $\oplus$ and the ordering $\leq$ form ordered commutative semigroups with neutral element 0, and are all isomorphic to the semigroup of nonnegative elements of the MYCIN group **MC**.

(iv) The mapping h is an ordered homomorphism of the ordered Dempster's semigroup onto its subgroup G (i.e. onto **PP***).*

(v) The mapping f is a homomorphism of Dempster's semigroup onto its subsemigroup S (but it is not an ordered homomorphism).

For proofs see [9, 11, 17].

Definition 5. A *dempsteroid* is an algebra $\mathbf{D} = (D, \oplus, \ominus, o, o', \leq)$ satisfying the following:

1. $(D, \oplus, o, \leq)$ is an ordered commutative semigroup with o as a neutral element,
2. $\ominus(\ominus x) = x$,
   $\ominus(x \oplus y) = (\ominus x) \oplus (\ominus y)$ for each $x, y \in D$,
3. $o \leq o'$,
4. for each $x \in D$: $o \leq x \leq o'$ iff $x \oplus o' = o'$ iff $x = \ominus x$, the set of all x satisfying any of these conditions is denoted by S.
5. for each $x, y \in S$ such that $x \leq y$ there is $z \in S$ such that $x \oplus y = z$ (subtraction in S).

Let us define mappings h and f on dempsteroid D: $h(x) = x \oplus o'$, $f(x) = x \oplus (\ominus x)$ and let us denote G the set $G = \{x \oplus o' : x \in D\}$.

Definition 6. The *standard dempsteroid* $\mathbf{D}_0 = (D_0, \oplus, -, 0, 0', \leq)$ is the dempsteroid defined by Dempster's semigroup, by the operation $-$, and by the ordering $\leq$, see definition 3.

In order to use dempsteroids in uncertain information processing, it is necessary to enrich the defined algebraic structures with extremal elements:

Definition 7. An *extended dempsteroid* $\mathbf{D}^+ = (D \cup \{\bot, \top\}, \oplus, \ominus, o, o', \leq)$ is an algebraic structure resulting from taking a dempsteroid and adding extremal elements $\bot, \top$ in the following way:

$$x \oplus \bot = \bot \quad \text{and} \quad x \oplus \top = \top \qquad \text{for all } x \in D,$$
$$\ominus\bot = \top \quad \text{and} \quad \ominus\top = \bot,$$
$$\bot \leq x \leq \top \qquad \text{for all } x \in D.$$

For the standard dempsteroid let us define $\bot = (0,1)$, $\top = (1,0)$.

4 Automorphisms of Dempster's semigroup

It is useful to summarize features of potential nontrivial automorphisms of Dempster's semigroup before attempting to find them.

In [1] it is proved for any automorphism g of Dempster's semigroup (as a semigroup in the simple pure algebraic sense $(D_0, \oplus)$) that g preserves d-pairs 0 and 0′, and that g commutes with the homomorphism h. Further, it is proved that g commutes with the operation $-$ (see 3) if and only if it commutes with the homomorphism f.

In order to stress that g commutes with $-$, we shall say that g is an automorphism of the standard dempsteroid:

Definition 8. An automorphism of the standard dempsteroid is a one-to-one mapping $\mathbf{D}_0$ onto $\mathbf{D}_0$ which preserves the elements 0 and 0′ and commutes with the operations $\oplus, -$. If it preserves the ordering $\leq$ of $\mathbf{D}_0$, we call it an o-automorphism of $\mathbf{D}_0$.

When looking for nontrivial automorphisms of $\mathbf{D}_0$ it is very important to show preserving of subalgebras G and S of $\mathbf{D}_0$ by all its automorphisms, and preserving of subgroups S_1 and S_2 of $\mathbf{D}_0$ by an arbitrary o-automorphism of the standard dempsteroid.

For description of o-automorphisms of the standard dempsteroid it is useful to describe all the o-automorphisms of its subgroup G and of its subsemigroups S, S_1, and S_2, too.

Proposition 9. *(i) Any automorphism g of $\mathbf{D}_0$ maps elements of subgroup G (subsemigroup S) onto the subgroup G (subsemigroup S).*

(ii) Any o-automorphism g of $\mathbf{D}_0$ maps elements of subsemigroups S_1 and S_2 onto the subsemigroups S_1 and S_2, respectively.

For proof of this proposition and for proofs of other following statements see [1].

Proposition 10. *O-automorphisms of the ordered Abelian group G and of the ordered Abelian semigroups S, S_1 and S_2 are the just following mappings:*

$$G : \quad g_G(x) = \frac{x^k}{x^k+(1-x)^k}, \quad k > 0 \qquad \text{(G)}$$
$$S : \quad g_S(x) = \frac{(1-2x)^l-(1-x)^l}{(1-2x)^l-2(1-x)^l}, \quad l > 0 \qquad \text{(S)}$$
$$S_1, S_2 : \quad g_M(x) = 1-(1-x)^m, \quad m > 0 \qquad \text{(M).}$$

From proposition 9.(i) follows that the restriction of any o-automorphism of the standard dempsteroid to its subgroup G (subsemigroup S, S_1, S_2) is an o-automorphism of the subgroup G (subsemigroup S, S_1, S_2); i.e. from preservation of the subalgebras G, S, S_1 and S_2 of $\mathbf{D}_0$ it is seen that any o-automorphism g of $\mathbf{D}_0$ induces o-automorphisms of these subalgebras of $\mathbf{D}_0$.

On the other hand we can try to use o-automorphisms of these subalgebras to express o-automorphism g of the standard dempsteroid. For this reason we shall use the following lemmata.

Lemma 11. *Any o-automorphism g of $\mathbf{D}_0$ has the following property:*

$$\begin{aligned} g(a,b) \in f^{-1}(g(f(a,b))) \cap h^{-1}(g(h(a,b))) = & \qquad\qquad (\mathrm{g}\cap) \\ = \{(x,y) : \ f(x,y) = (g_S(p_i(f(a,b))), g_S(p_i(f(a,b)))) & \\ \& \ h(x,y) = (g_G(p_1(h(a,b))), g_G(p_2(h(a,b))))\}, & \end{aligned}$$

where g_G and g_S are o-automorphisms of the subalgebras G and S induced by g and for $i = 1,2$ p_i is projection to the i-th coordinate.

The proof of the lemma follows from commutativity of o-automorphisms of the standard dempsteroid with the homomorphisms h and f. h^{-1} denotes a set of preimages of a point (a,b) with respect to the homomorphism h, similarly for f and $f^{-1}(a,b)$. For particularities see [1].

According to lemma 11 we can express any o-automorphism of $\mathbf{D}_0$ by a pair of o-automorphisms g_G and g_M of subalgebras G and S. Now, we can ask the principial question: "Which pairs of o-automorphisms g_G and g_S define an o-automorphism of $\mathbf{D}_0$?". It would be easy in the case of looking for the o-automorphisms of the structure $G \times S$, where every such a pair defines by (g$\cap$) an o-automorphisms of the global structure. The situation is more complicated in the case of $\mathbf{D}_0$.
E.g.: Let $g_S = 1_S$ and g_G in the form **(G)** (i.e., g_S is in the form **(S)**, where $l = 1$). If $k > 1$ then g defined by (g$\cap$) maps elements of S_i out of $\mathbf{D}_0$. In the case of $k < 1$, g defined by (g$\cap$) maps elements of S_i into a proper subset of $\mathbf{D}_0$. So neither the first nor the second example of pair of o-automorphisms does not define by (g$\cap$) any o-automorphism of $\mathbf{D}_0$, because they both violate 9.(ii).

We have the biparametrical description of potential o-automorphisms of $\mathbf{D}_0$ with parameters k and l, $k, l > 0$, see **(G)**, **(S)**. Now, there is a question to solve: "Is there any pair $(k,l) \neq (1,1)$ which defines an o-automorphism of $\mathbf{D}_0$?", and if Yes: "What is the set of all such a pairs ?" The answer follows from the next lemmata.

Lemma 12. *(Necessary conditions for g_G and g_S)*

(o) A pair of o-automorphisms g_G and g_M induced by the same o-automorphism g of the standard dempsteroid $\mathbf{D}_0$ satisfies: $k = m$.

(o) A pair of o-automorphisms g_S and g_M induced by the same o-automorphism g of the standard dempsteroid $\mathbf{D}_0$ satisfies: $l = m$.

Here k, l, m are constants from expressions **(G)**, **(S)**, **(M)**.

Conclusion:

(•) *A pair of o-automorphisms g_G and g_S induced by the same o-automorphism g of the standard dempsteroid $\mathbf{D}_0$ satisfies: $k = l$.*

Lemma 12 says the very important fact that all the o-automorphisms g_G, g_S and g_M induced by the same o-automorphism of $\mathbf{D}_0$ are defined by the same constant from the expressions **(G)**, **(S)**, **(M)**.

To prove it, one heavily uses the preservation of subsemigroups S_1 and S_2 by any o-automorphism of $\mathbf{D}_0$, see proposition 9.(ii).

Lemma 13. *(Sufficient condition for g_G and g_S) If the mapping defined according to ($\mathbf{g}\cap$) satisfies $k = l$ then it maps elements of algebras S_1 and S_2 onto S_1 and S_2 respectively.*

Now we can formulate the theorem on o-automorphisms of the standard dempsteroid.

Theorem 14. (On o-automorphisms) *O-automorphisms of the standard dempsteroid are just the mappings defined according to ($\mathbf{g}\cap$) by pairs of o-automorphisms of the subgroup G and of the subsemigroup S such that*

$$g_G(x) = \frac{x^k}{x^k + (1-x)^k},$$

$$g_S(x) = \frac{(1-2x)^l - (1-x)^l}{(1-2x)^l - 2(1-x)^l},$$

where $k = l > 0$.

Corollary 15. *There is just one extension of any o-automorphism of the subgroup G to an o-automorphism of the whole $\mathbf{D}_0$.*

The theorem on o-automorphisms both says that there exist nonidentical o-automorphisms of the standard dempsteroid and describes the method for determining and of constructing of all of them. The explicit form of o-automorphisms of $\mathbf{D}_0$ is expressed in the following theorem, according to which it is possible to compute all the functional values for any o-automorphism g of $\mathbf{D}_0$.

Theorem 16. (On explicit o-automorphisms) *O-automorphisms of the standard dempsteroid are just mappings in the form*

$$g(a, b) = (A, B),$$

where

$$A = \frac{P_0^{2c} - (1-b)^c Q + (1-a)^c P_0^c}{P_0^{2c} - Q^2},$$

$$B = 1 - (1-A) \cdot \frac{(1-b)^c}{(1-a)^c} = 1 - \frac{(1-b)^c}{(1-a)^c} \cdot \frac{-Q^2 + (1-b)^c Q - (1-a)^c P_0^c}{P_0^{2c} - Q^2},$$

$$P_0 = 1 - (a+b),$$

$$Q = (1-a)^c + (1-b)^c, \qquad c > 0.$$

It is possible to make a non-trivial generalization of the theorem 14 to general (nonordered) automorphisms of $\mathbf{D_0}$ as 17. From the theorems 16 and 17 we can derive the explicit form of general automorphisms of $\mathbf{D_0}$. For all the proofs see [1].

Theorem 17. (On automorphisms) *Automorphisms of the standard dempsteroid are just mappings $g(a,b) = g_0(a,b)$ or $g(a,b) = -g_0(a,b)$, where $g_0(a,b)$ is an o-automorphism of $\mathbf{D_0}$.*

The proof is similar but a little bit more complicated than in the case of *o*-automorphisms because an automorphism of $\mathbf{D_0}$ generally does not preserve the semigroups S_i. So it is necessary to prove that a general automorphism of $\mathbf{D_0}$ maps S_1 (S_2) either onto itself or onto S_2 (S_1).

In this case we get for constants k and l determining g_G and g_S the following condition $k = l \vee k = -l, l > 0$.

Theorem 18. (On fixed points of *o*-automorphisms) *Every non-identical o-automorphism of the standard dempsteroid has just the four fixed points, namely (0,0), (0,1), $(\frac{1}{2},\frac{1}{2})$ and (1,0).*

Corollary 19. *A general automorphism of $\mathbf{D_0}$ has either four fixed points (as o-automorphisms) or two fixed points (0,0) and $(\frac{1}{2},\frac{1}{2})$ (in the case of the violation of the ordering).*

5 Endomorphisms of Dempster's semigroup

Let us investigate the case of endomorphisms now. We can use similar construction as in the case of automorphisms, thus we generalize the expression (g$\cap$) to (g$\cap'$) by admitting endomorphisms g_G and g_S of the subalgebras G and S.

So we get either a pair of automorphisms g_G and g_S, or a pair of two nonbijective endomorphisms, or a pair formed by one automorphism (g_G or g_S) and one nonbijective endomorphism (the remaining one). Similarly, as in the case of automorphisms, such a pair does not generally define an endomorphism of $\mathbf{D_0}$. For example, endomorphisms g_1 or g_2 of the subgroup S (where $g_1(x) \equiv 0$, $g_2(0') = 0'$, $g_2(x) = 0$ otherwise) with any automorphism g_G of the subgroup G does not define it.

The situation is fully described by the theorems "On endomorphisms" and "On *o*-endomorphisms. We mention only the latter one here, for particulars see [1].

Theorem 20. (On *o*-endomorphisms) *An o-endomorphism of the standard dempsteroid is defined according to* (g$\cap'$) *by one of the following possibilities:*

1. *by a pair of o-automorphisms g_G and g_S of the subgroup G and the subsemigroup S*

   $g(x) = \frac{x^k}{x^k+(1-x)^k}$, $\quad g(x) = \frac{(1-2x)^l-(1-x)^l}{(1-2x)^l-2(1-x)^l}$, $\quad$ *where* $l \geq k > 0$.

2. by an o-automorphism g_G of the subgroup G and by an non bijective o-endomorphism g_S of the subsemigroup S satisfying:

$$g_S(x) = 0' \qquad \forall x \in S$$

or $g_S(0) = 0$ and $g_S(x) = 0'$ for $x \in S - \{0\}$.

3. following trivial o-endomorphisms of Dempster's semigroup (defined by a pair of non-bijective endomorphisms):

$$g : \mathbf{D_0} \to \{0\},$$
$$g : \mathbf{D_0} \to \{0'\}.$$

Theorem 21. *Every o-endomorphism of the standard dempsteroid* $\mathbf{D_0}$ *defined by* ($\mathbf{g}\cap$), *i.e., by a pair of o-automorphisms, where* $l > k > 0$, *is a bijective mapping from* $\mathbf{D_0}$ *onto a proper subdempsteroid of* $\mathbf{D_0}$.

Corollary 22. (Theorem on subdempsteroids) *Every pair of numbers* k *and* l, *where* $l > k > 0$, *uniquely defines a proper subdempsteroid* $D_{[k,l]}$ *of the standard dempsteroid* $\mathbf{D_0} = D_{[1,1]}$ *and a bijective o-isomorphism between* $D_{[1,1]}$ *and* $D_{[k,l]}$.

6 Concluding remarks

6.1 Ordered morphisms of Dempster's semigroup and processing of uncertainty

A motivation of an algebraization of D-S theory came from rule-based systems managing uncertain rules. So the stated results and *o*-automorphisms of Dempster's semigroup are useful in rule-based knowledge systems [1], but their utilization is not at all restricted to field of rule-based systems.

We can utilize *o*-automorphisms of Dempster's semigroup everywhere we have discrete user scale of degrees of uncertainty and uncertainty management based on D-S theory. In that case, it is necessary to realize discrete degrees and intervals of them into continuous structure, into Dempster's semigroup. *O*-automorphisms of Dempster's semigroup give classes of equivalence of such realizations.

Discrete user scales of degrees of uncertainty is used, e.g., in expert systems AL/X [13] and EQUANT [8]. A discrete scale could be displayed to a user e.g. in the following form: False, almost false, a little false, maybe, a little true, almost true, true.

A realization of discrete degrees and their intervals into Dempster's semigroup is a special case of their realization into a general dempsteroid. Theoretical results on subdempsteroids and on endomorphisms of Dempster's semigroup are useful for such a realization.

6.2 Perspectives for future investigation

The following are *open problems* of the theory of morphisms of Dempster's semigroup:

The examining of the existence of an automorphism of the $(D_0, \oplus)$ which violates the operation $-$ and finding a form of a such an automorphism. Or proving of the theorem that every automorphism of $(D_0, \oplus)$ preserves the operation $-$, so that it is also an automorphism of the standard dempsteroid $(D_0, \oplus, -, 0, 0', \leq)$. It does mean to discover if it is necessary to distinguish between automorphisms of Dempster's semigroup and automorphisms of the standard dempsteroid.

Further it is possible to investigate which other subalgebras are preserved by *o*-automorphisms of the standard dempsteroid besides G, S and S_i.

Another interesting problem is the question whether $D_{[k,l]} = D_{[m,n]}$ for $\frac{k}{l} = \frac{m}{n}$, and what is the relation of isomorphisms from these dempsteroids to $\mathbf{D}_0$, etc.

The great challenge for future research is the generalization of the algebraic structure for D-S with a general finite frame of discernment. The general algebraic structure will be much more complicated. A complexity of the structure is dependent on the number of various focal elements $\sim 2^n - 1$, where n is a number of elements of a used frame.

After description of the structure it will be necessary to try to define reasonable ordering on it, and study its *o*-automorphisms. This will be a very interesting research task.

6.3 Conclusions

An algebraic analysis of morphisms of Dempster's semigroup were presented. It contributes to recognition of the fact that algebraic structures related to Dempster-Shafer (D-S) theory are mathematically very rich and interesting. A complete parametric description of (*o*-)automorphisms (and *o*-endomorphisms as well) of the standard dempsteroid has been given.

The results on ordered morphisms are related to the investigation of comparative properties of belief functions (i.e., comparison of beliefs, not assigning exact numerical values), since comparative properties are preserved by *o*-automorphisms (and some of them even by *o*-endomorphisms). This should be related to the work by Dubois [3].

D-S theory with a two-element frame of discernment is useful in the case of processing of uncertainty expressed by discrete finite scale, so called degrees of belief, degrees of plausibility. And described *o*-automorphisms are very principal within the proces of realization of these degrees of uncertainty (resp. intervals of them) in the continuous Dempster's semigroup. The structure on the space of these intervals is in some sense related to representation of uncertainty by bilattices in the works by Esteva-Garcia-Godo [4].

References

1. Daniel,M.: Dempster's Semigroup and Uncertainty Processing in Rule Based Expert Systems. Ph.D. thesis, Academy of Sciences of the Czech Republic, Prague, 1993, (in Czech).
2. Dempster,A.P.: A generalization of Bayesian inference. J. of Roy. Stat. Soc. Ser. B, 30, 1968, 205.
3. Dubois,D.: Belief structures, possibility theory and decomposable confidence measures. Computers and AI, 5, 1986, 403-416.
4. Esteva,F., Garcia-Calvés,P., Godo,L.: Enriched Interval Bilattices and Partial Many-Valued Logics: An Approach to Deal with Graded Truth and Imprecision. In: International Journal of Uncertainty, Fuzziness and Knowledge-based Systems, World Scientific, Vol.2, No.1, 1994, pp.37-54.
5. Fuchs,L.: Partially ordered algebraic systems. Pergamon Press, 1963.
6. Gordon,J., Shortliffe,E.H.: The Dempster-Shafer theory of evidence. In: Buchanan, B.G., Shortliffe, E.H., (eds): Rule based expert systems: Mycin experiments, Addison Wesley Reading, MA, 1984.
7. Hájek,P.: Deriving Dempster's Rule. In: Bouchon-Meunier,B., Valverde,L., Yager,R.R., (eds): Uncertainty in Intelligent Systems, Elsevier, 1993, 75-84.
8. Hájek,P., Hájková,M.: The Expert System Shell EQUANT-PC: Philosophy, Structure and Implementation. In: Computational Statistics Quarterly, Vol.4, 261-267, 1990.
9. Hájek,P., Havránek,T., Jiroušek,R.: Uncertain Information Processing in Expert Systems. CRC Press, Inc., Boca Raton, 1992.
10. Hájek,P., Valdés,J.J.: Algebraic foundations of uncertainty processing in rule-based expert systems (Group theoretic approach). Computers and Artificial Intell., Vol.9, No.4, 1990, 325-344.
11. Hájek,P., Valdés,J.J.: A generalized algebraic approach to uncertainty processing in rule-based expert systems (Dempsteroids). Computers and Artificial Intelligence, Vol.10, No.1, 1991, 29-42.
12. Klawonn,F., Schwecke, E.: On the axiomatic justification of Dempster's rule of combination. Int. Journal Intell. Systems, 7, 1992, 469-478.
13. Patterson,A.: AL/X - User Manual, Intelligent Terminals Ltd., Oxford, 1981.
14. Shafer,G.: A Mathematical Theory of Evidence. Princeton University Press, Princeton, 1976.
15. Shenoy,P.P., Shafer,G.: Propagating belief functions with local computations. IEEE Expert, 1, 1986, 43.
16. Smets,P.: The combination of evidence in the transferable belief model. IEEE Trans. Pattern. Anal. and Machine Int., 12, 1990, 447-458.
17. Valdés,J.J.: Algebraic and logical foundations of uncertainty processing in rule-based expert systems of Artificial Intelligence. Ph.D. thesis, Czechoslovak Academy of sciences, Prague, 1987.

Extension of Lower Probabilities and Coherence of Belief Measures

Zhenyuan Wang and Wei Wang

Department of Systems Science and Industrial Engineering
Thomas J. Watson School of Engineering and Applied Science
State University of New York at Binghamton
Binghamton, New York 13902-6000, U. S. A.

Abstract. Coherence is an important concept which is introduced and discussed in a new mathematics branch, Imprecise Probability Theory. By using the Choquet integral, belief measures can be extended to be coherent lower previsions on the linear space consisting of all bounded functions. As a special case, we establish that all belief measures are coherent imprecise probabilities.

1 Introduction

To deal with the uncertainty and discuss how to assign subjective probabilities in systems research, a new mathematical approach, imprecise probability, was established recently. A detailed discussion about imprecise probabilities, including lower probabilities and upper probabilities, and more general mathematical models of lower and upper previsions is given by Walley [5]. A principle of avoiding sure loss or a more stronger principle of coherence is used to establish reasonable lower previsions. Belief measures are one kind of lower probabilities. In this paper, the Choquet integral is used to extend a belief measure to be a lower prevision on the linear space of all bounded functions. We prove that such a lower prevision is coherent and, as a corollary, any belief measure is coherent.

2 Belief Measures

Let X be a nonempty set, $\mathcal{F}$ be a σ-algebra of subsets of X. A set function μ: $\mathcal{F} \rightarrow [0, \infty]$ is called a lower (upper) semicontinuous fuzzy measure iff it is monotone, continuous from below (above), and vanishing at ∅. Both lower and upper semicontinuous fuzzy measures are referred to as semicontinuous fuzzy measures (SC-fuzzy measures, for short). μ is called a fuzzy measure iff it is both lower and upper semicontinuous fuzzy measure. A fuzzy measure (or, SC-fuzzy measure) μ is called regular if $\mu(X) = 1$.

The power set $\mathcal{P}(X)$ is a σ-algebra. A set function m: $\mathcal{P}(X) \rightarrow [0,1]$ is called a basic probability assignment iff

(BPA 1) $m(\emptyset) = 0$;
(BPA 2) $\sum_{E \subset X} m(E) = 1$.

From this definition, we know that there exists a countable subset $\mathcal{D}$ of $\mathcal{P}(X)$ such that $m(E) = 0$ whenever $E \notin \mathcal{D}$.

If m is a basic probability assignment, then the set function Bel: $\mathcal{P}(X) \to [0,1]$ determined by

$$\mathrm{Bel}(E) = \sum_{E \subset F} m(F) \qquad \text{for any } E \in \mathcal{P}(X)$$

is called a belief measure on $(X, \mathcal{P}(X))$ or, more exactly, a belief measure induced from m.

A belief measure Bel is a regular upper semicontinuous fuzzy measure and satisfies the inequality

$$\text{(BM 1)} \qquad \mathrm{Bel}\left(\bigcup_{i=1} E_i\right) \geq \sum_{I \subset \{1,\ldots,n\},\ I \neq \emptyset} (-1)^{|I|+1} \mathrm{Bel}\left(\bigcap_{i \in I} E_i\right)$$

where $\{E_1, \ldots, E_n\}$ is any finite subset of $\mathcal{P}(X)$.

If X is finite, then any regular nonnegative monotone set function defined on $\mathcal{P}(X)$ which vanishes at $\emptyset$ and satisfies (BM 1) is a belief measure.

3 Choquet Integrals

Let $(X, \mathcal{F})$ be a measurable space, μ be a finite nonnegative monotone set function defined on $\mathcal{F}$ with $\mu(\emptyset) = 0$, and $R+ = [0,\infty)$. If f is a nonnegative measurable function defined on $(X, \mathcal{F})$, then the Choquet integral of f with respect to μ on any set $A \in \mathcal{F}$ is defined by

$$\int_{R+} \mu(F_\alpha \cap A)\, d\nu,$$

and denoted by

$$\int_A f\, d\mu,$$

where

$$F_\alpha = \{x: f(x) \geq \alpha\}, \quad \alpha \in R+,$$

and ν is the Lebesgue measure. When $A = X$, we simply replace $\int_A f\, d\mu$ with $\int f\, d\mu$.

Lemma 3.1 Let f be a nonnegative measurable function defined on $(X, \mathcal{F})$. Then

$$\int_A (f + c)\, d\mu = \int_A f\, d\mu + c\, \mu(A)$$

for any $c \in [0, \infty)$.

Proof. Denote $g = f + c$ and $G_\alpha = \{x: g(x) \geq \alpha\}$ for any $\alpha \in [0, \infty)$. Both of F_α and G_α are nonincreasing functions of α. Since $G_{\alpha+c} = F_\alpha$ for any $\alpha \in [0, \infty)$, we have

$$\begin{aligned}
\int_A (f + c)\, d\mu &= \int_A g\, d\mu \\
&= \int_{R+} \mu(G_\alpha \cap A)\, dv \\
&= \int_c^\infty \mu(G_\alpha \cap A)\, d\alpha + \int_0^c \mu(A)\, d\alpha \\
&= \int_0^\infty \mu(G_{\alpha+c} \cap A)\, d\alpha + \int_0^c \mu(A)\, d\alpha \\
&= \int_0^\infty \mu(F_\alpha \cap A)\, d\alpha + c\, \mu(A) \\
&= \int_A f\, d\mu + c\, \mu(A). \quad \square
\end{aligned}$$

When μ is regular, we can simply write

$$\int (f + c)\, d\mu = \int f\, d\mu + c, \quad \forall c \in [0, \infty).$$

Taking advantage of Lemma 3.1, we can generally define the Choquet integral for any bounded (not necessary to be nonnegative) measurable function f as follows.

Let f be a bounded measurable function defined on $(X, \mathcal{F})$. Then the Choquet integral of f with respect to μ on any set $A \in \mathcal{F}$ is defined by

$$\int_A f\, d\mu = \int_A (f - b)\, d\mu + b\, \mu(A),$$

where b is a lower bound of f on A. This definition is unambiguous.

The Choquet integral coincides with the Lebesgue integral when set function μ is a classical measure. The Choquet integral is a generalization of the Lebesgue integral. So we may us the same integral notation for it as used for the Lebesgue integral.

4 Properties of the Choquet Integral

The Choquet integral has following basic properties.

Theorem 4.1 Let f, f_1, and f_2 be any bounded measurable functions on $(X, \mathcal{F})$, c be a constant, and $A \in \mathcal{F}$.

(1) If $\mu(A) = 0$, then $\int_A f\, d\mu = 0$.

(2) If f is nonnegative, then

$$\int_A f\, d\mu = 0 \iff \mu(\{x: f(x) \neq 0\} \cap A) = 0.$$

(3) $\int_A c\, d\mu = c\, \mu(A)$.

(4) If $f_1 \le f_2$, then $\int_A f_1\, d\mu \le \int_A f_2\, d\mu$.

(5) $\int_A f d\mu = \int f \chi_A\, d\mu$, where χ_A is the characteristic function of set A.

(6) $\int_A (f + c)\, d\mu = \int_A f d\mu + c\, \mu(A)$.

(7) $\int_A c f d\mu = c\, (\int_A f d\mu)$ for any $c \ge 0$.

Proof. We only prove the result in (7). Due to (3), we can assume that $c > 0$. Let b be a lower bound of f. Then cb is a lower bound of cf, and we have

$$\begin{aligned}\int_A c f d\mu &= \int_A (cf - cb)\, d\mu + cb\, \mu(A) \\ &= \int_0^\infty \mu(\{x: c(f(x) - b) \ge \alpha\} \cap A)\, d\alpha + cb\, \mu(A) \\ &= \int_0^\infty \mu(\{x: f(x) - b \ge \alpha/c\} \cap A)\, d\alpha + cb\, \mu(A) \\ &= c \int_0^\infty \mu(\{x: f(x) - b \ge \alpha/c\} \cap A)\, d(\alpha/c) + cb\, \mu(A) \\ &= c \int_0^\infty \mu(\{x: f(x) - b \ge \alpha\} \cap A)\, d\alpha + cb\, \mu(A) \\ &= c \int_A (f - b)\, d\mu + cb\, \mu(A) \\ &= c \int_A f d\mu - cb\, \mu(A) + cb\, \mu(A) \\ &= c \int_A f d\mu. \quad \square\end{aligned}$$

Furthermore, if μ is a belief measure on $\mathcal{P}(X)$, then the Choquet integral is superlinear as shown below.

Theorem 4.2 Let Bel be a belief measure on $\mathcal{P}(X)$, and let f_1 and f_2 be bounded functions on X. Then

$$\int_A (f_1 + f_2)\, d\,\text{Bel} \ge \int_A f_1\, d\,\text{Bel} + \int_A f_2\, d\,\text{Bel}$$

for any $A \in \mathcal{P}(X)$.

Proof. $(X, \mathcal{P}(X))$ is a measurable space, and in this scope, any function defined on X is measurable. Due to the property given in Theorem 4.1,(5), we may only consider the situation when $A = X$.

Assume first that both f_1 and f_2 are nonnegative. Let m be the basic probability assignment from which belief measure Bel is induced. The focal sets of m are denoted by A_i, i = 1, 2, ... Also, we define three families of functions of a on $[0, \infty)$ as follows:

$$\beta_i(\alpha) = \begin{cases} m(A_i) & \text{if } \{x: f_1(x) + f_2(x) \ge \alpha\} \supset A_i \\ 0 & \text{otherwise,} \end{cases}$$

$$\beta_i'(\alpha) = \begin{cases} m(A_i) & \text{if } \{x: f_1(x) \geq \alpha\} \supset A_i \\ 0 & \text{otherwise ,} \end{cases}$$

and

$$\beta_i''(\alpha) = \begin{cases} m(A_i) & \text{if } \{x: f_2(x) \geq \alpha\} \supset A_i \\ 0 & \text{otherwise ,} \end{cases}$$

Arbitrarily fixing i, since $\{x: f_1(x) \geq \alpha_1\} \supset A_i$ and $\{x: f_2(x) \geq \alpha_2\} \supset A_i$ imply that $\{x: f_1(x) + f_2(x) \geq \alpha_1 + \alpha_2\} \supset A_i$ for any α_1 and α_2 , we obtain the inequality

$$\begin{aligned} &\sup\{\alpha: \{x: f_1(x) + f_2(x) \geq \alpha\} \supset A_i\} \\ &\qquad \geq \sup\{\alpha: \{x: f_1(x) \geq \alpha\} \supset A_i\} + \sup\{\alpha: \{x: f_2(x) \geq \alpha\} \supset A_i\}. \end{aligned}$$

Hence, by using the definition of the Choquet integral and the monotone convergence theorem of integral sequence (see Halmos [2], 27, Theorem B), we have

$$\begin{aligned} \int (f_1 + f_2)\, d\,\mathrm{Bel} &= \int_0^\infty \mathrm{Bel}(\{x: f_1(x) + f_2(x) \geq \alpha\})\, d\alpha \\ &= \int_0^\infty \sum_{i=1}^\infty \beta_i(\alpha)\, d\alpha \\ &= \sum_{i=1}^\infty \int_0^\infty \beta_i(\alpha)\, d\alpha \\ &= \sum_{i=1}^\infty \sup\{\alpha: \{x: f_1(x) + f_2(x) \geq \alpha\} \supset A_i\}\, m(A_i) \\ &\geq \sum_{i=1}^\infty [\sup\{\alpha: \{x: f_1(x) \geq \alpha\} \supset A_i\} + \sup\{\alpha: \{x: f_2(x) \geq \alpha\} \supset A_i\}]\, m(A_i) \\ &= \sum_{i=1}^\infty \sup\{\alpha: \{x: f_1(x) \geq \alpha\} \supset A_i\} m(A_i) + \sum_{i=1}^\infty \sup\{\alpha: \{x: f_2(x) \geq \alpha\} \supset A_i\} m(A_i) \\ &= \sum_{i=1}^\infty \int_0^\infty \beta_i'(\alpha)\, d\alpha + \sum_{i=1}^\infty \int_0^\infty \beta_i''(\alpha)\, d\alpha \\ &= \int_0^\infty \sum_{i=1}^\infty \beta_i'(\alpha)\, d\alpha + \int_0^\infty \sum_{i=1}^\infty \beta_i''(\alpha)\, d\alpha \\ &= \int_0^\infty \mathrm{Bel}(\{x: f_1(x) \geq \alpha\})\, d\alpha + \int_0^\infty \mathrm{Bel}(\{x: f_2(x) \geq \alpha\})\, d\alpha \\ &= \int f_1\, d\,\mathrm{Bel} + \int f_2\, d\,\mathrm{Bel}. \end{aligned}$$

Then, if at least one of f_1 and f_2 are not nonnegative, let $g_1 = f_1 - b_1$, and $g_2 = f_2 - b_2$, where b_1 and b_2 are the lower bounds of f_1 and f_2 respectively. Thus, both g_1 and g_2 are nonnegative. By using the property given in Theorem 4.1,(6), we have

$$\begin{aligned}\int (f_1 + f_2)\, d\,\mathrm{Bel} &= \int (g_1 + g_2)\, d\,\mathrm{Bel} + b_1 + b_2 \\ &\geq \int g_1\, d\,\mathrm{Bel} + \int g_2\, d\,\mathrm{Bel} + b_1 + b_2 \\ &= \int (g_1 + b_1)\, d\,\mathrm{Bel} + \int (g_2 + b_2)\, d\,\mathrm{Bel} \\ &= \int f_1\, d\,\mathrm{Bel} + \int f_2\, d\,\mathrm{Bel}.\end{aligned}$$

The proof now is complete. □

The superlinearity of the Choquet integral with respect to a belief measure is used in the next section to establish the main result of this paper.

5 Coherent Extension of Belief Measures

Let X be a nonempty set, $\mathcal{K}$ be a nonempty class of bounded functions defined on X, and $\underline{P}$ be a real valued functional defined on $\mathcal{K}$. Then, $\underline{P}$ (or in detail, $(X, \mathcal{K}, \underline{P})$) is called a lower prevision (Walley [5]).

When $\mathcal{K}$ is a class of characteristic functions of sets in $\mathcal{P}(X)$, $\underline{P}$ is called a lower probability. If $(X, \mathcal{F})$ is a measurable space, μ is a finite nonnegative monotone set function defined on $\mathcal{F}$ with $\mu(\emptyset) = 0$, and f is any bounded measurable function, then μ can be regarded as a lower probability, and the Choquet integral $\int f\, d\mu$ is a true extension of μ to be a lower prevision $\underline{P}$ on the linear space consisting of all bounded measurable functions on $(X, \mathcal{F})$. In fact, for any $A \in \mathcal{F}$, taking $f(x) = \chi_A(x)$, we have

$$\begin{aligned}\underline{P}(\chi_A) &= \int \chi_A\, d\mu \\ &= \int_A d\mu \\ &= \mu(A)\end{aligned}$$

by properties given in Theorem 4.1,(3) and (5). That is, $\underline{P}$ coincides with m on $\mathcal{F}$.

We say that a lower prevision $(X, \mathcal{K}, \underline{P})$ avoids sure loss if

$$\sup_{x \in X} \sum_{i=1}^{n} [f_i(x) - \underline{P}(f_i)] \geq 0,$$

whenever $n \leq 1$, and $f_1, \ldots, f_n \in \mathcal{K}$; a lower prevision $(X, \mathcal{K}, \underline{P})$ is coherent if

$$\sup_{x \in X} \sum_{i=1}^{n} [f_i(x) - \underline{P}(f_i)] - m \cdot [f_0(x) - \underline{P}(f_0)] \geq 0,$$

whenever $n \geq 1$, $m \geq 1$, and $f_0, f_1, \ldots, f_n \in \mathcal{K}$. It is evident that coherence implies avoiding sure loss.

When $\mathcal{K}$ is restricted to be a linear space (denoted by $\mathcal{L}$), a very useful equivalent definition of coherence was given by Walley [5] as follows.

Theorem 5.1 Suppose that $\mathcal{L}$ is a linear space. Then (X, $\mathcal{L}$, $\underline{P}$) is coherent if and only if it satisfies the following conditions:
(C1) $\underline{P}(f) \geq \inf_{x \in X} f(x) \quad \forall f \in \mathcal{L}$.
(C2) $\underline{P}(cf) = c\,\underline{P}(f) \quad \forall f \in \mathcal{L}$ and $c \geq 0$.
(C3) $\underline{P}(f_1 + f_2) \geq \underline{P}(f_1) + \underline{P}(f_2) \quad \forall f_1, f_2 \in \mathcal{L}$.

Now, we turn to consider the belief measures defined on $\mathcal{P}(X)$. When we use $\mathcal{P}(X)$ as the σ-algebra $\mathcal{F}$ in the measurable space (X, $\mathcal{F}$), any function defined on X is measurable. Let Bel be a belief measure on $\mathcal{P}(X)$. The class of all bounded functions defined on X is a linear space, and we use $\mathcal{L}$ to denote it. The Choquet integral of f on X with respect to Bel for all $f \in \mathcal{L}$ defines a real valued functional $\underline{\text{Bel}}$ on $\mathcal{L}$:

$$\underline{\text{Bel}}(f) = \int f \, d\,\text{Bel} \quad \forall f \in \mathcal{L}.$$

Then, (X, $\mathcal{L}$, $\underline{\text{Bel}}$) is a lower prevision. If we regard $\mathcal{P}(X)$ as $\mathcal{K}_0$, the class of all characteristic functions of set in $\mathcal{P}(X)$, that is, $\mathcal{K}_0 = \{\chi_A : A \in \mathcal{P}(X)\}$, then the lower prevision $\underline{\text{Bel}}$ is a true extension of Bel from $\mathcal{K}_0$ onto $\mathcal{L}$.

The main result of this paper is the following.

Theorem 5.2 Let Bel be a belief measure on $\mathcal{P}(X)$. Then its extension $\underline{\text{Bel}}$ onto $\mathcal{L}$ is coherent.

Proof. Since $\mathcal{L}$ is a linear space, we just need to verify that $\underline{\text{Bel}}$ satisfies the conditions (C1), (C2), and (C3).

Since $f \geq \inf_{x \in X} f(x)$ for any $f \in \mathcal{L}$, by using the properties given in Theorem 4.1,(3) and (4), we have

$$\begin{aligned}\underline{\text{Bel}}(f) &= \int f \, d\,\text{Bel} \\ &\geq \int \inf_{x \in X} f(x) \, d\,\text{Bel} \\ &= \inf_{x \in X} f(x)\, \text{Bel}(X) \\ &= \inf_{x \in X} f(x).\end{aligned}$$

That is, (C1) is satisfied. Furthermore, (C2) is guaranteed by the property given in Theorem 4.1,(7). Finally, (C3), the superlinearity of $\underline{\text{Bel}}$ is shown in Theorem 4.2 by taking A = X. □

From the definition of coherence directly, it is easy to show that a restriction of any coherent lower prevision within a subclass of $\mathcal{K}$ is still coherent. Since any belief measure Bel is a restriction of its extension $\underline{\text{Bel}}$ within $\mathcal{K}_0$, we obtain the following corollary.

Corollary 5.1 Any belief measure is coherent.

The result shown in Corollary 5.1 also can be obtained by constructing a family of probabilities such that the given belief measure is just their lower envelope (see Walley [5]).

6 Conclusions

By using the Choquet integral, any belief measure can be extended to be a coherent lower prevision on the linear space $\mathcal{L}$ consisting of all bounded functions. Walley [5] discusses the natural extension $\underline{E}$ of a lower prevision $\underline{P}$, and shows that $\underline{E}$ is the minimal coherent extension of $\underline{P}$ to $\mathcal{L}$ if $\underline{P}$ is coherent. So, we may conclude that $\underline{Bel}$ dominates the natural extension of Bel. An open question is: what is a further relation between $\underline{Bel}$ and the natural extension of Bel? More generally, for any given regular nonnegative monotone set function defined on $\mathcal{P}(X)$ that vanishes at the empty set, we can always use the Choquet integral to extend it to be a lower prevision on $\mathcal{L}$. This is a true extension, that is, the extension and the original set function coincide on $\mathcal{P}$ (X). But it is not coherent necessarily. Since this set function can be regarded as a lower prevision, we can also consider its natural extension, and such an extension is always coherent, even though it may not be a true extension, that is, it may be really larger than the original set function. Thus, we should look for a further relation between these two different kinds of extensions.

Acknowledgment

Work on this paper was supported in part by Rome Laboratory, Air Force Material Command, USAF, under grant number F30602-94-1-0011. The U.S. Government is authorized to reproduce and distribute reprints for Governmental purposes notwithstanding any copyright notation there on. The views and conclusions contained herein are those of the author and should not be interpreted as necessarily representing the official policies or endorsements, either expressed or implied, of Rome Laboratory or the U.S. Government.

References

1. G. Choquet: Theory of capacities. *Annales de l'Institut Fourier* , 5, 131-295 (1954).
2. P. R. Halmos: *Measure Theory* . Van Nostrand, New York (1967).
3. G. Shafer: *A Mathematical Theory of Evidence* . Princeton University Press, Princeton, New Jersey (1976).
4. P. Walley: Coherent lower (and upper) probabilities. *Statistics Research Report 22* , University of Warwick, Coventry (1981).
5. P. Walley: *Statistical Reasoning with Imprecise Probabilities* . Chapman and Hall, London (1991).
6. Z. Wang and G. J. Klir: *Fuzzy Measure Theory* . Plenum press, New York (1992).

A Note on Multi-objective Information Measures

Colette Padet
Laboratoire de Thermomécanique, S.U.E., GRSM
B.P. 347, 51062 Reims Cedex, France
Arthur Ramer
School of CSE, University of New South Wales
Sydney, 2052, Australia
Ronald Yager
Machine Intelligence Institute, Iona College
New Rochelle, NY 10801, USA

1 Introduction

Updating of an evidence assignment can be equated with finding a compromise between the prior distribution and the new evidence, with the resulting posterior a function these two distributions. In general evidence theory [11] such update is effected by Dempster rule of combination; in probability theory one can speak of Bayesian updating [6], and in possibility [1] one can view conjunction as updating operator.

We surmise that all these procedures try to locate a new assignment which, in some way, deviates least from all the inputs. This can be formalized by defining a measure of distance which is *multi-objective*—expresses proximity of a new assignment from all the inputs taken together. The assignment which minimizes sych measure would serve as the posterior distribution, Heuristic argument based on the known two-argument information distances [3, 5] leads to proposing related expressions in all three models. It is then fairly straightforward to verify that the standard rules of combination/updating follow,. Provenance of these functions (going back to Kullback-Leiber entropy-based information distance) and the fact that they 'arbitrate' among several inputs, motivate the term *multiobjective information measures.*

In the next section we desvribe briefly the combination rules we wish to model, and the known information distances, on which we base our construction.

The following section presents these measures separately in each framework (probability, evidence, possibility). In the last section we formulate the results about the minimization and the minimizing distributions.

We use term distribution to denote an assigment of uncertainty values. On several occasions, for stylisitc reasons, these two terms are used interchancheably.

Abbreviation *wrt* means: 'with respect to.'

The basis of logarithms is not stated, and only assumed fixed throughout. Changing it would simply result in multiplying all the formulae by a constant factor. We assume that $0 \log 0 = 0$.

2 Two-argument distances

These distances had been introduced to measures proximity of a given (prior) distribution and a new, 'proposed' (posterior) distribution. If not subject to additional constraints, the minimization of such distance (in every uncertainty model) is zero—simply copy the prior. The interest in such distances and their wide applicability come from their usefulness under the constraints. We usually impose certain *ex ante* conditions that any 'admissible' posterior must satisfy and seek one that is also as close to the prior as possible. Various considerations of consistency of such procedure, uniqueness of the results, and agreement with certain variational postulates led to establishing the 'prefered' distances in probability and in possibility theories.

In probability it is the *cross-entropy* [5], while in possibilty G-distance plays a similar role [3]. No attenpt, to the best of our knowledge, has been made to define such functuins in the general framework of evidence.

Given two probability distributions $\mathbf{p} = (p_1, \ldots, p_n)$ and $\mathbf{q} = (q_1, \ldots, q_n)$ we put

$$d(\mathbf{p}, \mathbf{q}) = \sum_i \log \frac{p_i}{q_i}.$$

This expression is defined whenever $\mathbf{p}$ is absolutely continuous wrt $\mathbf{q}$; as they are finite distributions, it means that $p_i = 0$ whenever $q_i = 0$.

The expresion is used to selct $\mathbf{p}_{post}$ given $\mathbf{p}_0 = \mathbf{p}_{prior}$ and a family of constraints on $\mathbf{p}_{post}$. The latter usually take form of linear equalities

$$\sum_i \alpha_i^{(k)} p_i = \beta^{(k)}, \; k = 1, \ldots, m$$

and inequalities

$$\sum_i \gamma_i^{(k)} p_i > \delta^{(k)}, \ k = 1, \ldots, l.$$

The optimizing strategy consists of selecting that $\mathbf{p}_{post}$ (from all $\mathbf{p}$ satisfying the constraints) for which

$$d(\mathbf{p}_0, \mathbf{p}_{post}) = \min_{\mathbf{p}} d(\mathbf{p}_0, \mathbf{p}).$$

This strategy can be justified axiomatically as a unique *consistent* strategy based on minimization of a single functional expression of $\mathbf{p}_0$ and $\mathbf{p}$. It has also been demonstrated, on ample examples, thea it is a very useful *practical* strategy, leading to the results which might have been ususally prefered on other grounds as well.

The structure of Kullback-Leiber distance makes apparent is relatioship with the basic (probabilistic) information measure—Shannon entropy. The latter is a single-distribution functional

$$H(\mathbf{p}) = \sum_i p_i \log \frac{1}{p_i};$$

locating its *constrained* extrema correspondds in many ways to minimization of $d(\mathbf{p}, \mathbf{q})$.

In possibility theory a similar role is played by G-distance, closely related to U-uncertainty [4, 9]. The latter is an uncertainty measure of a single distribution $\mathbf{p} = (p_1, \ldots, p_n)$

$$U(\mathbf{p}) = \sum_i (\tilde{p}_i - \tilde{p}_{i+1}) \log i,$$

where $(\tilde{p}_i)$ form a descending rearrangement of (p_i). Given any $\mathbf{p}$ and $\mathbf{q}$ which is uniformly greater ($q_i \geq p_i$ for each i) we define

$$G(\mathbf{p}, \mathbf{q}) = U(\mathbf{q}) - U(\mathbf{p}).$$

Although the definitions can be extended to arbitrary $\mathbf{p}$ and $\mathbf{q}$, it will not be needed here.

Again the method of minimizing the distance $G(\mathbf{p}_0, \mathbf{p}_{post})$ applies, although the constraints should, in general, be posited in form of maxima [7]

$$\max_i \alpha_i^{(k)} p_i = \beta^{(k)}, \ k = 1, \ldots, m$$

$$\max_i \gamma_i^{(k)} p_i > \delta^{(k)}, \; k = 1, \ldots, l.$$

In Dempster-Shafer model various measures of uncertainty (for a single assignment) have been proposed [2, 12]. Most notable are

$$\begin{aligned} C(\mu) &= -\sum_A m_A \log Bel(A) \\ E(\mu) &= -\sum_A m_A \log Pl(A) \\ N(\mu) &= \sum_A m_A \log |A| \end{aligned}$$

each summation extending over all the focal subsets A. The first is called *conflict* and is defined on the basis of belief $Bel(B) = \sum_{A \subset B} m_A$. The next one is *dissonance*, based on plausibility $Pl(B) = \sum_{A \cap B \neq \emptyset} m_A$. The last expression, nonspecificity, corresponds more to the possibilistic uncertainty.

However, the first two generalize Shannon entropy and we suggest they can be used as a basis of *evidence* distances

$$c(\pi, \rho) = -\sum m_A^{\rho} \log \frac{Bel^{\pi}(A)}{Bel^{\rho}(A)},$$

$$e(\pi, \rho) = -\sum m_A^{\rho} \log \frac{Pl^{\pi}(A)}{Pl^{\rho}(A)},$$

subject to condition that the numerators vanish whenever the demoninators do. We shall not pursue discussing their properties, as we are primarily interested in multi-objective distances.

3 Multi-objective distances

Another application of the notion of proximity (to be defined below) is to consider a case of several 'priors'. We intend that a few distributions $\mathbf{p}_1, \ldots, \mathbf{p}_k$ are given, and we wish to find a single distribution $\mathbf{p}_{post}$ which, in some way, is the closest possible to all $\mathbf{p}_1, \ldots, \mathbf{p}_k$. The problem, though stated without any prior constraints on $\mathbf{p}_{post}$, is highly nontrivial. While for a single prior, the answer would simply reproduce it, now we have to find areasonable *compromise* among the competing priors. An answer, if computed 'reasonably', would have the effect of composing the initial distributions $\mathbf{p}_1, \ldots, \mathbf{p}_k$.

We wish to propose forming a multi-distance function

$$d(\mathbf{q}; \mathbf{p}_1, \ldots, \mathbf{p}_k)$$

and request that $\mathbf{p}_{post}$ is defined as that distribution $\mathbf{q}$ for which d is minimized. The definitions are proposed on a heuristic basis, thropugh formation of expressions somewhat paralleling those known for the two-argument distances.

It turns out that the resulting functions are well designed, as they lead to some very well known rules of updating uncertianty values.

In probability model we propose

$$d(\mathbf{r}; \pi^{(1)}, \ldots, \mathbf{p}^{(k)}) = -\sum_{i=1}^{n} r_i(\log p_i^{(1)} + \cdots + \log p_i^{(k)} - \log r_i)$$

When applied with two 'prior' distributions $p_1, \ldots, p_n$ and $q_1, \ldots, q_n$, it leads to

$$\mathbf{r} = \alpha(p_1 q_1, \ldots, p_n q_n).$$

To prove it, it is enough to apply Gibbs inequality to $\mathbf{r}$ and $\mathbf{p} \times \mathbf{q}$ (coordinatewise). With a suitable normalizing factor α, it corresponds to the Bayesian updating [6].

In evidence model we propose

$$d(\rho; \pi^{(1)}, \ldots, \pi^{(k)}) = -\sum_A m_A^\rho(\log Bel^1(A) + \cdots + \log Bel^k(A) - \log Bel^\rho(A))$$

In the last expression we write, for simplicity, $Bel^j(A)$ instead of $Bel^{\pi^{(j)}}(A)$.

When applied to two assignments $\mu = (m_A)$ and $\nu = (n_B)$, whose focal sets form antichains, the effet is

$$\rho(C) = \frac{1}{1-k} \sum_{A \cap B = C} m_A n_B,$$

which gives Dempster rule of combination [11]. (Normalizing constant is $k = \sum_{A \cap B = \emptyset} m_A n_B$.)

The antichain property means that there are no proper containments among any two focal sets. In a sense it represents a completely dispersed evidence, where no single item fully reinforces another. We postulate (as an open problem) that the antichain condition is not necessary, or at least can be significantly weakened.

The very opposite is the case of fully nested focal sets [13]. This is exactly a possibilistic model (when presented in evidence setting). Our proposed formula is

$$d(\rho; \pi^{(1)}, \ldots, \pi^{(k)}) = U(\pi^{(1)}) - \ldots - U(\tilde{p}^{(k)}) - U(\rho)$$

Its application will lead, given two priors **p** and **q**, to their pointwise supremum **r**

$$r_i = \max(p_i, q_i), \; i = 1, \ldots, n.$$

Whether this is the best method of updating possibilities is a somewhat moot question. Proper discussion of updating must await resolving various questions of computing conditional possibility. We can only remark that taking maxima seem to accord with the spirit of several proposed conditioning methods.

References

[1] D. Dubois, H. Prade, 1988, *Possibility Theory*, Plenum Press, New York.

[2] D. Dubois, H. Prade, 1987, Properties of measures of information in evidence and possiblity theories. *Fuzzy Sets Syst.*, 24.

[3] M. Higashi, G. Klir, 1983, On the notion of distance representing information closeness, *Int. J. Gen. Sys.*, 9.

[4] M. Higashi, G. Klir, 1982, Measures of uncertainty and information based on possibility distributions, *Int. J. Gen. Sys.*, 8.

[5] J. N. Kapur, H. K. Kesavan, 1992, *Entropy Optimization Principles with Applications*, Academic Press, New York.

[6] J. Pearl, 1988, *Probabilistic Reasoning in Intelligent Systems*, Morgan Kaufmann, San Mateo, CA.

[7] A. Ramer, C. Padet, 1995, Principle of maximizing fuzzy uncertainty, *Int. J. General Systems*, (to appear).

[8] A. Ramer, 1994, Information Semantics of Possibilistic Uncertainty, *Fuzzy Sets Syst.*, 23.

[9] A. Ramer, 1992, Information and decisions in non-probabilistic uncertainty, *Proc. 2nd IIZUKA Int. Conf. on Fuzzy Logic and Neural Networks*, Iizuka, Japan.

[10] A. Ramer, 1990, Structure of possibilistic information metrics and distances, *Int. J. Gen. Sys.*, 17.

[11] G. Shafer, 1976, *A Mathematical Theory of Evidence.* Priceton University Press, Princeton.

[12] R. R. Yager, 1983, Entropy and specificity in a mathematical theory of evidence, *Int. J. Gen. Sys.* 9.

[13] R. R. Yager, 1980, Aspects of possibilistic uncertainty, *Int. Man-Machine Stud.*, 12.

Decision Making Using Belief Functions Evaluation of Information

Khaled Mellouli
Professor, Institut Supérieur de Gestion de Tunis
41 Rue de la liberté LeBardo Tunisie 2000

Abstract. In this paper, we use the belief functions framework to represent the available information in a decision making problem. We start by presenting the related decision process. Then, we define and caracterize the supporting knowledge of a decision. Finally, we give an evaluation of the confidence in a decision that is supported by a given knowledge.

1 Introduction

Decision making under uncertainty and incomplete information is a major problem facing decision makers. Several approaches to this problem are developped in the litterature especially in the work of Jaffray [2], Smets [7], Strat [8], Yager [9] and Weber [10]. In this paper, we consider that knowledge about the state of nature on which depends the payoffs related to a decision is given by a belief function. Following Strat's approach [8], we have a criteria for decision making based on expected value of payoffs using belief functions. We define and caracterize, in this paper, the supporting knowledge to a given decision. We then define the degree of precision of a belief function. This definition will help us associate to a piece of evidence supporting a given decision a degree of confidence in that decision

2 Belief function decision making model.

In this section, we will start by presenting some elements of the theory of belief functions and then we present the belief function decision making model.

2.1 Elements of the theory of belief functions

A belief function model is defined by a set Θ called a frame of discernement and by a basic probability assignment (b.p.a.) function. A basic probability assignment function assigns to every subset A of Θ a number m(A) such that m(A)≥0 for all $A \subseteq \Theta$, m(∅)=0, and $\Sigma\{m(A) \mid A \subseteq \Theta\}=1$. The basic probability assignment function is a generalization of the probability mass function in probability theory. Shafer [4] defines the belief function Bel corresponding to a basic probability assignment m as the function such that for any subset B of Θ, $Bel(B)=\Sigma\{m(A) \mid A \subseteq B\}$. The notation used for summation denotes the number obtained by adding numbers m(A) for all subsets of B. A subset A of Θ such that m(A)>0 is called a focal element of Bel. m(A) is called a basic probability number and it measures the belief that one commits exactly to A. The total belief commited to A is given by Bel(A). We say that Bel is vacuous if the only focal element is Θ, such belief function represents ignorance in the theory of belief functions. We say that a belief

function Bel is a bayesian belief function if all its focal elements are singletons hence Bel becomes a probability distribution.
In order to combine independent pieces of evidence, Shafer [4] uses the direct sum operation introduced by Dempster [1] and calls it Dempster's rule of combination. Let Bel_1 and Bel_2 be two belief functions over a frame of discernment Θ, and suppose Bel_1 and Bel_2 correspond to two independent bodies of evidence. Let m_1, m_2 be their respective basic probability assignment functions. The direct sum of Bel_1 and Bel_2 is denoted by $Bel=Bel_1 \oplus Bel_2$ and is given by its basic probability assignment function m:
$m(A)=K\Sigma\{m_1(A_1)m_2(A_2) \mid A_1 \subseteq \Theta, A_2 \subseteq \Theta, A_1 \cap A_2 = A\}$ for $\varnothing \neq A \subseteq \Theta$, where
$K^{-1}=1-\Sigma\{m_1(A_1)m_2(A_2) \mid A_1 \subseteq \Theta, A_2 \subseteq \Theta, A_1 \cap A_2 = \varnothing\}$.
The belief function $Bel=Bel_1 \oplus Bel_2$ is not defined whenever $K^{-1}=0$. In this case, the two belief functions totally contradict each other.

2.2 The Belief function decision model.

We suppose that we are facing a decision situation where we have a set of possible decisions $\mathfrak{a}=\{ai \mid i \in I\}$ and $\Theta=\{\Theta_j \mid j \in J\}$ representing the set of possible states of nature. Let $V(ai, \Theta j)$ the payoff received when we decide ai and Θj is the state of nature that prevails. Let $V(\mathfrak{a}, \Theta)$ the matrix of payoffs. Let us suppose that our knowledge about the state of nature is given by a belief function Bel over the frame of discernemnt Θ. We define for every decision ai element of A and every subset A of Θ
$Max(ai,A)= max\{V(ai,\Theta j) \mid \Theta j \in A\}$ and $Min(ai,A)= min\{V(ai,\Theta j) \mid \Theta j \in A\}$.
Following Yager [9] and adopting Strat [8] approach, we define for every decision ai element of $\mathfrak{a}$, the expected value
$E_{(\alpha,Bel)}(ai)=\Sigma\{m(A)[\alpha max(ai,A)+(1-\alpha)min(ai,A)] \mid A \subseteq \Theta\}$ where α is hurwitz coefficient of optimism, $0 \leq \alpha \leq 1$. By using this definition we suppose that we have complete ignorance inside the set of states of nature corresponding to a focal element A of the belief fonction Bel and facing this complete ignorance inside this focal element we use Hurwitz criteria for decision making $\alpha max(ai,A)+(1-\alpha)min(ai,A)$. We write $ai \geq_{(\alpha,Bel)} aj$ and say that ai αdominates aj whenever $E_{(\alpha,Bel)}(ai) \geq E_{(\alpha,Bel)}(aj)$.
Then. for $0 \leq \alpha \leq 1$, we select the best decision $a^{*(\alpha,Bel)}$ such that:
$E_{(\alpha,Bel)}(a^{*(\alpha,Bel)})=max\{E_{(\alpha,Bel)}(ai) \mid a_i \in \mathfrak{a}\}$. It is easy to prove that if Bel is a bayesian belief function, we get the bayesian decision model and if Bel is a vaccuous belief function we get the Hurwitz criteria for decision making. It is also easy to show that: $E_{(\alpha,Bel)}(ai) = \alpha E_{(1,Bel)}(ai)+(1-\alpha)E_{(0,Bel)}(ai)$. We denote by $ai \geq_{(Bel)} aj$ and say that ai strictly dominates aj whenever $ai \geq_{(\alpha,Bel)} aj$ for every α such that $0 \leq \alpha \leq 1$. It is quite clear that $ai \geq_{(Bel)} aj$ if and only if $ai \geq_{(1,Bel)} aj$ and $ai \geq_{(0,Bel)} aj$. We also have: $ai \geq_{(1,Bel)} aj$ if and only if $\Sigma\{m(A)[max(ai,A)-max(aj,A)] \mid A \subseteq \Theta\} \geq 0$.

ai $\geq_{(0,Bel)}$ aj if and only if $\Sigma\{m(A)[\min(ai,A)-\min(aj,A)] \mid A\subseteq\Theta\}\geq 0$.

If ai $\geq_{(Bel)}$ aj for every aj≠ai we select the decision ai and denote it by $a^{*(Bel)}$.

Example 1: let us consider the following decision problem:

Let A={a_1,a_2,a_3} and Θ={Θ_1,Θ_2,Θ_3}.

Let V(a, Θ) =

	Θ_1	Θ_2	Θ_3
a_1	40	-20	10
a_2	-5	30	20
a_3	20	5	15

let us suppose that our knowledge about the state of nature could be represented by a belief function Bel1 over Θ given by its basic probability assignment m1 such that m1({Θ1,Θ3})=0.7 and m1(Θ)=0.3, We want to find $a^{*(\alpha,Bel1)}$ for every $0\leq\alpha\leq 1$, and eventually $a^{*(Bel1)}$. Let us consider that the Hurwitz coefficent of optimism is α=1. The following tableau gives us for each ai the value $E_{(1,Bel1)}(ai)$. We first compute for each ai and for each subset A of Θ the value of max(ai,A), then we compute $E1=E_{(1,Bel1)}(ai)=\Sigma\{m1(A)\max(ai,A) \mid A\subseteq\Theta\}$.

m1(A)					0,7		0,3	
A	Θ_1	Θ_2	Θ_3	{Θ1,Θ2}	{Θ1,Θ3}	{Θ2,Θ3}	Θ	E1
a_1	40	-20	10	40	40	10	40	40
a_2	-5	30	20	30	20	30	30	23
a_3	20	5	15	20	20	15	20	20

Thus if our belief about the state of nature is given by Bel1, we find that a1 $\geq_{(1,Bel1)}$a2 and a1 $\geq_{(1,Bel1)}$ a3 hence the best decision $a^{*(1,Bel1)}$ is a_1. Similarly for α=0 the best decision $a^{*(0,Bel1)}$ is a_3. Now we compute $E_{(\alpha,Bel)}(ai)$ for each ai$\in$A. Since $E_{(\alpha,Bel)}(ai)=\alpha E_{(1,Bel)}(ai)+(1-\alpha)E_{(0,Bel)}(ai)=\alpha(E_{(1,Bel)}(ai)-E_{(0,Bel)}(ai))+E_{(0,Bel)}(ai)$ we have:

	$E_{(1,Bel1)}(ai)$	$E_{(0,Bel1)}(ai)$	$E_{(\alpha,Bel1)}(ai)$
a1	40	1	39α+1
a2	23	-5	28α-5
a3	20	12	8α+12

and we get for $0\leq\alpha\leq 11/31$, $a^{*(\alpha,Bel1)}$=a3, and for $11/31\leq\alpha\leq 1$, $a^{*(\alpha,Bel1)}$=a1, $a^{*(Bel1)}$ does not exist.

3. Supporting Knowledge.

Again we suppose that we are facing the decision situation described in the last section. For a given decision ai, we want to caracterize the supporting knowledge for this decision. Hence, we want to find the set of all belief functions over Θ that enable us to get $a^{*(Bel)}$=ai. Such belief functions are said to support decision ai. Let us denote this set by B(ai). Thus

$B(ai)=\cap\{\{Bel \mid ai \geq_{(Bel)} aj\} \mid aj\neq ai\}=\cap\{\{Bel \mid ai \geq_{(1,Bel)} aj \text{ and } ai \geq_{(0,Bel)} aj\} \mid aj\neq ai\}$

$B(ai)=\{Bel \mid [\Sigma\{m(A)[max(ai,A)-max(aj,A)] \mid A\subseteq\Theta\}\geq 0, \text{ for all } aj\neq ai] \text{ and }$
$[\Sigma\{m(A)[min(ai,A)-min(aj,A)] \mid A\subseteq\Theta\}\geq 0, \text{ for all } aj\neq ai]\}$

The set B(ai) gives us a caracterization of the information (belief functions) supporting decision ai.

Example 2: Let us consider the decision problem given in example 1 and let us caracterize the supporting knowledge for a decision ai. We suppose that our knowledge is given by a belief function Bel1 over Θ and let us denote its basic probability assignment function by $m1(\{\Theta_1\})=x1$, $m1(\{\Theta_2\})=x2$, $m1(\{\Theta_3\})=x3$, $m1(\{\Theta_1,\Theta_2\})=x12$, $m1(\{\Theta_1,\Theta_3\})=x13$, $m1(\{\Theta_2,\Theta_3\})=x23$, $m1(\Theta)=x123$ and $x1+x2+x3+x12+x13+x23+x123=1$. A belief function will then be noted as a 7uple $(x1,x2,x3,x12,x13,x23,x123)\in[0,1]^7$ such that $x1+x2+x3+x12+x13+x23+x123=1$.

We will first caracterize B(a1). We know that B(a1) is the set of belief function over Θ such that $\Sigma\{m(A)[max(a1,A)-max(aj,A)] \mid A\subseteq\Theta\}\geq 0$, for all $aj\neq a1$ and $\Sigma\{m(A)[min(a1,A)-min(aj,A)] \mid A\subseteq\Theta\}\geq 0$, for all $aj\neq a1$. The following tableau gives the values of max(ai,A) for all $ai\in A$ and all $A\subseteq\Theta$.

m1(A)	x1	x2	x3	x12	x13	x23	x123
	Θ_1	Θ_2	Θ_3	$\{\Theta_1,\Theta_2\}$	$\{\Theta_1,\Theta_3\}$	$\{\Theta_2,\Theta_3\}$	Θ
a_1	40	-20	10	40	40	10	40
a_2	-5	30	20	30	20	30	30
a_3	20	5	15	20	20	15	20

Hence $a1 \geq_{(1,Bel)} a2$ could be expressed by the following inequality:

$45x1-50x2-10x3+10x12+20x13-20x23+10x123\geq 0$.

Also we express $a1 \geq_{(1,Bel)} a3$ by: $20x1-25x2-5x3+20x12+20x13-25x23+20x123\geq 0$

Similarly we express $a1 \geq_{(0,Bel)} a2$ by: $45x1-50x2-10x3-15x12+15x13-40x23-15x123\geq 0$

and $a1 \geq_{(0,Bel)} a3$ by: $20x1-25x2-5x3-25x12-5x13-25x23-25x123\geq 0$.

We now can write that: $B(a1)=\{(x1,x2,x3,x12,x13,x23,x123)\in[0,1]^7 \mid$

$45x1-50x2-10x3+10x12+20x13-20x23+10x123\geq 0$
$45x1-50x2-10x3-15x12+15x13-40x23-15x123\geq 0$
$20x1-25x2-5x3+20x12+20x13-5x23+20x123\geq 0$
$20x1-25x2-5x3-25x12-5x13-25x23-25x123\geq 0$.
$x1+x2+x3+x12+x13+x23+x123=1\}$.

Similarly we can find B(a2) and B(a3). We note that for every ai, B(ai) is a convex subset of R^7, that gives all belief functions supporting decision ai

Let us suppose that the cardinal of the frame of discernement Θ is n. We define the degree of precision of a belief function Bel over Θ by:

$dp(Bel)=\Sigma\{(n-|A|)m(A) \mid A\subseteq\Theta\}/(n-1)$ where |A| is the cardinal of A. We notice that if Bel is a vaccuous belief function then dp(Bel)=0, and if Bel is a Bayesian belief function then

$dp(Bel)=1$. We also notice that for any belief function Bel over Θ we have $0 \le dp(Bel) \le 1$. We will say that Bel1 is more precise then Bel2 whenever $dp(Bel1) \ge dp(Bel2)$. In fact we will be looking for the less precise belief function supporting a given decision. Hence, we have to find the elements of B(ai) with the minimal value of dp(Bel), these elements are called minimal elements of B(ai) and could be found by solving the following linear programming problem:

$$\text{Min } C=dp(Bel)=\Sigma\{(n-|A|)m(A) \mid A \subseteq \Theta\}/(n-1)$$
$$\text{subject to } ai \ge_{(1,Bel)} aj \text{ for all } aj \ne ai \text{ and } ai \ge_{(0,Bel)} aj \text{ for all } aj \ne ai$$

This program could also be written as follows:

$$\text{Min } C=dp(Bel)=\Sigma\{(n-|A|)m(A) \mid A \subseteq \Theta\}/(n-1)$$
$$\text{subject to } \Sigma\{m(A)[\max(ai,A)-\max(aj,A)] \mid A \subseteq \Theta\} \ge 0, \text{for all } aj \ne ai$$
$$\text{and } \Sigma\{m(A)[\min(ai,A)-\min(aj,A)] \mid A \subseteq \Theta\} \ge 0, \text{for all } aj \ne ai$$

Let us denote C*(ai) the optimal value of the objective function of the preceding linear program. Using this value, the degree of precision of minimal elements of B(ai), we get a necessary condition for a belief function to support the decision ai. Thus, if a belief function Bel has a degree of precision dp(Bel) such that $dp(Bel) < C^*(ai)$ then it is impossible for such belief function to support decision ai.

Example 3: Let us consider the decision problem given in example 2. The degree of precision of a given belief function Bel can be computed as $dp(Bel)=x1+x2+x3+0.5x12+0.5x13+0.5x23$. We can find C*(a1) and caracterize the minimal elements of B(a1) by solving the following linear program:

$$\text{Min } dp(Bel)=x1+x2+x3+0.5x12+0.5x13+0.5x23$$
$$\text{subject to: } 45x1-50x2-10x3+10x12+20x13-20x23+10x123 \ge 0$$
$$45x1-50x2-10x3-15x12+15x13-40x23-15x123 \ge 0$$
$$20x1-25x2-5x3+20x12+20x13-5x23+20x123 \ge 0$$
$$20x1-25x2-5x3-25x12-5x13-25x23-25x123 \ge 0.$$
$$x1+x2+x3+x12+x13+x23+x123=1$$

All variables are positifs between zero and one.

Similarly we can find C*(a2) and caracterize the minimal elements of B(a2) by solving the following liear program:

$$\text{Min } dp(Bel)=x1+x2+x3+0.5x12+0.5x13+0.5x23$$
$$\text{subject to: } -45x1+50x2+10x3-10x12-20x13+20x23-10x123 \ge 0$$
$$-45x1+50x2+10x3+15x12-15x13+40x23+15x123 \ge 0$$
$$-25x1+25x2+5x3+10x12+15x23+10x123 \ge 0$$
$$-25x1+25x2+5x3-10x12-20x13+15x23-10x123 \ge 0.$$
$$x1+x2+x3+x12+x13+x23+x123=1$$

All variables are positifs between zero and one.

Similarly by using a linear program we can find C*(a3) and caracterize the minimal elements of B(a3). By solving these linear programs we get $C^*(a1)=0.555$, $C^*(a2)=0.2$ and $C^*(a3)=0.886$. Hence a belief function Bel with a degree of precision $dp(Bel) < 0.2$ is unable to support any decision, also if dp(Bel) is between 0.2 and 0.555 such a belief function is unable to support decision a1 or a3 but could support decision a2, if dp(Bel) is between 0.555 and 0.886 such a belief function is unable to support decision a3 but could support a1 or a2, finally if $dp(Bel)>0.886$ such a belief function could support any of the three decisions.

4. Confidence in a decision.

We now suppose that we have a piece of evidence given by Bel1 that gives $a^{*(Bel1)}$=ai (Bel1∈B(ai)). Let us consider the set of belief function Bel2 such that $a^{*(Bel1 \oplus Bel2)}$=aj (Bel1⊕Bel2∈B(aj)). We denote this set B(Bel1,→aj). We define the degree of transfert from ai to aj to be the degree of precision of the minimal elements of B(Bel1,→aj). We denote it by dt(Bel1,→aj). We define and denote the confidence in our decision ai as follows Conf(ai)= min{dt(Bel1,→aj)/ aj≠ai}.

Example 4: Let us consider the setting of example 3 and let us consider that our knowledge is given by a belief function Bel1 element of B(a1) (given by $m1(\{\Theta_1\})=0.6$, $m1(\Theta)=0.4$). Let us caracterize the set B(Bel1,→a2). Let us suppose that Bel2 is given by its basic probability assignment function m2: $m2(\{\Theta_1\})=y1$, $m2(\{\Theta_2\})=y2$, $m2(\{\Theta_3\})=y3$, $m2(\{\Theta_1,\Theta_2\})=y12$, $m2(\{\Theta_1,\Theta_3\})=y13$, $m2(\{\Theta_2,\Theta_3\})=y23$ and $m2(\Theta)=y123$ withy1+y2+y3+y12+y13+y23+y123=1 $Bel=Bel_1 \oplus Bel_2$ is given by its basic assignment function m. First we compute

$K^{-1}=1-\Sigma\{m_1(A_1)m_2(A_2)|A_1\subseteq\Theta,A_2\subseteq\Theta,A_1\cap A_2=\varnothing\}=1-0.6y2-0.6y3-0.6y23$

$m(\{\Theta_1\})=k(y1+0.6y12+0.6y13+0.6y123)$, $m(\{\Theta_2\})=0.4ky2$, $m(\{\Theta_3\})=0.4ky3$, $m(\{\Theta_1,\Theta_2\})=0.4ky12$, $m(\{\Theta_1,\Theta_3\})=04ky13$, $m(\{\Theta_2,\Theta_3\})=0.4ky23$ and $m(\Theta)=0.4ky123$. We now can compute dt(Bel1,→a2) by solving the linear programming problem caracterizing the minimal elements of B(Bel1,→a2):

Min dp(Bel)=y1+y2+y3+0.5y12+0.5y13+0.5y23

subject to: -45(y1+0.6y12+0.6y13+0.6y123)+50(0.4y2)+10(0.4y3)-10(0.4y12) -20(0.4y13)+20(0.4y23)-10(0.4y123) ≥0

-45(y1+0.6y12+0.6y13+0.6y123)+50(0.4y2)+10(0.4y3)+15(0.4y12) -15(0.4y13)+40(0.4y23)+15(0.4y123) ≥0

-25(y1+0.6y12+0.6y13+0.6y123)+25(0.4y2)+5(0.4y3)+10(0.4y12) +15(0.4y23)+10(0.4y123) ≥0

-25(y1+0.6y12+0.6y13+0.6y123)+25(0.4y2)+5(0.4y3)-10(0.4y12) -20(0.4y13)+15(0.4y23)-10(0.4y123) ≥0.

y1+y2+y3+y12+y13+y23+y123=1

All variables are positifs between zero and one.

By simplifying this program we get:

Min dp(Bel)=y1+y2+y3+0.5y12+0.5y13+0.5y23

subject to:
-45y1+20y2+4y3-31y12-35y13+8y23-31y123≥0
-45y1+20y2+4y3-21y12-33y13-16y23-21y123≥0
-25y1+10y2+2y3-11y12-15y13+6y23-11y123≥0
-25y1+10y2+2y3-19y12-23y13+6y23-19y123≥0.
y1+y2+y3+y12+y13+y23+y123=1
All variables are positifs between zero and one.

Solving this linear program we find dt(Bel1,→a2)=0.397. Similarly we compute dt(Bel1,→a3)= 0.455 . Thus the confidence in our decision a1 is:
Conf(a1)= min{dt(Bel1,→aj)/ aj≠a1}=0.397.

5. Conclusion

In this paper, we adopt strat's approach to decision making under uncertainty using belief functions. We first present this approach, then we define the supporting knowledge to a given decision as the set of belief functions that give this decision as optimal independently of the degree of optimism of the decision maker. We then give a caracterization of this set of belief functions. We also define a degree of precision of a belief function. This definition gives us a necessary condition for a belief function to support a given decision, Then, we define a degree of confidence in a decision when our knowedge is given by a belief function supporting that decision. By using the degree of confidence in a decision and the propagation of belief functions in Markov tree setting, see Mellouli [3], Shafer et al.[5], we could define a value of the information given by a given belief function. Also we could extend this work to valuation based systems see Shenoy [6].

References

1. A.P. Dempster: A generalization of Bayesian inference (with discussion), *Journal of the Royal Statistical Society, Series B,* vol. 30, 205-247 (1968).
2. J. Y. Jaffray: Linear Utility Theory for Belief Functions, *Operations Research Letters 8* North-Holland, 107-112 (1989).
3. K Mellouli: On the Combination of Beliefs in Networks using the Dempster-Shafer Theory of Evidence, Doctoral dissertation, school of Business, University of Kansas (1987).
4. G. Shafer: *A Mathematical Theory of Evidence*, Princeton University Press, NJ (1976).
5. G. Shafer, P.P. Shenoy and K. Mellouli: Propagating belief functions in qualitative Markov trees, *International Journal of Approximate Reasoning.* vol1, 349-400 (1987)
6. P.P. Shenoy: Valuation based Systems for Bayesian Decision Analysis, *Working Paper 220.* School of Business University of Kansas K.S (1990).
7. P. Smets: Decisions and belief functions. Technical report 90-10 Institut de Recherches Interdisciplinaires et de développements en Intélligence Artificielle. Université Libre de Bruxelles (1990).
8. T. M. Strat: Decision making using belief functions, *International Journal of Approximate Reasoning.* North-Holland, vol4, 391-417 (1990).
9. R. R. Yager: Decision Making Under Dempster-Shafer Uncertainties, *Tech. Report No MII-915,* Machine Intelligence Institute, Iona College, New Rochelle, N.Y. 10801. (1989).
10. M. Weber: Decision making with incomplete information, *European Journal of Operational Research.* North-Holland, vol 4, 44-57 (1987).

3. NETWORKS

Hybrid Propagation in Junction Trees

A. Philip Dawid[1], Uffe Kjærulff[2], Steffen L. Lauritzen[2]

[1] University College London, Gower Street, London WC1E 6BT, England
[2] Aalborg University, Fredrik Bajers Vej 7E, DK-9220 Aalborg, Denmark

Abstract. We introduce a methodology for performing approximate computations in complex probabilistic expert systems, when some components can be handled exactly and others require approximation or simulation. This is illustrated by means of a modified version of the familiar 'chest-clinic' problem.

1 Introduction

Markov Chain Monte-Carlo (MCMC) methods have over the last decade become increasingly popular as a computational tool in complex stochastic systems of various type (Gelfand and Smith 1990, Thomas et al. 1992, Gelman and Rubin 1992, Geyer 1992, Smith and Roberts 1993). The methods are flexible, easy to implement, and computing time tends to scale manageably with the size of the problem under consideration. Their main disadvantages are associated with their reliance on pseudorandom number generators (Ripley 1987), some difficulty in deciding whether the necessary equilibrium has been reached, as well as the fact that even moderately sized problems compute slowly.

Alternatively, fast and exact methods have been developed for computation in particular types of graphical models (Lauritzen and Spiegelhalter 1988, Shenoy and Shafer 1990, Jensen et al. 1990, Dawid 1992, Lauritzen 1992, Spiegelhalter et al. 1993) and implemented, for example, in the shell HUGIN (Andersen et al. 1989). Whenever exact methods apply, they are usually preferable, but the weakness of exact methods is precisely the limitations on the problems that they are currently able to deal with appropriately. Even though the development of exact methods has not stopped nor found its limit, there will always be interesting problems that cannot be appropriately dealt with by means of an exact analysis.

Currently the situation is that if just a single variable in a Bayesian network has a continuous distribution, then the exact analysis cannot be performed and the entire network must be handled with Monte-Carlo methods – except in special cases (Lauritzen 1992).

This paper is concerned with establishing the basic ideas behind a scheme that enables the combination of virtues of the two types of methods. The idea is to split up a graphical model into submodels (universes) that each use their favourite method of computation, organise the universes in a junction tree as also done in the case of exact methods, and letting them communicate with each other. The hope is that this will produce a scheme which has the flexibility of the

MCMC methods but exploits speed and correctness of exact methods whenever feasible. In the research horizon, more general schemes involving combination with belief functions and objects from other uncertainty formalisms appear as interesting possibilities.

2 A Basic Example

To illustrate our scheme we have chosen a modified version of the chest-clinic problem (Lauritzen and Spiegelhalter 1988). The directed acyclic graph in Fig. 1a shows the independence graph of the associated model with variables 'Visit to Asia?' (A), 'Smoker?' (S), 'Tuberculosis?' (T), 'Lung cancer?' (L), 'Bronchitis?' (B), 'Either tuberculosis or lung cancer?' (E), 'Positive X-ray?' (X), and 'Dyspnoea?' (D). All variables have states 'yes' and 'no', in the following denoted by 1 and 0, respectively. For convenience, we shall also use the notation $A = a$ for $A = 1$, $A = \bar{a}$ for $A = 0$, etc.

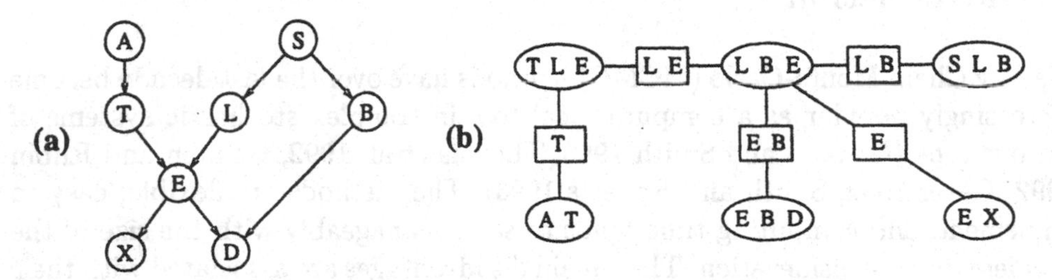

Fig. 1. Independence graph (a) and a junction tree (b) for the original chest-clinic model

The original conditional probabilities associated with the model are reproduced in Table 1.

Table 1. Conditional probabilities for the chest-clinic model

$P(A=1) = 0.01$	$P(E=1 \mid L=1, T=1) = 1$
	$P(E=1 \mid L=1, T=0) = 1$
$P(T=1 \mid A=1) = 0.05$	$P(E=1 \mid L=0, T=1) = 1$
$P(T=1 \mid A=0) = 0.01$	$P(E=1 \mid L=0, T=0) = 0$
$P(S=1) = 0.50$	$P(X=1 \mid E=1) = 0.98$
	$P(X=1 \mid E=0) = 0.05$
$P(L=1 \mid S=1) = 0.10$	
$P(L=1 \mid S=0) = 0.01$	$P(D=1 \mid B=1, E=1) = 0.90$
	$P(D=1 \mid B=1, E=0) = 0.80$
$P(B=1 \mid S=1) = 0.60$	$P(D=1 \mid B=0, E=1) = 0.70$
$P(B=1 \mid S=0) = 0.30$	$P(D=1 \mid B=0, E=0) = 0.10$

A junction tree corresponding to the graph of Fig. 1a is constructed via the operations of moralization (edges between E and B and between T and L) and triangulation (edge between L and B) (Lauritzen and Spiegelhalter 1988) and is shown in Fig. 1b, with the belief universes (cliques) marked by ovals, and the separators by rectangles. The initial (i.e., with no evidence) marginal probabilities for the individual variables can be calculated by appropriate marginalisations of the clique potentials after a propagation has been conducted. The result of these operations appear in Table 2.

Table 2. Marginal probabilities for the individual variables of the chest-clinic model

$P(A) = (0.0100, 0.9900)$	$P(B) = (0.4500, 0.5500)$
$P(T) = (0.0104, 0.9896)$	$P(E) = (0.0648, 0.9352)$
$P(S) = (0.5000, 0.5000)$	$P(X) = (0.1103, 0.8897)$
$P(L) = (0.0550, 0.9450)$	$P(D) = (0.4361, 0.5639)$

The modified chest-clinic model illustrating our hybrid propagation scheme is as follows. First, assume that we want to relax the constraint that bronchitis is either present or absent, and instead be able to express degrees of bronchitis on a continuous scale. We could accomplish this by defining the conditional distribution of acquiring bronchitis given smoking habits by some standard continuous distribution. If a beta distribution (Johnson and Kotz 1970) is found appropriate, we could define

$$B \mid S \sim \mathrm{Be}(3 + 3S, 4 + 3(1 - S)) \ , \tag{1}$$

which yields mean values $3/(3 + 7) = 0.3$ and $6/(6 + 4) = 0.6$ for $S = 0$ and $S = 1$, respectively, complying with the original specification of $P(B \mid S)$ (see Table 1).

Next, when $B \mid S$ has been specified as a beta distribution, we need to respecify the conditional probabilities of suffering from shortness of breath (dyspnoea) given bronchitis and lung cancer or tuberculosis. This could be accomplished through

$$\text{logit } P(D = 1 \mid B, E) = \alpha B + \beta E + \gamma \ , \tag{2}$$

where logit $x = \log x/(1 - x)$. For $\alpha = 2.5$, $\beta = 1.9$ and $\gamma = -1.65$ we get

$$\begin{aligned}
\text{logit } P(D = 1 \mid B = 1, E = 1) &= 2.75\ (0.94) \\
\text{logit } P(D = 1 \mid B = 1, E = 0) &= 0.85\ (0.70) \\
\text{logit } P(D = 1 \mid B = 0, E = 1) &= 0.25\ (0.56) \\
\text{logit } P(D = 1 \mid B = 0, E = 0) &= -1.65\ (0.16)
\end{aligned}$$

where the numbers in parentheses are the corresponding conditional probabilities obtained by $\text{logit}^{-1}(2.75) = 0.94$, etc, where $\text{logit}^{-1}(x) = e^x/(1 + e^x)$. Note that these probabilities are somewhat different from those of Table 1, but the closest we can get when $P(D \mid B, E)$ is given by an expression as (2).

The modification of the model which we have introduced is sufficient to destroy the possibility of computing marginal probabilities at all nodes by exact methods. This example is small enough that direct Monte-Carlo or other numerical methods can be exploited to obtain the marginal probabilities. In this paper we exploit this fact to check our calculations. The example is supposed only to illustrate a scheme that has its practical interest in problems of larger size.

With the above modifications, the mean of B is $E(B) = E(B \mid S=0)P(S=0)+E(B \mid S=1)P(S=1) = 0.5 \cdot 0.3 + 0.5 \cdot 0.6 = 0.45$. The marginal distribution for D can be calculated to be $(0.4042, 0.5958)$. The last result has been found by numerical integration.

3 Propagation by Message Passing

Here we briefly recall the general idea of propagation by message passing in a junction tree as described by e.g. Dawid (1992). We use notation and terminology from this reference.

Each belief universe (clique) C and separator S in the junction tree is initially equipped with a *potential* such that the joint density satisfies

$$f(x) \propto \frac{\prod_{C \in \mathcal{C}} a_C(x_C)}{\prod_{S \in \mathcal{S}} b_S(x_S)} \quad . \tag{3}$$

When evidence has been received, the potentials might have changed in some universes, but an expression of the form (3) continues to be valid.

Messages can now be sent from one universe C to another universe D via their separator S in the following way:

1. C calculates the *marginal* to S, $a_C^{\downarrow S}$, of its potential.
2. The potential on D *receives* the message by changing to

$$a_D^* = a_D \frac{a_C^{\downarrow S}}{b_S} = \frac{a_D}{b_S} a_C^{\downarrow S} \quad . \tag{4}$$

3. The separator potential changes to $b_S^* = a_C^{\downarrow S}$.

Two important points should be noted here. Firstly, whenever a message has been sent, the correctness of (3) has not been affected. Secondly, if the messages are scheduled appropriately, then after a finite number of message passes, all messages are neutral, i.e., they do not change anything. We then say that the junction tree is in *equilibrium*. When this happens, marginals to all subsets of a universe calculated from its potential are identical to marginals calculated from the joint expression (3).

The basic idea behind this paper is to consider the potentials and (3) as purely formal objects, allowing the computer to represent these either directly as (discrete) probability tables (as in the current implementation of HUGIN), or as programs such as numerical integrators or Monte-Carlo machines that can draw at random from a distribution with density proportional to a given potential.

For this to be of any practical interest we must also think in terms of the universes being large networks and other more complex objects than just cliques in a graph. If we think of such objects as sets, they are organised in a junction tree of usual type.

The difficulties encountered in such a general approach is concerned with an elaboration of the abstract theory, as well as the specific description of types of universes and the scheduling of messages. The abstract theory is to be discussed in a subsequent paper and the other aspects in the following sections. One aspect of the abstract theory must be briefly mentioned to explain the following developments: The potentials should be replaced by measures rather than functions, and ratios interpreted as Radon-Nikodym derivatives. We avoid the technical details here.

However, we point out that in general it may happen in our scheme that a marginalised measure $\mu^{\downarrow S}$ is not absolutely continuous with respect to the separator measure μ_S. Then the second version of (4) must be used, rather than the first. In the first case, we say that D receives its message *as an update ratio*, and in the second case *as a measure.*

4 Universe Types

This section describes in general terms the main types of universes dealt with in our scheme and how they communicate with each other. We distinguish between the following three main types.

DE (discrete exact) The archetype of a belief universe. A DE-universe is simply given by a list of possible configurations and their associated weights (real numbers). The term 'exact' refers to the fact that it knows to do whatever computation it needs in an exact way.

E (other exact) There are other types of universes that can do exact computations. Examples of such E-universes are
- E-hugin: A universe of discrete variables represented in the HUGIN sense, i.e., by a junction tree of DE-universes.
- E-cghugin: A universe with possibly both discrete and continuous conditional Gaussian variables in the HUGIN sense. DE-universes and E-hugin universes become special cases.
- E-formula: A universe of discrete variables with a belief potential expressed through exact formulae. Computations can for example be handled through some form of computer algebra program.

N (numerical) This is a type of universe that in some way calculates only approximately. An N-universe is typically one of
- N-int: A universe where computations are made by numerical integration; such universes will not be discussed in the present paper, but we find it important to point at the possibility.
- N-mc: This is a universe that makes its calculations through some form of a Monte-Carlo method, for example Gibbs sampling in a moral graph, forward sampling in a directed, acyclic graph, or direct sampling without exploiting any graphical structure. The calculations can involve importance weights.

The different kinds of universes and the way they send and receive messages will be described in detail in Sects. 4.1–4.3.

Returning to our example, we choose to implement AT, TLE and EX as universes of type DE, $SLBE$ as a universe of type N-mc, and EBD as a universe of type E-formula, where we use AT as shorthand for $\{A, T\}$, etc. The universes are connected in the junction tree displayed in Fig. 2.

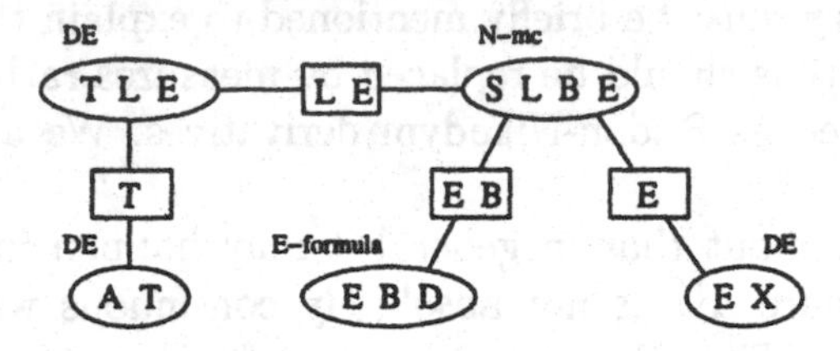

Fig. 2. Junction tree for the modified chest-clinic model with three DE-universes, one N-mc universe, and one E-formula universe

The belief potentials of the universes are instantiated as shown in Table 3, where $\phi(B) \sim U(0,1)$ in universe EBD, $\phi(E) = (1,1)$ in the universes $SLBE$ and EBD, and where $\phi(A) = P(A)$, $\phi(T \mid A) = P(T \mid A)$, etc. No evidence is assumed present. We shall later explain the precise meaning of these potentials in the specific cases.

Table 3. Initial instantiation of the belief potentials of the universes

Univ.	Kind	Initial potential(s)
AT	DE	$\phi(A), \phi(T \mid A)$
TLE	DE	$\phi(E \mid T, L)$
$SLBE$	N-mc	$\phi(S)$, $\phi(B \mid S)$, $\phi(L \mid S)$; $\phi(E)$
EBD	E-formula	$\phi(E)$, $\phi(B)$, $\phi(D \mid E, B)$
EX	DE	$\phi(X \mid E)$

4.1 Discrete Exact Universes

A DE-universe, U, is represented by a finite subset

$$\mathcal{X}'_U = \{x_U^1, \ldots, x_U^n\} \subseteq \mathcal{X}_U = \times_{v \in U} \mathcal{X}_v$$

of n possible states in the product space $\mathcal{X}_U$, and a set of associated weights $w_1, \ldots, w_n$, where w_i is the weight associated with x_U^i. As for other universes, its potential is given by a positive measure μ_U. For a function f, the integral of f with respect to μ_U is thus represented as

$$\int f(x) \mu_U(dx_U) = \sum_{i=1}^{n} w_i f(x_U^i) \ .$$

The universe marginalises to the space

$$\mathcal{X}'_A = \{x_A, x_U \in \mathcal{X}'_U\} \subseteq \mathcal{X}_A = \times_{v \in A} \mathcal{X}_v$$

by just adding the weights of points with the same A-marginal. That is,

$$w_i^{\downarrow A} = \sum_{j:x_A^j = x_A^i} w_j .$$

Messages are received as update ratios – messages sent from neighbouring universes as measures are transformed to update ratios upon division by separator measures. When receiving an update ratio ϕ_S, the revised weights are obtained as

$$w_i^* = w_i \cdot \phi_S(x_S^i) .$$

DE-universes can always receive and send messages when they are requested to do so.

In our basic example, the AT-universe has states $\{(a,t),(\overline{a},t),(a,\overline{t}),(\overline{a},\overline{t})\}$. The weights (potentials) are obtained by multiplying the appropriate numbers from Table 1. Similarly with the other DE-universes in our example.

4.2 Other Exact Universes

An E-universe, U, has some representation of a probability measure μ_U on $\mathcal{X}_U$ such that $\mu_U^{\downarrow S}$ can be computed in exact form for all separators S associated with U. We assume that if $\mu_U^{\downarrow S}$ is computable, so is $\mu_{U\setminus S|S}$. This seems no severe restriction.

As described above, U can be one of sub-types E-hugin, E-cghugin, or E-formula. Any kind of E-universe can send exact messages and receive any message. Upon receiving an update ratio, ϕ_S, the revised measure, μ_U^*, is obtained as

$$\mu_U^* = \mu_U \cdot \phi_S , \tag{5}$$

and upon receiving a measure μ_S, μ_U^* is obtained as

$$\mu_U^* = \mu_{U\setminus S|S} \cdot \mu_S , \tag{6}$$

but U may have to change to some other type such as N-type, since μ_U^* may not allow exact calculations.

In our basic example, the potential on the universe EBD is represented as a table (over DE) of algebraic expressions involving standard functions of B such as displayed in Table 4.

4.3 Numerical Universes

An N-universe, U, is equipped with some representation of a probability measure μ_U defined on $\mathcal{X}_U = \times_{v \in U} \mathcal{X}_v$, but it cannot calculate separator marginals exactly and must therefore make some approximate calculation.

If U is of type N-int, this approximation is obtained via numerical integration, possibly by approximating the entire universe and measure $(\mathcal{X}_U, \mu_U)$ by a discrete DE-universe $\mathcal{X}'_U$ with suitable weights. Thus, when integration has been performed, U may have changed type from N-int to DE.

Table 4. Initial exact potentials for the universe EBD

	$E=1$	$E=0$
$D=1$	$\frac{e^{2.5B+.25}}{1+e^{2.5B+.25}}$	$\frac{e^{2.5B-1.65}}{1+e^{2.5B-1.65}}$
$D=0$	$\frac{1}{1+e^{2.5B+.25}}$	$\frac{1}{1+e^{2.5B-1.65}}$

If U is of type N-mc, the approximation is obtained by using a program $\pi(\nu_U, \lambda_U, n)$ which performs forward sampling in a DAG on U drawing from ν_U, Gibbs sampling in an undirected graph on U drawing from ν_U, or Monte-Carlo sampling directly from ν_U without exploiting a graphical structure. Here we assume that $\mu_U = \lambda_U \nu_U$. The program returns a discrete DE-universe $\mathcal{X}'_U$ with points equal to those in the sample. If the function λ_U differs from unity, the samples are weighted according to λ_U. That is, each sample x^i_U is given importance weight $\lambda_U(x^i_U)$. The third argument determines the sample size. Thus, when the program terminates, U has changed type from N-mc to DE.

An N-universe can receive any kind of message (update ratio or measure), and the revised measure is obtained as described in (5) and (6). In any particular instance it must be specified whether the program representing the universe takes in the message by changing ν or by changing the importance weights λ. It may also have to change its Monte-Carlo sampling procedure from forward to Gibbs etc.

Since U becomes a DE-universe before sending, messages from U are sent as described in Sect. 4.1.

In our basic example we have implemented the universe $SLBE$ as an N-mc universe with forward sampling that receives its messages as importance factors. Hence, with the initial potential, it samples by first choosing S uniformly, then L from the distribution of L given S, B from a beta distribution with the correct parameters determined by the sampled value of S, and finally E uniformly. This gives a first point (s, l, b, e) in the approximating space, and the procedure is then repeated n times. If messages have been received from neighbouring universes, the points are weighted appropriately.

Note that this is not in any way an optimal implementation of our network, but it is chosen to illustrate the flexibility of the methods.

5 Network Architecture

In a certain sense the possibilities for scheduling of messages are like in the general case. The novel element is the fact that in the propagation process, universes may change type, in particular they may change into approximating universes, and these can be drastically different from the original ones. This happens for example when some universe with continuous state space changes to a discrete universe. This may destroy the property that the marginal of the universe is absolutely continuous with respect to a separator measure. When it has sent its message, this property will be restored in the separator, but then it may have been destroyed in the separators of the neighbouring universe etc., thus requiring further messages until the property is restored everywhere. We refer to this phenomenon as a *cascade* of messages.

Noting that messages never destroy correctness of the representation, we see that essentially the usual message schedule carries over, with the above modification. We shall illustrate a possible correct scheduling in our example. Recall that a message is termed *active* if it is sent from a universe that has received active messages from all its other neighbours. When active messages have been passed between any pair of universes, equilibrium has been reached. The scheduling in our example is as follows:

1. $AT \to TLE$;	5. $EBD \to SLBE$;
2. $EX \to SLBE$;	6. $SLBE \to TLE$;
3. $TLE \to SLBE$;	7. $TLE \to AT$;
4. $SLBE \to EBD$;	8. $SLBE \to EX$.

All messages are active and equilibrium is therefore reached. Below we indicate in more detail how the messages are sent.

AT → TLE This is a standard message. The potential is marginalised to T by summing the AT-table of numbers over the values of A. This is then placed as the new potential in the separator and is also multiplied onto the initial TLE-table. No division is needed because the separator initially contained a neutral potential.

EX → SLBE This message is almost a standard message. The universe EX sends the potential $a^*_E(e) = \sum_x a_{XE}(x,e)$ to the N-mc universe $SLBE$. The result is stored as an importance factor to be used when $SLBE$ is forward sampling. In our particular case the factor turns out to be unity, but this would not have been so if a chest X-ray had been taken.

TLE → SLBE This is a message of the same type as above. The universe TLE sends the separator potential $a^*_{LE}(l,e) = \sum_t a_{TLE}(t,l,e)$ to the N-mc universe $SLBE$ and this function is multiplied onto the existing importance factor.

SLBE → EBD For this message to be sent, we must first forward sample from $SLBE$ as described in Sect. 4.3, obtaining values $(s_i, l_i, b_i, e_i), i = 1, \ldots, n$ with weights w_i calculated from the importance factor potential.

Since B is a continuous variable, we cannot arrange our samples in a contingency table and we need to store all samples (see Table 5) and send the information pertaining to B and E to universe EBD.

When the sampling has ended, the universe has changed type to DE. It can now send to EBD in the usual way. Thus Table 5 is marginalised to BE by ignoring the S and L values and EBD receives the message by becoming of

Table 5. Sample values for universe *SLBE*

i	s_i	l_i	b_i	e_i	w_i
1	1	0	0.613862	0	0.989600
2	1	0	0.744140	0	0.989600
3	0	0	0.480592	1	0.010400
⋮	⋮	⋮	⋮	⋮	⋮
$n-1$	1	0	0.468109	0	0.989600
n	1	0	0.707468	1	0.010400

type DE with weights obtained from those in Table 5 by multiplication with appropriate values of the functions in Table 4.

When this message has been sent, all universes have become standard DE-universes and messages are sent in the usual way.

The hybrid propagation method given by the scheduling of messages described above has been implemented in the C programming language. Also, a forward sampling procedure has been implemented and run on our example. The marginal distributions on the individual variables obtained by running these two methods with a sample size of 1000 are displayed in Table 6 together with the exact values obtained by direct calculation (variables A, T, S, L, E, and X) and numerical integration (variables B and D). We observe no significant difference between the precision of the distributions provided by forward sampling and those provided by the hybrid propagation method. Since we have implemented the hybrid method in a very inefficient way in this example, this gives good promise for further developments.

Table 6. Exact marginal distributions and approximate ones obtained by forward sampling and hybrid propagation with a sample size of 1000. For variable B, the first number displayed is the mean and the second is the variance.

Variable	Forward	Hybrid	Exact
A	(0.0110, 0.9890)	(0.0100, 0.9900)	(0.01, 0.99)
T	(0.0140, 0.9860)	(0.0107, 0.9893)	(0.0104, 0.9896)
S	(0.5190, 0.4810)	(0.4816, 0.5184)	(0.5, 0.5)
L	(0.0520, 0.9480)	(0.0611, 0.9389)	(0.055, 0.945)
B	(0.4579, 0.0416)	(0.4430, 0.0426)	(0.45, 0.04300)
E	(0.0660, 0.9340)	(0.0712, 0.9288)	(0.0648, 0.9352)
X	(0.1020, 0.8980)	(0.1162, 0.8838)	(0.110, 0.890)
D	(0.4080, 0.5920)	(0.3908, 0.6092)	(0.4042, 0.5958)

Acknowledgements

This research was supported by ESPRIT Basic Research Action through project no. 6156 (DRUMS II) as well as the PIFT programme of the Danish Research

Councils. We are grateful to Ross Curds for providing the numerical integrations using *Mathematica*.

References

Andersen, S. K., Olesen, K. G., Jensen, F. V., Jensen, F.: HUGIN - A shell for building Bayesian belief universes for expert systems, *Proceedings of the 11th International Joint Conference on Artificial Intelligence* (1989) 1080–1085. Also reprinted in Shafer & Pearl (1990)

Dawid, A. P.: Applications of a general propagation algorithm for probabilistic expert systems, *Statistics and Computing* **2** (1992) 25–36

Gelfand, A. E., Smith, A. F. M.: Sampling-based approaches to calculating marginal densities, *Journal of the American Statistical Association* **85** (1990) 398–409

Gelman, A., Rubin, D. B.: Inference from iterative simulation using single and multiple sequences (with discussion), *Statistical Science* **7** (1992) 457–511

Geyer, C. J.: Practical Markov Chain Monte Carlo (with discussion), *Statistical Science* **7** (1992) 473–511

Jensen, F. V., Lauritzen, S. L., Olesen, K. G.: Bayesian updating in causal probabilistic networks by local computation, *Computational Statistics Quarterly* **4** (1990) 269–282

Johnson, N. L., Kotz, S.: *Distributions in Statistics. Continuous Univariate Distributions*, Vol. 2, Wiley & Sons, New York (1970)

Lauritzen, S. L.: Propagation of probabilities, means and variances in mixed graphical association models, *Journal of the American Statistical Association* **86** (1992) 1098–1108

Lauritzen, S. L., Spiegelhalter, D. J.: Local computations with probabilities on graphical structures and their application to expert systems (with discussion), *Journal of the Royal Statistical Society, Series B* **50** (1988) 157–224

Ripley, B. D.: *Stochastic Simulation*, Wiley & Sons (1987)

Shafer, G. R., Pearl, J. (eds): *Readings in Uncertain Reasoning*, Morgan Kaufmann, San Mateo, California (1990)

Shenoy, P. P., Shafer, G. R.: Axioms for probability and belief-function propagation, *in* R. D. Shachter, T. S. Levitt, L. N. Kanal and J. F. Lemmer (eds), *Uncertainty in Artificial Intelligence IV*, North-Holland, Amsterdam, (1990) 169–198

Smith, A. F. M., Roberts, G. O.: Bayesian computation via the Gibbs sampler and related Markov chain Monte Carlo methods, *Journal of the Royal Statistical Society, Series B* **55**(1) (1993) 5–23

Spiegelhalter, D. J., Dawid, A. P., Lauritzen, S. L., Cowell, R. G.: Bayesian analysis in expert systems (with discussion), *Statistical Science* **8** (1993) 219–247 and 247–283

Thomas, A., Spiegelhalter, D. J., Gilks, W. R.: BUGS: A program to perform Bayesian inference using Gibbs sampling, *in* J. M. Bernardo, J. O. Berger, A. P. Dawid and A. F. M. Smith (eds), *Bayesian Statistics 4*, Clarendon Press, Oxford, UK, (1992) 837–842

Heuristic Algorithms for the Triangulation of Graphs*

Andrés Cano and Serafín Moral

Departamento de Ciencias de la Computación e I.A.
Universidad de Granada
18071 - Granada - Spain

Abstract. Different uncertainty propagation algorithms in graphical structures can be viewed as a particular case of propagation in a joint tree, which can be obtained from different triangulations of the original graph. The complexity of the resulting propagation algorithms depends on the size of the resulting triangulated graph. The problem of obtaining an optimum graph triangulation is known to be NP-complete. Thus approximate algorithms which find a *good* triangulation in *reasonable* time are of particular interest. This work describes and compares several heuristic algorithms developed for this purpose.

1 Introduction

A number of different algorithms for the propagation of uncertainty in graphical structures have been developed in recent years. The main effort has been devoted to the study of probabilistic propagation. Original algorithms were proposed on a directed acyclic graph without loops by Kim and Pearl [12, 17]. Different propagation algorithms have been described for the general case [13, 18, 19, 6, 21, 15, 16].

Sachter, Andersen and Szolovits [20] have shown that the different exact algorithms described in the literature are all particular cases of a single general algorithm, called the clustering algorithm. This algorithm shows that the essence of the efficiency achieved in the original propagation algorithms, comes from the factorization of the global probability distribution which is obtained from the independence relationships among the variables of the problem. This factorization gives rise to the cluster tree, where all the computations are carried out. If we define the size of the cluster tree like the sum of the sizes of the cliques then the efficiency of the computations depends on the size of the associated cluster tree; in other words, efficiency is a polynomial function of size.

The key point in the construction of the cluster tree is the triangulation of the undirected graph expressing the independences of the problem. Through this

* This work was supported by the Commission of the European Communities under project DRUMS2, BRA 6156

process, the most efficient exact algorithms can be obtained by considering the optimum triangulation (the triangulation with a minimum size of the associated clusters). However, the construction of the optimal triangulation is known to be an NP-complete problem [1].

Kjærulff [9] has studied different heuristic methods and a simulated annealing algorithm in order to obtain efficient triangulations within a reasonable amount of time. One heuristic was shown to be the best, with similar results to the simulated annealing algorithm, which is much more time consuming.

In this work we propose new heuristics and compare them with Kjærulff's best heuristic, by using different randomly generated graphs. It is shown that these heuristics can improve the optimality of the resulting cluster trees.

The obtained results are applicable to other uncertainty formalisms. Shafer and Shenoy [21] and Cano, Delgado Moral [2] have shown that propagation algorithms can be applied in order to propagate other uncertainties represented by other methodologies in which combination and marginalization are defined as elementary operations, verifying a set of axioms. The efficiency of the calculations will be also much related to the size of the associated cluster tree.

This work is organized in the following way: in the second section, we briefly give the fundamental concepts related to the problem of triangulation of a graph and the construction of the associated cluster tree. In the third section, we describe the different heuristics, that will be compared and evaluated in the fourth section. Finally, in the fifth section, we present the conclusions.

2 Graph Triangulation

Let us assume an n-dimensional variable $(X_1, \ldots, X_n)$, each variable, X_i taking values on a finite set U_i. A usual way of expressing independences among variables is by means of a directed acyclic graph in which each node represents a variable, X_i (see Fig. 2) [16].

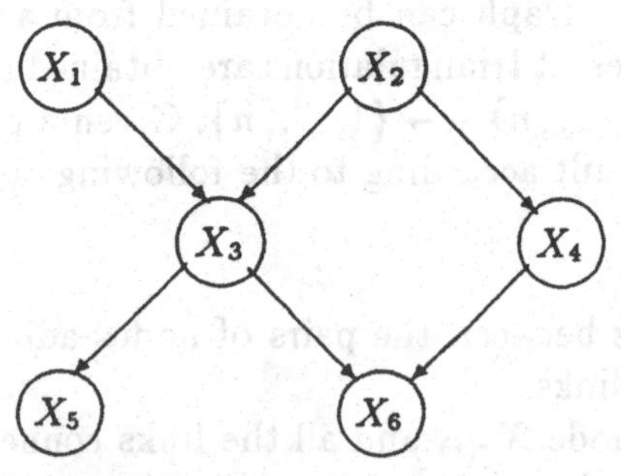

Fig. 1. A Directed Acyclic Graph

From a directed acyclic graph we can build its associated undirected graph, also called the moral graph. This graph is constructed by adding undirected links between two parents of the same node and ignoring the direction of the links in the directed graph (see Fig. 2) [13].

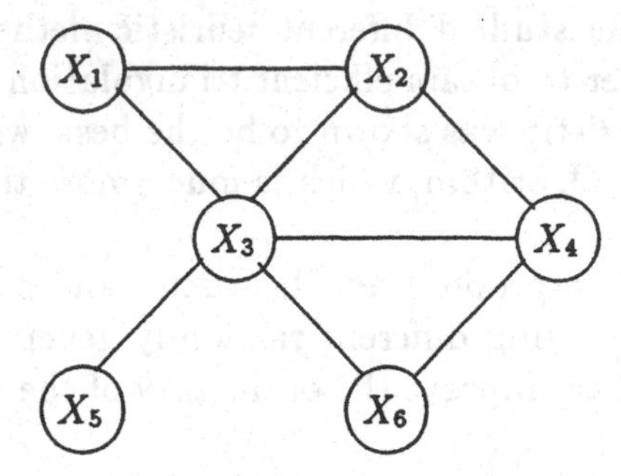

Fig. 2. The Moral Graph

An undirected graph is said to be chordal or triangulated if, and only if, every cycle of length four or more has an arc between a pair of nonadjacent nodes. To build a tree of clusters in which to carry out the computations we need a triangulated undirected graph. Graphs in Fig. 2 and 3 are triangulated and non triangulated graphs, respectively.

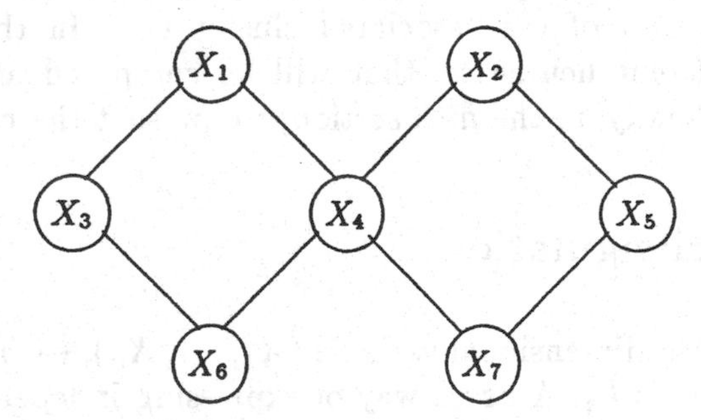

Fig. 3. A non chordal undirected graph

A triangulated graph can be obtained from a general undirected graph by adding links. Different triangulations are obtained from different permutations of the nodes, $\sigma : \{1, \ldots, n\} \longrightarrow \{1, \ldots, n\}$. Given a permutation, σ the associated triangulation is built according to the following procedure.

- For i=1 to n
  - Add links between the pairs of nodes adjacent to $X_{\sigma(i)}$. Let L_i the set of added links.
  - Remove node $X_{\sigma(i)}$ and all the links connecting $X_{\sigma(i)}$ with other nodes of the graph.

- To obtain the triangulated graph from the original graph, add all the links in the sets L_i $(i = 1, \ldots, n)$.

The permutation σ is called a deletion sequence. If we consider the deletion sequence $(\sigma(1), \sigma(2), \sigma(3), \sigma(4), \sigma(5), \sigma(6), \sigma(7)) = (3, 2, 7, 4, 1, 5, 6)$ in the graph in Fig. 3, we obtain the triangulated graph in Fig. 4.

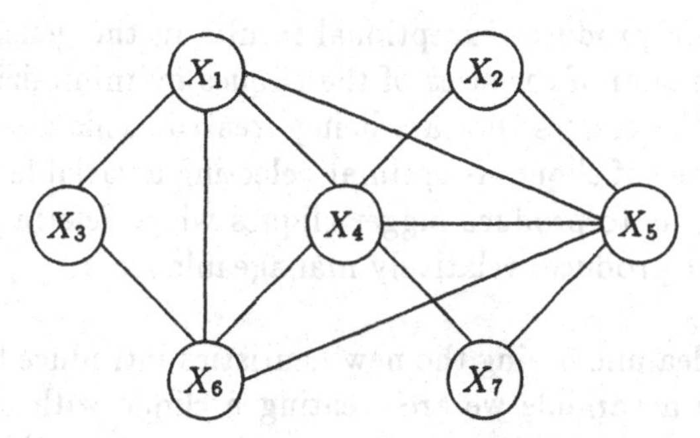

Fig. 4. Associated triangulated graph under σ

The clusters of a triangulated graph are cliques: maximal complete (all the nodes are adjacent) subgraphs. In the graph in Fig. 4 the cliques are $\{X_1, X_3, X_6\}$, $\{X_1, X_4, X_5, X_6\}$, $\{X_2, X_4, X_5\}$, $\{X_4, X_5, X_7\}$.

The size of a clique $\{X_i\}_{i \in I}$ is the number of elements of the cartesian product of the sets in which each X_i takes its values. That is the number of elements of $\Pi_{i \in I} U_i$, which is equal to the product of the number of elements of each U_i.

From a triangulated graph we can construct the cluster tree, also called the junction tree [14]. This can be done by the maximum cardinality search algorithm [23]. The graph in Fig. 5 is one of the possible tree of clusters associated to the triangulated graph in Fig. 4.

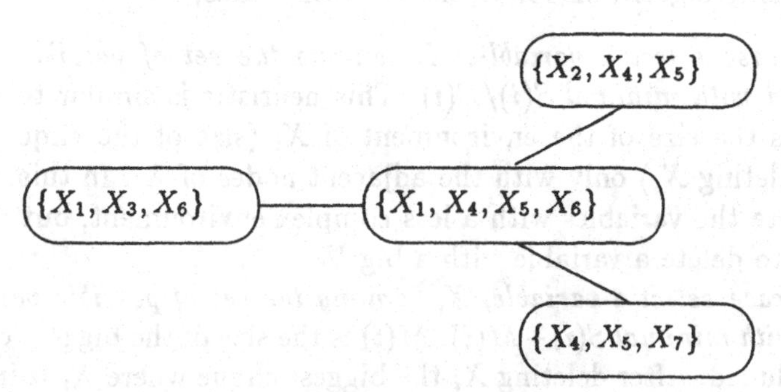

Fig. 5. The tree of cliques

The size of a tree of cliques is the sum of the sizes of each one of its cliques. To obtain efficient computations of this tree it is essential that its size be as small as possible.

3 Heuristic Algorithms

Olmsted [10], Kong [11], Kjærulff [9] proposed the following heuristic in order to obtain a good deletion sequence:

H1 *In each case, select the variable among the set of non-deleted variables producing a clique of minimal size and then delete this variable.*

This heuristic produces exceptional results in the general case. It attempts to minimize the sum of the sizes of the cliques by minimizing, in each step, the size of each of the cliques that are being created. This does not guarantee that the size of the tree of cliques is optimal; selecting a variable producing a minimal clique can force us to produce bigger cliques when deleting the other variables, but in general it produces relatively manageable trees.

The main idea underlying the new heuristics introduced in this work, is that when we delete a variable we are creating a clique with a size that should be minimized. However, at the same time, we are removing this variable and all the corresponding links, thereby simplifying the resulting graph. The simplification obtained in the resulting graph can also be a hepful guide in order to obtain efficient triangulations.

Following this idea we introduce the next heuristic algorithms to find a deletion sequence:

If X_i is a possible variable to be deleted then,

- $S(i)$ will be the size of the clique created by deleting this variable.
- $E(i)$ will be the number of elements of U_i.
- $M(i)$ will be the maximum size of the cliques of the subgraph given by X_i and its adjacents nodes.
- $C(i)$ will be the sum of the size cliques of the subgraph given by X_i and its adjacent nodes.

The heuristic algorithms follow the following rules,

H2 *In each case select a variable, X_i, among the set of possible variables to be deleted with minimal $S(i)/E(i)$.* This heuristic is similar to H1, but H2 computes the size of the environment of X_i (size of the clique to be produced deleting X_i) only with the adjacent nodes of X_i. In this way, we not only delete the variables with a less complex environment, but also make it possible to delete a variable with a big U_i.

H3 *In each case select a variable, X_i, among the set of possible variables to be deleted with minimal $S(i)-M(i)$.* $M(i)$ is the size of the biggest clique where X_i is included. After deleting X_i the biggest clique where X_i is included will have the size $S(i)$. With H3 we try to minimize the increment in the size of the biggest clique where each variable is included.

H4 *In each case select a variable, X_i, among the set of possible variables to be deleted with minimal $S(i) - C(i)$.* This heuristic is similar to H3, but the difference is computed with $C(i)$, that is the sum of the sizes of the cliques where X_i is included.

H5 *In each case select a variable, X_i, among the set of possible variables to be deleted with minimal $S(i)/M(i)$.*

H6 *In each case select a variable, X_i, among the set of possible variables to be deleted with minimal $S(i)/C(i)$.*

When we have two or more equally good variables to delete the next, we choose randomly one of that variables. We think that it would be possible to find a better tie-breaking.

The heuristics H5 and H6 are similar to H3 and H4 respectively, but rather than computing a difference, we compute the factor in the increment of the cliques. The complexity of H2 is equal to the complexity of H1. The idea of simplying the resulting graph is only partially considered by H2: the cause of deleting a variable X_i is the size of the created clique, $S(i)$. It is considered that the resulting graph is simpler if we delete a variable with more cases. By dividing $S(i)$ by the number of cases, $E(i)$, we balance the cost of deleting a variable with the simplicity of the obtained graph. However, $E(i)$ is not a very precise indicator of the simplicity of the graph; more precise indicators of this simplicity are considered in the heuristics H3, H4, H5 and H6. The intuitive basis of this rules are the following: if we delete variable X_i, the arcs of the old graph, and not present in the new graph, are the arcs linking X_i with its adjacents nodes. $C(i)$ and $M(i)$ aim to measure the complexity of the subgraph given by these arcs. But these rules are more complex than H1 and H2 because we must compute the cliques of a graph, although this graph will generally be small because it is the graph given by a node and its adjacent nodes.

4 Evaluation of the Heuristic Algorithms

We have implemented the different heuristic algorithms in C language in a SUN Workstation. To evaluate them, we have generated 500 graphs of each one of the following groups, and we are triangulated each one of the graphs with the six heuristics.

a) The graphs have 50 nodes which are are enumerated from 1 to 50. Each node has 2 cases, a number of parents chosen from the set $\{0, \ldots, 5\}$, and generated according to an uniform distribution of mean 2.5, rounding to the closest integer. The parents are the nodes immediately preceding the given node (this produces chain-like graphs).
b) The same as a), but now the parents are selected randomly among the preceding nodes. The nearest nodes have a higher probability of being chosen as parents.

c) The same as a), but now the parents are selected randomly among all of the preceding nodes. All the nodes have the same probability of being chosen as parents.
d) The same as a), but now the number of cases for each node is selected according to a Poisson distribution of mean 2 with a minimum of 2.
e) The same as b), but now the number of cases for each node is selected according to a Poisson distribution of mean 2 with a minimum of 2.
f) The same as c), but now the number of cases for each node is selected according to a Poisson distribution of mean 2 with a minimum of 2.
g) The same as a), but now the number of cases for each node is selected according to a Poisson distribution of mean 3 with a minimum of 2.
h) The same as b), but now the number of cases for each node is selected according to a Poisson distribution of mean 3 with a minimum of 2.
i) The same as c), but now the number of cases for each node is selected according to a Poisson distribution of mean 3 with a minimum of 2.
j) The same as a), but now the number of cases for each node is selected according to a Poisson distribution of mean 4 with a minimum of 2.
k) The same as b), but now the number of cases for each node is selected according to a Poisson distribution of mean 4 with a minimum of 2.
l) Same as c), but now the number of cases for each node is selected according to a Poisson distribution of mean 4 with a minimum of 2.

The mean size and standard deviation of sizes of the tree of cliques obtained by using each of the heuristic algorithms are given in tables 1 and 2.

Method	H1	H2	H3	H4	H5	H6
a)	489	489	428	483	428	483
b)	19.965	19.965	18.027	17.564	20.153	15.613
c)	23.378	23.378	22.826	22.660	26.046	22.152
d)	2.700	2.613	2.278	2.301	2.278	2.301
e)	2.158.184	1.606.454	2.199.538	2.091.678	1.691.858	1.052.726
f)	4.457.325	2.420.105	2.890.733	2.872.668	1.820.238	1.423.942
g)	6.606	6.407	5.548	5.558	5.548	5.558
h)	8.377.588	7.981.114	7.943.450	8.002.658	6.290.989	5.197.635
i)	65.147.965	32.137.124	53.160.398	53.028.113	28.502.411	24.144.396
j)	16.825	16.275	14.189	14.196	14.189	14.196
k)	130.974.850	128.772.625	131.136.043	133.913.541	123.910.850	77.734.851
l)	1.627.641.198	1.506.177.237	1.605.723.127	1.606.452.599	528.890.374	475.232.064

Table 1. Mean sizes of the tree of cliques

The results are the following:

– The new algorithms, with the exceptions of H3 and H4, are generally better than the previous known best algorithm, H1.

Method	H1	H2	H3	H4	H5	H6
a)	64	64	54	71	54	64
b)	34.332	34.332	27.561	28.929	30.047	23.179
c)	39.452	39.452	38.911	31.731	39.188	32.630
d)	1.097	1.063	956	954	956	954
e)	7.423.401	5.080.951	8.543.176	8.367.031	7.670.390	3.276.971
f)	50.441.890	13.401.449	18.944.797	18.946.028	7.086.978	5.472.380
g)	2.630	2.546	2.188	2.187	2.188	2.187
h)	21.490.256	23.676.980	21.069.018	21.380.543	19.737.130	16.726.049
i)	705.106.339	288.762.073	466.932.003	466.839.764	201.002.001	209.095.702
j)	7.940	7.636	6.916	6.914	6.916	6.914
k)	344.446.921	451.214.399	360.803.452	380.197.254	500.892.991	570.893.175
l)	25.081.485.346	24.760.156.861	24.797.026.066	24.796.998.549	4.180.926.330	4.280.997.524

Table 2. Standard Deviations of the sizes of the tree of cliques

- The relative improvement increases as a function of the complexity of graphs. The mean size applying H1 divided by the mean size applying H6 is close to one in the case of the simplest graphs (type a). In the case of the more complicated graphs (type l) it is greater than 3.
- The best heuristic is H6. This heuristic has a greater computer cost than H1. However, H2 has the same cost as H1 and produces, in general, better results than H1 (sometimes it produces a factor of 2).
- When the parents of a node are chosen among its preceding nodes (cases a, d, g, and j) the graphs are relatively simple and all of the heuristic algorithms produce similar results.
- When the nearest nodes have a higher probability of being chosen as parents, (b, e, h, k), the graphs have a smaller size than when the nodes have the same probability (c, f, i, l), but we obtain larger sizes if we compare with the sizes of the simple cases (a, d, g, j).

5 Conclusions

In this work, we have presented new heuristic algorithms to triangulate a graph. The problem of triangulation of graphs is the key point for the efficiency of propagation algorithms in graphical structures.

We have presented new heuristic algorithms which have been tested with regard to that proposed by Kjærulff [9]. For this we have used 500 randomly generated graphs for each one of 12 different types of graphs. We have obtained better triangulations than with Kjærulff's heuristic.

Since the problem of obtaining an optimal triangulation is NP-hard, some new heuristics could be considered. More complex procedures could introduce further improvements, as in the case of simulated annealing algorithms proposed in [9].

The main problem is what amount of calculation should be devoted to the triangulation. That is, should we choose very complex triangulation procedures, giving rise to good trees of cliques, or very fast triangulation procedures, giving rise to poor triangulations?. In general, the answer will depend on the particular problem we are trying to solve, but in general, we can say that the good trees of cliques can save time for a lot of different inference problems: the same tree will be used for several propagation cases, so, in most of the situations it will be worthy to spend some reasonable time on the triangulation.

The results of this work are useful for not only probabilistic propagation. The size of the associated tree of cliques is also fundamental for the propagation of uncertainty represented by other formalisms, such as belief functions and convex sets of probabilities.

References

1. Arnborg S., D.G. Corneil, A. Proskurowski (1987) Complexity of finding embeddings in a k-tree. *SIAM Jour. Alg. Discr. Meth.* **8**, 277-284.
2. Cano J.E., M. Delgado, S. Moral (1993) An axiomatic framework for the propagation of uncertainty in directed acyclic graphs. *International Journal of Approximate reasoning* 8 253-280.
3. Chin H.L., G.F. Cooper (1989) Bayesian network inference using simulation. In: *Uncertainty in Artificial Intelligence, 3* (Kanal, Levitt, Lemmer, eds.) North-Holland, 129-147.
4. Cooper G.F. (1988) Probabilistic inference using belief networks is NP-hard. Technical Report KSL-87-27, Stanford University, Stanford, California.
5. Cooper G.F. The computational complexity of probabilistic inference using bayesian belief networks is NP-hard. *Artificial Intelligence* **42**, 393-405.
6. D'Ambrosio B. (1991) Symbolic probabilistic inference in belief nets. Department of Computer Science, Oregon State University.
7. Geman S., D. Geman (1984) Stochastic relaxation, Gibbs distributions, and the bayesian restoration of images. *IEEE Transactions on Pattern Analysis and Machine Intelligence* **6**, 721-741.
8. Henrion M. (1986) Propagating uncertainty by logic sampling in Bayes' networks. Technical Report, Department of Engineering and Public Policy, Carnegie-Mellon University.
9. Kjærulff U. (1990) Triangulation of graphs-algorithms giving total state space. R 90-09, Department of Mathematics and Computer Science, Institute for Electronic Systems, Aalborg University.
10. Olmsted, S.M.(1983). On representing and solving decision problems. Ph.D. thesis, Department of Engineering-Economic Systems, Stanford University, Stanford, CA.
11. Kong, A. (1986). Multivariate belief functions and graphical models. Ph.D. dissertation, Department of Statistics, Harvard University, Cambridge, MA.
12. Kim J.H., J. Pearl (1983) A computational model for causal and diagnostic reasoning in inference engines, *Proceedings 8th IJCAI*, Karlsruhe, Germany.

13. Lauritzen S.L., D.J. Spiegelharter (1988) Local computation with probabilities on graphical structures and their application to expert systems. J. of the Royal Statistical Society, B 50, 157-224.
14. Jensen F. V. Junction trees and decomposable hypergraphs, *Research report*, Judex Datasystemer A/S, Aalborg, Denmark.
15. Pearl J. (1986) A constraint-propagation approach to probabilistic reasoning. In: *Uncertainty in Artificial Intelligence* (L.N. Kanal, J.F. Lemmer, eds.) North-Holland, 357-370.
16. Pearl J. (1986) Fusion, propagation and structuring in belief networks. *Artificial Intelligence* **29** 241-288.
17. Pearl J. (1988) *Probabilistic Reasoning in Intelligent Systems.* Morgan & Kaufman, San Mateo.
18. Shachter R.D. (1986) Evaluating influence diagrams. *Operations Research* **34**, 871-882.
19. Shachter R.D. (1988) Probabilistic inference and influence diagrams. *Operations Research* **36**, 589-605.
20. Shachter R.D., S.K. Andersen, P. Szlovits (1991) The equivalence of exact methods for probabilistic inference on belief networks. Submitted to *Artificial Intelligence.*
21. Shafer G., P.P. Shenoy (1990) Probability Propagation. *Annals of Mathematical and Artificial Intelligence* **2**, 327-351.
22. Shenoy P.P., G. Shafer (1990) Axioms for probability and belief-functions propagation. In: *Uncertainty in Artificial Intelligence, 4* (R.D. Shachter, T.S. Levitt, L.N. Kanal, J.F. Lemmer, eds.) North-Holland, Amsterdam, 169-198.
23. Tarjan R.E., M. Yannakakis (1984) Simple linear-time algorithm to test chordality of graphs, test acyclicity of hypergraphs, and selectively reduce acyclic hypergraphs. *SIAM Journal of Computing* **13** 566-579.

Computing Marginals from the Marginal Representation in Markov Trees

Hong Xu

IRIDIA and Service d'Automatique, Université libre de Bruxelles
50, Ave. F. Roosevelt, CP194/6, 1050-Bruxelles, Belgique

Abstract. Local computational techniques have been proposed to compute marginals for the variables in belief networks or valuation networks, based on the secondary structures called clique trees or Markov trees. However, these techniques only compute the marginal on the subset of variables contained in one node of the secondary structure. This paper presents a method for computing the marginal on the subset that may not be contained in one node. The proposed method allows us to change the structure of the Markov tree without changing any information contained in the nodes, thus avoids the possible repeated computations. Moreover, it can compute the marginal on any subset from the marginal representation already obtained. An efficient implementation of this method is also proposed.

1 Introduction

Belief networks (BN) and valuation networks (VN) are two well-known frameworks for the graphical representations of uncertain knowledge. BN is implemented for the probabilistic inference [4], while VN can represent several uncertainty formalisms in a unified framework [9]. Shenoy [12] has also shown that each BN has a corresponding VN which can represent the same knowledge. Thus, without loss of generality, we use the language of VN in this paper.

We can perform inferences on VN by: First, combining all the valuations in the network, yielding *the joint valuation*; Second, computing the marginals of the joint valuation for each variables or on some subsets of the variables. This method is not feasible when the number of variables is too large to compute the joint valuation. Many researchers have studied the methods for computing marginals without computing the joint valuation. A well-known approache is the local propagation scheme in Markov trees (also known as clique trees or join trees. See [1, 3, 4, 8]). Other methods are the arc reversal node reduction approach [5], the fusion algorithm [10] and the belief propagation by conditional belief functions [13]. Shachter et al. [6] and Shafer [7] have shown the similarities and the differences among the Markov tree propagation approaches. All these approaches can be regarded as two steps: 1. find a secondary structure called Markov tree[1] representation, where each node of the Markov tree is a subset of the variables; 2. use a so-called message-passing scheme to propagate the uncertainty in the Markov tree.

The difference among these approaches is the forms of the messages sent among the nodes. After propagation being done, what we obtain is the marginals of the joint valuation for each nodes in the Markov tree. We call this *a marginal representation*. Then, the marginal on the subset of variables contained in one node of the Markov tree

[1]Markov trees, clique trees, join trees and cluster trees have the same properties. They are the secondary representatives for belief networks or VN constructed from different techniques.

(we call such subsets *existing subsets*) can be easily computed, but from these approaches, we can't directly compute the marginal on the subset not contained in one node (we call such subsets *non-existing subsets*). However, it might be useful to know the marginals on some subset of variables that happens to be a non-existing one. In this paper, we will present a method for computing the marginals for such non-existing subsets. The proposed method is based on the local propagation techniques. It allows us to change the Markov tree structure without changing any information already contained in the nodes of the existing Markov tree. It can also compute the marginal on any subset from the marginal representation. In this way, we can avoid the reconstruction of new Markov tree and recomputation.

The rest of this paper is organized as follows: In section 2, we will review the local computation techniques for propagating information in Markov trees; In section 3, we will present the method for computing marginal on any subset from the marginal representation; In section 4, we propose an efficient implementation for the method presented in sections 3; Finally some conclusions are given in section 5.

2 Propagating Uncertainty in Markov Trees

A valuation network (VN) is a hypergraph with valuations defined on the hyperedges of the hypergraph. Formally, a *hypergraph* is a pair $(\mathbf{U}, \mathbf{H})$, where $\mathbf{U}$ is a finite set of nodes, and $\mathbf{H}$ is a set of non-empty subsets of $\mathbf{U}$, called *hyperedges*. Each node represents a variable u_i, associated with its *frame* W_i, the set of all its possible values. For each hyperedge h, its *frame* W_h is the Cartesian product of the frames of the variables it includes, and a function called *a valuation*, denoted by V_h, is defined on this frame. A valuation V_h represents some knowledge of the variables in h. One of the tasks of valuation network inference is to compute the marginals of the joint valuation, the combination of all the valuations in the network: $\otimes\{V_h \mid h\in \mathbf{H}\}$, for the variables in $\mathbf{U}$ or on some subsets of $\mathbf{U}$. The *marginal on a subset* denoted by R_S where $\mathbf{S}\subseteq\mathbf{U}$ is defined by:

$$R_S = (\otimes\{V_h \mid h\in \mathbf{H}\})^{\downarrow S} \tag{2.1}$$

$\otimes$ and $\downarrow$ denote two operations combination and marginalization respectively. Abstractly, combi-nation is a mapping: $V_{h_1}\times V_{h_2}\rightarrow V_{h_1\cup h_2}$, which corresponds to the aggregation of knowledge; marginalization is a mapping: $V_{h_1}\rightarrow V_{h_2}$ where $h_1\supseteq h_2$, which corresponds to the coarsening of knowledge. In the later discussion, we will also use an operation removal, denoted by ®. Removal is a mapping: $V_{h_1}\times V_{h_2}\rightarrow V_{h_1\cup h_2}$, that can be regarded as an "inverse" of combination in the sense that division is an inverse of multiplication [12]. If V_1 and V_2 represent some knowledge, then V_1®V_2 represents the remaining knowledge after V_2 is removed from V_1[2]. In probability theory, combination is multiplication, marginalization is addition, and removal is division. Details of these operations and the axioms that should be satisfied for any uncertainty calculus to be represented in the framework of VN and to make inference using these operators are described in [11, 12].

Local computational techniques aim to avoid computing the joint valuation when computing the marginals. Shafer and Shenoy [8] have shown that if the valuation network

---

[2] the removal operation for the case of belief function might result in a non-belief function if V_2 has not been combined in V_1 before being removed. In this paper, this would not happen since in the later discussion, all the valuations to be removed are those having been combined.

can be represented by a secondary structure, called Markov tree, the valuations can be "propagated" in the Markov tree by a local message-passing scheme, producing as a result the marginals of the joint valuation for each variable. (See [3] for the case of belief networks).

Let G=(**M**, **E**) be a tree where each $X_i \in \mathbf{M}$ is a non-empty subset of **U**, and **E** is the set of edges (X_i, X_j) in G. We say G is *Markov* if for any two nodes X_i and X_j, $X_i \cap X_j \subseteq X_t$ where X_t is any node on the path between X_i and X_j. We let **a**, **b**, **c** denote the elements in the nodes, V_i denote the valuation of node X_i, $S_{i,j}$ denote the intersection of two adjacent nodes X_i and X_j. X_i and X_j are adjacent if $(X_i, X_j) \in \mathbf{E}$, and we call them neighbors of each other. For a given hypergraph (**U**, **H**), we can always find a corresponding Markov tree (**M**, **E**) where $\mathbf{H} \subseteq \mathbf{M}$ and $X_i \subseteq \mathbf{U}$ for $X_i \in \mathbf{M}$. For $X_i \in \mathbf{M}$, if $X_i \in \mathbf{H}$, then its valuation is the one defined on the corresponding hyperedge, otherwise its valuation is an identity valuation. A valuation V is an identity valuation if $V \otimes V' = V'$ for any V' on the same frame of V. Algorithms for constructing Markov trees for hypergraphs can be found in [2, 16]. See also [3, 14] for finding clique trees for belief networks. Fig.1 illustrates an example for a hypergraph (left hand side) and its corresponding Markov tree (right hand side).

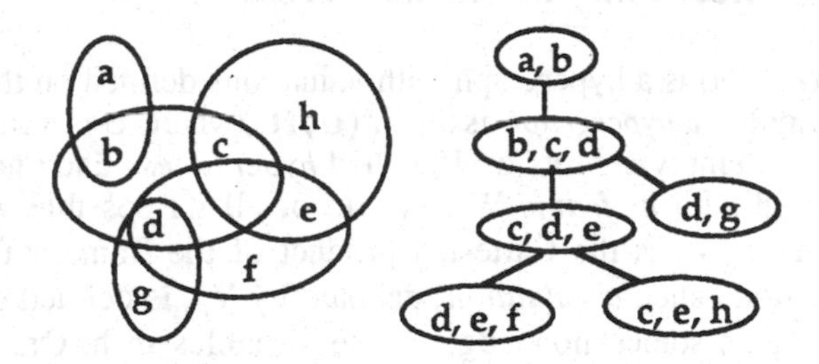

Fig. 1. An example for a hypergraph and its Markov tree $\mathbf{U} = \{a, b, c, d, e, f, g, h\}$

Given a Markov tree, we use a local message-passing scheme to compute the marginal for any node X_i in the Markov tree. The marginal R_i for X_i is the combination of its own valuation and the messages sent by all its neighbors. i.e.,

$$R_i = V_i \otimes (\otimes\{M_{X_j \to X_i} | (X_i, X_j) \in \mathbf{E}\}) \tag{2.2}$$

where $M_{X_j \to X_i}$ is the message received by X_i from its neighbor X_j. The message sent by X_j to its neighbor X_i is computed by projecting on $S_{i,j}$ the combination of the valuation of X_j and the messages sent by the neighbors of X_j except X_i:

$$M_{X_j \to X_i} = (V_j \otimes (\otimes\{M_{X_k \to X_j} | (X_k, X_j) \in \mathbf{E}, k \neq i\}))^{\downarrow X_i \cap X_j} \tag{2.3}$$

After computing the marginals for all the nodes in the Markov tree, we have a marginal representation. Then we can compute the marginal for a subset, say **S**, that is contained in some X_i, i.e., $\mathbf{S} \subseteq X_i$. The computation is obvious:

$$R_{\mathbf{S}} = R_i^{\downarrow \mathbf{S}} \tag{2.4}$$

The above technique yields the marginals of the joint valuation for each node avoiding the computation on the frame of the set of all variables, thus it drastically reduces the complexity of computation. However, for the marginal on a non-existing subset, the computation is not as easy as eq. (2.4). The most straightforward way is to add a hyperedge on the original network, reconstruct the Markov tree and repropagate the valuation in the new Markov tree. Generally, this method is inefficient since we have to recompute everything. In the next section, we will give a method to compute such marginals without changing any information we have already obtained.

3 Marginal Computation on Any Subset

In this section, we always assume we've already got the Markov tree G=(**M**, **E**) for a given VN represented by (**U**, **H**). Suppose $\mathbf{S}(\subseteq \mathbf{U})$ is a non-existing subset, i.e., there doesn't exist $X_i \in \mathbf{M}$ such that $\mathbf{S} \subseteq X_i$. To compute the marginal on **S** in G, we need to add a node containing **S**. In order that the local computational technique can be used, the joint valuation and the Markov property of the modified network shouldn't be changed. One method proposed by Shachter et. al. [6] is called Synthesis. The basic idea is: First, chose a node as a so-called pivot node; Next, add the elements of **S** in all the nodes on the path between where they already appear and the pivot node. Then collecting messages at the pivot node can get the desired result. This method has changed the structures of some original nodes, thus changed the valuations in those nodes. It also needs some recomputation. In this paper, we propose another method for modifying the Markov tree without changing any information contained in nodes. Moreover, from such new structure, we can compute the marginal on **S** directly from the marginal representation. First, let's look at the algorithm for modifying G=(**M**, **E**).

Algorithm MTM (Markov Tree Modification)

1. Find a set $\mathfrak{b}=\{X_{t_1}, X_{t_2}, ..., X_{t_k}\}$ in G such that $X_{t_i} \cap \mathbf{S} \neq \emptyset$, $\cup\{X_{t_i}\} \supseteq \mathbf{S}$, i=1, 2, ..., k;
2. Create a set $\mathfrak{a}$ such that it contains $\mathfrak{b}$ and all the nodes on the paths among the nodes in $\mathfrak{b}$:
   $$\mathfrak{a} = \mathfrak{b} \cup \{X_k \mid X_k \text{ is on the path between any } X_{t_i} \text{ and } X_{t_j} \text{ where } X_{t_i}, X_{t_j} \in \mathfrak{b}\} \quad (3.1)$$
3. Let $\mathbf{M}' = \mathbf{M} \cup \{X^*\}$, where $X^* = \mathbf{S} \cup \{\cup S_{i,j} \mid X_i, X_j \in \mathfrak{a}\}$ and the valuation V* be an identity valuation ($S_{i,j} = X_i \cap X_j$, $(X_i, X_j) \in \mathbf{E}$).
4. Remove all the edges among the nodes in $\mathfrak{a}$ and add the edges between X* and the nodes in $\mathfrak{a}$. i.e., $\mathbf{E}' = (\mathbf{E} - \{(X_i, X_j) \mid X_i, X_j \in \mathfrak{a}\}) \cup \{(X^*, X_i) \mid X_i \in \mathfrak{a}\}$.

The algorithm can be illustrated by the example shown in Fig. 2. Given a Markov tree(Fig. 2a), suppose we need to include a node containing **S**={**a**, **g**}. First we find a set of nodes such that each node has a non-empty intersection with **S**, and the union of the nodes contains **S**: $\mathfrak{b}$={{**a**,**b**} {**d**,**g**}}; Next, create set $\mathfrak{a}$ whose elements compose the path between {**a**,**b**} and {**d**,**g**}: $\mathfrak{a}$={{**a**,**b**} {**b**,**c**,**d**} {**d**,**g**}}. According to step3 in the algorithm, we have

$$X^* = \{a,g\} \cup (\{a,b\} \cap \{b,c,d\}) \cup (\{b,c,d\} \cap \{d,g\} = \{a,b,d,g\}.$$

Adding X* and the edges between X* and every elements in $\mathfrak{a}$ to the network and removing the edges among the elements in $\mathfrak{a}$, we have the final result shown in Fig. 2b.

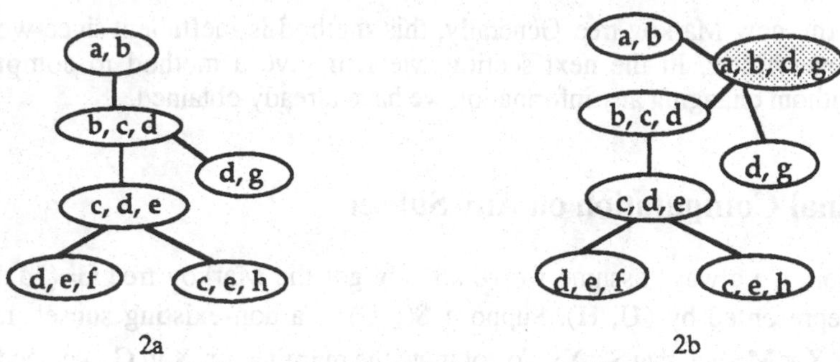

Fig. 2. An example of Markov tree modification

We modify the Markov tree to use it to compute the marginal on $\mathbb{S}$. The following theorem implies that the desired result can be computed from the resulting graph by using local propagation scheme (A proof can be found in [15]).

Theorem 1: Given a Markov tree G=(M, E) for a valuation network where $\mathbf{U}=\cup\{X_i \mid X_i\in \mathbf{M}\}$ is a finite set of variables concerned. Let $\mathbb{S}$ be a non-existing subset. By applying Algorithm MTM, the resulting graph G'=(M', E') is also a Markov tree for the original network.

Note that in Algorithm MTM, set $\mathfrak{a}$ is not unique. Lemma 1 states we can find the smallest one, thus the frame of X* is also the smallest.

Lemma 1: Let $\mathfrak{b}$ and $\mathfrak{a}$ be obtained from step 1 and 2 of Algorithm MTM for subset $\mathbb{S}$. There always exists the smallest set $\mathfrak{a}$ for $\mathbb{S}$. We say $\mathfrak{a}$ is *the smallest* for $\mathbb{S}$ if removing any element X_r from $\mathfrak{a}$ will yield $\mathbb{S}\not\subset\cup\{X_i \mid X_i\in\mathfrak{a}\ X_i\neq X_r\}$, or will yield the nodes in $\mathfrak{a}$ disconnected. If $\mathfrak{a}$ is the smallest, then the frame of X* constructed from $\mathfrak{a}$ is the smallest (A proof can be found in [15]).

For example, given a Markov tree in Fig. 2a, suppose we need to compute the marginal on $\mathbb{S}$={a,d}. We can modify Fig. 2a to Fig. 3a where $\mathfrak{a}$={{a,b}, {b,c,d}} and X*={a,b,d}, or to Fig. 3b where $\mathfrak{a}$={{a,b}, {b,c,d}, {c,d,e}} and X*={a,b,c,d}. We find that Fig. 3a and 3b are both the Markov trees that we can use to compute our desired result. But it's obvious that the Markov tree in Fig. 3a involves less computation than that in Fig. 3b.

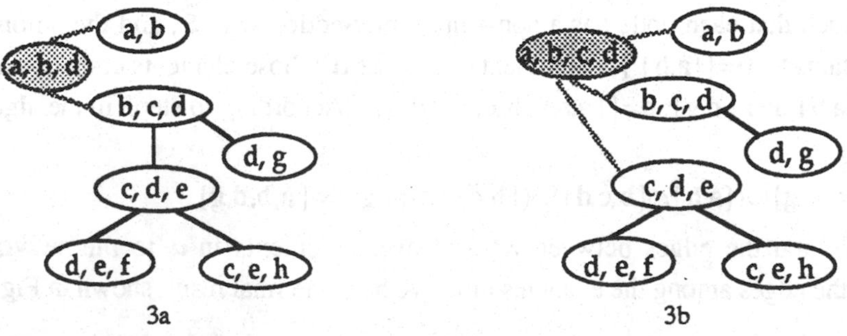

Fig. 3. An example showing the different result when $\mathfrak{a}$ is different for $\mathbb{S}$={a,d}.
In 3.a, $\mathfrak{a}$={{a,b},{b,c,d}}. In 3.b, $\mathfrak{a}$={{a,b},{b,c,d},{c,d,e}}.

Up to now, we've shown we may compute the marginal on $\mathbf{S}$ from the modified structure G'. The rest of this section will show that the marginal on $\mathbf{S}$ can be computed directly from the marginal representation in G.

Theorem 2: Given G and $\mathbf{S}$ as described in Theorem 1, let G' be the graph obtained by Algorithm MTM. Then R*, the marginal of the joint valuation for X*, denoted by $R^{\downarrow X^*}$, is computed by [15]:

$$R^* = (\otimes \{R_i^{\downarrow X_i \cap X^*} | X_i \in \alpha \}) \otimes (\otimes \{R_{S_{i,j}} | (X_i, X_j) \in E, X_i, X_j \in \alpha \}) \tag{3.2}$$

After applying Theorem 2, the marginal on $\mathbf{S}$ can be computed by marginalizing R* to the subset $\mathbf{S}$. i.e., $R_{\mathbf{S}} = (R^*)^{\downarrow \mathbf{S}}$.

Theorem 2 implies that we can manage computing the marginal on any subset without changing any information we have already obtained. i.e., we get a marginal representation for the new Markov tree G' by Theorem 1 and 2. In practice, if we need $\mathbf{S}$ for the further computation, we can change the structure as described above, and update the information using the same technique we used before. If we don't want to change the structure, we can keep the original structure G and compute the marginal only using eq. (3.2). In this case, we can use a two-level structure for the data management. Let's consider the example in Fig. 2. Suppose we have propagated all the valuations and got the marginal representation. Suppose now we need to compute the marginal on {a, g}. Instead of modifying the original structure as shown in Fig. 2, we can construct the problem as two levels, shown in Fig. 4. We find α={{**a**,**b**}, {**b**,**c**,**d**}, {**d**,**g**}}, create a subtree with nodes in α at a different level, add X*={**a**,**b**,**d**,**g**}, remove and add edges as described in Algorithm MTM. Then, the marginal for X* is computed by:

$$R^* = R^{\downarrow \{a,b,d,g\}} = (R^{\downarrow \{a,b\}} \otimes (R^{\downarrow \{b,c,d\}})^{\downarrow \{b,d\}} \otimes R^{\downarrow \{d,g\}}) \otimes (R^{\downarrow \{b\}} \otimes R^{\downarrow \{d\}}) \tag{3.3a}$$

and

$$R^{\downarrow \{a,g\}} = (R^{\downarrow \{a,b,d,g\}})^{\downarrow \{a,g\}}. \tag{3.3b}$$

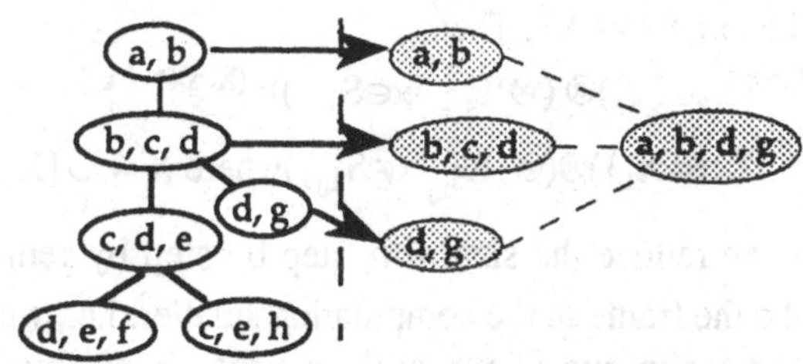

Fig. 4. Two-level structure for computing a marginal from marginal representation

To summarize, we find that our approach provides a simple way to modify the existing Markov tree for computing the marginals which could not be obtained directly from the propagation. It can be used at any time during the computation: before propagation has been done or after the marginals for the nodes have been computed. One distinct advantage of our approach is that we can always use the information we have got without any recomputation. In fact, our approach can be regarded as a fully general solution for "marginalizing in marginals" [3]. However, from the construction process of X*, it seems that the frame of X* might be very large in some cases. The next section will give a local computation technique to solve this problem.

4 Local Computation Techniques

The previous section have posed the problem that the frame in the involved computation is Ω_{X^*}, which might be much larger than $\Omega_{\mathbf{S}}$. The question is: can we find a way to compute $R_{\mathbf{S}}$ involving a smaller frame? The answer is yes if we use the idea of fusion algorithm proposed by Shenoy [11]. In this section, we will present how to apply the fusion algorithm to make the computation for $R_{\mathbf{S}}$ more efficient.

The fusion algorithm is an application of local computation to the inference on VN. The core of the algorithm is the fusion operation defined as follows [10]: Consider a set of valuations $V_1,...,V_k$, V_i is a valuation for h_i. Let $Fus_X\{V_1,...,V_k\}$ denote the collection of valuations after deleting a variable X, then

$$Fus_X\{V_1,...V_k\}=\{V^{\downarrow(h-\{X\})}\}\cup\{V_i|X\notin h_i\} \quad (4.1)$$

where $V=\otimes\{V_i|X\in h_i\}$, and $h=\cup\{h_i|X\in h_i\}$. Equation (4.1) implies the following equation:

$$(\otimes\{V_1,...V_k\})^{\downarrow(\cup\{h_i|i=1,..,k\}-\{X\})}=\{V^{\downarrow(h-\{X\})}\}\otimes\{V_i|X\notin h_i\} \quad (4.2)$$

To apply the fusion algorithm in our problem, we have to modify it since the removal operation is involved. Note that the valuations concerned are those for the nodes in $\boldsymbol{\alpha}$. The variables we need to delete are those contained in X*, but not in $\mathbf{S}$. We state the following theorem for this computation (A proof can be found in [15]).

Theorem 3: Suppose $\boldsymbol{\alpha}=\{X_1, ..., X_k\}$ and X* are obtained by Algorithm MTM. Let the marginal for each X_i be denoted by R_i, and the intersection of two adjacent node X_i and X_j ($X_i, X_j\in\boldsymbol{\alpha}$) by $S_{i,j}$. Suppose **x** is a variable that we need to delete for computing the marginal on $\mathbf{S}$, i.e., $\mathbf{x}\notin\mathbf{S}$, $\mathbf{x}\in X^*$. Then,

$$(R^*)^{\downarrow X^*-\{x\}}=((\otimes\{R_i^{\downarrow X_i\cap X^*}|x\in X_i\})\circledR(\otimes\{R_{S_{i,j}}|x\in S_{i,j}\}))^{\downarrow(X\cap X^*-\{x\})}$$
$$\otimes((\otimes\{R_i^{\downarrow X_i\cap X^*}|x\notin X_i\})\circledR(\otimes\{R_{S_{i,j}}|x\notin S_{i,j}\})\text{where } X=\cup\{X_i|x\in X_i\}). \quad (4.3)$$

Given Theorem 3, we can reduce the size of $\boldsymbol{\alpha}$ step by step by removing the variables irrelevant to $\mathbf{S}$, and make the frame in the computation smaller. i.e., we compute our result by a local computation technique. To show the process of this local computation technique [15], let's consider the example shown in Fig. 4. Instead of computing $R^{\downarrow\{a,g\}}$ using (3.3), where $X^*=\{\mathbf{a,b,d,g}\}$ we do as follows (Fig. 5):

Step 1: compute the marginal for $\{\mathbf{a,b,d}\}=(\{\mathbf{a,b}\}\cup\{\mathbf{b,c,d}\})\cap X^*$ from $\{\mathbf{a,b}\}$ and $\{\mathbf{b,c,d}\}$ as eq. (4.4a), and reconstruct the network as shown in Fig. 5b. Then $X^*=X^*-((\{\mathbf{a,b}\}\cap\{\mathbf{b,c,d}\})-\mathbf{S})=\{\mathbf{a,d,g}\}$

$$R^{\downarrow\{a,b,d\}}=(R^{\downarrow\{a,b\}}\otimes(R^{\downarrow\{b,c,d\}})^{\downarrow\{b,d\}})\circledR R^{\downarrow\{b\}} \quad (4.4a)$$

Step 2: compute the marginal for $\{\mathbf{a,d,g}\}=(\{\mathbf{a,b,d}\}\cup\{\mathbf{d,g}\})\cap X^*$ from $\{\mathbf{a,b,d}\}$ and $\{\mathbf{d,g}\}$ by:

$$R^{\downarrow\{a,d,g\}} = ((R^{\downarrow\{a,b,d\}})^{\downarrow\{a,d\}} \otimes R^{\downarrow\{d,g\}}) \otimes R^{\downarrow\{d\}} \quad (4.4b)$$

Step 3: compute the marginal on {a,g} by $(R^{\downarrow\{a,d,g\}})^{\downarrow\{a,g\}}$ from the result of eq. (4.4b).

It's obvious that the computation from the structure in Fig. 5 is more efficient than that in Fig. 4 since the size of the frame involved in eqs. (4.4) is $\max(|\Omega_{\{a,b,d\}}|, |\Omega_{\{a,d,g\}}|)$ which is smaller than $|\Omega_{\{a,b,d,g\}}|$, the frame involved in eq. (3.3a).

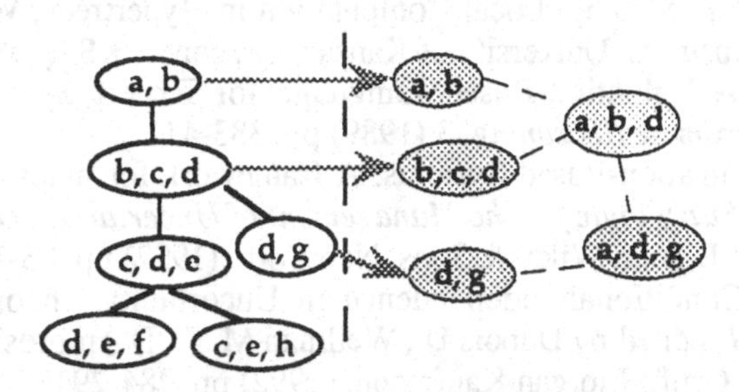

Fig. 5. Computing marginal on {a, g} using local computation

5 Conclusions

We have presented a method for computing the marginal on a subset which is not included in any node of the existing Markov tree. The proposed method modifies the structure of the existing Markov tree without changing any information already contained in the nodes. Thus we compute our desire result directly from the marginal representation without any recomputation. This method can be regarded as a general solution for computing marginal on any subset given a marginal representation. We have also proposed a local computation technique for implementing the presented method in an efficient way. Generally, it can prevent the computation from the exponential explosion.

Acknowledgments

The author is grateful to Philippe Smets, Alessandro Saffiotti, Prakash Shenoy, Liping Liu and anonymous reviewers for their comments and discussion. This work has been supported by a grant of IRIDIA, Université libre de Bruxelles.

References

1. F. V. Jensen, K. G. Olesen and K. Anderson: An Algebra of Bayesian belief Universes for Knowledge-based Systems, *Networks*, 20, (1990) pp. 637-659
2. A. Kong: Multivariate Belief Functions & Graphical Models, Ph.D dissertation, Department of Statistics, Harvard University, Cambridge, MA (1986).
3. S. L. Lauritzen and D. J. Spiegelhalter: Local Computation with Probabilities on Graphical Structures and Their Application to Expert Systems, *Journal of the Royal Statistical Society*, Series B, 50, No 2, (1988) pp. 157-224.
4. J. Pearl: *Probabilistic Reasoning in Intelligence Systems: Networks of Plausible Inference*, Morgan Kaufmann, Los Altos, CA (1988).

5. R. D. Shachter: Evaluating Influence Diagrams, *Operations Research* 34, (1986) pp. 871-882
6. R. D. Shachter, S. K. Andersen and P. Szolovits: The Equivalence of Exact Methods for Probabilistic Inference in Belief Networks, Department of Engineering-Economic Systems, Stanford University (1992).
7. G. Shafer: *Probabilistic Expert Systems* Society for Industrial and Applied Mathematics, Philadelphia, PA, (1993) to appear.
8. G. Shafer and P. P. Shenoy: Local Computation in Hypertrees, Working Paper No. 201, School of Business, University of Kansas, Lawrence, KS (1988).
9. P. P. Shenoy: A Valuation-Based Language for Expert Systems, *International Journal of Approximate Reasoning*, 3 (1989) pp. 383-411.
10. P. P. Shenoy: Valuation-Based Systems: A framework for managing uncertainty in expert systems, *Fuzzy logic for the Management of Uncertainty* edited by Zadeh, L. A. and Kacprzyk J., John Wiley & Sons, NewYork. (1992) pp. 83-104.
11 P. P. Shenoy: Conditional Independence in Uncertainty Theories, in *Proc. 8th Uncertainty in AI* edited by Dubois D., Wellman M. P., D'Ambrosio B. D. and Smets Ph. , San Mateo, Calif.: Morgan Kaufmann, (1992) pp. 284-291.
12. P. P. Shenoy: Valuation Networks and Conditional Independence, in *Proc. 9th Uncertainty in AI* edited by Wellman M. P., Mamdani A. and Heckerman D., San Mateo, Calif.: Morgan Kaufmann (1993).
13. Ph. Smets: Belief Functions: the Disjunctive Rule of Combination and the Generalized Bayesian Theorem, *International Journal of Approximate Reasoning*,, Vol. 9, No. 1 (1993) pp1-35.
14. R. E. Tarjan and M. Yannakakis: Simple Linear-time Algorithms to Test Chordality of Graphs, Test Acyclicity of Hypergraphs, and Selectivity Reduce Acyclic Hypergraphs, *SIAM J. Comput.*, 13, (1984) pp. 566-579.
15. H. Xu: Computing Marginals from the Marginal Representation in Markov Trees, Technical Report TR/IRIDIA/93-17 (1993).
16. L. Zhang: Studies on Finding Hypertree Covers of Hypergraphs, Working Paper No. 198, School of Business, University of Kansas, Lawrence, KS, (1988).

Representation of Bayesian Networks as Relational Databases

S.K.M. Wong, Y. Xiang, and X. Nie
Department of Computer Science, University of Regina
Regina, Saskatchewan, Canada S4S 0A2
wong@cs.uregina.ca fax:(306)585-4745

Abstract. This paper suggests a representation of Bayesian networks based on a generalized relational database model. The main advantage of this representation is that it takes full advantage of the capabilities of conventional relational database systems for probabilistic inference. Belief update, for example, can be processed as an ordinary query, and the techniques for query optimization are directly applicable to updating beliefs. The results of this paper also establish a link between knowledge-based systems for probabilistic reasoning and relational databases.

1 Introduction

A Bayesian network can be regarded as a summary of an expert's experience with an implicit population. A database can be viewed as a detailed documentation of such knowledge with an explicit population. In this view, databases are treated as information resources and used as a tool for knowledge acquisition [2, 3, 13]. That is, databases facilitate the automatic generation of Bayesian networks. However, once Bayesian networks have been generated, databases are regarded as being no longer relevant to the uncertain reasoning process.

We believe that there exists a deeper connection between the concepts of a Bayesian network and those of a relational database. There are many similarities between these two knowledge representation systems. Relational databases manipulate tables of tuples, whereas Bayesian networks manipulate tables of probabilities. Relational databases answer queries that involve attributes in different relations by joining the appropriate relations and then projecting the result onto the set of target attributes. In a Bayesian network, the joint probability distribution of many variables is defined by a product of local distributions. Each local distribution is defined on a smaller number of variables. Belief update [10] computes the marginalization of the joint distribution on a particular subset of variables. The notion of a join-tree (an acyclic hypergraph) plays an important role in schema design [1, 8], as it provides many desirable properties in database applications. On the other hand, to achieve inference efficiency, multiple-connected Bayesian networks are transformed into join (junction) trees [6] or junction forests [15].

This paper explores the connection between Bayesian networks and relational databases. We suggest a framework to represent a Bayesian network as a generalized relational database. Our method is based on a join-tree representation of Bayesian networks [11]. The proposed representation facilitates the use of Bayesian networks for probabilistic inference. From the theoretical point of view, our approach reveals the existence of a close relationship between Bayesian networks and relational databases. On the practical side, our method allows conventional relational database management systems to be directly used for probabilistic reasoning, and thus facilitates the development and implementation of knowledge-based systems based on Bayesian networks.

2 Bayesian Networks

A Bayesian network [10, 9, 7, 6] is a triplet (N, E, P). N is a set of nodes. Each node is labeled by a random variable (an attribute). Since the labeling is unique, we shall use 'node' and 'variable' interchangeably. E is a set of arcs such that $D = (N, E)$ is a directed acyclic graph (DAG). The arcs signify the existence of direct causal influences between the linked variables. For each node x_i, the strengths of the causal influences from its parent nodes π_i (a set of variables) are quantified by a conditional probability distribution $p(x_i|\pi_i)$ conditioned on the values of x_i's parents. The basic dependency assumption embedded in a Bayesian network is that a variable is independent of its non-descendants given its parents. For example, the joint probability distribution defined by a Bayesian network shown in Figure 1 can be expressed as:

$$p(x_1, x_2, x_3, x_4, x_5, x_6) \\ = p(x_1)p(x_2|x_1)p(x_3|x_1)p(x_4|x_1, x_2)p(x_5|x_2, x_3)p(x_6|x_5). \quad (1)$$

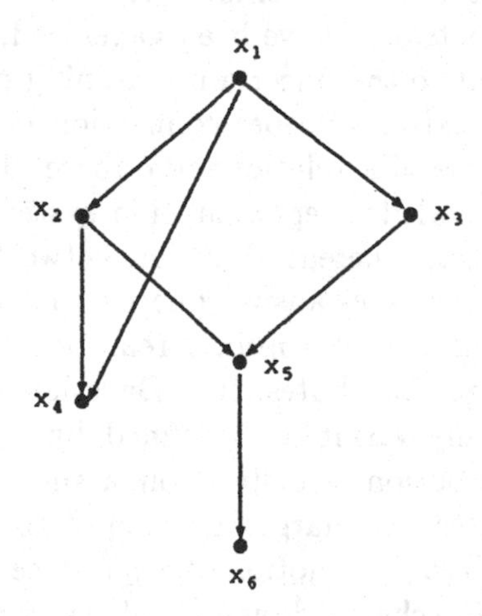

Fig. 1. The DAG of a Bayesian network defined by equation (1).

The joint probability distribution of a Bayesian network can be equivalently factored based on an undirected graph derived from the original DAG [7, 6]. For example, the above joint distribution $p(x_1, x_2, ..., x_6)$ can be written as a product of the distributions of the cliques of the graph G (depicted in Figure 2) divided by a product of the distributions of some of their pairwise intersections, namely:

$$\begin{aligned} &p(x_1, x_2, x_3, x_4, x_5, x_6) \\ &= (p(x_1, x_2, x_3)p(x_1, x_2, x_4)p(x_2, x_3, x_5)p(x_5, x_6)) \\ &\quad /(p(x_1, x_2)p(x_2, x_3)p(x_5)). \end{aligned} \tag{2}$$

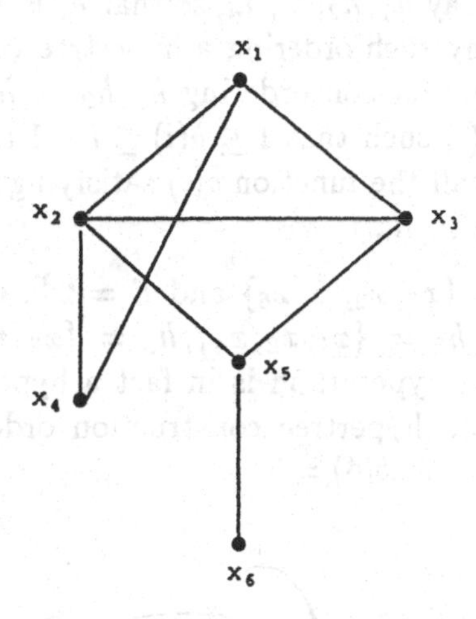

Fig. 2. An undirected graph G derived from the DAG in Figure 1.

Note that the conditional independence, $p(x_2, x_3|x_1) = p(x_2|x_1)p(x_3|x_1)$, in equation (1) is not explicitly represented in equation (2). It will be shown, however, that a tree organization of the cliques of G as shown in Figure 2 provides a convenient way of representing a Bayesian network as a relational database and of performing probabilistic inference with such a network.

3 Representation of a Factored Probability Distribution as a Product on a Hypertree

In order to facilitate the development of a relational representation of a Bayesian network and to use it for answering queries that involve marginal distributions, we first demonstrate how to express a joint probability distribution as a product on a *hypertree*[1]. The ideas developed in this section are based on the work of Shafer [11].

[1] Other terms like 'junction tree' and 'join tree' [10, 5, 6, 8] have been used to describe a hypertree in the literature. We will use 'hypertree' for the rest of the paper.

3.1 Hypergraphs and Hypertrees

Let $\mathcal{L}$ denote a lattice. We say that $\mathcal{H}$ is a *hypergraph*, if $\mathcal{H}$ is a finite subset of $\mathcal{L}$. Consider, for example, the power set $2^{\mathcal{X}}$, where $\mathcal{X} = \{x_1, x_2, ..., x_n\}$ is a set of variables. The power set $2^{\mathcal{X}}$ is a lattice of all subsets of $\mathcal{X}$. Any subset of $2^{\mathcal{X}}$ is a hypergraph on $2^{\mathcal{X}}$. We say that an element t in a hypergraph $\mathcal{H}$ is a *twig* if there exists another element b in $\mathcal{H}$, distinct from t, such that $t \cap (\cup(\mathcal{H} - \{t\})) = t \cap b$. We call any such b a *branch* for the twig t. A hypergraph $\mathcal{H}$ is a *hypertree* if its elements can be ordered, say $h_1, h_2, ..., h_n$, so that h_i is a twig in $\{h_1, h_2, ..., h_i\}$, for $i = 2, ..., n$. We call any such ordering a hypertree *construction ordering* for $\mathcal{H}$. Given a hypertree construction ordering $h_1, h_2, ..., h_n$, we can choose, for i from 2 to n, an integer $b(i)$ such that $1 \leq b(i) \leq i - 1$ and $h_{b(i)}$ is a branch for h_i in $\{h_1, h_2, ..., h_i\}$. We call the function $b(i)$ satisfying this condition a branch function for $\mathcal{H}$ and $h_1, h_2, ..., h_n$.

For example, let $\mathcal{X} = \{x_1, x_2, ..., x_6\}$ and $\mathcal{L} = 2^{\mathcal{X}}$. Consider a hypergraph, $\mathcal{H} = \{h_1 = \{x_1, x_2, x_3\}, h_2 = \{x_1, x_2, x_4\}, h_3 = \{x_2, x_3, x_5\}, h_4 = \{x_5, x_6\}\}$, depicted in Figure 3. This hypergraph is in fact a hypertree; the ordering, for example, h_1, h_2, h_3, h_4, is a hypertree construction ordering. Furthermore, we obtain: $b(2) = 1, b(3) = 1$, and $b(4) = 3$.

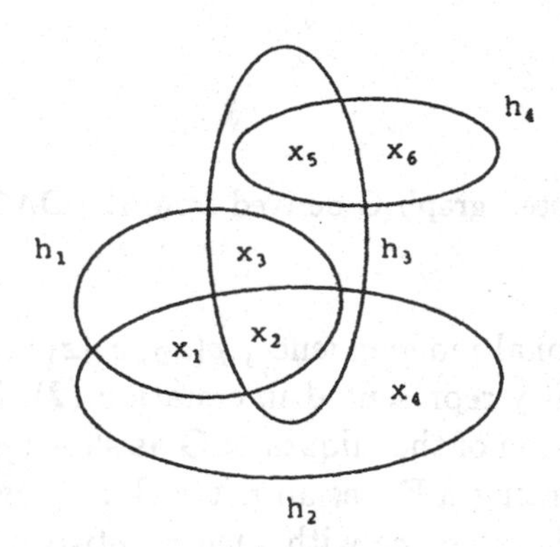

Fig. 3. A graphical representation of the hypergraph $\mathcal{H} = \{h_1, h_2, h_3, h_4\}$.

A hypertree $\mathcal{K}$ on $\mathcal{L}$ is called a *hypertree cover* for a given hypergraph $\mathcal{H}$ on $\mathcal{L}$ if for every element h of $\mathcal{H}$, there exists an element $k(h)$ of $\mathcal{K}$ such that $h \subseteq k(h)$. In general, a hypergraph $\mathcal{H}$ may have many hypertree covers. For example, the hypertree depicted in Figure 3 is a hypertree cover of the hypergraph, $\{\{x_1, x_2\}, \{x_1, x_3\}, \{x_1, x_2, x_4\}, \{x_2, x_5\}, \{x_3, x_5\}, \{x_5, x_6\}\}$.

3.2 Factored Probability Distributions

Let $\mathcal{X} = \{x_1, x_2, ..., x_n\}$ denote a set of variables. A factored probability distribution $p(x_1, x_2, ..., x_n)$ can be written as:

$$p(x_1, x_2, ..., x_n) = \phi = \phi_{h_1}\phi_{h_2}...\phi_{h_n},$$

where each h_i is a subset of $\mathcal{X}$, i.e., $h_i \in 2^{\mathcal{X}}$, and ϕ_{h_i} is a real-valued function on h_i. Moreover, $\mathcal{X} = h_1 \cup h_2 \cup ... \cup h_n = \bigcup_{i=1}^{n} h_i$. By definition, $\mathcal{H} = \{h_1, h_2, ..., h_n\}$ is a hypergraph on the lattice $2^{\mathcal{X}}$. Thus, a factored probability distribution can be viewed as a product on a hypergraph $\mathcal{H}$, namely:

$$\phi = \prod_{h \in \mathcal{H}} \phi_h.$$

Let v_x denote the discrete frame (state space) of the variable $x \in \mathcal{X}$. We call an element of v_x a *configuration* of x. We define v_h to be the Cartesian product of the frames of the variables in a hyperedge $h \in 2^{\mathcal{X}}$:

$$v_h = \times_{x \in h} v_x.$$

We call v_h the frame of h, and we call its elements configurations of h.

Let $h, k \in 2^{\mathcal{X}}$, and $h \subseteq k$. If c is a configuration of k, i.e., $c \in v_k$, we write $c^{\downarrow h}$ for the configuration of h obtained by deleting the values of the variables in k and not in h. For example, let $h = \{x_1, x_2\}, k = \{x_1, x_2, x_3, x_4\}$, and $c = (c_1, c_2, c_3, c_4)$, where $c_i \in v_{x_i}$. Then, $c^{\downarrow h} = (c_1, c_2)$.

If h and k are disjoint subsets of $\mathcal{X}$, c_h is a configuration of h, and c_k is a configuration of k, then we write $(c_h * c_k)$ for the configuration of $h \cup k$ obtained by *concatenating* c_h and c_k. In other words, $(c_h * c_k)$ is the unique configuration of $h \cup k$ such that $(c_h * c_k)^{\downarrow h} = c_h$ and $(c_h * c_k)^{\downarrow k} = c_k$. Using the above notation, a factored probability distribution ϕ on $\cup\mathcal{H}$ can be defined as follows:

$$\phi(c) = (\prod_{h \in \mathcal{H}} \phi_h)(c) = \prod_{h \in \mathcal{H}} \phi_h(c^{\downarrow h}),$$

where $c \in v_{\mathcal{X}}$ is an arbitrary configuration and $\mathcal{X} = \cup\mathcal{H}$.

3.3 Marginalization

Consider a function ϕ_k on a set k of variables. If $h \subseteq k$, then $\phi_k^{\downarrow h}$ denotes the function on h defined as follows:

$$\phi_k^{\downarrow h}(c_h) = \sum_{c_{k-h}} \phi_k(c_h * c_{k-h}),$$

where c_h is a configuration of h, c_{k-h} is a configuration of $k - h$, and $c_h \cdot c_{k-h}$ is a configuration of k. We call $\phi_k^{\downarrow h}$ the *marginal* of ϕ_k on h.

A major task in probabilistic reasoning with Bayesian networks is to compute marginals as new evidence becomes available.

4 Representation of a Factored Probability Distribution as a Generalized Acyclic Join Dependency

Let c be a configuration of $\mathcal{X} = \{x_1, x_2, ..., x_n\}$. Consider a factored probability distribution ϕ on $\mathcal{H}$:

$$\phi(c) = \prod_{h \in \mathcal{H}} \phi_h(c^{\downarrow h}).$$

We can conveniently express each function ϕ_h in the above product as a *relation* Φ_h. Suppose $h = \{x_1, x_2, .., x_l\}$. The function ϕ_h can be expressed as a relation on the set $\{x_1, x_2, ..., x_l, f_{\phi_h}\}$ of *attributes* as shown in Figure 4. A configuration $c_i = (c_{i1}, c_{i2}, ..., c_{il})$ in the above table denotes a row excluding the last element in the row, and s is the cardinality of v_h.

$\Phi_h =$

x_1	x_2	$\cdots$	x_l	f_{ϕ_h}
c_{11}	c_{12}	$\cdots$	c_{1l}	$\phi_h(c_1)$
c_{21}	c_{22}	$\cdots$	c_{2l}	$\phi_h(c_2)$
$\vdots$	$\vdots$		$\vdots$	$\vdots$
c_{s1}	c_{s2}	$\cdots$	c_{sl}	$\phi_h(c_s)$

Fig. 4. The function ϕ_h expressed as a relation.

By definition, the product $\phi_h \cdot \phi_k$ of any two function ϕ_h and ϕ_k is given by:

$$(\phi_h \cdot \phi_k)(c) = \phi_h(c^{\downarrow h}) \cdot \phi_k(c^{\downarrow k}),$$

where $c \in v_{h \cup k}$. We can therefore express the product $\phi_h \cdot \phi_k$ equivalently as a *join* of the relations Φ_h and Φ_k, written $\Phi_h \otimes \Phi_k$, which is defined as follows:

(i) Compute the *natural join*, $\Phi_h \bowtie \Phi_k$, of the two relations of Φ_h and Φ_k.
(ii) Add a new column with attribute $f_{\phi_h \cdot \phi_k}$ to the relation $\Phi_h \bowtie \Phi_k$ on $h \cup k$. Each value of $f_{\phi_h \cdot \phi_k}$ is given by $\phi_h(c^{\downarrow h}) \cdot \phi_k(c^{\downarrow k})$, where $c \in v_{h \cup k}$.
(iii) Obtain the resultant relation $\Phi_h \otimes \Phi_k$ by projecting the relation obtained in Step (ii) on the set of attributes $h \cup k \cup \{f_{\phi_h \cdot \phi_k}\}$.

For example, let $h = \{x_1, x_2\}$, $k = \{x_2, x_3\}$, and $v_h = v_k = \{0, 1\}$. The join $\Phi_h \otimes \Phi_k$ is illustrated in Figure 5.

Since the operator $\otimes$ is both commutative and associative, we can express a factored probability distribution as a join of relations:

$$\phi = \prod_{h \in \mathcal{H}} \phi_h = \bigotimes_{h \in \mathcal{H}} \Phi_h = \bigotimes \{\Phi_h | h \in \mathcal{H}\}.$$

$\Phi_h \bowtie \Phi_k \overset{(i)}{=}$

x_1	x_2	f_{ϕ_h}
0	0	a_1
0	1	a_2
1	0	a_3
1	1	a_4

$\bowtie$

x_2	x_3	f_{ϕ_k}
0	0	b_1
0	1	b_2
1	0	b_3
1	1	b_4

$=$

x_1	x_2	x_3	f_{ϕ_h}	f_{ϕ_k}
0	0	0	a_1	b_1
0	0	1	a_1	b_2
0	1	0	a_2	b_3
0	1	1	a_2	b_4
1	0	0	a_3	b_1
1	0	1	a_3	b_2
1	1	0	a_4	b_3
1	1	1	a_4	b_4

$\xrightarrow{(ii)}$

x_1	x_2	x_3	f_{ϕ_h}	f_{ϕ_k}	$f_{\phi_h \cdot \phi_k}$
0	0	0	a_1	b_1	$a_1 \cdot b_1$
0	0	1	a_1	b_2	$a_1 \cdot b_2$
0	1	0	a_2	b_3	$a_2 \cdot b_3$
0	1	1	a_2	b_4	$a_2 \cdot b_4$
1	0	0	a_3	b_1	$a_3 \cdot b_1$
1	0	1	a_3	b_2	$a_3 \cdot b_2$
1	1	0	a_4	b_3	$a_4 \cdot b_3$
1	1	1	a_4	b_4	$a_4 \cdot b_4$

$\xrightarrow{(iii)}$

x_1	x_2	x_3	$f_{\phi_h \cdot \phi_k}$
0	0	0	$a_1 \cdot b_1$
0	0	1	$a_1 \cdot b_2$
0	1	0	$a_2 \cdot b_3$
0	1	1	$a_2 \cdot b_4$
1	0	0	$a_3 \cdot b_1$
1	0	1	$a_3 \cdot b_2$
1	1	0	$a_4 \cdot b_3$
1	1	1	$a_4 \cdot b_4$

$= \Phi_h \otimes \Phi_k$

Fig. 5. The join of two relations: Φ_h and Φ_k.

We can also define marginalization as a relational operation. Let $\Phi_k^{\downarrow h}$ denote the relation obtained by marginalizing the function ϕ_k on $h \subseteq k$. We can construct the relation $\Phi_k^{\downarrow h}$ in two steps:

(a) Project the relation Φ_k on the set of attributes $h \cup \{f_{\phi_k}\}$, without eliminating identical configurations.
(b) For every configuration $c_h \in v_h$, replace the set of configurations of $h \cup \{f_{\phi_k}\}$ in the relation obtained from Step (a) by the singleton configuration $c_h * (\sum_{c_{k-h}} \phi_k(c_h * c_{k-h}))$.

Consider, for example, the relation Φ_k with $k = \{x_1, x_2, x_3\}$ as shown in Figure 6. Suppose we want to compute $\Phi_k^{\downarrow h}$ for $h = \{x_1, x_2\}$. From Step (a), we obtain the relation in Figure 7 by projecting Φ_k on $h \cup \{f_{\phi_k}\}$. The final result is shown in Figure 8.

Two important properties are satisfied by the operator $\downarrow$ of marginalization:

Lemma 1

(1) If Φ_k is a relation on k, and $h \subseteq g \subseteq k$, then

$$(\Phi_k^{\downarrow g})^{\downarrow h} = \Phi_k^{\downarrow h}.$$

$\Phi_k =$

x_1	x_2	x_3	f_{ϕ_k}
0	0	0	d_1
0	0	1	d_2
0	1	0	d_3
0	1	1	d_3
1	0	0	d_4
1	0	1	d_4
1	1	0	d_5
1	1	1	d_6

Fig. 6. A relation Φ_k with attributes x_1, x_2, x_3 ,f_{ϕ_k}, and $k = \{x_1, x_2, x_3\}$.

(2) If Φ_h and Φ_k are relations on h and k, respectively, then

$$(\Phi_h \otimes \Phi_k)^{\downarrow h} = \Phi_h \otimes (\Phi_k^{\downarrow h \cap k}).$$

x_1	x_2	f_{ϕ_k}
0	0	d_1
0	0	d_2
0	1	d_3
0	1	d_3
1	0	d_4
1	0	d_4
1	1	d_5
1	1	d_6

Fig. 7. The projection of the relation Φ_k in Figure 6 onto $\{x_1, x_2\} \cup \{f_{\phi_k}\}$.

Proof:

(1) By definition, a configuration of the set $g \cup \{f_{\phi_g}\} = g \cup \{f_{\phi_k^{\downarrow g}}\}$ of attributes in the relation $\Phi_g = \Phi_k^{\downarrow g}$ is

$$c_g * \phi_g(c_g) = c_g * \phi_k^{\downarrow g}(c_g) = c_g * (\sum_{c_{k-g}} \phi_k(c_g * c_{k-g})),$$

where $c_g \in v_g$. Similarly, a configuration of the set of attributes $h \cup \{f_{\phi_g^{\downarrow h}}\}$ in the relation $\Phi_g^{\downarrow h} = (\Phi_k^{\downarrow g})^{\downarrow h}$ is

$$c_h * \phi_g^{\downarrow h}(c_h)$$

$$= c_h * \sum_{c_{g-h}} \phi_g(c_h * c_{g-h})$$

$$= c_h * \sum_{c_{g-h}} \phi_k^{\downarrow g}(c_h * c_{g-h})$$

$$= c_h * (\sum_{c_{g-h}} \sum_{c_{k-g}} \phi_k(c_h * c_{g-h} * c_{k-g}))$$

$$= c_h * (\sum_{c_{k-h}} \phi_k(c_h * c_{k-h})),$$

which is a configuration of the set of attributes in the relation $\Phi_k^{\downarrow h}$.

(2) A configuration of the set of attributes $h \cup k \cup \{f_{\phi_h \cdot \phi_k}\}$ in the relation $\Phi_h \otimes \Phi_k$ is

$$c * (\phi_h(c^{\downarrow h}) \cdot \phi_k(c^{\downarrow k})),$$

where $c \in v_{h \cup k}$. Thus a configuration of the set of attributes $h \cup f_{\phi_h \cdot \phi_k}$ in the relation $(\Phi_h \otimes \Phi_k)^{\downarrow h}$ is

$$c^{\downarrow h} * (\phi_h(c^{\downarrow h}) \cdot \sum_{c_{k-h}} \phi_k(c^{\downarrow h \cap k} * c_{k-h})).$$

$\Phi_k^{\downarrow h} =$

x_1	x_2	$f_{\phi_k^{\downarrow h}}$
0	0	d_1+d_2
0	1	d_3+d_3
1	0	d_4+d_4
1	1	d_5+d_6

Fig. 8. The marginalization $\Phi_k^{\downarrow h}$ of the relation Φ_k in Figure 6 onto $h = \{x_1, x_2\}$.

On the other hand, by definition, a configuration of the set of attributes $(h \cap k) \cup \{f_{\phi_k^{\downarrow h \cap k}}\}$ in the relation $\Phi_k^{\downarrow h \cap k}$ is

$$(c^{\downarrow h \cap k} * \sum_{c_{k-h}} \phi_k(c^{\downarrow h \cap k} * c_{k-h})).$$

Also, a configuration of the set of attributes $h \cup \{f_{\phi_h}\}$ in the relation Φ_h is

$$(c^{\downarrow h} * \phi_h(c^{\downarrow h})) = (c^{\downarrow h-k} * c^{\downarrow h \cap k} * \phi_h(c^{\downarrow h})).$$

By the definition of the join operation $\otimes$, we can therefore conclude that

$$(\Phi_h \otimes \Phi_k)^{\downarrow h} = \Phi_h \otimes \Phi_k^{\downarrow h \cap k}. \quad \square$$

Before discussing the computation of marginals of a factored distribution, let us first state the notion of computational feasibility introduced by Shafer [11]. We call a set of attributes *feasible* if it is feasible to represent relations on these attributes, join them, and marginalize on them. We assume that any subset of feasible attributes is also feasible. Furthermore, we assume that the factored distribution is represented on a hypertree and every element in $\mathcal{H}$ is feasible.

Lemma 2 *Let $\Phi = \otimes\{\Phi_h | h \in \mathcal{H}\}$ be a factored probability distribution on a hypergraph $\mathcal{H}$. Let t be a twig in $\mathcal{H}$ and b be a branch for t. Then,*

(i) $(\otimes\{\Phi_h | h \in \mathcal{H}\})^{\downarrow \cup \mathcal{H}^{-t}}$
$= (\otimes\{\Phi_h | h \in \mathcal{H}^{-t}\}) \otimes \Phi_t^{\downarrow t \cap b}$.

(ii) If $k \subseteq \cup\mathcal{H}^{-t}$, then $(\otimes\{\Phi_h | h \in \mathcal{H}\})^{\downarrow k} = (\otimes\{\Phi_h^{-t} | h \in \mathcal{H}^{-t}\})^{\downarrow k}$, where $\mathcal{H}^{-t}$ denotes the set of hyperedges $\mathcal{H} - \{t\}$, $\Phi_b^{-t} = \Phi_b \otimes \Phi_t^{\downarrow t \cap b}$, and $\Phi_h^{-t} = \Phi_h$ for all other h in $\mathcal{H}^{-t}$.

Proof:

Since $\otimes\{\Phi_h | h \in \mathcal{H}^{-t}\}$ is a relation on $\cup\mathcal{H}^{-t}$ and $t \cap (\cup\mathcal{H}^{-t}) = t \cap b$, we obtain from property *(2)* of Lemma 1:

$$
\begin{aligned}
&(\otimes\{\Phi_h | h \in \mathcal{H}\})^{\downarrow \cup \mathcal{H}^{-t}} \\
&= ((\otimes\{\Phi_h | h \in \mathcal{H}^{-t}\}) \otimes \Phi_t)^{\downarrow \cup \mathcal{H}^{-t}} \\
&= (\otimes\{\Phi_h | h \in \mathcal{H}^{-t}\}) \otimes \Phi_t^{\downarrow t \cap (\cup \mathcal{H}^{-t})} \\
&= (\otimes\{\Phi_h | h \in \mathcal{H}^{-t}\}) \otimes \Phi_t^{\downarrow t \cap b}.
\end{aligned}
$$

The right-hand side of the above equation can be rewritten as:

$$(\otimes\{\Phi_h | h \in \mathcal{H}^{-t}\}) \otimes \Phi_t^{\downarrow t \cap b} = \otimes\{\Phi_h^{-t} | h \in \mathcal{H}^{-t}\}.$$

Since $k \subseteq \mathcal{H}^{-t}$, from property *(1)* of Lemma 1, it follows:

$$
\begin{aligned}
&(\otimes\{\Phi_h | h \in \mathcal{H}\})^{\downarrow k} \\
&= ((\otimes\{\Phi_h | h \in \mathcal{H}\})^{\downarrow \cup \mathcal{H}^{-t}})^{\downarrow k} \\
&= (\otimes\{\Phi_h^{-t} | h \in \mathcal{H}^{-t}\})^{\downarrow k}. \quad \square
\end{aligned}
$$

We now describe an algorithm for computing $\Phi^{\downarrow k}$ for $k \in \mathcal{H}$, where $\Phi = \otimes\{\Phi_h | h \in \mathcal{H}\}$ and $\mathcal{H}$ is a hypertree. Choose a hypertree construction ordering for $\mathcal{H}$ that begins with $h_1 = k$ as the root, say $h_1, h_2, ..., h_n$, and choose a branching $b(i)$ function for this particular ordering. For $i = 1, 2, ..., n$, let

$$\mathcal{H}^i = \{h_1, h_2, ..., h_i\}.$$

This is a sequence of sub-hypertrees, each larger than the last; $\mathcal{H}^1 = \{h_1\}$ and $\mathcal{H}^n = \mathcal{H}$. The element h_i is a twig in $\mathcal{H}^i$. To compute $\Phi^{\downarrow k}$, we start with $\mathcal{H}^n$ going backwards in this sequence. We use Lemma 2 each time to perform the *reduction*. At the step from $\mathcal{H}^i$ to $\mathcal{H}^{i-1}$, we go from $\Phi^{\downarrow\cup\mathcal{H}^i}$ to $\Phi^{\downarrow\cup\mathcal{H}^{i-1}}$. We omit h_i in $\mathcal{H}^i$ and change the relation on $h_{b(i)}$ in $\mathcal{H}^{i-1}$ from $\Phi^i_{h_{b(i)}}$ to

$$\Phi^{i-1}_{h_{b(i)}} = \Phi^i_{h_{b(i)}} \otimes (\Phi^i_{h_i})^{\downarrow h_i \cap h_{b(i)}},$$

and the other relations in $\mathcal{H}^{i-1}$ are not changed. The collection of relations with which we begin, $\{\Phi^n_h | h \in \mathcal{H}^n\}$, is simply $\{\Phi_h | h \in \mathcal{H}\}$, and the collection with which we end, $\{\Phi^1_h | h \in \mathcal{H}^1\}$, consists of the single relation $\Phi^1_h = \Phi^{\downarrow h_1}$.

Consider a factored probability distribution $\Phi = \otimes\{\Phi_h | h \in \mathcal{H}\}$ on a hypertree $\mathcal{H} = \{h_1, h_2, ..., h_n\}$. We say that Φ satisfies the acyclic join dependency (AJD), $*[h_1, h_2, ..., h_n]$, if Φ decomposes losslessly onto a hypertree construction ordering $h_1, h_2, ..., h_n$, i.e., Φ can be expressed as:

$$\Phi = (...((\Phi^{\downarrow h_1} \otimes' \Phi^{\downarrow h_2}) \otimes' \Phi^{\downarrow h_3})... \otimes' \Phi^{\downarrow h_n}),$$

where $\otimes'$ is a generalized join operator defined by:

$$\Phi^{\downarrow h_i} \otimes' \Phi^{\downarrow h_j} = \Phi^{\downarrow h_i} \otimes \Phi^{\downarrow h_j} \otimes (\Phi^{\downarrow h_i \cap h_j})^{-1}.$$

The relation $(\Phi^{\downarrow h})^{-1}$ is defined as follows. First, let us define the inverse function $(\phi^{\downarrow h})^{-1}$ of ϕ. That is,

$$(\phi^{\downarrow h})^{-1}(c) = \left(\sum_{c'} \phi(c * c') \right)^{-1}, \quad \text{for} \sum_{c'} \phi(c * c') > 0,$$

where c is a configuration of $h \subseteq \cup\mathcal{H}$, and c' is a configuration of $\cup\mathcal{H} - h$. We call the function $(\phi^{\downarrow h})^{-1}$ the *inverse marginal* of ϕ on h. The *inverse relation* $(\Phi^{\downarrow h})^{-1}$ is the relation constructed from the inverse function $(\phi^{\downarrow h})^{-1}$. Obviously, the product $(\phi^{\downarrow h})^{-1} \cdot \phi^{\downarrow h}$ is a unit function on h, and $(\Phi^{\downarrow h})^{-1} \otimes \Phi^{\downarrow h}$ is an identity relation on h.

Theorem 1 *Any factored probability distribution $\Phi = \otimes\{\Phi_h | h \in \mathcal{H}\}$ on a hypertree, $\mathcal{H} = \{h_1, h_2, ..., h_n\}$, decomposes losslessly onto a hypertree construction ordering $h_1, h_2, ..., h_n$. That is, Φ satisfies the AJD, $*[h_1, h_2, ..., h_n]$.*

Proof:

Suppose $t \in \mathcal{H}$ is a twig. By Lemma 2,

$$\begin{aligned}
\Phi^{\downarrow\cup\mathcal{H}^{-t}} &= ((\otimes\{\Phi_h | h \in \mathcal{H}^{-t}\}) \otimes \Phi_t)^{\downarrow\cup\mathcal{H}^{-t}} \\
&= (\otimes\{\Phi_h | h \in \mathcal{H}^{-t}\}) \otimes \Phi_t^{\downarrow t \cap (\cup\mathcal{H}^{-t})} \\
&= (\otimes\{\Phi_h | h \in \mathcal{H}^{-t}\}) \otimes \Phi_t^{\downarrow t \cap b}.
\end{aligned}$$

Note that $(\Phi_t^{\downarrow t\cap b})^{-1} \otimes \Phi_t^{\downarrow t\cap b} \otimes \Phi_t = \Phi_t$, as $(\Phi_t^{\downarrow t\cap b})^{-1} \otimes \Phi_t^{\downarrow t\cap b}$ is an identity relation on $t \cap b$. Thus,

$$\begin{aligned} &\Phi^{\downarrow \cup \mathcal{H}^{-t}} \otimes (\Phi_t^{\downarrow t\cap b})^{-1} \otimes \Phi_t \\ &= (\otimes\{\Phi_h | h \in \mathcal{H}^{-t}\}) \otimes \Phi_t^{\downarrow t\cap b} \otimes (\Phi_t^{\downarrow t\cap b})^{-1} \otimes \Phi_t \\ &= (\otimes\{\Phi_h | h \in \mathcal{H}^{-t}\}) \otimes \Phi_t \\ &= \Phi. \end{aligned}$$

Now we want to show that:

$$(\Phi_t^{\downarrow t\cap b})^{-1} \otimes \Phi_t = (\Phi^{\downarrow t\cap b})^{-1} \otimes \Phi^{\downarrow t}.$$

Note that by property *(2)* of Lemma 1, we obtain:

$$\begin{aligned} &\Phi_t \otimes (\otimes\{\Phi_h | h \in \mathcal{H}^{-t}\})^{\downarrow t\cap(\cup \mathcal{H}^{-t})} \\ &= (\Phi_t \otimes (\otimes\{\Phi_h | h \in \mathcal{H}^{-t}\}))^{\downarrow t} \\ &= \Phi^{\downarrow t}. \end{aligned}$$

On the other hand, we have:

$$\begin{aligned} &(\Phi_t^{\downarrow t\cap b})^{-1} \otimes ((\otimes\{\Phi_h | h \in \mathcal{H}^{-t}\})^{\downarrow t\cap(\cup \mathcal{H}^{-t})})^{-1} \\ &= (\Phi_t^{\downarrow t\cap b} \otimes (\otimes\{\Phi_h | h \in \mathcal{H}^{-t}\})^{\downarrow t\cap(\cup \mathcal{H}^{-t})})^{-1} \\ &= ((\Phi_t \otimes (\otimes\{\Phi_h | h \in \mathcal{H}^{-t})^{\downarrow t\cap(\cup \mathcal{H}^{-t})})^{\downarrow t\cap b})^{-1} \\ &= (((\Phi_t \otimes (\otimes\{\Phi_h | h \in \mathcal{H}^{-t}\}))^{\downarrow t})^{\downarrow t\cap b})^{-1} \\ &= ((\Phi^{\downarrow t})^{\downarrow t\cap b})^{-1} \\ &= (\Phi^{\downarrow t\cap b})^{-1}. \end{aligned}$$

Hence,

$$\Phi_t \otimes (\Phi_t^{\downarrow t\cap b})^{-1} = \Phi^{\downarrow t} \otimes (\Phi^{\downarrow t\cap b})^{-1}.$$

The relation Φ can therefore be expressed as:

$$\begin{aligned} \Phi &= \Phi^{\downarrow \cup \mathcal{H}^{-t}} \otimes \Phi^{\downarrow t} \otimes (\Phi^{\downarrow(\cup \mathcal{H}^{-t})\cap t})^{-1} \\ &= \Phi^{\downarrow \cup \mathcal{H}^{-t}} \otimes' \Phi^{\downarrow t}. \end{aligned}$$

Moreover,

$$\begin{aligned} \Phi^{\downarrow \cup \mathcal{H}^{-t}} &= \otimes\{\Phi_h | h \in \mathcal{H}^{-t}\} \otimes \Phi_t{}^{\downarrow t\cap b} \\ &= \otimes\{\Phi_h^{-t} | h \in \mathcal{H}^{-t}\}. \end{aligned}$$

We can immediately apply the same procedure to $\Phi^{\downarrow \cup \mathcal{H}^{-t}}$ for further reduction. Thus, by applying this algorithm recursively, the desired result is obtained. □

5 Conclusion

The previous section demonstrates how a factored join probability distribution can be conveniently expressed as a generalized acyclic join dependency. By adding two new relational operations, namely, the marginalization operator $\downarrow$ and the generalized join operator $\otimes'$ to conventional database management systems, belief updates can be processed as ordinary queries. All the techniques developed for query optimization are applicable to perform belief updates. In other words, this new representation of Bayesian networks enables us to take full advantage of the capabilities of conventional relational database systems for probabilistic inference. Clearly, our approach considerably shortens the time required for implementing a Bayesian network.

The preliminary results of this paper have also established a link between Bayesian networks and relational databases. This unified representation of knowledge may benefit both database design and reasoning with probabilities.

References

1. C. Beeri, R. Fagin, D. Maier and M. Yannakakis. On the desirability of acyclic database schemes. *Journal of the Association for Computing Machinery*, 30(3):479–513, 1983.
2. C.K. Chow and C.N. Liu. Approximating discrete probability distributions with dependence trees. *IEEE Transactions on Information Theory*, 14:462–467, 1968.
3. G.F. Cooper and E. Herskovits. A Bayesian method for the induction of probabilistic networks from data. *Machine Learning*, 9:309–347, 1992.
4. M. Henrion and M.J. Druzdzel. Qualitative propagation and scenario-based approaches to explanation of probabilistic reasoning. *Proc. Sixth Conference on Uncertainty in Artificial Intelligence*, pages 10–20, Cambridge, Mass., 1990.
5. F.V. Jensen. Junction tree and decomposable hypergraphs. Technical report, JUDEX, Aalborg, Denmark, February 1988.
6. F.V. Jensen, S.L. Lauritzen, and K.G. Olesen. Bayesian updating in causal probabilistic networks by local computations. *Computational Statistics Quarterly*, 4:269–282, 1990.
7. S.L. Lauritzen and D.J. Spiegelhalter. Local computation with probabilities on graphical structures and their application to expert systems. *Journal of the Royal Statistical Society, Series B*, 50:157–244, 1988.
8. D. Maier. *The Theory of Relational Databases*. Computer Science Press, 1983.
9. R.E. Neapolitan. *Probabilistic Reasoning in Expert Systems*. John Wiley and Sons, 1990.
10. J. Pearl. *Probabilistic Reasoning in Intelligent Systems: Networks of Plausible Inference*. Morgan Kaufmann, 1988.
11. G. Shafer. An axiomatic study of computation in hypertrees. School of Business Working Paper Series, (232), University of Kansas, Lawrence, 1991.
12. D.J. Spiegelhalter, R.C.G. Franklin, and K. Bull. Assessment, criticism and improvement of imprecise subjective probabilities for a medical expert system. *Proc. Fifth Workshop on Uncertainty in Artificial Intelligence*, pages 335–342, Windsor, Ontario, 1989.

13. W.X. Wen. From relational databases to belief networks. B. D'Ambrosio, P. Smets, and P.P. Bonissone, editors, *Proc. Seventh Conference on Uncertainty in Artificial Intelligence*, pages 406–413. Morgan Kaufmann, 1991.
14. Y. Xiang, M.P. Beddoes, and D. Poole. Sequential updating conditional probability in Bayesian networks by posterior probability. *Proc. 8th Biennial Conf. Canadian Society for Computational Studies of Intelligence*, pages 21–27, Ottawa, 1990.
15. Y. Xiang, D. Poole, and M. P. Beddoes. Multiply sectioned Bayesian networks and junction forests for large knowledge based systems. *Computational Intelligence*, 9(2):171–220, 1993.

Decision Influence Diagrams with Fuzzy Utilities

Miguel López

Departamento de Matemáticas, Universidad de Oviedo
33071 Oviedo, Spain

Abstract In this paper, decision influence diagrams are studied when the assessment of utilities with real numerical values is considered to be too restrictive, and the use of fuzzy sets to model the problem in terms of fuzzy utilities seems appropiate. An algorithm to solve decision influence diagrams with fuzzy utilities is suggested.

Keywords: Kolodiejczyk coefficient, decision influence diagram, fuzzy random variable, fuzzy set.

1 Introduction

In decision studies, the utility function is assumed to take on numerical values, though this assumption can be sometimes viewed as too retrictive. In many problems, the use of fuzzy sets to describe utilities in a more realistic way is convenient. To justify this assertion we can consider the following example, taken from [1], which is based on an introductory text of Statistics [10]:

A neurologist has to classify his patients as requiring exploratory brain surgery (action a_1) or not (action a_2). From previous experiences with this type of patients, he knows that 60% of the people examined need the operation, while 40% do not. The problem can be regarded as a decision problem in a Bayesian context with state space $\Theta = \{\theta_1, \theta_2\}$, where $\theta_1 \equiv$ the patient requires surgery, $\theta_2 \equiv$ the patient does not require surgery, action space $A = \{a_1, a_2\}$, prior distribution ξ, with $\xi(\theta_1) = 0.6$ and $\xi(\theta_2) = 0.4$ and utility function $u(\theta_1, a_1) = u(\theta_2, a_2) = 0$, $u(\theta_1, a_2) = 5u(\theta_2, a_1)$ with $u(\theta_2, a_1) < 0$, and utilities are intended in this example as opposed to losses.

The assessment of utilities seems to be extremely precise for the nature of actions and states in the problem, so we could express the decision maker's "preferences" in a more imprecise way, saying: $u(\theta_1, a_1) = u(\theta_2, a_2) = 0$, $u(\theta_2, a_1) =$ "inconvenient", $u(\theta_1, a_2) =$"dangerous", being $u(\theta_2, a_1)$ and $u(\theta_1, a_2)$ described by means of fuzzy sets on $\mathbb{R}$.

2 Preliminary concepts

In this section we introduce concepts required thorought this paper.
Let $(\Theta, \mathcal{C}, \xi)$ be a probability space and let $\mathcal{F}_o(\mathbb{R})$ be the collection of all fuzzy subsets

$\widetilde{V}$ of $\mathbf{R}$, characterized by membership functions $\mu_{\widetilde{V}} : \mathbf{R} \mapsto [0,1]$, having the following properties:

i) support $\widetilde{V}$ = closure of $\{x \in \mathbf{R} | \mu_{\widetilde{V}}(x) > 0\}$ is compact,

ii) $\widetilde{V}_\alpha = \alpha$-level set of $\widetilde{V} = \{x \in \mathbf{R} | \mu_{\widetilde{V}}(x) \geq \alpha\}$ is closed for each $0 \leq \alpha \leq 1$,

iii) $\widetilde{V}_1 = \{x \in \mathbf{R} | \mu_{\widetilde{V}}(x) = 1\} \neq \emptyset$.

Definition 2.1. *A fuzzy random variable is a function $\phi : \Theta \mapsto \mathcal{F}_o(\mathbf{R})$ satisfying the condition: $\{(\theta, w) | w \in (\phi(\theta))_\alpha\} \in \mathcal{C} \times \mathcal{B}_R$, for all $\alpha \in [0,1]$, where $\mathcal{B}_R$ is the Borel $\sigma-$field.*

To define the expected value of a fuzzy random variable, we first define that concept for simple fuzzy random variables.

Definition 2.2. *Let $\phi : \Theta \mapsto \mathcal{F}_o(\mathbf{R})$ be a simple fuzzy random variable, what means a fuzzy random variable taking on the "fuzzy values" $\widetilde{V}_1, \widetilde{V}_2, \ldots, \widetilde{V}_k \in \mathcal{F}_o(\mathbf{R})$, on the pairwise disjoint subsets of Θ, $C_1, C_2, \ldots, C_k \in \mathcal{C}$, respectively. So, we can write this variable by*

$$\phi = \sum_{j=1}^{k} \widetilde{V}_j \aleph_{C_j}$$

$\aleph_C$ being the indicator function of C. The expected value of ϕ, with respect to the probability measure ξ on $(\Theta, \mathcal{C})$, is the fuzzy set denoted by $\widetilde{E}(\phi/\xi) = \widetilde{\int}_\Theta \phi(\theta) d\xi(\theta) \in \mathcal{F}_o(\mathbf{R})$ given by

$$\widetilde{\sum_{j=1}^{k}} V_j \otimes \xi(C_j)$$

where $\otimes \equiv$ fuzzy product and $\widetilde{\sum} \equiv$ fuzzy addition (operations based on Zadeh's extension principle, cf. [11]).

Definition 2.3. *If ϕ is a fuzzy random variable, it is possible to obtain a sequence of simple fuzzy random variables, $\{\phi_n\}$, such that $lim_{n \to \infty} d_\infty(\phi_n(\theta), \phi(\theta)) = 0$, for almost $\theta \in \Theta$, where*

$$d_\infty(\phi_n(\theta), \phi(\theta)) = \sup_{\alpha \in [0,1]} d_H((\phi_n(\theta))_\alpha, (\phi(\theta))_\alpha)$$

is a metric on $\mathcal{F}_o(\mathbf{R})$, and d_H is the Hausdorff metric defined on the set of all non-empty compact subsets of $\mathbf{R}$, given by

$$d_H(\Omega, \Omega') = max\{\sup_{\omega \in \Omega} \inf_{\omega' \in \Omega'} | \omega - \omega' |, \sup_{\omega' \in \Omega'} \inf_{\omega \in \Omega} | \omega - \omega' |\}$$

Then, the expected value of ϕ, with respect to the probability measure ξ on $(\Theta, \mathcal{C})$, is the unique fuzzy set $\widetilde{E}(\phi/\xi) \in \mathcal{F}_o(\mathbf{R})$ such that $lim_{n \to \infty} d_\infty(\widetilde{E}(\phi/\xi), \widetilde{E}(\phi_n/\xi)) = 0$.

A rigorous and detailed justification of the definition of the above expected value, is given by Puri and Ralescu [6], [7], and by Negoita and Ralescu, [3].

It is also worth pointing out that, in practice, the computation of $\widetilde{E}(\phi/\xi)$ for non-simple fuzzy random variables, obtained from a limiting process, usually becomes complicated.

We need now to consider another element: the comparison between expected values. As these expected values will be fuzzy numbers, that comparison has to be a ranking of fuzzy numbers. Several procedures have been proposed in the literature of fuzzy numbers. Kolodziejczyk [2], following ideas from Orlovsky [5], analyzed different fuzzy relations, satisfying some properties that confirm their suitability to rank fuzzy numbers. In particular, some of them were suggested so that calculations in the set of fuzzy numbers with respect to the fuzzy addition and product by a positive real number could be performed in a manner analogous to those for real numbers (and, consequently, so that calculations through the expected value for a fuzzy random variable could be performed in an analogous way as those for random variables, which is very convenient and plausible for our purposes).

That relation, introduced by Kolodziejczyk, could be expressed as follows:
Let $\widetilde{V} \in \mathcal{F}_o(\mathbb{R})$ be a fuzzy number and let $\geq \widetilde{V}$ and $\leq \widetilde{V}$ denote the fuzzy sets of $\mathbb{R}$ "more than or equal to $\widetilde{V}$" and "less than or equal to $\widetilde{V}$" whose membership functions are given by

$$\mu_{\geq \widetilde{V}}(w) = \begin{cases} \mu_{\widetilde{V}}(w), & \text{si } w \leq z; \\ 1, & \text{si } w > z. \end{cases}$$

where z is the biggest real number such that $\mu_{\widetilde{V}}(z) = 1$ and

$$\mu_{\leq \widetilde{V}}(w) = \begin{cases} \mu_{\widetilde{V}}(w), & \text{si } w \geq z; \\ 1, & \text{si } w < z. \end{cases}$$

where z is the smallest real number such that $\mu_{\widetilde{V}}(z) = 1$.

Definition 2.4. *Let $\widetilde{U}, \widetilde{V} \in \mathcal{F}_o(\mathbb{R})$, be two fuzzy numbers. The following coefficient represents "the degree of truth for the expression ≪$\widetilde{U}$ is not higher than $\widetilde{V}$≫"*

$$R(\widetilde{U}, \widetilde{V}) = \frac{d(\geq \widetilde{U} \,\widetilde{\vee} \geq \widetilde{V}, \geq \widetilde{U}) + d(\leq \widetilde{U} \,\widetilde{\vee} \leq \widetilde{V}, \leq \widetilde{U}) + d(\widetilde{U} \widetilde{\cap} \widetilde{V}, \widetilde{0})}{d(\geq \widetilde{U}, \geq \widetilde{V}) + d(\leq \widetilde{U}, \leq \widetilde{V}) + 2d(\widetilde{U} \widetilde{\cap} \widetilde{V}, \widetilde{0})}$$

where d is the Hamming distance between fuzzy sets given by

$$d(\widetilde{U}, \widetilde{V}) = \int_{R} | \mu_{\widetilde{U}}(z) - \mu_{\widetilde{V}}(z) | \, dz$$

$\widetilde{\vee}$ is the fuzzy maximun operator, characterized by

$$\mu_{\widetilde{U} \widetilde{\vee} \widetilde{V}}(z) = \sup_{x \vee y = z} [min\{\mu_{\widetilde{U}}(x), \mu_{\widetilde{V}}(y)\}], \forall z \in R \text{ with } \vee = maximun$$

$\widetilde{\cap}$ is the fuzzy intersection given by

$$\mu_{\widetilde{U} \widetilde{\cap} \widetilde{V}}(z) = min\{\mu_{\widetilde{U}}(z), \mu_{\widetilde{V}}(z)\} \ \forall z \in R,$$

and $\widetilde{0}$ is the special fuzzy set whose membership function is

$$\mu_{\widetilde{0}}(w) = \begin{cases} 1, & \text{if } w = 0; \\ 0, & \text{otherwise.} \end{cases}$$

On the basis of coefficient R, we can establish the following criterion to order fuzzy numbers.

Definition 2.5. *Let $\widetilde{U}, \widetilde{V} \in \mathcal{F}_o(\mathbb{R})$, two fuzzy numbers; $\widetilde{U}$ is said to be greater than or equal to $\widetilde{V}$ if and only if $R(\widetilde{U}, \widetilde{V}) \leq 0.5$.*

Kolodziejczyk [2] proved the following properties:
Let $\widetilde{T}$, $\widetilde{U}$, $\widetilde{V}$, $\widetilde{W} \in \mathcal{F}_o(\mathbb{R})$ be four fuzzy numbers, then

- **if $R(\widetilde{T}, \widetilde{U}) \leq 0.5$ and $R(\widetilde{V}, \widetilde{W}) \leq 0.5$, then $R(\widetilde{T} \oplus \widetilde{V}, \widetilde{U} \oplus \widetilde{W}) \leq 0.5$, where $\oplus$ means the fuzzy addition;**
- **if λ is a positive real number, then $R(\widetilde{T}, \widetilde{U}) = R(\lambda \otimes \widetilde{T}, \lambda \otimes \widetilde{U})$;**
- **if $R(\widetilde{T}, \widetilde{U}) \leq 0.5$, then $R(\widetilde{T} \oplus ((-1) \otimes \widetilde{U}, \widetilde{0}) \leq 0.5$**

In general, any suitable comparison satisfying these properties would also be useful for our purposes.

On the basis of the preceding concepts we can now model the fuzzy utilities as follows:

Definition 2.6. *Consider a decison problem involving n probability spaces denoted by (Θ_i, C_i, ξ_i), $i = 1, 2, ..., n$ (Θ_i being either a state space or a sample space associated with a random variable), and m action spaces D_j, $j = 1, ..., m$. A fuzzy utility function for that problem is a fuzzy numbered-valued function U on $(\Theta_1 \times ... \times \Theta_n) \times (D_1 \times ... \times D_m)$ such that*

i) for each $i \in \{1, ..., n\}$, if $\theta_l \in \Theta_l$ for all $l \neq i$ and $a_j \in D_j$, $j \in \{1, ..., m\}$, the function $U(\theta_1, ..., \theta_{i-1}, \bullet, \theta_{i+1}, ..., \theta_n; a_1, ..., a_m)$ defined on Θ_i is a fuzzy random variable associated with the measurable space (Θ_i, C_i), whose expected value with respect to ξ_i is a fuzzy number that will be denoted by

$$\widetilde{E}(U(\theta_1, ..., \theta_{i-1}, \bullet, \theta_{i+1}, ..., \theta_n; a_1, ..., a_m)/\xi_i)$$

or alternatively (when ξ_i is not required to be specified), by

$$U(\theta_1, ..., \theta_{i-1}, \theta_{i+1}, ..., \theta_n; a_1, ..., a_m)$$

(and this process could be repeated and extended for a joint distribution on the cartesian product of various Θ_i).

ii) if $\theta_{i_1}, ..., \theta_{i_p}, \{i_1, ..., i_p\} \subset \{1, ..., n\}$, are such that $\theta_{i_1} \in \Theta_{i_1}, ..., \theta_{i_p} \in \Theta_{i_p}$ and they are assumed to be known at the time of making the decision in D_j, then $a_j \in D_j$ is said to be preferred or indifferent to $a'_j \in D_j$ with respect to $\theta_{i_1}, ..., \theta_{i_p}$ and actions $a_1, ..., a_{j-1}, a_{j+1}, ..., a_m$ if and only if

$$R(U(\theta_{i_1}, ..., \theta_{i_p}; a_1, ..., a_j, ..., a_m), U(\theta_{i_1}, ..., \theta_{i_p}; a_1, ..., a'_j, ..., a_m)) \leq 0.5$$

3 Decision influence diagrams

A decision influence diagram for a Bayesian decision problem is a directed acyclic graph involving three types of nodes: ***chance nodes***, ***decision nodes*** and the ***value node***.

Chance nodes are associated with random variables and they are represented by circles, ***decision nodes*** are associated with action spaces or decision rules and they

are represented by rectangles, and the *value node* is associated with the criterion to choose decisions and it is commonly represented by a diamond.
Arcs between nodes in the diagram reveal the presence of *conditional*, *functional* or *informational* influences. Thus, an *arc to a chance node* implies the associated variable can be probabilistically conditioned by the input variable or decision, although the probabilities themselves are not explicitely shown in the diagram. An *arc to a decision node* means that the information from the input variable value or decision made is known at the time of making the output decision.
The value node has no successors (there are no arcs from the value node to other nodes) in the diagram and the criterion for choosing decisions is usually formalized in terms of a measurable utility function depending on the direct predecessors of the value node (nodes with an arc to the value node) and whose expected value is to be maximized for an optimal policy.

The general procedure for solving a decision problem by means of the associated influence diagram is the following:

- Describing the initial information of the problem by means of an influence diagram.
- Describing the objective and allocate to it a node (value node) in the diagram.
- Using elementary graphical transformation to gradually reduce the diagram to the value node.
- Performing the statistical rules, for each transformation which reduces the diagram.

The elementary graphical transformations to reduce an influence diagram are some value preserving transformations. A *value preserving transformation* of an influence diagram is a graphical representation of an operation that transforms any diagram to another one having the same optimal policy and expected value as the first one. The elementary value preserving transformations used in the literature of influence diagrams with decisions (see, for instance, Olmsted, [4], Shachter, [8], Smith [9]) are arc reversal, chance node removal, and decision node removal. *Arc reversal* is the influence diagram representation for *Bayes'rule*. *Chance node removal* is the representation for *conditional expectation*. *Decision node removal* is the representation for *maximization* of a conditional expected utility.

4 Transformations in decision influence diagrams with fuzzy utilities

In this section, we analyze the extension of the preceding transformations to influence diagrams for decision problems in which the utility function is not a real-valued one but a fuzzy utility function.

We will employ the usual notations: if B is a node, we will denote by P_B the set of direct predecessors of that node, which are the nodes with an arc to node B, and by S_B the set of direct successors of B, which are the nodes with an arc from B to them.

Definition 4.1. *A node B is said to be a barren node if it has no sucessors in the diagram.*

Definition 4.2. *A path from X_{i_1} to X_{i_n} is a succession of nodes $X_{i_2}, X_{i_3},...,X_{i_{n-1}}$ such that there is an arc from X_{i_j} to $X_{i_{j+1}}$, for $j = 1,...,n-1$.*

We are now going to examine the effects of fuzziness on the value preserving transformations.

Transformation 4.1. *(Arc reversal) If there is an arc from X_i to X_j (chance nodes) and there is not another path from X_i to X_j, then the arc from X_i to X_j can be replaced by an arc from X_j to X_i and arcs from direct predecessors of X_i to X_j, and arcs from direct predecessors of X_j to X_j.*

Proof: We denote by $P_{X_iX_j} = P_{X_i} \cup P_{X_j} - \{X_i\}$. In accordance with the meaning of arcs to chance nodes, we must obtain the conditional distribution $P(X_i/X_j, P_{X_i} \cup P_{X_j} - \{X_i\})$.
Then $P(X_i, X_j/P_{X_iX_j}) = P(X_j/P_{X_j})P(X_i/P_{X_i})$, so we can obtain the marginal distribution $P(X_j/P_{X_iX_j})$, and the conditional one $P(X_i/Y, P_{X_iX_j})$, as well as confirm that $P(X_j/P_{X_j})P(X_i/P_{X_i}) = P(X_i/X_j, P_{X_iX_j})P(X_j/P_{X_iX_j})$.
Then we must replace arcs corresponding to $P(X_j/P_{X_j})$ and $P(X_i/P_{X_i})$ by arcs corresponding to $P(X_i/X_j, P_{X_iX_j})$ and $P(X_j/P_{X_iX_j})$. •

Transformation 4.2. *A barren node (excepting the value node) may be simply removed from the influence diagram.*

Proof: If a chance node or a decision node does not have successors in the diagram, then the associated random variable or decision space either does not condition other variables or its value is not known at the time of making a decision, so we can remove it from the diagram.

Other nodes may become barren nodes after this transformation and then they may be removed. •

Transformation 4.3. *(Chance node removal) If node X_i, associated with a random variable, is a direct predecessor of the value node and nothing else, then it may be removed by conditional expectation, and all direct predecessors of node X_i will be direct predecessors of the value node.*

Proof: Let P_V be the set of direct predecessors of the value node in the influence diagram, let P'_V be the set of direct predecessors of the value node after applying the transformation ($P'_V = P_V \cup P_{X_i} - \{X_i\}$).
We must obtain the expected utility given P'_V. This can be done by calculating the expected value of $U(P_V - \{X_i\}, \bullet)$ with respect to $P(X_i/P_{X_i})$, so $U(P'_V) = \widetilde{E}(U(P_V - \{X_i\}, \bullet)/P(X_i/P_{X_i})$. •

According to the last result, conditions and studies for the case of fuzzy utilities do not essentially differ from those in the real-valued case. However, there is a transformation explicitly involving those fuzzy utilities, in which conclusions strongly depend on the choice of the ranking of fuzzy numbers to be used. The result we are going to present allows us to conclude that when using the ranking in Definition 2.5, conclusions are not affected by the incorporation of fuzziness in the utility assessment.

Transformation 4.4. *(Decision node removal) If node D_i, associated with a decision rule is a direct predecessor of the value node and $P_V \subset P_{D_i} \cup D_i$, that node may be removed by "maximizing" the expected utility. It is possible that the process creates barren nodes, which can be removed according to Transformation 4.1.*

Proof: In this case the only successor in the diagram of D_i is the value node.
Let P_V be the set of direct predecessors of the value node in the influence diagram, let P'_V be the set of direct predecessors of the value node after applying the transformation (i.e.,$P'_V = P_V - \{D_i\}$).
We must look for the action(s) of D_i which "maximizes" the expected utility.
For that purpose, we will use the criterion to rank fuzzy numbers. We can consider the following steps. (n_i being the number of possible actions in D_i)
Let $a* = a_{il_1}$;
For $j = 2$ to n_i do:
if $R(U(P_V - \{D_i\} = p_v - a*, D_i = a*), U(P_V - \{D_i\} = p_v - a_{il_j}, D_i = a_{il_j})) > 0.5$, then $a* = a_j$.
With this criterion, we consider $a*$ as an optimal action in D_i, and $U(P'_V = p'_v) = U(P_V - \{D_i\} = p'_v, D_i = a*)$. •

Proposition 4.1. *If all barren nodes have been deleted and there is no decision node to be removed, then there is a chance node that is a direct predecessor of the value node and not of a decision node which can be removed, perhaps after some arc reversal.*

Proof: See, for instance, Shachter [8]. •

5 Algorithm for decision problems

In this section we present an algorithm based on the preceding transformations, to obtain optimal policies and the final expected utility value.

Step 1.- Eliminate all barren nodes from the diagram until it no longer contains barren (except the value node) (Transformation 4.2).

Step 2.- While $P_V \neq \emptyset$, do:

- If X_i is a chance node with $X_i \in P_V$ and $S_{X_i} = \{V\}$, then remove X_i; go to Step 2. (Transformation 4.3.)
- If D_i is a decision node with $D_i \in P_V$, $P_V - \{D_i\} \subset P_{D_i}$, then remove node D_i and remove barren nodes; go to Step 2 (Transformation 4.4.)
- Find X_i, chance node with $X_i \in P_V$, $S_{X_i} \cap \{D_1, ..., D_1\} = \emptyset$, while $S_{X_i} \neq \{V\}$, do:
  - find X_j chance node with $X_j \in S_{X_i}$ without another path from X_i to X_j
  - reverse the arc from X_i to X_j (Transformation 4.1.)
  - remove X_i; go to Step 2 (Transformation 4.3.)

Step 3.- End.

6 Conclusions

From the above discussion, we can conclude that extending decisions influence diagrams to fuzzy utilities is easy from a theoretical viewpoint. All transformations are identical, except the decision and chance node removal; the former needs a criterion to rank fuzzy numbers and the latter needs the use of fuzzy random variables and their expected value. From an algorithmic viewpoint, decision influence diagrams with fuzzy utilities can be handled in analogous form to those without fuzzy utilities. We should remark that in the preceding algorithm, the number of possible actions in each decision space has been considered finite in order to use the ranking in Definition 2.5. However, that is not a necessary condition, since it is often possible to order fuzzy numbers depending on some parameter.

Acknowledgments

This research has been financially supported by DGICYT Grant PB92-1014. I am also grateful to María Angeles Gil for her helpful comments.

References

1 Gil, M.A. & Jain, P. (1992). Comparison of Experiments in Statistical Decision problems with fuzzy utilities. *IEEE Trans. Syst., Man and Cybern.* **22**, 662-670.

2 Kolodziejczyk, W. (1986). Orlovsky's concept of Decision Making with fuzzy preference relation-further results. *Fuzzy Sets and Systems* **19**, 11-20.

3 Negoita, C.V. & Ralescu, D.A. (1987). *Simulation, Knowledge-based Computing, and Fuzzy Statistics.* New York, NY: Van Nostrand Reinhold Co.

4 Olmsted, S.M. (1983). On representing and solving decision problems. Ph. D. Thesis, Stanford University.

5 Orlovsky, S.A. (1980). On formalization of a general fuzzy mathematical problem. *Fuzzy Sets and Systems* **3**, 311-321.

6 Puri, M.L. & Ralescu, D.A. (1985). The concept of Normality for Fuzzy Random Variables. *Ann. Probab.* **13**,1373-1379.

7 Puri, M.L. & Ralescu, D.A. (1986). Fuzzy Random Variables. *J.Math.Anal.Appl.* **114**, 409-422.

8 Shachter, R. (1986). Evaluating Influence Diagrams. *Opns. Res.* **34**, 871-882.

9 Smith, J.Q. (1988). *Decision Analysis. A Bayesian Approach.* Chapman and Hall, London.

10 Wonnacott, R.J. & Wonnacott, T.H. (1985). *Introductory Statistics.* New York, NY:Wiley.

11 Zadeh, L.A. (1975). The concept of a linguistic variable and its application to approximate reasoning. *Inf. Sci.*, Part 1: Vol.**8**, 199-249; Part 2: **8**, 301-353; Part 3: **9**, 43-80.

Causal Networks and Their Toolkit in KSE*

Jianming Liang[1,2], Qinliang Ren[3], Zhuoqun Xu[4], Jiaqing Fang[2]

[1] Department of Computer Science, University of Turku, DataCity, Lemminkäisenkatu 14 A, FIN-20520 Turku, Finland
[2] North China Institute of Computing Technology, Beijing, P. R. China
[3] SRIFFI, Beijing, P. R. China
[4] Beijing University, P. R. China

Abstract. Causal Networks have recently received much attention in AI, and have been used in many areas as a knowledge representation. First, from the knowledge engineering point of view, we present causal networks, introduce a concept of network parameters, propose some principles for construction and use of causal networks, and indicate the advantages of the knowledge bases with the form of causal networks. In order to put them into practice and incorporate with other techniques in AI, we have introduced the idea of causal networks into a Knowledge System Environment (KSE) as a toolkit (Bent). Then we give an overview of KSE, followed by a detailed discussion about Bent.

1 Introduction

Causal models have recently received much attention in AI, and have been used in many areas [2]. This is because the core of human knowledge and experience is causal relationships among things. The knowledge and experience, however, are always incomplete and inexact, thus uncertainty management techniques should be fused into causal models. In a sense, a rule-based system can represent causal relationships, and there are many approaches to uncertainty management, but they are not ideal. The reason is that the distinguished feature of rule-based systems is *modularity,* i.e. detachment and locality. It is the modularity that constitutes the assumption of independence, i.e. all rules are independent in a knowledge base. In fact, rules are highly connected, and it is very difficult to satisfy the assumption in many domains, and it is this that causes the semantic deficiencies [9] and makes knowledge acquisition and knowledge base maintenance a difficult task. From the strictest sense, a rule-based system only is a shallow model, can not qualify as a causal model. It combines control knowledge with domain knowledge [2], can not express the underlying domain principle, and lacks the deep domain knowledge, makes the brittleness of expert systems and the difficulties in understanding the behavior of its inference and explanation. However, it possesses efficient reasoning mechanisms. A causal model is a deep model. It aims to capture the behavior and structure of the modeled system, can explicitly represent causal relationships among all components, separates control knowledge from domain knowledge, and makes the resulting knowledge bases usable for different applications. From the expert's point of view, a knowledge base organized as a causal model, whose structure may

* This work has been supported in part by national 8:5 programs and in part by national 863 program.

correspond to the domain structure, makes knowledge acquisition easier. From the developer's point of view, the causal model representation imposes an organization on the knowledge base which simplifies its maintenance. However, it generally requires sophisticated reasoning mechanisms, and lacks the mechanism for plausible inference. Thus it is important to combine the advantages of shallow models and deep models. A particular type of causal models, called the causal network, has been proposed, which provides a natural, efficient method for representing probabilistic causal relationships among a set of variables, can be used to express both shallow and deep knowledge, is a desirable knowledge representation with plausible inference.

2 Causal Networks

2.1 Overview of Theoretical Foundation

A causal network consists of a graphical structure that is augmented by a set of probabilities. The graphical structure is a directed, acyclic graph in which nodes represent domain variables that may be events, states, objects, propositions, or other entities, arcs are interpreted frequently causal relationships among the variables. Prior probabilities are assigned to source nodes, and conditional probabilities are associated with arcs. In particular, for each source node X_i, there is a prior probability function $P(X_i)$; for each node X_i with one or more direct predecessors Π_i, there is a conditional probability function $P(X_i|\Pi_i)$. A causal network, in fact, represents a full-joint probability space over the n domain variables in the network, for $P(X_1, X_2, ..., X_n) = \prod_i P(X_i|\Pi_i)$. Generally, this representation greatly reduces the number of probabilities that must be assessed and stored (relative to the full joint-probability space), and the probability of any sample point in the space can be computed from the probabilities in the network, and further based on the concepts of *conditional independence* and *graph separability*[9, 10], it can convey the pattern of dependency: two events do not become relevant to each other merely by virtue of predicting a common consequence, but they do become relevant when the consequence is actually observed; two consequences do not become irrelevant to each other merely by virtue of being caused by a common cause, but they do become irrelevant when the cause is actually observed.

Causal networks have the following properties. First, they represent probabilistic relationships concisely. It is necessary to consider only the known dependencies among variables in a domain, rather than to assume that all variables are dependent on each other. This provides an efficient and expressive language for acquiring and representing knowledge in many domains. Second, they manage probabilistic relationships efficiently. It is necessary to reason only on the current dependencies among variables, for dependencies are dynamic relationships, which are created and destroyed as new evidence is obtained. This lays a theoretical foundation for qualitative reasoning, which only involves the related variables, and ignores all unrelated variables for the current task. Finally, numerous approximation and special-case algorithms have been developed for efficient probabilistic inference using causal networks [1,4,6,7,8,9].

2.2 Construction Method for Causal Networks

We propose some principles here for constructing causal networks, supported by Bent.

Choosing Modeled Domain. In principle, application areas of causal networks include diagnosis, forecasting, image understanding, multi-sensor fusion, decision support systems, plan recognition, planning and control, speech recognition – in short, almost any task requiring that conclusions be drawn from uncertain clues and incomplete information [9], however the objective world is so complicated that a causal network like other knowledge representations can only express one part of the world, forms an incomplete knowledge base, the brittleness problem is postponed rather than avoided, thus we must understand our application and let it determine the exact choice of behaviors to model, and further, we must consider the model conciseness, for there is a tradeoff between conciseness on one hand, and understandability and complexity of the model on the other. The more concise the model, the more information it contains, and therefore the more complex, less understandable, and less efficient it will be. Thus it is important to select appropriate domains and model conciseness.

Choosing Abstract Objects. After choosing the modeled domain, we can draw abstract objects from the domain, which constitute nodes in causal networks. An abstract object is generally correspondent to a particular object in the domain, however, at some times, an abstract object may be correspondent to several particular objects, several abstract objects to one particular object, in order to improve the understandability and efficiency of inference. For example, in medical diagnosis, mediate states may be introduced, while in electric circuit diagnosis, several circuit components may be aggregated to one abstract object. Having chosen the abstract objects, we can determine their value sets.

Determining Network Parameters. In order to improve the degree of abstraction and shareability of causal networks, we permit one or more members in value sets of abstract objects to be variables besides normal values, and different abstract objects can take the same variable. These variables are termed the *network parameters,* which establish special relationships among abstract objects, and improve the power of inference of causal networks. They are handled as normal values, but they need instantiating when the causal network with them being used.

Determining Qualitative Relationships. This step is to determine the causal relationships among the abstract objects obtained above, i.e. to gain the structure of causal networks. In practice, it has been proved to be easy for a domain expert to judge the direct influence relationships among domain concepts. If abstract object A causes abstract object B, then draw an arrow from A to B, otherwise, there is no arrow between them. The direction of the arrow is generally from *cause* to *effect,* however, it can be adjusted according to the convenience of assessment of quantitative relationships (see next step), and this does not influence the inference using the causal network constructed. In this step, it is possible that there is no arrow between two sets of abstract objects, forming more than one causal network, but we may consider them as one causal network to manage in Bent.

Determining Quantitative Relationships. In this step, we only determine the local quantitative relationships, i.e. for an abstract object X_i, if its parents set is Π_i, we only assess the probabilities $P(X_i|\Pi_i)$. According to the features of causal networks, as long as we guarantee the local consistency, i.e. let the sum of $P(X_i|\Pi_i)$ be a unit, we can gain the global consistency of the whole causal network. If the quantitative relationships satisfy *"or model"* or *"and model"*, the correspondent causal networks may become simpler, i.e. a few of probabilities need assessing and storing.

The main sources of qualitative relationships and quantitative relationships above are domain experts, textbooks, and statistic material. For man-made systems, e.g. electric systems, the knowledge can come directly from their structures and behaviors, exists the possibility of constructing causal networks automatically; for natural systems, e.g. medical systems, it mostly comes from the expert experience, statistic material, or other related records. In many applications the exact values of the quantitative relationships play a minor role; most of the knowledge needed for reasoning plausibly about a domain lies in its structure [9], thus we should grasp consciously the qualitative relationships in the domain. When constructing a causal network, we could update it repeatedly, while the related abstract objects are linked with arrows, this makes it easy to check the consistency of the causal network.

When constructing a causal network, we should make it hierarchical, modular, and contain not very lots of abstract objects. For a large scale system, we can incorporate several causal networks, even other appropriate methods, to model it. This idea is reflected in our Knowledge System Environment KSE.

2.3 Inference Mechanisms & Explanation Mechanism

When finishing the construction of a causal network, we can use it as inference engines and get two kinds of results: *beliefs* of every abstract object and *most-probable explanation.* We can also derive beliefs of composite sets of abstract objects, and second-best interpretation, and circumscribing explanation from above two kinds of results. Correspondingly, causal networks have two inference methods: *belief updating* and *belief revision.* The belief updating is to assign each hypothesis in a network a degree of belief, BEF(.), consistent with all observations. The function BEL(.) changes smoothly and incrementally with each new item of evidence. The resulting output is beliefs of every abstract object. The belief revision is to find the tentative categorical acceptance of a subset of hypothesis which constitute the most satisfactory explanation of the evidence at hand. The resulting output is an optimal list of jointly accepted propositions - most-probable explanation, which may change abruptly as more evidence is obtained. In general, complexity of inference using causal networks is NP-hard [1], but numerous approximation and special-case algorithms have been developed for efficient inference. For example, the propagation procedures reported by Professor Judea Pearl [9]. For *singly* connected causal networks, there are known algorithms for belief updating and belief revision that can distributively propagate impacts of every item of evidence throughout in time that is linear as a function of the diameter of the causal network. For *multiply* connected causal networks, we can employ any one method of *clustering, conditioning, stochastic simulation,* and *preprocessing* to cope with *loops.* Both the algorithms for belief updating and belief revision are very similar. We can achieve them in one propagation mechanism having been implemented in Bent. There are several *belief parameters,* by which interacting abstract objects communicate in Bent. The definitions of them can be found in [9].

All abstract objects and arcs in causal networks have clear semantics, thus, in the propagation procedures, it is easy that tracing the most influential messages back to the generating evidence would yield a verbal explanation.

2.4 Use Method for Causal Networks

In this subsection, we give the use method supported by Bent/KSE.

1) First, point out the causal network to be used. If it is used at the first time, you should initialize the causal network, and instantiate its network parameters, otherwise, you can directly use it.
2) Obtain the evidence. It can integrate evidence from several blackboard levels, which may be input by users, or retrieved from data bases, or inferred using other methods.
3) Use the inference mechanism to reason with the current evidence. This can be automatically finished in Bent.
4) Obtain the results inferred. They can be stored in specific blackboard levels so as to be used by other methods, or output directly to users.
5) Store the causal network, if you want to use it later on. For example, in medical diagnosis, the stored causal networks can be as the clinical records.

2.5 Summary for Causal Networks

Causal networks provide a natural, efficient method for representing and managing probabilistic dependencies among a set of variables, can explicitly express the causal relationships, possess a firm theoretical foundation, combine qualitative with quantitative relationships, can be used to represent both shallow and deep knowledge, efficiently manage uncertainty in expert systems, support object-oriented design, separate control knowledge from domain knowledge, and the resulting knowledge base structure can correspond to domain structure. Thus they make it easier knowledge acquisition and knowledge base maintenance, improve the reusability of knowledge bases and increase acceptance of expert systems in the user community. Causal networks also possess numerous inference mechanisms, can integrate evidence from multiple sources, generate a coherent interpretation of the evidence. Causal networks have proved to be a useful knowledge representation tool, which can be applied to a number of areas. But, we do not think they are all-powerful. In order to incorporate with other techniques in AI, we introduced them into KSE.

3 KSE: Knowledge System Environment

The KSE is a Knowledge System Environment under development. We took the ***blackboard architecture*** as its core and the ***comprehensive knowledge model*** as its basis so that it can integrate the abilities of logic reasoning, numerical computing and data processing. Naturally, the environment is so complicated that we must balance between performance and flexibility, i.e. avoid sacrificing performance for flexibility or flexibility for performance. The essential notion is that instead of having one large system with an abundance of features, one should be able to construct systems which have just the set of features which the application calls for. Thus we adopted a unique software engineering methodology incorporating techniques from several well-known methodologies, specially, function decomposition, Object-Oriented design and Jackson method, and made it with a reconfigurable architecture. The following is a brief description about KSE (see [5] for details).

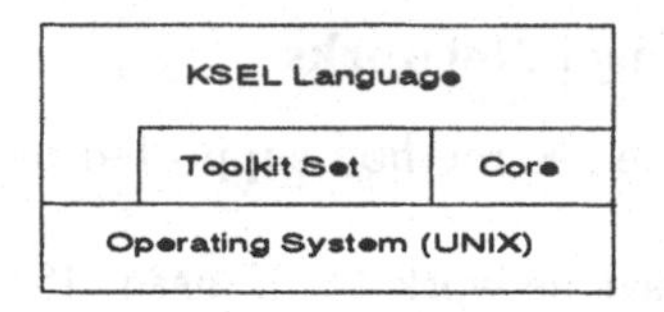

Fig. 1. The Architecture

3.1 Architecture

The KSE consists of the *blackboard kernel system* (the *core*), *toolkit set* and *KSEL language,* which are based on the operating system. The architecture is shown in Figure 1. When installing it, one can choose some toolkits in the toolkit set to integrate a configuration with the core: blackboard kernel system. The interfaces of the toolkits chosen and blackboard kernel system constitute the *interface* for knowledge engineers called KSEL language, which is a non-procedural language, can be used interactively or embedded in Ada and C languages.

3.2 Multiple Level Interfaces

Knowledge engineers can use KSEL to develop knowledge systems with the *multiple level interfaces* including the interface for domain experts, interface for knowledge system administrators and interface for operators (end users). These interfaces improve the security of knowledge systems. Their relationships are indicated in Figure 2.

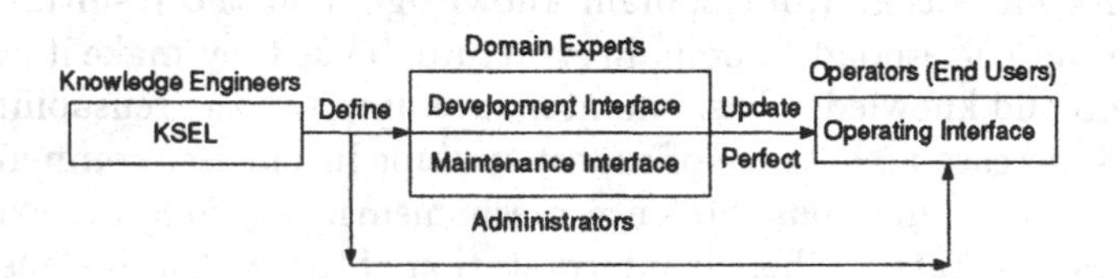

Fig. 2. Multiple Level Interfaces

3.3 Comprehensive Knowledge Model

Building knowledge systems, we should fully utilize the current achievements, for in many domains a lot of data bases (including graphical data bases, terrain data bases), model bases, and way bases have been set up. Various knowledge representations should be comprehensively utilized, for human knowledge possesses various forms. Thus we take the comprehensive knowledge model [11] as the basis of KSE, which is intensively reflected in knowledge bodies of knowledge sources as shown in Figure 3. An application system can incorporate several knowledge sources. Each knowledge source can selectively reference data bases, model bases, way bases and knowledge bases, which are regarded as *objects* in KSE. Each knowledge base possesses a unique KB structure. The KB structure may contain various knowledge representations, including *frames, rules, causal networks,* etc. For each knowledge representation, there is a correspondent management mechanism, e.g. Rult, Frat, Bent, etc.

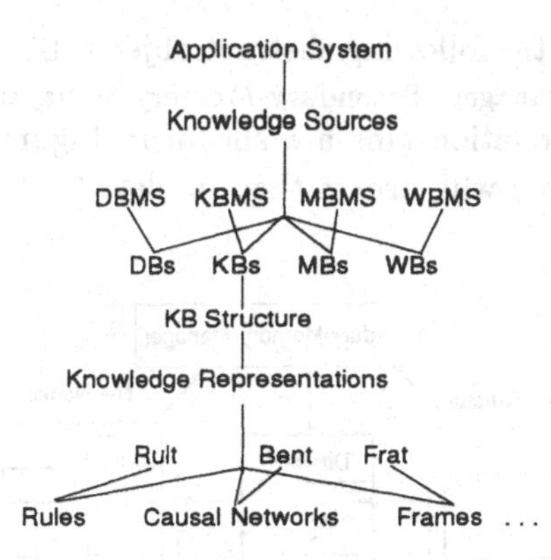

Fig. 3. Comprehensive Knowledge Model

3.4 Blackboard Kernel System

The implementations of blackboard systems vary considerably, but they all exhibit the same major architectural constructs: the blackboard (an explicit global data base), knowledge sources (effecting and reacting to changes to the blackboard) and control mechanism (determining when knowledge sources are executable). Thus the blackboard system design can be broken up into the design and representation of the blackboard itself, design and representation of the knowledge sources, design and representation of the control mechanism, and the design of a shell. The blackboard kernel system is a reconfigurable blackboard skeletal system, which is based on the experience of the Erasmus version of Boeing blackboard system [3].

3.5 Toolkit Set

Every toolkit in KSE is independent of each other, but they can constitute a configuration with the blackboard kernel system. In addition to generator for interfaces, editor for knowledge sources and toolkits for various knowledge representations (Rult, Frat, Bent, etc.), there are toolkits for connecting computer networks, data bases, model bases and way bases. Users can also combine their toolkits with the environment through related facilities. This makes our environment with an open architecture to suit the rapid development of artificial intelligence.

4 Bent: Toolkit for Causal Networks

4.1 Logical Structure

In general, an application system always needs several causal networks to model. In order to manage them efficiently, we introduced a Directory. We divided each causal network into two parts: Net-Structure and Net-Nodes so as to suit the qualitative and quantitative steps in the construction of causal networks. All operations on a causal network can be performed only when it has been in the primary memory, thus we designed a Primary-Memory-Manager. Causal networks may be updated and used repeatedly, thus we must keep them, the task is finished by the Secondary-Memory-Manager. In order to propagate the evidence, achieve belief updating and belief revision, we need a Propagating-Controller. As a toolkit, it should have a User-Interface so that users can conveniently utilize it. Thus from the Object-Oriented design viewpoint, we have the following abstract objects: Directory, Net-Structure, Net-Nodes, Primary-Memory-Manager, Secondary-Memory-Manager, Propagating-Controller and User-Interface. Their relationships are shown in Figure 4 (User-Interface is not shown). In the following, we will discuss them in detail.

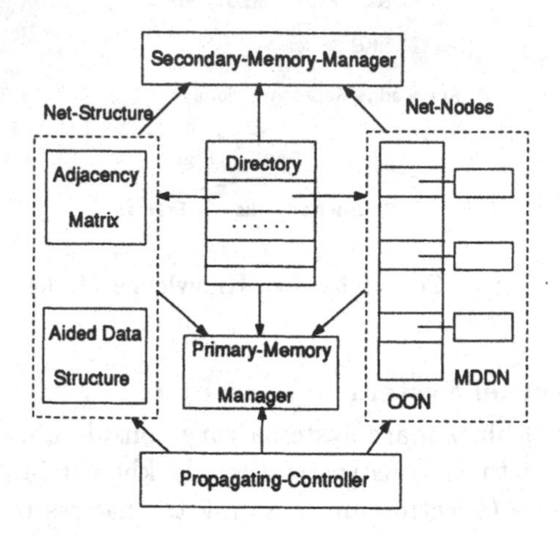

Fig. 4. The Logical Structure

4.2 Directory

The function of the Directory is to manage all causal networks in Bent. All operations on causal networks are based on it. In Bent, the name of a causal network is unique, thus users can access it with its name. In the Directory, for each causal network, there is a correspondent register, which contains the basic information about the causal network, including net-name, creator, date-created, date-last-used, Net-Structure-pointer, Nodes-pointer, number-of-nodes, etc. These registers are stored in a file, called Directory-File. There is a manager to manage the registers in the Directory-File. When a causal network is created, the manager creats a correspondent register in the Directory-File. When a causal network is deleted, the manager deletes its register in the Directory-File. The primitive operators on the registers are to *read* and to *write*.

4.3 Net-Structure

In general, there are two principle representations to describe a graph: adjacency matrix and adjacency list. However, the adjacency list is not suitable to causal networks, because many operations on causal networks can not be implemented easily. Therefore, we choose the adjacency matrix to represent the structure of causal networks. The adjacency matrix for a causal network with n nodes is an $n*n$ matrix A of bits. We set $A_{ij}=1$ if there is an edge from node i to node j and $A_{ij}=0$ if there is no such an edge. Since causal networks can not contain self-loops, the adjacency matrix always has zeros on its main diagonal. From the representation, we can easily obtain the direct parents and successors of every node (abstract object). In order to support propagating scheme effectively, we introduced two aided data structures: Father-Pointer and Successor-Pointer. In order to check the *loops* in a causal network, we introduced another aided data structure: *underlying graph,* which is undirected. Loops can be detected through matrix multiplication. All these aided data structures are automatically generated when the correspondent causal network is referenced.

4.4 Net-Nodes

Every node contains the information as to its value set, quantitative relationships, and belief parameters (its parents and successors sending). Since different nodes may have different value sets, different types, and different numbers of direct parents and successors, the sizes of space for different nodes may be different. In order to manage the information efficiently, we divided it into two parts. One is Overview-Of-Nodes (OON) with fixed length. The other is More-Detailed-Description-about-Nodes (MDDN) with variable length. OON records each node's Node-Name, Identifier, The-Number-of-Values, Type-of-Values, Position-of-its-item-in-MDDN, Heads-of-Belief-Parameters, Head-of-Conditional/Prior-Probabilities, etc. When a causal network is being created or used, this part is in the primary memory. For each node in a causal network, there is a correspondent item in MDDN, which mainly records Value-Set, Best-Value, Second-Best-Value, Belief Parameters, and Conditional/Prior-Probabilities. When a causal network is being used, not all its items are in the primary memory.

4.5 Primary-Memory-Manager

The function of the Primary-Memory-Manager is to manage the primary memory. Once a causal network is referenced, it distributes irreplaceable memory cells for its Structure and OON, and if needed, it allocates replaceable memory cells for the items involved in MDDN. When a causal network is not used, it collects all memory cells occupied by the network. Thus main operators are to *distribute* and to *collect.* Another operator is to *replace,* which is only for the items in MDDN, where we adopted the least recently used (LRU) algorithm.

4.6 Secondary-Memory-Manager

The function of the Secondary-Memory-Manager is to keep causal networks in the secondary storage. For a causal network created, the number of nodes, its structure, value set of every node are all fixed, thus all information about it can be stored in a file, called Net-File. When a causal network is kept, it creats such a net-file, and firstly stores its structure, then OON, finally MDDN. Before storing OON, it must compute the positions of every item in MDDN to be stored, and fill them in OON. When reading the causal network stored in a net-file, it firstly reads its structure and lets the Primary-Memory-Manager allocate the number of memory cells required, then OON. Only when some node is referenced, it reads its item in MDDN to replaceable memory cells. When a causal network is deleted, it deletes the correspondent net-file.

4.7 Propagating-Controller

The function of the Propagating-Controller is to finish belief updating and belief revision. In order to simulate the distributed procedures, we introduced the following control data structures: Pending Nodes Queue and Current Node Parameters List. The Pending Nodes Queue records all of the active nodes whose belief parameters can not be recomputed. For each such a node, there is a register, which contains Node-Identifier, Activator and their relationship. The Current Node Parameters List records all belief parameters which need recomputing for the current node. The recomputing formulae are defined in [9]. In fact, therefore, the procedures are implemented in series. We will implement them with an electric circuit, so as to take advantage of the distributed propagation procedures.

4.8 User-Interface

In Bent, we provided two kinds of user interfaces. One is the language binding interface, which is a non-procedure language, can be embedded in Ada and C languages. The other is the interactive interface, which is a set of commands used interactively. The effect of each command can be shown in an audio-visual way. For example, it can show how the belief parameters in belief updating and belief revision are being changed, and the paths of belief parameters propagated. Their functions mainly include three aspects: causal network *description* (to construct causal networks), causal network *manipulation* (to update causal networks, obtain evidence and propagate belief parameters) and causal network *control* (to manage all causal networks in Bent). The syntax is given in [5].

5 Conclusion

As a useful knowledge representation tool, causal networks can be used in many areas. We introduced the idea into KSE as a toolkit called Bent, which consists of seven abstract objects (modules), has been programmed in C. Since we adopted the unique software engineering methodology, they possess moderate sizes, low couping, high cohesion, and well information hiding.

References

1. G. F. Cooper. The computational complexity of probabilistic inference using Bayesian belief networks. *Artificial Intelligence*, 42(2–3):393–405, 1990.
2. E. Hudicka. Construction and use of a causal model for diagnosis. *International Journal of Intelligent Systems*, 3:315–349, 1988.
3. V. Jagannathan, R. T. Dodhiawala, and L. S. Baum. Boeing blackboard system: The erasmus version. *International Journal of Intelligent Systems*, 3:281–293, 1988.
4. S. L. Lauritzen and D. J. Spiegelhalter. Local computations with probabilities on graphical structures and their application to expert systems. *Journal of the Royal Statistical Society, Series B (Methodological)*, 50(2):157–224, 1988.
5. J. Liang. Probabilistic reasoning: Belief networks and their toolkit in KSE. Technical Report (M.Sc. Thesis), North China Institute of Computing Technology, P. O. Box 619 Ext. 70, Beijing 100083, P. R. China, 1989.
6. W. Long. Medical diagnosis using a probabilistic causal network. *Applied Artificial Intelligence*, 3(2–3):367–383, 1989.
7. K. C. Ng and B. Abramson. Uncertainty management in expert systems. *IEEE EXPERT*, 5(2):29–48, 1990.
8. K. G. Olesen, U. Kjærulff, F. Jensen, F. V. Jensen, B. Falck, S. Andreassen, and S. K. Andersen. A MUNIN network for the median nerve—a case study on loops. *Applied Artificial Intelligence*, 3(2–3):385–403, 1989.
9. J. Pearl. *Probabilistic Reasoning in Intelligent Systems.* Morgan Kaufmann, San Mateo, California, 1988.
10. J. Pearl. Belief networks revisited. *Artificial Intelligence*, 59:49–56, 1993.
11. X. Y. Tu. Methodology for design of large exper systems. In *Proceedings of the First Chinese Joint Conference on Artificial Intelligence*, pages 1–6, 1990.

Approximations of Causal Networks by Polytrees: an Empirical Study *

Silvia Acid and Luis M. de Campos

Universidad de Granada, Departamento de Ciencias de la Computación e I.A., 18071-Granada, Spain

Abstract. Once causal networks have been chosen as the model of knowledge representation of our interest, the aim of this work is to assess the performance of polytrees or Singly connected Causal Networks (SCNs) as approximations of general Multiply connected Causal Networks (MCNs). To do that we have carried out a simulation experiment in which we generated a number of MCNs, simulated them to get samples and used these samples to learn the SCNs that approximated the original MCNs, reporting the results.

1 Introduction

Causal networks have became common knowledge representation tools able to efficiently represent and manipulate dependence relationships by means of directed acyclic graphs (dags). However, when the dag corresponding to a given domain of knowledge is very dense, the processes necessary to estimate it from empirical data (learning) and to use it for inference tasks (propagation) may be very expensive.

Some simplified models, as trees and polytrees can alleviate these problems at the expense of losing some representation capabilities. Although an axiomatic characterization of dependency models isomorphic to trees and polytrees has been done in ([1]), in this paper we try to empirically assess the performance of polytrees or Singly connected Causal Networks (SCNs) as approximations of Multiply connected Causal Networks (MCNs). We have designed and carried out a simulation experiment to compare MCNs and SCNs from a perspective of classification. Basically, we randomly generated a number of MCNs, simulated them to get samples and used these samples to learn the SCNs that approximated the original MCNs. Next we compared the results, on the basis of the success rates of classification obtained by the original MCNs and their approximations.

The paper is organized as follows: in Section 2 we briefly describe MCNs and SCNs, together with their advantages and disadvantages. Section 3 contains a detailed description of the experiment carried out. The results of the experiment, as well as their analysis and interpretation, can be found in Section 4. Finally, Section 5 is devoted to the conclusions and future research.

* This work has been supported by the DGICYT under Project PB92-0939

2 Causal Networks

Causal networks [8] are directed acyclic graphs in which the nodes represent variables, the arcs signify the existence of direct causal dependencies between the linked variables, and the strengths of these dependencies are quantified by conditional probabilities (or any other measure of uncertainty, although here we restrict ourselves to probability measures).

Given the topology of the network, that establishes for each variable x_i in the graph the set (possibly empty) of variables $\pi(x_i)$ which are designated as parent variables of x_i, and given the probabilities $P(x_i|\pi(x_i))$ of each x_i conditioned to every combination of its parent set $\pi(x_i)$, we have specified a joint probability distribution

$$P(x_1, x_2, ..., x_n) = \prod_{i=1}^{n} P(x_i|\pi(x_i))$$

which includes all the independence relationships displayed by the graph.

Having a joint probability we can perform probabilistic inferences as, for example, to compute the posterior probability of a variable given that we know the values of some other variables. Propagation algorithms try to efficiently perform this computation by using local methods that take advantage of the independencies represented in the dag. However when the dag is multiply connected, that is, when there is more than one (undirected) path connecting two variables, the purely local procedures fail, although several methods retaining locality partially are available (see [6, 8]). Trees and polytrees constitute simplified causal models, where no more than one (undirected) path connects every two variables (singly connected networks). In polytrees a variable may have several (independent) parents and therefore they have more representation capabilities than trees. The absence of undirected cycles in the polytree permits the use of local methods of propagation.

On the other hand, if we do not have available the network and we need to learn it from examples, the algorithms for learning polytrees see ([2, 12]) are easier and more efficient than the algorithms for learning MCNs ([11, 13, 14]). Such algorithms are based on estimating the marginal and several first conditional independencies between pairs of variables in order to get independence measures or independency tests which are then used to constuc a polytree structure preserving the stronger dependencies.

So, using polytrees, we gain efficiency and simplicity in the procedures to learn the network as well as to propagate information through it. The price we have to pay is a less expressive power, because the kind of independence relationships that may be represented is more restricted for SNCs than for MCNs ([2]). Figures 1 and 2 display the topological differences between a dag with 7 nodes and its possible polytree approximation.

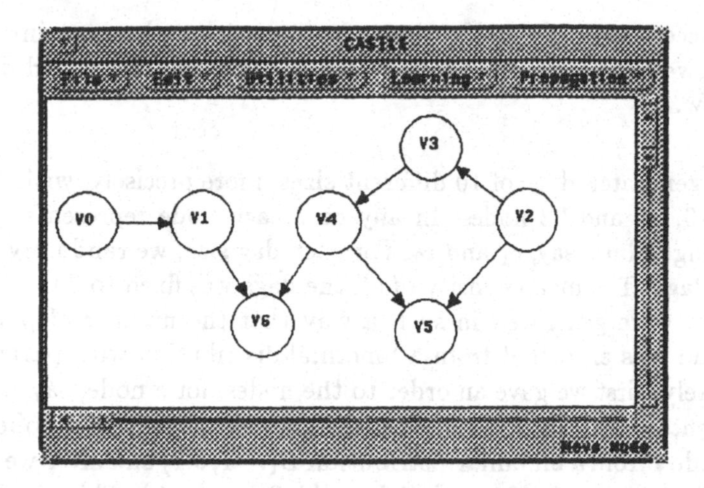

Fig. 1. A multiply connected network

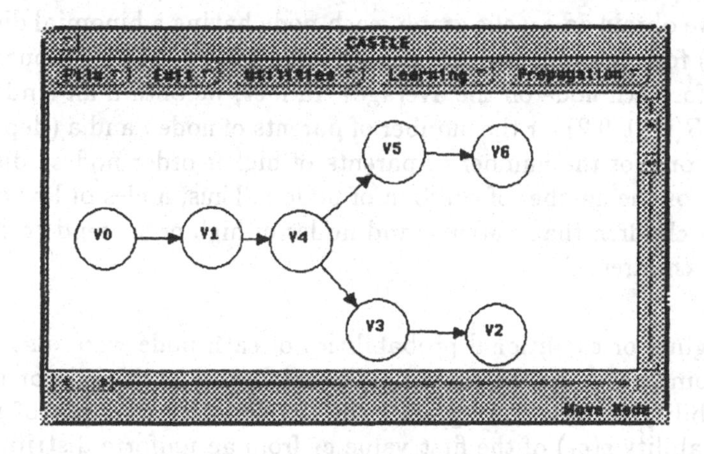

Fig. 2. An approximation by a singly connected network

The important question is the following: Is this price too high or are there practical situations where the replacement of a MCN by a SCN is acceptable? The next sections try to give answer to this question.

3 Empirical Comparison of Causal Networks and Polytrees

Our objective is to test whether SNCs can replace MCNs in practice. The criterion used to evaluate the quality of the approximation is neither whether the topologies of the MCN and its SCN approximation are more o less similar, as the independencies can be directly read off the graphical structure ([10, 8]), nor whether the joint probabilities obtained from the MCN and the SCN specify the same or a more restricted set of independency statements ([9]), but the capacity of SNCs to give results similar to that of MCNs for classification problems, measuring the success rates of classification in both models and comparing them. For this purpose we have carried out a large experiment, whose detailed description is given below.

We have generated dags of 10 different sizes, more precisely, with 4, 7, 9, 10, 12, 13, 15, 16, 19 and 22 nodes. In any case each node represents a bivalued variable taking values, say, c_1 and c_2. For each dag size, we randomly generated 50 different dags. The mean density of all the dags was fixed to 20%; this means that the dags were generated in such a way that the number of parents that each node had was extracted from a binomial distribution with parameter 0.2. More concretely: first we gave an order to the nodes; for a node, say i, the set of possible parents of i is $\{1, 2, \ldots, i-1\}$; next we extracted the number $pa(i)$ of parents of node i from a binomial distribution $B(i-1, 0.2)$ and after we randomly selected $pa(i)$ parent nodes for node i from $\{1, 2, \ldots, i-1\}$. This procedure was sequentially repeated (in the reverse order) for each node, starting from node n. In this way we obtain an acyclic graph, each node having a binomial distribution $B(n-1, 0.2)$ for the number of its neighbours ($0.2(n-1)$ neighbours (parents or children) for each node on the average). In fact, we obtain an (independent) distribution $B(i-1, 0.2)$ for the number of parents of node i and a (dependent on the distributions for the number of parents of higher order nodes) distribution $B(n-i, 0.2)$ for the number of children of node i. Thus, nodes of low order tend to have more children than parents, and nodes of high order tend to have more parents than children.

The marginal or conditional probabilities of each node were also randomly generated, computing, for the marginal probability on each node or the conditional probability of each node given each combination of values of its parent set, the probability $p(c_1)$ of the first value c_1 from an uniform distribution, and letting $p(c_2) = 1 - p(c_1)$.

For each one of these 50 dags of a given size we obtained 10 samples of 3000 examples each, using Montecarlo simulation (or logic sampling, see [5]). For each sample, we divided it in a learning set of 2000 examples and a test set of size 1000. Then we used the learning set to obtain a polytree approximation of the dag, using a batch version of the software CASTLE [2]. Next (and selecting one variable as the classification variable, and the other as attribute variables, and repeating this process for each variable), each example in the test set was classified using the original dag and its polytree approximation, and the success rates with respect to the true classification were collected. For propagation in dags we used the software ENTORNO [3], that was adapted to batch mode. Figure 3 depicts a schema of the experiment. It is worth noticing the enormous volume of data generated and processed: 500 dags generated, 5000 polytrees estimated, 15000000 examples generated, 127000000 classifications performed,... Running the overall experiment in batch mode needed 911 cpu hours in a Sparc Station 2.

The complete procedure to run the experiment is the following:

For each dag size n do
• For $i = 1$ to 50 do
⋆ generate dag(i)
⋆ For $j = 1$ to 10 do
∘ For $k = 1$ to 3000 do generate example(i,j,k)
∘ split the 3000 examples in learning_set(i,j) and test_set(i,j)
∘ learn polytree(i,j) using learning_set(i,j)
∘ For $h = 1$ to n do
· For $m = 1$ to 1000 do classify the example number m in the test_set(i,j) considering that the variable number h is the classification variable, using both dag(i) and polytree(i,j)
· calculate the success rates for classifying variable h using dag(i) and polytree(i,j), succ_dag(i,j,h) and succ_poly(i,j,h), respectively
⋆ For $h = 1$ to n do
∘ calculate the mean success rates of the 10 simulations for variable h using dag(i) and polytree(i,j), succ_dag(i,h) and succ_poly(i,h), respectively:
· $succ\_dag(i,h) = \sum_{j=1}^{10} succ\_dag(i,j,h)/10$
· $succ\_poly(i,h) = \sum_{j=1}^{10} succ\_poly(i,j,h)/10$
⋆ calculate the means and the deviations of the mean success rates of the n variables, succ_dag(i), succ_poly(i)), σ_dag(i) and σ_poly(i):
∘ $succ\_dag(i) = \sum_{h=1}^{n} succ\_dag(i,h)/n$
∘ $succ\_poly(i) = \sum_{h=1}^{n} succ\_poly(i,h)/n$
∘ $\sigma\_dag(i) = \sqrt{(}\sum_{h=1}^{n}(succ\_dag(i,h) - succ\_dag(i))^2/(n-1))$
∘ $\sigma\_poly(i) = \sqrt{(}\sum_{h=1}^{n}(succ\_poly(i,h) - succ\_poly(i))^2/(n-1))$
• calculate the means and deviations of the mean success rates per dag and its polytree approximation:
⋆ $succ\_dag = \sum_{i=1}^{50} succ\_dag(i)/50$
⋆ $succ\_poly = \sum_{i=1}^{50} succ\_poly(i)/50$
⋆ $\sigma\_dag = \sqrt{(}\sum_{i=1}^{50}(succ\_dag(i) - succ\_dag)^2/49)$
⋆ $\sigma\_poly = \sqrt{(}\sum_{i=1}^{50}(succ\_poly(i) - succ\_poly)^2/49)$
• estimate an approximate confidence interval for the difference between succ_dag and succ_poly
• calculate also the means of the deviations per variable, and their deviations:
⋆ $mean\_\sigma\_dag = \sum_{i=1}^{50} \sigma\_dag(i)/50$
⋆ $mean\_\sigma\_poly = \sum_{i=1}^{50} \sigma\_poly(i)/50$
⋆ $\sigma\_\sigma\_dag = \sqrt{(}(\sum_{i=1}^{50}(\sigma\_dag(i) - mean\_\sigma\_dag)^2)/49)$
⋆ $\sigma\_\sigma\_poly = \sqrt{(}(\sum_{i=1}^{50}(\sigma\_poly(i) - mean\_\sigma\_poly)^2)/49)$

Now, we are going to briefly describe the principles on which the software CASTLE is based. In its learning mode, CASTLE basically estimates, from a

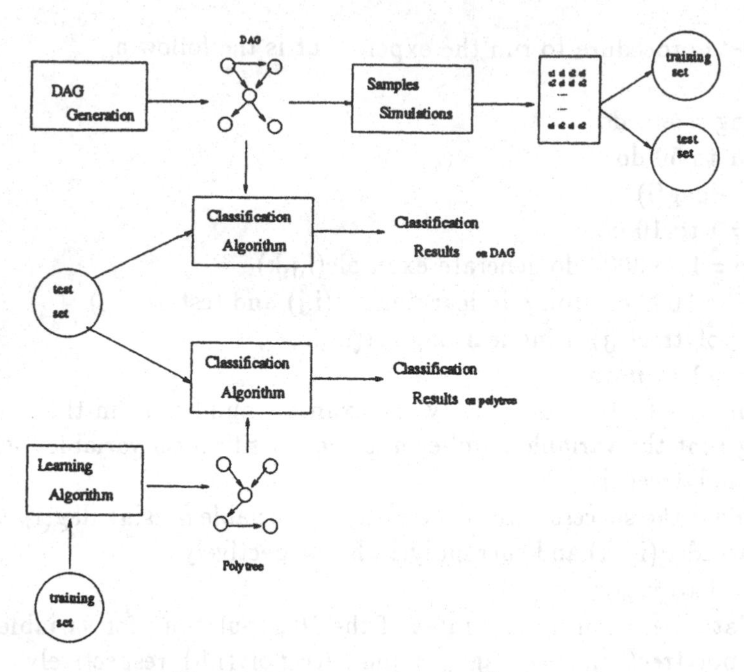

Fig. 3. The experiment diagram

file of examples, the independencies among the variables involved in the examples, in order to build a polytree displaying these independencies. The process of building the network is not focussed on any distinguished variable (like a classification variable, if we are interested on classification problems); therefore we obtain as a result a singly connected directed graph which approximates the underlying distribution of the data, without syntactic differences between the variables.

The first task in estimating the polytree is to learn the skeleton, the graph stripped of the arrows. To build the skeleton we know (see [8]) that if a positive distribution is representable by a polytree then a maximum weight spanning tree algorithm will recover its skeleton, where the weight of the branch connecting any two variables has the form of the Kullback-Leiber information measure. The underlying distribution of the polytree obtained in this way is an approximation of the true underlying distribution ([4]). CASTLE also incorporates other measures of dependency between variables, which have similar properties as the already mentionned Kullback measure, such as some distance measures between a bidimensional probability and the product of its marginals.

The second task in the learning procedure of CASTLE is to direct the skeleton, that is, to give direction to the branches. This is done by using certain independence measures or independency tests, which try to detect some patterns (such as head-to-head) in the graph [12]. In this process the algorithm also takes care partially of the consistence of the graph. Having estimated the structure of the polytree, the last step is to estimate the numerical parameters (marginal or conditional probabilities) which are necessary for the established network topology, from the data set. Once the network has been constructed, CASTLE can also exploit the net by fixing observed cases on some nodes, propagating these pieces of information through out the network, and consulting the posterior probabilities of the nodes. Finally, CASTLE also has simulation and graphical edition capabilities (see [2] for more details about CASTLE, and [7] for an application of CASTLE to classification problems).

Remark: The reason to use 10 polytrees for each dag in the experiment is that CASTLE can learn only from examples. So, the polytrees obtained by CASTLE depend on the samples generated, and therefore they are not necessarily identical to the theoretical polytree approximation that would be obtained by using the true joint probability distribution associated to the dag. By averaging the results of 10 polytrees per dag we aimed to obtain a behaviour more similar to this theoretical polytree.

4 Experimental Results

After having run the experiment described before, (the simulation, learning and classification processes), in this section we show in a summarised way some statistical results in tables 1, 2 and 3.

For each one of the sizes which appear in the first column, n, in table 1 we have *succ_dag* and *succ_poly* which are the mean success percentages obtained by classifying the test set using dag and polytree respectively. The columns σ*_dag* and σ*_poly* show the corresponding standard deviations.

The columns *mean_*σ*_dag* and *mean_*σ*_poly* in table 2 show the mean of the standard deviations (for each one of the dags and polytrees of a given size) with respect to the mean success percentages of classification for all the variables. They are measures that give information relative to the impact that the designation of the different variables as the classification variable has on the success rate of classification. The columns σ*_*σ*_dag* and σ*_*σ*_poly* show the standard deviation with respect to the previous ones.

Table 3 displays $diff\_med = succ\_dag - succ\_poly$, that is the means of the differences between the success percentages of classification using the original dags and their approximations, the standard deviations for these differences and approximate 95% confidence intervals.

In the light of the results shown in the tables above, some conclusions may be drawn:

First, when using the original dags, there is an increment in the success percentages of classification as the dag size increases (going from 77.8% for size 4 to

Table 1. Results on classification

n	*succ_dag*	*succ_poly*	σ*_dag*	σ*_poly*
4	77.844	77.342	6.422	6.659
7	78.014	76.954	4.991	5.109
9	78.537	76.518	3.531	3.897
10	79.931	77.430	3.810	4.381
12	79.899	76.817	2.950	3.812
13	80.535	77.330	2.685	3.382
15	80.692	76.461	3.253	4.008
16	80.787	76.226	2.409	2.510
19	81.826	75.161	1.783	2.538
22	82.341	75.140	2.000	2.749

Table 2. Results on classification

n	*mean_*σ*_dag*	*mean_*σ*_poly*	σ*_*σ*_dag*	σ*_*σ*_poly*
4	9.922	10.071	4.912	4.927
7	10.060	10.264	3.166	3.316
9	10.259	10.774	2.523	2.621
10	9.518	10.261	2.223	2.278
12	9.430	10.319	2.076	2.068
13	9.207	10.369	2.268	2.130
15	9.327	10.922	1.819	1.728
16	9.332	10.579	1.826	1.774
19	8.746	10.782	1.160	1.304
22	8.856	10.869	1.295	1.402

82.3% for size 22). The explanation for this at first sight surprising result is that the more variables there are in the dags, the more discriminating power the dags have because the joint probability distributions are more restricted (remember that the dags in our experiment are rather sparse).

Second, when we use the polytree approximations, the success percentage decreases as the size increases (going from 77.3% for size 4 to 75.1% for size 22), that is to say, the quality of polytrees as approximations of dags decreases. These two observations are more clearly appreciated in table 3: there are significant differences between *succ_dag* and *succ_poly* in all the cases, and these differences increase as the dag size increases. However, we want to emphasize that the error is not very high, at least in the studied cases (no more than 8% for size 22, and less than 1% for size 4). The linear regression model relating the difference in the success percentage for dags and polytrees and the dag size gives

$$diff\_med = 0.397 * n - 1.539$$

Third, the values of *mean_*σ*_dag* and *mean_*σ*_poly* (columns 1 and 2 in table 2) show that the success percentage of classification varies depending on what

Table 3. Differences between a dag and its polytree approximation

n	$diff\_med$	$diff\_\sigma$	[ ]
4	0.502	1.315	0.125 0.879
7	1.060	1.495	0.631 1.489
9	2.019	1.762	1.513 2.525
10	2.501	2.294	1.843 3.159
12	3.082	2.049	2.494 3.670
13	3.205	1.802	2.688 3.722
15	4.230	2.028	3.648 4.812
16	4.561	2.165	3.940 5.182
19	6.665	2.500	5.948 7.382
22	7.201	2.331	6.532 7.870

variable is designated as the classification variable, but the differences are not very high (around 9.5% and 10.5% for dags and polytrees respectively). Moreover $mean\_\sigma\_dag$ and $mean\_\sigma\_poly$ do not change very much when we change the dag size (they are nearly independent on the dag size), although the corresponding standard deviations, σ_σ_dag and σ_σ_poly decrease as the dag size increases.

So, we believe that the experimental results support the conclusion that polytrees are suitable approximations for dags with a degree of connectivity (or density) moderate, and a reduced number of variables. Moreover, it is interesting to remark that the conditional and marginal probability distributions were simulated from uniform distributions; this means that every probability distribution $(p(x_i = c_1), p(x_i = c_2)) = (p, 1-p))$ had the same probability of being selected. In real situations, we think that the probability distributions will be more restricted (possibly with sharper probability values). In that cases the polytree approximations could be better.

5 Concluding Remarks

We have tried to assess the quality of polytrees as approximations of multiply connected causal networks from a practical point of view. The results obtained after performing a big simulation experiment show that the inability of polytrees to represent independencies of order greater than two does not affect very much their capacity as classifiers: they can replace causal networks which neither are highly connected nor have a great number of variables, at the expense of a moderate decrease in the success rates of classification, but gaining in simplicity and efficiency.

However, in our experiment the only factor influencing the success rates of classification which has been investigated is the size of the network, and there are other potentially influencing factors that have been fixed in our study. So, in forthcoming studies we aim to experiment how the success rates of classification for dags and polytrees vary depending on the density of the graph (although we can expect a poor performance of polytrees approximating causal networks highly connected), the number of cases per variables, as well as biased probability distributions generated when constructing the dags. Another different research will lie in elaborating a criterion of quality of the approximation based on the similarity between the joint distributions.

References

1. L.M. de Campos, Independecy relationships in singly connected networks, Technical Report DECSAI-94214, Universidad de Granada, (1994).
2. S. Acid, L.M. de Campos, A. González, R. Molina, N. Pérez de la Blanca, Learning with CASTLE, in Symbolic and Quantitative Approaches to Uncertainty, Lecture Notes in Computer Science 548, R. Kruse, P. Siegel (Eds.), Springer Verlag, 99–106 (1991).
3. J. E. Cano, Propagación de probabilidades inferiores y superiores en grafos, Ph.D Thesis, Universidad de Granada (1993).
4. C.K. Chow, C.N. Liu, Approximating discrete probability distribution with dependence trees, IEEE Transactions on Information Theory 14, 462–467 (1968).
5. M. Henrion, Propagating uncertainty in bayesian networks by probabilistic logic sampling, Uncertainty in Artificial Intelligence 2, J.F. Lemmer and L.N. Kanal, Eds., North-Holland, 149–163 (1988).
6. S.L. Lauritzen, D.J. Spiegelhalter, Local computations with probabilities on graphical structures and their application to expert systems, J.R. Statist. Soc. Ser. B 50, 157–224 (1988).
7. R. Molina, L.M. de Campos, J. Mateos, Using bayesian algorithms for learning causal networks in classification problems, Uncertainty in Intelligent Systems, B. Bouchon-Meunier, L. Valverde, R.R. Yager (Eds.), North-Holland, 49–59 (1993).
8. J. Pearl, Probabilistic reasoning in intelligent systems: Networks of plausible inference, Morgan and Kaufmann (1988).
9. J. Pearl, T.S. Verma, Equivalence and Synthesis of causal models: Proceedings of the 6th International Conference, J.A. Allen, R. Fikes and E. Sandewall, Eds, Morgan and Kaufmann, 220–227 (1990).
10. T.S. Verma, Causal networks: Semantics and Expressiveness. Proceedings of 4th Uncertainty in Artificial Intelligence, R. Shachter, T.S. Levitt and L.N. Kanal, Elsevier Science Publisher, 325–359 (1989)
11. J. Pearl, T.S. Verma, A theory of inferred causation, in Principles of Knowledge Representation and Reasoning: Proceedings of the Second International Conference, J.A. Allen, R. Fikes and E. Sandewall, Eds, Morgan and Kaufmann, 441–452 (1991).
12. G. Rebane, J. Pearl, The recovery of causal polytrees from statistical data, Uncertainty in Artificial Intelligence 3, L.N. Kanal, T.S. Levitt and J.F. Lemmer, Eds, North-Holland, 175–182 (1989).
13. P. Spirtes, Detecting causal relations in the presence of unmeasured variables, Uncertainty in Artificial Intelligence, Proceedings of the Seventh Conference, 392–397 (1991).
14. G.F. Cooper, E. Herskovits, A bayesian method for the induction of probabilistic networks from data, Machine Learning 9, 309–347 (1992).

Evidence Propagation on Influence Diagrams and Value of Evidence

Kazuo J. Ezawa

AT&T Bell Laboratories
600 Mountain Avenue,
Murray Hill, New Jersey 07974, USA

Abstract. In this paper, we introduce evidence propagation operations on influence diagrams and a concept of value of evidence, which measures the value of experimentation. Evidence propagation operations are critical for the computation of the value of evidence, general update and inference operations in normative expert systems which are based on the influence diagram (Bayesian Network) paradigm. The value of evidence allows us to compute directly a value of perfect information and a value of control which are used in decision analysis (the science of decision making under uncertainty).

1. Introduction

Probabilistic expert systems that can interact with and interpret real world quantitative data and uncertainty have been developed thanks to algorithmic advances on evidence propagation (probabilistic inference) on Bayesian networks [3, 5, 6]. It replaces a rule-based knowledge representation with the Bayesian network based knowledge representation. For the development of *normative expert systems*, the use of influence diagrams [1, 2, 7] is a logical way to replace a heuristic based inference process of expert systems with a full fledged decision theoretic process. Hence evidence propagation operations on the influence diagrams are critical for the realization of the normative expert systems.

Evidence propagation on probabilistic influence diagrams which contain only chance (probabilistic) nodes has been previously discussed [6]. In this introduction, we briefly describe influence diagrams. We discuss evidence propagation in the presence of deterministic nodes, decision nodes, and a value node in addition to chance nodes in the regular influence diagrams in section 2. In section 3, we define and discuss the value of evidence in conjunction with the value of perfect information and the value of control.

An *influence diagram* is a graphical representation of a decision problem under uncertainty, explicitly revealing probabilistic dependence and the flow of information. It is an intuitive framework in which to formulate problems as perceived by decision makers and to incorporate the knowledge of experts. It is also a mathematically precise description of the problem that can be directly evaluated. An influence diagram is an acyclic directed graph with four types of nodes which represent different type of variables and two types of arcs which represent relationships between nodes.

A circle symbolizes a *chance* node which represents uncertain "event" and contains mutually exclusive potential outcomes and associated probabilities. A double circle symbolizes a *deterministic* node which represents functional relationships between nodes (variables) and contains a deterministic function that describes the relationships. A square symbolizes a *decision* node which represents a decision variable for the decision maker and contains decision alternatives. A diamond symbolizes a *value* node (V) which represents the goal of the decision problem and contains the value function which measures the "value." An arc into a chance node indicates there is a probabilistic dependency between the node and its predecessor(s), and it is called a *conditional arc.* An arc into a decision node is an *informational arc*, i.e., before you make a decision, you have information related to its predecessor(s). It represents time precedence between the decision node and its predecessor(s).

A *successor* of node i is a node on a directed path emanating from node i. A successor node that is adjacent to node i is called a *direct successor* of the node i and denoted as S(i). A *predecessor* of node i is a node on a directed path terminating at node i. A predecessor that is adjacent to node i is called a *direct predecessor* of node i and is denoted as C(i) (as *conditioning* nodes).

Each node i has an associated variable Xi, outcome space Ωi, and xi which represents a particular outcome of Ωi. A subset of Ωi is denoted by xi. **X** denotes all variables, and **D** denotes all decision variables which is a subset of **X** in the influence diagram. P{Xi} represents the probability distribution of the conditionally independent variable Xi. P{Xi|Xj} represents the probability distribution of conditionally dependent variable Xi given Xj.

There are three regular operations in the influence diagram evaluation algorithms. One is "chance node removal" which is to take the expectation of the joint probability given the chance node. The second operation is "decision node removal" which is to take the maximum (minimum) of expected value given the alternatives of the decision node. The last operation is "arc reversal" that is to change the direction of the arc which represents an application of Bayes rule.

2. Evidence Propagation on Influence Diagrams

In this section, we discuss three types of evidence propagation, 1) chance to chance evidence propagation and evidence reversal, 2) evidence propagation with a deterministic node, and 3) evidence propagation with a decision node.

The instantiation of evidence on a chance node and propagation of evidence among chance nodes involve the following operations depending on the network structure. These definitions are slightly different from the reference [6]:

- Evidence absorption: instantiation of evidence x_j on node Xj which is just the table lookup of the observed outcome, i.e., $P\{X_J=x_j \mid X_{C(X_J)}\}$
- Evidence propagation: propagation of evidence x_j to its successor node i, which is the identification of still valid potential outcomes, i.e., $P\{X_i \mid X_{C(X_I)} \,\&\, X_J=x_j\}*P\{X_J=x_j \mid X_{C(X_J)}\}$
- Evidence reversal: evidence absorption of x_j on X_J and arc reversal between X_J and its predecessor X_K and the propagation of evidence x_j to X_K.

• *Evidence Propagation to a Deterministic Node :* The propagation of evidence x_j to its successor node I, which is a function (F) of X_J and others is to set $X_J = x_j$ in the function F., i.e., $P\{X_I = F(X_{C(I)\backslash J} \& X_J{=}x_j)\}$ where " \ " indicates exclusion of J.

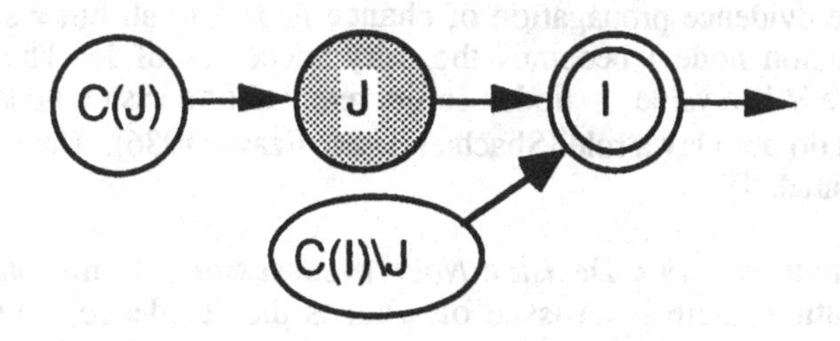

Figure 1. Evidence Propagation to a Deterministic Node

Note: We can first convert a deterministic node to a chance node, and then perform standard evidence propagation operations. An advantage of keeping it as a deterministic node is that we can save computational space better in this form in the influence diagram.

• *Evidence Reversal to a Deterministic Node:* Evidence reversal to a deterministic node requires the deterministic node to be converted to a chance node first, and then perform regular (chance to chance) evidence reversal.

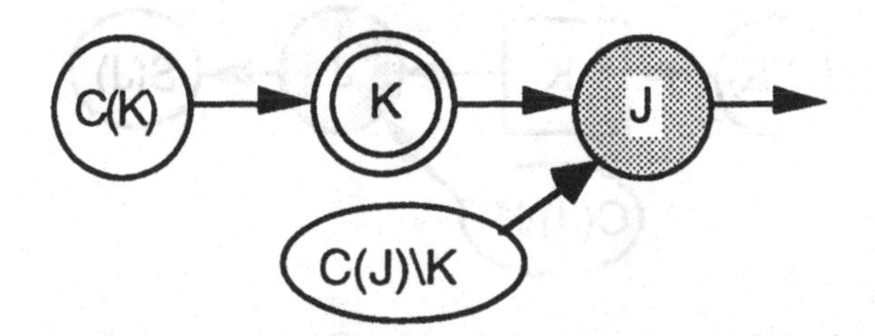

Figure 2. Evidence Reversal to a Deterministic Node

• *Evidence Propagation to a Decision Node:* It involves standard evidence propagation and an elimination of an arc from J to I (decision node).

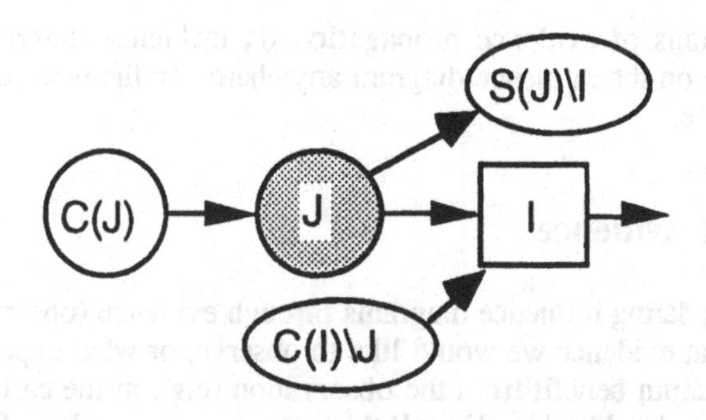

Figure 3. Evidence Propagation to a Decision Node

Proposition 1. After the evidence propagation of a chance node J to its successors S(J) \ I, if the chance node has a decision successor I, we can simply delete the arc from the chance node to the decision node.

Proof: After the evidence propagation of chance node J to all other successor nodes S(J) \ I, the decision node I becomes the only successor of J. Thus J belongs to $X_{C(I)\backslash C(V)}$ where V is a value node. In the removal of decision node I, $X_{C(I)\backslash C(V)}$ are irrelevant and do not play a role [Shachter 1986, Ezawa 1986]. Hence the arc from J to I can be eliminated. []

• *Evidence propagation with a Decision Node predecessor* : In this particular case of evidence propagation, there is an issue of what is the "evidence," i.e., whether we observe an outcome of a node or conditional outcomes of a node. We will discuss this issue in section 3 - the value of evidence. Here we assume we observe conditional outcomes of the node, and observation of an unconditional outcome of the node is a special case of observation of conditional outcomes.

Proposition 2. A chance node with decision node predecessors can propagate evidence with observed outcomes identified as conditional i.e., J|K. With the full evidence, it contains outcomes of J given all decision alternatives of K. After the propagation of evidence of {J|K}, decision node K inherits J's successors as successors.

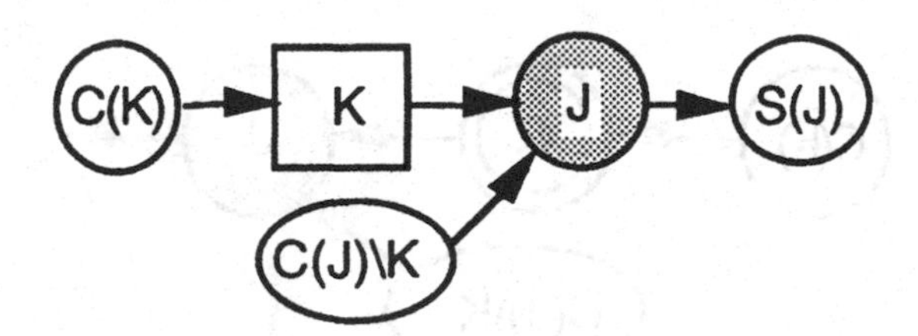

Figure 4. Evidence Propagation with a Decision Node Predecessor

Proof: The joint probabilities, P{S(J) U J U C(J)} = P{S'(J) U J U K} where S'(J) represents successors of J after the arc reversals to all chance nodes of C(J). If we apply evidence propagation of $J = x_j^*|x_k$, we get $S'(J = x_j^*|x_k)$ U K). []

With these operations of evidence propagation on influence diagrams, we can now propagate evidence on the influence diagram anywhere. In the next section, we discuss the value of evidence.

3. Value of Evidence

In the process of updating influence diagrams through evidence (observation), it is often useful to know what evidence we would like to observe, or what experiment we should do to receive maximum benefit from the observation (e.g., in the case of inquiry, what specific question we should ask). We call this measure of the value of an experiment as the *value of evidence (VOE)* defined below.

$$VOE (X_J = x_j) = EV(\mathbf{X} \setminus X_J, X_J = x_j) - EV(\mathbf{X})$$

i.e., the value of evidence of outcome x_j of variable X_J is the expected value (EV) of the influence diagram with the evidence x_j minus the expected value of the influence diagram.

In the normative expert systems which are based on influence diagram paradigm, the evidence propagation operation is the most often used functionality. It would be convenient if we could compute the value of perfect information and the value of control from the measure based on evidence propagation. The value of evidence is the measure which allows us to compute the value of perfect information and the value of control.

The value of perfect information is the difference between the expected values of knowing the outcomes of a node before making a decision and not knowing the outcome of the node of a given influence diagram in question. The *value of perfect information (VOPI)* is defined as

$$\mathrm{VOPI}(X_J) = \mathrm{EV}(\mathbf{X}\backslash\{\mathbf{D}, X_J\}, \mathbf{D}|X_J, X_J) - \mathrm{EV}(\mathbf{X}).$$

In the influence diagram operation, to obtain the expected value with perfect information, we add an arc from the chance node in question to the related decision node. When a decision node is a predecessor, we perform a special operation which we will discuss later. We can also define the value of perfect information as the expected value of the value of evidence of the evidence node J, i.e.,

Proposition 3.

$$\mathrm{VOPI}(X_J) = \Sigma\, \mathrm{VOE}(X_J = x_j) * P\{x_j\} \text{ for } \Omega_J \text{ of evidence node j.}$$

Proof: $\mathrm{VOPI}(X_J)=\mathrm{EV}(\mathbf{X}\backslash\{\mathbf{D},X_J\}, \mathbf{D}|X_J, X_J)-\mathrm{EV}(\mathbf{X})$
$=(\Sigma \mathrm{EV}(\mathbf{X}\backslash\{\mathbf{D},X_J\},\mathbf{D}|X_J,X_J=x_j)*P\{x_j\})- \mathrm{EV}(\mathbf{X})$
$=(\Sigma \mathrm{EV}(\mathbf{X}\backslash X_J,X_J=x_j)*P\{x_j\})-\mathrm{EV}(\mathbf{X})$
$=\Sigma(\mathrm{EV}(\mathbf{X}\backslash X_J, X_J=x_j)-\mathrm{EV}(\mathbf{X}))*P\{x_j\}$
$= \Sigma\, \mathrm{VOE}(X_J = x_j) * P\{x_j\}$
for Ω_J of evidence node j. []

In other words, once the evidence x_j is propagated, when we make the next decision (reduce decision node), this information is already incorporated(i.e., given the evidence of x_j.) Hence by weighing the value of evidence for each x_j with $P\{x_j\}$, we can compute the value of perfect information. The unconditional probability $P\{X_J\}$ can always be obtained by applying arc reversals between its predecessors as long as they are not decision nodes. Note that VOPI computed from VOE is the VOPI for overall decisions. Only when a decision predecessor is involved, we can compute VOPI for the decision. Note also a value of evidence could be negative, but the value of perfect information is always greater than or equal to 0.

The value of control is the difference between the expected values of controlling the outcomes of the node in question and not controlling it. The *value of control (VOC)* is defined by

$$\mathrm{VOC}(X_J) = \mathrm{Max}\ \mathrm{EV}(\mathbf{X}\backslash X_J, X_J=x_J) - \mathrm{EV}(\mathbf{X}) \text{ for all } \Omega_J.$$

In the influence diagram operation, to obtain the expected value with control, we change the chance node in question to a decision node. The value of control can be directly computed from the value of evidence.

Proposition 4.

VOC (X_J) = Max VOE ($X_J = x_j$) for Ω_J of evidence node J, if we are maximizing the value function.

Proof: $VOC(X_J)$=Max $EV(X\backslash X_J, X_J=x_j)$ - $EV(X)$
= Max ($EV(X\backslash X_J, X_J=x_j)$ - $EV(X)$)
= Max VOE ($X_J = x_j$) for Ω_J of evidence node J []

In other words, the evidence absorption is controlling of the event, i.e., assuming certainty of the outcome. Choosing the best outcome is the value of control. Hence by choosing the value of evidence which optimizes the value function, we get the value of control.

Now let's discuss the issue of evidence/observation of conditional outcomes and unconditional outcomes in the case of evidence propagation with a decision node predecessor. We discuss this in conjunction with the value of evidence, the value of perfect information, and the value of control using an example, Mars vs Venus [Matheson, 1990].

Mars vs Venus. Consider an hypothetical case of sending a landing craft to Mars or Venus. As a decision maker, we have a choice of sending the craft to Mars or Venus. The probability of success of the mission is 60% regardless of the destination. The values we receive from the mission is as follows:

Destination	Mission (Probability)	Value
Mars	Success (60%)	50
Mars	Failure (40%)	10
Venus	Success (60%)	100
Venus	Failure (40%)	-10

Table 1. Mars vs Venus

Figure 5 shows an influence diagram representation of Mars vs Venus problem.

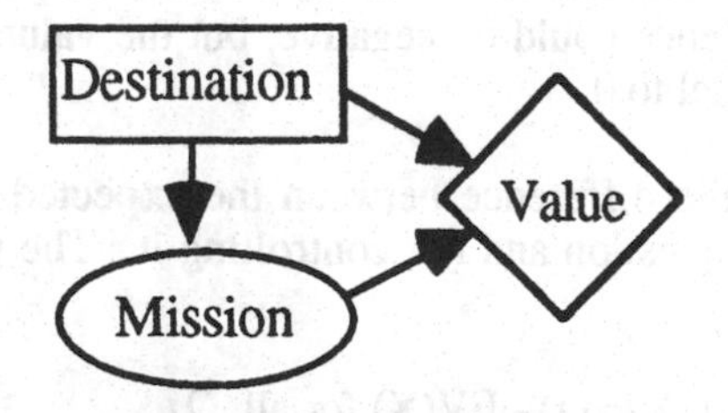

Figure 5. Mars vs Venus

The value of evidence for the mission is as follows:

Evidence	Destination	Value	VOE
Failure	Mars	10	- 46
Success	Venus	100	+ 44

Table 2. Value of Evidence

The value of perfect information is 8. It is computed from the value of evidence using information from table 1 and 2 (i.e., -46 * 0.40 + 44 * 0.60 = 8.) The value of control is 44. This is a "naive" computation of value of perfect information (i.e., based on observation of unconditional outcomes.) Next, we discuss a more sophisticated interpretations of the value of perfect information (i.e., based on an observation of conditional outcomes.)

3.2 With Full Evidence

In the case of value of perfect information, if we don't assume conditional independence between probability of success in Mars landing and Venus landing, we need to reassess these conditional probabilities for the computation of the value of perfect information for both using and not using value of evidence. For the computation of the value of perfect information, since we cannot directly reverse the arc between destination and mission in Figure 5, we need to modify the influence diagram to the one in Figure 6, which allows us to add an arc to Destination from Mission given Destination. Note that as shown in section 2, the value of evidence doesn't require this modification of the influence diagram.

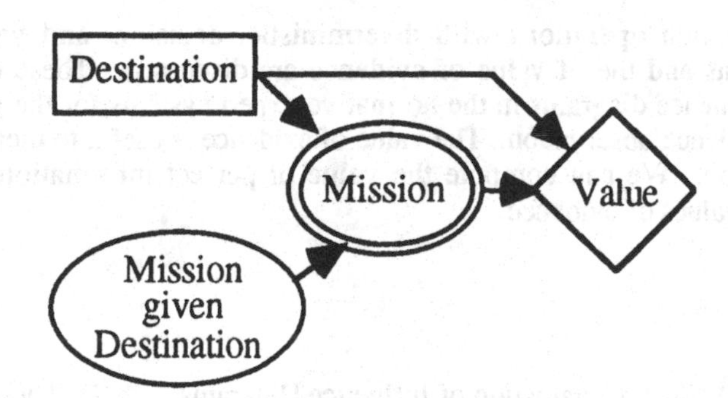

Figure 6. Modified Mars vs Venus Influence Diagram

	Venus: Failure	Venus: Success	
Mars: Failure	0.354	0.046	0.40
Mars: Success	0.046	0.554	0.60
	0.40	0.60	

Table 3. The joint probability distribution for the conditional outcomes of Mission given Destination

For the value of evidence for the conditional observation, is as follows:

Evidence (Observation)	Desti-nation	Value	VOE
Mars: Failure Venus: Failure	Mars	10	- 46
Mars: Success Venus: Failure	Mars	50	- 6
Mars: Failure Venus: Success	Venus	100	+ 44
Mars: Success Venus: Success	Venus	100	+ 44

Table 4. Value of Evidence

The value of perfect information is 9.84. As in the previous example, it is directly computed from table 3 and 4. The direction of the value of evidence, i.e., + or -, is also informative, it shows the direction of expected value's change as we observe more evidence. The value of control remains 44.

The advantage of the use of value of evidence is that we can avoid the modification of influence diagrams like the one we discussed here. Also it allows us to compute both the value of perfect information and the value of control directly from the value of evidence.

4 Summary

Evidence propagation operations with deterministic, decision, and value nodes on influence diagrams and the of value of evidence are discussed. These operations are crucial to use influence diagrams in the normative expert systems for the general update based on new evidence/observation. The value of evidence is useful to measure the value of experimentation. We can compute the value of perfect information and value of control from the values of evidence.

References

1. Ezawa, K. J., "Efficient Evaluation of Influence Diagrams," Ph. D. Thesis, Dept. of Engineering-Economic Systems, Stanford University, Palo Alto, CA, 1986
2. Ezawa, K. J., and Scherer, J. B., "Technology Planning For Advanced Telecommunication Services: A Computer-Aided Approach," *Telematics and Informatics*, pp. 101 - 112, 1992.
3. Lauritzen, S. L., and Spiegelhalter, D. J., "Local Computations with Probabilities on Graphical Structures and their Application to Expert Systems," *J. R. Statist. Soc.* B, 50, No.2 pp 157-224, 1988.
4. Matheson, James E., "Using Influence Diagrams to Value of Information and Control," Influence Diagrams, Belief Nets and Decision Analysis, John Wiley & Sons, 1990.

5. Pearl, Judea, "*Probabilistic Reasoning in Intelligent Systems: Networks of Plausible Inference*," Morgan Kaufmann, 1988.
6. Shachter, R. D., "Evidence Absorption and Propagation through Evidence Reversals," *Uncertainty in Artificial Intelligence,* Vol. 5, pp. 173-190, North-Holland, 1990.
7. Shachter, R. D., "Evaluating Influence Diagrams," *Operations Research,* Vol. 34, No. 6, 1986.

4. PROBABILISTIC, STATISTICAL AND INFORMATIONAL METHODS

Some Varieties of Qualitative Probability

Michael P. Wellman

University of Michigan
Dept of Electrical Engineering and Computer Science
Ann Arbor, MI 48109 USA
wellman@umich.edu

Abstract. In this essay I present a general characterization of *qualitative probability*, defining the concept of a qualitative probability *language* and proposing some bases for comparison. In particular, enumerating some of the distinctions that can be supported by a qualitative probability language induces a partial taxonomy of possible approaches. I discuss some of these in further depth, identify central issues, and suggest some general comparisons.

1. Introduction

In the standard theory of probability, degrees of belief for events or *propositions* take values in the real interval [0,1]. From degrees of belief on the primitive propositions, the theory dictates degrees of belief for various compound and conditional propositions, and vice versa. Computational schemes for *probabilistic reasoning* apply this theory to the automated derivation of degrees of belief for designated propositions of interest given prespecified degrees of belief over some other propositions and some particular conditioning propositions observed or hypothesized. This approach has, among other advantages, those accruing to a well understood and powerful underlying theory.

Despite these virtues, many have objected to the straightforward application of probability theory to uncertain reasoning because it appears to require undue precision in the specified degrees of belief. This and other supposed drawbacks has led researchers to invent innumerable variations on probability, as well as many other "non-probabilistic" uncertainty calculi. Among other distinctions, the variant approaches typically support the specification of degrees of belief that are less precise in some way or another. Motivations for these features include reduction of specification burden in representing uncertain knowledge, robustness of inference with respect to slight changes in degree of belief, and sometimes, the attempt to capture "ignorance" in a state of uncertain belief.

It is not my intent here to evaluate or survey the numerous mechanisms for incomplete or imprecise degree-of-belief specification in variant uncertainty calculi. Rather, I aim to identify and delineate some general classes of approaches that relax precision in representation and reasoning *within* the probabilistic framework. In doing so, I maintain a Bayesian perspective on probability, interpreting the imprecise specifications as *partial descriptions* of an uncertain belief state, not the belief state itself.[1] The various approaches have (legitimately) been termed "qualitative probability", though they are distinct enough to warrant an examination of their relative merits and limitations.

---

[1]Nevertheless, most of the technical points should be ideologically neutral enough to avoid putting off non-Bayesian readers, if I still have any.

The term "qualitative probability" was perhaps first introduced by de Finetti. In this paper, I am concerned with a broader class of schemes (explained below), of which de Finetti's approach (as developed by Savage and others) is a particular instance.

2. Qualitative Probability

Generally stated, *qualitative reasoning* is reasoning directly in terms of the qualities of interest for a particular problem or class of problems. The idea of qualitative reasoning is present (often implicitly) in many technical disciplines, where the aim of research is to identify exactly those qualities that are specific enough to characterize the phenomena of interest, yet general enough to comprise all situations where the phenomena apply. As a general methodology for representation and computation, qualitative reasoning has been investigated most explicitly and thoroughly within the field of Artificial Intelligence, especially by researchers interested in capturing commonsense or expert knowledge about the physical world (Weld and de Kleer 1989). These researchers have tended to focus on particular sorts of qualitative properties: signs of physical quantities and rates of change, monotone relations among parameters, and topological descriptions of space. But this emphasis can be attributed to the particular problems they have addressed, and we can consider qualitative reasoning more broadly as the approach to knowledge representation that shuns excess precision wherever possible in the effort to match the levels of description and objects of inference to those of the problem being addressed.

So how does the idea of qualitative reasoning apply to probabilistic or uncertain reasoning? Since few if any real problems in uncertain reasoning (specifically, decisions under uncertainty) depend on the full precision of real-valued degrees of belief, the principle of qualitative reasoning suggests that we should not require that degrees of belief be specified to this full precision, nor that the reasoning method manipulate degrees of belief at this precision. The usual (Bayesian, at least) objection at this point is that we cannot generally tell in advance which level of precision will be required, so it will be irrational to accept anything but the fullest precision. But it really does not follow from the lack of an assured precision bound that we need apply unbounded precision *in all cases*. Rather the qualitative dictum just calls for our reasoner to be flexible in its use of precision, and says that in those cases where precision can be avoided, it should be.

What does it mean to be less than fully precise about probability? There are several possible interpretation approaches, but the most straightforward is to regard an imprecise statement involving degrees of belief as a statement that the degrees of belief satisfy some constraint. Thus, whereas a fully precise description of an uncertain belief state will assign a unique number to the probability of each proposition, an imprecise description will merely constrain the probability to belong to some set. Or more generally, it will constrain the set of *joint* probability assignments to a subset of the joint space of probabilities.

Now, there are many potential ways that we can relax precision in degrees of belief. For example, we could simply replace our scalar representation of degrees of belief with a pair, where the two numbers are interpreted as lower and upper bounds on degrees of belief. This is, of course, standard in several of the variant uncertainty calculi.[2] But it is

---

[2]One representative attempt to apply probability bounds in uncertain reasoning was INFERNO (Quinlan 1983). There have been many others, before and since. Concepts of lower probability are commonplace in the statistical literature, and the notion of belief intervals (not necessarily interpreted as probability) is prominent in both Dempster-Shafer theory and Possibilistic Logic.

only the most obvious of the possible relaxations of precision, and not obviously the most advantageous for purposes of representation and reasoning. In order to make this judgment, we first need be more explicit about our desiderata for representation language and inference calculus. These will depend on the task we have in mind for qualitative probabilistic reasoning, and (once again) on the qualitative reasoning principle that reasoning be directly in terms of the qualities with which we are concerned.

3. Qualitative Probabilistic Reasoning

3.1. Semantics

Given our constraint-based interpretation of qualitative probability, we specify the reasoning process in the conventional way. Let L be a qualitative probability language, and Q a sentence of L. The meaning of Q is defined by the set of probability distributions that satisfy it. We write $\mu \models Q$ to say that the probability distribution μ satisfies the constraints Q. The set

$$\mathrm{M}(Q) = \{\mu \mid \mu \models Q\}$$

directly characterizes the extension of Q. From this we can define entailment in the qualitative probability language:

$$Q_1 \models Q_2 \text{ iff } \mathrm{M}(Q_1) \subseteq \mathrm{M}(Q_2).$$

Typically, systems of qualitative probability specify some inference regime, denoted $\vdash$, such that $Q_1 \vdash Q_2$ means that Q_2 is derivable from Q_1. This inference regime is *sound* if $Q_1 \vdash Q_2$ implies $Q_1 \models Q_2$, and *complete* if the converse holds. Most of the qualitative probability schemes we are concerned with are sound but not complete.

The expressive power of a qualitative probability language is defined by the sets of probability distributions that can be expressed. A set M is *expressible* in L if $\mathrm{M} = \mathrm{M}(Q)$ for some Q in L. A qualitative probability language is *closed under conjunction* if, for any two sentences Q_1 and Q_2, $\mathrm{M}(Q_1) \cap \mathrm{M}(Q_2)$ is expressible. Closure under disjunction and negation can be defined similarly. A perhaps more interesting property is closure under marginalization, which requires that the set $\mathrm{M} = \mathrm{M}(Q)$ remain expressible after averaging over some proposition or random variable.

3.2. Tasks

Qualitative probabilistic reasoning is the process of deriving some qualitative properties of a probability distribution from other ones. Which qualitative properties are of interest depends on the objectives of the reasoner. Two qualitative reasoning tasks have been featured in the literature: *decision making* and *acceptance*.

In decision making, the objective is to determine which action or strategy maximizes expected utility. This task requires that we have information about the available actions and preferences in addition to probability. The imprecision of qualitative probability means that we cannot in general determine a unique optimal decision, but the constraints may allow us to eliminate many of the possibilities based on qualitative dominance (Wellman 1990).

The aim of the acceptance task is to establish that the likelihood of particular propositions is sufficiently high that they should be accepted as beliefs. To accept a proposition is essentially to act as though it were certainly the case. Although a strict Bayesian would deny that it is ever rational to accept uncertain propositions,[3] some have advocated it on the grounds that it simplifies reasoning. Indeed, this is one of the central motivations for nonmonotonic logics. See Kyburg and respondents (Kyburg 1994) for a discussion of acceptance based on probabilistic thresholds. In any case, once an acceptance criterion is stated, the problem becomes a well-posed exercise in qualitative probability.

Irrespective of the overall task, another aim of qualitative probability has been to capture common inference patterns. That is, qualitative probabilistic reasoning schemes are often judged by whether they sanction particular inference schemas or how well they match the expected results on canonical examples. Examples include probabilistic theories of causality (Suppes 1970), and accounts of the phenomenon of "explaining away" in evidential reasoning (Pearl 1988; Wellman and Henrion 1993).

4. A Taxonomy of Qualitative Probability

Having described the general idea and formal framework, we are ready to classify some different forms of qualitative probability. The various distinctions considered in this paper are laid out in the taxonomy below.

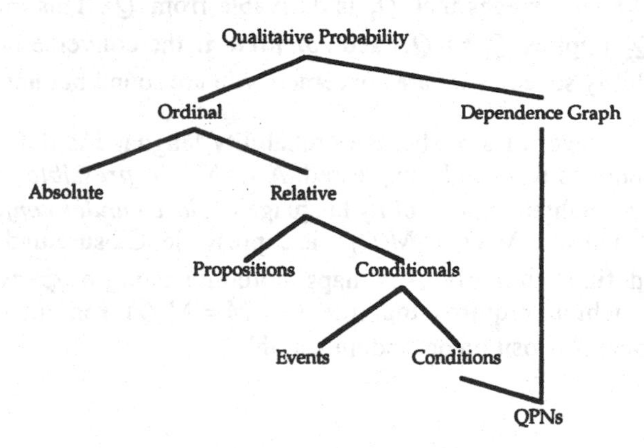

Fig. 1. A taxonomy of qualitative probability.

The two main categories of qualitative probability concepts, ordinal and dependence graph, are described in the sections below. Ordinal comparisons constitute the most

---

[3]Or at least the Bayesian would argue against separating acceptance of beliefs from decision making based on them. Nevertheless, the separation between so-called *theoretical* and *practical* reasoning is widely adopted in philosophy, and much research in uncertain reasoning presumes that the tasks are indeed separable. The case for categorical acceptance of uncertain beliefs can be made (in some cases) on computational grounds, but this is rarely explicit in treatments of the acceptance task.

common form of qualitative probability, and can be distinguished along several dimensions. First, we can compare probabilities against an absolute scale, or relative to each other. Second, the comparisons can involve unconditional propositions or conditional expressions (this distinction makes sense for absolute as well as relative comparisons, but is shown in the taxonomy only for the latter). Finally, in comparing conditional expressions, we can vary the base event part of the expression, or the conditioning events (or both). The qualitative probability language I have developed, QPNs (Wellman 1990), consists of relative ordinal comparisons among conditional expressions varying only the conditions, embedded within a dependence graph formalism.

5. Ordinal Qualitative Probability

Several forms of qualitative probability are defined by introducing an ordering relation $\not\prec$ over events signifying "not less likely". Its negation, $\prec$, denotes "less likely", and its symmetric restriction, $\sim$, "equally likely". In some foundational accounts of probability (Koopman 1940; Savage 1972), the numeric representation is derived from axioms on $\not\prec$. However, even if we accept the axioms, we are not *obliged* to use the numeric representation, and can design schemes that reason in terms of $\not\prec$ directly.

Specifying partial rankings among degrees of belief rather than assessing their magnitudes on a cardinal scale is perhaps the most prevalent form of qualitative probability. Many languages based on this idea have been proposed; the following sections outline some of the major distinctions.

5.1. Absolute vs. Relative

Often, ordinal qualitative probability schemes are augmented with a set of reference events measured on a cardinal scale. When all comparisons in the language are between a propositional expression and one of these reference events, we say that the language supports statements of *absolute* qualitative probability. All of the various interval calculi fall in this category.

In what cases will an absolute scheme achieve our goal of capturing the central qualities that we care about in uncertain reasoning? For example, suppose we fix a set of reference events corresponding to degrees of belief, such as 0.01, 0.05, 0.5, etc. These points may correspond to intuitive landmarks on the probability interval, but in reality it will be only by coincidence that they line up with significant thresholds in any decision or acceptance problem. (Kyburg 1994) proposes that in limited domains only a few threshold acceptance values will be required, but this claim is not really supported. The reason to be skeptical is that such thresholds are generally functions of a complex of other degrees of belief and preferences, and there is no a priori reason they should cluster around particular values. Except for the endpoints, no values of the probability interval seem to have general significance outside of some particular context.

Even if there exist some qualitatively meaningful reference probabilities, it is not clear that typical operations will respect the qualitative boundaries. For example, Bacchus (1989) has investigated an inheritance scheme based on properties holding with probability $\geq 1/2$, finding that it did not support chaining of inferences about typicality. Analysis of the idea of using fixed totally-ordered (though not cardinally scaled) reference events (Xiang et al. 1990) suggests that the problem is quite general, although

Dubois et al. (1992) have found that a scheme with parametric thresholds can be surprisingly robust.

An alternate approach is to invoke the ordering relation among the domain events rather than special reference events. We refer to such approaches as *relative* qualitative probability. The advantage of relative approaches is that the objects of comparison are known to be meaningful in the domain, and hence their relative likelihood is more apt to be a qualitatively significant property.

5.2. Conditional vs. Unconditional

Another significant distinction among ordinal approaches is whether the objects of comparison are event probabilities or conditional event probabilities. The use of conditionals is possible whether the comparison is relative or absolute. For example, the scheme of (Dubois et al. 1992) is based on classifying within parametric interval ranges conditional expressions corresponding to terms in a syllogism. Order-of-magnitude schemes (e.g., ε-semantics (Pearl 1988) or System-Z^+ (Goldszmidt and Pearl 1992)) rank conditional probabilities corresponding to default rules against an infinitesimal scale.

When comparing conditional expressions in a relative scheme, it also makes sense to distinguish whether the comparison varies the base proposition, the conditioning events in the expression, or both. For example, the logic of (Gärdenfors 1975) seems to allow arbitrary comparisons, whereas Neufeld's qualitative account of defaults (Neufeld 1989) makes use only of comparisons where the conditioning event is negated from one term to the other.

6. Dependence Graphs

Probabilistic independence, conditional or unconditional, is a basic qualitative concept exploited implicitly or explicitly in many probabilistic reasoning schemes. Independence imposes a constraint on probability distributions, and can be directly incorporated in languages of qualitative probability.

The logic of probabilistic independence has been studied in depth by Pearl and colleagues (1989), among others. In particular, they have proposed a set of axioms for conditional independence. Although these axioms could be used directly in a logic for reasoning about independence, much more common are representations based on *dependence graphs*, where dependence among propositions is encoded via connectivity in a graph. The most widely applied representation is that of directed acyclic graphs, variously called Bayesian, belief, causal, or probabilistic networks.

The independence relations encoded in a probabilistic network are that each node x is independent of its non-descendants given its immediate predecessors in the graph. That is,

$$Y \cap succ^*(x) = \varnothing \rightarrow I(x, pred(x), Y),$$

where $pred(x)$ denotes the direct predecessors of x, $succ^*(x)$ denotes the direct and indirect successors (i.e., nodes with a directed path from x, including x), and $I(X,Z,Y)$ means that the nodes in X are conditionally independent of those in Y given those in Z. From these explicit I relations, all that follow can be determined via *d-separation*, a sound, complete, and efficiently testable graphical connectivity criterion for conditional independence (Geiger et al. 1990).

How can we view dependence graphs as a qualitative probability logic? Although the language is not sentential in form, we can cast it sententially in several ways. First, we can treat the whole graph as a sentence, asserting the conditional independence statements above. This can be decomposed into a set of sentences $I(x, pred(x), \overline{succ^*(x)})$, one for each node x. However, in this decomposition the conditions are not local, as the I relation at each node depends on the structure of the rest of the graph.

Alternately, we could treat each link as a sentence, yet here we cannot associate a constraint with sentences, because the presence of a link by itself does not disallow any probability distributions. This suggests the final approach, taking the *absence* of links as sentences. The absence of a link does indeed express independence (if there is no link from y to x then $I(x, pred(x), y)$, assuming $y \notin succ^*(x)$), however, this too is a non-local decomposition.

Therefore, if we desire a logic with a modular semantics, that is, one where we can give a meaning to each sentence separately, it appears that we need to treat the entire graph as one sentence, or accept sentences with non-local scope. (Alternately, we could give up modularity and describe dependence graphs in a *nonmonotonic* logic (Grosof 1988).) The resulting logic is well characterized, and has a sound, complete, and efficient inference procedure based on d-separation (i.e., we can compute efficiently whether the independence conditions expressed in one graph are entailed by those of another). We could incorporate conjunction by allowing collections of graphs represent the union of associated independence conditions. This would strictly increase the expressive power of the language, though I am not sure if it would permit expression of arbitrary patterns of conditional independence.

Marginalization of a dependence graph can be accomplished via the graphical operations of node reduction and arc reversal (Shachter 1988). Although this preserves as many independencies as possible, the result may still lose information about conditional independence among the original variables. Just as adding auxiliary variables can enhance the expressive power of dependence graphs (even with respect to the original nodes) (Pearl 1988), eliminating variables can decrease it. Thus, dependence graphs are not closed under marginalization.

7. Qualitative Probabilistic Networks

Qualitative probabilistic networks (QPNs) are dependence graphs augmented with a specific form of ordinal, conditional comparison where the conditioning events vary *ceteris paribus*. Space does not permit a full account of QPNs here, but happily, there exist several available descriptions (Wellman 1990; Wellman and Henrion 1993). The main constructs in QPNs are *qualitative relations*, which specify the direction of interactions among random variables. For instance, a positive qualitative influence from node a to binary node b means that

$$a' > a'' \rightarrow \Pr(B \mid a'x) \geq \Pr(B \mid a''x),$$

for any assignment x to the nodes in $pred(b) - \{a\}$. There are also two varieties of qualitative *synergy*, which express ternary constraints capturing the interactions among qualitative influences.

As we can see above, QPNs impose constraints on probability distributions beyond those inherent in the dependence graph. However, just as for dependence graphs, it is difficult to interpret the qualitative relations sententially, as their meaning depends on the rest of the graph (in the case above, the other predecessors of node *b*). However, we can locally interpret all of the incoming links on a given node as a single sentence constraining the conditional distribution of a node given its predecessors. These, in conjunction with the sentence(s) describing the entire graph structure, constitute a complete description of the constraints imposed by a QPN.

Given this sketch, we can note some of the technical properties of QPNs as a qualitative probability logic. QPNs are trivially more expressive than dependence graphs, as any dependence graph can be represented by a QPN where all links are of sign ?. There are also sound inference algorithms, based on graph transformations (Wellman 1990) or qualitative sign propagation (Druzdzel and Henrion 1993). Both are efficient, and the latter is conjectured to be complete, at least with respect to queries about the qualitative influence relations among pairs of nodes. The transformation operations (node reduction and arc reversal) lose information; this is reflected in a lack of closure under marginalization and Bayesian revision.

8. This is not a Comprehensive Survey

I have described a general framework for characterizing schemes for qualitative probability, and identified some of the major distinctions. The analysis is not nearly exhaustive, having omitted discussion of several lines of work that are rightfully considered qualitative. Significant omissions include order-of-magnitude systems based on infinitesimal probabilities (Goldszmidt and Pearl 1992), and dependence models based on abstract degrees of belief (Darwiche 1993). Finally, I have also slighted non-probabilistic approaches for qualitative uncertain reasoning (Parsons 1993). See the report of a recent workshop on qualitative probability (Goldszmidt 1993) for further pointers and discussion of current research topics.

Acknowledgment

This work was supported in part by grant F49620-94-1-0027 from the US Air Force Office of Scientific Research. I have benefited from discussions of these concepts with Moisés Goldszmidt and Simon Parsons, among others.

References

Bacchus, F. (1989). A modest, but semantically well founded, inheritance reasoner. *Eleventh Int'l Joint Conference on Artificial Intelligence*, Detroit, MI, Morgan Kaufmann.

Darwiche, A. Y. (1993). *A Symbolic Generalization of Probability Theory*. PhD Thesis, Stanford University.

Druzdzel, M. J., and M. Henrion (1993). Efficient Reasoning in qualitative probabilistic networks. *Proceedings of the Eleventh National Conference on Artificial Intelligence*, Washington, DC, AAAI Press.

Dubois, D., H. Prade, L. Godo, et al. (1992). A symbolic approach to reasoning with linguistic quantifiers. *Eighth Conference on Uncertainty in Artificial Intelligence*, Palo Alto, CA, Morgan Kaufmann.

Gärdenfors, P. (1975). Qualitative probability as an intensional logic. *Journal of Philosophical Logic* **4**: 171-185.

Geiger, D., T. Verma, and J. Pearl (1990). *d*-separation: From theorems to algorithms. *Uncertainty in Artificial Intelligence 5* Ed. M. Henrion et al. North-Holland.

Goldszmidt, M., and J. Pearl (1992). Reasoning with qualitative probabilities can be tractable. *Eighth Conference on Uncertainty in Artificial Intelligence*, Palo Alto, CA, Morgan Kaufmann.

Goldszmidt, M. (1993). Research issues in qualitative and abstract probability. *AI Magazine* **15**(4): 63–66.

Grosof, B. N. (1988). Non-monotonicity in probabilistic reasoning. *Uncertainty in Artificial Intelligence 2* Ed. J. F. Lemmer, and L. N. Kanal. North-Holland. 237-249.

Koopman, B. O. (1940). The axioms and algebra of intuitive probability. *Annals of Mathematics* **42**: 269-292.

Kyburg, Jr., H. E. (1994). Believing on the basis of the evidence. *Computational Intelligence* **10**(1).

Neufeld, E. (1989). Defaults and probabilities; Extensions and coherence. *First Int'l Conf. on Principles of Knowledge Representation and Reasoning*, Toronto, Morgan Kaufmann.

Parsons, S. D. (1993). *Qualitative Methods for Reasoning under Uncertainty*. PhD Thesis, Queen Mary and Westfield College.

Pearl, J. (1988). *Probabilistic Reasoning in Intelligent Systems: Networks of Plausible Inference*. San Mateo, CA, Morgan Kaufmann.

Pearl, J., D. Geiger, and T. Verma (1989). Conditional independence and its representations. *Kybernetika* **25**: 33-44.

Quinlan, J. R. (1983). Inferno: A cautious approach to uncertain inference. *Computer Journal* **26**: 255-269.

Savage, L. J. (1972). *The Foundations of Statistics*. New York, Dover Publications.

Shachter, R. D. (1988). Probabilistic inference and influence diagrams. *Operations Research* **36**: 589-604.

Suppes, P. (1970). *A Probabilistic Theory of Causality*. Amsterdam, North-Holland.

Weld, D. S., and J. de Kleer, Ed. (1989). Readings in Qualitative Reasoning About Physical Systems. Morgan Kaufmann.

Wellman, M. P. (1990). *Formulation of Tradeoffs in Planning Under Uncertainty*. London, Pitman.

Wellman, M. P. (1990). Fundamental Concepts of Qualitative Probabilistic Networks. *Artificial Intelligence* **44**: 257-303.

Wellman, M. P., and M. Henrion (1993). Explaining "explaining away". *IEEE Transactions on Pattern Analysis and Machine Intelligence* **15**: 287-292.

Xiang, Y., M. P. Beddoes, and D. Poole (1990). Can uncertainty management be realized in a finite totally ordered probability algebra? Uncertainty in Artificial Intelligence 5 Ed. M. Henrion et al. North-Holland. 41-57.

The Qualitative Verification of Quantitative Uncertainty

Simon Parsons[1,2] and Alessandro Saffiotti[3]

[1] Advanced Computation Laboratory, Imperial Cancer Research Fund,
P.O. Box 123, Lincoln's Inn Fields, London WC2A 3PX, United Kingdom.
[2] Department of Electronic Engineering, Queen Mary and Westfield College,
Mile End Road, London, E1 4NS, United Kingdom.
[3] IRIDIA, Université Libre de Bruxelles, 50 av. Roosevelt, CP 194–6
B-1050 Bruxelles, Belgium

Abstract. We introduce a new application of qualitative models of uncertainty. The qualitative analysis of a numerical model of uncertainty reveals the qualitative behaviour of that model when new evidence is obtained. This behaviour can be compared with an expert's specifications to identify those situations in which the model does not behave as expected. We report the result of experiments performed using a probability model, and a model based on the Dempster-Shafer theory of evidence.

1 Introduction

Recently, there has been considerable interest in the qualitative representation of reasoning under uncertainty, including qualitative probabilistic networks [2, 13] as well as qualitative possibilistic and evidential networks [3, 4]. Such work aims to determine the impact of new evidence in situations in which full numerical results may not be obtained due to the incompleteness and imprecision of the available knowledge. In this paper we suggest a different use for qualitative methods. Since the qualitative behaviour of a system may be established from numerical knowledge, we can qualitatively analyse any numerical model. This analysis may then be used as a simple means of verifying that a system behaves as intended by the knowledge engineer who built it. It also provides a means for guiding the correction of any faults that may be found.

The basic method of our analysis is as follows. When we find new evidence about the state of a variable we update our prior values to take account of the evidence. When using the model, we are interested in the new value obtained after updating. When verifying the behaviour of the model, however, we are interested in checking that this updating corresponds to that described by the domain expert whose knowledge is captured in the model. Since the expert's knowledge is often expressed in the form "If we observe e then it is more likely that h is the case", we may be more interested in knowing the way in which the values change than in knowing the values themselves. Given the equations that relate two uncertainty values val_1 and val_2, expressed in some formalism, we can establish an expression, in terms of numerical uncertainty values, for

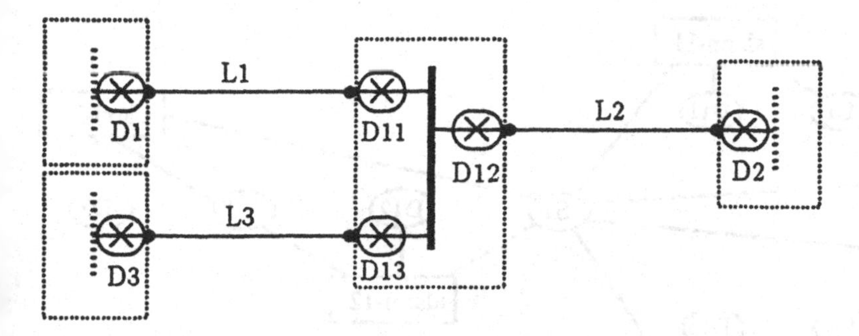

Fig. 1. A fragment of an electricity distribution network.

the derivative $\frac{dval_1}{dval_2}$ that relates the two quantities.This expression allows us to determine the qualitative value of the derivative, written as $[\frac{dval_1}{dval_2}]$, that is whether the derivative is positive, written as [+], negative, [−], or zero, [0]. The sign of the derivative indicates the direction of change of val_1 when val_2 increases. As $\Delta x = \Delta y\frac{dx}{dy} + \Delta z\frac{dx}{dz}$, we can get the effect of several successive pieces of evidence by combining the effect of each alone. To validate a given model, we establish $[\frac{dval(h)}{dval(e)}]$ for every interesting hypothesis h and piece of evidence e, and then compare these values with the knowledge expressed by the expert.

In the rest of this paper we demonstrate our qualitative analysis technique for a probabilistic and a Dempster-Shafer model on a small example extracted from a real application. A longer report [5] includes all the computations, extends the analysis to possibilistic models, and describes a debugging procedure.

2 Problem description

The problem under study is a simplified version of fault diagnosis in electricity networks, originally used by Saffiotti and Umkehrer [8] to investigate the use of different formalisms to model uncertainty. We consider the fragment of an electricity network shown in Fig 1. This fragment comprises four substations, linked by three lines $L1$, $L2$ and $L3$. The substation in the middle includes $S1$, a big conductive bar, known as a busbar, used for connecting more lines together. The $D$is are circuit breakers, that is devices which watch the part of the network on their "hot" side, marked by a dot in the picture, for overloads. If an overload occurs, a circuit breaker generates an alarm, either instantaneous or delayed, and transmits it to the control room. Talking to domain experts revealed what the qualitative behaviour of the modelled fragment should be (no matter what formalism is used to model the uncertainty):

- an instantaneous alarm from an outer circuit breaker should increase belief in the occurrence of a fault in the line that the breaker is on;
- a delayed alarm from an outer breaker should increase belief in the occurrence of a fault in either the line the breaker is on, or in the busbar;

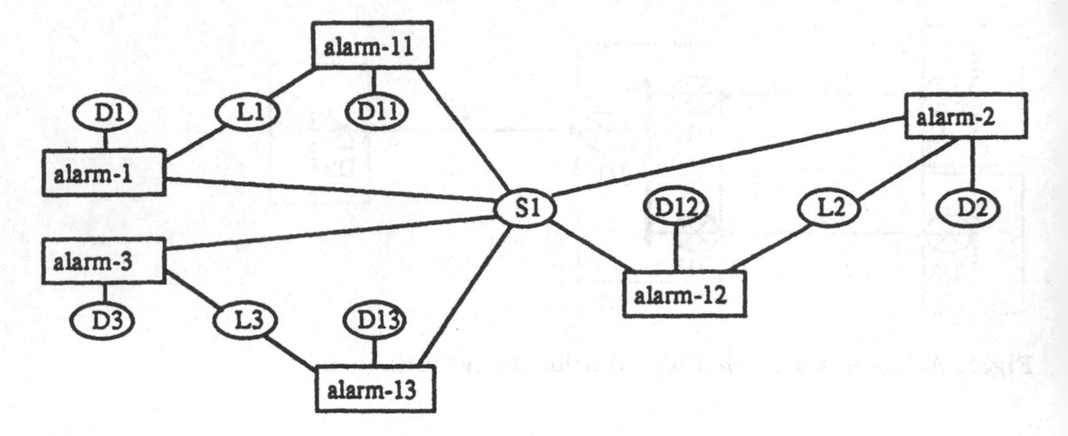

Fig. 2. The valuation system for the distribution network

– an alarm (of any kind) from an inner breaker should only increase our belief in the occurrence of a fault in the line the breaker is on.

The experts were also able to provide rough quantitative estimates of the uncertainty, for instance, in roughly 10% of the cases alarms are generated without faults or faults occur without alarms.

We have modelled our problem in both probability and Dempster-Shafer theory by using Shenoy and Shafer's valuation system formalism [10, 11]. The tool we have used for our experiment, Pulcinella [7, 8], is an implementation of valuation systems in which many uncertainty handling formalisms can be embedded. As according to the valuation system formalism, we model our problem through a set of variables, and a set of valuations linking sets of related variables. Fig 2 is a graphical representation of the model, where circles stand for variables, and rectangles for valuations.

A valuation over a set of variables expresses information about the values taken by the variables in that set, and defines the relations between the variables. Here, the *Di*s are variables representing circuit breaker states, with possible values 'ok' (no alarm), 'del' (delayed alarm), and 'inst' (instantaneous alarm); the *Li*s and *S*1 represent line and busbar states, respectively, with possible values 'ok' and 'fault'; and the *alarm-i*s relate generation of alarms by breakers with states of neighbour lines. New information can be propagated through the *alarm-i* relations to produce updated estimates of the states of the elements of the network. In order to build the *alarm-i* valuations so that they behave as described above, we first split them into two groups: those refering to outer circuit breakers (*alarm-1*, *alarm-2*, *alarm-3*), and those refering to inner circuit breakers (*alarm-11*, *alarm-12*, *alarm-13*). Table 1 shows the values for the two classes of valuations when probability values are used—valuations in this case are joint probability distributions.

To build a Dempster-Shafer model for our problem [9, 12], we we use basic belief assignments that correspond to the conditional belief functions shown in

Fig 3[1]. A more detailed analysis of this example is given in [8]. In the rest of this paper we will investigate whether these models of uncertainty correctly encode the behaviour described above.

Inner breakers ($D11$)			$P(\cdot)$		Outer breakers ($D1$)		
ok	del	inst	$L1$	$S1$	ok	del	inst
0.45	0.05	0.05	ok	ok	0.89	0.1	0.05
0.45	0.05	0.05	ok	fault	0.05	0.6	0.05
0.05	0.89	0.89	fault	ok	0.05	0.2	0.89
0.05	0.01	0.01	fault	fault	0.001	0.1	0.01

Table 1. The joint probability distributions for the alarm valuations

3 Qualitative analysis of the problem

It is helpful to reformulate the problem using a causal network representation similar to that of Pearl [6], where two nodes are joined by a directed arc if and only if the variable represented by the node at the end of the arc is directly dependent upon the variable represented by the node at the beginning of the arc. Thus the problem information of Section 2 may be represented by the network of Fig 4. Due to the rules of differential calculus, we need only consider changes in sub-networks of the form of Fig 5. The change at $S1$ is the sum of all the changes due to Di and $D1i$, as is the change at Li. In the qualitative analysis we look at changes in value of $S1$ and Li given changes in value of a single Di. If we wanted to assess the impact of several alarms we could sum the impact of the

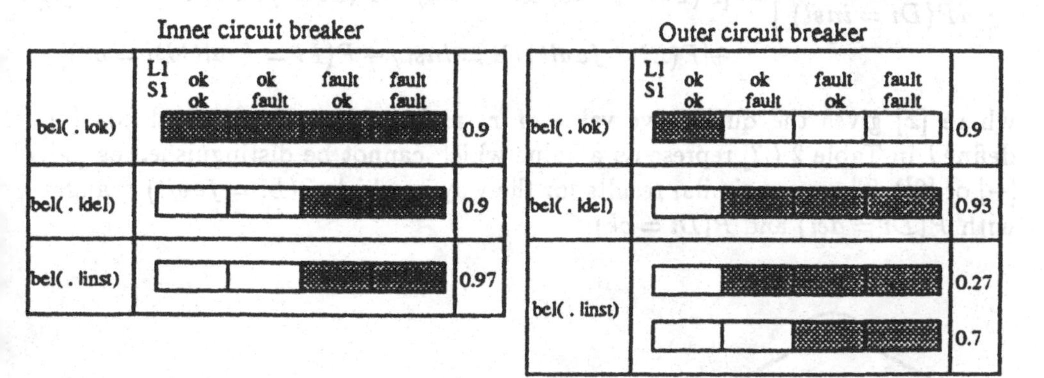

Fig. 3. Conditional belief functions for the "alarm" relations

[1] The combined assignments are fairly intricate, and are reported in the full paper [5] which also deals with the case in which possibility theory is used.

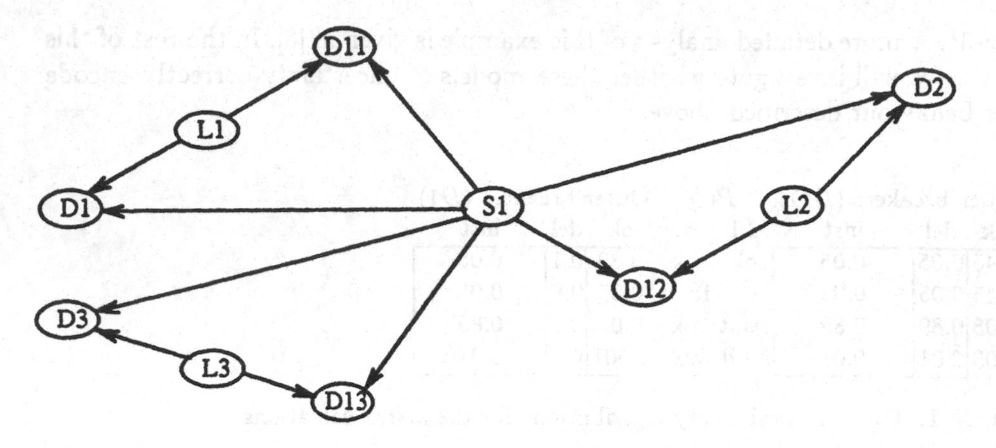

Fig. 4. The causal network representation of the electricity distribution problem.

individual alarms. We first analyze our problem in the probabilistic case. Every circuit breaker has three modes of operation, namely "send an instantaneous alarm", "send a delayed alarm", and "send no alarm". From Fig 5 we can write down the probability of failure of a given line as:

$$P(Li = fault) = \sum_{Di \in \{inst, del, ok\}} P(Li = fault|Di)p(Di).$$

for any Di. It is possible to determine how this probability changes as the probabilities of the various alarms change. Applying results proven in [3] we find that the qualitative value of the derivative relating $P(Li = fault)$ to $P(Di = inst)$ is:

$$\left[\frac{dP(Li = fault)}{dP(Di = inst)}\right] = [P(Li = fault\,|\,Di = inst) - P(Li = fault\,|\,Di = del)] \qquad (1)$$
$$\oplus [P(Li = fault\,|\,Di = inst) - P(Li = fault\,|\,Di = ok)]$$

where $[x]$ gives the qualitative value of x, and $\oplus$ is qualitative addition [1], defined in Table 2 ([?] represents a value which cannot be distinguished as [+], [−] or [0]). There are similar results for the way in which $P(Li = fault)$ changes with $P(Di = del)$ and $P(Di = ok)$.

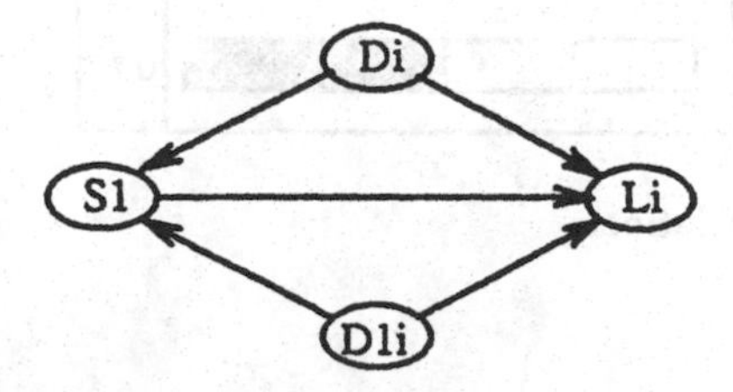

Fig. 5. A sub-network.

⊗	[+]	[0]	[−]	[?]
[+]	[+]	[0]	[−]	[?]
[0]	[0]	[0]	[0]	[0]
[−]	[−]	[0]	[+]	[?]
[?]	[?]	[?]	[?]	[?]

⊕	[+]	[0]	[−]	[?]
[+]	[+]	[+]	[?]	[?]
[0]	[+]	[0]	[−]	[?]
[−]	[?]	[−]	[−]	[?]
[?]	[?]	[?]	[?]	[?]

Table 2. Qualitative multiplication and addition

We also know [3] that:

$$\left[\frac{dP(Li = fault)}{dP(Di)}\right] = [-] \otimes \left[\frac{dP(Li = ok)}{dP(Di)}\right] \tag{2}$$

for all $P(Di)$, so when we know how $P(Li = fault)$ changes we can tell how $P(Li = ok)$ changes. These results are true for both inner and outer circuit breakers, and furthermore are true for any numerical values we put into the model. Similar calculations will give us the probabilities of a busbar failure.

We now turn to considering the belief function case. Once again we start the qualitative analysis with an equation relating L to D, this time expressed in terms of belief functions:

$$bel(Li = fault) = \sum_{Di \subseteq \{inst, del, ok\}} bel(Li = fault|Di)m(Di) \tag{3}$$

Once again we can obtain the relevant qualitative derivative directly by applying the results of [3], getting:

$$\left[\frac{dbel(Li = fault)}{dbel(Di = inst)}\right] = \tag{4}$$

$$[bel(Li = fault \,|\, Di = inst) - \min_{X \subseteq \{inst, del, ok\}, del \in X} bel(Li = fault \,|\, X)]$$

$$\oplus [bel(Li = fault \,|\, Di = inst) - \max_{X \subseteq \{inst, del, ok\}, del \in X} bel(Li = fault \,|\, X)]]$$

which tells us how $bel(Li = fault)$ varies with $bel(Di = inst)$, and we can establish similar results for the delayed and no alarm conditions. The analysis for $bel(Li = ok)$, $bel(S1 = fault)$, and $bel(S1 = ok)$ are similar.

4 Validating the models

Having analysed the way in which qualitative uncertainty values are propagated through the network structure of our test case, we can use the numerical information in Section 2 to examine the qualitative behaviour of the models we have proposed for our testbed.

The information that a particular alarm has arrived from a breaker is typically introduced in the model by increasing the value for the associated state, at the expense of the values of the alternative states—for example, a report of an instantaneous alarm is encoded by forcing $\Delta val(Di = inst) = [+]$, $\Delta val(Di = del) = [-]$, and $\Delta val(Di = ok) = [-]$, where $val(x)$ stands for both $P(x)$ and $bel(x)$. These changes are related to the change in the uncertainty value of a particular fault hypothesis, say a line fault, by:

$$\begin{aligned} \Delta val(Li = fault) = & \left[\frac{dval(Li = fault)}{dval(Di = inst)}\right] \otimes \Delta val(Di = inst) \\ & \oplus \left[\frac{dval(Li = fault)}{dval(Di = del)}\right] \otimes \Delta val(Di = del) \\ & \oplus \left[\frac{dval(Li = fault)}{dval(Di = ok)}\right] \otimes \Delta val(Di = ok) \end{aligned} \quad (5)$$

From the quantitative knowledge of Section 2 we can establish the qualitative derivatives using the results of Section 3. These can then be used along with (5) to establish the qualitative behaviour of the various models. In the case of probability theory, we have, for the outer circuit breakers:

$$\begin{aligned} P(Li = fault, S1 = fault \mid Di = inst) &= 0.01 \\ P(Li = fault, S1 = ok \mid Di = inst) &= 0.89 \\ P(Li = fault, S1 = fault \mid Di = del) &= 0.1 \\ P(Li = fault, S1 = ok \mid Di = del) &= 0.2 \\ P(Li = fault, S1 = fault \mid Di = ok) &= 0.001 \\ P(Li = fault, S1 = ok \mid Di = ok) &= 0.05 \end{aligned}$$

so that: $P(Li = fault|Di = inst) = 0.90$, $P(Li = fault|Di = del) = 0.30$, and $P(Li = fault|Di = ok) = 0.051$. Using these values in (1), we find the qualitative values of the derivatives that link the probability of a line fault to that of an alarm for the outer breakers:

$$\frac{dP(Li{=}fault)}{dP(Di{=}inst)} = [+] \quad \frac{dP(Li{=}fault)}{dP(Di{=}del)} = [-] \quad \frac{dP(Li{=}fault)}{dP(Di{=}ok)} = [-]$$

Using these values in (5), we can predict how the probability of a line fault, $P(Li = fault)$, changes qualitatively after different alarm reports:

Report	none	inst	delayed	ok
$[\Delta P(Li = fault)]$	[0]	[+]	[?]	[?]

We can refine this prediction for a particular model by exploiting some order of magnitude information. For instance, we know that

$$\left[\frac{dP(Li{=}fault)}{dP(Di{=}inst)}\right] > \left[\frac{dP(Li{=}fault)}{dP(Di{=}del)}\right] \approx \left[\frac{dP(Li{=}fault)}{dP(Di{=}ok)}\right]$$

and, from the probabilities in our model, we have $|\Delta P(Di = ok)| \approx |\Delta P(Di = del)| \gg |\Delta P(Di = inst)|$ when a delayed alarm is reported. Thus, when evaluating (5) to establish the change in the probability of a line fault given a delayed alarm, the third term dominates and $\Delta P(Li = fault) = [+]$. Similar reasoning tells us that $\Delta P(Li = fault) = [-]$ for no alarm. We can obtain the same results for the inner breakers.

Thus for both inner and outer breakers, the probability of the line being faulty increases with both instantaneous and delayed alarms, and decreases when we know that there is no alarm, and the model responds to the specifications in Section 2. We can use Pulcinella to confirm that these predictions are borne out in practice. The following table shows the values of $P(Li = fault)$ after a given report has been received by an outer or an inner circuit breaker.

Report	none	inst	delayed	ok
Outer breaker	0.0065	0.5	0.18	0.0064
Inner breaker	0.0065	0.67	0.5	0.0063

A similar verification can be performed for the probability of busbar faults given alarms from the outer breakers, which shows that the model behaves according to the specifications in Section 2. The inner breakers, however, constitute something of a surprise. We obtain:

Report	none	inst	delayed	ok
$[\Delta P(S1 = fault)]$	[0]	[−]	[−]	[+]

which means that if we have a report of any kind of alarm in the inner breakers then the probability of a busbar fault decreases, while knowing for sure that there is no alarm means that the probability of failure increases. This behaviour is rather odd, and marks a departure of our model from the specifications. Once an anomaly like this has been spotted, we can run Pulcinella over the specific data in order to get the quantitative data and evaluate the impact of the discrepancy.

Report	none	inst	delayed	ok
$P(S1 = fault)$	0.000175	0.000088	0.000088	0.000176

In this case, the changes in values are very small, and we may choose to ignore the anomaly.

We now turn to considering the belief function model. From the set of conditional assignments over $L1$ and $S1$ shown in Fig 3 we can establish, for instance, that for the inner circuit breakers $bel(Li = ok \mid Di = ok) = 0.9$, $bel(Li = fault \mid Di = del) = 0.9$, and $bel(Li = fault \mid Di = inst) = 0.97$, while for the outer breaker, $bel(Li = ok \cap S1 = ok \mid Di = ok) = 0.9$, $bel(Li = fault \cup S1 = fault \mid Di = del) = 0.93$, $bel(Li = fault \cup S1 = fault \mid Di = inst) = 0.27$, $bel(Li = fault \mid Di = inst) = 0.7$ and all other conditional beliefs are zero. From $bel(Li = ok \cap S1 = ok \mid Di = ok) = 0.9$ we know that $bel(Li = ok \mid Di = ok) \geq 0.9$ and $bel(S1 = ok \mid Di = ok) \geq 0.9$. From (4) we learn that for the outer circuit breakers we have:

$$\left[\frac{dbel(Li{=}fault)}{dbel(Di{=}inst)}\right] = [+] \left[\frac{dbel(Li{=}fault)}{dbel(Di{=}del)}\right] = [0] \left[\frac{dbel(Li{=}fault)}{dbel(Di{=}ok)}\right] = [0]$$

while for the outer circuit breakers:

$$\left[\frac{dbel(Li{=}fault)}{dbel(Di{=}inst)}\right] = [+] \left[\frac{dbel(Li{=}fault)}{dbel(Di{=}del)}\right] = [+] \left[\frac{dbel(Li{=}fault)}{dbel(Di{=}ok)}\right] = [0]$$

showing that our model behaves as it should. This prediction is once again verified by running Pulcinella on some sample data.

Report	none	inst	delayed	ok
Outer breaker	0	0.98	0	0
Inner breaker	0	0.97	0.9	0

Similar results may be established for the other cases, and these show that the qualitative behaviour of the belief function model always agrees with the domain expert's specification.

5 Conclusions

We have shown that our qualitative analysis may be used to validate a numerical model since the qualitative predictions may be compared to the opinion of a domain expert to determine whether the model has captured the expert's knowledge. In our example, this validation exposed an anomaly in the probabilistic model—the corresponding valuation system, with values as ascertained by the knowledge engineer, does not behave quite as might be expected from the description of its intended behaviour that is supplied in Section 2. The fact that the predictions are observed to be true when Pulcinella is run on the numerical models does not mean that the qualitative analysis is redundant. The qualitative equations focus on one aspect of the model—the way some values are influenced by some observations—while abstracting away from the numerical configurations of input/output values. This allows us to spot those points where the behaviour of the model does not meet the specifications without having to go through an empirical sequence of numerical tests. After the qualitative analysis has spotted an unexpected behavior, numerical tests can be carried out to assess the performance of the system at these points, and the quantitative impact of the discrepancy can be evaluated. In our case, Pulcinella has shown small quirks in the probabilistic model in the direction predicted by the qualitative analysis. It is the task of the model designer to decide whether these quirks are important enough to require a correction to the model. If so, the qualitative analysis can guide us, by indicating which numerical values we can freely change without affecting other "healthy" behaviour. In the full report [5], we describe a systematic technique for debugging the quantitative values based on the results of the qualitative analysis.

The qualitative analysis is a way to analyse the behaviour of a system at a high level of abstraction. As such, it relies on weak information and produces results that, although correct, may at times be too weak to be useful. We have shown in the above how we can enrich a purely qualitative analysis by introducing some informal order of magnitude arguments.

Acknowledgements

Work by the first author was partially supported by Esprit Basic Research Action 3085 DRUMS, and a grant from BT. The second author was partially supported by the Action de Recherches Concertées BELON founded by the Communauté Francaise de Belgique. Comments from Moisés Goldszmidt, Paul Krause, Inaki Laresgoiti, Philippe Smets, Elisabeth Umkehrer, Hong Xu and an anonymous reviewer have been very helpful.

References

1. Bobrow, D. G. *Qualitative Reasoning about Physical Systems*, North-Holland, Amsterdam, 1984.
2. Druzdzel, M. and Henrion, M. Belief propagation in qualitative probabilistic networks, in: N. Piera Carreté and M. G. Singh (Eds.) *Qualitative reasoning and decision technologies, CIMNE*, Barcelona, 1993.
3. Parsons, S. *Qualitative methods for reasoning under uncertainty*, MIT Press, (to appear).
4. Parsons, S. and Mamdani E. H. On reasoning in networks with qualitative uncertainty, *Proc of the 9th Conf. on Uncertainty in AI*, Washington DC, 1993.
5. Parsons, S. and Saffiotti, A. The qualitative verification and debugging of qualitative uncertainty: a case study, Technical Report TR/IRIDIA/93-1, Université Libre de Bruxelles, Brussels, B, 1993.
6. Pearl, J. *Probabilistic reasoning in intelligent systems: networks of plausible inference*, Morgan Kaufman, San Mateo, CA, 1988.
7. Saffiotti, A., and Umkehrer, E. Pulcinella User's Manual, Technical Report TR/IRIDIA/91-5, Université Libre de Bruxelles, Brussels, B, 1991.
8. Saffiotti, A. and Umkehrer, E. Pulcinella: a tool for propagating uncertainty by local computation. *Proc. of the 6th Conf. on Uncertainty in AI*, Los Angeles, CA, 1991.
9. Shafer, G. *A mathematical theory of evidence*, Princeton University Press, Princeton, NJ, 1976.
10. Shenoy, P. P. Valuation Based Systems, Working Paper 226, School of Business, University of Kansas, 1991.
11. Shenoy, P., P. and Shafer, G. Axioms for probability and belief function propagation, in: R. D. Shachter, T. S. Levitt, L. N. Kanal, and J. F. Lemmer, (Eds.) *Uncertainty in AI 4*, (Elsevier, North-Holland, 1990.
12. Smets, P and Kennes, R. The transferable belief model, *Artificial Intelligence*, 66, 1994.
13. Wellman, M. P. Fundamental concepts of qualitative probabilistic networks, *Artificial Intelligence*, 44, 1990.

Uncertainty Management Using Probability Intervals*

Luis M. de CAMPOS, Juan F. HUETE, Serafín MORAL

Departamento de Ciencias de la Computación e I.A.
Universidad de Granada
18071-Granada, Spain

Abstract. We study probability intervals as a interesting tool to represent uncertain information. Basic concepts for the management of uncertain information, as combination, marginalization, conditioning and integration are considered for probability intervals. Moreover, the relationships of this theory with some others, as lower and upper probabilities and Choquet capacities of order two are also clarified. The advantages of probability intervals with respect to these formalisms in computational efficiency are highlighted too.

1 Introduction

There are several theories to deal with numerical uncertainty in Artificial Intelligence. Some of them, at least in a formal sense, are hierarchically ordered, going from the general to the particular: general fuzzy measures, lower and upper probabilities, Choquet capacities of order two and belief/plausibility functions, which include both necessity/possibility and probability measures. Usually, the more general the theory is, the more expressive capabilities it has, and the lesser efficiently the computations whitin this theory can be done.

In this paper we study other formalism for representing uncertain information, namely probability intervals. This formalism is easy to understand and it combines a reasonable expressive power and efficient computation. The main concepts and tools necessary for the development of a theory of uncertain information, as precision (inclusion), combination, marginalization, conditioning and integration are studied for probability intervals. Moreover, the place of probability intervals in the hierarchy above is also analized.

The paper is divided in 6 sections. In section 2 we introduce formally probability intervals, and study their relationships with lower and upper probabilities and convex sets of probabilities. Section 3 is devoted to the combination of probabilit intervals, and the associated problem of inclusion. The basic concepts of marginalization and conditioning of probability intervals are analized in section 4. Section 5 studies methods of integration with respect to probability intervals. Finally, section 6 contains the concluding remarks and some proposals for future work.

---

* This work has been supported by the DGICYT under Project n. PB92-0939

2 Probability Intervals

Let us consider a variable X taking its values in a finite set $D_X = \{x_1, x_2, \ldots, x_n\}$ and a family of intervals $L = \{[l_i, u_i],\ i = 1, \ldots, n\}$, verifying $0 \leq l_i \leq u_i \leq 1\, \forall i$. We will interpret these intervals as a set of bounds of probability by defining the set $\mathcal{P}$ of probability distributions on D_X as

$$\mathcal{P} = \{P \mid l_i \leq p(x_i) \leq u_i\ \forall i\} \tag{1}$$

So, we say that L is a set of probability intervals, and $\mathcal{P}$ is the set of possible probabilities associated to L. As $\mathcal{P}$ is obviously a convex set, we can consider a set of probability intervals as a particular case of convex set of probabilities with a finite set of extreme points ([7, 10]).

In order to avoid for the set $\mathcal{P}$ to be empty, it is necessary to impose to the intervals $[l_i, u_i]$ the condition

$$\sum_{i=1}^{n} l_i \leq 1 \leq \sum_{i=1}^{n} u_i \tag{2}$$

A set of probability intervals verifying condition (2) will be called proper. We always use proper probability intervals, because non proper intervals, associated to the empty set, are useless.

In addition to a convex set $\mathcal{P}$, we can also associate with the proper intervals $[l_i, u_i]$ a pair (l, u) of lower and upper probabilities (also called a pair of representable measures, or a probability envelope [1]) through $\mathcal{P}$ as follows:

$$l(A) = \min_{P \in \mathcal{P}} P(A),\ u(A) = \max_{P \in \mathcal{P}} P(A), \tag{3}$$

So, probability intervals can be also considered as particular cases of lower and upper probabilities, where the set of associated probabilities is defined by restrictions affecting only the individual probabilities $p(x_i)$. Moreover, it can be proved that probability intervals always belong to a well-known subclass of lower and upper measures, namely, Choquet capacities of order two ([8]).

In order to maintain consistency between both views of probability intervals, it is important that the restriction of $l(.)$ and $u(.)$ to the singletons (sets with only one element) be equal to the original bounds, that is to say, that

$$l(\{x_i\}) = l_i,\ u(\{x_i\}) = u_i\ \ \forall i \tag{4}$$

However, these equalities are not always true. We will call reachable to a set L of probability intervals verifying conditions (4). In order to be reachable, the conditions that L must verify can be characterized as follows:

$$\sum_{j \neq i} l_j + u_i \leq 1 \text{ and } \sum_{j \neq i} u_j + l_i \geq 1\ \ \forall i \tag{5}$$

We can always obtain reachable probability intervals from the original ones by modifying the bounds l_i and u_i but without altering the set $\mathcal{P}$, that is to say, without changing the set of possible probabilities, in the following way: define a new set of probability intervals $L' = \{[l'_i, u'_i], i = 1, \ldots, n\}$ by means of

$$l'_i = l_i \vee (1 - \sum_{j \neq i} u_j), \; u'_i = u_i \wedge (1 - \sum_{j \neq i} l_j) \; \forall i \tag{6}$$

Then the set of probabilities associated to L' is also $\mathcal{P}$; moreover L' is a reachable set of probability intervals. As the replacement of the original set of probability intervals L by the narrower set L' does not change the possible probabilities, and L' constitutes a more accurate representation of these probabilities, we will perform the substitution when L does not satisfy the conditions (5), thus always using reachable probability intervals.

For reachable sets of probability intervals we have guarantee that the values $l(\{x_i\})$ and $u(\{x_i\})$ of the associated lower and upper probabilities, (l, u), coincide with the initial probability bounds l_i and u_i. But how can we calculate the values $l(A)$ and $u(A)$ for any other subset A of D_X? These values can be computed directly from the values l_i and u_i, according to the following expressions:

$$l(A) = (\sum_{x_i \in A} l_i) \vee (1 - \sum_{x_i \notin A} u_i) \text{ and } u(A) = (\sum_{x_i \in A} u_i) \wedge (1 - \sum_{x_i \notin A} l_i) \tag{7}$$

For general lower and upper probability measures we need to give all the values $l(A)$ or $u(A)$ in order to have a complete specification of these measures, that is, we need $2^{|D_X|}$ values ($|D_X|$ stands for the cardinal of the set D_X). For several distinguished kinds of measures, as probabilities or possibilities [15], it suffices to have the $|D_X|$ values of these measures for singletons. For probability intervals, we need to specify only $2|D_X|$ values (l_i and u_i) instead of $2^{|D_X|}$. This fact makes probability intervals to be much easier to manage than lower and upper probability measures or even belief and plausibility functions.

As we have mentioned, a probability interval may be interpreted as a convex set of probabilities, $\mathcal{P}$. We conclude this section presenting an algorithm that calculates the extreme probabilities of $\mathcal{P}$. This is interesting because the extreme probabilities provide an alternative representation for $\mathcal{P}$. Moreover, in some aplications may be necessary to calculate the extreme probabilities [7].

In the method below, we mantain a 'partial probability' P (this means a set of values p_i, $i = 1, \ldots, n$ verifying the restrictions $l_i \leq p_i \leq u_i$, $\forall i$ but not necessarily the restriction $\sum_i p_i = 1$) where each p_i is initializated to l_i. The others parameters are: a list *Expl* of explored indices, initializated to the empty set; a real value λ, whose initial value is $1 - \sum_i l_i$; and a record *Current* of the current index being explored, initializated to 0. After these initialization steps, the algorithm is the following:

If $\lambda = 0$ then output$(p_1, \ldots, p_n)$ else *Getprob*$(P, \lambda, \mathit{Expl}, \mathit{Current})$

where *Getprob* is a recursive procedure that uses an implicit tree search, where each node is a partial probability and a child node represents a refinement of its parent node by increasing one component p_i. The leaf nodes in this tree are the extreme probabilities. The algorithm could duplicate extreme probabilities when it tries to refine a partial probability by increasing the same components p_i and p_j in a different order (this may only happen when these components p_i and p_j can be increased to their maximum values u_i and u_j). The easiest way to avoid this behaviour is to keep in mind the order used in the refinement, so that if $i < j$ and both components can be increased to their maximum values, u_i and u_j, the algorithm will branch first from i and after for j, but when branching first from j, then the branch corresponding to i will be pruned. The recursive procedure *Getprob* is defined as follows:

Getprob$(P, \lambda, Expl, Current)$
For $i = 1$ to n do
  If not belong(i,*Expl*)
  then if $\lambda < u_i - l_i$
    then output$(p_1, \ldots, p_{i-1}, p_i + \lambda, p_{i+1}, \ldots, p_n)$
    else if $i >$ *Current* and $u_i - l_i > 0$
      then if $\lambda - u_i + l_i > 0$
        then
          $v \leftarrow p_i; p_i \leftarrow u_i;$
          Getprob$(P, \lambda - u_i + l_i, Expl \cup \{i\}, i)$;
          $p_i \leftarrow v$;
        else output$(p_1, \ldots, p_{i-1}, p_i + \lambda, p_{i+1}, \ldots, p_n)$

Using this algorithm we obtained a drastic decreasing in computing time with respect to the algorithm given in [6], which maintains a list of extreme probabilitites found so far; when the algorithm reaches a new probability P, it is included in the list only if P does not belong to the list. So, the new algorithm avoids the necessity of searching P in the list of probabilities. The next table summarizes the results of a simulation experiment where, for each size n, we randomly generated 100 probability intervals and then applied the algorithms.

n	m	t_1	t_2
5	13.46	0.0066	0.0006
10	83.01	0.4101	0.0057
15	312.18	15.0981	0.0169
20	784.76	189.7078	0.0481

Here m is the the mean number of extreme probabilities, t_1 (for the algorithm given in [6]) and t_2 (for the new algorithm) represent the mean times (in seconds) needed to calculate these extreme probabilities.

Remark: The proofs of all the results in this paper can be found in [6]

3 Inclusion and Combination of Probability Intervals

Two important issues when dealing with uncertain information are those of precision of a piece of information and aggregation of several pieces of information. With respect to the first issue, we are going to study the concept of inclusion of probability intervals, which tries to clarify when a set of probability intervals is more precise or contains more information than another set. In relation to the issue of aggregation, we will study methods to combine two (or more) sets of probability intervals in conjunctive and disyunctive ways. To do that, we will take advantage of the interpretation of probability intervals as particular cases of lower and upper probability measures, because the concepts of inclusion and combination are defined within this theory ([1, 9]).

Given two pairs of lower and upper probability measures (l_1, u_1) and (l_2, u_2), defined on the same domain D_X, (l_1, u_1) is said to be included in (l_2, u_2), and it is denoted by $(l_1, u_1) \subseteq (l_2, u_2)$, if and only if (see [1, 9])

$$[l_1(A), u_1(A)] \subseteq [l_2(A), u_2(A)],\ \forall A \subseteq D_X \tag{8}$$

which is equivalent to any of the following inequalities:

$$l_1(A) \geq l_2(A)\ \ \forall A \subseteq D_X \text{ or } u_1(A) \leq u_2(A)\ \ \forall A \subseteq D_X \tag{9}$$

Inclusion of (l_1, u_1) in (l_2, u_2) means that (l_1, u_1) is a more precise assessment of the information about the values of one variable than (l_2, u_2). Moreover, (8) is also equivalent to the inclusion of the set $\mathcal{P}_1$ of probabilities associated to (l_1, u_1) in the corresponding set $\mathcal{P}_2$ associated to (l_2, u_2), $\mathcal{P}_1 \subseteq \mathcal{P}_2$.

We will say that a set L of probability intervals is included in another set L' if the pair of lower and upper measures (l, u) associated to L is included in the corresponding pair (l', u') associated to L'. It can be proved that this definition is equivalent to the inclusions

$$[l_i, u_i] \subseteq [l'_i, u'_i],\ \forall i = 1, \ldots, n \tag{10}$$

Therefore, as we could expect, to check the inclusion between two sets of probability intervals, only the single values l_i, l'_i, u_i and u'_i need to be considered.

With respect to the combination of lower and upper probability measures, the conjunctive and disjunctive combinations of these measures, corresponding to the logical operators 'and' and 'or' respectively, were defined in [1]. The idea is simple: the relation of inclusion defines a partial order relation on the family of lower and upper probability pairs. The conjunction of two pairs (l, u) and (l', u'), denoted by $(l \otimes l', u \otimes u')$, is defined as the infimum of (l, u) and (l', u'), if there exist common lower bounds, that is, it is the greatest pair included in both (l, u) and (l', u'). Analogously, the disjunction of (l, u) and (l', u'), denoted by $(l \oplus l', u \oplus u')$, is the supremum of (l, u) and (l', u'), the least pair including both (l, u) and (l', u'). The conjunction is the pair of lower and upper measures associated to the intersection $\mathcal{P} \cap \mathcal{P}'$ of the sets of probabilities $\mathcal{P}$ and $\mathcal{P}'$ associated to the initial lower and upper measures. Similarly, the disjunction is the pair associated to the set of probabilities $\mathcal{P} \cup \mathcal{P}'$. The semantics of these operators is clear: the conjunction represents the conclusion we can obtain if we suppose that the two initial pieces of information are true; the disjunction is that we can infer if at least one piece of information is considered to be true.

The calculus of the disjunction $(l \oplus l', u \oplus u')$ is very easy: it can be shown (see [1]) that for all $A \subseteq D_X$

$$(l \oplus l')(A) = \min(l(A), l'(A)) \text{ and } (u \oplus u')(A) = \max(u(A), u'(A)) \qquad (11)$$

However, the calculus of the conjunction $(l \otimes l', u \otimes u')$ is not so easy. In general, we need to solve a linear programming problem for each value $(l \otimes l')(A)$. Moreover the conjunction does not always exist. In these cases we say that the two pairs are not compatible: the information they represent cannot be simultaneously true. Clearly, compatibility holds if and only if the set $\mathcal{P} \cap \mathcal{P}'$ is not empty.

Now, we are in position to define the combination of two sets of probability intervals L and L' as the combination of their associated lower and upper probability pairs.

Compatibility between L and L' holds if and only if the following conditions are true:

$$\begin{gathered} l_i \leq u'_i \text{ and } l'_i \leq u_i \ \forall i = 1, \ldots, n, \\ \sum_{i=1}^{n} (l_i \vee l'_i) \leq 1 \leq \sum_{i=1}^{n} (u_i \wedge u'_i) \end{gathered} \qquad (12)$$

The specific formulas for the conjunction $L \otimes L'$ of L and L' are very simple:

$$\begin{aligned} (l \otimes l')_i &= \max\{l_i, l'_i, 1 - \sum_{j \neq i} \min(u_j, u'_j)\} \\ (u \otimes u')_i &= \min\{u_i, u'_i, 1 - \sum_{j \neq i} \max(l_j, l'_j)\} \end{aligned} \qquad (13)$$

With respect to the disjunction $L \oplus L'$ of L and L', although it is very easy to calculate it, the problem is that this operation is not closed for probability intervals: the disjunction of two sets of probability intervals is always a pair of lower and upper probability measures, but it is not necessarily a set of probability intervals. In order to get a set of probability intervals as result of the disjunction, we can try to find the set of probability intervals that is the best approximation of $L \oplus L'$. So, we look for the set of probability intervals, say $(L \oplus L')^a$, such that, first, $L \oplus L'$ is included in $(L \oplus L')^a$ (in order to not to add information), and second, every other set of probability intervals including $L \oplus L'$ must also include $(L \oplus L')^a$ (we should lose as least information as possible). We can always find a set of probability intervals verifying these conditions, which is

$$(L \oplus L')^a = \{[l_i \wedge l'_i, u_i \vee u'_i], i = 1, \ldots, n\} \qquad (14)$$

So, if we want to have a disjunctive combination closed for probability intervals, the best choice is to define it as $(L \oplus L')^a$ in (14).

4 Marginalization and Conditioning of Probability Intervals

Usually, in most of the problems, our interest is not restricted to only one variable but we are dealing with several variables defined on different domains that exhibit some relationships among each other. In these cases we have one joint piece of information on the set of variables (or a number of pieces of information relative to several subsets of variables). In such situations, we need a tool to obtain information on one variable or a subset of variables from the joint information. Such a tool is the marginalization operator. Moreover, it is also necessary to have available a mechanism to update our information about one or several variables once we know for sure the values taken by other variables. This is a conditioning operator. In this section we define and study the concepts of marginalization and conditioning for probability intervals. We will study the simple case in which we have only two variables, but the generalization to deal with more variables is straightforward.

So, let us consider two variables X and Y taking values in the sets $D_X = \{x_1, x_2, \ldots, x_n\}$ and $D_Y = \{y_1, y_2, \ldots, y_m\}$ respectively, and a set of reachable bidimensional probability intervals $L = \{[l_{ij}, u_{ij}], i = 1, \ldots, n, j = 1, \ldots, m\}$, defined on the cartesian product $D_X \times D_Y$, representing the joint available information on these two variables.

First, we want to define the marginals of these probability intervals. To do that, we can use the interpretation of a set of probability intervals as a pair of lower and upper probabilities (l, u). Given (l, u), the marginal measures (l_X, u_X) on D_X are defined for all $A \subseteq D_X$ as (see [5]):

$$l_X(A) = l(A \times D_Y),\ u_X(A) = u(A \times D_Y) \tag{15}$$

Alternatively, we could use the interpretation of probability intervals as convex sets of probabilities, and define the marginal of L on D_X as the set $\mathcal{P}_X$ of marginal probabilities of the probabilities in the convex set $\mathcal{P}$, being $\mathcal{P}$ the set of probabilities associated to L, that is to say,

$$\mathcal{P}_X = \{P \mid \exists Q \in \mathcal{P}, p(x_i) = \sum_{j=1}^{m} q(x_i, y_j) \forall i\} \tag{16}$$

Both definitions are equivalent, in the sense that $\mathcal{P}_X$ is just the set of probabilities associated to (l_X, u_X), that is, for all $A \subseteq D_X$

$$l_X(A) = \min_{P \in \mathcal{P}_X} P(A), \text{ and } u_X(A) = \max_{P \in \mathcal{P}_X} P(A), \tag{17}$$

Moreover, it can be seen that these marginals are in fact reachable probability intervals, $L_X = \{[l_i, u_i], i = 1, \ldots, n\}$, whose expressions can be easily calculated in the following way:

$$l_i = \sum_{j=1}^{m} l_{ij} \vee (1 - \sum_{k \neq i} \sum_{j=1}^{m} u_{kj}),\ i = 1, \ldots, n$$

$$u_i = \sum_{j=1}^{m} u_{ij} \wedge (1 - \sum_{k \neq i} \sum_{j=1}^{m} l_{kj}), \ i = 1, .., n \tag{18}$$

In order to define the conditioning of probability intervals we will again use their interpretation as lower and upper probabilities, because there are available several definitions of conditioning in this framework (see [11] for a review). We will use the definition of conditioning proposed in [3]: Given a pair of lower and upper probabilities (l, u) defined on a domain D, and a subset $B \subseteq D$, the conditional lower and upper measures given that we know B, $(l(.|B), u(.|B))$ are defined as

$$l(A|B) = \frac{l(A \cap B)}{l(A \cap B) + u(\overline{A} \cap B)} \ \forall A \subseteq D$$

$$u(A|B) = \frac{u(A \cap B)}{u(A \cap B) + l(\overline{A} \cap B)} \ \forall A \subseteq D \tag{19}$$

In our case, we have a set of bidimensional probability intervals $L = \{[l_{ij}, u_{ij}], i = 1, \ldots, n, j = 1, \ldots, m\}$, and we want to calculate the conditional probability intervals for one variable, say X, given that we know the value of the other variable, $Y = y_j$, $L(X|Y = y_j) = \{[l_{i|j}, u_{i|j}], i = 1, \ldots, n\}$. Then the previous expressions (19) become

$$l_{i|j} = \frac{l_{ij}}{l_{ij} + (\sum_{k \neq i} u_{kj} \wedge (1 - \sum_k \sum_{h \neq j} l_{kh} - l_{ij}))}$$

$$u_{i|j} = \frac{u_{ij}}{u_{ij} + (\sum_{k \neq i} l_{kj} \vee (1 - \sum_k \sum_{h \neq j} u_{kh} - u_{ij}))} \tag{20}$$

Note that the calculus of the conditional probability intervals is very simple. Moreover, these intervals are always reachable, and therefore it is not necessary to transform them in reachable intervals.

5 Integration with Respect to Probability Intervals

In probability theory, the concept of mathematical expectation or integral with respect to a probability measure plays an important role from both a theoretical and a practical point of view. Indeed, integration is useful, for example, to derive the probability of an event A, $P(A)$, from the conditional probabilities $P(A|B_i)$ of that event given a set of mutually exhaustive and exclusive events $B_1, \ldots, B_m$, and the probabilities of these events $P(B_i)$. Concepts as the entropy of a probability distribution or the quantity of information about one variable that another variable contains are defined with the help of an integral. Integration is also essential in decision-making problems with uncertainty. In this section we are going to study the concept of integration when the underlying uncertainty measure is not a probability measure but a set of probability intervals.

As it is usual in this paper, we will use the interpretation of probability intervals as particular cases of pairs of lower and upper probability measures, which in turn are particular cases of fuzzy measures ([13]), since we have available several methods of integration for fuzzy measures (fuzzy integrals). The two main fuzzy integrals are Sugeno integral [13] and Choquet integral [8]. We will use Choquet integral, because it is closer in spirit to the mathematical expectation than Sugeno integral, and therefore it seems us more appropriate for probability intervals (see [4] for an in-deepth study of Choquet and Sugeno integrals).

In our case, we have a set L of probability intervals, and the associated pair of lower and upper measures (l, u). So, we can define the Choquet integral with respect to the two fuzzy measures $g = l$ or $g = u$. We will denote them as the lower $E_l(h)$ and the upper $E_u(h)$ Choquet integrals, and they form an interval $[E_l(h), E_u(h)]$. This interpretation is justified because of the holding of the following equalities,

$$E_l(h) = \min_{P \in \mathcal{P}} E_P(h) \text{ and } E_u(h) = \max_{P \in \mathcal{P}} E_P(h) \tag{21}$$

which relate the values $E_l(h)$ and $E_u(h)$ with the mathematical expectations $E_P(h)$ with respect to probabilities P that belong to the set of probabilities $\mathcal{P}$ associated to L.

The specific expressions for $E_l(h)$ and $E_u(h)$ for the particular case of probability intervals (assuming that $h(x_1) \leq h(x_2) \leq \ldots \leq h(x_n)$) are

$$E_l(h) = \sum_{i=1}^{n} p_i h(x_i) \text{ and } E_u(h) = \sum_{i=1}^{n} q_i h(x_i) \tag{22}$$

where $(p_1, p_2, \ldots, p_n) = (u_1, u_2, \ldots, u_{k-1}, 1 - L_{k+1} - U_{k-1}, l_{k+1}, \ldots, l_n)$ and k is the index such that $l_k \leq 1 - L_{k+1} - U_{k-1} \leq u_k$, and $L_i = \sum_{j=i}^{n} l_j$, $U_i = \sum_{j=1}^{i} u_j \; \forall i$, $(q_1, q_2, \ldots, q_n) = (l_1, l_2, \ldots, l_{h-1}, 1 - L^{h-1} - U^{h+1}, u_{h+1}, \ldots, u_n)$ and h is the index such that $l_h \leq 1 - L^{h-1} - U^{h+1} \leq u_h$, and $L^i = \sum_{j=1}^{i} l_j$, $U^i = \sum_{j=i}^{n} u_j \; \forall i$.

An easy algorithm to calculate the weights p_i in (22) is the following:

$S \leftarrow 0$;
For $i = 1$ to $n - 1$ do $S \leftarrow S + u_i$;
$S \leftarrow S + l_n$;
$k \leftarrow n$;
While $S \geq 1$ do
    $S \leftarrow S - u_{k-1} + l_{k-1}$;
    $p_k \leftarrow l_k$;
    $k \leftarrow k - 1$;
For $i = 1$ to $k - 1$ do $p_i \leftarrow u_i$;
$p_k \leftarrow 1 - S + l_k$;

An analogous algorithm would obtain the weights q_i in (22).

6 Concluding Remarks

In this paper we have studied probability intervals as an interesting formalism to represent uncertain information. We think that probability intervals, because of their computational simplicity and expressive power, are promising tools for uncertain reasoning.

Further work should include the study of the concept of independence [5] and estimation methods for probability intervals from empirical data, and their use for learning and inference in belief networks [12, 14]. For this task, the study of ways to obtain bidimensional distributions from marginal and conditional ones, and the related problems of generalizing total probability and Bayes theorems to probability intervals will also be necessary.

References

1. L.M. de Campos, M.T. Lamata, S. Moral, Logical connectives for combining fuzzy measures, in Methodologies for Intelligent Systems **3**, Z.W. Ras, L. Saitta (Eds.) North-Holland, 11–18 (1988).
2. L.M. de Campos, M.J. Bolaños, Representation of fuzzy measures through probabilities, Fuzzy Sets and Systems **31**, 23–36 (1989).
3. L.M. de Campos, M.T. Lamata, S. Moral, The concept of conditional fuzzy measure, International Journal of Intelligent Systems **5**, 237–246 (1990).
4. L.M. de Campos, M.J. Bolaños, Characterization and comparison of Sugeno and Choquet integrals, Fuzzy Sets and Systems **52**, 61–67 (1992).
5. L.M. de Campos, J.F. Huete, Independence concepts in upper and lower probabilities, in Uncertainty in Intelligent Systems, B. Bouchon-Meunier, L. Valverde, R.R. Yager (Eds.), North-Holland, 85–96 (1993).
6. L.M. de Campos, J.F. Huete, S. Moral, Probability intervals: a tool for uncertain reasoning, DECSAI Technical Report 93205, Universidad de Granada (1993).
7. J.E. Cano, S. Moral, J.F. Verdegay, Propagation of convex sets of probabilities in directed acyclic networks, in Uncertainty in Intelligent Systems, B. Bouchon-Meunier, L. Valverde, R.R. Yager (Eds.), North-Holland, 15–26 (1993)
8. G. Choquet, Theory of capacities, Ann. Inst. Fourier **5**, 131–295 (1953).
9. D. Dubois, H. Prade, A set-theoretic view of belief functions, International Journal of General Systems **12**, 193–226 (1986).
10. H.E. Kyburg, Bayesian and non-bayesian evidential updating, Artificial Intelligence **31**, 271–293 (1987).
11. S. Moral, L.M. de Campos, Updating uncertain information, in Uncertainty in Knowledge Bases, Lecture Notes in Computer Science 521, B. Bouchon-Meunier, R.R. Yager, L.A. Zadeh (Eds.), Springer Verlag, 58–67 (1991).
12. J. Pearl, Probabilistic reasoning in intelligent systems: networks of plausible inference, Morgan and Kaufmann, San Mateo (1988).
13. M. Sugeno, Theory of fuzzy integrals and its application, Ph.D. Thesis, Tokio Inst. of Technology (1974).
14. B. Tessem, Interval representation on uncertainty in Artificial Intelligence, Ph.D. Thesis, Department of Informatics, University of Bergen, Norway (1989).
15. L.A. Zadeh, Fuzzy sets as a basis for a theory of possibility, Fuzzy Sets and Systems **1**, 3–28 (1978).

Probabilistic Consistency of Conditional Probability Bounds

Angelo Gilio

Dipartimento di Matematica, Cittá Universitaria
Viale A. Doria, 6 - 95125 Catania, Italy

Abstract. In this paper, given an arbitrary finite family of conditional events $\mathcal{F}$, a generalized probabilistic knowledge base represented by a set of conditional probability bounds defined on $\mathcal{F}$ is considered. Following the approach of de Finetti we define the concept of coherence for the given set of bounds. Then, some results on the probabilistic consistency of the knowledge base are obtained. Finally, an algorithm to check the coherence of the set of bounds is described and some examples are examined.

1 Introduction

In many applications of expert systems the probabilistic treatment of uncertainty entails practical difficulties. In fact, in a situation of partial knowledge the assignment of precise probability values may be not easy. In these cases it can be preferable an approach based on qualitative judgements or on interval - valued probability assessments. General results within this approach have been obtained in [3]. Adopting a numerical approach, the methodology consists in assigning some conditional probability bounds on the family of conditional events of interest. See also [6, 8, 9, 17, 20]. To check the probabilistic consistency we can rely on the coherence principle of de Finetti [7]. This methodology has been adopted in [2, 4, 5, 11, 13, 14, 15, 18, 19]. An approach to represent imprecision of probabilistic knowledge is proposed in [16]. In this paper we consider a generalized probabilistic knowledge base $(\mathcal{F}, \mathcal{A})$, where $\mathcal{F}$ is a family of n conditional events $\{E_1 \mid H_1, E_2 \mid H_2, \ldots, E_n \mid H_n\}$ and $\mathcal{A}$ is a set of conditional probability bounds $(\alpha_1, \alpha_2, \ldots, \alpha_n)$ on $\mathcal{F}$. In Sect. 2 we give a review of some preliminary results. Then, in Sect. 3 some results on coherence of the knowledge base $(\mathcal{F}, \mathcal{A})$ are examined and an algorithm to check coherence of the set of bounds $\mathcal{A}$ is described. Finally, in Sect. 4 some examples are examined.

2 Some preliminary results

We denote, respectively, by $\vee$ and $\wedge$ the operations of logical union and logical product of events; by Ω the certain event and by $\emptyset$ the impossible one; by E^c the contrary of the event E and by EH the logical product of E and H. Given a family of n conditional events $\mathcal{F} = \{E_1 \mid H_1, E_2 \mid H_2, \ldots, E_n \mid H_n\}$, let Π be the partition $\{C_0, C_1, \ldots, C_m\}$ of Ω obtained by developing the expression

$$(E_1H_1 \vee E_1^cH_1 \vee H_1^c) \wedge (E_2H_2 \vee E_2^cH_2 \vee H_2^c) \wedge \cdots \wedge (E_nH_n \vee E_n^cH_n \vee H_n^c)$$

where the atoms C_h are all the (not impossible) events $A_1A_2 \cdots A_n$, with $A_i \in \{E_iH_i, E_i^cH_i, H_i^c\}$ and $C_0 = H_1^cH_2^c \cdots H_n^c$. The atoms $C_1, \ldots, C_m$ are contained in the event $H_0 = H_1 \vee \cdots \vee H_n$. Based on a vector of real numbers $\mathcal{V} = (\nu_1, \nu_2, \ldots, \nu_n)$ and on the atoms $C_1, C_2, \ldots, C_m$ we introduce m vectors $V_1, V_2, \ldots, V_m$, defined as $V_h = (v_{h1}, v_{h2}, \ldots, v_{hn})$, where for each pair (i, h) it is

$$v_{hi} = \begin{cases} 1, & \text{if } C_h \subseteq E_iH_i \\ 0, & \text{if } C_h \subseteq E_i^cH_i \\ \nu_i, & \text{if } C_h \subseteq H_i^c \end{cases} . \tag{1}$$

Consider a conditional probability assessment $\mathcal{P} = (p_1, p_2, \ldots, p_n)$ on $\mathcal{F}$. We observe that applying (1) with $\mathcal{V} = \mathcal{P}$ we obtain the generalized atoms $Q_1, Q_2, \ldots, Q_m$ (see [10]) relative to the probabilistic knowledge base $(\mathcal{F}, \mathcal{P})$. Given a subset J of the set of integers $J_0 = \{1, 2, \ldots, n\}$ we put

$$\mathcal{F}_J = \{E_j \mid H_j \,;\, j \in J\}, \;\; \mathcal{P}_J = (p_j \,;\, j \in J) \;.$$

We denote by $\mathcal{I}_J$ the convex hull of the generalized atoms $Q_1^J, \ldots, Q_{m_J}^J$ relative to the probabilistic knowledge base $(\mathcal{F}_J, \mathcal{P}_J)$.
We observe that, for $J = J_0$, it is $\mathcal{F}_J = \mathcal{F}$, $\mathcal{P}_J = \mathcal{P}$ and in this case we put $\mathcal{I}_J = \mathcal{I}$. Then we have

Theorem 2.1 The conditional probability assessment $\mathcal{P}$ on $\mathcal{F}$ is coherent if and only if $\mathcal{P}_J \in \mathcal{I}_J$ for every $J \subseteq J_0$.

We observe that the condition $\mathcal{P}_J \in \mathcal{I}_J$ is equivalent to

$$\mathcal{P}_J = \sum_{t=1}^{m_J} \lambda_t^J Q_t^J \,, \;\; \sum_{t=1}^{m_J} \lambda_t^J = 1 \,, \;\; \lambda_t^J \geq 0 \,, \; t = 1, \ldots, m_J \;.$$

Given a vector $\mathcal{V} = (\nu_1, \nu_2, \ldots, \nu_n)$, the inequality

$$\sum_{t=1}^{m} \lambda_t V_t \geq \mathcal{V} \,, \;\; \text{with} \;\; \sum_{t=1}^{m} \lambda_t = 1 \,, \;\; \lambda_t \geq 0 \,, \; t = 1, \ldots, m,$$

amounts to compatibility of the following system $(\mathcal{S})$, with vector of unknowns $\Lambda = (\lambda_1, \lambda_2, \ldots, \lambda_m)$

$$\sum_{t=1}^{m} \lambda_t v_{ti} \geq \nu_i, \; i = 1, 2, \ldots, n, \;\; \sum_{t=1}^{m} \lambda_t = 1 \,, \;\; \lambda_t \geq 0 \,, \; t = 1, \ldots, m \;. \tag{2}$$

Denote by S the set of solutions of (2) and, for each subscript $j \in J_0$, consider the (non- negative linear) function

$$\Phi_j(\Lambda) = \Phi_j(\lambda_1, \lambda_2, \ldots, \lambda_m) = \sum_{t: C_t \subseteq H_j} \lambda_t$$

defined for $\Lambda \in S$. Then, introduce the set I_0 defined as

$$I_0 = \{j \in J_0 : Max\, \Phi_j = 0\} \ . \tag{3}$$

I_0 has been introduced in a previous paper ([12]), with the functions $\Phi_j(\Lambda), j \in J_0$, defined on the set of solutions of the following system

$$\mathcal{P} = \sum_{t=1}^{m} \lambda_t Q_t \,, \ \sum_{t=1}^{m} \lambda_t = 1 \,, \ \ \lambda_t \geq 0 \,, \, t = 1,2,...,m,$$

which corresponds to condition $\mathcal{P} \in \mathcal{I}$. Then, in the quoted paper the following result has been obtained

Theorem 2.2 The assessment $\mathcal{P}$ is coherent if and only if the following conditions are satisfied: $(i)\, \mathcal{P} \in \mathcal{I}$; $(ii)\, If\, I_o \neq \emptyset$, then $\mathcal{P}_{I_o}$ is coherent.

3 Coherent conditional probability bounds

Given a family of n conditional events $\mathcal{F} = \{E_1 \mid H_1, E_2 \mid H_2, \ldots, E_n \mid H_n\}$, consider the following vector of probability lower bounds $\mathcal{A} = (\alpha_1, \alpha_2, \ldots, \alpha_n)$ on $\mathcal{F}$:

$$(\mathcal{A}) : \qquad P(E_i \mid H_i) \geq \alpha_i \ , \quad i = 1,2,...,n.$$

We observe that an upper bound on the probability of a conditional event $E \mid H$, say $P(E \mid H) \leq \beta$, is equivalent to the lower bound $P(E^c \mid H) \geq 1 - \beta$. Therefore to consider only lower bounds is not restrictive.

Definition 3.1 A vector of lower bounds $\mathcal{A} = (\alpha_1, \alpha_2, \ldots, \alpha_n)$ defined on a family $\mathcal{F} = \{E_1 \mid H_1, E_2 \mid H_2, \ldots, E_n \mid H_n\}$ is coherent if and only if there exists a coherent conditional probability assessment on $\mathcal{F}$, $\mathcal{P} = (p_1, p_2, \ldots, p_n)$, with $p_i = P(E_i \mid H_i)$, such that $p_i \geq \alpha_i, i = 1, 2, \ldots, n$.

Given a subset $J \subseteq J_0 = \{1, 2, \ldots, n\}$, consider the corresponding subfamily $\mathcal{F}_J = \{E_j \mid H_j \ ; \ j \in J\} \subseteq \mathcal{F}$. Then, denote by Π_J the partition $\{C_0^J, C_1^J, \ldots, C_{m_J}^J\}$ of Ω obtained developing the expression $\bigwedge_{j \in J} (E_j H_j \vee E_j^c H_j \vee H_j^c)$. Based on the vector of bounds $\mathcal{A}_J = (\alpha_j \ ; \ j \in J)$ and on the atoms $C_1^J, \ldots, C_{m_J}^J$ relative to the subfamily $\mathcal{F}_J$, as shown in (1), we can define m_J vectors $V_1^J, \ldots, V_{m_J}^J$, with$V_h^J = (v_{hj}^J \ ; \ j \in J)$. Then, consider the inequality

$$\sum_{t=1}^{m_J} \lambda_t^J V_t^J \geq \mathcal{A}_J, \ \text{ with } \sum_{t=1}^{m_J} \lambda_t^J = 1 \,, \ \ \lambda_t^J \geq 0 \,, \, t = 1, ..., m_J,$$

i.e. the following system $(\mathcal{S}_{\mathcal{A}_J})$

$$\sum_{t=1}^{m_J} \lambda_t^J v_{tj}^J \geq \alpha_j \,, \ \ j \in J \,, \ \sum_{t=1}^{m_J} \lambda_t^J = 1 \,, \ \ \lambda_t^J \geq 0 \,, \, t = 1, ..., m_J, \tag{4}$$

with vector of unknowns $\Lambda_J = (\lambda_1^J, \lambda_2^J, \ldots, \lambda_{m_J}^J)$, and denote by S_J the set of its solutions. From (1) we have that the inequality $\sum_{t=1}^{m_J} \lambda_t^J v_{tj}^J \geq \alpha_j$ is equivalent to

$$\sum_{t:C_t^J \subseteq E_j H_j} \lambda_t^J \geq \alpha_j \sum_{t:C_t^J \subseteq H_j} \lambda_t^J . \qquad (5)$$

Then we have the following result which generalizes Theorem (2.1) to the case of conditional probability bounds.

Theorem 3.2 The vector of bounds $\mathcal{A}$ relative to the family of conditional events $\mathcal{F} = \{E_1 \mid H_1, E_2 \mid H_2, \ldots, E_n \mid H_n\}$ is coherent if and only if, for every $J \subseteq J_0$, the system (4) is compatible; that is there exist m_J non-negative numbers $\lambda_1^J, \ldots, \lambda_{m_J}^J$, with $\sum_{t=1}^{m_J} \lambda_t^J = 1$, such that $\sum_{t=1}^{m_J} \lambda_t^J V_t^J \geq \mathcal{A}_J$.

Proof. Assume $\mathcal{A}$ coherent, so that there exists a coherent probability assessment $\mathcal{P} = (p_1, p_2, \ldots, p_n)$ on $\mathcal{F}$ such that $\mathcal{P} \geq \mathcal{A}$. Then, for every $J \subseteq J_0$, the assessment $\mathcal{P}_J = (p_j \; ; \; j \in J)$ on $\mathcal{F}_J$ is coherent and it is $\mathcal{P}_J \geq \mathcal{A}_J$. Moreover $\mathcal{P}_J \in \mathcal{I}_J$, that is, putting $Q_t^J = (q_{tj}^J \; ; \; j \in J)$, there exists a non-negative vector $\Lambda_J = (\lambda_1^J, \lambda_2^J, \ldots, \lambda_{m_J}^J)$ such that the system

$$\sum_{t=1}^{m_J} \lambda_t^J q_{tj}^J = p_j \, , \; j \in J, \qquad (6)$$

with $\sum_{t=1}^{m_J} \lambda_t^J = 1$, is compatible. We observe that (6) amounts to

$$\sum_{t:C_t^J \subseteq E_j H_j} \lambda_t^J \geq p_j \sum_{t:C_t^J \subseteq H_j} \lambda_t^J \, , \; j \in J,$$

so that, being $p_j \geq \alpha_j$, we obtain

$$\sum_{t:C_t^J \subseteq E_j H_j} \lambda_t^J \geq \alpha_j \sum_{t:C_t^J \subseteq H_J} \lambda_t^J \, , \; j \in J,$$

that is the inequality (5), therefore the system (4) is compatible.
Conversely, assume that for every $J \subseteq J_0$ the system (4) is compatible. Then, considering the set I_0 defined in (3) and denoting by $\setminus$ the *difference* between sets, for each $j \in I_0^c = J_0 \setminus I_0$ there exists a vector Λ_j, solution of the system (2), with $\mathcal{V} = \mathcal{A}$, such that $\Phi_j(\Lambda_j) > 0$. Defining a vector

$$\Lambda = (\lambda_1, \lambda_2, \ldots, \lambda_m) = \sum_{i \in I_0^c} x_i \Lambda_i \, ,$$

with $\sum_{i \in I_0^c} x_i = 1$ and $x_i > 0$ for every i, we have that Λ is a solution of (2) and, for every $j \in I_0^c$, it is

$$\Phi_j(\Lambda) = \Phi_j \left(\sum_{i \in I_0^c} x_i \Lambda_i \right) = \sum_{i \in I_0^c} x_i \Phi_j(\Lambda_i) \geq x_j \Phi_j(\Lambda_j) > 0 \; .$$

On the subfamily $\mathcal{F}_{I_0^c} = \{E_j \mid H_j ; j \in I_0^c\}$ it can be introduced an assessment $\mathcal{P}_{I_0^c} = (p_j ; j \in I_0^c)$, with $p_j = P(E_j \mid H_j)$ defined by

$$p_j = \left(\sum_{t:C_t \subseteq E_j H_j} \lambda_t \right) / \left(\sum_{t:C_t \subseteq H_j} \lambda_t \right) , \quad j \in I_0^c . \tag{7}$$

We have $p_j \geq \alpha_j$ for every $j \in I_0^c$.
Denote by $(\mathcal{S}_0)$ the system obtained from (4) putting $J = I_0$ and by S_0 the set of its solutions. Then, for each $j \in I_0$ introduce the (non - negative linear) functions

$$\Phi_j^0(\Lambda^0) = \Phi_j^0(\lambda_1^0, \lambda_2^0, \ldots, \lambda_{m_0}^0) = \sum_{t:C_t^0 \subseteq H_j} \lambda_t^0 ,$$

with $\Lambda^0 \in S_0$, and consider the set

$$I_1 = \{j \in I_0 : Max\, \Phi_j^0 = 0\} . \tag{8}$$

Denoting by I_1^c the set $I_0 \setminus I_1$, for each $j \in I_1^c$ there exists a vector $\Lambda_j^0 \in S_0$ such that $\Phi_j^0(\Lambda_j^0) > 0$ and using a convex linear combination of these vectors, with positive coefficients, we can define a vector $\Lambda^0 \in S_0$ such that $\Phi_j^0(\Lambda^0) > 0$, for every $j \in I_1^c$. Then, as shown in (7), using the vector Λ^0 we can define an assessment $\mathcal{P}_{I_1^c} = (p_j ; j \in I_1^c)$ on the subfamily $\mathcal{F}_{I_1^c} = \{E_j \mid H_j ; j \in I_1^c\}$. Notice that

$$J_0 = I_0^c \cup I_0 = I_0^c \cup I_1^c \cup I_1 .$$

Then, iterating the above procedure, after t steps we obtain

$$J_0 = I_0^c \cup I_1^c \cup \cdots \cup I_t^c \cup I_t$$

and we can define a conditional probability assessment $(\mathcal{P}_{I_0^c}, \mathcal{P}_{I_1^c}, \ldots, \mathcal{P}_{I_t^c})$ on the family $\{E_j \mid H_j ; j \in I_0^c \cup I_1^c \cup \cdots \cup I_t^c\}$.
Clearly, there exists a number k, with $k \leq n$, such that $I_k = \emptyset$. Then we have

$$J_0 = I_0^c \cup I_1^c \cup \cdots \cup I_k^c$$

and, after k steps, an assessment $\mathcal{P} = (\mathcal{P}_{I_0^c}, \mathcal{P}_{I_1^c}, \ldots, \mathcal{P}_{I_k^c})$, such that $\mathcal{P} \geq \mathcal{A}$, can be defined on the family $\mathcal{F}$. As one can verify, the assessment $\mathcal{P}$ is coherent (see [12], *Theorem*(3.5)), therefore the vector of lower bounds $\mathcal{A}$ is coherent too.
Theorem (3.2) seems difficult to be applied for checking coherence of the vector of bounds $\mathcal{A}$. In order to construct a more practical algorithm to check coherence of $\mathcal{A}$, we examine another theoretical result which generalizes Theorem (2.2) to the case of conditional probability bounds.
Denote by $(\mathcal{S}_{\mathcal{A}})$ the system obtained from (2) putting $\mathcal{V} = \mathcal{A}$ and consider the sub–vector $\mathcal{A}_0 = (\alpha_i ; i \in I_0)$ of lower bounds

$$(\mathcal{A}_0) : \qquad P(E_i \mid H_i) \geq \alpha_i , \quad i \in I_0 .$$

Then we have

Theorem 3.3 The vector of bounds $\mathcal{A}$ is coherent if and only if the following conditions are satisfied: (*i*) the system $(\mathcal{S}_{\mathcal{A}})$ is compatible; (*ii*) If $I_0 \neq \emptyset$, then the sub–vector of bounds $\mathcal{A}_0$ is coherent.

Proof. Assume $\mathcal{A}$ coherent. Then there exists a coherent assessment $\mathcal{P}$ on $\mathcal{F}$ such that $\mathcal{P} \geq \mathcal{A}$. Since $\mathcal{P}$ is coherent there exists a non-negative vector $\Lambda = (\lambda_1, \lambda_2, \ldots, \lambda_m)$, with $\sum_{t=1}^{m} \lambda_t = 1$, extending $\mathcal{P}$ to the conditional events $C_1 \mid H_0$, $C_2 \mid H_0, \ldots, C_m \mid H_0$, with $\lambda_t = P(C_t \mid H_0)$. For each j, from the well known relation $P(E_j H_j \mid H_0) = P(E_j \mid H_j) P(H_j \mid H_0)$ we obtain

$$\sum_{t: C_t \subseteq E_j H_j} \lambda_t = p_j \sum_{t: C_t \subseteq H_j} \lambda_t \geq \alpha_j \sum_{t: C_t \subseteq H_j} \lambda_t ,$$

and using (1) with $\mathcal{V} = \mathcal{A}$ it follows $\sum_{t=1}^{m} \lambda_t v_{hj} \geq \alpha_j$, so that the system $(\mathcal{S}_{\mathcal{A}})$ is compatible.
Moreover, if $I_0 \neq \emptyset$, the sub–vector $\mathcal{P}_{I_0} = (p_i \, ; \, i \in I_0)$ is coherent and from $\mathcal{P}_{I_0} \geq \mathcal{A}_0$ it follows that $\mathcal{A}_0$ is coherent too.
Conversely, assume $(\mathcal{S}_{\mathcal{A}})$ compatible and (if I_0 is not empty) $\mathcal{A}_0$ coherent.
We first assume $I_0 = \emptyset$. In this case it is

$$Max\, \Phi_j > 0 , \;\; j = 1, 2, \ldots, n,$$

so that there exist n solutions $\Lambda_1, \Lambda_2, \ldots, \Lambda_n$ of $(\mathcal{S}_{\mathcal{A}})$ such that

$$\Phi_j(\Lambda_j) > 0 , \;\; j = 1, 2, \ldots, n \; .$$

Then, given a positive vector $X = (x_1, x_2, \ldots, x_n)$, with $\sum_{i=1}^{n} x_i = 1$, to the vector X there corresponds a solution $\Lambda = \sum_{i=1}^{n} x_i \Lambda_i$ of $(\mathcal{S}_{\mathcal{A}})$ such that, for every $j = 1, 2, \ldots, n$, it is

$$\Phi_j(\Lambda) = \Phi_j \left(\sum_{i=1}^{n} x_i \Lambda_i \right) = \sum_{i=1}^{n} x_i \Phi_j(\Lambda_i) \geq x_j \Phi_j(\Lambda_j) > 0 \; . \tag{9}$$

Moreover, putting $\Lambda = (\lambda_1, \lambda_2, \ldots, \lambda_m)$ and defining the probability p_j of each conditional event $E_j \mid H_j$ as

$$p_j = \left(\sum_{t: C_t \subseteq E_j H_j} \lambda_t \right) / \left(\sum_{t: C_t \subseteq H_j} \lambda_t \right) , \;\; j = 1, 2, \ldots, n,$$

we have $p_j \geq \alpha_j$, $j = 1, 2, \ldots, n$. Then, using (1) with $\mathcal{V} = \mathcal{P}$, it can be verified that $\mathcal{P} = \sum_{t=1}^{m} \lambda_t Q_t$, i.e. $\mathcal{P} \in \mathcal{I}$, and from (9) and Theorem (2.2) we obtain that the probability assessment $\mathcal{P} = (p_1, p_2, \ldots, p_n)$ is coherent. Then, from $\mathcal{P} \geq \mathcal{A}$ it follows that the vector of lower bounds $\mathcal{A}$ is coherent too.

Assume now $I_0 \neq \emptyset$. In this case for every $j \in I_0^c$ it is $Max\,\Phi_j > 0$ and there exists a solution Λ_j of $(\mathcal{S}_{\mathcal{A}})$ such that $\Phi_j(\Lambda_j) > 0$. Then, defining a vector $\Lambda = \sum_{i \in I_0^c} x_i \Lambda_i$, with $\sum_{i \in I_0^c} x_i = 1$ and $x_i > 0$ for every $i \in I_0^c$, we have that Λ is a solution of $(\mathcal{S}_{\mathcal{A}})$ and for every $j \in I_0^c$

$$\Phi_j(\Lambda) = \Phi_j\left(\sum_{i \in I_0^c} x_i \Lambda_i\right) = \sum_{i \in I_0^c} x_i \Phi_j(\Lambda_i) \geq x_j \Phi_j(\Lambda_j) > 0 \,. \tag{10}$$

Moreover, putting $\Lambda = (\lambda_1, \lambda_2, \ldots, \lambda_m)$ and defining

$$P(E_j \mid H_j) = p_j = \left(\sum_{t: C_t \subseteq E_j H_j} \lambda_t\right) / \left(\sum_{t: C_t \subseteq H_j} \lambda_t\right), \; j \in I_0^c,$$

for each $j \in I_0^c$ we have $p_j \geq \alpha_j$, so that using (1), with $\nu_j = p_j$ for $j \in I_0^c$, it follows

$$\sum_{t=1}^{m} \lambda_t v_{tj} = p_j \,, \; j \in I_0^c \,.$$

Then, based on Theorem (2.2) and ([12], *Theorem*(3.1)), we can verify that the assessment $\mathcal{P}_{I_0^c} = (p_j \,;\, j \in I_0^c)$, on the subfamily $\{E_j \mid H_j \,;\, j \in I_0^c\}$ is coherent.
Since the vector of bounds $\mathcal{A}_0$ is coherent there exists a coherent probability assessment $\mathcal{P}_{I_0} = (p_j \,;\, j \in I_0)$, defined on the subfamily $\{E_j \mid H_j \,;\, j \in I_0\}$ such that $\mathcal{P}_{I_0} \geq \mathcal{A}_0$.
Then, the assessment $\mathcal{P} = (\mathcal{P}_{I_0}, \mathcal{P}_{I_0^c})$ is such that $\mathcal{P} \geq \mathcal{A}$.
To prove that $\mathcal{A}$ is coherent we need to verify coherence of $\mathcal{P}$. We observe that, using (1) with $\mathcal{V} = \mathcal{P}$ we obtain

$$\sum_{t=1}^{m} \lambda_t q_{tj} = p_j \,, \; j = 1, 2, \ldots, n, \tag{11}$$

that is $\mathcal{P}$ belongs to the convex hull $\mathcal{I}$ of the generalized atoms $Q_1, \ldots, Q_m$ relative to the knowledge base $(\mathcal{F}, \mathcal{P})$. Then, from (10), (11) and coherence of $\mathcal{P}_{I_0}$, using Theorem (2.2), it follows that $\mathcal{P}$ is coherent, so that the vector of lower bounds $\mathcal{A}$ is coherent too.
Using the previous result we can construct the following algorithm to check the coherence of $\mathcal{A}$.

3.1 An algorithm for checking coherence of $\mathcal{A}$

Step 0. We check the compatibility of the system $(\mathcal{S}_{\mathcal{A}})$. If $(\mathcal{S}_{\mathcal{A}})$ is not compatible, then $\mathcal{A}$ is not coherent. If $(\mathcal{S}_{\mathcal{A}})$ is compatible, we determine the set I_0 defined in (3). If $I_0 = \emptyset$, then $\mathcal{A}$ is coherent and the procedure stops.
Step 1. If $I_0 \neq \emptyset$, we check the compatibility of the system $(\mathcal{S}_0)$, obtained by system (4) choosing $J = I_0$. If $(\mathcal{S}_0)$ is not compatible, then $\mathcal{A}$ is not coherent. If $(\mathcal{S}_0)$ is

compatible, we determine the set I_1 defined in (8). If $I_1 = \emptyset$, then $\mathcal{A}$ is coherent and the procedure stops.
Step 2. If $I_1 \neq \emptyset$, we apply the same procedure to the system $(\mathcal{S}_1)$, obtained by system (4) choosing $J = I_1$, determining the set I_2, and so on.
It is clear that the algorithm stops after k steps, with $k \leq n$. Then the vector of lower bounds A is coherent if the system $(\mathcal{S}_{k-1})$ examined at step k is compatible and $I_k = \emptyset$. Finally, $\mathcal{A}$ is not coherent if the system $(\mathcal{S}_{k-1})$ is not compatible. In this case, to achieve the coherence of $\mathcal{A}$, we need to make coherent the sub-vector $\mathcal{A}_{I_{k-1}} = (\alpha_j \,;\, j \in I_{k-1})$ by suitably changing some of its components.
As one can see, the algorithm just described is easy to be applied. Concerning the computational complexity, the only aspect to be considered is the number m of atoms of the partition Π. If m is too large it becomes difficult to determine the atoms $C_1, C_2, \ldots, C_m$. Then, we have problems in starting the algorithm with the checking of the compatibility of the system $(\mathcal{S}_{\mathcal{A}})$.

4 Some examples

To illustrate the algorithm (3.1) we examine some examples.
Example 1. Consider the vector of lower bounds $\mathcal{A} = (0.9, 0.9, 0.9, 0.5)$ defined on the family $\mathcal{F} = \{A \mid H,\, B \mid K, (A^cH \vee B^cK) \mid (H \vee K), H^cK^c \mid \Omega\}$. The atoms of the partition Π are

$$C_1 = AHBK,\ C_2 = A^cHBK,\ C_3 = H^cBK,\ C_4 = AHB^cK,\ C_5 = A^cHB^cK,$$

$$C_6 = H^cB^cK,\ C_7 = AHK^c,\ C_8 = A^cHK^c,\ C_9 = H^cK^c,$$

and, using (1), the corresponding vectors are

$$V_1 = (1,1,0,0),\ V_2 = (0,1,1,0),\ V_3 = (0.9,1,0,0),\ V_4 = (1,0,1,0),\ V_5 = (0,0,1,0),$$

$$V_6 = (0.9,0,1,0),\ V_7 = (1,0.9,0,0),\ V_8 = (0,0.9,1,0),\ V_9 = (0.9,0.9,0.9,1)\ .$$

It can be verified that: (i) the system $(\mathcal{S}_{\mathcal{A}})$ is compatible; $(ii) I_0 = \{1,2,3\}$; (iii) the system $(\mathcal{S}_0)$ is not compatible. Therefore, the vector of bounds $\mathcal{A}$ is not coherent.
Example 2. Consider the family $\mathcal{F} = \{A \mid B, C \mid B, C \mid A\}$, with $A \subseteq B, C \subseteq B$, and the vector of lower bounds $\mathcal{A} = (0.8, 0.5, 0.3)$. The atoms and the corresponding vectors are

$$C_0 = B^c,\ C_1 = ABC,\ C_2 = ABC^c,\ C_3 = A^cBC,\ C_4 = A^cBC^c,$$

$$V_1 = (1,1,1),\ V_2 = (1,0,0),\ V_3 = (0,1,0.3),\ V_4 = (0,0,0.3)\ .$$

Applying the algorithm (3.1) it can be verified that the vector $\mathcal{A}$ is coherent. Notice that in our case $C \mid A = C \mid AB$. Then, based on a geometrical approach described in [11], it can be verified that the assessment $P(A \mid B) = \alpha, P(C \mid B) = \beta$ propagates to $P(C \mid AB) \geq (\alpha + \beta - 1)/\alpha$ and that the *imprecise* assessment $P(A \mid B) \geq a$, $P(C \mid B) \geq b$ propagates to $P(C \mid AB) \geq (a + b - 1)/a$. Therefore, the probability bounds $P(A \mid B) \geq 0.8, P(C \mid B) \geq 0.5$ *exactly* propagate to $P(C \mid AB) \geq 0.375$.

Example 3. (Bayes' rule) Consider the family $\mathcal{F} = \{B \mid A, C \mid AB, C \mid A\}$ and the vector of lower bounds $\mathcal{A} = (a, b, c)$. The atoms and the corresponding vectors are

$$C_0 = A^c,\ C_1 = ABC,\ C_2 = ABC^c,\ C_3 = AB^cC,\ C_4 = AB^cC^c,$$

$$V_1 = (1,1,1),\ V_2 = (1,0,0),\ V_3 = (0,b,1),\ V_4 = (0,b,0)\ .$$

It could be verified that the precise assessment $P(B \mid A) = \alpha, P(C \mid AB) = \beta$ propagates to $P(C \mid A) \geq \alpha\beta$, so that the bounds $P(B \mid A) \geq a$, $P(C \mid AB) \geq b$ exactly propagate to $P(C \mid A) \geq ab$. Therefore the vector of bounds $\mathcal{A} = (a, b, c)$ is coherent for every a, b, c. We also observe that, choosing $a = b = 1 - \varepsilon$, we obtain

$$P(B \mid A) \geq 1 - \varepsilon, P(C \mid AB) \geq 1 - \varepsilon \Rightarrow P(C \mid A) \geq 1 - 2\varepsilon + \varepsilon^2,$$

that is, $C \mid A$ is ***probabilistically entailed*** (see [1]) by $B \mid A$ and $C \mid AB$.

References

1. E. W. Adams: The logic of conditionals. Dordrecht: Reidel, 1975
2. G. Coletti: Coherent numerical and ordinal probabilistic assessments, IEEE Transactions on Systems, Man, and Cybernetics 24, n. 12, 1747-1754 (1994)
3. G. Coletti: Numerical and qualitative judgements in probabilistic expert systems. In: R. Scozzafava (ed.): Probabilistic Methods in Expert Systems. Roma: S.I.S. 1993, pp. 37-55
4. G. Coletti, A. Gilio, R. Scozzafava: Conditional events with vague information in expert systems. In: B. Bouchon- Meunier, R. R. Yager, L. A. Zadeh (eds.): Uncertainty in Knowledge Bases. Lecture Notes in Computer Science 521. Berlin: Springer 1991, pp. 106-114
5. G. Coletti, A. Gilio, R. Scozzafava: Comparative probability for conditional events: a new look through coherence. Theory and Decision 35 , 237-258 (1993)
6. L. M. De Campos, J. F. Huete: Learning non probabilistic belief networks. In: (M. Clarke, R. Kruse, S. Moral (eds.): Symbolic and Quantitative Approaches to Reasoning and Uncertainty. Lecture Notes in Computer Science 747. Berlin: Springer 1993, pp. 57- 64
7. B. de Finetti: Theory of probability, 2 Volumes (A.F.M. Smith and A. Machí trs.). New York: John Wiley 1974, 1975
8. D. Dubois, H. Prade: Probability in automated reasoning: from numerical to symbolic approaches. In: R. Scozzafava (ed.): Probabilistic Methods in Expert Systems. Roma: S.I.S. 1993, pp. 79-104.
9. D. Dubois, H. Prade, L. Godo, R. L. De Mántaras: Qualitative reasoning with imprecise probabilities. Journal of Intelligent Information Systems 2, 319-362 (1993)

10. A. Gilio: Criterio di penalizzazione e condizioni di coerenza nella valutazione soggettiva della probabilita'. Boll. Un. Matem. Ital.(7) 4-B, 645-660 (1990)

11. A. Gilio: Algorithms for precise and imprecise conditional probability assessments. In: G. Coletti, R. Scozzafava, D. Dubois (eds.): "Mathematical Models for Handling Partial Knowledge in Artificial Intelligence", London: Plenum Publ. Co. 1995 (to appear)

12. A. Gilio: Probabilistic consistency of knowledge bases in inference systems. In: M. Clarke, R. Kruse, S. Moral (eds.): Symbolic and Quantitative Approaches to Reasoning and Uncertainty. Lecture Notes in Computer Science 747. Berlin: Springer 1993, pp. 160-167

13. A. Gilio, R. Scozzafava: Le probabilita' condizionate coerenti nei sistemi esperti. In: Atti Giornate di lavoro A.I.R.O. su "Ricerca Operativa e Intelligenza Artificiale", Pisa: IBM 1988, pp. 317-330

14. A. Gilio, R. Scozzafava: Conditional events in probability assessment and revision. IEEE Transactions on Systems, Man, and Cybernetics 24, n. 12, 1741-1746 (1994)

15. A. Gilio, F. Spezzaferri: Knowledge integration for conditional probability assessments. In: D. Dubois, M. P. Wellman, B. D'Ambrosio, P. Smets (eds.): Uncertainty in Artificial Intelligence, San Mateo, California: Morgan Kaufmann Publishers 1992, pp. 98-103.

16. G.D. Kleiter: Expressing imprecision in probabilistic knowledge. In: Scozzafava (ed.): Probabilistic Methods in Expert Systems. Roma: S.I.S. 1993, pp. 139-158

17. S. Moral: A formal language for convex sets of probabilities. In: M. Clarke, R. Kruse, S. Moral (eds.): Symbolic and Quantitative Approaches to Reasoning and Uncertainty. Lecture Notes in Computer Science 747 Berlin: Springer 1993, pp. 274-281

18. R. Scozzafava: Subjective probability versus belief functions in artificial intelligence. International Journal of General Systems 22, 197-206 (1994)

19. R. Scozzafava: How to solve some critical examples by a proper use of coherent probability. In: B. Bouchon-Meunier, L. Valverde, R. R. Yager (eds.): Uncertainty in Intelligent Systems. North-Holland: Elsevier Science Publ. B. V. 1993, 121-132

20. H. Thöne, U. Güntzer, W. Kießling: Towards precision of probabilistic bounds propagation. In: D. Dubois, M. P. Wellman, B. D'Ambrosio, P. Smets (eds.): Uncertainty in Artificial Intelligence, San Mateo, California: Morgan Kaufmann Publishers 1992, pp.315-322

Testing the convenience of a variate for stratification in estimating the Gini-Simpson diversity

María Carmen Alonso, María Angeles Gil

Departamento de Matemáticas, Universidad de Oviedo
33071 Oviedo, Spain

Abstract In the literature on the quantification of the diversity in a population, one of the indices that has been proven to be useful in dealing with uncensused populations is the Gini-Simpson one. On the other hand, uncensused populations whose diversity with respect to a certain aspect is required to be quantified , often arise naturally stratified and large samples from them are available. Targets of this paper are: to approximate the optimum allocation in estimating the Gini-Simpson diversity in stratified sampling so that the asymptotic precision is maximized; to introduce the relative gain in precision from stratified sampling (with optimum allocation) over random one as a measure of the adequacy of a variate for stratification in approximating diversity; to construct a procedure to test the convenience of a given variate for stratification in estimating diversity from large samples.

1 Introduction

There are several aspects or attributes many populations vary with respect to, that is, they are diverse. In this sense, ***diversity*** can be regarded as the population variation with respect to a qualitative aspect (or a quantitative one, but the magnitude of its values is irrelevant for purposes of computing that variation). More precisely, two compounds may be distinguished in diversity: the number of categories of the considered aspect in the population, and the associated distribution. Diversity and its numerical quantification is a topic having many interesting applications in fields like Ecology (cf. Pielou, 1975, Ludwig & Reynolds, 1988), Sociology (cf., Agresti & Agresti, 1978), Linguistic (cf. Greenberg, 1956), and others. Several measures have been suggested in the literature as weighted diversity indices (cf., Patil & Taille, 1982). In that way, indices based on entropy measures have been often considered. If the population is finite but too large to be censused, the diversity may be estimated from a sample drawn at random from it. The problem of estimating diversity in simple random sampling has been discussed in previous papers (cf. Rao, 1982, Nayak, 1985, and others).
To improve the estimation precision we can take advantage of the fact that most of real-life populations this estimation is applied to (like ecological, sociological, or linguistic communities) arise naturally stratified. Consequently, it is useful to estimate diversity from a stratified sample independently chosen from different strata, and for that purpose the asymptotic behaviour of Shannon' s and Havrda-Charvát' s

(in particular, Gini-Simpson' s) diversity indices in stratified random sampling with proportional allocation have been examined (1989a).
This paper is first focussed on the extension of our previous study (1989a) to an arbitrary allocation and on the obtainment of the (asymptotically) optimum one, intended as that maximizing the asymptotic precision. On the basis of the variance for the asymptotically optimum allocation, we then define the relative gain in precision from stratified random sampling (with that allocation) over the simple random one, and we suggest to employ it as a measure of the convenience of a variate for stratification of a population in estimating the value of its diversity. A procedure to test such a convenience is finally constructed in terms of the measure above defined.

2 Basic concepts and results; asymptotically optimum allocation

Consider a finite population of N individuals which is classified according to a given variable or aspect X into M values or classes $x_1, \dots, x_M$. Suppose that the population can be divided into r non-overlapping subpopulations, called ***strata*** (as homogeneous as possible with respect to the diversity associated with X). Let $N_k =$ number of individuals in the kth stratum, $W_k = N_k/N$, and $p_k(x_i) =$ probability that an individual randomly selected from the kth stratum belongs to x_i. From those elements we obtain: $p_{ik} =$ probability that an individual randomly selected from the whole population simultaneously belongs to the kth stratum and to x_i, and $p_{i.} =$ probability that a randomly selected individual in the whole population belongs to x_i ($\sum_k N_k = N, \sum_i p_k(x_i) = 1, \sum_i p_{ik} = W_k, \sum_i \sum_k p_{ik} = 1, p_{i.} = \sum_k p_{ik}, i = 1, \dots, M, k = 1, \dots, r$).
Then, the diversity in the population with respect to X may be quantified by means of the measure below.

Definition 2.1. *The value $D(X)$ defined by*

$$D(X) = 1 - \sum_{i=1}^{M} p_{i.}^2 = 1 - \sum_{i=1}^{M} (\sum_{k=1}^{r} W_k p_k(x_i))^2$$

*is called **the Gini-Simpson population diversity associated with** X.*

Assume that a stratified sample of size n is drawn at random from the population independently from different strata. We first suppose that the sample is chosen by a specified allocation, say $\{w_k, k = 1, \dots, r\}$, so that a sample of size n_k is drawn at random and with replacement from the kth stratum, where $n_k/n = w_k$. Let $f_k(x_i)$ = relative frequency of individuals in the sample from the kth stratum belonging to x_i. From those elements we obtain: $f_{ik} =$ relative frequency of individuals in the sample from the whole population simultaneously belonging to the kth stratum and to x_i, and $f_{i.} =$ relative frequency of individuals in the sample from the whole population belonging to x_i ($\sum_k n_k = n, \sum_i f_k(x_i) = 1, \sum_i f_{ik} = w_k, \sum_i \sum_k f_{ik} = 1, f_{i.} = \sum_k f_{ik}, i = 1, \dots, M, k = 1, \dots, r$.
Then, according to some well-known results in large sample theory (cf. Rao, 1973, Serfling, 1980) we can conclude that

Theorem 2.1. *In the stratified random sampling, if* $\delta_{(n)}(X)$ *given by*

$$\delta_{(n)}(X) = 1 - \sum_{i=1}^{M} (\sum_{k=1}^{r} W_k f_k(x_i))^2$$

then, the sequence $\{n^{\frac{1}{2}}(\delta_{(n)}(X) - D(X))\}_n$ *converges in law (as* $n_k \to \infty, k = 1, \ldots, r$*) to a univariate normal distribution with mean zero and variance equal to*

$$(\sigma^2)^{st} = \sum_{k=1}^{r} \frac{W_k^2}{w_k} [\sum_{i=1}^{M} p_k(x_i) V_i^2 - (\sum_{i=1}^{M} p_k(x_i) V_i)^2]$$

(where $V_i = -2p_{i.} = -2\sum_k W_k p_k(x_i)$*) whenever* $(s^2)^{st} > 0$.

On the basis of Theorem 2.1, and assuming that large samples are available, we are going to approximate the optimum allocation so that sample sizes n_k in the strata are chosen to minimize the asymptotic variance (coinciding in this case with the asymptotic mean square error).
When diversity is estimated, stratification is usually appropriate to be considered since it often produces a gain in estimates precision. We are now looking for the best choice of sample sizes n_k to get the maximum precision, when strata are assumed previously constructed, the size of the whole sample is already specified, and the cost per unit is the same for all strata. Under those assumptions and by using the Lagrange multipliers method, we obtain the result in Theorem 2.2, which will provide us with the asymptotically optimum allocation when the population diversity is approximated by the estimators in Theorem 2.1.

Theorem 2.2. *In the stratified random sampling, the asymptotic variance of the estimator* $\delta_{(n)}(X)$ *is minimized for a fixed total size of sample, n, if*

$$n_k = \frac{W_k (T_k)^{1/2}}{\sum_{k=1}^{r} W_k (T_k)^{1/2}} n$$

where

$$T_k = \sum_{i=1}^{M} p_k(x_i) V_i^2 - (\sum_{i=1}^{M} p_k(x_i) V_i)^2$$

whenever $\sum_k W_k (T_k)^{1/2} > 0$. *The minimum asymptotic variance is then given by*

$$[\frac{(\sigma^2)^{st}}{n}]_{min} = \frac{(\sum_{k=1}^{r} W_k (T_k)^{1/2})^2}{n}$$

Remark. Theorem 2.2 suggests that it is advisable choosing the sample size in each stratum so that the larger the stratum is, or the higher the variance of the random variable $V(X)$ (that takes on the value V_i with probability $p_k(x_i)$ in stratum k, and $p_{i.}$ in the whole population, $i = 1, \ldots, M$), the larger the corresponding "optimum" sample size.

3 The relative gain in precision due to the stratification: a measure of the degree of adequacy of a variate for stratification

In this section we are going to theoretically verify advantages of the optimum allocation over the proportional one in stratified random sampling, and advantages of the last sampling technique over the simple random one, when large samples are available.
If we denote by V_{ran}, V_{prop} and V_{opt} the generic asymptotic variances of $\delta_{(n)}(X)$ in simple random sampling, stratified random sampling with proportional allocation and stratified random sampling with optimum allocation, respectively, we have that

Theorem 3.1. *For fixed total size of the sample, n, we have*

$$V_{ran} \geq V_{prop} \geq V_{opt}$$

In addition,

$$V_{ran} - V_{prop} = \frac{1}{n}[\sum_{k=1}^{r} W_k(\sum_{i=1}^{M} p_k(x_i)V_i)^2 - (\sum_{k=1}^{r} W_k \sum_{i=1}^{M} p_k(x_i)V_i)^2]$$

and

$$V_{prop} - V_{opt} = \frac{1}{n}[\sum_{k=1}^{r} W_k T_k - (\sum_{k=1}^{r} W_k (T_k)^{1/2})^2]$$

Obviously, the gain in precision from stratified random sampling with proportional allocation over simple random one can be considerable when the mean of the variable $V(X)$ greatly varies from stratum to stratum. In the same way, the gain in precision from stratified random sampling with optimum allocation over the stratified one with proportional allocation can be considerable when the variance of V(X) greatly varies from stratum to stratum.
The preceding result shows that a natural way to quantify the relative gain in precision due to the stratification by a given variate is the following:

Definition 3.1. *Given a variate for stratification in approximating the Gini-Simpson population diversity asssociated with X, the quantity RGP[D(X)] defined by*

$$RGP[D(X)] = 1 - \frac{(\sum_{k=1}^{r} W_k (T_k)^{1/2})^2}{\sum_{i=1}^{M} p_{i.} V_i^2 - (\sum_{i=1}^{M} p_{i.} V_i)^2}$$

if $V_{ran} > 0$, and $RGP[D(X)] = 0$ if $V_{ran} = 0$, will be called ***the relative gain in precision due to the stratification*** *in estimating $D(X)$.*

Remark. **Obviously, whenever $V_{opt} > 0$, the value of $RGP[D(X)]$ equals that for the quotient $(V_{ran} - V_{opt})/V_{ran}$.**

The measure above defined satisfies several properties being useful for posterior purposes of comparing different variates for stratification, or making decisions about the convenience of a specified variate. In that sense, we have that

Theorem 3.2. *Given a variate for stratification in approximating the Gini-Simpson population diversity asssociated with X, then*

i) $0 \leq RGP[D(X)] \leq 1$.

ii) $RGP[D(X)] = 0$ *if and only if the mean and the variance of the variable $V(X)$ does not depend on the stratum. In particular, if X and the variate for stratification are stochastically independent, then* $RGP[D(X)] = 0$.

iii) $RGP[D(X)] = 1$ *if and only if V_i does not depend on i. In particular, if a mapping from the set of strata of the variate of stratification to the set of values or classes of X can be defined, then* $RGP[D(X)] = 1$.

To later construct a procedure to test the convenience of a variate or criterion, S, for stratification in approximating the value of $D(X)$, we have to examine the asymptotic behaviour of the relative gain in precision on stratified populations. That can be formally done by considering a simple random sampling from the whole population and examining the bivariate distribution of $(X, S), x_1, \ldots, x_M$ being the values or classes of X, and $s_1, \ldots, s_r$ being the strata determined by S.
In that case, assume that a sample of size n is drawn at random and with replacement from the population. Let f_{ik} = relative frequency of individuals in the sample belonging to x_i and s_k (with $\sum_i \sum_k f_{ik} = 1$). Then, we have that

Theorem 3.3. *In simple random sampling, if $\gamma_{(n)}(X)$ denotes the estimator*

$$\gamma_{(n)}(X) = 1 - [A_{(n)}]^2 / B_{(n)},$$

where

$A_{(n)} = \sum_{k=1}^{r} [\sum_{j=1}^{M} f_{jk} \sum_{i=1}^{M} f_{ik} \widehat{V}_i^2 - (\sum_{i=1}^{M} f_{ik} \widehat{V}_i)^2]^{1/2}$

$B_{(n)} = \sum_{i=1}^{M} \sum_{k=1}^{r} f_{ik} \widehat{V}_i^2 - (\sum_{i=1}^{M} \sum_{k=1}^{r} f_{ik} \widehat{V}_i)^2$

(with $\widehat{V}_i = -2 \sum_k f_{ik}$), then the sequence $\{n^{1/2}(\gamma_{(n)}(X) - RGP[D(X)]/\tau_n\}_n$, where

$(\tau_n)^2 = \sum_{i=1}^{M} \sum_{k=1}^{r} f_{ik} \widehat{V}_{ik})^2 - (\sum_{i=1}^{M} \sum_{k=1}^{r} f_{ik} \widehat{V}_{ik}))^2$

with $\widehat{V}_{ik} = \{A_{(n)} B_{(n)} C_{ik(n)} + [A_{(n)}]^2 D_{ik(n)}\} / [B_{(n)}]^2$

and

$$C_{ik(n)} = -\frac{\sum_{j=1}^{M} f_{ik} [(\widehat{V}_i^2 - \widehat{V}_j^2)^2 - 4 f_{ik} (\widehat{V}_i - \widehat{V}_j)]}{[\sum_{h=1}^{M} f_{hk} \sum_{j=1}^{M} f_{jk} \widehat{V}_j^2 - (\sum_{j=1}^{M} f_{jk} \widehat{V}_j)^2]^{1/2}}$$

$$D_{ik(n)} = \widehat{V}_i(\widehat{V}_i - 4 f_{ik}) - 2 \sum_{j=1}^{M} \sum_{l=1}^{r} f_{jl} \widehat{V}_j (\widehat{V}_i - 2 f_{ik})$$

converges in law (as $n \to \infty$) to a normal distribution with mean zero and variance equal to one, whenever $\tau_{(n)}^2 > 0$ and $\tau^2 > 0$ (where τ^2 is obtained by replacing f_{ik} by p_{ik} in the expression for $\tau_{(n)}^2$.

Remark. It is worth emphasizing that whenever $RGP[D(X)] = 1$ we have that $\tau^2 = 0$, so that the result in Theorem 3.3 cannot be applied. Consequently, inferences on the hypothesis $RGP[D(X)] = 1$ (indicating that the considered variate for stratification would be ideal to estimate the Gini-Simpson diversity associated with X) cannot be obtained on the basis of that theorem.

4 A procedure to test the convenience of a variate for stratification in estimating Gini-Simpson's diversity

From Theorem 3.3 we can now state a method to test the hypothesis "$RGP[D(X)] \geq \varepsilon$" (under the assumption that $RGP[D(X)] < 1$), $\varepsilon < 1$ being the desired degree of relative gain in precision for the stratification by the given variate. Assume that a sample of size n is drawn at random and with replacement from the population. Then, the steps to test the preceding hypothesis are the following:

Step 1. Compute the sample value $\gamma_{(n)}(X)$.

Step 2. Compute the sample value $\tau_{(n)}$ and check its positiveness.

Step 3. Compute the sample value $z = n^{1/2}[\gamma_{(n)}(X) - \varepsilon]/\tau_{(n)}$.

Step 4. Compute the p-value, $p = \Phi(z)$, Φ being the cumulative function of the standard normal distribution.

Step 5. The hypothesis must be rejected at any significance level $\alpha \in (0,1)$ higher than p.

5 Concluding remarks

Similar studies to that in this paper could be developed for other indices of diversity, for indices of *divergence* between two probability distributions (see, Kullback and Leibler, 1951, Csiszár, 1967), or for indices of *inequality* with respect to a quantitative attribute (e.g., income, wealth, cf. Bourguignon, 1979). Preliminary analyses of those indices in sampling can be found in Gil *et al.* (1989b), and Zografos *et al.* (1990).

On the other hand, it would be useful to state tests to compare two stratification variates in estimating the diversity of a common aspect.

Finally, we want to remark the interest of discussing in a near future, and by means of simulation studies, the advantages of using the procedure in Section 4 in comparison with that based on the mutual information measure (cf., Gil & Gil, 1991), which involves more combersome computations.

Acknowledgements

The research in this paper is supported in part by DGICYT Grant No. PB92-1014. Its financial support is gratefully acknowledged.

References

1 Agresti, A. & Agresti, B.F. (1978). Statistical analysis of qualitative variation. *Social Methodology*, 204-237.

2 Bourguignon, F. (1979). Decomposable Income Inequality Measures. *Econometrica* 47, 901-920.

3 Csiszár, I. (1968). Information-type Measures of Difference of Probability Distributions and Indirect Observations. *Studia Sci. Math. Hungar.* 2, 299-318.

4 Gil, M.A., (1989a). A note on the Stratification and Gain in Precision in Estimating Diversity from Large Samples. *Commun. Statist. Theory and Methods* 18, 1521-1526.

5 Gil, M.A. and Gil, P. (1991). A procedure to test the suitability of a factor for stratification in estimating diversity. *Appl. Math. Comp.* 43, 221-229.

6 Gil, M.A., Pérez, R. & Gil, P. (1989b). A family of measures of uncertainty involving utilities: definitions, properties, applications and statistical inferences. *Metrika* 36, 129-147.

7 Greenberg, J.H. (1956). The measurement of linguistic diversity. *Language* 32, 109-115.

8 Kullback, S. & Leibler, A. (1951). On the information and sufficiency. *Ann. Math. Statist.* 27, 986-1005.

9 Ludwig, J.A. & Reynolds, J.F. (1988). *Statistical Ecology.* Wiley Int., New York.

10 Nayak, T.K. (1985). On Diversity Measures based on Entropy Functions. *Commun. Statist. Theory and Methods* 14, 203-215.

11 Patil, G.P. & Taille, C. (1982). Diversity as a concept and its Measurement. *J. Am. Stat. Assoc.* 77, 548-567.

12 Pielou, E.C. (1975). *Ecological Diversity.* Wiley Int., New York.

13 Rao, C.R. (1982). Diversity: its measurement, decomposition, apportionment and analysis. *Sankhyā, Ser. A* 44, 1-22.

14 Rao, C.R. (1973). *Linear Statistical Inference and its applications.* Wiley, New York.

15 Serfling, R.J. (1980). *Approximation Theorems of Mathematical Statistics.* Wiley, New York.

16 Zografos, K., Ferentinos, K. and Papaioannou, T. (1990). ϕ-divergence statistics: sampling properties and multinomial goodness of fit and divergence tests. *Commun. Statist. Theory and Methods* 19, 1785-1802.

Fuzzy Probabilities and their Applications to Statistical Inference

Dan Ralescu[1,2]

Department of Systems Science
Tokyo Institute of Technology
4259 Nagatsuta, Midori-ku, Yokohama 227, Japan
and
The Institute of Statistical Mathematics
4-6-7 Minami-Azabu, Minato-ku, Tokyo 106, Japan

Abstract. We discuss the issue of probability of a fuzzy event. The value of this probability is a fuzzy set rather than a number between zero and one. We compare this concept with the non fuzzy probability of a fuzzy event and we investigate some possible shortcomings of the latter. Explicit formulae for the fuzzy probability are given in both the discrete and continuous cases. In particular, in the discrete uniform case we relate the concepts of fuzzy probability and fuzzy cardinality. Our concept of fuzzy probability is applied to statistical inference and in particular to testing of hypotheses of the form "θ is F", where F is a fuzzy set. The level of significance and the power function of testing procedures are evaluated as fuzzy probabilities.

1 Introduction

Consider the following simple examples:

(a) A student in a class is chosen at random. What is the probability that the student is *tall*?.

(b) It is known that the score on a particular test follows the normal distribution with mean 65 and standard deviation 2.4. For a student chosen at random, what is the probability that he/she had a *very high score* on that test?.

(c) Among 20 light bulbs *a few* are defective. What is the probability that among 3 light bulbs chosen at random, at least 2 are defective?

All these examples have in common a mixture of randomness and fuzziness. The experiment is random, giving rise to a (real-valued) random variable, X. However, fuzziness is present in terms of the range of the random variable. More specifically, we want to calculate a probability of the form P(X is M) where M is a fuzzy set. To do this,

---

[1] On leave from the Department of Mathematical Sciences, University of Cincinnati, Cincinnati, Ohio 45221, U. S. A.

[2] Research supported by the National Science Foundation Grant INT-9303202.

we have to address the basic problem of a probability of a fuzzy event. The first to recognize the importance of this problem was Zadeh [7] who defined the probability of a fuzzy event as a real, (non fuzzy) number between 0 and 1. To review this concept briefly, let $(\Omega, \wp, P)$ be a probability space: Ω is the sample space of all outcomes of some random experiment; $\wp$ is a σ-algebra of subsets of Ω (called events), and $P: \wp \to [0, 1]$ is a (σ-additive) probability measure in the usual sense. A *fuzzy event* in Ω is a fuzzy subset A of Ω, given by its membership function $\mu_A: \Omega \to [0, 1]$. For the concepts which follow we need to restrict attention to measurable μ_A's thus to fuzzy events A whose membership function is a random variable.

Then the *probability of a fuzzy event* A (Zadeh [7]) is defined as

$$P(A) = E(\mu_A) = \int_\Omega \mu_A \, dP \tag{1}$$

where E() stands for the expected value.

Example 1: If Ω is a finite set with n elements and if P is the uniform measure assigning probability 1/n to each outcome of Ω, then

$$P(A) = (1/n) \Sigma \mu_A(x) ; x \in \Omega \tag{2}$$

It is only natural, in this case, to think of the sum $\Sigma \mu_A(x)$ as the number of elements in A; this has been done by De Luca and Termini [2] and their concept, later called σ-count is

$$\sigma\text{-count } A = \Sigma \mu_A(x) ; x \in \Omega \tag{3}$$

With this notation, (2) becomes

$$P(A) = \frac{\sigma\text{-count } A}{n} \tag{4}$$

However, one of the drawbacks to the concept of cardinality of a fuzzy set obtained by summing up all the membership degrees of all the elements in the set is as follows: a fuzzy set with many elements with small degrees may have the same cardinality as a fuzzy set with fewer elements but with larger degrees: Imagine a fuzzy set with 10^6 elements each with degree 0.001. In many respects this fuzzy set is "empty"; however, its σ-count is 1,000. Thus as an answer to the question "how many elements are there in A?" the σ-count gives "A has 1,000 elements". Clearly, this is a counter intuitive answer (for more on the cardinality of a fuzzy set see Ralescu [5]). Similar problems also arise with the (non fuzzy) probability of a fuzzy event (1), or in particular with (4). We will discuss this in more detail later; a simple example here will make the point. Consider a binomial experiment where Ω has two elements: S(success) and F(failure), and $P(S) = P(F) = \frac{1}{2}$.

Consider the fuzzy event A given by its membership function $\mu_A(S) = \mu_A(F) = \frac{1}{2}$. It is clear that in this set-up, the event A is a rare event; however, its probability as calculated from (1) is $P(A) = (\frac{1}{2})(\frac{1}{2}) + (\frac{1}{2})(\frac{1}{2}) = \frac{1}{2}$. This would imply that if this experiment is repeated, in the long run event A will occur 50% of the times. This again, is a very high probability for such a rare event and is, therefore, counterintuitive.

Our purpose is to discuss a probability of a fuzzy event which is a *fuzzy set* rather than a number. This idea itself is not new; it has been suggested by Zadeh as more realistic to assume that cardinality, probability, etc. of a fuzzy set should be fuzzy. Actually, for finite sample space Ω, Yager [6] has defined a concept of fuzzy probability along the lines of the fuzzy cardinality of Zadeh [8] (the latter based on an earlier concept of fuzzy cardinality of Blanchard [1]).

We will discuss our concept of fuzzy probability, study its properties, and compare it with the other concept that has appeared in literature. We will also introduce another concept of non fuzzy probability and show that it is a point of maximum of the fuzzy probability.

In both discrete and continuos space we provide explicit formulae for the fuzzy probability. In particular, in a finite uniform space, we relate the concept of fuzzy probability to that of fuzzy cardinality (see Anca Ralescu [4]) for an explicit formula of fuzzy cardinality).

It is natural to look for applications of fuzzy probabilities to various problems of statistical inference. As an explicit situation we will discuss testing of fuzzy hypotheses of the form H_0: θ is F, where θ is an unknown parameter and F is a fuzzy set.

A testing procedure for such a hypothesis is of the form: Reject H_0 if $\delta(X_1, ..., X_n)$ is M, where $\delta(X_1, ..., X_n)$ is a test statistic and M is another fuzzy set. We will evaluate the level of significance and the power function of such a testing procedure by using our concept of fuzzy probability.

2 The Concept of Fuzzy Probability

Let $(\Omega, \wp, P)$ be a probability space, and let A be a fuzzy event, i.e. its membership function μ_A: $\Omega \rightarrow [0,1]$ is measurable. We want to define the probability $\Pi(A)$ as a fuzzy subset of [0, 1], thus $\Pi(A) : [0, 1] \rightarrow [0, 1]$.

Intuitively, $\Pi(A)(p)$ will measure the extent to which the number p is the "probability" of A. More specifically, we want to make precise the following statement: "probability of A is p and probability of A' is 1-p" where A' stands for the complement of A with membership function $\mu_{A'} = 1 - \mu_A$.

Our definition is based on the concept of quantile of a probability distribution.

Definition 1. The *fuzzy probability* of A is

$$\Pi(A)(p) = \sup\{\alpha \mid \alpha \text{ is a (1-p)-quantile of } \mu_A\} \wedge \sup\{\alpha \mid \alpha \text{ is a p-quantile of } \mu_{A'}\} \quad (5)$$

The first and very important example we will consider, is that of a finite sample space and uniform probability P. More specifically, $\Omega = \{x_1, x_2, ..., x_n\}$ and $P\{x_1\} = ... = P\{x_n\} = \frac{1}{n}$.
In this case A is a fuzzy subset of a finite space. The explicit formula for fuzzy cardinality of A as given in Anca Ralescu [4], is

$$\text{card}A(k) = \mu_{(k)} \wedge (1 - \mu_{(k+1)}), \; k=0,...,n \quad (6)$$

where

$1 = \mu_{(0)} \geq \mu_{(1)} \geq ... \geq \mu_{(n)} \geq \mu_{(n+1)} = 0$ are the ordered values of the membership degrees $\mu_A(x_1), ..., \mu_A(x_n)$.

It is possible to prove the following.

Theorem 1: If Ω is finite and P is uniform, then

$$\Pi(A)(\frac{k}{n}) = \frac{1}{n}\,\mathrm{card}A(k) = \frac{1}{n}\,[\mu_{(k)} \wedge (1-\mu_{(k+1)})] \tag{7}$$

for k=0,1,...,n.

Essentially this result says that the probability of A is equal to $\frac{\mathrm{card}\,A}{n}$ just like the non fuzzy case.

Next we assume the case of a finite space $\Omega = \{x_1, x_2,...,x_n\}$ but P is no longer uniform: $P\{x_i\}=p_i$, i=1,..., n. The following result generalizes Theorem 1 and therefore also extends the fuzzy cardinality formula (6).

Theorem 2: If Ω is finite and A is a fuzzy event then

$$\Pi\,(A)[\,P(\mu_A \leq \mu_r)\,] = \mu_r \wedge (1-\mu_{r+1}) \tag{8}$$

where $\mu_1>\mu_2>...>\mu_s$ are the distinct ordered membership values $\mu_A(x_1), ..., \mu_A(x_n)$.

The next important case is when fuzzy event A has a membership function μ_A (which, let us recall, is also a random variable) of the continuous type. Let us, in this case, introduce the (cumulative) probability distribution function F of μ_A

$$F(x) = P(\mu_A \leq x),\ x \in R \tag{9}$$

It is also possible to define the "inverse" distribution function $F^{-1}(y) = \sup\{x \mid F(x) \leq y\}$. The explicit formula for the fuzzy probability in the continuous case is given in the following

Theorem 3: If fuzzy event A has a membership function μ_A of the continuous type, then

$$\Pi(A)(p) = F^{-1}(1-p) \wedge (1 - F^{-1}(1-p)) \tag{10}$$

Concerning a concept of non fuzzy probability of A we can introduce

Definition 2 : The non fuzzy probability of A is

$$NP(A) = P(\,\mu_A \geq 0.5\,) \tag{11}$$

In other words, NP(A) is the probability of the 0.5-level set of A, $A_{0.5} = \{\,\mu_A \geq 0.5\}$.

Just like the case of fuzzy cardinality (see Ralescu [5] for more details) it is possible to prove that NP(A) is a point of maximum of $\Pi(A)$:

Proposition 1: The following relationship holds

$$\Pi(A)[NP(A)] = \sup \Pi(A)(p),\ 0 \le p \le 1 \tag{12}$$

To end this section, consider again one of the examples briefly discussed in Section 1: $\Omega = \{S, F\}$, $P(S) = P(F) = \frac{1}{2}$ and A is the fuzzy event with $\mu_A(S) = \mu_A(F) = \frac{1}{2}$. It is possible to show that $\Pi(A) = \frac{1}{2}$.

3 Testing of fuzzy hypotheses

Consider $X_1, ..., X_n$ to be a random sample from some density function $f_\theta(x)$ where θ is an unknown parameter of interest. We want to test a null hypothesis of the form H_0: θ is F, where F is a fuzzy set. It seems natural to consider procedures of the form: Reject H_0 is $\delta(X_1, ..., X_n)$is M, where $\delta(X_1, ..., X_n)$ is a statistic and M is a fuzzy subset of the sample space.

The probability of type I error (level of significance) of such a test is

$$\sup_F P_{H_0}[\delta(X_1, ..., X_n) \text{ is } M] \tag{13}$$

and it is clear that such a probability should be fuzzy rather than exact; it can be calculated by using Definition 1 and the results presented in Section 2.

More importantly, still, is the power function of the test $P_{H_1}[\delta(X_1, ..., X_n)$ is M] where H_1 stands for the alternative hypothesis. This power function is now fuzzy-valued and it is definitely a nontrivial matter that of comparing two testing procedures on the basis of their (fuzzy) power functions.

Concerning (13) (where the supremum is taken over a fuzzy set) we can set the level of significance at a fixed (but fuzzy) value: $\sup_F P_{H_0}[\delta(X_1, ..., X_n)$ is M] is π.

In specific cases π stands for *small, very small,* and so on. Let us also mention that a similar approach is possible for confidence estimation; in that case we deal with statements of the form $P(\theta$ is in $\phi(X_1, ..., X_n))$ is π where $\phi(X_1, ..., X_n)$ is a fuzzy random variable (Puri and Ralescu [3]) and π is a fuzzy probability.

4 Conclusions

We have defined a concept of fuzzy probability of a fuzzy event which includes many interesting particular concepts, such as the fuzzy the cardinality. Of course this concept can be defined in a more general set-up, for a measure which is not necessarily a probability.

Our concepts seem very well suited for statistical decision analysis with fuzzy information and, in particular, for testing of fuzzy hypotheses and confidence estimation based on fuzzy random variables.

Acknowledgment: This work benefited from frequent discussions with Anca Ralescu.

References

1. Blanchard, N. Cardinal and ordinal theories about fuzzy sets, in Fuzzy Information and Decision Processes (M. M. Gupta and E. Sanchez, eds.) pp. 149-157 (North Holland 1982).

2. De Luca, A. and Termini S. A definition of a non probabilistic entropy in the setting of fuzzy sets theory, Information and Control 20 (1972) pp 301-312.

3. Puri, M. L. and Ralescu, D. A. Fuzzy random variables, J. Math. Analysis and Applications 114(1986) 409-422.

4. Ralescu, Anca L. A note on rule representation in expert systems, Information Sciences 38(1986) 193-203.

5. Ralescu, D. Cardinality, quantifiers, and the aggregation of fuzzy criteria, to appear in Fuzzy Sets and Systems.

6. Yager, R. R. A representation of the probability of a fuzzy subset, Fuzzy Sets and Systems 13(1984) 273-283.

7. Zadeh, L. A. Probability measures of fuzzy events, J. Math. Analysis and Applications 23(1968) 421-427.

8. Zadeh, L. A. A computational approach to fuzzy quantifiers in natural languages, Computers and Mathematics 9(1983) 149-184.

A Bayesian Functional Approach to Fuzzy System Representation

Cristian GEORGESCU †, Afshin AFSHARI †, Guy BORNARD *

† ETL - Schlumberger Industries,
B.P. 620-05, 50, av. Jean-Jaurès,
92542 Montrouge, France.

* Laboratoire d'Automatique de Grenoble,
B.P. 46, ENSIEG - INPG,
38402 Saint Martin d'Herès, France.

Abstract. In the present contribution, we develop a fuzzy function representation based on a probabilistic approach to fuzzy sets – the likelihood sets. Fuzzy functions, rather than fuzzy sets, are placed in the center of the fuzzy paradigm. Fuzzification, inference, defuzzification stages are naturally established as results deriving from bayesian estimation theory. Some important problems such as fuzzy system prediction and model inversion are addressed in this framework and some results are presented. The input-output behavior of a fuzzy system is an interpolating scheme with a symbolic specification given in terms of fuzzy logic. The formulation of the semantic framework for the fuzzy systems we develop here provides suitable way to deal with the introduction of a priori information – expressed in a qualitative way.

1 Introduction

In the present article we introduce a generalization of the fuzzy sets in terms of conditional probabilities of a class of attributes or properties over a class of objects. In this framework we rebuild some standard constructions and results of the fuzzy sets in terms of the new introduced likelihood sets, accentuating on the different aspects in which the two approaches differ.

The fuzzy function representation is placed in the center of the fuzzy paradigm and the likelihood sets allows handling the uncertainty associated with the problem of approximating an unknown function based on qualitative and symbolic information. The representation of a fuzzy function requires some constructions and some preliminary results of estimation theory that we present in the first sections. All these results and constructions makes obvious that the alternative approach we consider here is going further than the usual interpretation of the membership function in terms of probabilities [7] (Dubois & Prade [93] for an exhaustive presentation of this topic). Fuzzy set membership interpretation in terms of conditional probability can be found in early research of Loginov [8], however likelihood sets are more general and allows the direct use of estimation theory in the derivation of our results. Other interesting results obtained in the direction of generalizing the fuzzy sets in the framework of probability calculus are obtained by Guiasu [9], based this time on joint probability rather than conditional probability. The fuzzy representation of a crisp function is obtained step-by-step from the definition of likelihood sets. We propose a standard decomposition for the fuzzy representation of a function, that incorporates our approach as well as the classic approach. We obtain some new results, especially in the defuzzification, model inversion and system prediction, that show that the approach we considered here is promising.

The results obtained here were developed in the scope of addressing concrete problems encountered in the case of fault detection, fuzzy optimal control and pattern recognition. Some of them are presented in [14-17] leading to successful industrial applications that were based on the fuzzy system representation and the algorithms that presented here which proves that the presented approach is interesting not only from a theoretical point of view.

The results presented in this paper are intended to create a link between the fuzzy set theory and probability theory, and we don't claim that the bayesian fuzzy system representation we have introduced here may be advocated in all cases and for all applications. Some criteria allowing a suitable choice between the classical fuzzy system representation ([1-4], [13]) and our representation are discussed in the last section.

2 A Probabilistic Approach to Fuzzy Sets

Let $\mathbf{X}$ and Ξ be two universes and (Ω, K, Π) a probability space, with Ω the sample space (or the space of point events), K a σ-algebra of the family $P(\Omega)$ of subsets of Ω, and Π a probability measure on Ω. We consider X and Ξ two random variables defined on $\mathbf{X}$ and Ξ respectively:

$$x = X(\omega): \Omega \rightarrow \mathbf{X} \tag{1}$$

$$\xi = \Xi(\omega): \Omega \rightarrow \Xi \tag{1'}$$

with the conditional distribution of probability $p_{\Xi|X}(\xi|x)$ and the probability density functions:

$$\Pi(\omega / X(\omega) \in X_0 \subset X) = \int_{t \in X_0} p_X(t)\, dt \tag{2}$$

$$\Pi(\omega / \Xi(\omega) \in \Xi_0 \subset \Xi) = \int_{t \in \Xi_0} p_\Xi(t)\, dt \tag{2'}$$

If we fix the variable x in the expression $p_{\Xi|X}(\xi|x)$ we obtain a family of probability distribution functions $p_{\Xi|X}(\bullet|x)$ indexed by x, that respects the normalizing relation:

$$\int_{\xi \in \Xi} p_{\Xi|X}(\xi|x)\, d\xi = 1 \tag{3}$$

and for which the bayesian inference rule holds true:

$$p_\Xi(\xi) = \int_{x \in X} p_{\Xi|X}(\xi|x) \cdot p_X(x)\, dx \tag{4}$$

If we fix instead the variable x we obtain a function: $p_{\Xi|X}(\xi|\bullet)$ that, in fact, does not depend on the distribution of the random variable X . This is a likelihood function (or sometimes defined as the logarithm of this function, but what is essential in our discussion is that the it is a function of the cause events x rather than a function of the effect events ξ). This notion is rather intuitive (and can be subject to more than one mathematical formalization - f.i. the likelihood is considered to be logarithm of the conditional probability) but we use here the concept that was introduced in [11-12] by Fisher, in the framework of the estimation by the maximum likelihood method. The likelihood is not a probability. It exists in the case of discrete random variables as well as in the case of continuous variables.

We may well remark that the role of variables X and Ξ is not symmetric, the variable Ξ being conditioned by X. To underlie this non symmetry we shall call **X** the space of *objects* and Ξ the space of *attributes*. The fact that the random variable Ξ is conditioned by the random variable X will be referred as Ξ being a *property* of X. The fact that $p(\xi|x)$ is particularly addressed is due to the fact that there is an intrinsic causal link from **X** to **Y**. This means that, for a given "property" $X \rightarrow \Xi$ the conditional probability $p(\xi|x)$ is invariant with respect with the distribution $p(x)$, whereas the reciprocal conditional probability $p(x|\xi)$ depends on theparticularly specified distributions for either ξ or x. A likelihood set is defined by the likelihood of a certain attribute given by a certain property over the space of objects.

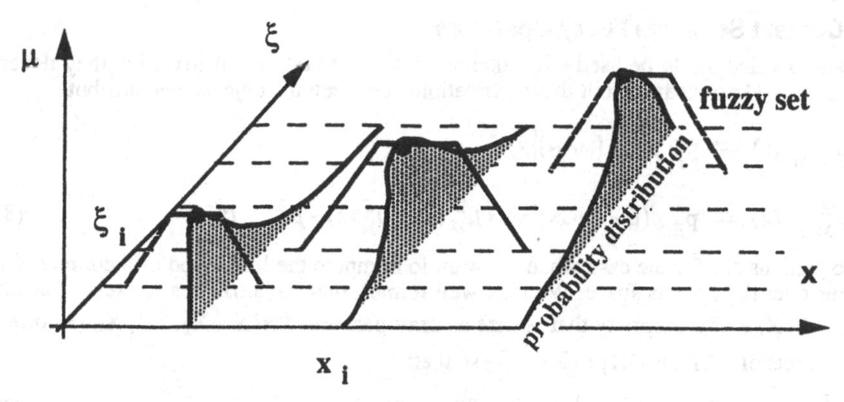

Fig. 1. Fuzzy sets issued as likelihood sets from the conditional distribution of probability $p_{\Xi|X}(\xi|x)$ that links objects x to attributes ξ.

Once the random variable Ξ defining a particular set of attributes was fixed, the value $p_{\Xi|X}(\xi|x)$ can be interpreted as the density of probability of object x having the attribute ξ. The value $\mu_{\xi}^{\Xi}(x)$ can be interpreted as the likelihood of the attribute ξ over the space of objects x The value $\mu_{M}^{\Xi}(x)$ can be interpreted as the likelihood of the set of attributes ξ∈ M over the space of objects x. We have the obvious relation:

$$\mu_{M}^{\Xi}(x) \stackrel{def}{=} p_{\Xi|X}(\xi \in M|x) = \int_{\xi \in M} p_{\Xi|X}(\xi|x)\, d\xi \tag{5}$$

This introduces the fuzzy sets as a natural particularization of likelihood sets. The intuitive interpretation of this formula is that the degree of membership $\mu_{M}^{\Xi}(x)$ of the objects $x \in \mathbf{X}$ to a certain set of attributes $M \subset \Xi$ or the truth value $t_{(\xi(x)\in M)}$ of the proposition "x has an attribute from M" or simply "x is M" can be defined as the probability of the attribute $\xi \in M$ conditioned by the object x or "the measure of x having the attribute $\xi \in M$" (as probability is a measure!).

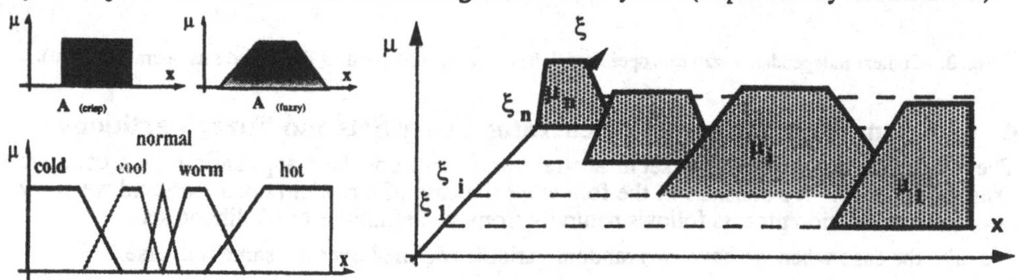

Fig. 2. Fuzzy partition over the universe of attributes Ξ.

When the space of attributes consists of only of two values that are the two logical values {0, 1} associated with the pair {FALSE, TRUE}, then we obtain the definition of a simple *fuzzy set* as a particular case. We have the following formulae:

$$\mu_{X_1}(x) = \mu_{1}^{\Xi}(x) = p_{\Xi|X}(1|x) \tag{6}$$

$$\mu_{X_0}(x) = \mu_{\bar{X}_1}(x) = \mu_{0}^{\Xi}(x) = 1 - \mu_{1}^{\Xi}(x) = p_{\Xi|X}(0|x) = 1 - p_{\Xi|X}(1|x) \tag{7}$$

3 Context Sensitive Fuzzy Operators

The operators that are to be used with likelihood sets are contex sensitive, i.e. they depend on the background knowledge about the interrelations between the objects and attributes:

$$\mu^{\Xi}_{M\cap N}(x) \stackrel{def}{=} p_{\Xi|X}(\xi \in (M\cap N)|x) \tag{8}$$

$$\mu^{\Xi}_{M\cup N}(x) \stackrel{def}{=} p_{\Xi|X}(\xi \in (M\cup N)|x) = \mu^{\Xi}_{M}(x) + \mu^{\Xi}_{N}(x) - \mu^{\Xi}_{M\cap N}(x) \tag{8'}$$

These equations are for the case when we want to compute the likelihood of a composed set of attributes over the objects space. We may well remark that the attributes (or sets of attributes) $X_1, X_2, \ldots, X_{nx}$ of a propriety that realize a fuzzy partition $\tilde{X}=\{X_1, X_2, \ldots, X_{nx}\}$ correspond to disjoint sets of attributes: $X_1 \cap X_1 = \varnothing$ so that:

$$\mu^{\Xi}_{X_i\cap X_j}(x) \stackrel{def}{=} p_{\Xi|X}(\xi \in \varnothing|x) = 0 \quad (\forall)\, i \neq j \tag{9}$$

$$\mu^{\Xi}_{X_i\cup X_j}(x) = \mu^{\Xi}_{X_i}(x) + \mu^{\Xi}_{X_j}(x) \quad (\forall)\, i \neq j \tag{9'}$$

Context dependent operators may be seen as a drawback but they are instead a more refined way to handle uncertainty. In the case of properties that are defined by independent random variables (i.e. independent properties) we have a simple formula:

$$\mu^{A\text{ and }B}_{\xi}(x) \stackrel{def}{=} p_{A\text{ and }B|X}(\xi|x) = p_{A|X}(\xi|x)\cdot p_{B|X}(\xi|x) = \mu^{A}_{\xi}(x)\cdot\mu^{B}_{\xi}(x) \tag{10}$$

$$\begin{aligned}\mu^{A\text{ or }B}_{\xi}(x) &= p_{A|X}(\xi|x) + p_{B|X}(\xi|x) - p_{A\text{ and }B|X}(\xi|x) = p_{A|X}(\xi|x) + \\ &+ p_{B|X}(\xi|x) - p_{A|X}(\xi|x)\cdot p_{B|X}(\xi|x) = \mu^{A}_{\xi}(x) + \mu^{B}_{\xi}(x) - \mu^{A}_{\xi}(x)\cdot\mu^{B}_{\xi}(x)\end{aligned} \tag{10'}$$

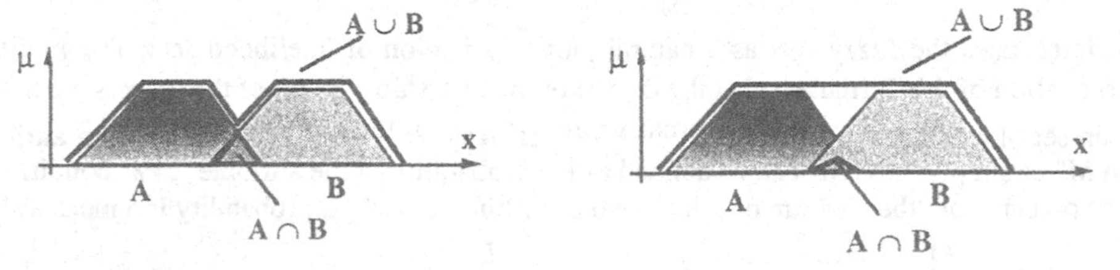

Fig. 3. Context independent fuzzy sets operators (left) and context dependent likelihood sets operators (right).

4 Random Quantization as a Generator Fuzzy Sets and Fuzzy Partitions

The precedent definitions may seem abstract and therefore their application on a concrete example is useful. We consider in the following the case of a random quantizor, and we show how the fuzzification process follows naturally from the definitions of likelihood sets.

We take the case when we have two random variables defined over the same universe **X**:

$$x = X(\omega): \Omega \to \mathbf{X} \tag{11}$$

$$\xi = \Xi(\omega): \Omega \to \mathbf{X} \tag{11'}$$

Suppose that $\xi = \Xi(\omega): \Omega \to \mathbf{X}$ is a random variable that is conditioned by the values of the random variable $x = X(\omega): \Omega \to \mathbf{X}$. Suppose also that we know the conditional probability $p(\xi\,|\,x)$ of the real value ξ given the value x. We partition the set **X** into disjoint sets $\Xi_i : \Xi_j \cap \Xi_i = \varnothing \ (\forall) i \neq j$, so that $\bigcup_i \Xi_i = \mathbf{X}$. These sets can be disjoint intervals that cover the real line, for instance. The usual "crisp" *quantization* consists in determining for each value x the corresponding set Ξ_i from the partition, for which we have $x \in \Xi_i$. We can formally introduce a discrete set of values: $\Xi = \{\xi_1,\ldots,\xi_l\}$ in a bijective correspondence with the sets: $\tilde{\Xi} = \{\Xi_1,\ldots,\Xi_n\}$. These values can be chosen as representative values in the sets Ξ_i or as a set of nominal values identified by linguistic labels. The quantization function $\tilde{q}: X \to \Xi$ is then, defined as: $\tilde{q}(\xi) = \xi_i$ iff $\xi_i \in \Xi_i$. Then $\tilde{q}^{-1}(\xi_i) = \Xi_i$ and the characteristic function of the sets Ξ_i is:

$$\mu_{\Xi_i}(x) = \begin{cases} 1 & \text{if } x \in \Xi_i \\ 0 & \text{if } x \notin \Xi_i \end{cases} \tag{12}$$

The *random quantization* can be defined as the process of passing from the conditional probability $p(\xi \mid x)$ to the conditional probabilities:

$$\mu_{X_i}(x) \overset{\text{def}}{=} p(\xi \in \Xi_i \mid x) = \int_{\xi \in \Xi_i} p(\xi \mid x)\, d\xi = \tag{13}$$

$$= \int_{\xi \in X} \mu_{\Xi_i}(\xi) \cdot p(\xi \mid x)\, d\xi = p(\tilde{q}(\xi) = \xi_i \mid x) = \int_{\xi \in X} p(\xi \in \Xi_i) \cdot p(\xi \mid x)\, d\xi$$

The above formula obviously defines a fuzzy partition $\partial X = \{X_1,\ldots,X_n\}$ associated to the crisp partition $\partial \Xi = \{\Xi_1,\ldots,\Xi_n\}$. The characteristic functions of the sets in the fuzzy partition are $\mu_{X_i}(x)$. The sets Ξ_i can be thought as cells that cover the space and that can have a name associated to their order of magnitude: *"big"*, *"medium"*, *"small"*. In the case where we have a uniform distribution of the real value ξ around the measured value x :

$$p(\xi \mid x) = \begin{cases} \dfrac{1}{2d} & \text{if } \xi \in [x-d, x+d] \\ 0 & \text{elsewhere} \end{cases} \tag{14}$$

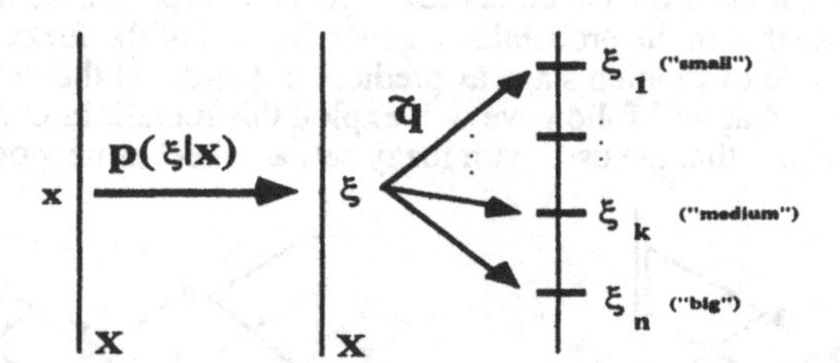

Fig. 4. Random quantization

with a dispersion equal to the cells radium, then we obtain the usual triangular fuzzy sets. When the dispersion d is smaller than the cells radium we obtain trapezoidal fuzzy sets.

The limit case when the probability distribution of ξ is the Dirac distribution $p(\xi|x) = \delta_x(\xi)$ reduces the process of random quantization to the usual quantization. In this case we have:

$$\mu_{X_i}(x) \overset{\text{def}}{=} p(\xi \in \Xi_i \mid x) = \int_{\xi \in \Xi_i} p(\xi \mid x)\, d\xi = \int_{\xi \in \Xi_i} \delta_x(\xi)\, d\xi = \begin{cases} 1 & \text{if } x \in \Xi_i \\ 0 & \text{if } x \notin \Xi_i \end{cases} \tag{15}$$

The process of associating to every value of x a likelihood function over the space of attributes, $x \longrightarrow p(\xi \mid x) \longrightarrow p(\xi \in \Xi_o \mid x) = \mu_{X_o}(x)$ can be equivalated to the fuzzification stage.

5 Probabilistic Estimation, Likelihood Sets and Fuzzification Methods

We have seen that the fuzzification stage can be equivalated to a random quantization associating to each object x the fuzzy degree of possesing a crisp attribute ξ:

$$\mu_{X_i}(x) = p(\xi \in \Xi_i \mid x) = \int_{\xi \in \Xi_i} p(\xi \mid x)\, d\xi \tag{16}$$

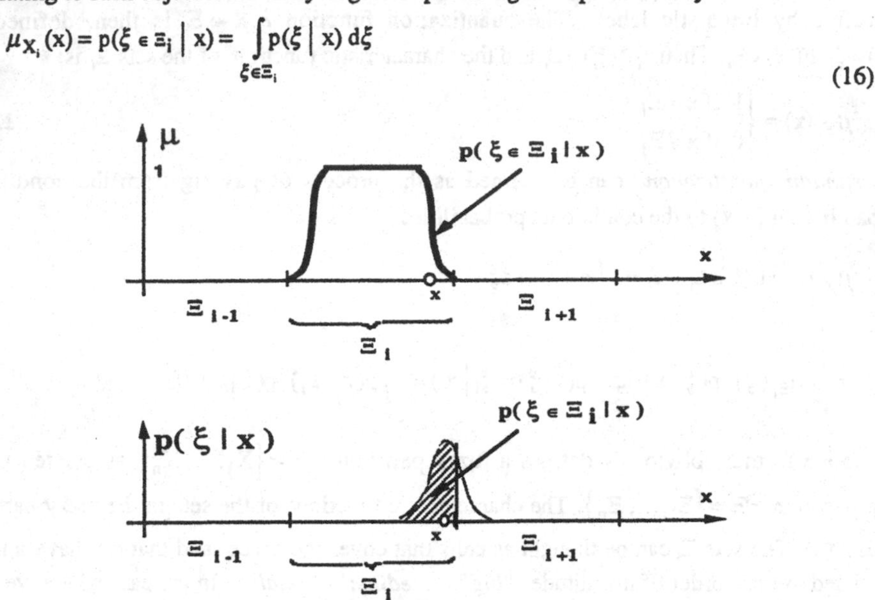

Fig.5. Random quantization as a way to obtain fuzzy sets

This corresponds to a simplified blurring formula that consists in mapping the crisp input values into the corresponding singletons. The following stage of the fuzzification, that is the projection, is also simplified to direct evaluation of the membership values of the fuzzy sets in the fuzzy partition. If we view the canonical fuzzy function representation as a block diagram fig. (5), then we can see that in the probabilistic generalization of the fuzzy sets we can interpret that fuzzification and rule evaluation stage as prediction stages and the defuzzification stage as a retrodiction stage. In what will follow we will exploit this remark in order to deduce most of the computational schemes that are used with fuzzy sets as wheel as new or modified formulas.

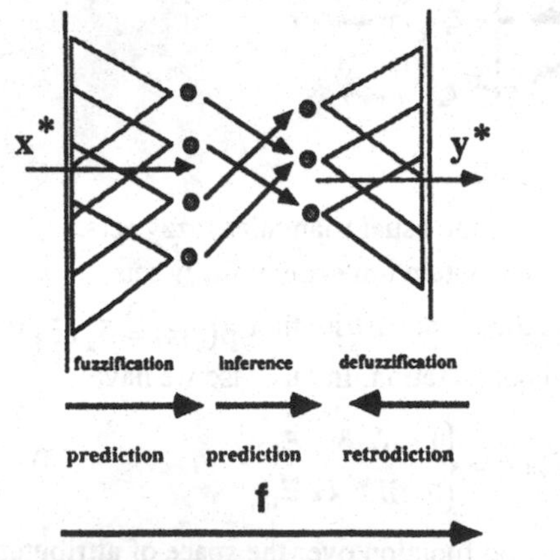

Fig. 6a. The canonical fuzzy function representation as a combined estimation scheme.

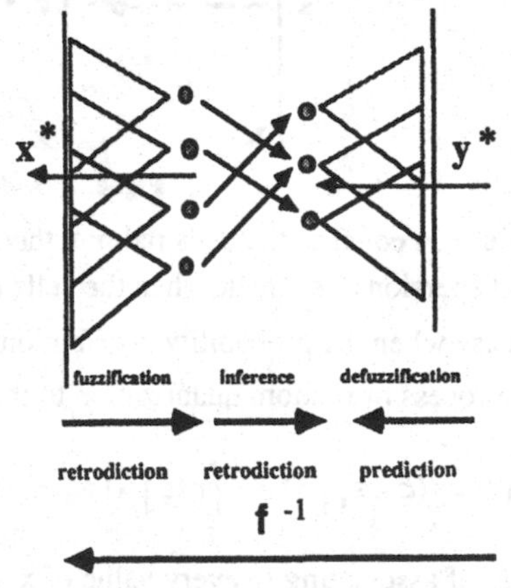

Fig. 6b. Model inversion as an inverse estimation scheme of the fuzzy function.

6 Model inversion

Model inversion often arise in control problems and it can sometimes be posed as the inversion of the function describing the evolution of the system. In this section we define the fuzzy inverse of a function f (given by the canonical fuzzy function representation) as the function inv(f) (given by its corresponding canonical fuzzy function representation) that naturally approximates the function f. The sense of this "natural approximation" will be detailled below.

The problem of inverting a function f given by its canonical fuzzy function representation $f = B \circ P \circ H \circ R \circ C = F(x) \circ H \circ D(y)$ can be addressed with the remark that fuzzification and inference are predictions as for the defuzzification is a retrodiction. As the retrodiction is the inverse of the prediction we can obtain $\mathrm{inv}(f) = D^{-1} \circ H^{-1} \circ F^{-1}$ by inverting the prediction steps, that is by solving the associated retrodiction problems, and by replacing the retrodiction step with the corresponding prediction. As defuzzification is formally the inverse operation of the fuzzification, we can write that $\mathrm{inv}(f) = F(y) \circ H^{-1} \circ D(x)$ so that finally we need to know only how to inverse the inference operation H.

$$\begin{array}{ccc} q & \xrightarrow{\;H\;} & r \\ F(x)\uparrow\;\downarrow D(x) & & F(y)\uparrow\;\downarrow D(y) \\ x & \xrightarrow{\;f\;} & y \end{array}$$

We can see that the inversion of f is not exact, that is we do not have the relation:

$$f \circ inv(f) = id.$$

This is due to loss of information in the specification of function f. The exact nature of this loss is firstly due to the process of quantization between the continuos universes **X** and **Y** and the discrete universes $\tilde{\mathbf{X}}$ and $\tilde{\mathbf{Y}}$ (i.e. the corresponding fuzzy partitions) and secondly to the probabilistic nature of the inference operation H. When the function f is injective we can achieve an exact inversion $f^{-1} : f(\mathbf{X}) \to \mathbf{Y}$ by analyzing the input output behavior of the function f.

7 Probabilistic Estimation and Fuzzy Rule Evaluation Methods

The rule evaluation corresponds to the prediction of a model given by the conditional probabilities h_{ij} starting from the initial probabilities q_i and is given by the bayesian formula:

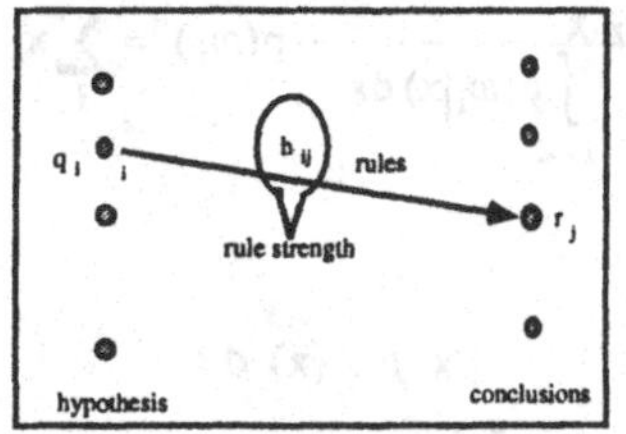

Fig. 7. Rule evaluation or fuzzy inference.

$$r_j = \sum_{i=1:nx} h_{ij} \cdot q_i \qquad (17)$$

Therefore the rule strengths (in the probabilistic generalization of fuzzy sets) must satisfy the normalization relation:

$$1 = \sum_{j=1:ny} h_{ij} \qquad (18)$$

The Entropy of a Fuzzy Rule Base.

By identifying the membership function of a fuzzy set with the likelihood of a certain attribute ξ over the space of objects x: $\mu_{\xi}^{\Xi}(x) \triangleq p_{\Xi|X}(\xi|x)$ we can naturally introduce the entropy of a fuzzy partition:

$$H(\Xi|X)(x) = \int_{\xi\in\Xi} p_{\Xi|X}(\xi|x)\ \ln p_{\Xi|X}(\xi|x)\, d\xi \qquad (19)$$

For a discrete set of attributes and using the fuzzy set notation, the entropy of a fuzzy partition becomes: $H(\Xi|X)(x) = \sum_i \mu_{\xi_i}^{\Xi}(x) \ln \mu_{\xi_i}^{\Xi}(x)$.

For a fuzzy inference rule-base, we extend the same formula as: $H(x \mapsto y) = \sum_j h_{ij} \ln h_{ij}$ (19')

As the entropy is maximal for the uniform distribution of probability, we see that the entropy of a fuzzy partition is maximal at the cross-overs of the fuzzy sets and is minimal on the kernels of the fuzzy sets, taking a null value. In the case of a fuzzy rule-base the information is maximum (entropy is minimum) when the inference is crisp.

8 Probabilistic Estimation and Defuzzification Methods

We have seen that a fuzzy partition $\tilde{X} = \{X_1, \ldots, X_n\}$ is defined by :

$$\mu_{X_i}(x) = p(\omega_i \mid x) = p(\xi \in \Xi_i \mid x) = \int_{\xi \in \Xi_i} p(\xi \mid x)\, d\xi$$

Suppose now that we have also the probabilities: $\rho_{X_i} = p(\omega_i) = p(\xi \in \Xi_i)$. The problem is to compute the probability distribution of x and its estimated value $\hat{x}$. as this is a retrodiction problem we follow the steps of the Bayesian algorithm (where $u(x)$ is the uniform distribution):

$$p(\omega_i, x) = p(\omega_i | x) \cdot u(x) \qquad (19)$$

$$p(x|\omega_i) = \frac{p(\omega_i, x)}{\int_{x \in X} p(\omega_i, x)\, dx} = \frac{u(x) \cdot p(\omega_i|x)}{u(x) \cdot \int_{x \in X} p(\omega_i|x)\, dx} = \frac{p(\omega_i|x)}{\int_{x \in X} p(\omega_i|x)\, dx}$$

$$p(x) = \sum_i p(x|\omega_i) \cdot p(\omega_i) = \sum_i \frac{p(\omega_i|x)}{\int_{x \in X} p(\omega_i|x)\, dx} \cdot p(\omega_i)$$

If we want the mean punctual estimator:

$$\hat{x} = \int_{x \in X} x \cdot p(x)\, dx = \int_{x \in X} x \cdot \left(\sum_i \frac{p(\omega_i|x)}{\int_{x \in X} p(\omega_i|x)\, dx} \cdot p(\omega_i) \right) dx = \sum_i \frac{\int_{x \in X} x \cdot p(\omega_i|x)\, dx}{\int_{x \in X} p(\omega_i|x)\, dx} \cdot p(\omega_i) = \sum_i x_i \cdot p(\omega_i)$$

or with standard fuzzy notations:

$$\hat{x} = \sum_i \frac{\int_{x \in X} x \cdot \mu_{X_i}(x)\, dx}{\int_{x \in X} \mu_{X_i}(x)\, dx} \cdot \rho_{X_i} = \sum_i x_i \cdot \rho_{X_i} \quad \text{where } x_i = \frac{\int_{x \in X} x \cdot \mu_{X_i}(x)\, dx}{\int_{x \in X} \mu_{X_i}(x)\, dx} \qquad (20)$$

This is a simplified defuzzification method that we call the *revisited center of gravity* (RCOG).

Remark. The usual defuzzification method would correspond to the formula:

$$\hat{x} = \frac{\int_{x \in X} x \cdot \sum_i p(\omega_i|x) \cdot p(\omega_i)\, dx}{\int_{x \in X} \sum_i p(\omega_i|x) \cdot p(\omega_i)\, dx} \qquad \text{(COG)}$$

We can see that RCOG method acts firstly by condensing the fuzzy sets into their centers of gravity and secondly by combining these points (as singletons) into the final output, each point carrying a weight equal to the truth value of the corresponding output. This method is equivalent to a defuzzification algorithm using singletons as output fuzzy sets. These singletons are in fact the centers of gravity of the corresponding fuzzy sets.

The conventional COG method, on the other side, firstly aggregates the membership functions in a single output (each membership function carrying a weight equal to the truth value of the corresponding output) and thereafter the resulting membership function is condensed into a single point thát is the center of gravity of the graphic of the function. We can well remark that the COG method risks to give more importance to conclusions that come from fuzzy output sets that has a relatively larger area than the other sets. However a flat fuzzy set expresses a bigger incertitude than a sharp pointed fuzzy set, so that it seems that no justification exist to give to such uncertain sets a higher weight than to other sets.

In Fig. 7 we present an application for the classification of two electrical loads that are sampled in the active (P) and reactive (Q) power. The fuzzy sets represent the projection of the likelihood sets derived from the joint probability distribution obtained from the sample of measures displayed in the figure. The two rules issues the decision on the nature of the load.

9 Conclusions

The approach of deriving fuzzy set membership functions from likelihood functions leading to the introduction of so called likelihood sets (for semantic compatibility) constitute a rigorous way of introducing fuzzy sets and offers as well a sound semantic interpretation framework of the corresponding membership functions. As a first bonus, we remark that the fuzzy partitions implicitly respect the otherwise ad hod normalizing formula for the sum of membership functions. The normalizing relation corresponds to the sum of probabilities of independent events. Another benefice is that operators between fuzzy sets can be deduced from the application context..

Fuzzy sets are the projection of the likelihood sets on the space of attributes for a specific value of the attributes and represent for an object, the measure (or the degree) of possesing an attribute. A fuzzy partition is associated with a property which is a specific dependence between the objects and the attributes.

One way to naturally visualise the fuzzification process is to apply these concepts to the random quantifier. The random quantifier is the composition of a conventional quantifier with a random variable defined on the underlying universe of discourse. The resulting conditional probability may be an appropriate way of introducing the definition of fuzzy sets by identifying the likelihood functions with the membership functions of the corresponding sets. The fact that we associate for each value x a conditional probability $p(\xi \mid x)$ corresponds to the implicit hypothesis that the "real" value ξ is always affected by a basic a priori incertitude and we refer it by some representative associated value x (f.i. its mean or the measured value). The difference between the conventional quantization and the random quantization is that in the case of random quantization we work with the probability distributions instead of precise numerical values as is the case of conventional quantization.

Handling a priori information is the most interesting feature of the proposed approach. Therefore the proposed approach is suited for problems where the loss of information is critical, due to lack of redundant information. It is the case of control problems where identification must be achieved with few data that can be affected by a high level of noise and the use of a priori information can highly improve the performances of the system.

Another advantage of the formalism we developed here is that analitical results are simpler to obtain and to extend. We have seen how natural the entropy of fuzzy partitions is introduced and more generally how general estimation theory can be used in the derivation of results. In [17] we derive, based on analog ideas, the adjoint of a dynamical system described by a fuzzy logic table and the adjoint system leads directly to an optimal control iterative solution.

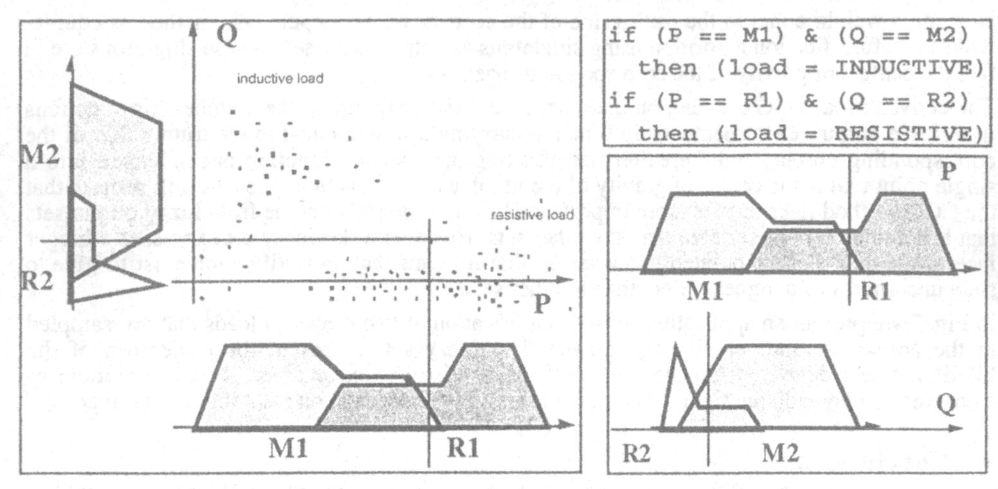

Fig. 8 Fuzzy rule based decision on an inductive / resistive load classification problem

10 References

[1] Zadeh. L.A. (1965) "Fuzzy Sets", Information and Control, no.8, pp.330-353, 1965.

[2] Zadeh L.A. (1968) "Fuzzy Algorithms", Information and Control, no.12, pp.94-102, 1968.

[3] Zadeh L.A. (1973) "Outline of a New Approach to the Analysis of Complex Systems and Decision Process". IEEE Trans. on Systems, Man and Cybernetics, 3, 28-44, 1973.

[4] Zadeh L.A. (1978) "Fuzzy sets as a basis for a theory of possibility". Fuzzy sets and systems, 1, 3-28, 1978.

[5] Black M. (1937) "Vagueness: An exercise in Logical Analysis", Philosophy of science, vol. 4, 427-455, 1937.

[6] Shafer G. (1976) "A Mathematical Theory of Evidence", Princeton University Press, Princeton, N.J., 1976.

[7] Dubois D., Prade H. (1993) "Fuzzy Sets and Probability - Misunderstanding, Bridges and Gaps", pp. 1091-1098 in the Proceedings of The II-nd IEEE International Conference on Fuzzy Systems, March 28 - April 1, 1993, San Francisco.

[8] Loginov V.I. (1966) "Probability treatment of Zadeh membership functions and their use in pattern recognition" Eng. Cyber.

[9] Guiasu S. (1994) "Stochastic logic" Proceedings of International Conference on Information Processing and Management of Uncertainty in Knowledge-Based Systems IPMU '94, July 4-8, 1994, Paris, France.

[10] Kendall M.G. (1956) "The Beginnings of a Probability Calculus", Biometrika, vol. 43, p. 1-14, 1956.

[11] Fisher R.A. (1922) "On the Mathematical Foundations of Theoretical Statistics", Phil.Trans.Royal.Soc., A, 222, pp.368-390, 1922.

[12] Fisher R.A. (1925) "Theory of Statistical Estimation", Proc.Comb.Phil.Soc., 22, pp.700-725, 1925.

[13] Lee.C.C. (1990) "Fuzzy Logic in Control Systems, Part I and II", in IEEE Transactions on Systems, Man and Cybernetics, vol. 20, no. 2, March-April, 1990.

[14] Georgescu C., A. Afshari and G. Bornard, (1993) "Fuzzy Predictive PID Controllers", pp. 1091-1098 in the Proceedings of The II-nd IEEE International Conference on Fuzzy Systems, March 28 - April 1, 1993, San Francisco.

[15] Georgescu C., A. Afshari and G. Bornard, (1993) "Fuzzy Model Based Predictive Control and Its Application to Building Energy Management Systems", pp. 487-492 in the Proceedings of The II-nd European Control Conference, June 28 - July 1, 1993, Groningen, The Netherlands.

[16] Georgescu C., A. Afshari and G. Bornard, (1993) "Qualitative Fuzzy Model-Based Fault Detection and Process Diagnosis Applied to Building Heating, Ventilating and Air Conditioning", pp. 683-692 in Proceedings of Quardet '93, III-rd IMACS International Workshop on Qualitative Reasoning and Decision Technologies, 16-18 June, 1993, Barcelona, Spain.

[17] Georgescu C., A. Afshari and G. Bornard, "A Generalised Backpropagation Method for Fuzzy Model-Based Predictive Control" in the Proceedings of The 2-nd European Congress on Intelligent Techniques and Soft Computing, Aachen, Germany, September 20-23, 1994.

Information Matrices Associated to $(\underline{h}, \underline{\phi})$-Divergence Measures: Applications to Testing Hypotheses *

D. Morales, L. Pardo
Dep. de Estadística e I.O., Univ. Complutense de Madrid
Madrid 28040 (Spain)

M. Salicrú
Dep. de Estadística, Univ. de Barcelona
Barcelona 08028 (Spain)

M.L. Menéndez
Dep. de Matemática Aplicada, Univ. Politécnica de Madrid
Madrid 28040 (Spain)

Abstract

In this work, parametric measures of information (information matrices) are defined from the nonparametric information measures called $(\underline{h}, \underline{\phi})$– divergences. Asymptotic distributions for information matrices are obtained when the parameter is replaced by its maximum likelihood estimator. On the basis of these results, tests of hypotheses are constructed.

1 Introduction

Divergence measures play an important role in statistical theory, specially in large sample theories of estimation and testing. The underlying reason is that they are indices of statistical distance between probability distributions P and Q; the smaller these indices are the harder it is to discriminate between P and Q. Many divergence measures have been proposed since the publication of the paper of Kullback and Leibler [8]. For this reason Menéndez et al. [10] introduced a generalized divergence, called $(\underline{h}, \underline{\phi})$-divergence, which includes most of the more important divergence measures. Under different assumptions, in Menéndez et al. [10] and Morales

*The research in this paper was supported in part by DGICYT Grants N. PB93-0068 and N. PB93-0022. Their financial support is gratefully acknowledged

et al. [11], it is shown that the asymptotic distribution of the $(\underline{h},\underline{\phi})$ -divergence statistic, in a stratified random sampling, are either normal or a linear combination of chi square distributions, and applications to testing statistical hypotheses in multinomial populations are given. More concretely, tests for goodness of fit, homogeneity and independence, among others, are constructed. If we consider random sampling the asymptotic distribution are either normal or chi square.

In Salicrú et al. [13] the $(\underline{h},\underline{\phi})$ -divergence measures are studied in general populations and their asymptotic distribution is obtained when the parameter is replaced by its maximum likelihood estimator.

Let $(\mathfrak{X}, \mathfrak{B}_{\mathfrak{X}}, P_\theta)_{\theta\in\Theta}$ be a statistical space, where Θ is an open subset of $\mathbb{R}^M$. We shall assume that there exists a generalized probability density function (p.d.f.) $f_\theta(x)$ for the distribution P_θ with respect to a σ-finite measure μ. In this context, the $(\underline{h},\underline{\phi})$-divergence between the generalized probability densities $f_{\theta_1}(x)$ and $f_{\theta_2}(x)$ is given by

$$D_{\underline{\phi}}^{\underline{h}}(\theta_1,\theta_2) = \int_\Lambda h_\alpha \left(\int f_{\theta_2}(x)\phi_\alpha \left(\frac{f_{\theta_1}(x)}{f_{\theta_2}(x)} \right) d\mu(x) - \phi_\alpha(1) \right) d\eta(\alpha), \qquad (1)$$

where $\underline{h} = (h_\alpha)_{\alpha\in\Lambda}$, $\underline{\phi} = (\phi_\alpha)_{\alpha\in\Lambda}$, ϕ_α and h_α are real valued C^2 functions, with $h_\alpha(0) = 0$ and η being a σ-finite measure on the measurable space (Λ, B). Furthermore, we suppose that

$$\lim_{x\to 0+} \phi_\alpha(x) \qquad \lim_{x\to 0+} \frac{\phi_\alpha(x)}{\alpha}$$

exist, in the extended sense, for all α, and $\underline{h}$ is a family of functions in order that the integral makes sense, i.e.,

$$\int_\Lambda h_\alpha d\eta = \int_\Lambda h_\alpha^+ d\eta - \int_\Lambda h_\alpha^- d\eta$$

provided not both integrals are ∞, where $h_\alpha^+ = \max\{h_\alpha, 0\}$ and $h_\alpha^- = \max\{-h_\alpha, 0\}$.

For various functions h_α and ϕ_α, well known divergences such as the measures of (1) Csiszar [5], (2) Kullback-Leibler [8], (3) Variational or Statistical, (4) χ^2-Divergence or Kagan [7], (5) Matusita [9], (6) Balakrishman and Sanghvi [1], (7) Havrda and Charvat [6], (8) Cressie and Read [4], (9) Renyi [12], (10) and (11) Sharma and Mittal [14], (12) Battacharyya [2], (13) Trigonometric, (14), (15), (16) and (17) Taneja [15] among others are obtained. Next table gives the corresponding expressions of their functions h_α and ϕ_α. We will take $\Lambda = \{1,2\}$ and $\eta(1) = \eta(2) = 1$. The function

$$h(x) = \frac{1}{s-1}\left((x+1)^{\frac{s-1}{r-1}} - 1 \right),$$

plays an important role in what follows.

N	Divergence	h_α	ϕ_α
(1)	$D_\varphi(\theta_1,\theta_2)$	$h_1(x)=x, h_2(x)=0$	$\phi_1(x)=\varphi(x)+\varphi(1)$
(2)	$D(\theta_1,\theta_2)$	$h_1(x)=x, h_2(x)=0$	$\phi_1(x)=x\ln x$
(3)	$D^{VA}(\theta_1,\theta_2)$	$h_1(x)=x, h_2(x)=0$	$\phi_1(x)=\mid x-1\mid$
(4)	$D^{KA}(\theta_1,\theta_2)$	$h_1(x)=x, h_2(x)=0$	$\phi_1(x)=(1-x)^2$
(5)	$D^{MA}(\theta_1,\theta_2)$	$h_1(x)=x, h_2(x)=0$	$\phi_1(x)=(1-x^a)^{1/a}, 0<a\leq 1$
(6)	$D^{BS}(\theta_1,\theta_2)$	$h_1(x)=x, h_2(x)=0$	$\phi_1(x)=\frac{(x-1)^2}{(x+1)^2}$
(7)	$D^{HC}(\theta_1,\theta_2)$	$h_1(x)=x, h_2(x)=0$	$\phi_1(x)=\frac{(x-x^a)}{1-a}, a>0$
(8)	$D^{CR}(\theta_1,\theta_2)$	$h_1(x)=x, h_2(x)=0$	$\phi_1(x)=\frac{x^{a+1}-1}{a(a+1)}$
(9)	$D_r^1(\theta_1,\theta_2)$	$h_1(x)=\frac{1+\ln x}{r-1}, h_2(x)=0$	$\phi_1(x)=x^r$
(10)	$D_1^s(\theta_1,\theta_2)$	$h_1(x)=\frac{e^{(s-1)x}-1}{s-1}, h_2(x)=0$	$\phi_1(x)=x\ln x$
(11)	$D_r^s(\theta_1,\theta_2)$	$h_1(x)=h(x), h_2(x)=0$	$\phi_1(x)=x^r$
(12)	$D^{BA}(\theta_1,\theta_2)$	$h_1(x)=-\ln(x+1), h_2(x)=0$	$\phi_1(x)=x^{1/2}$
(13)	$D^{\mathrm{Tri}}(\theta_1,\theta_2)$	$h_1(x)=\tan^{-1}(x), h_2(x)=0$	$\phi_1(x)=x^\alpha, 0<\alpha<1$
(14)	${}^1J_r^s(\theta_1,\theta_2)$	$h_1(x)=h_2(x)=h(x)$	$\phi_1(x)=x^r, \phi_2(x)=x^{1-r}$
(15)	${}^2J_r^s(\theta_1,\theta_2)$	$h_1(x)=2h(x), h_2(x)=0$	$\phi_1(x)=\frac{1}{2}(x^r+x^{1-r})$
(16)	${}^1R_r^s(\theta_1,\theta_2)$	$h_1(x)=h_2(x)=\frac{1}{2}h(x)$	$\phi_1(x)=\left(\frac{1+x}{2}\right)^{1-r}$
			$\phi_2(x)=x^r\left(\frac{1+x}{2}\right)^{1-r}$
(17)	${}^2R_r^s(\theta_1,\theta_2)$	$h_1(x)=h(x), h_2(x)=0$	$\phi_1(x)=2^{r-2}(1+x^r)(1+x)^{1-r}$

Our aim is not to introduce new measures. We have tried to give a very general functional, which can be used to do global studies, instead of measure to measure individualized studies. Therefore, we would like to consider the formulae (1) as a simple way to summarize previously defined divergence measures, in such a way that when we talk about $(\underline{h}, \underline{\phi})$-divergences is because we have in mind an already existing an studied divergence. The final purpose is to save time and work.

In this paper, on the basis of the $(\underline{h}, \underline{\phi})$ -divergence measures, the information matrices associated to a differential metric in direction to the tangent space are obtained. Finally, we get the asymptotic distribution of the information matrices obtained when the parameter is replaced by its maximum likelihood estimator.

2 Information matrix associated to $(\underline{h}, \underline{\phi})$ -divergences

In what follows,the following regularity assumptions hold:

(i) The set $A = \{x \in X / f(x, \theta) > 0\}$ does not depend on θ and for all $x \in A, \theta \in \Theta$

$$\frac{\partial f_\theta(x)}{\partial \theta_i}, \frac{\partial^2 f_\theta(x)}{\partial \theta_i \partial \theta_j}, \frac{\partial^3 f_\theta(x)}{\partial \theta_i \partial \theta_j \partial \theta_k}, \quad i, j, k = 1, \ldots, M,$$

exist and are finite.

(ii) There exist real valued functions $F(x)$ and $H(x)$ such that

$$\left| \frac{\partial f_\theta(x)}{\partial \theta_i} \right| < F(x), \left| \frac{\partial^2 f_\theta(x)}{\partial \theta_i \partial \theta_j} \right| < F(x), \left| \frac{\partial^3 f_\theta(x)}{\partial \theta_i \partial \theta_j \partial \theta_k} \right| < H(x),$$

where F is finitely integrable and $E[H(X)] < K$, with K independent of θ.

(iii) $\left(E \left(\frac{\partial \ln f_\theta(x)}{\partial \theta_i} \frac{\partial \ln f_\theta(x)}{\partial \theta_j} \right) \right)_{i,j=1,\ldots,M}$ is finite and positive definite.

Taking in account that each population can be characterized by a particular point θ of Θ, we may interpret P_θ as a manifold and consider $\theta = (\theta_1, \ldots, \theta_M)$ as a coordinate system. In general, it is also assumed that for a fixed $\theta \in \Theta$ the M functions

$$\frac{\partial f_\theta(x)}{\partial \theta_i}, \quad i = 1, \ldots, M$$

are linearly independent. Thus, the tangent space T_θ at point θ is the span of

$$\left\{ \frac{\partial f_\theta(x)}{\theta_i} \right\}_{i=1,\ldots,M}$$

Now consider any divergence measure $\mathcal{D}$ and define the arc differential element in direction to the tangent space T_θ and with respect to the divergence measure $\mathcal{D}$ as follows:

$$ds^2(\theta) = \lim_{t\to 0} \frac{\mathcal{D}(f_\theta(x), f_\theta(x) + t\, df_\theta(x)) - \mathcal{D}(f_\theta(x), f_\theta(x))}{t^2}. \tag{2}$$

In the following theorem we obtain the arc differential element in direction to the tangent space T_θ and with respect to the $(\underline{h}, \underline{\phi})$-divergence measures.

Theorem 2.1 *The arc differential element in direction to the tangent space T_θ and with respect to the $(\underline{h}, \underline{\phi})$-divergence measures is given by*

$$ds^2(\theta)^{\underline{h}}_{\underline{\phi}} = (d\theta_1, ..., d\theta_M) \left\| \int h'_\alpha(0) \frac{\phi''_\alpha(1)}{2} d\eta I^F_{ij}(X,\theta) \right\|_{M\times M} (d\theta_1, ..., d\theta_M)^t.$$

Proof
Considering Mc-Laurin expansion of the function

$$g(t) = D^{\underline{h}}_{\underline{\phi}}(f_\theta(x), f_\theta(x) + tdf_\theta(x))$$

we have

$$g(0) = 0$$

$$g'(0) = \int h'_\alpha(0)(\phi_\alpha(1) - \phi'_\alpha(1))d\eta \left(\sum_{i=1}^M \frac{\partial}{\partial\theta_i} \int f_\theta(x) d\mu d\theta_i \right) = 0$$

and

$$g''(0) = \int h'_\alpha(0)\phi''_\alpha(1)d\eta \sum_{i,j=1}^M \left(\int \frac{1}{f_\theta(x)} \frac{\partial f_\theta(x)}{\partial\theta_i} \frac{\partial f_\theta(x)}{\partial\theta_j} d\mu \right) d\theta_i d\theta_j.$$

Then, the arc differential element is given by

$$\begin{aligned} ds^2(\theta)^{\underline{h}}_{\underline{\phi}} &= \liminf_{t\to\infty} \frac{g(0) + g'(0) + \frac{1}{2}g''(0)t^2 + O(t^3)}{t^2} \\ &= (d\theta_1, \ldots, d\theta_M) \left\| \int h'_\alpha(0) \frac{\phi''_\alpha(1)}{2} d\eta I^F_{ij}(X,\theta) \right\|_{M\times M} (d\theta_1, \ldots, d\theta_M)^t \end{aligned}$$

Remark 2.1

(a) We can observe that

$$ds^2(\theta)^{\underline{h}}_{\underline{\phi}} = \sum_{i=1}^M \sum_{j=1}^M g_{ij}(\theta) d\theta_i d\theta_j.$$

The matrix $IM^{\underline{h}}_{\underline{\phi}}(\theta) = (g_{ij}(\theta))_{ij}$ is positive-definite for every $\theta \in \Theta$, then $ds^2(\theta)^{\underline{h}}_{\underline{\phi}}$ gives a Riemannian metric, because $IM^{\underline{h}}_{\underline{\phi}}(\theta)$ defines a covariant symmetric tensor of second order for all $\theta \in \Theta$.

(b) In the literature of Information Theory, the metric $ds^2(\theta)^{\underline{h}}_{\underline{\phi}}$ and the matrix $IM^{\underline{h}}_{\underline{\phi}}(\theta) = (g_{ij}(\theta))_{ij}$ are called the "$(\underline{h}, \underline{\phi})$ -information metric" and the "$(\underline{h}, \underline{\phi})$ -information matrix", Burbea [3].

3 Asymptotic distribution of information matrices

Now we obtain the asymptotic distribution of $IM^{\underline{h}}_{\underline{\phi}}(\hat{\theta}) = (g_{ij}(\hat{\theta}))_{ij}$ where $\hat{\theta} = (\hat{\theta}_1, \ldots, \hat{\theta}_M)$ is the maximum likelihood estimator of θ based on a random sample of size n.

Theorem 3.1 *If we suppose the usual conditions to get the asymptotic normality of the maximum likelihood estimator, then we obtain*

$$n^{1/2}\left(VecIM^{\underline{h}}_{\underline{\phi}}(\hat{\theta}) - VecIM^{\underline{h}}_{\underline{\phi}}(\theta)\right) \xrightarrow[n\uparrow\infty]{L} N(0, A^t I^F_X(\theta)^{-1} A)$$

where

$$VecIM^{\underline{h}}_{\underline{\phi}}(\hat{\theta}) = (g_{11}(\hat{\theta}), ..., g_{MM}(\hat{\theta}))$$

$$VecIM^{\underline{h}}_{\underline{\phi}}(\theta) = (g_{11}(\theta), ..., g_{MM}(\theta))$$

$$A = \begin{pmatrix} \frac{\partial g_{11}(\theta)}{\partial \theta_1} & \cdots & \frac{\partial g_{11}(\theta)}{\partial \theta_M} \\ \cdots & \cdots & \cdots \\ \cdots & \cdots & \cdots \\ \frac{\partial g_{MM}(\theta)}{\partial \theta_1} & \cdots & \frac{\partial g_{MM}(\theta)}{\partial \theta_M} \end{pmatrix}$$

Proof

The Taylor expansion of $IM^{\underline{h}}_{\underline{\phi}}(\hat{\theta}) = (g_{ij}(\hat{\theta}))_{ij}$ at the point θ is:

$$VecIM^{\underline{h}}_{\underline{\phi}}(\hat{\theta}) = VecIM^{\underline{h}}_{\underline{\phi}}(\theta) + \left(\sum_{i=1}^{M} \frac{\partial g_{11}}{\partial \theta_i}(\hat{\theta}_i - \theta_i), \ldots, \sum_{i=1}^{M} \frac{\partial g_{MM}}{\partial \theta_i}(\hat{\theta}_i - \theta_i)\right)^t + R_n.$$

As $n^{1/2} R_n \xrightarrow[n\uparrow\infty]{L} 0$, we have that the random variables

$$n^{1/2}\left(VecIM^{\underline{h}}_{\underline{\phi}}(\hat{\theta}) - VecIM^{\underline{h}}_{\underline{\phi}}(\theta)\right) \quad \text{and} \quad An^{1/2}(\hat{\theta} - \theta)$$

have asymptotically the same distribution, where the matrix A is given above. Then we get

$$n^{1/2}\left(VecIM_{\underline{\phi}}^{\underline{h}}(\hat{\theta}) - VecIM_{\underline{\phi}}^{\underline{h}}(\theta)\right) \xrightarrow[n\uparrow\infty]{L} N(0, A^t I_X^F(\theta)^{-1} A)$$

Now on the basis of this result we can test for goodness of fit. We want to test the null hypothesis:

$$H_0 : IM_{\underline{\phi}}^{\underline{h}}(\theta) = IM_{\underline{\phi}}^{\underline{h}}(\theta_0).$$

For this purpose we consider the statistic

$$T_1 = n\left(VecIM_{\underline{\phi}}^{\underline{h}}(\hat{\theta}) - VecIM_{\underline{\phi}}^{\underline{h}}(\theta_0)\right)^t \left(A^t I_X^F(\theta)^{-1} A\right)^{-1} \left(VecIM_{\underline{\phi}}^{\underline{h}}(\hat{\theta}) - VecIM_{\underline{\phi}}^{\underline{h}}(\theta_0)\right).$$

This statistic is asymptotically distributed as a Chi square distribution with s degrees of freedom being $s = \text{rank}(A^t I_X^F(\theta)^{-1} A)$. We reject the null hypothesis if $T_1 \geq \chi^2_{s,\alpha}$.

References

[1] V. Balakrishman, L.D. Sanghvi: Distance between populations on the basis of attribute data. Biometrics, 24, 859-865 (1986)

[2] A. Battacharyya: On a measure of divergence between two statistical populations defined by their probability distributions. Bull. Cal. Math. Soc.,35, 99-109 (1946)

[3] J. Burbea: Informative geometry of probability spaces. Expositions Mathematicae, 4, 347-378 (1986)

[4] N. Cressie, T.R.C. Read: Multinomial goodness of fit tests. J. Royal Statist. Soc. B, 46, 440-464 (1984)

[5] I. Csiszar: Information type measures of difference of probability distributions and indirect observations. Studia Sci. Mat. Hung., 2, 299-318 (1967)

[6] J. Havrda, F. Charvát: Quantification of classification processes. Concept of structural α-entropy. Kybernetika, 2, 30-35 (1967)

[7] M. Kagan: On the theory of Fisher's amount of information. Sov. Math. Stat., 27, 986-1005 (1963)

[8] S. Kullback, A. Leibler: On the information and sufficiency. Ann. Math. Statist., 27, 986-1005 (1951)

[9] K. Matusita: Decision rules, based on the distance for problems of fit, two samples and estimation. Ann. Math. Statist. 26, 631-640 (1964)

[10] M.L. Menéndez, D. Morales, L. Pardo, M. Salicrú: Asymptotic distribution of the generalized distance measure in a random sampling. Proceeding of Distance'92. 337-340 (1992)

[11] D. Morales, L. Pardo, M. Salicrú, M.L. Menéndez: Asymptotic properties of divergence statistics in a stratified random sampling and its applications to test statistical hypotheses. Journal of Statistical Planning and Inference, 38, 201-221 (1994)

[12] A. Renyi: On measures of entropy and information. Proceedings of 4th Berkeley Symp. Math. Statist. and Prob., 1, 547-561 (1961)

[13] M. Salicrú, M.L. Menéndez, D. Morales, L. Pardo: On the applications of divergence type measures in testing statistical hypotheses. Journal of Multivariate Analysis, 51, 372-391 (1994)

[14] B.D. Sharma, D.P. Mittal: New nonadditive measures of entropy for discrete probability distributions. J. Math. Sci., 10, 28-40 (1977)

[15] I.J. Taneja: On generalized information measures and their applications. Adv. Elect. and Elect. Phis. 76, 327-413 (1989)

Informational Energy Test for Equality of Variances

Pardo, J.A., Pardo, M.C., Vicente, M.L., Esteban, M.D.

Departamento de Estadística e I.O.
Universidad Complutense de Madrid
28040 - Madrid (Spain).

Abstract

On the basis of the informational energy an alternative test of homogeneity of variances in samples from normal populations is developed. In addition, several of the more well known statistics for performing this type of analysis and the new test proposed here are compared in a Monte Carlo study.

1 Introduction

Tests for equality of variances are of interest in many situations such as analysis of variance or quality control. From Neyman and Pearson [6], who proposed a test based on the likelihood ratio statistic, to Conover, Johnson and Johnson [3] who performed an interesting study about robustness and power for many of the existing parametric and nonparametric test for homogeneity of variances, there exist many authors who had studied this problem. Recently, one important application of the statistical information theory is to develop tests of hypothesis based on entropy and divergence type statistics. See. e.g. Salicrú et al. [7] and Morales et al. [5].

In this work, we present an alternative test for equality of variances based on the informational energy. This test is compared with several of the more well known statistics for performing this type of analysis. The comparisons are made for power and for stability of error rates using Monte Carlo simulations. We find that teh overall performance of the informational energy procedure is as good as or better than any of the other test statistics.

2 Test for homogeneity of variances

Consider the statistical space $(\mathbb{R}, \beta_{\mathbb{R}}, P_\theta);\ \theta \in \Theta)$, where $\theta = (\mu, \sigma)$, $\Theta = \mathbb{R} \times \mathbb{R}_+$ and $\{P_\theta; \theta \in \Theta\}$ is the family of normal probability functions. Let $f(x, \theta)$ be the probability density function associated to P_θ with respect to a σ-finite measure λ. In this context the informational energy associated to $f(x, \theta)$ is given by

$$e(\theta) = \int_{\mathbb{R}} f(x, \theta)^2 d\lambda(x) = \frac{\pi^{-1/2}}{2\sigma}.$$

We get the following result:

Theorem 2.1 *If $T^t I_F(\theta)^{-1} T > 0$, then*

$$n^{1/2}\left(e(\widehat{\theta}) - e(\theta)\right) \xrightarrow[n\uparrow\infty]{L} \mathcal{N}(0, \sigma^2 = T^t I_F(\theta)^{-1} T)$$

where $\widehat{\theta}$ is the maximum likelihood estimator of θ, $T^t = (t_1, t_2)$, $t_1 = \dfrac{\partial e(\theta)}{\partial \mu} = 0$, $t_2 = \dfrac{\partial e(\theta)}{\partial \sigma^2} = -\dfrac{1}{4\pi^{1/2}\sigma^3}$ and $I_F(\theta) = \begin{pmatrix} \sigma^2 & 0 \\ 0 & 2\sigma^4 \end{pmatrix}^{-1}$ is the Fisher information matrix.

Proof: The proof of the theorem follows from the asymptotic normality of $n^{1/2}(\widehat{\theta} - \theta)$ and the Taylor expansion of $e(\widehat{\theta})$ around θ.

The results obtained in theorem 2.1 can be used in statistical inference and, in particular, to construct a test for homogeneity of variance based on information energy. Previously we consider two tests which are useful to construct it. These tests are:

(i) Test for a predicted value of the population certainty; i.e., $H_0 : e(\theta) = e^*$. In this case we can use the statistic

$$T_1 = \frac{n^{1/2}(e(\widehat{\theta}) - e^*)}{\sigma^2(\widehat{\theta})},$$

which has approximately the standard normal distribution under H_0 for sufficiently large n and $\sigma^2(\widehat{\theta})$ is obtained from theorem 2.1 by replacing θ by its maximum likelihood estimator $\widehat{\theta}$ in $\sigma^2(\theta)$. We can use this procedure for testing $H_0 : \sigma = \sigma_*$ because $e(\sigma)$ is a bijective function of σ.

(ii) Test for a common predicted value of r population certainties; i.e., $H_0 : e(\theta_1) = \ldots = e(\theta_r) = e^*$. In this case we can use the statistic

$$T_2 = \sum_{i=1}^{r} \frac{n_j(e(\theta_j) - e^*)^2}{\sigma^2(\widehat{\theta}_j)},$$

which is asymptotically chi-square distributed with r degrees of freedom. We can use this procedure for testing $H_0 : \sigma_1 = \ldots = \sigma_r = \sigma_*$ because $e(\sigma)$ is a biyective function of σ.

The test for homogeneity of variances is equivalent to testing the equality of r population certainties; i.e., $H_0 : e(\theta_1) = \ldots = e(\theta_r)$. In this case we can use the statistic

$$T_3 = \sum_{j=1}^{r} \frac{n_j (e(\theta_j) - \overline{e})^2}{\sigma^2(\widehat{\theta}_j)},$$

where

$$\overline{e} = \frac{\sum_{i=1}^{r} \frac{n_j e(\widehat{\theta}_j)}{\sigma^2(\widehat{\theta}_j)}}{\sum_{i=1}^{r} \frac{n_j}{\sigma^2(\widehat{\theta}_j)}}.$$

This statistic is asymptotically chi-squared distributed with $r-1$ degrees of freedom. The proof follows from Cochran's theorem and from the T_2 distribution.

So

$$\sigma^2(\widehat{\mu}_i, \widehat{\sigma}_i) = T^t \left(I_F(\widehat{\mu}_i, \widehat{\sigma}_i^2)\right)^{-1} T = (8\pi\widehat{\sigma}_i^2)^{-1},$$

we have

$$\overline{e} = \frac{1}{2\pi^{1/2}} \frac{\sum_{i=1}^{r} n_i \widehat{\sigma}_i}{\sum_{i=1}^{r} n_i \widehat{\sigma}_i^2}$$

therefore the test statistic is given by

$$T_3 = 2 \sum_{i=1}^{r} n_i \widehat{\sigma}_i^2 \left[\frac{1}{\widehat{\sigma}_i} - \frac{\sum_{i=1}^{r} n_i \widehat{\sigma}_i}{\sum_{i=1}^{r} n_i \widehat{\sigma}_i^2} \right]^2.$$

Finally the null hypothesis of equality of variances should be rejected at a level α if $T_3 > \chi^2_{r-1,\alpha}$.

3 An example

As an illustration of the application of the test for equality of variances given in the above section we consider the following example.

Example (Bickel and Doksum, [2])

Consider the following data giving blood cholesterol levels of men in three different socioeconomic groups labeled I, II and III with I being the high end.

Table 1

I	403	311	269	336	259					
II	312	222	302	420	420	286	353	210	286	290
III	403	244	353	235	319	260				

We want to test whether all the variances are equal.

From observed data given before, we have

Group	n_i	$\widehat{\sigma}_i$
I	5	51.867523
II	10	91.750536
III	6	75.248477

and

$$T_3 = 1.7061454.$$

The critical point for a significance level $\alpha = 0.05$ is $\chi^2_{2,0.05} = 5.991$ and as

$$T_3 = 1.7061454 < \chi^2_{2,0.05} = 5.991$$

there is no evidence to indicate that variance blood cholesterol is different for the three socioeconomic groups.

4 A comparative study of the informational energy test

Many of the existing parametric and nonparametric tests for homogeneity of variances, and some variations of these tests are examined by Conover, Johnson and Johnson [3]. The purpose of their study is to provide a list of tests that have a stable Type I error rate when the normality assumption may not be true, when the sample sizes may be small and/or unequal, and when distributions may be skewed and heavy-tailed. We have used their method of comparing tests in order to evaluate the test statistic given in section 2.

To study the information energy test robustness and power we used samples from several distributions, using several sample sizes and various combinations of variances. The normal and double exponential were chosen for symmetric distributions. Uniform random numbers were generated using a multiplicative congruential generator. The normal variates were obtained by the Box-Müller method. The double exponential variates were obtained from the inverse cumulative distribution function. Four samples were drawn with respective sample sizes (n_1, n_2, n_3, n_4) = $(5,5,5,5)$, $(10,10,10,10)$, $(20,20,20,20)$ and $(5,5,20,20)$. The null hypothesis of equal variances (all equal to 1) was examined along with four alternatives $(\sigma_1^2, \sigma_2^2, \sigma_3^2, \sigma_4^2)$=$(1,1,1,2)$, $(1,1,1,4)$, $(1,1,1,8)$ and $(1,2,4,8)$. The means were set equal to the standard deviation in each population under the alternative hypothesis. Zero means were used for H_0. Each of these 40 combinations of distribution type, sample size and variances was repeated 10,000 times (1,000 in Conover et al [3]), so the informational energy test statistic, as well as Neyman–Pearson [6], Barlett [1] and Hartley [4] statistics, were computed and compared with their 5 percent nominal critical values 400,000 times each. The observed frequency of rejection is reported in table 2 for normal distributions and in table 3 for double exponential distributions. The figures in parentheses in those tables represent the averages over the four variance combinations under the alternative hypothesis.

For the asymmetric case, we obtain the corresponding figures doing $\sigma X_i^2 + \mu$, where X_i represents the null distributed random variable. The two distributions (normal)^2 and (double exponential)^2, the two sample sizes $(10, 10, 10, 10)$ and $(5, 5, 20, 20)$ and the five variance combinations gave a total of 20 combinations. For each combination, 10,000 repetitions were run for each of the 4 statistics. With the same structure as tables 2 and 3, the observed frequencies of rejection are reported in table 4 for these asymmetric distributions.

To interpret their simulation results, Conover et al [3] define a test to be robust if the maximum Type I error rate is less than 0.10 for a 5 per cent test. In this sense, if we compare the type I error rate values of the tables 2 and 3, the four tests considered are sensitive to departures from normality. On the other hand, if we observe these values of table 4, we can conclude that the tests considered are rather sensitive to the highly skewed and extremely leptokurtic distributions too. Among these tests, the informational energy test appear to have slightly more power than two of the other three although the third one have a type I error rate greater than the informational energy test.

If we look at the results in table 5, pp. 357, of Conover et al [3], then we obtain that for the columns 1, 2, 3 and 4 there are only 10, 8, 2 and 14 power values (in parentheses) respectively greater than the corresponding power values of the informational energy statistic. So we conclude that the informational energy gives a "good" test, among the 56 tests statistics considered, when the normality assumptions hold.

Table 2 **Normal Distribution**

(n_1, n_2, n_3, n_4)	(5,5,5,5)	(10,10,10,10)	(20,20,20,20)	(5,5,20,20)
T_c	.056 (.369)	.051 (.623)	.053 (.812)	.058 (.662)
Neyman-Pearson	.123 (.460)	.073 (.653)	.064 (.816)	.103 (.761)
Barlett	.041 (.303)	.044 (.591)	.049 (.796)	.044 (.646)
Hartley	.042 (.234)	.044 (.546)	.051 (.781)	.247 (.669)

Table 3 **Double exponential distribution**

(n_1, n_2, n_3, n_4)	(5,5,5,5)	(10,10,10,10)	(20,20,20,20)	(5,5,20,20)
T_c	.219 (.462)	.280 (.672)	.318 (.830)	.268 (.713)
Neyman-Pearson	.330 (.564)	.333 (.707)	.341 (.839)	.344 (.801)
Barlett	.179 (.410)	.259 (.653)	.309 (.824)	.237 (.705)
Hartley	.157 (.355)	.237 (.625)	.288 (.811)	.462 (.828)

Table 4

	$(\text{Normal})^2$ distribution		$(\text{Double exp.})^2$ distribution	
(n_1, n_2, n_3, n_4)	(10,10,10,10)	(5,5,20,20)	(10,10,10,10)	(5,5,20,20)
T_c	.637 (.786)	.615 (.812)	.865 (.893)	.850 (.892)
Neyman-Pearson	.679 (.815)	.697 (.877)	.887 (.916)	.892 (.938)
Barlett	.621 (.777)	.595 (.813)	.858 (.896)	.848 (.908)
Hartley	.684 (.764)	.872 (.951)	.849 (.879)	.926 (.971)

References

[1] Bartlett, M.S.: "Properties of Sufficiency and Statistical Tests". *Proc. Roy. Soc.*, **Ser. A, 160**, 268-282 (1937).

[2] Bickel, P.J. and Doksum, K.A.: *"Mathematical statistics: Basic ideas and selected topics"*. Holden-Day (1977).

[3] Conover, W.J., Johnson, M.E. and Johnson, M.M.: "A Comparative Study of Tests for Homogeneity of Variances, with Applications to the Outer Continental Shelf Bidding Data". *Technometrics*, **23, 4**, 351-361 (1981).

[4] Hartley, H.O.: "The Maximum F-Ratio as a Short-Cut Test for Heterogeneity of Variance". *Biometrika*, **37**, 308-312 (1950).

[5] Morales D., Menéndez M.L., Salicrú M. and Pardo L.: "Asymptotic properties of divergence statistics in a stratified random sampling and its applications to test statistical hypotheses". *the Journal of Statistical Planning and Inference*, **38**, 201-222 (1994).

[6] Neyman, J. and Pearson, E.S.: "On the Problem of k Samples". *Bull. Acad. Polon. Sci. et Lettres*, **Ser. A**, 460-481 (1931).

[7] Salicrú M., Pardo L., Morales D. and Menéndez M.L.: "Asymptotic distributions of (h, ϕ)-entropies". *Communications in Statistics (theory and methods)*. **22**, **7**, 2015-2031 (1993).

Compositive Information Measure of a Fuzzy Set

Carlo Bertoluzza[a] - Teofilo Brezmes[b] - Gloria Naval[b]

[a]Dip.Informatica e Sistemistica - Universitá di Pavia
Corso Strada Nuova 65 - 27100 Pavia - ITALY

[b]Departamento de Matemáticas - Universidad de Oviedo
calle Calvo Sotelo s/n - 33071 Oviedo - Spain

Abstract In this paper the infomation associated to an event $\widetilde{A}$ (crisp or fuzzy) of a space $(\Omega, \mathcal{S})$ measures the degree of surprise which arises when this event occurs. J.Kampé de Fèriet and B.Forte gave in 1967 ([5]) an axiomatic definition of this concept, in which the information measure J is associated directly to the crisp event A of a measurable space $(\Omega, \mathcal{S})$. We will propose here a possible way to measure the information in the fuzzy context. We proceed by supposing that an axiomatic information measure is defined on the crisp events (crisp subsets): the information of a fuzzy subset will be constructed by means of the informations of their α-cuts.

1. Preliminaries

Fuzzy Sets. Let $L = \{0 = \alpha_0 < \alpha_1 < \ldots < \alpha_n = 1\}$ a finite subset of the interval $[0,1]$ such that $0 \in L$ and $\alpha \in L \Longrightarrow 1 - \alpha \in L$. An *L-fuzzy subset* of a given universe Ω is a map $\widetilde{A}$ from Ω to L. The family of the fuzzy subsets of Ω is equipped with union and intersection operations and a partial ordering (inclusion) $\sqsubseteq$ defined by

$$(\widetilde{A} \cup \widetilde{B})(\omega) = \sup\{\widetilde{A}(\omega), \widetilde{B}(\omega)\} \quad (1.1.a)$$

$$(\widetilde{A} \cap \widetilde{B})(\omega) = \inf\{\widetilde{A}(\omega), \widetilde{B}(\omega)\} \quad (1.1.b)$$

$$\widetilde{A} \sqsubseteq \widetilde{B} \iff \widetilde{A}(\omega) \leq \widetilde{B}(\omega) \quad (1.1.d)$$

$$\forall \omega \in \Omega$$

For each value $\alpha_i \in L$ the $\alpha_i - cut$ of $\widetilde{A}$ is the crisp set A_i defined by

$$A_i = \{\omega \in \Omega | \widetilde{A}(\omega) \geq \alpha_i\} \quad (1.2)$$

Information measure Let $(\mathcal{T}, \preceq)$ a lattice with minimum (m) and maximum (M). An *information measure* over $(\mathcal{T}, \preceq)$ is a map

$$J^* : \mathcal{T} \longrightarrow \mathbb{R}^+$$

with the following properties

$$J^*(M) = 0 \quad (1.4)$$

$$J^*(m) = +\infty \quad (1.5)$$

$$x \preceq y \Longrightarrow J^*(x) \geq J^*(y) \quad (1.6)$$

$$\forall (x,y) \in \mathcal{T} \times \mathcal{T}$$

The lattice underlying the Forte-Kampé de Fériet measure (J) is an algebra of crisp subsets of an universe Ω, whereas the fuzzy information $\widetilde{J}$ which we try to introduce is defined over an algebra $\widetilde{\mathcal{S}}$ of fuzzy subsets. The maximum and tha minimum are Ω and $\emptyset$ in both the cases.

The family of the α-cutcompletely determines the fuzzy set $\widetilde{A}$: then any information measure associated to $\widetilde{A}$ may be viewed as depending on the family $\{A_1 \dots A_n\}$. In particular we suppose that a set theoretic information J is defined over the crisp space (Ω, S) and that the information $\widetilde{J}(\widetilde{A})$ depends on the crisp informations $J_i = J(A_i)$ of their α-cut:

$$\widetilde{J}(\widetilde{A}) = G_n(J_1 \dots J_n) \tag{1.9}$$

2. Basic Properties

We can easily recognize that the function G_n is defined on a suitable subset of $\Delta_n = \{(x_1 \dots x_n) \mid 0 \leq x_1 \leq x_2 \leq \dots \leq x_n\}$, and that it satisfies the following properties

$$G_n(0 \dots 0) = 0 \tag{2.3}$$

$$G_n(+\infty \dots +\infty) = +\infty \tag{2.4}$$

$$x_i \leq y_i \quad i = 1 \dots n \implies G_n(x_1 \dots x_n) \leq G_n(y_1 \dots y_n) \tag{2.5}$$

We reserve a particular attention to the two classes of fuzzy subsets whose membership functionassumes only one value α_k or only two values α_r, α_S (with s¿r) different from zero.Let us consider the two following functions

$$g_k(u) = G_n(\underbrace{u \dots u}_{k\ times}, +\infty \dots +\infty) \tag{2.6}$$

$$g_{rs}(x,y) = G(\underbrace{x \dots x}_{r\ times}, \overbrace{y \dots y}^{t\ times} + \infty \dots + \infty) \tag{2.7}$$

where $t = s - r$. If we pose $u = J(A)$, $x = J(A' \cup A'')$, $y = J(A'')$, then$g_k(u)$ and $g_{rs}(x,y)$ represent respectively the informations of the fuzzy subsets $\widetilde{B} = \alpha_k A$ and $\widetilde{C} = \alpha_r A' \cup \alpha_s A''$, whose membership functions assume respectively only one (α_k0 or only two (α_r, α_s) values different from zero. The functions g_k and g_{rs} have the following properties

$$g_n(0) = 0 \tag{2.8}$$

$$g_k(+\infty) = +\infty \tag{2.9}$$

$$x' < x'' \implies g_k(x') \leq g_k(x'') \tag{2.10}$$

$$k < m \implies g_k(x) \geq g_m(x) \tag{2.11}$$

$$g_{rn}(0,0) = 0 \tag{2.12}$$

$$g_{rs}(+\infty, +\infty) = +\infty \tag{2.13}$$

$$x' \leq x'', y' \leq y'' \implies g_{rs}(x', y') \leq g_{rs}(x'', y'') \tag{2.14}$$

$$g_{rs}(x,x) = g_s(x) \tag{2.15}$$

$$g_{rs}(x, +\infty) = g_r(x) \tag{2.16}$$

$$g_{rs}(x,x) \leq g_{rs}(x,y) \leq g_{rs}(y,y) \tag{2.17}$$

The study of these functions will result the key point in order to determine the function G_n, and consequently the fuzzy information $\widetilde{J}$.

3. Compositive Measures

Definition 3.1. A set theoretic information measure is said to be *compositive*, if there exists a function of two variables

$$F \,:\, \mathbb{R}^+ \times \mathbb{R}^+ \longrightarrow \mathbb{R}^+ \tag{3.1}$$

such that

$$A \cap B = \emptyset \implies J(A \cup B) = F[J(A), J(B)] \tag{3.2}$$

P.Benvenuti, B.Forte and J.Kampé de Fériet gave a complete characterization of the compositive measures. They proved that the general form of F is, roughly speaking, a suitable combination of the two extremal forms

$$F_i(x,y) = \inf(x,y) \tag{3.3}$$

$$F_S(x,y) = f^{-1}[f(x)+f(y)] \tag{3.4}$$

where $f(x)$ is any continuous and strictly decreasing function with $f(+\infty) = 0$. The compositive measure are the best known and the most utilized in the field of the set theoretic information theory; therefore it seems quite natural textend this notion to the fuzzy ones.

Definition 3.2. Formally we say that a fuzzy information is *compositive* if there exists a function

$$\widetilde{F} \,:\, \mathbb{R}^+ \times \mathbb{R}^+ \longrightarrow \mathbb{R}^+ \tag{3.5}$$

such that

$$\widetilde{A} \cap \widetilde{B} = \emptyset \implies \widetilde{J}(\widetilde{A} \cup \widetilde{B}) = \widetilde{F}[\widetilde{J}(\widetilde{A}), \widetilde{J}(\widetilde{B})] \tag{3.6}$$

It is easy to recognize that the function $\widetilde{F}$ presents the same characteristics of the function F of definition 4.1, and therefore its general form coincides with the one given in [7]. Moreover, if we express the fuzzy information $\widetilde{J}$ in terms of the information functions G , then it is easy to recognize that the following compatibility relations must hold

$$\widetilde{F}[G(x_1 \dots x_n), G(y_1 \dots y_n)] = G[F(x_1, y_1) \dots F(x_n, y_n)] \tag{3.7}$$

$$\widetilde{F}[g_k(x_k), g_k(y_k)] = g_k[F(x_k, y_k)] \tag{3.8}$$

$$\widetilde{F}[g_{rs}(x_r, x_s), g_{rs}(y_r, y_s)] = g_{rs}[F(x_r, y_r), F(x_s, y_s)] \tag{3.9}$$

The next two sections are devoted to the study of the two classes of compositive fuzzy information which correspond to the two extremal cases specified by the expression (3.3) and (3.4). In particular we suppose that a composition law F (of the form (3.3) and (3.4) respectively) is defined over the crisp information space $(\Omega, \mathcal{S}, J)$ and also that the fuzzy information space is compositive. Starting from these hipotesis, we will determine the information measure $\widetilde{J}(\widetilde{A})$. We use the one level subsets in order to determine the form of the fuzzy composition law $\widetilde{F}$. The two levels subsets are used in the case of the type-inf composition law, in order to build up a recursive process which allows us to construct the function G_n. In the case of a type-M composition law the function G_n is determined directly by solving a suitable functional equation.

4. Type-inf composition law

Suppose that the crisp information space $(\Omega, \mathcal{S}, J)$ is compositive with composition law given by

$$F(x,y) = \inf(x,y) \tag{4.1}$$

It is easy to recognize, using (3.8) that the function $\widetilde{F}$ is of the same type, that is

$$\widetilde{F}(u,v) = \inf(u,v) \tag{4.2}$$

for all u, v in the domain $\Gamma = [g_k(0), +\infty)$ .Then the compatibility equations (3.7) and (3.9) assume the form

$$\inf[G(x_1 \dots x_n), G(y_1 \dots y_n)] = G[\inf(x_1, y_1) \dots \inf(x_n, y_n)] \tag{4.3}$$
$$\inf[g_{rn}(x_r, x_n), g_{rn}(y_r, y_n)] = g_{rn}[\inf(x_r, y_r), \inf(x_n, y_n)] \tag{4.4}$$

We start by determining the form of the function g_{rs} , that is the function which determines the information of the two-level subsets.

norTheorem 4.1 Let $\tau(y)$ a (i) not decreasing, (ii) lower semicontinuous function with (iii) $\tau(0) = 0$, (iiii) $\tau(t) \leq t$ and let $g(t)$ a (e) continuous (ee) not decreasing function , and let us pose $\overline{\tau}(x) = \sup\{y \mid \tau(y) < x\}$. Then any function $g_{rs}(x,y)$ defined by

$$g_{rs}(x,y) = \begin{cases} g[\overline{\tau}(x)] & \text{if } x \leq \tau(y) \\ g(y) & \text{if } x > \tau(y) \end{cases} \tag{4.5}$$

is a type-inf information function, and on the contrary any type-inf information function has the form (4.5), with a suitable choice of the functions τ, g .

It is easy to verify that a function of type (4.5) satisfy the compatibility equation (4.4), as well as the characteristic properties (2.12)—(2.17) of the information functions g_{rs} . So the "if" part of the theorem is proved.

To prove the "only if" part let us suppose that an information function g_{rs} is given satisfying the properties (2.12)—(2.17) and the compatibility equation (4.4). The functions g and τ defined by

$$g(y) = g_{rs}(y,y) \tag{4.6.1}$$
$$\tau(y) = \inf\{t \mid g_{rs}(t,y) = g(y)\} \tag{4.6.2}$$

fulfill the properties (e)—(eee) and (i)—(iiii) and the compatibility equation (4.4) forces g_{rn} to assume the form (6.1). The details of the proof may be found in [9], pp. 14-17.

The two extremal cases correspond to the two extremal choices of the function $\tau(t)$, namely $\tau(t)=t$ and $\tau(t)=0$, which correspond to

$$g_{rs}(x,y) \;=\; g(x) \;= g[\inf(x,y)] \tag{4.7}$$

$$\widetilde{J}(\widetilde{A}) \;=\; \widetilde{J}(\{A_1,A_2\}) \;= g[\widetilde{J}(A_1)] \tag{4.8}$$

$$g_{rs}(x,y) \;=\; g(y) \;= g[\sup(x,y)] \tag{4.9}$$

$$\widetilde{J}(\widetilde{A}) \;=\; \widetilde{J}(\{A_1,A_2\}) \;= g[\widetilde{J}(A_2)] \tag{4.10}$$

In particular, if we choose $g(t)=t$, which corresponds to impose the condition $\widetilde{J}(A)=J(A)$, then we have

$$\widetilde{J}(\widetilde{A}) \;=\; \inf\{J(A'),J(A'')\} \;=\; \widetilde{J}(A') \tag{4.11}$$

$$\widetilde{J}(\widetilde{A}) \;=\; \sup\{J(A'),J(A'')\} \;=\; \widetilde{J}(A'') \tag{4.12}$$

We can start from the result just obtained to deduce the form of the information function G in the case where we have an arbitrary number n of different α–cuts. We describe in particular the case where $|L|=3$, that is the case where the membership function may assume only four distinct values $\alpha_0=0<\alpha_1<\alpha_2<\alpha_3=1$.

Let $G(x,y,z)$ $(x\le y\le z)$ the information function, and let us consider its restriction to the set of the points of type (y,y,z) . It is easy to prove that the function $\gamma(y,z)=G(y,y,z)$ have all the properties of the function g_{rn} which we analized previously. Then we can establish the following

Lemma 1. For each $z\in \mathbb{R}^+$, let $\tau(z)=\inf\{y|G(y,y,z)=G(z,z,z)\}$. Then, by posing $G(z,z,z)=g(z)$, we have

$$G(y,y,z) \;=\; \begin{cases} g[\overline{\tau}(y)] & \text{if } y\le\tau(z) \\ g(z) & \text{if } y>\tau(z)\end{cases} \tag{4.13}$$

We can also recognize that the function $\gamma_z(x,y)=G(x,y,z)$ shows all the properties of the functions g_{rs} , and therefore the following lemma holds:

Lemma 2. Let $\tau_z(y)=\inf\{x|G(x,y,z)=G(y,y,z)\}$. Then we have

$$G(x,y,z) \;=\; \begin{cases} g[\overline{\tau}_z(x)] & \text{if } x\le\tau_z(y) \\ G(y,y,z) & \text{if } x>\tau_z(y)\end{cases} \tag{4.14}$$

where $\overline{\tau}_z$ is defined like to $\overline{\tau}$ in theorem 4.1

We can observe that the function τ_z and τ are strongly related. In fact we have

$$\tau_z(y) \;=\; \inf\{x \mid G(x,y,z)=\begin{cases} g(y) & \text{if } y\le\tau(z) \\ g(z) & \text{if } y>\tau(z)\end{cases} \tag{4.15}$$

Combining (4.13),(4.14) and (4.15) we obtain the general form of the information function G , by means of the following procedure:

(a) fix three functions $\tau(z)$, $\tau_z(y)$ and $g(t)$,

(b) perform $G(y,y,z)$ by means of (4.13),

(c) verify the compatibility between τ_z , τ and g using (4.15),
(d) construct $G(x,y,z)$ according to (4.14).

The general case. The method used in the case $|L| = 3$ is also allowable in the general case. The form of G may be obtained by means of a recurrent procedure of the type described in the previous case. We determine firstly $G(x_{n-1} \dots x_{n-1}, x_n)$, then we determine $G(x_{n-2} \dots x_{n-2}, x_{n-1}, x_n)$, and so on, till to reach $G(x_1 \dots x_n)$. At each step i we have to introduce a function τ_i , depending on the parameters $x_n \dots x_{n-i+1}$ which plays the role of the functions τ and τ_z in the case $n = 3$. The family τ_i , together with the function $g(t) = G(t \dots t)$, determines completely the information function G .

5. Type-M composition law

Suppose that the crisp information space (Ω, S, J) has a type-M composition law with additive generator f , that is

$$F(x,y) = f^{-1}[f(x) + f(y)] \tag{5.1}$$

Then the fuzzy composition law $\tilde{F}$ is also of type-M (possibly with a different generator $\tilde{f}$), that is

$$\tilde{F}(x,y) = \tilde{f}^{-1}[\tilde{f}(x) + \tilde{f}(y)] \tag{5.2}$$

We proved this result (see [9], p.23) by recognizing that the set of the elements which are idempotent for the law $\tilde{F}$, that is the set $\tilde{\Delta} = \{\alpha \in \mathbb{R}^+ \mid \tilde{F}(\alpha,\alpha) = \alpha\}$ contains only the values 0 and $+\infty$. The compatibility equation becomes in this case

$$\begin{aligned} \tilde{f}^{-1}[\tilde{f}(G(x_1 \dots x_n)) + \tilde{f}(G(y_1 \dots y_n)) &= \\ = G[f^{-1}(f(x_1 + f(y_1)) \dots f^{-1}(f(x_n) + f(y_n))] \end{aligned} \tag{5.3}$$

which may be written in the form

$$\theta(\xi_1 \dots \xi_n) + \theta(\eta_1 \dots \eta_n) = \theta(\xi_1 + \eta_1 \dots \xi_n + \eta_n) \tag{5.4}$$

where $\xi_i = f(x_i), \eta_i = f(y)i)$, $\theta(t_1 \dots t_n) = \tilde{f} \circ G[f(t_1) \dots f(t_n)]$.
Using a well known rwesult of functional equation, we recognize that the unique monotonic solution of this equation is ths function

$$G(x_1 \dots x_n) = \tilde{f}^{-1}[\sum_{i=1}^{n} a_i f(x_i)] \tag{5.5}$$

with a_i are non-negative fixed values.

We can use, in particular, (5.5) to deduce the form of the fuzzy information of the crisp subsets, thus obtaining

$$\tilde{J}(A) = \tilde{f}^{-1}[a \cdot f(J(A))] \tag{5.6}$$

where $a = \sum a_i$. We deduce immediatly, from the above relation, that the restriction of the fuzzy information to the class S of the crisp subsets reduces to the crisp infomation iff $\tilde{f}(t) = a \cdot f(t)$:

$$\tilde{J}(A) = J(A), \ \forall A \in S \iff \tilde{f}(t) = a \cdot f(t) \tag{5.7}$$

References

[1] J.Kampé de Fériet: Note di teoria dell'informazione. Quaderni dei Gruppi di ricerca del CNR, Ist.Mat.Appl. Roma 1972.

[2] N.Wiener: Cybernetics. Herrmann, Paris 1948.

[3] C.F.Picard: Aspects informatiques de l'information hyperbolique. Symposia Mathematica, vol 10, pp. 55-82 1975.

[4] B.Forte, J.Kampé de Fèriet: Information et Probabilité. C.R.A.S Paris, 265A, pp.110 and 142 1967.

[5] L.A.Zadeh: Fuzzy sets. Information and Control, 8 pp 338-353.

[6] G.Comyn and J.Losfeld: Définition d'une information composable sur un treillis. C.R.A.S. Paris, t.278 (25 février 1974), Série A, pp.633- 636.

[7] P.Benvenuti, B.Forte and J.Kampé de Fériet: Forme générale de l'operation de composition .... C.R.A.S Paris, 265 Serie A, 1967.

[8] J.Aczel:Lectures on functional equations and thei characterizations. A.P., New York, 1966, pg.348.

[9] C.Bertoluzza, T.Brezmes, M.L.Capodieci and G.Naval: Compositive information measures of a fuzzy set: two equivalent approaches. Manuscript.

Informational Time

Robert Vallée
Wosc, 2, rue de Vouillé
F-75015 Paris, France

Abstract. We propose an "informational time" concerning some stochastic dynamical systems where there is an uncertainty on the state of the system or the space localization of a phenomenon. In each case the informational time is defined by the increase of a Shannon's informational entropy, formulated either exactly or asymptotically or in the case of Laplace-Gauss probability densities. After giving some remarks on Shannon's informational entropy, three examples are considered : a linear differential system with uncertain initial state, a "stochastic system with diffusion", and a quantum system. The informational times proposed are expressed, in terms of classical time t, by the integral of a positive function of t, the logarithm of the square root of t and the logarithm of t.

1 About Shannon's Informational Entropy

We shall use Shannon's informational entropy in the continuous case, that is to say for a probability density r, or more explicitely r(x), defined on R^n. It is given by the integral extended to R^n

$$H_r = - \int r(x) \log r(x)\, dx$$

where dx is the differential volume element of R^n and log the natural logarithm for the sake of simplicity. This informational entropy is also the opposite of Neumann-Wiener information. It is well known that this entropy has the inconvenient not to be invariant in a general change of coordinates. To avoid this difficulty a generalized informational entropy $H_{r,m}$ has been proposed

$$H_{r,m} = - \int r(x) \log(r(x)/m(x))\, dx$$

Where m is a referential probability density which makes $H_{r,m}$ invariant in every change of coordinates. If we choose for m a centered LaplaceGauss function defined on R^n by

$$L_g(x) = 1 / (g\sqrt{2\pi})^n \exp(-(N(x))^2/2g^2)\ ,$$

N(x) being the euclidian norm of x, we have

$$H_{r,m} = H_r + \int r(x) \log L_g(x)\, dx = H_r - n \log(g\sqrt{2\pi}) - M_{2,r}/2g^2$$

Where $M_{2,r}$ is the moment of order 2 of r relative to the origin of coordi nates. So if we consider the variation of the generalized information en tropy corresponding to the change from r_0 to r, we have

$$H_{r,m} - H_{r_0,m} = H_r - H_{r_0} + (M_{2,r_0} - M_{2,r})/2g^2.$$

But since we can choose g very large, it is possible to write, with as go od an approximation we like,

$$H_{r,m} - H_{r_0,m} = H_r - H_{r_0},$$

an equality which becomes perfectly rigorous if we accept to have for m what we call the "epsilon distribution" [9] . In the particular case of La place-Gauss probability density, defined on R, with standard deviation s, we have classically

$$H_r = \log(s\sqrt{2\pi e}) = \log s + \log\sqrt{2\pi e}$$

and so, if m is the epsilon distribution, or ε ,

$$H_{r,\varepsilon} - H_{r_0,\varepsilon} = H_r - H_{r_0} = \log(s/s_0).$$

If we consider now a dynamical system involving a probability density, as we shall see in the coming examples, the variation of the associated inf ormational entropy from its value at initial instant defines a time intr insic to the system. This time may be said to be informational because it varies as the Shannonian uncertainty linked to the evolution of the sy stem.

2 Linear Differential Systems with Uncertain Initial Conditions

We consider a linear differential system

$$dx(t)/dt = A(t)\, x(t) + B(t)\, u(t)$$

where $x(t)$ belongs or R^n, $A(t)$ being a real square matrix of order n. We suppose that the only information we have about the initial state $x(t_0)$ is the probability density r_{t_0}, or more precisely $r_{t_0}(x)$, concerning its localization. We have shown (1979) that the Shannon's informational en tropy concerning the probability density r_t, or $r_t(x)$, of state $x(t)$ is independant of function u and given by [5,6,7]

$$H_{r_t} = H_{r_{t_0}} + \int_{t_0}^{t} \mathrm{tr}\, A(v)\, dv,$$

a formulation proposed again later (1991) by another author [2] .

We make now the hypothesis that the trace, tr $A(t)$, of matrix $A(t)$, which is equal to the sum of its eigen-elements, is positive or null without being equal to zero on an interval of non zero length. We can tell that we have a "globally exploding system" : we may have eigen-values with negative real parts (implosion) and eigen-values with positive real parts (explosion) but, in a way, the last ones overcome the first. Consequently the integral , in the second member of the expression giving H_{r_t} is strictly increasing with t.

Then it is possible to propose an "informational time" $T(t)$ associated to classical time t and given by

$$T(t) = \int_{t_0}^{t} \operatorname{tr} A(v)\, dv = H_{r_t} - H_{r_{t_0}} = H_{r_t,\varepsilon} - H_{r_{t_0},\varepsilon} \; .$$

This intrinsic time $T(t)$ increases according to the variation of the information entropy concerning the probability density of the Shannon uncertainty about the localization of the state. In the special case where $A(t)$ reduces to a constant matrix A, with strictly positive trace, we have more simply

$$T(t) = (t-t_0) \operatorname{tr} A$$

and informational time $T(t)$ is proportional to classical time elapsed since initial instant t_0.

3 Stochastic System with Diffusion

We shall call stochastic system with diffusion a system whose state at t is only known by a probability density function r_t, or more explicitly $r_t(x)$ or even $r(x,t)$, where x belongs to R^n and satisfies an evolution equation of the diffusion type. For the sake of simplification we shall consider only the case of dimension one. The diffusion equation is

$$\partial r(x,t)/\partial t - \partial^2 r(x,t)/\partial x^2 = o.$$

We know that the solution of this equation is given by

$$r(x,t) = \int_{-\infty}^{+\infty} (1/2\sqrt{\pi t}\;\; \exp(-(x-y)^2/4t)\; r(y,o)\, dy,$$

that is to say the convolution $\circledast$ of density $r(x,o)$, at instant o, by a Laplace-Gauss kernel. So we can write

$$r_t = r_o \circledast L_{\sqrt{2t}} \, ,$$

with, for the Laplace-Gauss function, a notation already used

We suppose now that r_o is a centered Laplace-Gauss function L_{s_o}, that is to say with standard deviation s_o, so we have

$$r_t = L_{s_o} * L_{\sqrt{2t}} = L_{\sqrt{s_o^2+2t}},$$

because the convolution of two Laplace-Gauss functions is a Laplace-Gauss function whose standard deviation is the square root of the sum of squares of the standard deviations of the two functions. Knowing that as seen before the informational entropy of a Laplace-Gauss density with standard deviation s is given by

$$\log s + \log \sqrt{2\pi e},$$

we have, with t positive or equal to zero, a possible informational time

$$H_{r_t,\varepsilon} - H_{r_o,\varepsilon} = H_{r_t} - H_{r_o} = \log \sqrt{1+ 2t/s_o^2} = T(t)$$

or, asymptotically for great values of t,

$$T(t) = \log (\sqrt{t}/s_o) .$$

4 Quantum System

We consider a system whose state, at instant t, is defined by a complex valued function $G(x,t)$, where x belongs to R^3, and which satisfies Schrödinger's equation for a free particle. In the case of dimension one this equation, with suitable units, is

$$\partial G(x,t)/\partial t - i\partial^2 G(x,t)/\partial x^2 = 0.$$

The classical interpretation of $G(x,t)$, as a function of x, which we shall represent by G_t, is that $|G_t|^2$ is the probability density about the space localization, at instant t, of the quantum particle, as it can be revealed by a measurement process. The Fourier transform respective to x, with well chosen factors, of $G(x,t)$ is a function $F(p,t)$, or F_t if we consider it as a function of momentum p at instant t. Then classically $|F_t|^2$ is interpreted as the probability density about the localization of the particle in momentum space, at instant t. If $|G_o|^2$ is a Laplace-Gauss function, with standard deviation s_o, it is well known that $|F_o|^2$ is also a Laplace-Gauss function. If moreover the product of the standard deviations of the two probability densities has its sma llest value,according to Heisenberg's inequality, the initial wave pack et is said to be minimal. Then $|G_t|^2$ and $|F_t|^2$ are Laplace-Gauss functi ons at any ulterior instant, the wave packet being no longer minimal.Ac cording to a known result [3] the standard deviation of the Laplace-Gau ss probability density $|G_t|^2$ is

$$\sqrt{s_o^2 + t^2/s_o^2} \, .$$

Then, if we represent $|G_t|^2$ by r_t, the informational entropy of this pr obability density is given by

$$H_{r_t} = \log \sqrt{s_o^2 + t^2/s_o^2} + \log \sqrt{2\pi e}$$

and so

$$H_{r_t,\varepsilon} - H_{r_o,\varepsilon} = H_{r_t} - H_{r_o} \quad \log \sqrt{1+t^2/s_o^4}$$

or, asymptotically for great values of t, $\log (t/s_o^2)$.So we can propose on the basis of this particular case an informational time

$$T(t) = \log \sqrt{1+t^2/s_o^4} \, ,$$

or for great values of t

$$T(t) = \log (t/s_o^2).$$

We have studied [4,8] the general case where space has dimension 3 and the wave packet is not initially minimal.Considering $G(x,t)$ given by a convolution implying $G(x,o)$ and $F(p,t)$ expressed as the Fourier transfo rm of $G(x,t)$, it is possible to use the so called method of the statio nary phase [1] and obtain, asymptotically for great values of t , the same formulation as above and so propose the same informational time.

References

1. A.Erdélyi: Asymptotic Expansion.New York:Dover 1956

2. G.Jumarie: Extension of quantum information by using entropy of de terministic functions.Mathematical and Computer Modelling, 15 , 103-116 (1991)

3. A.Messiah: Mécanique Quantique I.Paris:Dunod 1959

4. R.Vallée:Expression asymptotique,pour les grandes valeurs du temps de l'information associée à la fonction d'ondes dans le cas d'un corpuscule libre.Comptes rendus de l'Académie des Sciences, B , 267, 529-532 (1968)

5. R.Vallée: Aspect informationnel du problème de la prévision dans le cas d'une observation initiale imparfaite.Economie Appliquée,32,2.3 221-227 (1979)

6. R.Vallée:Evolution of a dynamical linear system with random initial conditions.In:R.Trappl (ed.): Cybernetics and Systems Research. Amsterdam:North Holland Publishing Company 1982,pp.163-164

7. R.Vallée: Information entropy and state observation of a dynamical system.In: B.Bouchon,R.R. Yager (eds.): Uncertainty in Knowledge Based Systems.Berlin:Springer-Verlag 1987,pp.403-405

8. R.Vallée: Information and Schrödinger's equation.SCIMA Systems and Cybernetics in Management,20,3,71-75 (1991)

9. R.Vallée: The "epsilon-distribution" or the antithesis of Dirac delta. In:R.Trappl (ed.): Cybernetics and Systems Research'92 . Singapore: World Scientific 1992,pp.97-102

5. POSSIBILITY THEORY

Updating, Transition Constraints and Possibilistic Markov Chains

Didier DUBOIS – Florence DUPIN de SAINT CYR – Henri PRADE

Institut de Recherche en Informatique de Toulouse (I.R.I.T.) – C.N.R.S.
Université Paul Sabatier, 118 route de Narbonne
31062 TOULOUSE Cedex – FRANCE

Abstract. Possibility theory is applied to the updating problem in a knowledge base that describes the state of an evolving system. The system evolution is described by a possibilistic Markov chain whose agreement with the axioms of updating is examined. Then it is explained how to recover a possibilistic Markov chain from a set of transition constraints, on the basis of a specificity ordering.

1. Introduction

One of the most challenging problems in databases and knowledge-based systems is that of modifying a knowledge base under the arrival of a new piece of information. Basically, if the new information contradicts the contents of the knowledge base, there exists several ways of restoring consistency, unless strict guidelines are supplied that lead to a unique solution. An important distinction has been drawn by Katsuno and Mendelzon [12] between revising a knowledge base and updating it. In revision, the new information is meant to improve our cognitive state regarding a given situation; what was plausibly thought as being true may actually be false. In updating, the new information is meant to inform the knowledge base that something has changed in the actual world; what was thought to be true may no longer be true because things have changed.

This paper is a preliminary investigation about a new approach to updating a knowledge base describing the behaviour of an evolving system. More specifically an attempt is made to unify three points of view on this problem: the point of view of the system analyst who describes the evolution of a system via a transition graph between states, the point of view of formal philosophers who have laid bare the postulates of rational updating, and the point of view of database research, where the update is achieved at the syntactic level by means of transition constraints [4]. It is suggested that such transition constraints partially determine a transition graph between states of the system. This graph can be viewed as a possibilistic Markov chain. The same observation can be derived from Katsuno and Mendelzon's postulates, except that the obtained transition graph underlies an inertia property. Some hints about how to derive Markovian transition graphs from transition constraints which are either strong or defeasible, are suggested, in the framework of possibility theory [16] [7]. Lastly, it should be pointed out that temporal aspects are not explicitly dealt with and, without loosing generality, actions are not formalized here, only their effects are accounted for (e.g., "shoot" is replaced by "shot" and the proposition "shot" is true when the action "shoot" has been executed).

2. Evolution and Uncertainty

There are at least two kinds of evolutive systems: *dynamical* systems and so-called "*inert* systems" [14]. The difference lies more in the point of view that is chosen for studying these systems than in their nature. One can be interested in modelling the transient behaviour of the system under its own dynamics, with a view to controlling its trajectory towards a stable state. On the contrary, in inert systems, the evolution is only due to some external event, and the system is supposed to stay in the same state unless such an event occurs. One is then only interested in computing the resulting state of the system. The main difficulty is then to find the parts of the system which change (this is called the ramification problem) and those which persist (this is called the frame problem, see, e.g., [3]). The well-known "world of blocks" in robotics, where blocks are moved by some manipulator, is a typical example of inert system. A knowledge base which maintains the available information about the world, and whose contents only changes when new information comes in about the world, is another example of inert system. This paper deals with some aspects of updating knowledge about inert systems.

Uncertainty pervades this kind of problem. First the current state of a system may be ill-known. This means that the knowledge base describing its state may be incomplete. Sometimes, probabilistic knowledge about the state of the system is available. But generally, one may only have some idea about what is/are the most plausible state(s) of the system, among possible ones. This type of incomplete knowledge can be described in terms of possibility distributions. Uncertainty may also affect the transitions between states, i.e., the mapping that computes the following state may be one-to-many. This form of uncertainty expresses a form of non-determinism in the behaviour of the system and can be represented by a Markov chain where the set $\Gamma(\omega)$ of potential successors of the state ω is equipped with a probability distribution, $P(\omega'|\omega)$ being the probability of reaching state $\omega' \in \Gamma(\omega)$ when the system is in state ω. Updating means computing the current state of the system, or the most plausible current state taking into account the knowledge on the previous state and receiving a piece of information about the current state.

Formally, let T be a set of time points, at each time point the real system is in a state. Let S be a set of symbols which can be associated to properties of the system. A fluent is a property of the system considered at a given time point. It is of the form p_t where p is a property of S and t a time point of T, and can be true or false. Let F be the set of fluents. The state ω of the system at time t is described by the set of fluents that are true at t. From a set of fluents of F indexed by the same time point t, sentences can be built using classical connectives. Let K_{to} describes our knowledge of the system at a certain time point $t_o \in T$, K_{to} contains three kinds of information:

- a set of sentences at time t_o, which describes *the state of the system* as it is known, maybe incompletely (K_{to} may have several models);
- a set of sentences of the form $\forall t \in T, X_t$ where X_t is a sentence at instant t, which are *static rules* that restrict the possible states of the world to a subset of interpretations of the language. They are true at any instant and called *domain constraints;*
- a set of sentences of the form $\forall t \in T, p_t \mapsto q_{t+1}$, where p_t and q_{t+1} are sentences, and $\mapsto$ is a non-classical arrow, which are *dynamic rules* that characterize the way the world can change. A rule $p_t \mapsto q_{t+1}$ means that if presently $K_t \vdash p_t$ then the world evolves in such a way that $K_{t+1} \vdash q_{t+1}$ eventually, where K_{t+1} is what is known of the resulting state. They are called *transition constraints.*

Let α_{to+1} be a sentence representing a new piece of information, and called *input*: it can be viewed as a constraint that the result of a change must satisfy.

Some transition constraints are called hard if they must always be respected. Other transition constraints are not hard in the sense that they can be defeated by harder transition constraints that contradict them. These constraints will be called defeasible. The strength of a transition constraint can be modelled by means of a priority ordering, as suggested by Cordier and Siegel [6]. Such priorities may have different meanings. They may express degrees of persistence through time (e.g., "dead" is more persistent than "asleep"); they may also reflect a specificity ordering among transition constraints.

In order to compute the resulting state of an update α_{t_0+1}, transition constraints are applied to the initial state and a resulting knowledge base K_{t_0+1} is obtained such that $K_{t_0+1} \vdash \alpha_{t_0+1}$ holds.

3. Possibilistic Markov Chains

The natural approach in systems analysis consists in enumerating all possible states of the system under study, and then all possible direct transitions between states, with a possible quantification of the likelihood of these transitions. Having built the state transition graph, it is possible to simulate the system behaviour. The transition graph is said to be Markovian if the next state only depends on the current state, not on the past earlier states. The most widely known type of such transition graphs are probabilistic Markov chains where transitions $\omega \rightarrow \omega'$ are quantified by a conditional probability $P(\omega'|\omega)$, such that $\Sigma_{\omega'} P(\omega'|\omega)=1$.

In the following we consider *possibilistic* Markov chains, whose transitions $\omega \rightarrow \omega'$ are quantified by conditional possibility distributions $\Pi(\omega'|\omega)$ such that $\max_{\omega'} \Pi(\omega'|\omega)=1$. Possibilistic Markov chains describe for each state ω the set of more or less plausible next states ω'. $\Pi(\omega'|\omega)=1$ means that nothing prevents ω' from being the state succeeding to ω. $\Pi(\omega'|\omega)=0$ means that the direct transition from ω to ω' is strictly forbidden. $\Pi(\omega'|\omega)>\Pi(\omega''|\omega)$ means that in the absence of further information, one prefers to consider ω' as more plausible successor state to ω than ω''. Compared to a probabilistic Markov chain, a possibilistic Markov chain is a much more qualitative description of a system behaviour. A possibilistic Markov chain can be viewed

i) either as a set of possibility distributions π_ω, one per state $\omega \in \Omega$, where $\Pi(\omega'|\omega)=\pi_\omega(\omega')$, $\forall \omega, \omega' \in \Omega$;

ii) or as a ternary relation expressing "closeness" of the form $\omega' <_\omega \omega''$ on Ω^3 where $\omega' <_\omega \omega'' \Leftrightarrow \Pi(\omega'|\omega)>\Pi(\omega''|\omega)$. It rank-orders all states $\omega' \in \Omega$ according to their proximity (here understood as: plausibility of being the next state) to ω, for all ω.

The meaning of such possibilistic Markov chains differs from probabilistic ones in the way simulation will take place. In the case of probabilistic Markov chains, the next state is chosen at random following the given probabilities. Simulation of a possibilistic Markov chain is performed differently. The chosen next state is the most plausible one, i.e., ω' such that $\Pi(\omega'|\omega)$ is maximal, given current knowledge about what this next state is. When the most plausible next state is not unique, the simulation becomes non-deterministic. In the absence of knowledge about the next state, all ω' such that $\Pi(\omega'|\omega)=1$ are selected. However we may know from some external source that the next state lies in A. Then the most plausible next state is $\omega' \in A$ such that $\Pi(\omega'|\omega)=\Pi_\omega(A)$, where $\Pi_\omega(A)= \sup\{\Pi(\omega''|\omega), \omega'' \in A\}$. This is how the updating problem is addressed. Note that this mode of simulation could be applied to probabilistic Markov chains (always choosing the most probable next state).

Example: The door-to-door shuttle

Assume a taxicab is running according to customer calls, and a central broadcast station communicates with it, either to receive information about what the cab is doing, or to transmit the call of a new customer. We denote by H the customer house, and D the destination of the customer.

The meanings of the literals are the following. I: idle, C: a call has been received, G_H: taxicab going to H, H: taxicab is at H, G_D: taxicab going to D. G_D, G_H, H, I are mutually exclusive, and in this example it is assumed that no new call is registered by the taxicab before taking the current customer. Moreover if a call has not been received, the taxi cannot go to the client's home nor be there, so the states $H\,\bar{C}$ and $G_H\,\bar{C}$ are impossible.

The transition graph modelling the normal behaviour of a taxi is on Fig. 1. $G_H\,C$, $G_D\,C$, $G_D\,\bar{C}$, $I\bar{C}$ are persistent states. For instance, when a taxicab is idle, without any customer, it remains so until new information comes in; if a customer calls then the state changes to IC. On the contrary IC and HC are transient states. The difference between transient and persistent states here is chosen so as to distinguish between short activities (letting a customer into the car) and long ones (a drive). A persistent state ω is such that $\Pi(\omega|\omega)=1>\Pi(\omega'|\omega)$ if $\omega\neq\omega'$. A transient state is such that $\Pi(\omega|\omega)<1$. Using this kind of modelling, update takes place upon receiving new information as follows: assume the current state is $G_D\,\bar{C}$. If a new information comes in saying the cab is now idle, then the possible current states are IC and $I\bar{C}$, and most plausibly $I\bar{C}$.

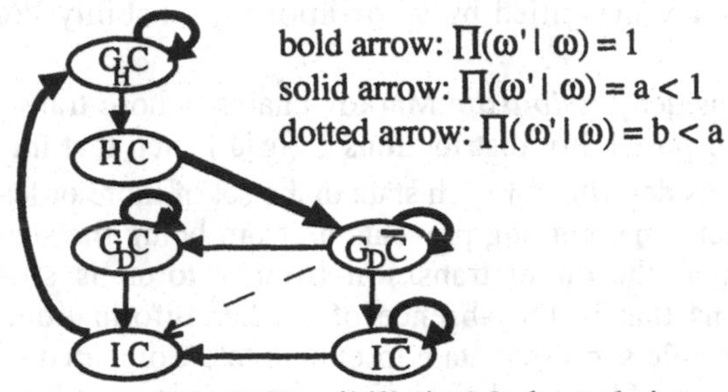

Fig. 1. Possibilistic Markov chain

4. A Possibilistic Analysis of Update Postulates

Katsuno and Mendelzon [12] have suggested 8 postulates that should be satisfied when updating a knowledge base K with a new information α. The arrival of α is supposed to be caused by an evolution in the system whose state is (incompletely) described in K: now things have changed in such a way that α is true. In the following K is a set of propositional sentences whose conjunction is ϕ; α is a propositional sentence. ϕ & α denotes the result of updating K by α. The set of postulates are as follows:

U1 $\phi\&\alpha\vdash\alpha$; U2 $\phi\vdash\alpha\Rightarrow\phi\&\alpha=\phi$

U3 if ϕ and α are satisfiable then so is $\phi\&\alpha$

U4 if $\phi=\phi'$ then $\phi\&\alpha=\phi'\&\alpha$

U5 $(\phi\&\alpha)\wedge\beta\vdash\phi\&(\alpha\wedge\beta)$

U6 if $\phi\&\alpha\vdash\beta$ and $\phi\&\beta\vdash\alpha$ then $\phi\&\alpha=\phi\&\beta$

U7 if ϕ is complete then $(\phi\&\alpha)\wedge(\phi\&\beta)\vdash\phi\&(\alpha\vee\beta)$

U8 $(\phi_1\vee\phi_2)\&\alpha=(\phi_1\&\alpha)\vee(\phi_2\&\alpha)$.

Katsuno and Mendelzon have proved that an update operation satisfies U1-U8 if and only if there exists a partial ordering that satisfies $\omega<_\omega\omega'$, $\forall\omega'\neq\omega$, such that, the set of models of $\phi\&\alpha$ is $\{\omega' \mid \omega\models\alpha, \exists\omega\models\phi, \omega'\leq_\omega\omega'', \forall\omega''\models\alpha\}$. In other words, $\phi\&\alpha$ is obtained by selecting for each model ω of ϕ the models of α closest to ω in the sense of the partial ordering $\leq_\omega$. It is clear that the set of partial orderings $\{\leq_\omega,\omega\in\Omega\}$ that underlie the postulates U1-U8 are very similar to possibilistic Markov chains on Ω (set of interpretations of the language on which K and α are built). However there are some differences: first, the orderings $\{\leq_\omega,\omega\in\Omega\}$ induced by a possibilistic Markov chain are complete, while the ordering relations in Katsuno and Mendelzon are partial. A complete preorder is derived from the postulates U1-U8 and the following additional weak converse of U5:

U9 if ϕ is complete and $(\phi\&\alpha)\wedge\beta$ is satisfiable then $\phi\&(\alpha\wedge\beta)\vdash(\phi\&\alpha)\wedge\beta$.

When ϕ is complete, ϕ corresponds to a single state ω; let $\alpha(\omega)$ be the set of most plausible next states after ω, that satisfy α. Axioms U5 and U9 say that $\alpha(\omega)\cap[\beta]=(\alpha\wedge\beta)(\omega)$, i.e., the set of most plausible next states that satisfy α, which also satisfy β should coincide with the best next states in $[\alpha\wedge\beta]$, where $[\beta]$ is the set of models of β. This is due to the fact that if one maximises a function f over a set A, then if the set $A^*{}_f\cap B$ of maxima of f over A that belong to B is non empty then it is equal to $(A\cap B)^*{}_f$.

There are other more noticeable differences between possibilistic Markov chains and the type of graph obtained by means of the postulates. U2 is an inertia postulate: if $\phi\vdash\alpha$, then the truth of α is already known and the new information causes no change. Hence as a particular case, letting $[\phi]=\{\omega\}$ and $\alpha=T$, we get $\omega<_\omega\omega'$, $\forall\omega'\neq\omega$, which in possibilistic terminology reads $\Pi(\omega|\omega)=1>\Pi(\omega'|\omega)$, $\forall\omega'\neq\omega$. In other words, all states are persistent. Note that this assumption may look debatable, e.g., in the previous example. Besides, U3 implies that $\forall\omega'\in\Omega$, $\Pi(\omega'|\omega)\neq0$. Indeed, if it is not the case, the result of updating ϕ such that $[\phi]=\{\omega\}$ by α such that $[\alpha]=\{\omega'\}$ leads to an impossible situation. Hence according to Katsuno and Mendelzon any state can be attained from any other state, so that any constraint on the next state can be satisfied via one transition only. Again this assumption may look too strong. As proved by Dubois and Prade [9], axiom U8, cast in the possibilistic setting, is nothing but Zadeh's extension principle which considers the function f that maps each ω to the next plausible states that satisfy α, and extends this function to the set of models of ϕ. More generally, we can define the models of $\phi\&\alpha$ as follows:

$$[\phi\&\alpha] = f([\phi]) \quad \text{with } f(\omega)=\mathrm{ArgMax}_{\omega'}\ \mu_{[\alpha]\cap R(\omega)}(\omega'),$$

where "ArgMax" retrieves the arguments ω' which here maximize the considered expression, μ denotes membership functions, and $R(\omega)$ is the fuzzy set of states that are successors of ω (whose membership function is π_ω) and $\mu_{[\alpha]\cap R(\omega)}=\min(\mu_{[\alpha]},\mu_{R(\omega)})$. Updating at the semantic level using the possibilistic Markov chain Π means predicting the states ω' that plausibly follows each model ω of ϕ, deleting those which fail to satisfy α, and keeping only the most plausible ones in α. For instance suppose that the current state of the taxicab is that it is either letting a customer into the car, or going to its destination, or idle, with no new customer call. Hence $[\phi]=\{HC, G_D\bar{C}, I\bar{C}\}$. Assume the new information is that it is idle. The result of the update by α such that $[\alpha]=\{IC, I\bar{C}\}$ is $\{I\bar{C}\}$. But the other state IC is reachable (from $I\bar{C}$ and from $G_D\bar{C}$) with lower plausibility.

Note that if ϕ is complete, i.e., $[\phi]=\{\omega\}$, we have $\phi\&\alpha\vdash\beta$ if and only if $\Pi_\omega(\beta\wedge\alpha)>\Pi_\omega(\neg\beta\wedge\alpha)$ where Π_ω is the possibility measure based on the possibility distribution $\pi_\omega=\Pi(\cdot|\omega)$, i.e., $\Pi_\omega(\gamma)=\sup_{\omega'\models\gamma}\pi_\omega(\omega')$. Equivalently $\phi\&\alpha\vdash\beta$ if and only if $N_\omega(\beta|\alpha)>0$ where N_ω is the conditional necessity associated with π_ω. See [8] for the semantics of preferential models expressed in terms of conditional necessity. More generally, we conclude that $\phi\&\alpha\vdash\beta$ if and only if $\forall\omega\in[\phi]$, $N_\omega(\beta|\alpha)>0$, i.e., $N(\beta|\alpha)=\inf_{\omega\models\phi}N_\omega(\beta|\alpha)>0$. $N(\beta|\alpha)$ is still a necessity measure which is based on the possibility distribution $\max_{\omega\models\phi}\pi_\omega(\cdot|\alpha)$ where

$$\pi_\omega(\omega'|\alpha)=\begin{cases}1 & \text{if } \pi_\omega(\omega')=\Pi_\omega(\alpha),\ \omega'\models\alpha\\ \pi_\omega(\omega') & \text{if } \pi_\omega(\omega')<\pi_\omega(\alpha),\omega'\models\alpha\\ 0 & \text{if } \omega'\not\models\alpha\end{cases}$$

The above results point out that the decision $\phi\&\alpha\vdash\beta$ cannot be computed by a simple aggregation of possibility distributions π_ω for $\omega\models\phi$. Each possibility distribution π_ω

must be revised and only the $\pi_{\omega}(\cdot|\alpha)$ can be combined. The need for reasoning state by state is typical of the updating problem and is embodied by axiom U8.

5. From Transition Constraints to Possibilistic Markov Chains

Where do possibilistic Markov chains come from? How can the possibilities of the transitions be computed? A way for rank-ordering the transitions between states is to take advantage of the relative specificity of the condition parts of the transition constraints. This can be done either by computing directly a possibilistic ordering between pairs of states or by computing a rank-ordering between transition constraints viewed as default rules [1] and then deducing the ordering between pairs of states.

- the first method consists in considering each transition constraint $p_t \mapsto q_{t+1}$ as a constraint specifying *feasible possibility distributions on $\Omega_t \times \Omega_{t+1}$* ($\Omega_t$ represents the set of the system states at time t). Each transition constraint $p_t \mapsto q_{t+1}$ means implicitly that when p is known at time t, the next state more plausibly satisfies q than ¬q. The chosen ordering on transitions is then encoded by the least specific possibility distribution π on $\Omega_t \times \Omega_{t+1}$ that satisfies the set of inequality constraints $\{\Pi(p_t \wedge q_{t+1}) > \Pi(p_t \wedge \neg q_{t+1})$ for $p_t \mapsto q_{t+1} \in C\}$ and which assigns the greatest possible possibility degrees to pairs of interpretations in $\Omega_t \times \Omega_{t+1}$.

- the second method consists, first, in rank-ordering the transition constraints (considered as classical default rules) in priority levels, using Pearl's algorithm [13]. Namely, the idea is to partition C into an ordered set $\{C_0, C_1, \ldots, C_k\}$ such that transition constraints in C_i are tolerated by all transition constraints in $C_{i+1} \cup \ldots \cup C_k$. A transition constraint $p_t \mapsto q_{t+1}$ is tolerated by a set $\{pi_t \mapsto qi_{t+1}, i=1 \ldots m\}$ if and only if $\{p_t \wedge q_{t+1}, \neg p1_t \vee q1_{t+1}, \ldots, \neg pm_t \vee qm_{t+1}\}$ is consistent. The level C_0 contains the less specific transition constraints and C_k the most specific ones. From this partition, transition between pairs of states can be compared :

$$\pi(\omega_1,\omega_2) \geq \pi(\omega'_1,\omega'_2) \Leftrightarrow \{p_t \rightarrow q_{t+1} | \omega_1 \models \neg p \text{ or } \omega_2 \models q\} \geq_{bo} \{p'_t \rightarrow q'_{t+1} | \omega'_1 \models \neg p' \text{ or } \omega_2' \models q'\}$$

where $\geq_{bo}$ is the "best out" ordering defined in [2]. More precisely if A and B are two sets of default rules : $A \geq_{bo} B$ iff $a(A) \geq a(B)$ where $a(A) = \min_i\{p_t \rightarrow q_{t+1} \notin A \cap C_i\}$. In other words a transition (ω_1,ω_2) is preferred to (ω'_1,ω'_2), if (ω_1,ω_2) violates transition constraints that are less specific (i.e., less important) than those violated by (ω'_1,ω'_2).

The two methods give the same rank-ordering between states. The proof has been made by Benferhat and al. [1] with usual default rules and can easily be generalized to our type of default rule.

Example: The dead battery (Goldszmidt and Pearl [11])

Let $C = \{lo_t \mapsto db_{t+1}, db_t \wedge tk_t \mapsto \neg cs_{t+1}, tk_t \mapsto cs_{t+1}\}$, where lo=lights on all night long, db=dead battery, tk=ignition key had been turned, cs=car starting. We restrict the set of states to the models of lo ∨ tk so that at least one rule applies. There are thus 12 states of interest

	lo tk	lo ¬tk	¬lo tk
cs db	ω_1	ω_5	ω_9
cs ¬db	ω_2	ω_6	ω_{10}
¬cs db	ω_3	ω_7	ω_{11}
¬cs ¬db	ω_4	ω_8	ω_{12}

The 3 transition constraints are modelled as follows: $\Pi(lo_t \wedge db_{t+1}) > \Pi(lo_t \wedge \neg db_{t+1})$, $\Pi(db_t \wedge tk_t \wedge \neg cs_{t+1}) > \Pi(db_t \wedge tk_t \wedge cs_{t+1})$, $\Pi(tk_t \wedge cs_{t+1}) > \Pi(tk_t \wedge \neg cs_{t+1})$.
Maximizing the possibility degrees, one easily finds that in $\Omega_t \times \Omega_{t+1}$

- the models (ω_t, ω_{t+1}) of $(\neg lo_t \vee db_{t+1}) \wedge (\neg db_t \vee \neg tk_t \vee \neg cs_{t+1}) \wedge (\neg tk_t \vee cs_{t+1})$ have maximal possibility
- the models of $db_t \wedge tk_t \wedge cs_{t+1}$ have minimal possibility
- the other models have intermediary possibility.

Taking into account inertia, it yields the following ranking of the transitions:

π	Transitions
high (= 1)	$(\omega_2,\omega_1),(\omega_4,\omega_1),(\omega_5,\omega_5),(\omega_6,\omega_5),$ $(\omega_{10},\omega_{10}),(\omega_{12},\omega_{10}),(\omega_8,\omega_7),(\omega_7,\omega_7)$
medium	$(\omega_1,\omega_3),(\omega_3,\omega_3),(\omega_9,\omega_{11}),(\omega_{11},\omega_{11})$
low	$\{\omega_1,\omega_9,\omega_3,\omega_{11}\} \times \{\omega_1,\omega_2,\omega_5,\omega_6,\omega_9,\omega_{10}\}$

Note that if the transition constraint $tk_t \mapsto ks_{t+1}$ is added to the knowledge base whith ks meaning that the key is in the starter then from an initial state where $tk \wedge db$ is true, this method can conclude that in the next state $\neg cs$ will be true but cannot conclude that ks will also be true. It is the *drowning effect* [2].

6. Prediction and updating

Once the transitions are rank-ordered, updating and prediction can be envisaged according to two different ways : either on the basis of a state by state analysis à la Katsuno and Mendelzon in which all the most plausible transitions are considered from *each* possible initial state ω, or by means of a default analysis where the most plausible transitions are applied to the most normal state corresponding with what is known of the initial state.

6.1. The state by state analysis

Updating in this method consists, as it is suggested by Katsuno and Mendelzon, in considering each possible initial state, then computing each best final state satisfying the input, and finally doing the union of all final states obtained. But in order to compute each best final state, a ternary relation $\{<_\omega, \omega \in \Omega\}$ is needed. Where does it come from? This problem is left open by these authors. Some researchers have tried to give an answer to that problem, and noticeably Winslett [15]. She derives the set of orderings $\{<_\omega, \omega \in \Omega\}$ right away from the language. Namely assume that the language is made of n propositional letters $x_1,\ldots, x_n$; let x_i^* be a literal defined from x_i, i.e., $x_i^* \in \{x_i, \neg x_i\}$. A state ω is equated to an interpretation $x_1^* \wedge x_2^* \wedge \ldots \wedge x_n^*$. Let $D(\omega,\omega') = \{$literals whose sign differs in ω and $\omega'\}$. Then the ternary relation proposed by Winslett is such that $\omega' <_\omega \omega''$ iff $D(\omega,\omega') \subset D(\omega,\omega'')$. In other words the most plausible transitions are those which tend not to change the signs of literals. This proposal has the drawback of being entirely language-driven. It presupposes that distinct propositional letters correspond to independent properties. In this paper, it has been suggested that the state transitions can also be syntactically specified by means of transition constraints attached to the knowledge base and we use the rank-ordering defined above. Winslett's method can then be used to complete the possibilistic Markov chain obtained by the method described above using persistence assumptions attached to literals.

Let K_{t0} be a knowledge base. A set C of transition constraints is supposed to be available. It contains assertions of the form $p_t \mapsto q_{t+1}$, as already said (those transitions can be hard or defeasible). For each state ω, we can compute the set of successor states as follows. Let $C(\omega)=\{p_t \mapsto q_{t+1}, \omega \models p\}$ be the set of transition constraints that are enabled in state ω. Then the set of successor state of ω is defined as $S(\omega)=\{q \mid p_t \mapsto q_{t+1} \in C_i(\omega)\ \not\exists\ p'_t \mapsto q'_{t+1}$, s.t. $p'_t \mapsto q'_{t+1} \in C_j(\omega), j<i\}$, where $C_i(\omega) = C(\omega) \cap C_i$, and $\{C_0, C_1, ..., C_k\}$ is the partition computed in Section 5. Namely we delete rules $p_t \mapsto q_{t+1}$ such that more specific rules $p'_t \mapsto q'_{t+1}$ can be applied to ω. We further assume that $S(\omega)$ will be consistent (otherwise possible successors of ω are undetermined). The set of models of $S(\omega)$, $[S(\omega)]$ is the set of possible states succeeding to ω. The most plausible among these states will be the ones which maximize inertia, i.e., $\{\omega' \in [S(\omega)], D(\omega,\omega')$ is minimal$\}$. If $S(\omega)$ is inconsistent then we can introduce a priority ordering among conflicting rules and select the most prioritary ones. Such an ordering can discriminate among more or less persistent literals as in Cordier and Siegel [6].

Example (continued) We now compute the most plausible transitions from complete states of knowledge obeying C and maximizing inertia. For instance, if we consider state ω_1 where the lights have been on, the battery is dead, the key has been turned and the car is starting, it is a transient state which is necessarily followed by ω_3 where the information cs is turned into $\neg$cs due to the transition constraints. Indeed, $S(\omega_1)=\{db_{t+1}, \neg cs_{t+1}\}$, which is completed by adding lo_{t+1} and tk_{t+1} by inertia. Using the same principle, ω_3 is clearly a persistent state.
On the whole the reader can check that the following transition graph is obtained:

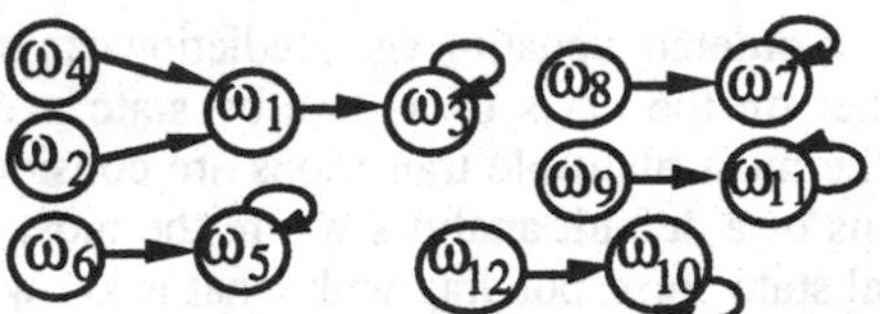

Fig. 2. The most plausible transitions

The arrows express the most plausible transitions (according to inertia and specificity). There are 5 persistent states: $\omega_3=\{lo,db,tk,\neg cs\}$, $\omega_5=\{lo,db,\neg tk,cs\}$, $\omega_7=\{lo,db,\neg tk,\neg cs\}$, $\omega_{10}=\{\neg lo,\neg db,tk,cs\}$, $\omega_{11}=\{\neg lo,db,tk,\neg cs\}$.

Note that ω_5 acknowledges the possibility of starting a car with a dead battery by other means than simply turning the key. This possibility is not ruled out by the set of transition constraints.

6.2 The default method

The above method has a drawback, it gives rather imprecise results (even if only the most plausible transitions are taken into account). For instance, if the initial knowledge base contains $lo \wedge tk$ it is clear that the resulting persistent state is ω_3. However if it contains tk only, then the final state is any of ω_3, ω_{10} or ω_{11}, hence it is not clear if the car starts. But from the transition constraints one should by default conclude that cs follows from tk, hence only ω_{10} should remain.

The default method consists in considering the most plausible transition constraints which can be applied to the initial state as it is known. In other words, the implicit assumption that the system is in *the most normal case* is made, so it gives us a precise but not absolutely sure result (because if the initial state were known more precisely then possibly more specific but opposite transition constraints should have been applied). For instance, if all that is known of the current state at time t_0 is that tk holds then the only possible persistent states are $\{\omega_3,\omega_{10},\omega_{11}\}$, and ω_{10} is the most likely to be reached, whereby we can conclude by default $\neg lo_{to+1}$, $\neg db_{to+1}$, cs_{to+1}, as expected.

In the case of *updating* this method must be applied more cautiously than in the case of prediction, the result must be checked in order to know if it verifies the input. If the result does, then there is no reason to question it, otherwise we must consider that the implicit asumption made about the initial state might be not true and that we should use the state by state method.

As already said, we might want to add persistence transitions à la Winslett, like $db_t \mapsto db_{t+1}$, $lo_t \mapsto lo_{t+1}$, $\neg db_t \mapsto \neg db_{t+1}$, $\neg lo_t \mapsto \neg lo_{t+1}$. This corresponds to new constraints which would restrict the transitions having the highest possibility level. Note that in the case of a complete knowledge of the initial state the two methods presented in this part give the same result.

7. Link with Other Approaches

Cordier and Siegel [5] handle transition constraints in the framework of default logic. Hence they cannot cope with rule specificity in a simple way. Moreover they assume that the initial state is complete. There are some connections between our proposal and the more recent semantic transition model of Cordier and Siegel [6]. In their model the transition constraints are rank-ordered right away in terms of persistence, and are then encoded in a circumscriptive logic with priorities. On the contrary, here we automatically rank the transition constraints so as to respect the specificity ordering, and we can then rank the transitions between states accordingly. The two approaches are complementary since some information about the relative strength of persistence of some litterals may be available (for instance for a regularly maintained car $\neg db$ should persist more than db). Note that this ranking procedure fails if some transition constraints bluntly contradict one another. Our approach also has same connections with Goldszmidt and Pearl's [11] stratified ranking approaches to updating, and especially the idea of going from transition constraints to a ranking of models. However in their approach Goldszmidt and Pearl consider only causal problems, such as the dead battery above, for which the transition constraints form a circuit-free graph. Clearly the above approach is meant to work if there are circuits, since once each transition constraint $p \rightarrow q$ has been written $p_t \mapsto q_{t+1}$ all possible circuits have been broken. On the contrary, examples that can be dealt with using stratified rankings are ones where the system evolves once for all and then "stops".

Our analysis of Katsuno and Mendelzon's postulates has led us to define a new set of update postulates see [10] which are exactly describing the behaviour of every possibilistic Markovian graph (where inertia principles are not taken for granted).

8. Further Research

A direction of study is to find an economic way of expressing the persistent constraints of the form $p_t \mapsto p_{t+1}$ and $\neg p_t \mapsto \neg p_{t+1}$ for each p of the set of symbols S. More work is needed to assess the level of generality and the limitations of the proposed treatment of transition constraints, specially their rank-ordering : dynamic constraints expressing change have clearly a higher priority degree than persistence constraints. Two automatic methods to rank-order dynamic change constraints have been presented in this paper. The rank-ordering of persistent constraints is to be studied, since as in Cordier and Siegel [6], we should take into account differences between degrees of persistence. Moreover, computing the transition graph is the semantic aspect of updating, there is another aspect to develop:

the syntactic one, i.e., how to compute updating without being obliged to develop all the possible states of the system...

Acknowledgements

This study has been partially supported by the French Computer Science Inter-P.R.C. project "Gestion de l'Evolutif et de l'Incertain" of the M.E.S.R. The authors are indebted to Marie-Odile Cordier for many comments and discussions which help them to prepare the final version of this paper.

References

1. Benferhat S., Dubois D., Prade H. (1992) Representing default rules in possibilistic logic. Proc. 3rd Conf. on Principles of Knowledge Representation and Reasoning (KR'92) (B. Nebel et al., eds.), Cambridge, MA, 673-684.
2. Benferhat S., Cayrol C., Dubois D., Lang J., Prade H. (1993) Inconsistency management and prioritized syntax-based entailment. Proc. 13th Int. Joint Conf. on Artificial Intelligence (IJCAI'93), 640-645, Chambéry.
3. Brown F. (Ed.) (1988) The Frame Problem in Artificial Intelligence. Morgan & Kaufmann, San Mateo, CA.
4. Cholvy L. (1994) Database updates and transition constraints. Int. J. Intelligent Systems, 9, 169-180.
5. Cordier M.O., Siegel P. (1992) A temporal revision model for reasoning about change. Proc. 3rd Conf. on Principles of Knowledge Representation and Reasoning (KR'92) (B. Nebel and al., eds.), Cambridge, MA, 732-739.
6. Cordier M.O., Siegel P. (1993) Prioritized transitions for updates. Proc. AI'93 Workshop on "Belief Change: Bridging the Gap Between Theory and Practice", Melbourne and Technical report IRISA, n 884, 1994, Rennes, France.
7. Dubois D., Prade H. (1988) Possibility Theory. Plenum Press, New York.
8. Dubois D., Prade H. (1991) Possibilistic logic, preferential models, non-monotonicity and related issues. Proc. 12th Int. Joint Conf. on Artificial Intelligence (IJCAI'91), Sydney, Australia, Aug. 24-30, 419-424.
9. Dubois D., Prade H.(1993) Belief revision and updates in numerical formalisms. Proc. 13th Int. Joint Conf. on Artificial Intelligence (IJCAI'90), 620-625, Chambéry.
10. Dubois D., Dupin de Saint-Cyr F., Prade H. (1995) Update postulates without inertia. Inter. Rep. IRIT, Toulouse University, France.
11. Goldszmidt M., Pearl J. (1992) Rank-based systems: A simple approach to belief revision belief update and reasoning about evidence and actions. Proc. 3rd Conf. on Principles of Knowledge Representation and Reasoning (KR'92) (B. Nebel and al., eds.), Cambridge, MA, 661-672.
12. Katsuno H., Mendelzon A.O. (1991) On the difference between updating a knowledge base and revising it. Proc. 2nd Conf. on Principles of Knowledge Representation and Reasoning (KR'91) (J. Allen et al., eds.), Cambridge, MA, 387-394.
13. Pearl J. (1990) System Z: a natural ordering of defaults with tractable applications to default reasoning. Proc 3rd Conf. on Theoretical Aspects of Reasoning About Knowledge (M. Vardi, ed.), Morgan Kaufman, San Mateo, CA, 121-135.
14. Sandewall E. (1993) The range of applicability of nonmonotonic logics for the inertia problem. Proc. 13th Int. Joint Conf. on Artificial Intelligence (IJCAI'93), 738-743, Chambéry.
15. Winslett M. (1990) Updating Logical Databases. Cambridge University Press, Cambridge, UK.
16. Zadeh L.A. (1978) Fuzzy sets as a basis for a theory of possibility. Fuzzy Sets and Systems, 1, 3-28.

Possibilistic Logic as Interpretability Logic *

Petr Hájek

Institute of Computer Science, Academy of Sciences of the Czech Republic
Pod vodárenskou věží 2, 182 07 Prague 8, Czech Republic

Abstract. It is shown that a variant of qualitative (comparative) possibilistic logic is closely related to modal interpretability logic, as studied in the metamathematics of first-order arithmetic. This contributes to our knowledge on the relations of logics of uncertainty to classical systems of modal logic.

1 Introduction

Possibilistic logic, as developed by Zadeh, Dubois, Prade and others (see e.g. [4] or [5]), deals with formulas and their possibilities, the possibility $\Pi(A)$ of a formula A being a real number from the unit interval, and the following axioms are assumed: $\Pi(\text{true}) = 1$, $\Pi(\text{false}) = 0$, equivalent formulas have equal possibilities, $\Pi(A \vee B) = max(\Pi(A), \Pi(B))$. It is very natural to ask how possibilistic logic relates to known systems of modal logics. This question was discussed in [1, 3, 6, 7]; in the last paper, possibilistic logic was related to tense (temporal) logic with finite linearly preordered time. One deals with Kripke models $\langle W, \Vdash \pi\rangle$ where W is a finite non-empty set of possible worlds, $\Vdash$ maps $Atoms \times W$ into $\{0, 1\}$ (truth evaluation), and π maps W into the unit interval [0,1]; sets $X \subseteq W$ have possibilities $\Pi(X) = max\{\pi(w) \mid w \in X\}$ and the possibility $\Pi(A)$ of a formula A is the possibility of the set of all worlds satisfying A. The mentioned three papers study the binary modality $\lhd$ defined as follows: $A \lhd B$ iff $\Pi(A) \leq \Pi(B)$. Classical Kripke models have the form $\langle W, \Vdash, R\rangle$ where R is a binary relation. In particular, each possibilistic model $\langle W, \Vdash, \pi\rangle$ determines a model $\langle W, \Vdash, R\rangle$ where $w_1 R w_2$ iff $\pi(w_1) \leq \pi(w_2)$; clearly, R is a linear preorder. [7] formulate an axiom system QPL sound for this semantics. The axioms are tautologies, transitivity $((A \lhd B) \& (B \lhd C)) \rightarrow (A \lhd C)$, linearity $(A \lhd B) \vee (B \lhd A)$, monotonicity $(A \lhd B) \rightarrow (A \vee C \lhd B \vee C)$; $0 \lhd A$ and $\neg(1 \lhd 0)$ (non-triviality); 1 is *true*. Deduction rules are modus ponens and the following necessitation: from $A \rightarrow B$ infer $A \lhd B$. It was proved in [1] that this axiom system is incomplete (even if complete for formulas with non-nested modalities) and an axiom scheme was exhibited making the system complete. A suggestion of Herzig has lead to the observation that the following pair of axioms sufficies:
(P) $(A \lhd B) \rightarrow \Box(A \lhd B)$,
(P^-) $\neg(A \lhd B) \rightarrow \Box\neg(A \lhd B)$,
where $\Box C$ is $\neg C \lhd 0$. The related system of tense logic of [1] has three necessity-like modalities G, H, I (meaning "in all future worlds", "in all past worlds", "in

* Partial support by the COPERNICUS grant No 10053 (MUM) is acknowledged.

all present worlds", respectively); $\Box A$ is $HA\&IA\&GA$ and $A \lhd B$ is defined as $\Box(A \rightarrow \neg(I(\neg B)\&G(\neg B))$; equivalently, $\Box(A \rightarrow (JB \vee FB))$ where J is $\neg I \neg$ and F is $\neg G \neg$ (dual modalities). Details will not be repeated here.

In this paper, we are going to relate possibilistic logic to *interpretability logic*, as developed by Smoryński, Hájek, Švejdar, de Jongh, Veltman, Visser and others. Interpretability logic extends *provability logic* L, and we comment first on the latter. In provability logic, necessity (box, $\Box$) is understood as provability in a fixed axiomatic arithmetic T (e.g. Peano arithmetic). As Gödel discovered, in T we can define a formula $Pr(x)$ formalizing the notion of provability in T, e.g. $\neg Pr$ (false) is the formula Con expressing the consistency of T in T. (Gödel's second incompleteness theorem says that under reasonable assumptions on T, T does not prove its own consistency, i.e. $T \nvdash Con$.) Gödel also invented the method of self-reference in arithmetic, by constucting a formula ν such that $T \vdash \nu \equiv \neg Pr(\bar{\nu})$ (ν says "I am unprovable"; $\bar{\nu}$ is the numerical code of ν) and showed that under reasonable assumptions on T, ν is an independent formula ($T \nvdash \nu, T \nvdash \neg\nu$). (This is Gödel's first incompleteness theorem). An *arithmetical translation* of modal logic is a mapping $*$ associating with each formula A of propositional modal logic (whose only modality is $\Box$) a sentence A^* of T in such a way that $*$ commutes with connectives (e.g. $(A\&B)^*$ is $A^*\&B^*$, etc.) and $(\Box A)^*$ is $Pr(\bar{A}^*)$ (this is how necessity is understood as provability). The *arithmetical completeness theorem* (cf. [12, 13]) says that a propositional modal formula A is provable in the provability logic L iff for each arithmetical translation $*$, $T \vdash A^*$. Provability logic also has its Kripke semantics and there is a corresponding completeness theorem (the same references).

For general theories T_1, T_2, T_1 is *interpretable* in T_2 if primitive notions of T_1 can be defined on T_2 in such a way that axioms of T_1 become provable in T_2. Consider the extension of our arithmetic $T: T_1 = (T+\phi)$, $T_2 = (T+\psi)$. In this case we can formalize the notion of interpretability, i.e., produce a formula $Intp(x, y)$ of T saying "$(T+\phi)$ interprets $(T+\psi)$" or "$(T+\psi)$ is interpretable in $(T+\phi)$"; T proves reasonable properties of this notion. This leads to a modal propositional logic with one unary modality $\Box$ and one binary modality $\lhd$; one has arithmetical interpretations ($(\Box A^*)$ is $Pr(\bar{A}^*)$, $(A \lhd B)^*$ is $Intp(\bar{A}^*, \bar{B}^*)$), arithmetical completeness, Kripke models, Kripke-style completeness [2, 8, 9, 10, 11, 14]. The double semantics of provability and interpretability logic (arithmetical and Kripke-like) gives interpretability logic its beauty; nevertheless, arithmetical interpretations will be disregarded here. We relate interpretability logic to a variant of qualitative possibilistic logic which we call the logic of sufficiently big possibilities (or the logic of future possibilities). In Section 1 we survey most basic facts on interpretability logic; in Section 2 we introduce our comparative logic of sufficiently big possibilities, in Section 3 we develop a tense logic with finite linearly preordered time and relate it to both preceding systems. Section 4 contains some remarks and Section 5 is an appendix giving indications for a completeness proof.

2 Preliminaries: Interpretability logics

Axioms are as follows:

axioms of L:
(L1) tautologies,
(L2) $\Box(A \rightarrow B) \rightarrow (\Box A \rightarrow \Box B)$,
(L3) $\Box A \rightarrow \Box\Box A$,
(L4) $\Box(\Box A \rightarrow A) \rightarrow \Box A$ (Löb's axiom),
additional axioms:
(J1) $\Box(A \rightarrow B) \rightarrow A \lhd B$
(J2) $((A \lhd B) \& (B \lhd C)) \rightarrow (A \lhd C)$
(J3) $((A \lhd C) \& (B \lhd C)) \lhd ((A \vee B) \lhd C)$
(J4) $A \lhd B \rightarrow (\Diamond A \rightarrow \Diamond B)$
(J5) $\Diamond A \lhd A$
(M) $(A \lhd B) \rightarrow ((A \& \Box C) \lhd (B \& \Box C))$
(P) $(A \lhd B) \rightarrow \Box(A \lhd B)$

Deduction rules: modus ponens and necessitation: from A infer $\Box A$. (Clearly, $\Diamond A$ is $\neg\Box\neg A$).

Remark Arithmetical validity of most axioms is easy to see; we comment on Löb's axiom. In fact, this is a variant of Gödel's second incompleteness theorem: by trivial manipulations, it can be written as $\Diamond\neg A \rightarrow \Diamond(\neg A \& \neg\Diamond\neg A)$, thus (replacing $\neg A$ by B) $\Diamond B \rightarrow \Diamond(B \& \neg\Diamond B)$ and hence $\Diamond B \rightarrow \neg\Box(B \rightarrow \Diamond B)$ which has the following arithmetical interpretation: if B is consistent (with T) then the formula $B \rightarrow Con(B)$ is unprovable (in T, thus: if $(T+B)$ is consistent then $(T+B)$ does not prove its own consistency; we disregard technical details.).

The following are important *axiom systems*: $IL = L + (J1) - (J5)$, $ILM = IL + (M)$, $ILP = IL + (P)$. See [2, 8, 9, 11, 14].

Note that in IL, box is definable from triangle:

$$IL \vdash \Box A \equiv (\neg A \lhd 0)$$

where 0 is *false*). A *Veltman model* has the form $\langle W, \Vdash, R, S\rangle$ where $\langle W, \Vdash, R\rangle$ is a Kripke model with R transitive and asymmetric (hence irreflexive) and S is a reflexive transitive relation containing R. (This is only a particular case; see [10] for the general case). One defines $w \Vdash \Box A$ iff for all $v \in W$, wRv implies $v \Vdash A$; $w \Vdash A \lhd B$ iff

$$(\forall v)(wRv \;\&\; v \Vdash A \;\rightarrow\; (\exists u)(wRu \;\&\; vSu \;\&\; u \Vdash B)).$$

The *completeness theorem* for ILP says that $ILP \vdash A$ iff A is true in each finite Veltman model $\langle W, \Vdash, R, S\rangle$ satisfying the following condition:

$$(wRv \;\&\; wRu \;\&\; vSu \;\&\; wRw' \;\&\; w'Rv) \rightarrow w'Ru$$

Another formulation is as follows: let $vS_w u$ mean $wRv \;\&\; wRu \;\&\; vSu$. Then

$$(vS_w u \;\&\; wRw' \;\&\; w'Rv) \rightarrow vS_{w'} u.$$

To get completeness for IL and ILM one needs a more complicated notion of a Veltman model.

3 The comparative logic of future possibilities

Comparing ILP with $QPL+(P)$ we see that $QPL+(P)$ proves (J1-J4) but not (J5) and clearly does not prove Löb's axiom. If one restricts oneself to positive models (for each $w \in W, \pi(w) > 0$) then $\Box$ of QPL becomes an (S5)-modality; in particular, $\Box A \to A$ is sound; $QPL+(P)+(\Box A \to A)$ axiomatizes completely $\lhd$ with respect to positive models.

Our aim is to relate possibilistic logic more closely to interpretability logic. This is done below.

To marry ILP with possibility theory, consider the world-dependent *future possibility*: $\Pi(A, w) = sup\{\pi(w') \mid w' > w \text{ and } w' \Vdash A\}$. Here $w' > w$ means $\pi(w') > \pi(w)$. Define $w \Vdash A \lhd B$ if $\Pi(A, w) \leq \Pi(B, w)$. Thus $A \lhd B$ is satisfied in the world w if either $\Pi(A)$, $\Pi(B) \leq \pi(w)$ or $\Pi(A) \leq \Pi(B)$. This suggests the following (fuzzy) reading of the new triangle-modality: the possibility of A is less-than-or-equal to the possibility of B, or neither A nor B are too much possible.

Our comparative *logic of future possibilities* (or, if the reader prefers, *logic of sufficiently big possibilities*) has formulas built from propositional variables using connectives and the modality $\lhd$; its models are finite possiblistic Kripke models $K = \langle W, \Vdash, \pi, \rangle$ and the semantics of $\lhd$ is given by comparison of future possibilities as above. We shall find a complete axiomatization.

Define $\Box$ to be the *future necessity*: $w \Vdash \Box A$ iff for all $w' > w$, $w' \Vdash A$. This relates possibilistic logic and its Kripke models to tense logic and its Kripke models; we shall investigate the corresponding tense logic in the next section. At this moment, let us stress that relating possibilistic logic to tense logics (and other logics, e.g. interpretability logic) should contribute to our understanding of what possibility theory is; one interpretation is that $\Pi(A)$ is the *last moment* in which A is possible (or zero). This temporal interpretation makes our future possibility natural: $\Pi(A, w)$ means (in the world w) the *last moment after now* in which A is possible (or zero). To elucidate this, let us verify the validity of the axiom (J5): $\Diamond A \lhd A$. Given w, if $\Pi(\Diamond A, w) > 0$, let w' be the last world after w satisfying $\Diamond A$; thus there is a last $w'' > w$ satisfying A; $\Pi(\Diamond A, w) = \Pi(w') < \Pi(w'') = \Pi(A, w)$. Clearly, each possibilistic model $\langle W, \Vdash, \pi\rangle$ determines a particular Veltman model called an LPO-model (linear preorder):

An *LPO-model* is a Veltman model $\langle W, \Vdash, R, S\rangle$ where S is a linear preorder of W, i.e. S is transitive and dichotomous (wSv or vSw for all $w, v \in W$) and R is the corresponding strict preorder: wRv if wSv and not vSw.

Observe that $w \Vdash A \lhd B$ in the possibilistic model $\langle W, \Vdash, \pi\rangle$ (with respect to the future-possibility semantics) iff $w \Vdash A \lhd B$ in the corresponding LPO model $\langle W, \Vdash, R, S\rangle$ (with respect to Veltman semantics). Let (D) be the axiom of dichotomy $(A \lhd B) \vee (B \lhd A)$. Then:

Fact. Axioms of ILPD are tautologies of LPO-models.

But we shall show that ILPD is not complete for LPO-models. In the sequel we develop a tense logic with finite linearly preordered time and one (future) necessity extending ILPD and complete for LPO -models.

4 A tense logic with finite linearly preordered time

In this section we shall investigate the modality of future necessity introduced above. Note that here we have only the future necessity (always in the future), no past necessity and no present necessity. This is in contrast to the system of [1] discussed in Section 1, with the same Kripke models (with finite linearly preordered time) but with three necessities mentioned. We show that our present tense logic is completely axiomatized by an axiom system extending the axioms of provability logic L by a single axiom of linear preorder:

Let (E) be the axiom

$$\Box(\Box A \rightarrow B) \vee \Box(\Box B \rightarrow \Box A)$$

(linEar prEordEr).

Theorem 1. *1. The logic L+(E) (provability logic plus (E)) is sound and complete for finite Kripke models $\langle W, \Vdash, R\rangle$ such that there is a linear preorder S on W (transitive and dichotomous) whose corresponding strict preorder is R (thus wRv iff wSv and not vSw; in other words, $\langle W, \Vdash, R, S\rangle$ is an LPO-model).*

2. In $L+(E)$ define $\lhd$ as follows (definition):
(F) $A \lhd B$ iff $\Box(\neg A \& \neg B) \vee \Diamond(B \& \Box \neg A)$.
Then $L+(E)+(F)$ proves all axioms of ILPD.

3. The definition (F) is true in each LPO-model if $\lhd$ means comparison of future possibilities.

4. ILPD does not prove (E); in particular, the model

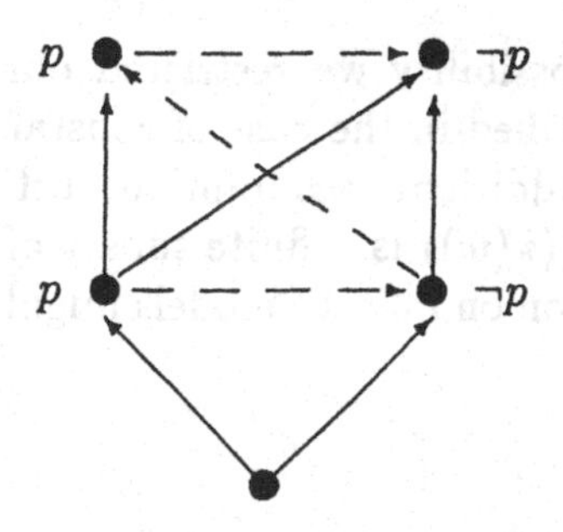

is a model of ILPD, but the formula $\Box(\Box p \rightarrow \neg p) \vee \Box(\Box \neg p \rightarrow \Box p)$ is false in the root.

The *proof* of (1) is a modification of the proof of the completeness theorem in [1]; (2)-(4) are easy.

Summarizing, $L+(E,F)$ is the logic of comparison of future possibilities (over finite models); it strictly extends the interpretability logic *ILPD*.

5 Remarks

We continue with a series of remarks.

Remark Over LPO models, $\Box$ and $\lhd$ are interdefinable: (F) defines $\lhd$ from $\Box$ and evidently $\Box A \equiv (\neg A \lhd 0)$ is a tautology. Thus the theory L + (E,F) may be presented as a theory with a single modality $\lhd$.

Remark We compare the modality of comparison of future possibilities (denoted $\lhd_f$ in the present remark) with the (world independent) modality of comparison of possibilities ($\lhd_c, c$ for "constant over worlds"). For a moment, let a CNN-formula (closed non nested) be a boolean combination of formulas of the form $A \lhd B$. If C is such a formula, C_f and C_c mean results of replacement of $\lhd$ by $\lhd_f$ and $\lhd_c$ respectively. The relation is as follows:

Fact. Let C be a CNN-formula. If C_f is a tautology (over LPO-models) then C_c is a tautology. On the other hand, for each axiom C of QPL except $\neg(1 \lhd 0), C_f$ is a tautology. (But $1 \lhd_f 0$ is satisfied in each maximal element of each LPO-model).

The first part is proved by adding to an arbitrary LPO-model a new least element.

Let QPL_0 result from QPL by removing the axiom $\neg(1 \lhd 0)$. It can be shown, in an analogy to [1], that QPL_0 axiomatizes all CNN formulas that are tautologies in the future semantics. Alternatively, it axiomatizes all CNN formulas that are tautologies in the constant semantics with respect to all LPO-models plus the *empty* model (no worlds, possibility of each formula is 0).

Remark The task remains to analyze the meaning of ILPD (which is weaker than $L+(E, F)$) for possibilistic logic. As the example above shows, ILPD admits models substantially different from (not elementarily equivalent to) any LPO-model. Does every model of ILPD have a possibilistic interpretation?

Remark Studying future possibility we restricted ourselves to finite models. Such a restriction is fully justified in the case of constant possibilities, as it was shown in [1]. Here this is an additional assumption, justified, e.g. by postulating that the scale of possibilities ($\pi(w)$) is a finite subset of [0,1]. Expressive power of future possibility comparison on infinite models might deserve additional consideration.

6 Appendix

A proof of the completeness theorem for $L + (E)$ may be obtained by an inspection of the proof of completeness of FLPOT in [1] Section 3. We briefly indicate the necessary changes; the reader is assumed to have a copy of [1] at his/her disposal.

In 3.5, change the definition of E:

$\Gamma_1 E \Gamma_2$ iff, for each C, $GC \in \Gamma_1$ iff $GC \in \Gamma_2$

(G replaces $\Box$; we have no H, I modalities). Delete 3.6, 3.7, 3.8(1) holds as it stands; 3.8(2) will read: If $\Gamma E \Gamma'$ then $\Gamma R \Delta$ implies $\Gamma' R \Delta$ (and also $\Delta R \Gamma$ implies $\Delta R \Gamma'$, but this will be proved later).

Proof. If $GA \in \Gamma'$ then $GA \in \Gamma$ and $A \in \Delta$.

Define irreflexive theories as in 3.9; 3.10 holds as it stands.
3.11(1) holds by a new proof:

Proof. Assume $\Theta R \Gamma_1$, $\Theta R \Gamma_2$, and let all of the following three conditions be false: $\Gamma_1 R \Gamma_2$, $\Gamma_2 R \Gamma_1$, $\Gamma_1 E \Gamma_2$. Then there are A, B, C witnessing this, i.e.

$$(GA \& \neg B \& GC) \in \Gamma_1$$

$$(GB \& A \& \neg GC) \in \Gamma_2$$

But $\Theta \vdash G(G(A\&C) \to B) \vee G(GB \to G(A\&C))$, thus Θ proves one of the disjuncts. If $\Theta \vdash G(G(A\&C) \to B)$ then $(G(A\&C) \to B) \in \Gamma_1$ and Γ_1 is inconsistent; if $\Theta \vdash G(GB \to G(A\&C)$ then $\Gamma_2 \vdash GB \to G(A\&C) \to GC$, thus Γ_2 is inconsistent. But Γ_1, Γ_2 are consistent.

In Def.3.12 of a critical formula delete H, I; define $S = \{\Gamma \mid \Delta R \Gamma \text{ or } \Delta E \Gamma\}$. 3.13 is O.K.

In 3.14 show that, if Ω is a set of irreflexive theories, then for $\Gamma_1, \Gamma_2 \in \Omega$ *exactly* one of the conditions $\Gamma_1 R \Gamma_2$, $\Gamma_1 E \Gamma_2$, $\Gamma_2 R \Gamma_1$ holds; i.e., that they are mutually exclusive; use the proof is it stands. For transitivity, we have to prove the second half of 3.8:

$$\Gamma_1 R \ \Gamma_2 E \ \Gamma_3 \text{ implies } \Gamma_1 R \ \Gamma_3.$$

If $\Gamma_1 R \ \Gamma_2 E \ \Gamma_3$ and not $\Gamma_1 R \ \Gamma_3$ then $\Gamma_3 R \ \Gamma_1$ or $\Gamma_1 E \ \Gamma_3$. But $\Gamma_3 R \ \Gamma_1$ implies $\Gamma_3 R \ \Gamma_2$ which is incompatible with $\Gamma_2 E \ \Gamma_3$, and $\Gamma_1 E \ \Gamma_3$ gives $\Gamma_1 E \ \Gamma_2$ which is incompatible with $\Gamma_1 R \ \Gamma_2$.

3.15(1) is O.K. as its stands and (2), (3) are deleted. Similarly for 3.16(1), (2), (3). This completes the proof; the rest is O.K.

References

1. Bendová K., Hájek P.: Possibilistic logic as tense logic, in: Proc. QUARDET'93, Barcelona 1993.
2. Berarducci A.: The interpretability logic of Peano arithmetic, Journal of Symb. Log. 55, 1990, 1059-1089.
3. Boutilier C.: Modal logics for qualitative possibility and beliefs, in: Uncertainty in Artificial Intelligence VIII, Morgan-Kaufmann Publ. 1992, 17-24.
4. Dubois D., Lang J., Prade H.: Possibilistic logic, Report IRIT-91-98/R, Decembre 1991, Toulouse.
5. Dubois D., Prade H.: Possibilistic and fuzzy logic, in: (Smets et al., eds): Non-standard logics for automated reasoning, Academic Press 1988, 287.
6. Dubois D., Prade H., Testemate C.: In search for a modal system for possibility theory, in: Proc. ECAI'89, Munich 1988, 581-586.

7. Fariñas del Cerro L., Herzig A.: A modal analysis of possibilistic logic, in: (Kruse et al.,eds): Symbolic and Quantitative Approaches to Uncertainty, Lect. Notes in Comp. Sci. 548, Springer-Verlag 1991, p.58, (cf. also: Jorrand and Kelemen, eds: Fundamentals of AI research, Lect. Notes in AI 535, Springer-Verlag 1991, 11).
8. Hájek P., Montagna F.: The logic of Π_1 - conservativity, Arch. Math. Logic 30, 1990, 113-123.
9. Hájek P., Montagna F.: The logic of Π_1 - conservativity continued, Arch. Math. Logic 32, 1992, 57-63.
10. de Jongh D., Veltman F.: Provability logics for relative interpretability, in: Mathematical Logic (Proceedings of the Heyting conference, Bulgaria 1988), Plenum Press 1990.
11. Shavrukov V.Yu.: The logic of relation interpretability over Peano arithmetic (in Russian) Moscow 1988, preprint.
12. Smoryński C.: Self-reference and modal logic, Springer-Verlag 1988.
13. Solovay R.: Provability interpretations of modal logic, Israel Journ. Math. 25, 1976, 287-304.
14. Visser A.: Interpretability logic, in: Mathematical Logic (Proc. Heyting conference Bulgaria 1988), Plenum Press 1988.

A Numerical Framework for Possibilistic Abduction *

Jörg Gebhardt and Rudolf Kruse

Department of Mathematics and Computer Science
University of Braunschweig
D-38106 Braunschweig, Germany
Email: gebhardt@ibr.cs.tu-bs.de

Abstract. We introduce a numerical framework for possibilistic abduction that is based on a relational setting for handling imprecise data and an information–compression view of possibility distributions as one–point coverages of random sets. Existing dependencies among disorders, manifestations, and intermediary characteristics are modelled with the aid of a hypergraph representation. The underlying reasoning concept of a possibilistic focusing system is outlined and compared with two alternative approaches in this field.

1 Introduction

In current research there are two communities working on model–based diagnosis, the first of which considers numerical settings, especially Bayesian networks [8] as the choice of their modelling, the second prefers logical approaches, characterized by the use of consistency–based diagnosis [13, 5]
Restricting ourselves to pure numerical environments, this paper introduces a framework for possibilistic abduction that incorporates Reggia's basic model of parsimonious covering theory [12] as well as a new approach to possibilistic abduction discussed in [2].
In particular, section 2 proposes a relational setting for imprecise reasoning in a multidimensional space of attributes, interpreted as disorders, manifestations, and intermediary characteristics, respectively. Considering a universe of discourse $\mathcal{U} = (\Omega_i)_{i \in N}$, where Ω_i denotes the domain of the i–th attribute, and $\Omega \stackrel{\text{Df}}{=} \bigtimes_{i \in N} \Omega_i$ the induced product space, we suppose that *general knowledge* R about relationships between attribute values as well as *evidential knowledge* E about a specific object state $\omega_0 \in \Omega$ under consideration is characterized in terms of relations $R \subseteq \Omega$ and $E \subseteq \Omega$. Then we introduce the concept of a *focusing system*, which specifies qualitative dependencies between attributes by means of a hypergraph $(N, \mathcal{I})$, whereas quantitative dependencies are expressed by a rule system $\mathcal{R}(\mathcal{U}, \mathcal{I})$, consisting of relations R^I, referred to the attributes identified by all indices contained in $I \in \mathcal{I}$.

* This work has partially been funded by CEC-ESPRIT III Basic Research Project 6156 (DRUMS II)

In section 3 the proposed model is compared with the two approaches to abduction mentioned above.
Section 4 deals with a specific problem of abductive reasoning in our model, which is calculating parsimonious covers of explanations for a given diagnostic problem.
Incorporating uncertainty aspects by generalizing from imprecise to possibilistic knowledge, section 5 briefly introduces the concept of a *possibilistic focusing system* and its semantic background for possibilistic abduction.

2 A Relational Setting for Imprecise Reasoning

Let $\mathcal{A} = \{a_1, \ldots, a_n\}$ be a set of attributes that characterize an object under consideration and $\Omega_i = Dom(a_i)$, $i = 1, \ldots, n$, their attached domains. Stating the *closed world assumption*, the current state of this object is expected to be specifiable in terms of a tuple $(\omega_0{}^{(1)}, \ldots, \omega_0{}^{(n)}) = \omega_0$, with $\omega_0{}^{(i)} \in \Omega_i$ and $\omega_0 \in \Omega \stackrel{\text{Df}}{=} \bigtimes_{i=1}^{n} \Omega_i$. From a semantic point of view we want to distinguish between different types of attributes, reflected by a partition $\{\mathcal{D}, \mathcal{M}, \mathcal{S}\}$ of $\mathcal{A}$, where $\mathcal{D} = \{d_1, \ldots, d_k\} = \{a_1, \ldots, a_k\}$ denotes a *set of causes (disorders)*, $\mathcal{M} = \{m_1, \ldots, m_l\} = \{a_{k+1}, \ldots, a_{k+l}\}$ a *set of effects (manifestations)*, and $\mathcal{S} = \{s_1, \ldots, s_m\} = \{a_{k+l+1}, \ldots, a_n\}$ a set of additional attributes that serve as intermediary characteristics, if no direct association between all disorders and manifestations is available. For applications it is often sufficient to suppose that $\Omega_i = \{\text{present}, \text{absent}\}$, $i = 1, \ldots, k + l$, if disorders and manifestations are assumed to be either present or absent, but from a more general point of view it is sufficient only to require the finite cardinality of all involved domains.
Independent from what kind of reasoning problem we intend to investigate, it is appropriate in our setting to presuppose that *general knowledge* about relationships between the chosen attributes (e.g. the presence/absence of disorders and manifestations) is formalized by means of a *relation* $R \subseteq \Omega$, whereas *evidential knowledge* (to be interpreted as additional information on the specific object state ω_0) is specified by another relation $E \subseteq \Omega$.
In standard cases of practical interest, R and E will of course not directly be available, but rather occur as the result from combining smaller pieces of knowledge R_j, $j = 1, \ldots, r$, that refer to *quantitative dependencies* between the possible values of subsets $\mathcal{A}_j \subseteq \mathcal{A}$ of attributes, where $\mathcal{A}_j$ reflects an existing *qualitative dependency* with respect to the attributes contained in $\mathcal{A}_j$.
To support a nice handling of R_j we choose the following formal environment:

Let $\mathcal{A}' \subseteq \mathcal{A}$ be a set of selected attributes and

$$I(\mathcal{A}') \stackrel{\text{Df}}{=} \{i \in \{1, \ldots, n\} \mid a_i \in \mathcal{A}'\}$$

its *identifying index set*.
For short, we introduce the index sets $D \stackrel{\text{Df}}{=} I(\mathcal{D})$, $M \stackrel{\text{Df}}{=} I(\mathcal{M})$, $S \stackrel{\text{Df}}{=} I(\mathcal{S})$, and $N \stackrel{\text{Df}}{=} \{1, \ldots, n\}$, respectively.

An arbitrary set $\mathcal{I} = \{I_1, \ldots, I_r\}$ of index sets is called a *modularization* of N, iff

$$(a)\ (\forall I \in \mathcal{I})\ (\emptyset \neq I \subseteq N),$$
$$(b)\ \bigcup_{I \in \mathcal{I}} I = \{1, \ldots, n\},$$
$$(c)\ (\forall I, I' \in \mathcal{I})\ (I \neq I' \text{ implies } I \not\subseteq I').$$

If $\mathcal{I} = \{I(\mathcal{A}_j) \mid j = 1, \ldots, r\}$ is a modularization of N, it can be used to define the qualitative dependencies among all attributes of $\mathcal{A}$, which are especially clarified in terms of the induced hypergraphs $(\mathcal{A}, \{\mathcal{A}_1, \ldots, \mathcal{A}_r\})$ and $(N, \mathcal{I})$, respectively.
In order to get a pleasant representation of R_j, $j = 1, \ldots, r$, we apply the concept of the general cartesian product:

Let $I \subseteq N$ be a non-empty index set and $(\Omega_i)_{i \in I}$ its attached set system of domains. Then,

$$\Omega^I = \bigtimes_{i \in I} \Omega_i = \Big\{ \varphi \mid \varphi : I \to \bigcup_{i \in I} \Omega_i \text{ and } (\forall i \in I)\ (\varphi(i) \in \Omega_i) \Big\}$$

is the product space of $(\Omega_i)_{i \in I}$.
Hence, $\Omega \stackrel{\mathrm{Df}}{=} \Omega^N$ denotes the product space of our universe of discourse $\mathcal{U} = (\Omega_i)_{i \in N}$.

The mentioned quantitative dependencies can now be formalized by a *rule system* $\mathcal{R}(\mathcal{U}, \mathcal{I}) = \{R^I \mid I \in \mathcal{I}\}$, $R^I \subseteq \Omega^I$, such that $R_j = R^{I(\mathcal{A}_j)}$, $j = 1, \ldots, r$.
In a corresponding way the evidential knowledge may be specified with the aid of an *evidential system* $\mathcal{E}(\mathcal{U}, \mathcal{N}) = \{E^J \mid J \in \mathcal{N}\}$, where $\mathcal{N}$ is a partition of (non-empty subsets of) N, expected to be *compatible with* $\mathcal{I}$, which means that for all $J \in \mathcal{N}$ there exists an $I \in \mathcal{I}$ such that $\emptyset \neq J \subseteq I$. If, for example, only single attribute values, but no relations between attribute values are to be observed, then $\mathcal{N} = \{\{1\}, \ldots, \{n\}\}$ is the right choice and always compatible with $\mathcal{I}$. Note that compatibility ensures that the specific qualitative dependencies in the evidential knowledge are consistent with the stated general qualitative dependencies (reflected by the chosen modularization $\mathcal{I}$).
In order to determine R from $\mathcal{R}(\mathcal{U}, \mathcal{I})$ and E from $\mathcal{E}(\mathcal{U}, \mathcal{N})$, respectively, and furthermore, to calculate the resulting restrictions for the $\omega_0^{(i)}$–values, induced by the assumption that imprecise general and evidential knowledge is *correct* w.r.t. ω_0 (which means that $\omega_0 \in R \cap E$), we need three elementary set–theoretical operations on product spaces, which are *cylindrical extension*, *intersection*, and *projection*.

Let the two non-empty index sets $I, J \subseteq N$ fulfil $I \subseteq J$. Additionally, let $\mathcal{P}(\Omega^I)$ and $\mathcal{P}(\Omega^J)$ be the power sets of Ω^I and Ω^J, respectively.

$$\widehat{\prod}_I^J : \mathcal{P}(\Omega^I) \rightarrow \mathcal{P}(\Omega^J),$$

$$\widehat{\prod}_I^J(\Phi) \stackrel{\text{Df}}{=} \left\{\psi \in \Omega^J \;\middle|\; (\exists \varphi \in \Phi)\,(\forall i \in I)\,(\psi(i) = \varphi(i))\right\}$$

denotes the *cylindrical extension of Ω^I onto Ω^J*.

$$\prod_I^J : \mathcal{P}(\Omega^J) \rightarrow \mathcal{P}(\Omega^I),$$

$$\prod_I^J(\Psi) \stackrel{\text{Df}}{=} \left\{\varphi \in \Omega^I \;\middle|\; (\exists \psi \in \Psi)\,(\forall i \in I)\,(\psi(i) = \varphi(i))\right\}$$

is the *projection of Ω^J onto Ω^I*.

Using this formalism, we are now in the position to define the exact interpretation of our rule system $\mathcal{R}(\mathcal{U},\mathcal{I})$ and the evidential system $\mathcal{E}(\mathcal{U},\mathcal{N})$ in the way that

$$R \stackrel{\text{Df}}{=} \bigcap_{I \in \mathcal{I}} \widehat{\prod}_I^N(R^I) \quad \text{is our } \textit{induced general knowledge base}$$

$$\text{and } E \stackrel{\text{Df}}{=} \bigcap_{J \in \mathcal{N}} \widehat{\prod}_J^N(E^J) \quad \text{our } \textit{induced total evidence w.r.t. } \omega_0 .$$

Note that these definitions consider both, the underlying qualitative dependency structure (in terms of the modularization $\mathcal{I}$ and the compatible partition $\mathcal{N}$), and the quantitative knowledge, given by the rule system $\mathcal{R}(\mathcal{U},\mathcal{I})$ and the evidential system $\mathcal{E}(\mathcal{U},\mathcal{N})$.

With respect to knowledge propagation, we call the tuple $\mathfrak{X} = (\mathcal{U},\mathcal{I},\mathcal{N},\mathcal{R}(\mathcal{U},\mathcal{I}))$ a *focusing system*, and $\sigma(\mathfrak{X},\mathcal{E}(\mathcal{U},\mathcal{N})) \stackrel{\text{Df}}{=} R \cap E$ the *current state of $\mathfrak{X}$*.
The name "focusing system" is due to the fact that we focus our general knowledge $\mathcal{R}(\mathcal{U},\mathcal{I})$ in the light of the available evidential knowledge $\mathcal{E}(\mathcal{U},\mathcal{N})$.
The resulting *projections* w.r.t. single attributes are

$$\pi^{(i)} \stackrel{\text{Df}}{=} \prod_{\{i\}}^N \Big(\sigma(\mathfrak{X},\mathcal{E}(\mathcal{U},\mathcal{N}))\Big),\ i = 1,\ldots,n.$$

Note that ${\omega_0}^{(i)} \in \pi^{(i)}$ are the most specific restrictions we can get for the attribute values of the object state $\omega_0 = ({\omega_0}^{(1)},\ldots,{\omega_0}^{(n)})$, assuming that our whole available knowledge on ω_0, represented by $\mathcal{R}(\mathcal{U},\mathcal{I})$ and $\mathcal{E}(\mathcal{U},\mathcal{N})$, is correct w.r.t. ω_0. Furthermore note that the projections $\pi^{(i)}, i = 1,\ldots,n$, are of course less informative than $\sigma(\mathfrak{X},\mathcal{E}(\mathcal{U},\mathcal{N}))$, since $\bigcap_{i=1}^{n} \widehat{\prod}_{\{i\}}^N \pi^{(i)} \supseteq R \cap E$, where the equality in general does not hold.

Efficient local propagation techniques that take advantage from the qualitative dependency structure (i.e. the semantics of the modularization $\mathcal{I}$), are, for example, presented in [6].

3 Relationships to Abductive Reasoning

Up to this point we introduced a formal environment for imprecise reasoning in a numerical framework.
Now we will clarify how specific aspects of abductive reasoning are captured in this setting.
Note that abductive reasoning refers to finding plausible explanations (in terms of disorders) for observed manifestations, which are entailed, caused, or produced by disorders. In this connection general knowledge is often described in terms of *imprecise inference rules* of the form

$$R_j : \ if\, \xi^{A_j} \ in\ X_j \ then\ \xi^{B_j} \ in\ Y_j \ ,$$

where ξ denotes a variable taking its values on $\mathcal{P}(\Omega)$, and $\emptyset \neq A_j \subseteq N$, $\emptyset \neq B_j \subseteq N$, $A_j \cap B_j = \emptyset$, $\emptyset \neq X_j \subseteq \Omega^{A_j}$, $\emptyset \neq Y_j \subseteq \Omega^{B_j}$ is supposed to be satisfied. For $\emptyset \neq A \subseteq N$, the value ξ^A denotes the Ω^A–projection of ξ.

R_j is interpreted in the way that

$$\begin{aligned} &\xi^{A_j} \in X_j \ \text{ implies } \xi^{B_j} \in Y_j, \quad \text{and} \\ &\xi^{A_j} \in \Omega^{A_j} \backslash X_j \ \text{ implies } \xi^{B_j} \in \Omega^{B_j}, \end{aligned}$$

which induces the representing relation

$$R_j = \left(\widehat{\prod}^{I_j}_{A_j}(X_j) \cap \widehat{\prod}^{I_j}_{B_j}(Y_j) \right) \cup \widehat{\prod}^{I_j}_{A_j}(\Omega^{A_j} \backslash X_j)$$

with respect to the common index set $I_j = A_j \cup B_j$.

The occuring system of index sets I_j induces a modularization $\mathcal{I}$ of N, and therefore, incorporating the relations R_j, a rule system $\mathcal{R}(\mathcal{U},\mathcal{I})$. Hence, for representing our general knowledge, we get a focusing system $(\mathcal{U},\mathcal{I},\mathcal{N},\mathcal{R}(\mathcal{U},\mathcal{I}))$, where $\mathcal{N} = \{\{m\} \mid m \in \mathcal{M}\}$, since only evidence on manifestations is expected to be available.
Realizing a first step of abductive reasoning means to calculate $\pi^{(i)}$ for all $i \in D$ with respect to a given evidential system $\mathcal{E}(\mathcal{U},\mathcal{N})$. Further steps of abduction, like finding parsimonious covers in the sense of [10], will be discussed in the next section.

In the following, we present two examples which show that our framework incorporates less general relational approaches to abduction, as they are proposed in other publications.
From this point (in order to reach formal simplicity), we prefer to relate the whole discussion to identifying index sets rather than to sets of underlying attributes. Hence, the index sets D and M are now themselves addressed as sets of disorders and manifestations, respectively.

Example 1

Following the relational setting for the classical diagnostic problem as suggested in [2], let $M(d)^+$ denote the set of all manifestations which are certainly produced by disorder d alone, and $M(d)^-$ the set of manifestations that certainly cannot be caused by d alone. Furthermore, let $M(d) \stackrel{\text{Df}}{=} M \setminus (M(d)^+ \cup M(d)^-)$, $d \in D$. If $M(d) = \emptyset$, then $M(d)^+ \cup M(d)^- = M$, which is the completely informed case, such that for each considered disorder $d \in D$ we can decide, whether its presence implies the presence of a specific manifestation or not. Hence, $M(d) \neq \emptyset$ reflects the case of incomplete information.

For all $m \in M$ let

$$
\begin{aligned}
D(m)^+ &\stackrel{\text{Df}}{=} \{d \in D \mid m \in M(d)^+\},\\
D(m)^- &\stackrel{\text{Df}}{=} \{d \in D \mid m \in M(d)^-\}, \text{ and}\\
D(m) &\stackrel{\text{Df}}{=} D \setminus (D(m)^+ \cup D(m)^-).
\end{aligned}
$$

Furthermore we define for arbitrary index sets $I \subseteq N$, $I \neq \emptyset$, the relations

$$
\begin{aligned}
P^I &\stackrel{\text{Df}}{=} \{\varphi \mid \varphi : I \rightarrow \bigcup_{i \in I} \Omega_i \wedge (\forall i \in I)\,(\varphi(i) = \text{present})\},\\
A^I &\stackrel{\text{Df}}{=} \{\varphi \mid \varphi : I \rightarrow \bigcup_{i \in I} \Omega_i \wedge (\forall i \in I)\,(\varphi(i) = \text{absent})\},\\
\overline{P}^I &\stackrel{\text{Df}}{=} \Omega^I \setminus P^I, \text{ and}\\
\overline{A}^I &\stackrel{\text{Df}}{=} \Omega^I \setminus A^I.
\end{aligned}
$$

In our setting, the knowledge of $M(d)^+$ and $M(d)^-$ for all disorders $d \in D$, and the interpretation of these sets of manifestations in the sense of [2], corresponds to stating the following imprecise inference rules:

$$
\begin{aligned}
&(1)\ \textit{if } \xi^{\{d\}} \textit{ in } P^{\{d\}} \textit{ then } \xi^{\{m\}} \textit{ in } P^{\{m\}},\\
&(2)\ \textit{if } \xi^{\{m\}} \textit{ in } P^{\{m\}} \textit{ then } \xi^{D^*(m)} \textit{ in } \overline{A}^{D^*(m)},\\
&\text{where } D^*(m) = D(m)^+ \cup D(m)\ .
\end{aligned}
$$

Rules of type (1) relate to all $m \in M(d)^+$ and $d \in D$ such that $M(d)^+ \neq \emptyset$, whereas rules of type (2) refer to all $m \in M$ such that $D(m)^+ \cup D(m) \neq \emptyset$.

Note that the whole general knowledge in this approach concerns only direct relationships among disorders and manifestations, but does not include intermediary chracteristics.

Turning over to the evidential knowledge, Dubois and Prade [2] refer to the set M^+ of manifestations that are certainly present, and the set M^- of manifestations that are certainly absent, where in the case of incomplete information $M^+ \cup M^- \neq M$ holds.

From our viewpoint this induces the evidential system $\mathcal{E}(\mathcal{U}, \mathcal{N}) = \{E^J \mid J \in \mathcal{N}\}$, defined by

$$E^{\{m\}} = \begin{cases} \{\text{absent}\} & \text{, iff } m \in M^- \\ \{\text{present}\} & \text{, iff } m \in M^+ \\ \{\text{absent,present}\} & \text{, iff } m \in M \setminus (M^+ \cup M^-) . \end{cases}$$

□

Example 2

In abductive reasoning research, another relation-based formulation of diagnostic inference has been studied by Peng and Reggia [12], [10]. In their basic model, they assume the knowledge of a causation $\mathcal{C} \subseteq D \times M$ between disorders and manifestations, such that $(d, m) \in \mathcal{C}$ means that disorder d may directly cause m. Hence, effects$(d) \stackrel{\text{Df}}{=} \{m \mid (d, m) \in \mathcal{C}\}$ is the set of all objects directly caused by d, whereas causes$(m) \stackrel{\text{Df}}{=} \{d \mid (d, m) \in \mathcal{C}\}$ is the set of all objects which *might* directly cause m.
In our setting, $\mathcal{C}$ with its specific interpretation corresponds to stating the imprecise inference rules

$$(1') \; \textit{if } \xi^{\{d\}} \textit{ in } P^{\{d\}} \textit{ then } \xi^{\textit{effects}(d)} \textit{ in } \overline{A}^{\textit{effects}(d)},$$

$$(2') \; \textit{if } \xi^{\{m\}} \textit{ in } P^{\{m\}} \textit{ then } \xi^{\textit{causes}(m)} \textit{ in } \overline{A}^{\textit{causes}(m)}.$$

for all $d \in D$ and $m \in M$, respectively, presupposing that

$$\text{domain}(\mathcal{C}) \stackrel{\text{Df}}{=} \left\{d \in D \mid (\exists m \in M)((d, m) \in \mathcal{C})\right\} = D$$

and

$$\text{range}(\mathcal{C}) \stackrel{\text{Df}}{=} \left\{m \in M \mid (\exists d \in D)((d, m) \in \mathcal{C})\right\} = M .$$

Evidential knowledge in the referenced approach specifies the set M^+ of manifestations that are present with respect to ω_0.
Consequently, the induced evidential system $\mathcal{E}(\mathcal{U}, \mathcal{N}) = \{E^J \mid J \in \mathcal{N}\}$, is defined by

$$E^{\{m\}} = \begin{cases} \{\text{present}\} & \text{, iff } m \in M^+ \\ \{\text{absent,present}\} & \text{, iff } m \in M \setminus M^+ . \end{cases}$$

□

Concerning the two considered examples, it should be mentioned that the concepts of a focusing system and an evidential system are general enough to specify arbitrary dependencies between the values of all considered attributes, whether they are disorders, manifestations, or intermediary characteristics, respectively. For this reason it is no problem to generalize example 1 to the treatment of dependencies between sets of disorders and sets of manifestations, as suggested, for instance, in [2]. Finally note that in our approach the underlying domains Ω_i, $i = 1, \ldots, n$, need not to be restricted to the special case $\Omega_i = \{\text{present, absent}\}$, which is of major relevance for many types of diagnostic systems.

4 Parsimonious Covers

In the previous sections we concentrated our interest on a first step of abductive reasoning, which in our framework is characterized by the determination of the current state $\sigma(\mathfrak{X}, \mathcal{E}(\mathcal{U}, \mathcal{N}))$ of a focusing system $\mathfrak{X}$, induced by the evidential system $\mathcal{E}(\mathcal{U}, \mathcal{N})$ that serves as the specification of an observation of the object state $\omega_0 \in \Omega$ under consideration and strongly relates to the presence or absence of selected manifestations (i.e. $\mathcal{N} = \{\{m\} \mid m \in M\}$).
The second step of abductive reasoning refers to finding ***plausible explanations*** (in terms of disorders) for the set of present manifestations. With respect to [9], the tuple $P = (D, M, C, M^+)$ with components as defined above, is called a ***classical diagnostic problem***. A set $D^* \subseteq D$ is said to be an ***explanation*** of M^+ for the problem P, iff D^* ***covers*** M^+(which means that $M^+ \subseteq \text{effects}(D^*)$), and satisfies given ***parsimony criteria*** (relevancy, irredundancy, and minimum cardinality being just three of them). More particularly, D^* is ***relevant***, iff $D^* \subseteq \text{causes}(M^+)$, ***irredundant***, iff none of the proper subsets of D^* also covers M^+, and ***minimum***, iff D^* has the smallest cardinality among all covers of M^+.
Note that relevancy is a looser criterion than irredundancy, and irredundancy itself is looser than minimum cardinality.
Transferred to our setting, using a focusing system $\mathfrak{X}$ and an evidential system $\mathcal{E}(\mathcal{U}, \mathcal{N})$, propagation yields for any subset D^* of disorders:

$$D^* \text{ covers } M^+, \text{ iff } (\forall d \in D^*)\,(\text{present} \in \pi^{(d)}),$$

$$D^* \text{ is a relevant cover of } M^+, \text{ iff } (\forall d \in D^*)\,(\pi^{(d)} \not\subseteq \{\text{absent}\}).$$

Note that each propagation algorithm that calculates $\sigma(\mathfrak{X}, \mathcal{E}(\mathcal{U}, \mathcal{N}))$ and the projections $\pi^{(d)}$, $d \in D$, induces the relevant cover $D^* = \{d \mid \pi^{(d)} \not\subseteq \{\text{absent}\}\}$. This is the best result one can expect from a propagation algorithm, since irredundancy as well as minimality are additional, separate concepts that go further than applying the available information $\mathcal{R}(\mathcal{U}, \mathcal{I})$ and $\mathcal{E}(\mathcal{U}, \mathcal{N})$ on ω_0 to compute $\sigma(\mathfrak{X}, \mathcal{E}(\mathcal{U}, \mathcal{N}))$ as the most specific relation that guarantees the truth of $\omega_0 \in \sigma(\mathfrak{X}, \mathcal{E}(\mathcal{U}, \mathcal{N}))$.

A quite simple way of using the propagation algorithm to obtain an irredundant cover of M^+ is the following:
Determine the mentioned relevant cover D^* of M^+, select non–empty subsets D_j of D^*, and start new propagations with respect to the modified evidential system $\mathcal{E}_j = \mathcal{E}(\mathcal{U}, \mathcal{N}_j)$, $\mathcal{N}_j = \{\{i\} \mid i \in M \cup D_j\}$, where $\mathcal{E}_j = \{E_j^{\{i\}} \mid i \in M \cup D_j\}$ is defined by

$$E_j^{\{i\}} = \begin{cases} \{\text{present}\} & \text{, iff } i \in M^+ \\ \{\text{absent,present}\} & \text{, iff } i \in M \setminus M^+ \\ \{\text{absent}\} & \text{, iff } i \in D_j \,. \end{cases}$$

The subsets D_j are chosen to fulfil $D_0 = D^*$ and $D_{j+1} \subseteq D_j$, $|D_j| = |D_{j-1}| - 1$ for $j = 1, \ldots, |D^*|$. Let k be the smallest index such that

$$\left(\exists m \in M^+\right)\left(\prod_{\{m\}}^{N}(\sigma(\mathfrak{X},\mathcal{E}_k)) = \emptyset\right).$$

Then, D_{k-1} is an irredundant cover of M^+.
Note that the presented iteration proposes to find irredundant covers by means of the propagation algorithm itself.
A reasonable alternative approach is that of using algebras of generator–sets as investigated in parsimonious covering theory [10].
Nevertheless there are crucial complexity problems in applying algorithms for the determination of irredundant or even minimal covers. The problem of finding all minimal covers of M^+ is NP–hard [12]. A combinatorial explosion with an increasing cardinality of M^+ presumably also occurs when one tries to calculate all irredundant covers for a given M^+. Approximative algorithms may circumvent such difficulties.

5 Possibilistic Abduction

So far we have restricted ourselves to consider a formal environment for abductive reasoning that is capable of handling imprecise, but certain knowledge. On the other hand, our model can of course be generalized to the treatment of uncertain information. Among the well–known uncertainty calculi for numerical settings such as Bayes theory, Dempster–Shafer theory, possibility theory, and fuzzy set theory, a promising and straight forward extension of our approach refers to possibility theory in the sense of Zadeh [17], where possibility distributions are introduced as the epistemic counterparts of fuzzy sets rather than via the definition of possibility measures and necessity measures, respectively. Diagnosis methods based on a fuzzy relational model were already developed more than ten years ago, e.g. [14, 16, 15], but an existing problem concerning those approaches is the fact that it is not clear how to exactly interpret degrees of association in fuzzy relations between disorders and manifestations or degrees of membership in a fuzzy set of present manifestations. For this reason we prefer to consider *possibility functions* $\pi : \Omega \rightarrow [0,1]$, where $\pi(\omega)$, applied to our reasoning process, quantifies the degree of possibility with which $\omega \in \Omega$ equals the current object state ω_0 of interest. Note that we talk about possibility functions rather than possibility distributions, since in our setting it is more appropriate to reject the normalization assumption and to view π as the one–point coverage of a random set (γ, P), $\gamma : C \rightarrow 2^{\Omega}$ [7]. C is a finite set of consideration contexts for ω_0, $\gamma(c)$ the specification of ω_0 in context $c \in C$, saying that $\omega_0 \in \gamma(c)$ is true, if c is the true context (which is an event that occurs with probability $P(\{c\})$). With this semantic background π can be defined as $\pi \equiv \pi_\Gamma$, where $\pi_\Gamma : \Omega \rightarrow [0,1]$, $\pi_\Gamma(\omega) \stackrel{\text{Df}}{=} P(\{c \in C \mid \omega \in \gamma(c)\})$ is fulfilled. Regarding a detailed discussion of possibility functions in a more general view of consideration contexts, including an investigation of reasonable operations on possibilistic data, we refer to [3].
Note that $\pi(\omega)$ is the mass of all contexts $c \in C$ that support the possibility of truth of "$\omega = \omega_0$" and therefore serves as a possibility degree. Furthermore,

note that π_Γ can be interpreted as an information–compressed representation of Γ, restricting to pure possibilistic aspects, avoiding the explicit treatment of underlying consideration contexts.

Based on the random set approach, π_Γ reflects the occurence of uncertainty (probabilistic background in form of the probability measure P) and imprecision (due to the set–valued function γ) for the imperfect specification of ω_0. In the special case of precision (i.e. $|\gamma(c)| = 1$ for all $c \in C$), π_Γ formally coincides with a probability distribution on Ω, but $\pi_\Gamma(\omega)$ should in general not be mixed up with a probability mass, since — from a semantic point of view — possibility degrees and probability masses are quite different concepts.

With respect to the realization of possibilistic abduction it has to be pointed out that the whole discussion on focusing systems and evidential systems presented in the previous sections can easily be adopted for the possibilistic case in the way that *possibilistic rule systems* $\mathcal{R}(\mathcal{U}, \mathcal{I}) = \{\rho^I \mid I \in \mathcal{I}\}$ and *possibilistic evidential systems* $\mathcal{E}(\mathcal{U}, \mathcal{N}) = \{\varepsilon^J \mid J \in \mathcal{N}\}$ with possibility functions $\rho^I : \Omega^I \to [0, 1]$ and $\varepsilon^J : \Omega^J \to [0, 1]$, respectively, are applied.

In this connection all operations on relations used for the propagation algorithm (which are intersection, cylindrical extension, and projection) have to be replaced by their possibilistic counterparts (e.g. the min–operation for intersection).

Furthermore it turns out that the mentioned context–related view of possibility functions requires to interpret possibilistic inference rules (as a generalization of imprecise inference rules discussed in section 3) with the aid of the corresponding *Gödel relation*. For a clarification of this fact, see, for example, [3].

6 Further Remarks

Possibilistic abduction as introduced in the previous sections, using the concept of a possibilistic focusing system $(\mathcal{U}, \mathcal{I}, \mathcal{N}, \mathcal{R}(\mathcal{U}, \mathcal{I}))$, $\mathcal{U} = (\Omega_i)_{i \in N}$, is going to be realized as an extension of the software tool POSSINFER (Possibilistic Inference) [6], which supports the calculation of relevant covers in a multidimensional space of disorders, manifestations, and intermediary characteristics. This means that qualitative dependencies are represented in terms of a hypergraph $(N, \mathcal{I})$, whereas quantitative dependencies are defined as possibility relations w.r.t. the subspaces Ω^I, $I \in \mathcal{I}$. More details on the semantic background of possibilistic focusing systems as well as quantitative and qualitative learning in such systems are out of the scope of this paper, but have been investigated in [4, 6].

An additional interesting issue to be investigated later on is to work out relationships between numerical and symbolic settings for abductive reasoning [11]. As Reiter's logical model [13] encodes the basis of parsimonious covering theory [12], it should be expected to find references between the concept of a possibilistic focusing system and methods related to possibilistic assumption–based truth–maintenance systems [1].

References

1. S. Benferhat, D. Dubois, J. Lang, and H. Prade. Hypothetical reasoning in possibilistic logic: Basic notions and implementation issues. In P.Z. Wang and K.F. Loe, editors, *Advances in Fuzzy Systems – Vol. 1.* World Scientific Publisher, Singapore, 1992.
2. D. Dubois and H. Prade. A fuzzy relation–based extension of Reggia's relational model for diagnosis handling uncertain and incomplete information. In *Proc. 9th Conf. on Uncertainty in Artificial Intelligence, Washington*, pages 106–113, 1993.
3. J. Gebhardt and R. Kruse. A new approach to semantic aspects of possibilistic reasoning. In M. Clarke, R. Kruse, and S. Moral, editors, *Symbolic and Quantitative Approaches to Reasoning and Uncertainty, Lecture Notes in Computer Science, 747,*, pages 151–160. Springer, Berlin, 1993.
4. J. Gebhardt and R. Kruse. Learning possibilistic networks from data. In *Proc. 5th Int. Workshop on Artificial Intelligence and Stati stics*, pages 233–244, Fort Lauderdale, 1995.
5. J. de Kleer and B.C. Williams. Diagnosing multiple faults. *Artificial Intelligence*, 32:97–130, 1987.
6. R. Kruse, J. Gebhardt, and F. Klawonn. *Foundations of Fuzzy Systems.* Wiley, Chichester, 1994.
7. H.T. Nguyen. On random sets and belief functions. *J. of Mathematical Analysis and Applications*, 65:531–542, 1978.
8. J. Pearl. *Probabilistic Reasoning in Intelligent Systems: Networks of Plausible Inference.* Morgan Kaufmann, New York, 1988.
9. Y. Peng and J.A. Reggia. Diagnostic problem solving with causal chaining. *Int. J. of Intelligent Systems*, 2:265–302, 1987.
10. Y. Peng and J.A. Reggia. *Abductive Inference Models for Diagnostic Problem-Solving.* Springer, New York, 1990.
11. D. Poole. Probabilistic horn anduction and bayesian networks. *Artificial Intelligence*, 64:81–129, 1993.
12. J.A. Reggia, D.S. Nau, P.Y. Wang, and H. Peng. A formal model of diagnostic inference. *Information Sciences*, 37:227–285, 1985.
13. R. Reiter. Nonmonotonic reasoning. *Annual Review of Computer Science*, 2:147–186, 1987.
14. E. Sanchez. Solutions in composite fuzzy relation equations: Application to medical diagnosis in Brouwerian logic. In M.M. Gupta, G.N. Saridis, and B.R. Gaines, editors, *Fuzzy Automata and Decision Processes*, pages 221–234. North Holland, Amsterdam, 1977.
15. E. Sanchez. Inverses of fuzzy relations, application to possibility distributions and medical diagnosis. *Fuzzy Sets and Systems*, 2:75–86, 1979.
16. Y. Tsukamoto and T. Terano. Failure diagnosis by using fuzzy logic. In *Proc. IEEE Conf. on Decision and Control, New Orleans*, pages 1390–1395, 1977.
17. L.A. Zadeh. Fuzzy sets as a basis for a theory of possibility. *Fuzzy Sets and Systems*, 1:3–28, 1978.

Possibility Theory and Independence

Luis Fariñas del Cerro, Andreas Herzig

I.R.I.T.- Université Paul Sabatier
118 Route de Narbonne, F-31062 Toulouse Cedex (France)
email: {farinas, herzig}@irit.fr, fax: (33) 61 55 62 58

Abstract. We study the links between uncertainty and the notion of dependence. We investigate three possible definitions of dependence that are constructed from possibility measures. One of them has been proposed be Zadeh, and the two others are based on the notion of conditional possibility. We show that the latter have enough expressive power to support the whole possibility theory and that complete axiomatizations can be given.

1 Introduction

The notion of epistemic independence has been studied in the framework of reasoning under uncertainty. It can be derived naturally using conditioning: "C is independent of A iff the uncertainty degree of C is equal to the uncertainty degree of B knowing A."

Traditionally, the formal basis of the dependence relation (that has also been called relevance relation) is probability theory, or more precisely conditional probability: Given two formulas A and C and a probability measure Prob, *A is (probabilistically) independent of C* iff Prob(A|C) = Prob(A). It follows from the axioms of probability theory that dependence is a symmetric relation, and that it is not sensitive to negation. In other words, if A depends on C then C depends on A, and A depends on ¬C. Unfortunately it turns out that we cannot study the formal properties of probabilistic dependence separately from the probabilistic framework: The axiomatisation of dependence involves qualitative probabilities (see (Kolmogorov 1956), cited in (Fine 1973)). We note that Gärdenfors (1978) has criticised such a definition, and has investigated several weaker ones.

In this paper we show that possibilistic independence based on possibility theory (developed by Zadeh, Dubois and Prade) has quite different properties. We present three definitions of dependence and study their formal properties. For two of them we show that they have enough expressive power to support the whole possibility theory, and that complete axiomatizations not involving the underlying uncertainty theory can be given. Finally we show that one must be careful in the use of dependence in the context of belief revision: We prove that we get an impossibility theorem if we naively employ a preservation principle based on dependence.

2 Possibility Theory

In this section we introduce possibility measures and possibilistic independence.

Let FOR be the set of formulas of a language of propositional logic. Possibility measures allow to associate an uncertainty degree to the elements of FOR. Following

(Dubois, Prade 1986), a function Π from FOR into the real interval [0,1] is a *possibility measure*[1] if it satisfies the three axioms $\Pi(\text{True}) > \Pi(\text{False})$, $\Pi(A \vee B) = \max(\Pi(A), \Pi(B))$, and If $A \leftrightarrow B$ then $\Pi(A) = \Pi(B)$.

Every possibility measure induces a relation "≥" defined by $A \geq B$ iff $\Pi(A) \geq \Pi(B)$. $A \geq B$ is read "A is at least as possible as B". For such relations we have:

(non triviality)	True > False
(transitivity)	If $A \geq B$ and $B \geq C$ then $A \geq C$
(disjunctiveness)	$A \vee C \leq A$ or $A \vee C \leq C$
(dominance)	If $A \rightarrow C$ then $A \leq C$

Relations satisfying these conditions are called *qualitative possibility relations*. There is the following formal relation between possibility theory and qualitative possibility relations (Dubois 1986):

Theorem (qualitative possibility relations). Let Π be any measure on FOR, and ≥ any binary relation on FOR. Then Π is a possibility measure iff ≥ is a qualitative possibility relation.

Following Hisdal (1978) and Dubois and Prade (1986, 1992) the *conditional possibility* $\Pi(A|C)$ is defined as the maximal solution of the equation $\Pi(A \wedge C) = \min(\Pi(A|C), \Pi(C))$. This definition is clearly inspired from Bayes' Rule, where min corresponds to the product. A solution of the equation is

$$\Pi(A|C) = 1 \qquad \text{if } \Pi(C) = \Pi(A \wedge C)$$

$$\Pi(A|C) = \Pi(A \wedge C) \qquad \text{if } \Pi(C) > \Pi(A \wedge C)$$

Facts.

1. If $\Pi(C) = 0$ then $\Pi(A|C) = 1$
2. If $\Pi(A \wedge C) = 1$ then $\Pi(A|C) = 1$
3. If $\Pi(A) = 0$ and $\Pi(C) > 0$ then $\Pi(A|C) = 0$
4. $\Pi(A|\neg A) = 1$ iff $\Pi(\neg A) = 0$
5. $\Pi(A|\neg A) = 0$ iff $\Pi(\neg A) > 0$

---

[1] The statement of the axioms is a little bit unusual for two reasons: First, we have adapted the original terminology and give the definition in terms of formulas (an not in terms of events being sets of elementary events). Therefore we had to add a further clause saying that logically equivalent formulas are equally possible. Second, the usual presentation contains a stronger version of the first axiom, namely $\Pi(\text{True}) = 1$ and $\Pi(\text{False}) = 0$. We prefer the weaker version, because the stronger one is just a convention (Dubois, Prade 1992), which obviously has no qualitative counterpart (which we need in the sequel).

In the following sections we present three different definitions of dependence. Two of them are based on the notion of conditional possibility. We show that both can express qualitative possibility, and that complete axiomatizations can be given for them. We conjecture that this is not possible for the third. In all three cases, a necessary condition for the independence of A and C is that the conjunction $A \wedge C$ is interpreted truth-functionally, in the sense that $\Pi(A \wedge C) = \min(\Pi(A), \Pi(C))$.

3 The Non-Symmetric Dependence Relation

In the sequel we suppose given some qualitative possibility relation $\leq$. The *dependence relation* can then be defined in the same way as for probabilities:

(Def ≈≈>) *A depends on C* (denoted by C ≈≈> A) iff $\Pi(A|C) \neq \Pi(A)$.

Dually, we say that A is *independent of* C (denoted by C ≈/≈> A) if it is not the case that A depends on C. A synonymous expression for "A depends on C" is "C is relevant for A".

Facts.

1. If A ≈≈> B∨C then A ≈≈> B or A ≈≈> C
2. If A∨B ≈≈> C then A ≈≈> C or B ≈≈> C
3. If A ≈≈> C and B ≈≈> C then A∨B ≈≈> C
4. If A ≈≈> B and A ≈≈> C then A ≈≈> B∨C
5. False ≈≈> False
6. True ≈/≈> A
7. C ≈/≈> True
8. C ≈/≈> False iff $\Pi(C) > 0$
9. A ≈/≈> A iff False ≈/≈> A iff A∧C ≈/≈> A iff $\Pi(A) = 1$
10. A ≈≈> A∨C iff $\Pi(A \vee C) < 1$
11. A∨C ≈≈> A iff $1 > \Pi(A) \geq \Pi(C)$
12. A∨C ≈/≈> A∨¬C iff $\Pi(A) \leq \Pi(C)$ and $\Pi(A) \leq \neg\Pi(C)$
13. ¬A ≈/≈> A iff A ≈/≈> ¬A iff ($\Pi(A) = 1$ or $\Pi(\neg A) = 0$)
14. If $\Pi(C) = 0$ then (C ≈/≈> A iff $\Pi(A) = 1$)
15. If $\Pi(A) = 1$ then (C ≈/≈> A iff $\Pi(C) > 0$)

Perhaps the only fact whose proof is not immediate is 12: A∨C ≈/≈> A∨¬C means that $\Pi(A) = \min(\Pi(A \vee C), \Pi(A \vee \neg C))$ and ($\Pi(A \vee C) = 1$ or $\Pi(A \vee C) < \Pi(A \vee \neg C)$). Now the second condition always holds, because $\Pi(A \vee C) = 1$ or $\Pi(A \vee \neg C) = 1$, and the first condition means that $\Pi(A) \geq \Pi(C)$ or $\Pi(A) \geq \Pi(\neg C)$.

A ≈≈> A means that "A depends on A" or "If I am told that A is true I may change my opinion on the possibility of A". This is the case only if the possibility of A is less than 1, i.e., A was not completely possible for me. (Else the information that A is true does not teach me anything.)

Clearly, probabilistic dependence and possibilistic dependence are quite different concepts. Probabilistic properties such as symmetry ("If B depends on A then A depends on B") or transparency w.r.t. negation ("If B depends on A then B depends on ¬A") do not hold in the possibilistic case. In other words, A ≈/≈> B neither implies B ≈/≈> A nor A ≈/≈> ¬B. On the other hand, possibilistic dependence has some "nice" properties such as 1., 2., 3., 4.. neither of which holds in the probabilistic case case.

In the next theorem we characterize our dependence relation without using conditional possibilities.

Theorem (from Π to ≈≈>). Let Π be a possibility measure, and let ≈≈> be defined from Π.

1. C ≈/≈> A iff $\Pi(A \wedge C) = \min(\Pi(A), \Pi(C))$ and ($\Pi(A) = 1$ or $\Pi(A) < \Pi(C)$).

2. C ≈≈> A iff $\Pi(A \wedge C) \neq \min(\Pi(A), \Pi(C))$ or $1 > \Pi(A) \geq \Pi(C)$.

Proof. $\Pi(A|C) = \Pi(A)$ means that either $\Pi(C) = \Pi(A \wedge C)$ and $\Pi(A|C) = \Pi(A) = 1$, or $\Pi(C) > \Pi(A \wedge C)$ and $\Pi(A|C) = \Pi(A \wedge C) = \Pi(A)$. The former is equivalent to $\Pi(A \wedge C) = \min(\Pi(A), \Pi(C))$ and $\Pi(A) = 1$. The latter case is equivalent to $\Pi(A \wedge C) = \min(\Pi(A), \Pi(C))$ and $\Pi(A) < \Pi(C)$.

It is interesting to observe that the dependence relation possesses the same expressive power as possibility theory itself. This follows from the next result.

Theorem (from ≈≈> to Π). Let Π be a possibility measure, and let ≈≈> be defined from Π.

1. $\Pi(A) < \Pi(C)$ iff (A∨C ≈/≈> A and A∧C ≈≈> A) iff (A∨C ≈/≈> A and A ≈≈> A)

2. $\Pi(A) \geq \Pi(C)$ iff (A∨C ≈≈> A or A∧C ≈/≈> A) iff (A∨C ≈≈> A or A ≈/≈> A)

Proof. Putting 1. and 8. of the previous fact together, A∨C ≈/≈> A and A∧C ≈≈> A is equivalent to $\Pi(A) < \Pi(C)$ and $\Pi(A) < 1$.

The theorem can be read as follows: A is strictly less possible than C if and only if learning that A∧C is true may make me change my mind on A, whereas learning that A∨C is true does not (because it is more possible that it was C that made A∨C true). It follows from the theorem that we are able to express qualitative possibilities by means of ≈≈>. In a trivial manner, this correspondence enables us to obtain an axiomatization of the dependence relation by translating the qualitative counterpart of possibility theory. Note that by contrary in probability theory, the independence relation can not capture completely qualitative probability (which on its turn characterises exactly the probability measure). Here we give a simpler *axiomatization* of ≈≈>:

(≈≈> 1) True ≈/≈> A

(≈≈> 2) False ≈≈> False

(≈≈> 3) If (A∨B) ≈≈> A and (B∨C) ≈≈> B then (A∨C) ≈≈> A

(≈≈> 4) If (A∨C) ≈≈> A then C ≈≈> C

(≈≈> 5) If (A ≈≈> A and C ≈≈> C) then (A∨C) ≈≈> (A∨C)

(≈≈> 6) If A ≈≈> B∨C then A ≈≈> B or A ≈≈> C

(≈≈> 7) If A ↔ B then (A ≈≈> C iff B ≈≈> C)

Theorem (soundness and completeness of the axiomatics of ≈≈>). Let ≈≈> be a relation on formulas, and Π a mapping from the set of formulas to $[0,1]$ such that C≈≈>A iff $\Pi(A|C) \neq \Pi(A)$. Then ≈≈> is a dependence relation iff Π is a possibility measure.

Proof. From the right to the left, it is sufficient to prove that the above axioms (rewritten as qualitative possibilities) are valid. Then we can use the soundness of qualitative possibility relations w.r.t. possibility theory.

From the left to the right, we prove that the axioms of qualitative possibility (expressed through ≈≈>) are derivable from the above axiomatics (and then use the completeness of qualitative possibility relations w.r.t. possibility theory).

1. (non triviality) True > False becomes False∨True ≈/≈> False and False ≈≈> False. Via (≈≈> 7), the former is an instance of (≈≈> 1) and the latter is (≈≈> 2).

2. (transitivity) if A ≥ B and B ≥ C then A ≥ C becomes "If (A≈/≈>A or A∨B≈≈>A) and (B≈/≈>B or B∨C≈≈>B) then (A≈/≈>A or A∨C≈≈>A)". This can be proved by establishing both "If A∨B ≈≈> A and B ≈/≈> B then (A ≈/≈> A or A∨C ≈≈> A)", and "If A∨B ≈≈> A and B∨C ≈≈> B then (A ≈/≈> A or A∨C ≈≈> A)". The antecedens of the first clause cannot be the case because of (≈≈> 4), and the second clause follows from (≈≈> 3).

3. (disjunctiveness) A≥A∨C or C≥A∨C is A∨C≈≈>A or A≈/≈>A or A∨C≈≈>C or C≈/≈>C. Suppose A ≈≈> A and C ≈≈> C. By (≈≈> 5), we have (A∨C) ≈≈> (A∨C). From that it follows by (≈≈> 6) that (A∨C) ≈≈> A or (A∨C) ≈≈> C.

5. (dominance) can be replaced by the two axioms If A ↔ C then C ≤ A (equivalence) and A∨C ≥ A (monotony). The latter is translated to (A∨C∨C) ≈≈> (A∨C) or (A∨C) ≈/≈> (A∨C), which (via ≈≈> 7) is trivially the case. Hence what remains to prove is "If A ↔ C then A∨C ≈≈>A or A ≈/≈> A". Now from A ↔ C we get A ↔ A∨C. From the latter we get (A ≈≈> A iff A∨C ≈≈> A) by (≈≈> 7). It follows that A ≈/≈> A or A∨C ≈≈> A.

4 The Symmetric Dependence Relation

Contrary to probability-based dependence, the fact that A depends on C does not imply that C depends on A. If we want to exclude the case of non-symmetric dependence relations, we get a stronger notion that we study in the present section.

(Def $\approx\approx$) *A and C are dependent* (denoted by $A \approx\approx C$) iff $A \approx\approx> C$ or $C \approx\approx> A$

As expected, A and C are *independent* (denoted by $C \approx/\approx A$) if A and C are not dependent.

Facts.

0. If $A \approx\approx> C$ then $A \approx\approx C$, but the converse is not true in general.
1. If $A \approx\approx B \vee C$ then $A \approx\approx B$ or $A \approx\approx C$
2. If $A \vee B \approx\approx C$ then $A \approx\approx C$ or $B \approx\approx C$
3. If $A \approx\approx C$ and $B \approx\approx C$ then $A \vee B \approx\approx C$
4. If $A \approx\approx B$ and $A \approx\approx C$ then $A \approx\approx B \vee C$
5. False $\approx\approx$ False
6. True $\approx/\approx$ A
7. C $\approx/\approx$ True
8. C $\approx/\approx$ False iff $\Pi(C) = 1$
9. $A \approx/\approx A$ iff False $\approx/\approx A$ iff $A \wedge C \approx/\approx A$ iff $\Pi(A) = 1$
10. $A \approx\approx A \vee C$ iff $\Pi(A \vee C) < 1$
11. $A \vee C \approx/\approx A \vee \neg C$ iff $\Pi(A) \geq \min(\Pi(C), \Pi(\neg C))$
12. $A \approx/\approx \neg A$ iff $\Pi(A) = 0$ or $\Pi(\neg A) = 0$

E.g. the first fact in not true for probabilistic independence. To make it true we need some supplementary proviso: The sentences A, C and B must be probabilistically independent. Again, we can characterize our dependence relation without using conditional possibilities as follows:

Theorem (from Π to $\approx\approx$). $C \approx/\approx A$ iff $(\Pi(A \wedge C) = \min(\Pi(A), \Pi(C))$ and $\Pi(A \vee C) = 1)$.

Proof. Acording to the definition, $C \approx/\approx A$ means that the following three conditions hold:

(a) $\Pi(A \wedge C) = \min(\Pi(A), \Pi(C))$

(b) $\Pi(A) = 1$ or $\Pi(A) < \Pi(C)$

(c) $\Pi(C) = 1$ or $\Pi(C) < \Pi(A)$

Now (b) and (c) are equivalent to $\Pi(A) = \Pi(C) = 1$, or $\Pi(A) < 1 = \Pi(C)$, or $\Pi(C) < 1 = \Pi(A)$, which is nothing else than the condition that $\Pi(A) = 1$ or $\Pi(C) = 1$.

Again we observe that $\approx\approx$ possesses the same expressive power as possibility theory itself. This follows from the following theorem, which is slightly more complicated than that for $\approx\approx>$.

Theorem (from $\approx\approx$ to Π). $\Pi(A) \geq \Pi(C)$ iff $\neg C \approx/\approx$ False and $A \vee C \approx/\approx A \vee \neg C$.

Proof. It suffices to note that $\neg C \approx/\approx$ False and $A \vee C \approx/\approx A \vee \neg C$ means $\Pi(\neg C) = 1$ and $\Pi(A) \geq \min(\Pi(C), \Pi(\neg C))$, which on its turn is equivalent to $\Pi(A) \geq \Pi(C)$.

Thus we are again able to express qualitative possibilities by means of $\approx\approx$. As before, via translation of the qualitative counterpart of possibility theory this correspondence enables us to obtain an axiomatization of $\approx\approx$, together with a soundness and completeness theorem.

5 Zadeh's Dependence Relation

Zadeh (1978) gave a definition of independence that is not based on conditional possibilities:

(Def $\approx\approx_Z$) *A and C are dependent in Zadeh's sense* (denoted by $A \approx\approx_Z C$)

iff $\Pi(A \wedge C) \neq \min(\Pi(A), \Pi(C))$.

Facts.

0. If $A \approx\approx C$ then $A \approx\approx_Z C$, but the converse is not true in general.
1. If $A \approx\approx_Z B \vee C$ then $A \approx\approx_Z B$ or $A \approx\approx_Z C$
2. If $A \vee B \approx\approx_Z C$ then $A \approx\approx_Z C$ or $B \approx\approx_Z C$
3. If $A \approx\approx_Z C$ and $B \approx\approx_Z C$ then $A \vee B \approx\approx_Z C$
4. If $A \approx\approx_Z B$ and $A \approx\approx_Z C$ then $A \approx\approx_Z B \vee C$
5. False $\approx\approx_Z A$
6. True $\approx/\approx_Z A$
7. $A \approx/\approx_Z A$
8. $A \approx/\approx_Z \neg A$ iff $\Pi(A) = 0$ or $\Pi(\neg A) = 0$
9. $A \approx\approx_Z C$ iff $C \approx\approx_Z A$
10. $A \vee C \approx/\approx_Z A$

Here, there seems to be no way to express Π by means of $\approx\approx_Z$, the reason being that we cannot express $\Pi(A) = 1$. Therefore, we conjecture that (just as in the case of probabilistic independence) $\approx\approx_Z$ cannot be axiomatized alone.

6 Possibility Theory, Belief Revision, and Dependence: An Impossibility Theorem

6.1 Possibility Theory and Revision

The notion of dependence is certainly a useful tool in the context of belief revision: If we suppose given a dependence relation ≈≈>, the following requirement seems to be natural:

(I*) For every A and C, if A≈/≈>C and $C \in K$ then $C \in K^*A$

Just as in the case of nonmonotonic reasoning, belief revision can be based on particular orderings on formulas called epistemic entrenchment orderings. A given ordering $\leq_{EE}$ determines a knowledge base K by K = {p: False $<_{EE}$ p}. There is a *representation theorem* saying that the revision operations satisfying the AGM postulates are exactly those generated by particular orderings on formulas called epistemic entrenchment orderings. Formally, $C \in K^*A$ iff $\models \neg A$ or $\neg A <_{EE} \neg A \vee C$. If we replace $\leq_{EE}$ by its dual relation $\leq$ defined by "$A \leq B$ iff $\neg A$ is at least as entrenched as $\neg B$", we get $C \in K^*A$ iff $\models \neg A$ or $A \wedge \neg C < A$, and it has been shown by Dubois and Prade (1991a) that the uncertainty theory associated to $\leq$ is possibility theory. Moreover the axioms "$C \in K$ iff $\Pi(\neg C)=0$" (maximality) and "If $A \leq C$ for all C, then A is inconsistent" (inconsistence) must be satisfied.

In the sequel we show that dependence cannot be based on epistemic entrenchment: We prove that if a possibility-based dependence relation ≈≈> satisfies the property stated in the preceeding section there is no AGM revision operation * that satisfies (I*).

6.2 The Impossibility Theorem

A common feature of all three notions of dependence that we have defined in possibility theory.is that the independence of A and B has as a necessary condition that the conjunction of A and B is interpreted truth-functionally. In other words, we have

$$\Pi(A \wedge B) = \min(\Pi(A), \Pi(B)) \text{ whenever } A \approx/\approx> B \text{ or } A \approx/\approx B \text{ or } A \approx/\approx_Z B$$

Note that this property is *a fortiori* satisfied by possibility measures satisfying (maximality) and (inconsistence). Then we have the following impossibility theorem.

Theorem. Suppose that there are two formulas A and C such that

1. $A \wedge C$, $A \wedge \neg C$, $\neg A \wedge C$, $\neg A \wedge \neg C$ are all consistent

2. ¬A ≈/≈> C, ¬C ≈/≈> A.

Then there is no possibility measure verifying (I*) and the AGM-postulates.

Proof. By hypothesis $A \wedge C$ is consistent. Hence there is a possibility measure Π such that $\Pi(\neg(A \wedge C)) = 0$. By (maximality), we have that $A \wedge C \in K$. By hypothesis ¬A ≈/≈> C and ¬C ≈/≈> A. Due to (I*) we would expect now that $K^*\neg A \vdash C$, and $K^*\neg C \vdash A$. By the representation theorem, this would amount to have both $\Pi(\neg A \wedge \neg C) < \Pi(\neg A)$ and $\Pi(\neg A \wedge \neg C) < \Pi(\neg C)$. (Our consistency hypotheses exclude the cases where A or C are theorems.) On the other hand, the definition of possibilistic independence says that if A and C are independent then the conjunction $\neg A \wedge \neg C$ should be interpreted truth-functionally by Pos. In other words we should have $\Pi(\neg A \wedge \neg C) \geq \Pi(\neg A)$ or $\Pi(\neg A \wedge \neg C) \geq \Pi(\neg C)$, which contradicts the above.

7 Discussion and Conclusion

In this paper we have analysed three notions of dependence that can be defined in possibility theory. A common feature of all of them is that the independence of A and B has as a necessary condition that the conjunction of A and B is interpreted truth-functionally. In other words, we have

$$\text{If } \Pi(A \wedge B) = \min(\Pi(A), \Pi(B)) \text{ then } A \approx\!/\!\approx> B \text{ and } A \approx\!/\!\approx B \text{ and } A \approx\!/\!\approx_Z B$$

Another common property of all the relations is their regular behaviour w.r.t. disjunction: The properties

1. If $A \approx\approx> B \vee C$ then $A \approx\approx> B$ or $A \approx\approx> C$
2. If $A \vee B \approx\approx> C$ then $A \approx\approx> C$ or $B \approx\approx> C$
3. If $A \approx\approx> C$ and $B \approx\approx> C$ then $A \vee B \approx\approx> C$
4. If $A \approx\approx> B$ and $A \approx\approx> C$ then $A \approx\approx> B \vee C$

hold not only for $\approx\approx>$, but also for $\approx\approx$ and $\approx\approx_Z$.

We are convinced that dependence is a fruitful notion in the study of belief change, and that a lot of things remain to be done in this area. In particular we think that it will be a useful tool in the practical implementation of contraction and revision operations.

We have shown that in the context of belief revision, if dependence is constructed from the same possiblility ordering as that underlying the revision operation, one cannot naively apply a dependence-based preservation principle such as (I*). We have obtained similar relsults for nonmonotonic reasoning, using the correspondence between the former and revision of (Makinson, Gärdenfors 1990).

We could have introduced as well a ternary dependence relation "B and C are independent, given A", as studied by Gärdenfors (1978, 1990) and Pearl et al. (Pearl et al. 1988). For reasons of simplicity we have restricted our analysis to binary dependence relations here, but it is clear that a ternary relation is certainly the most general one. This will be subject of further investigations.

Acknowledgements

This work has been supported by the project ESPRIT BRA DRUMS II and the projet inter-PRC Raisonnement sur l'évolutif et l'incertain dans une base de connaissances. The paper was partly written during a stay of the first author at the Universty of Konstanz that has been supported by CNRS and DAAD. He wishes to thank Hans Rott for the useful discussions he had there.

References

D. Dubois (1986), Belief Structures, Possibility Theory, Decomposable Confidence Measures on finite sets. Computer and Arti. Intell. vol 5, N° 5, pp 403-417.

D. Dubois, H. Prade (1986), Possibilistic inference under matrix form. In: *Fuzzy Logic in Knowledge Engeneering*, ed. H. Prade and C.V. Negoita, Velag TÜV Rheinland, Köln, pp. 112-126.

D. Dubois, H. Prade (1988), Possibility Theory: An Approach to Computerized Processing of Uncertainty. Plenum Press, New York.

D. Dubois, H. Prade (1992), Belief change and Possibility Theory. In: *Belief Revision*, ed. P. Gärdenfors, Cambridge University Press, pp. 142-182.

D. Dubois, L. Fariñas del Cerro, A. Herzig, H. Prade (1994), An ordinal view of independence with application to plausible reasoning. *Proc. Uncertainty in AI (UAI-94)*, Seattle.

L. Fariñas del Cerro, A. Herzig (1993), Interference logic = conditional logic + frame axiom. *Proc. Eur. Conf. on Symbolic and Quantitative Approaches to Reasoning and Uncertainty (ECSQARU-93)*, eds. M. Clarke, R. Kruse, S. Moral. Granada, Espagne, nov. 1993. Springer Verlag, LNCS 747, 1993, pp 105-112.

T. Fine (1973), Theories of probability. Academic Press.

P. Gärdenfors (1978), On the logic of relevance.Synthese 351-367.

P. Gärdenfors (1990), Belief revision and irrelevance. *PSA*, Vol. 2, pp 349-356.

E. Hisdal (1978), Conditional possibilities, independence and noninteraction. *Fuzzy Sets and Systems*, 1, 283-297.

A. Kolmogorov (1956), Foundations of the Theory of Probability. Bronx, New York:Chelsea.

J. Pearl (1988), Probabilistic reasoning in intelligent systems: Networks of plausible inference. Morgan Kauffman.

D. Scott (1964), Measurement structures and linear inequalities. *J. of Mathematical Psychology*, Vol. 1, 233-247.

K.Segerberg (1971), Qualitative probability in a modal setting. In: *Proc. of the Second Scandinavian Logic Symposium* (ed. J. E. Fenstad), North Holland.

Zadeh (1978), Fuzzy sets as a basis for a theory of possibility. *Fuzzy Sets and Systems*, Vol. 1, pp 3-28.

Handling hard rules and default rules in possibilistic logic

Salem Benferhat

I.R.I.T. (U.P.S.) 118, Route de Narbonne
Toulouse 31062 Cedex France
e.mail: Benferha@irit.fr

Abstract. This paper extends results done jointly with Dubois and Prade [3] about reasoning with generic knowledge in a possibilistic setting. We propose an approach to reasoning with both *hard* and *default* rules. We mean by a hard rule, the complete information "if we observe ϕ then we conclude (certainly) ψ", and by a default rule, the generic information "if we observe ϕ then generally we conclude ψ".

1. Introduction

In the last decade and since Reiter's paper [15] on default logic, there have been several proposals to reasoning with conditional knowledge bases. We mean by conditional (generic) knowledge bases a set of pieces of information of the type "generally, if α is believed then β is also believed". A typical example of a conditional information is "birds fly".

In this paper, we propose an approach to reasoning with generic information developed in the framework of possibilistic logic [5]. The connection between nonmonotonic logic and possibilistic logic has been laid bare by Dubois and Prade [6], and more recently a natural way to represent default rules in the framework of possibilistic logic has been proposed by Benferhat, Dubois and Prade [3]. In this paper we propose an extension of the system developed in [3], by considering both generic information and complete information. We propose a comparative study with other formalisms, and especially a comparison with Reiter's default logic and with some recent works of Yager [17].

The paper is organized as follows: next section recalls semantics of possibilistic logic and introduces the possibilistic inference relation. Section 3 shows how possibilistic logic can address the problem of exceptions in generic knowledge bases. Section 4 extends results of Section 3 and gives an algorithm to generate plausible results from a given set of beliefs. Section 5 gives some related works and compares our approach with the ones proposed in nonmonotonic reasoning literature.

2. Possibilistic logic

Issued from possibility theory of Zadeh, possibilistic logic has been proposed by Dubois and Prade and developed also by Lang to take account uncertain and imprecise information. In this section, we only recall the semantics of possibilistic logic (See [5] for a complete exposition of possibilistic logic).

Let us first recall elementary notions of classical logic. Throughout this paper, we denote by $\mathcal{L}$ a finite propositional language, namely a set of well founded formulas (represented by Greek letters $\alpha,\beta,\delta,\ldots,\phi,\psi,\ldots$) constructed over a finite set of propositional symbols (represented by small letters p,q,r,...) and the following logical operator $\wedge$ (conjunction), $\vee$ (disjunction), $\neg$ (negation). An interpretation ω for $\mathcal{L}$ is an affectation of a truth value {T or 1,F or 0} to each propositional symbols, and to each well founded formulas with respect to the traditional definitions of the logical operators. We denote by Ω the set of classical interpretations. We say that an interpretation ω satisfies a formula ϕ or ω is a model for ϕ, denoted by $\omega \models \phi$ iff $\omega(\phi)=T$ (namely the interpretation ω assigns the logical value T to the formula ϕ); we say that ψ is a logical consequence of ϕ, denoted by $\phi \models \psi$, iff each model of ϕ is also a model for ψ. Therefore a given formula ϕ can be seen as a set of its models, namely as sub-set of Ω, denoted by $[\phi]$, such that for each interpretation ω in Ω we have $\omega \models \phi$.

Default rules considered in this paper are all of the form "generally, if we have α then β". A base of default rules is denoted by $\Delta=\{\alpha_i \rightarrow \beta_i, i=1,n\}$ where the symbol "$\rightarrow$" is a *non-classical* arrow relating two Boolean propositions. In the whole paper the arrow $\rightarrow$ has this non-classical meaning, the material implication being written as a disjunction.

Semantics of possibilistic logic is based on the notion of *possibility distribution* π: $\Omega \rightarrow [0,1]$ which is a mapping from the set of classical interpretations Ω to the interval [0,1]. A possibility distribution gives a ranking on a set of models such that interpretations with the highest value (degree of possibility) are the most plausible ones. By convention, $\pi(\omega)=1$ means that ω is totally possible to be the real world, $\pi(\omega)>0$ means that ω is only somewhat possible, while $\pi(\omega)=0$ means that ω is not candidate to be the real world.

From now on, we denote by π some given possibility distribution on a set of classical interpretations Ω and we only consider a finite propositional language for the sake of simplicity. A possibility distribution π leads to evaluate subsets of possible worlds by mean of necessity (certainty):

$$N(\phi)=\text{Inf}\{1-\pi(\omega)/\ \omega \models \neg\phi\}$$

which evaluates the extent to which ϕ is entailed by the available knowledge; thus we have:

$$N(\phi \wedge \psi)=\min(N(\phi),\ N(\psi)).$$

Now, let us introduce semantics of possibilistic logic.

Definition 1: An interpretation ω is *preferred* to an interpretation ω' in the possibility distribution π iff $\pi(\omega)>\pi(\omega')$.

Definition 2: An interpretation ω is a *preferential interpretation* of a formula ϕ in the possibility distribution π iff

(i) $\omega \models \phi$,
(ii) $\pi(\omega)>0$, and
(iii) $\nexists \omega', \omega' \models \phi$ and $\pi(\omega')>\pi(\omega)$.

The notion of preferential entailment $\models_\pi$ can be defined in the spirit of Shoham [16]'s proposal:

Definition 3: A formula ψ is a *possibilistic consequence* of ϕ in the possibility distribution π, denoted by $\phi \models_\pi \psi$, iff each preferential interpretation of ϕ satisfies ψ.

In classical logic, a given set of beliefs Δ decomposes the set of interpretations Ω into two disjoint sets: the first one contains interpretations which are consistent with Δ (called models of Δ), and the second set contains interpretations which are inconsistent with Δ. The consequence relation in classical logic is defined in the following way: a formula ψ is a classical consequence of ϕ (wrt Δ) if and only if each model of Δ satisfying ϕ satisfies ψ. An interesting case is when the set of models of Δ satisfying ϕ is empty: in this case any formula is a classical consequence of ϕ; this behaviour is known as the principle "ex-falso quodlibet". The following example illustrates this particular case and shows informally how generic pieces of information can be handled in possibility framework:

Motivating example

Let Δ be the following set of beliefs:

$\Delta = \{$ $b \rightarrow f$ (birds fly)
$p \rightarrow \neg f$ (penguins do not fly)
$p \rightarrow b$ (penguins are birds$\}$

where b,p and f means respectively birds, penguins and fly.

And let Ω be the set of interpretations:

$\Omega = \{ \omega_0{:}\neg b \wedge \neg f \wedge \neg p,\ \omega_1{:}\neg b \wedge \neg f \wedge p,$
$\omega_2{:}\ \neg b \wedge f \wedge \neg p,\ \omega_3{:}\ \neg b \wedge f \wedge p,$
$\omega_4{:}\ b \wedge \neg f \wedge \neg p,\ \omega_5{:}\ b \wedge \neg f \wedge p,$
$\omega_6{:}\ b \wedge f \wedge \neg p,\ \omega_7{:}\ b \wedge f \wedge p\}.$

Intuitively, if we learn the fact "penguin" we would like to deduce from Δ that it does not fly, since a penguin is an exceptional bird.

In classical logic, the set of models of Δ is $M_\Delta=\{\omega_0, \omega_2, \omega_6\}$, therefore if we learn "penguin", we will deduce both it flies and it does not fly, since there exists no models of Δ satisfying p (i.e. $\Delta \cup \{p\}$ is inconsistent).

Now let us assume that we have some possibility distribution π, where $\{\omega_0, \omega_2, \omega_6\}$ are the most possible (for example their degree of possibility equals 1), $\{\omega_4, \omega_5\}$ are less possible (for example their degree of possibility is .5) and $\{\omega_1, \omega_3, \omega_7\}$are the least possible (for example their degree of possibility is 0).

Applying def. 2, we can check that ω_5 is the only preferential interpretation of p (ω_0, ω_2, ω_4 and ω_6 can not be preferential interpretations of p since they do not satisfy p, and ω_1, ω_3, ω_7 can not also be preferential interpretations of p since they are less possible than ω_5), and therefore, using def. 3, if we learn "penguin" we will just deduce that it does not fly ($\omega_5 \models \neg f$).

Now, the main question is how to built such a possibility distribution; this is the aim of next Section.

3. Possibilistic semantics and default reasoning

In the example described above, we have shown that if we choose an appropriate possibility distribution then possibilistic logic can address the problem of exceptions in generic knowledge.

In this Section, we try to find a procedure which selects automatically appropriate possibility distributions for a given set of beliefs. As in Reiter's formalism, we assume that our set of beliefs Δ is a couple $\Delta = (W, D)$ such that:

- $W = \{\alpha_i \Rightarrow \beta_i / i=1,n\}$ is a set of *complete pieces of information*, namely $\alpha_i \Rightarrow \beta_i$ means that "all α_i's are β_i's", and

- $D = \{\alpha_i \rightarrow \beta_i / i=1,m\}$ is a set of *generic pieces of information*, namely $\alpha_i \rightarrow \beta_i$ means only that "generally, α_i's are β_i's" (we may have some α_i's which are not β_i's).

The following definition selects a class of possibility distributions called *compatibles* with a set of beliefs:

Definition 4: A possibility distribution π is said to be *compatible* with $\Delta = (W, D)$ iff the following conditions are satisfied:

(i) for each complete information $\alpha_i \Rightarrow \beta_i$, and for any $\omega \models \alpha_i \wedge \neg\beta_i$ we have $\pi(\omega)=0$[1],

(ii) for each generic information $\alpha_i \rightarrow \beta_i$, we have $\alpha_i \models_\pi \beta_i$[2].

The first condition means that interpretations which falsify some complete information are not candidate to be the real world. The second requirement means that for each generic information $\alpha_i \rightarrow \beta_i$, in the context α_i, β_i is more possible than $\neg\beta_i$.

As in Reiter's formalism, for a given set of beliefs Δ, we may have one, none or many possibility distributions compatibles with Δ. We denote by Π_Δ the set of all possibility distributions compatibles with Δ. If Π_Δ is an empty set then Δ is said to be inconsistent. An example of an inconsistent set of beliefs is the following base is: Δ=(W=$\emptyset$, D={$a \rightarrow b$, $a \rightarrow \neg b$}).

From now on, we will consider only consistent sets of beliefs.

---

[1] i.e.$N(\neg\alpha_i \vee \beta_i)=1$, and N is a necessity measure.
[2] i.e. $N(\alpha_i \wedge \beta_i)>N(\alpha_i \wedge \neg\beta_i)$

To generate plausible results of the set of beliefs Δ, we may have several ways to define the inference relation. One possible definition, called universal (or cautious) consequence relation, takes account of all elements of Π_Δ, namely:

Definition 5 [8]: A formula ψ is said to be an universal consequence of ϕ, denoted by $\phi \models_\forall \psi$, iff ψ is a possibilistic consequence in each possibility distribution compatible with Δ, namely:

$$\phi \models_\forall \psi \text{ iff } \forall \pi \in \Pi_\Delta,\ \phi \models_\pi \psi$$

The universal consequence relation is conservative and suffers from a so-called irrelevance problem, since from a generic information "birds fly", we can not deduce that "red birds fly" (red birds can be exceptional birds).

To solve the irrelevance problem, we propose to choose one possibility distribution over Π_Δ given by *minimum specificity principle.*

4. Qualitative possibility distribution and minimum specificity principle

The idea in specificity principle is to select one (and unique) possibility distribution compatible which assign to each interpretation the highest possible value. Before introducing the inference relation which overcomes the problem of "Irrelevance" presented in the previous section, we need first to introduce the qualitative counterpart of a possibility distribution:

Definition 6 : Let π a possibility distribution. A qualitative counterpart of π, denoted by $Q\pi$, is a partition $E_0 \cup \ldots \cup E_n$ of Ω[1] such that:

$$\forall \omega \in E_i, \forall \omega' \in E_j, \qquad \pi(\omega) > \pi(\omega') \qquad \text{iff} \quad i<j.$$

Qualitative possibility distribution allows us to represent a possibility distribution in terms of classes of equally possible worlds.

We define now, the notion of the minimum specificity principle in qualitative case:

Definition 7: Let π, π' two possibility distributions, and let $Q\pi = E_0 \cup \ldots \cup E_n$ and $Q\pi' = E'_0 \cup \ldots \cup E'_m$ their qualitative counterpart. π is said to be *less specific* than π' iff:

$$\forall\, j = 1, \max(n,m) \quad , \bigcup_{i=1,j} E'_i \subseteq \bigcup_{i=1,j} E_i \text{ [2]}$$

Definition 8: A possibility distribution π is said to be the *least specific* possibility distribution over Π_Δ, iff for any possibility distribution π' of Π_Δ, we have π is less specific than π'.

---

[1] i.e. $\Omega = E_0 \cup \ldots \cup E_n$, and for $i \neq j$ we have $E_i \cap E_j = \emptyset$.

[2] for $j > \min(n,m)$ we use $E_j = \emptyset$ (resp. $E'_j = \emptyset$).

It is clear that the less specific π, the more numerous are the elements in the classes E_j of low rank j. Hence minimizing specificity comes down to minimize the number k of equivalence classes, so as to assign as many worlds as possible to classes of low rank. We denote by π_{spe} the least specific possibility distribution over Π_Δ[1].

Definition 9: A formula ψ is said to be a *Spe-consequence* of ϕ, denoted by $\phi \models_{\pi_{spe}} \psi$, iff ψ is a possibilistic consequence of ϕ in the possibility distribution π_{spe}.

Algorithm

We present in this Section the algorithm used to compute π_{spe}. This algorithm is an extension of the one developed in [3] since it takes account of both generic information and complete information. The construction of π_{spe} can be done by respecting the three following constraints:

(i) for each complete information $\alpha_i \Rightarrow \beta_i$, we have $\forall\omega$, $\omega \models \alpha_i \wedge \neg\beta_i$ then $\pi_{spe}(\omega)=0$,

(ii) for each generic information $\alpha_i \rightarrow \beta_i$, we have $\alpha_i \models_{\pi_{spe}} \beta_i$, and

(iii) Assign to each interpretation the highest possibility level (namely the minimal rank in the qualitative counterpart of π_{spe}) with respect to constraints (i) and (ii).

The constraints (i) and (ii) correspond to the compatibility of π_{spe} with the set of beliefs Δ, and the constraint (iii) corresponds to the minimum specificity principle.

<u>Procedure</u> Compute_Specificity(Δ; π_{spe})

a. Let Ω be the set of interpretations.

b. While W is not empty repeat b.1.-b.2.

b.1. Let $\alpha_i \Rightarrow \beta_i$ a complete information of W. remove $\alpha_i \Rightarrow \beta_i$ from W.

b.2. For each interpretation ω falsifying $\alpha_i \Rightarrow \beta_i$ put $\pi_{spe}(\omega)=0$ and remove ω from Ω.

c. Let i = 1.

d. While Ω is not empty repeat d.1.-d.3.

d.1. Let $E \subseteq \Omega$ the set of interpretations which do not falsify any rule of D.

d.2. Remove from D any rule $\alpha \rightarrow \beta$ such that there exists ω of E and $\omega \models \alpha \wedge \beta$.

d.3. for each element ω of E, put $\pi_{spe}(\omega)=1/i$, remove ω from Ω, and i=i+1.

<u>End</u>

---

[1] We can show that there exists exactly one least specific possibility distribution compatible with Δ.

5. Related work

• *An overview:*

In [2] a comparative study of different approaches developed in different formalisms has been established when we consider only default rules.

It has been shown that for a given set of beliefs containing only generic pieces of information (namely W=∅), the following results is valid:

$$\forall\phi,\psi \quad \phi\models_{\forall}\psi \text{ iff } \phi\vdash_{0} \text{ iff } \phi\vdash_{\varepsilon}\psi$$
$$\text{iff } \phi\models_{P}\psi \text{ iff } \phi\models_{CO}\psi.$$

Namely, the universal consequence relation $\models_{\forall}$ is equivalent to to 0-entailment of Pearl [14], ε–semantics of Adams [1][13], to System P of Kraus, Lehmann and Magidor [10], and also to a system developed by Dubois and Prade [7] based on the notions of conditional objects.

It has also been shown that when W=∅, the Spe-consequence relation is equivalent to 1-entailment in Pearl's System Z [14], to Lehmann's rational closure [12] and to the inference developed in conditional logic by Lamarre [11] based on the construction of a particular systems of spheres called "big model".

Besides, our approach to handle hard rules is slightly different from the one developed by Goldszmidt and Pearl [9] where "$\alpha\Rightarrow\beta$" has the following probabilistic interpretation $Pr(\beta|\alpha)=1$. Indeed, let us consider the following knowledge base $\Delta=\{W=\{\alpha\Rightarrow\beta, \alpha\Rightarrow\neg\beta\}, D=\emptyset\}$. In probabilistic approach this base is inconsistent since it is not possible to find a probability distribution Pr such that $Pr(\beta|\alpha)=1$ and $Pr(\neg\beta|\alpha)=1$. In possibilistic formalism, it is possible to work with such knowledge base and we will just deduce that α must be false.

• *Default logic vs Possibilistic approach:*

Default logic of Reiter [15] is more expressive than the possibilistic approach, since in this paper, we only consider normal default rules of the form: $\alpha:\beta/\beta$ which are a particular case of general default rules of the form: $\alpha:\xi/\beta$. Moreover Default logic can handle inconsistent set of beliefs while in possibility framework it is not possible[1].

On the other hand, the possibilistic approach has at least two advantages over Default logic. First, Default logic can not reason by case while the possibilistic approach does. Indeed, from the following set of beliefs: $\Delta=(W=\{b\}, D=\{b\wedge m\rightarrow f, b\wedge\neg m\rightarrow f\})$ (m, b and f mean respectively male, birds and fly) it is not possible that "birds fly" while in possibilistic approach we deduce it. And the second advantage is that the notion of

[1] As pointed by Pearl[13], in Reiter's formalism there is no differences between inconsistency (ex: $\Delta=(W=\{p\}, D=\{p\rightarrow f, p\rightarrow\neg f\})$ and exception (ex: $\Delta=(W=\{p\}, D=\{b\rightarrow b, b\rightarrow f, p\rightarrow\neg f\})$ in both cases we have two extensions: one contains "fly" and the other "not fly".

specificity is not captured by Reiter's formalism. Indeed, from the following set of beliefs: $\Delta=(W=\{b,p\}, D=\{b\rightarrow f, p\rightarrow\neg f, p\rightarrow b\})$ it is not possible to deduce the intuitive result that "penguins do not fly" while we have seen in the previous sections that it is possible to conclude such results in the possibility framework.

• *Maximal Buoyancy vs Minimum specificity:*

Yager [17] has proposed an approach to selecting preferred possibility distributions[1]. He defines a function called "Buoyancy measure" which associate to each possibility distribution a value in the following way:

Let π be a possibility distribution. We associate to each classical interpretation ω the following number:

$$g(\omega)=\left(\Sigma_{\omega_i} E(\omega_i)\right) / |\Omega|$$

where $n=|\Omega|$ is the cardinality of Ω, and $E(\omega_i)=1$ if $\pi(\omega)\geq\pi(\omega_i)$ and equal to 0 otherwise.

Let $a_0,\ldots,a_n$ be the ordered collection of the $g(\omega_i)$'s, that is, $a_0\geq\ldots\geq a_n$. Let $b_0,\ldots,b_n$ be a set of weights such that: (i) $b_i\in[0,1]$, (ii) $b_i\geq b_j$ if $i<j$, (iii) $\Sigma_i b_i=1$.

We define the *buoyancy* of the possibility distribution π as:

$$Buo(\pi)=\sum_i (b_i * a_i)$$

Given a set of possibility distributions, Yager proposes to select possibility distributions having the highest degree of Buoyancy.

Proposition 3: A possibility distribution π is the least specific over Π_Δ iff for each π' of Π_Δ we have $Buo(\pi)\geq Buo(\pi')$.

As a consequence of proposition 3, applying the Buoyancy measure or the minimum specificity principle to the set Π_Δ gives exactly the same results, namely π_{spe}.

6. Conclusions and future works

In this paper, we have proposed an approach to reasoning with both hard and default rules using possibilistic logic. We have seen that from the syntax (structure) of default rules we construct possibility distributions such that interpretations with high values represent defaults rules which are more specific. One future work is to extend the system developed in this paper to handle both hard, uncertain and generic information.

Acknowledgments:

The author is indebted to J. Lang for useful comments on first version of this paper.

---

[1]In fact, in Yager's paper, the selection is done over a set of total pre-orders.

7. References

[1] E.W. Adams (1975). The Logic of Conditionals. Dordrecht: D. Reidel.

[2] S. Benferhat (1994) Raisonnement non-monotone et traitement de l'inconsistance en logique possibiliste. Thèse de l'Université P. Sabatier, Toulouse.

[3] S. Benferhat, D. Dubois, H. Prade (1992) Representing default rules in possibilistic logic. KR'92, 673-684.

[4] S. Benferhat, C. Cayrol, D. Dubois, J. Lang, H. Prade (1993) Inconsistency management and prioritized syntax-based entailment. Proc. of IJCAI'93.

[5] D. Dubois, J. Lang, H. Prade (1994). Possibilistic logic. To appear in Handbook of Logic for Artificial Intelligence (D.M. Gabbay, ed.), vol. 3, 439-513.

[6] D.Dubois, H. Prade (1991) Possibilistic logic, preferential models, nonmonotonicity and related issues. IJCAI'91, 419-424.

[7] D. Dubois, H. Prade.(1993) Conditional Objects as Non-monotonic Consequence Relationships. Technical report, IRIT-93-08-R (University of Toulouse).

[8] L. Fariñas del Cerro, A. Herzig, and J. Lang (1992). From expectation-based nonmonotonic reasoning to conditional logics. *Working Notes of the 4th Inter. Workshop on Nonmonotonic Reasoning*, Plymouth, Vermont, May 28-31, 79-86.

[9] M. Goldszmidt, et J. Pearl (1991). On the consistency of defeasible databases. Artificial Intelligence 52:121-149.

[10] S. Kraus, D. Lehmann, and M. Magidor (1990). Nonmonotonic reasoning, preferential models and cumulative logics. Artificial Intelligence 44:167-207.

[11] P. Lamarre (1992), A Promenade from Monotonicity to Non-Monotonicity Following a Theorem Prover. Proc. of KR'92.

[12] D. Lehmann (1989), What does a conditional knowledge base entail? Proc. KR'89, pages 357-367.

[13] J. Pearl (1988), Probabilistic Reasoning in Intelligent Systems : Networks of Plausible Inference. Morgan Kaufmann Publ. Inc., San Mateo, Ca., 1988.

[14] J. Pearl (1990). System Z: a natural ordering of defaults with tractable applications to default reasoning. Proc.TARK, 121-135.

[15] R. Reiter (1980), A Logic for Default Reasoning; Artificial Intelligence N°13, 81-132.

[16] Y. Shoham, (1988), Reasoning About Change – Time and Causation from the Standpoint of Artificial Intelligence. Cambridge, Mass.: The MIT Press.

[17] R. R. Yager (1993), On the completion of priority orderings in nonmonotonic reasoning" Technical Report #1216, Iona College.

A General Possibilistic Framework for Reliability Theory

Bart Cappelle, Etienne E. Kerre

Department of Applied Mathematics and Computer Science
Krijgslaan 281-S9
B-9000 Gent, Belgium

Abstract. Recent developments in fuzzy set theory and the theory of multistate structure functions solved some important problems that arise in the classical reliability theory, based upon binary structure functions and probability theory. First we give an overview of some important non-probabilistic reliability models which are mainly based upon fuzzy set theory and possibility theory and we situate our approach based upon possibility theory and multistate structure functions among the existing models. Secondly, to gain more insight into our approach, we study a typical aspect of the possibilistic reliability functions: the structure function independence, s-independence for short.

1 A Non-Probabilistic Approach to Reliability

Reliability theory has been developed mainly during and after the Second World War. The experiences with weapon systems and the development of the aviation and space industry required a well-developed tool to analyse the reliability of these complex systems. The reliability of a system and its components, classically has been defined as the probability that the system or its components function properly during a time interval. Hence, probability theory served as the unifying tool to model the uncertainty w.r.t. the state a component or a system assumes. Since the introduction of fuzzy set theory and the development of possibility theory, alternative tools to model uncertainty became available. As has been pointed out throughout the "fuzzy" literature, possibility theory and fuzzy set theory can easily be applied to situations where a quantitative analysis is bearly possible. Hence, we can apply possibility and fuzzy set theory to reliability theory whenever qualitative information must be modeled.

There are several other important problems in the classical approach of reliability. In very large systems with a lot of components it is very difficult to obtain sufficient statistical information to derive the failure and functioning probability distribution within certain error bounds. We refer to the problems the NASA encountered when developing their Apollo Program in the sixties. Due to the lack of sufficient statistical information, the probability of failure of a space flight was almost 100%! Secondly, we must mention that the assumption that the components are probabilisticly independent in many cases does not hold. On the contrary, the probability that a component will fail, can be highly affected by the failure or functioning of other components. Finally, the assumption that a system or a component assumes either one of two possible states is a serious oversimplification of reality.

In order to solve these problems, since 1983 several alternative models are proposed to solve the main problems of the classical approach of reliability. Tanaka, Fan, Lai and Toguchi [15] introduced fuzzy probabilities to model the uncertainty about the exact value of the probability of failure. Based upon their work, Misra and Weber [8], and Singer [14] developed important properties and sophisticated models to apply fuzzy probabilities in a concrete reliability study. Singer, e.g. , fuzzified the classical reliability function by means of Zadeh's extension principle to allow a reliability analysis with fuzzy probabilities, and Misra and Weber introduced the basic concepts to perform a fuzzy fault-tree analysis, i.e. , a fault-tree analysis method that is based upon the properties of fuzzy probabilities.

One of the major steps towards a possibilistic approach of reliability is made by Onisawa. He introduces error possibilities, hence, he does not assume that the notion *reliability* is probabilistic by nature. His model applies whenever incomplete or linguistic information must be modeled [11, 13] and his approach is also very successful in modeling the human behaviour of, e.g. , operators in a control room. The strength of his ideas and notions has been stressed by his case study on the Chernobyl accident [12] which proved that the possibility of an accident at the Chernobyl nuclear power plant was much greater than generally thought and assumed.

However, till now all non-probabilistic reliability models are applied to binary systems and components, i.e., systems and components that may assume only one of two possible states. Hence, partial failures cannot be modeled under this assumption. Obviously, this assumption is an oversimplification of many real life components and systems since many of them are simply not binary by nature. From 1978 on, a general theory of multistate structure functions is developed by several reliability theoreticians [6, 10, 1, 9, 3]. The components and systems may assume intermediate levels, representing different levels of degradation of the component and system. A structure function is the deterministic relationship between the component states and the system state. Generally, a state space can be any complete lattice [3, 9]. In this paper, however, we assume that all state spaces equal the unit interval provided with its classical total order relation, denoted by $([0,1], \leq)$, and that a system has a finite number of components n.

We show, by means of a special property of the generalized reliability functions, how possibility theory can be combined with non-binary components and systems, and provide a general framework for possibilistic reliability theory. Therefore, we define two reliability functions: the possibilistic reliability function and the dual possibilistic reliability function. These reliability functions are based upon a general possibility measure which must not be a product possibility measure. Hence, we allow for possibilistic dependent components. Obviously, in a concrete reliability analysis, the possibility measure, can be determined by taking into account linguistic information [13] and the components in a system may be human operators in a plant that perform control actions.

First we introduce the notion structure function. We mentioned already that a structure function models the reliability aspects of a system with a finite number of components.

Definition 1 *Every* $[0,1]^n - [0,1]$ *mapping* ϕ *satisfying*

1. ϕ *is increasing in each component;*

2. $\phi(0,\ldots,0) = 0$ *and* $\phi(1,\ldots,1) = 1;$

is a n-ary structure function.

In the sequel, by **0**, **1** respectively, we denote $(0,\ldots,0)$, $(1,\ldots,1)$ respectively. The first condition mimics the fact that whenever a component functions better, the system itself cannot deteriorate. In the second condition some obvious boundary conditions are imposed: when every component functions perfectly, the system must function perfectly and when every component fails, the system must fail.

By $S(\alpha,\phi)$, $S^d(\alpha,\phi)$ respectively, we denote the α-level set, the dual α-level set respectively, of a structure function ϕ [3, 7], i.e.,

$$S(\alpha,\phi) = \{\mathbf{x} \mid \mathbf{x} \in [0,1]^n \text{ and } \phi(\mathbf{x}) \geq \alpha\}$$

and

$$S^d(\alpha,\phi) = \{\mathbf{x} \mid \mathbf{x} \in [0,1]^n \text{ and } \phi(\mathbf{x}) \leq \alpha\}.$$

In many real life situations it is not easy to carry out a probabilistic analysis of failure, degradation or functioning of a system and its components. Since a probabilistic analysis is rather quantitative by nature and in many cases a qualitative approach is more appropriate, alternative models are required. As pointed out above, we have chosen possibility theory to model the uncertainty about the failure or functioning of both components and systems. By Π we denote a possibility measure defined on $[0,1]^n$ and by π we denote the corresponding possibility distribution—for an exhaustive overview of possibility theory, we refer to [5].

Definition 2 *Let ϕ be a n-ary structure function. The possibilistic reliability function $\rho(\Pi,\phi)$ of ϕ is the* $[0,1]-[0,1]$ *mapping*

$$\rho(\Pi,\phi) : [0,1] \to [0,1] : \alpha \mapsto \Pi(S(\alpha,\phi));$$

the dual possibilistic reliability function $\rho^d(\Pi,\phi)$ of ϕ is the $[0,1]-[0,1]$ *mapping*

$$\rho^d(\Pi,\phi) : [0,1] \to [0,1] : \alpha \mapsto \Pi(S^d(\alpha,\phi)).$$

The value $\rho(\Pi,\phi)(\alpha)$ is the possibility that ϕ assumes at least the state α, while the value $\rho^d(\Pi,\phi)(\alpha)$ is the possibility that ϕ assumes at most the state α. In the next section, we give a concise overview of some important properties of the possibilistic reliability functions, while in section 3 we study the structure function independence, s-independence for short, of the possibilistic and the dual possibilistic reliability function.

2 Some Important Properties

In this section we give a concise overview of the main properties of the possibilistic reliability functions that we apply furtheron. For more detailed information we refer to [3, 4].

Property 1 *Let ϕ be a n-ary structure function, then*

$$\rho(\Pi,\phi)(0) = 1 \text{ and } \rho^d(\Pi,\phi)(1) = 1.$$

Property 2 *Let ϕ be a n-ary structure function, then $\rho(\Pi,\phi)$ is a decreasing mapping and $\rho^d(\Pi,\phi)$ is an increasing mapping.*

Theorem 1 *Let ϕ be a n-ary structure function, then*

$$\rho(\Pi,\phi):[0,1]\rightarrow[0,1]:\alpha\mapsto\sup_{\mathbf{x}\in S(\alpha,\phi)}\Pi([\mathbf{x},1])$$

and

$$\rho^d(\Pi,\phi):[0,1]\rightarrow[0,1]:\alpha\mapsto\sup_{\mathbf{x}\in S^d(\alpha,\phi)}\Pi([0,\mathbf{x}]).$$

From theorem 1 it follows that the reliability functions are completely determined by the events $[0,\mathbf{x}]$ and $[\mathbf{x},1]$, $\mathbf{x}$ an arbitrary element of $[0,1]^n$. In lattice theory, these sets are known as the principal ideals, the dual principal ideals respectively, of the complete lattice $([0,1]^n,\leq)$ [2]. Hence, it is not necessary to know the possibility distribution π to determine both the possibilistic reliability and the dual possibilistic reliability function, which is a remarkable result. If the possibility of every principal ideal and dual principal ideal of $([0,1]^n,\leq)$ is known, we easily obtain the following upper bound for $\pi(\mathbf{x})$, $\mathbf{x}\in[0,1]^n$:

$$\pi(\mathbf{x})\leq\min(\Pi([0,\mathbf{x}]),\Pi([\mathbf{x},1])).$$

Obviously we cannot derive the exact value of π, since the possibility of an intersection of two sets in most cases does not equal the minimum of the possibility of these two sets.

Theorem 2 *Let ϕ be a n-ary structure function and α an arbitrary element of $[0,1]$, then*

$$\max(\rho(\Pi,\phi)(\alpha),\rho^d(\Pi,\phi)(\alpha))=1.$$

This important relationship between $\rho(\Pi,\phi)$ and $\rho^d(\Pi,\phi)$ simplifies the calculation of the reliability functions considerably, since if either $\rho(\Pi,\phi)$ or $\rho^d(\Pi,\phi)$ takes a value strictly smaller than 1, the other reliability function assumes the value 1. Obviously, this theorem does not hold when we replace Π by a probability measure. Thus, it is a property typical for the possibilistic reliability functions.

3 S-Independence

In this section we develop an important potential property of the possibilistic and the dual possibilistic reliability functions. From this property we gain more insight into the meaning of the possibilistic reliability functions. By $\mathcal{M}([0,1]^n,[0,1])$ we denote the set of all n-ary structure functions.

Definition 3 *The possibilistic reliability function is s-independent if and only if for any two n-ary structure functions*

$$\rho(\Pi,\phi)=\rho(\Pi,\psi)$$

and the dual possibilistic reliability function is s-independent if and only if for any two n-ary structure functions

$$\rho^d(\Pi,\phi)=\rho^d(\Pi,\psi).$$

In the case of s-independence, the possibilistic reliability functions do not depend on the structure function. Hence, one can say that all systems are equally (possibilistic) reliable. We examine this remarkable situation in more detail and we characterize it completely.

Theorem 3 *The possibilistic reliability function is s-independent if and only if*

$$(\forall \mathbf{x} \in [0,1]^n \setminus \{\mathbf{0}\})(\pi(\mathbf{x}) \leq \pi(\mathbf{1})) \tag{1}$$

and the dual possibilistic reliability function is s-independent if and only if

$$(\forall \mathbf{x} \in [0,1]^n \setminus \{\mathbf{1}\})(\pi(\mathbf{x}) \leq \pi(\mathbf{0})). \tag{2}$$

Proof. Taking into account the analogy, we only prove the first statement. First, assume that (1) holds. Since for any structure function ϕ, $\rho(\Pi, \phi)(0) = 1$, we assume that α belongs to $]0,1]$. From the definition of a possibility measure, we find that

$$(\forall \mathbf{x} \in [0,1]^n \setminus \{\mathbf{0}\})(\Pi([\mathbf{x}, \mathbf{1}]) = \pi(\mathbf{1})).$$

Hence, taking into account theorem 1,

$$\rho(\Pi, \phi)(\alpha) = \pi(\mathbf{1}),$$

for any n-ary structure function ϕ, since $\mathbf{0}$ does not belong to $S(\alpha, \phi)$, $\alpha \in]0,1]$ [3]. The reverse follows immediately when the reliability functions of the structure functions

$$\mathbf{o} : [0,1]^n \to [0,1] : \mathbf{x} \mapsto \begin{cases} 1 & ; \quad \mathbf{x} = \mathbf{1} \\ 0 & ; \quad \text{elsewhere} \end{cases}$$

and

$$\mathbf{e} : [0,1]^n \to [0,1] : \mathbf{x} \mapsto \begin{cases} 0 & ; \quad \mathbf{x} = \mathbf{0} \\ 1 & ; \quad \text{elsewhere} \end{cases}$$

are calculated, taking into account the s-independence of the reliability function. □

Corollary 1 *If the possibilistic reliability function is s-independent, then for any structure function* ϕ

$$\rho(\Pi, \phi) : [0,1] \to [0,1] : \alpha \mapsto \begin{cases} 1 & ; \alpha = 0 \\ \pi(\mathbf{1}) & ; \textit{else} \end{cases}$$

and when the dual possibilistic reliability function is s-independent then for any structure function ϕ

$$\rho^d(\Pi, \phi) : [0,1] \to [0,1] : \alpha \mapsto \begin{cases} 1 & ; \alpha = 1 \\ \pi(\mathbf{0}) & ; \textit{else.} \end{cases}$$

Corollary 2 *Both the possibilistic and the dual possibilistic reliability function are s-independent if and only if for any* $\mathbf{x}$ *of* $[0,1]^n \setminus \{\mathbf{0}, \mathbf{1}\}$

$$\pi(\mathbf{x}) \leq \min(\pi(\mathbf{0}), \pi(\mathbf{1}))$$

holds.

The proofs are immediate from theorem 3.

Corollary 3 *If the possibilistic reliability function is s-independent, then*

$$\max(\pi(\mathbf{0}), \pi(\mathbf{1})) = 1.$$

Proof. Immediate from theorem 3 and the boundary conditions for possibility measures. □

Example 1 Consider the possibility distribution π on $[0,1]^2$,

$$\pi : [0,1]^2 \to [0,1] : (x_1, x_2) \mapsto x_1 x_2.$$

Since $\pi(1,1) = 1$, we find that the possibilistic reliability function is s-independent for any 2-ary structure function and equals to the constant mapping on 1. The dual possibilistic reliability function is not s-independent since $\pi(0,0) = 0$ and, e.g., $\pi(\frac{1}{2}, \frac{1}{2}) = \frac{1}{4} > 0$.

Remark 1 If the possibilistic reliability function is s-independent and $\pi(\mathbf{1}) < 1$, then $\pi(\mathbf{0})$ equals 1. Hence, the dual possibilistic reliability function is also s-independent. An analogous statement holds for the dual possibilistic reliability function. Obviously, the reverse does not hold, i.e., from $\pi(\mathbf{1}) = 1$, we cannot decide upon the s-independence of the dual possibilistic reliability function.

4 Conclusion

In this paper we have pointed out the importance of a non-probabilistic approach to reliability theory and we have stressed that the application of possibility theory is successful and indispensable, whenever there is a lack of sufficient statistical information or whenever human behaviour and linguistic information must be modeled. We have combined a possibilistic approach to reliability theory with the theory of multistate structure functions and we have given a concise overview of some of the basic properties of the possibilistic reliability functions. One rather unusual aspect of the possibilistic and the dual possibilistic reliability functions has been studied in more detail to gain insight into possibilistic reliability theory: the s-independence of the reliability functions.

At first glance, the s-independence property of a reliability function seems rather weird. However, taking into account theorem 3, we find that whenever the best (worst) state of the components is the most possible, the (dual) reliability function does not depend on the structure function. This is not surprising, since in these situations all systems behave similar according to the boundary conditions of structure functions. Hence, when all systems behave similar, they have equal possibilistic reliability functions.

We are convinced that possibility theory and the theory of multistate structure functions will play a major role in the future whenever studying the reliability aspects of complex and large systems, and whenever human behaviour and linguistic information must be modeled.

References

1. L.A. Baxter: Continuum Structures I. Journal of Applied Probability 21, 802-815 (1984)

2. G. Birkhoff: Lattice Theory. AMS Colloquium Publication Volume 24, Providence, Rhode Island 1967

3. B. Cappelle: Multistate Structure Functions and Possibility Theory: an Alternative Approach to Reliability. In: E.E. Kerre (ed.): Introduction to the Basic Principles of

Fuzzy Set Theory and Some of its Applications. Gent: Communication and Cognition 1991, pp. 252-293

4. B. Cappelle, E.E. Kerre: On a Possibilistic Approach to Reliability Theory. In: Proceedings of the Second International Symposium on Uncertainty Analysis (ISUMA 93), Maryland MD, 1993, pp. 415-418

5. D. Dubois, H. Prade: Théorie des possibilités. Masson, Paris (France) 1985

6. E. El–Neweihi, F. Proschan, J. Sethuraman: Multistate Coherent Systems. Journal of Applied Probability 15, 675-688 (1978)

7. E.E. Kerre: Basic Principles of Fuzzy Set Theory for the Representation and Manipulation of Imprecision and Uncertainty. In E.E. Kerre (ed.): Introduction to the Basic Principles of Fuzzy Set Theory and Some of its Applications. Gent: Communication and Cognition 1991, 1-158

8. K.B. Misra, G.G. Weber: Use of Fuzzy Set Theory for Level-I Studies in Probabilistic Risc Assesment. Fuzzy Sets and Systems 37, 139-160 (1991)

9. J. Montero, J. Tejada, J. Yáñez: General Structure Functions. In: Proceedings Workshop on Knowledge-Based Systems and Models of Logical Reasoning, Cairo (Egypt) 1988

10. B. Natvig: Two Suggestions of how to define Multistate Coherent Systems. Advances in Applied Probability 14, 434-455 (1982)

11. T. Onisawa: An Approach to Human Reliability in Man-Machine Systems using Error Possibility. Fuzzy Sets and Systems 27, 87-103 (1988)

12. T. Onisawa, Y. Nishiwaki: Fuzzy Human Reliability Analysis on the Chernobyl Accident. Fuzzy Sets and Systems 28, 115-127 (1988)

13. T. Onisawa: Use of Natural Language in System Reliability Analysis. In: R. Lowen, M. Roubens (eds.) Fuzzy Logic: State of the Art. Dordrecht: Kluwer Academic Publishers 1993, pp. 517-529

14. D. Singer: A Fuzzy Set Approach to Fault Tree and Reliability Analysis. Fuzzy Sets and Systems 34, 145-155 (1990)

15. H. Tanaka, L.T. Fan, F.S. Lai, K. Toguchi: Fault-Tree Analysis by Fuzzy Probability. IEEE Transactions on Reliability 32, 453-457 (1983)

Possibilistic Semantic Nets

Sandra Sandri, Guilherme Bittencourt

Instituto Nacional de Pesquisas Espaciais (INPE)
Cx Postal 515, 12201-370 S. J. Campos - SP - Brazil

Abstract. We investigate how possibilistic valuations can be introduced in the semantic net formalism. The model proposed here deals with multiple inheritance, exceptions and non-monotonicity by making use of necessity measures attached to the individual links in the net. We also define extensions for this formalism and compare them to other models in the literature.

1 Introduction

A *semantic net* [1] is a model for human associative memory, consisting of a set of nodes connected by directed edges called links. The formalism does not dispose of a general agreed-on semantics: the nodes can be used to represent concepts, predicates, objects, etc..., and the edges are associated to arbitrary binary relations between the nodes. The main inference mechanism in semantic nets is inheritance through the net and network matching.

The semantic nets formalism is adequate for taxonomically structured domains, but its main appeal is the expressivity obtained when the inheritance hierarchy is represented by a graph, allowing multiple inheritance, and the possibility of representing exceptions. This expressivity has its drawback in the complexity of the inheritance algorithms.

We explore here the potentiality of the use of possibility theory as a means of dealing with imperfect information in semantic nets. This framework has a great expressive power, in spite of its simple conception. We also define what are extensions in this model, providing a means of analysing the different lines of thought that can be taken in order to find the answer to a query.

This work has been motivated by the construction of an uncertainty module in a rule-based shell generator tool, called FASE, which furnishes logic, frames and semantic nets as knowledge representation models. All the pieces of knowledge given to an expert system constructed using a shell generated by the tool, — the rules extracted from experts and the facts furnished by the user —, may be inherently pervaded with uncertainty. On the other hand, several knowledge representation models may coexist in a single application. The uncertainty models provided by the tool should then be such as to deal with different knowledge representation models in a uniform manner. Moreover, these models should be of fairly easy comprehension by the average experts, knowledge engineers and expert system users. Possibility theory is such a model. In [2] we present a possibilistic frame formalism for system FASE.

In the next section, we give some of the definitions consacrated in this domain. In Section 3 we propose the model of possibilistic semantic nets, describing some of its features. In Section 4 we discuss what are extensions in this framework, and in Section 5 we verify how these extensions relate to some of its well-known definitions. Section 6 brings the conclusion.

2 Basic Definitions

The nodes in a semantic net can be of two types, classes and instances of classes, although not in all implementations this distinction is made explicit. Between each two nodes at most one link may exist. The links are generally of only two kinds, the positive links x -> y, called *"is-a"* and the negative links x -/-> y, called *"is-not-a"*. Link x -/-> y is called the *dual link* of x -> y and vice-versa. These links generally stand for *"x is usually a y"* and *"x is usually not a y"* respectively. There are subtle differences however in implementations and latter in this work we discuss this subject more deeply. In some of the implementations found in the literature a distiction is made between links connecting a class to another class from those connecting an instance to a class. In NETL, links may also connect nodes to links [3], as a means of treating exceptions. Extensive literature has established the relation between semantic nets and default logic (see for instance [4] and [5]).

In many works semantic nets are called inheritance systems, if multiple inheritance, exceptions and non-monotonicity are allowed. Here, we shall use both terms interchangebly. Touretzky and colleagues [6] give a classification of inheritance systems according to characteristics such as the treatment of exceptions, nonmonotonicity, etc...

Systems are said to be homogeneous if all links are either all strict or all defeasible, and heterogeneous otherwise. Some heterogeous systems have arised to deal with the concept of typicality (see [7] for a survey). In [6] and [8] mechanisms are proposed which deal with homogeneous systems (these works are compared to ours latter in this paper). [9] brings a mathematical treatment of defeasible reasoning, which is not however compared to our approach in the present work.

3 Possibilistic Semantic Nets

We propose here a semantic net formalism that deals with imperfect information through the use of possibility theory (see [10] for an introduction to this theory). The mechanisms proposed ressemble those of possibilistic logic PLI [11]. A Possibilistic Semantic Net (PSN) is characterized by a pair (K, L), where K denotes a set of nodes representing either classes or instances, and L a set of pairs (l, α), where l is either a positive link (x -> y), or a negative link (x -/-> y), and α, called a valuation, is a strictly positive necessity measure [10], defined as $N(l) \geq \alpha$, $\alpha > 0$. Between each two nodes at most one link, either positive or negative, may exist in the net.

The pair (l, α) is called a *(possibilistically) qualified link.* If l is a positive link x -> y then $N(l) \geq \alpha$, $\alpha > 0$ has the meaning "there is at least a necessity α that the x's are y's". In the same way, N(x -/-> y) $\geq \alpha$, $\alpha > 0$, means that "there is at least a necessity α that the x's are not y's".

Three possible interpretations for the necessity measure α in (x -> y, α) are:
(i) a measurement of the belief that all x's are y's,
(ii) a lower bound on an (unknown) frequency of x's that are y's,

(iii) a bound on how much of the characteristics of y's are shared by the x's, or in other words, to what extend are we allowed to say that the x's are typical y's.

Two interpretations for α in (x -/-> y, α) are directely derived by the suitable modification of i) and ii). In a third interpretation α represents the belief we have that x is an exception to y, when x is an instance of y.

The processing of the valutions proposed in this work is the same for all interpretations; the choice of interpretation by the user only influences the analysis of the results. It is of course, highly desirable that a single interpretation is adopted for all links.

Figure 1.a) shows a semantic net about an elephant named Clyde, of whom we would like to know the color. For instance, if no royal elephants are gray, and if Clyde is almost surely a royal elephant, then it will also be almost surely not gray. However, if there exists any gray royal elephant, then Clyde may be one of those exceptions. In the figures related to this example, the nodes C, E, R, and G stand for the concepts Clyde, elephant, royal elephants, and gray respectively.

From qualified links we derive paths called *annotated qualified links* (aql), characterized by pairs ((l, α), S), where l is a positive or negative link between two nodes, α a necessity measure, and S a sequence of $|S| = n$ nodes, $n \geq 0$. An original qualified link from the net is described by an aql of the form ((l, α), $\varnothing$), where $\varnothing$ denotes the sequence of size 0.

Let Z = (K, L) be a semantic net, with the links in L described by aql's ((l, α), $\varnothing$), and $S = <s_1,..., s_n>$ and $Q = <q_1,...,q_m>$ be sequences of nodes of K. The annotated qualified links representing the paths derivable from L are constructed using the following transitivity law :

If ((p -> r, α), S) and ((r -> q, β), Q) are aql's, then ((p -> q, δ), &(S, r, Q)) is an aql,
If ((p -> r, α), S) and ((r -/-> q, β), Q) are aql's, then ((p -/-> q, δ), &(S, r, Q)) is an aql,

where &(S, r, Q) denotes the concatenation of r between S and Q generating the sequence $<s_1,...,s_n,r,q_1,...,q_m>$, and $\delta = \min(\alpha, \beta)$ denotes de valuation calculated for the aql. The closure of a set of aql's A, denoted as C(A), is defined as all the aql's that can be derived from A with the recursive application of this transitivity law. It is important to note that no path ending with a negative link may be further augmented, i.e. "negative" information is not propagated in this formalism.

Figure 1.b) illustrates all the aql's that can be derived from the net shown in Figure l.a). Note that there will be as many aql's between two nodes as the number of possible paths between them; in the figures multiple paths are indicated by sets of valuations attached to the links.

The best answer that a PSN Z = (K, L) can give about the relation between two nodes x and y is the aql with highest credibility that is possible to derive between these nodes.

Formally, let C(L) be the closure of L, and N_Z(x -> y) = max {α / ((x -> y, α), S) $\in$ C(L)} and N_Z(x -/-> y) = max {α / ((x -/-> y, α), Q) $\in$ C(L)} be the highest credibilities that can be obtained for each positive and negative aql's in C(L) respectively. If N_Z(x->y) > N_Z(x-/->y) then x -> y is the most credible conclusion (with degree N_Z(x->y)) about the relation between x and y.

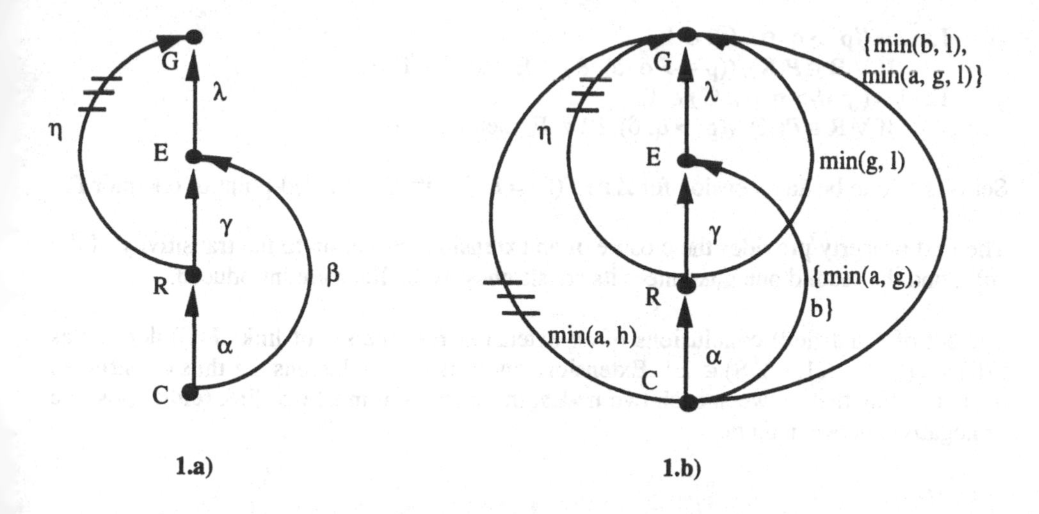

Fig. 1. a) A semantic net Z, **b)** Set C(L) of all aql's that can be derived from Z.

In the example shown in Figure 1, the best answer that the semantic net can give about Clyde's "grayness" depends on the maximum between $N_Z(C \text{->} G) = \max\{\min(\beta,\lambda), \min(\alpha,\gamma,\lambda)\}$ and $N_Z(C \text{-/->} G) = \min(\alpha, \eta)$. Let us suppose that $\eta = \beta = 1$, i.e. we know with certainty that no royal elephant is gray, and that Clyde is an elephant. We will then obtain $N_Z(C \text{-/->} G) = \alpha$, and $N_Z(C \text{->} G) = \lambda$, meaning that Clyde will be considered to be gray if the belief that elephants are gray is greater than the belief that Clyde is a royal elephant.

4 Possibilistic Semantic Nets Extensions

Extensions started being defined in the semantic nets formalism as non-monotonic models were introduced (see [6], [8]). Possibilistic semantic nets is closely related to possibilistic logic [11] and, as in that formalism, it can be considered monotonic if the information inside the base is consistent. Since this is not always the case, the definition of extensions in our framework is welcome. With the use of extensions, we are able to characterize complete lines of thought, which are internally consistent, and also to provide a powerful means for explanation. Last but not least, here their definition also serves to make it easier to compare PSN's with other inheritance systems.

Let Z = (K, L) be a PSN and *P*(K) the set of all possible sequences of nodes of K. Let S, Q and V sequences of nodes of K, E a set of aql's, and Γ(E) the smallest set satisfying the properties:

(i) Let $((p \text{->} r, \alpha), S) \in \Gamma(E)$.
If $\exists\, Q \in P(K)$, $((r \text{->} q, \beta), Q) \in \Gamma(E)$, then $((p \text{->} q, \delta), R) \in \Gamma(E)$,
If $\exists\, Q \in P(K)$, $((r \text{-/->} q, \beta), Q) \in \Gamma(E)$, then $((p \text{-/->} q, \delta), R) \in \Gamma(E)$,
where $\delta = \min(a, \beta)$, and $R = \&(S,r,Q)$;

(ii) Let $l = ((p \rightarrow q, \alpha), \emptyset) \in L$.
If $\forall R \in P(K)$, $((p \not\rightarrow q, \delta), R) \notin E$, then $l \in \Gamma(E)$.
Let $l = ((p \not\rightarrow q, \alpha), \emptyset) \in L$.
If $\forall R \in P(K)$, $((p \rightarrow q, \delta), R) \notin E$, then $l \in \Gamma(E)$.

Set E is said to be an extension for Z iff $\Gamma(E) = E$, i.e. iff E is a fixed-point of operator Γ.

The first property provides the closure of an extension in relation to the transitivity of the aql's, and the second one guarantees its consistency as the links are introduced.

The set of (qualified) conclusions of an extension E is the set of links F(E) defined as $F(E) = \{(l, \alpha) / ((l, \alpha), S) \in E\}$. Extensions and sets of conclusions are thus constructed in such a way that between each two nodes, there exists at most one link (either positive or negative) between them.

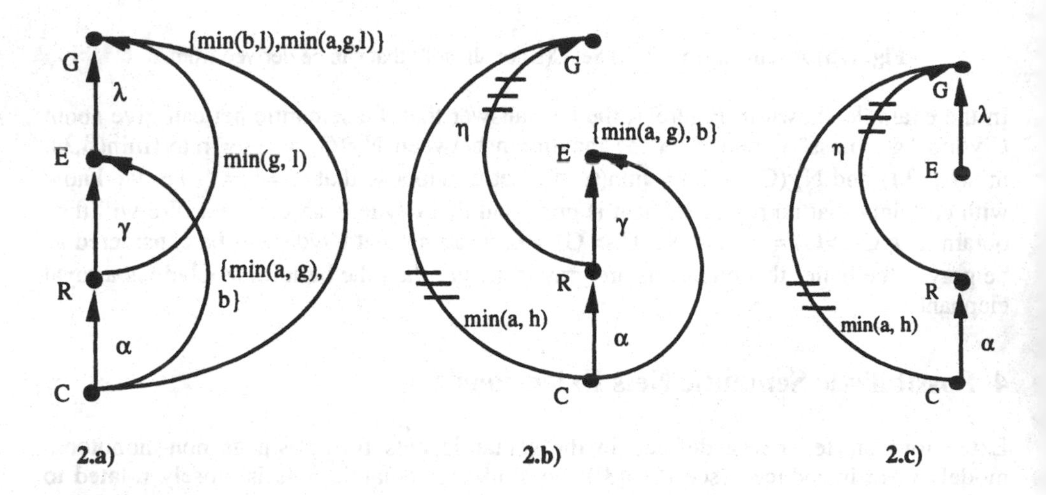

Fig. 2. Extensions of Z: **a)** A, **b)** B1, **c)** B2

Figure 2 shows graphically the extensions A, B1 and B2, that can be obtained for Clyde's problem. The paths are implicitly indicated by the valuations on the links. Note that in a given extension we may have more than one aql leading from a node x to a node y, if more than one path can be derived between them, provided they are either all positive or all negative.

Following extension A we conclude that Clyde is a gray royal elephant, and disconsider the information about royal elephants being not gray. From extensions B1 and B2 we conclude that Clyde is a royal elephant and that Clyde is not gray, and in order to guarantee the consistency of this line of thought, we either disconsider the information about elephants being gray, or of royal elephants being elephants.

5 Comparison to Some Seminal Works

In [6], system TMOIS is discussed, in which links x ->y and x -/->y are both defeasible. In this work, an extension of an inheritance system is described as a fixed point consisting of paths of form x_1->...->x_n or x_1->...->x_{n-1}-/->x_n that can be generated through a particular inheritance mechanism. The set of conclusions associated with an extension is the set of statements of the form x_1 -> x_n, if the extension contains a path of the form x_1->...->x_n; or x_1-/->x_n, if the extension contains a path of the form x_1->...->x_{n-1}-/->x_n.

In [8], an inheritance system I is a structure with simple nodes and two kinds of links; x->y means "any x is usually a y" and x -/-> y means "any x is usually not a y". Sandewall states that admissible extensions are the smallest supersets of I closed with reference to a particular inference mechanism, which is itself based upon the identification of some primitive structures, for each of which a particular treatement is furnished in a dictionary.

The example shown in Figure 1 is treated in these two seminal works (in [8] its structure is classified as Type 1b), and in both only one extension is accepted, namely the one that contains the paths Clyde->Royal Elephant, and Clyde->Royal Elephant-/->Gray. However, since in both systems the negative links are also considered to be defeasible, it is somewhat contradictory to exclude extensions containing positive paths linking Clyde and Gray. Indeed, if we cannot guarantee that all royal elephants are not gray, than Clyde could be that exceptional gray royal elephant.

In our approach, both situations can be modeled. Let us suppose that $\alpha = \beta = 1$, i.e. that the concept "Clyde" is in complete accordance with the concepts "elephants" and "royal elephants". If $\eta = 1$, then the best answer is surely that Clyde is not gray, coming from extensions Bl and B2. However, if η is a value close to 1 but $\eta < \lambda$, then Clyde is gray (with certainty λ) and it represents an exception in the class of royal elephants.

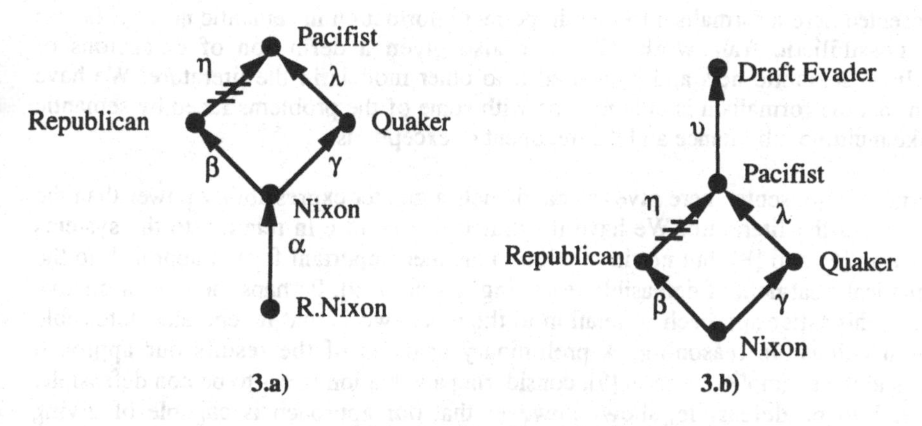

Fig. 3. Extended versions of Nixon's diamond

An extended Nixon's diamond is proposed in [8], where it is classified as a type 2 structure. This net is represented in our formalism as depicted in Figure 3.a) (the concept Nixon represents the Nixon family, and R.Nixon its member Richard Nixon). In our framework 4 extensions are yielded, two yielding the conclusions Nixon-/->Pacifist and R.Nixon-/->Pacifist, and two yielding the conclusions Nixon->Pacifist and

R.Nixon-> Pacifist. However, the best answers the PSN gives about the relations between Nixon and Pacifist, and between R.Nixon and Pacifist are independent. We might for instance obtain as best responses that Richard Nixon is a pacifist, whereas the Nixons in general are not. This is called *decoupling* in [6], and regarded as a undesirable feature in this work and in [8]. However, nothing really prevents a person to have different ideas from most of his family, and the situation shown here appears more often than not. Decoupling, in this case, is a way to accomodate different views. Moreover, if we had (R.Nixon->Nixon, 1), i.e. if the element Richard Nixon strictly agrees with all the characteristics of the class Nixon, then the conclusions about the pacifism of the element and the class would agree with the same strength, and decoupling would be avoided.

Figure 3.b) brings a structure classified as type 3 in [8]. In our formalism, 3 extensions can be derived therefrom; one in which Nixon is both a pacifist and a draft evader, and two in which Nixon is not a pacifist, and no conclusion can produced about draft evasion.

Formally decoupling would not occur here, even if we had Nixon-/->Pacifist and Nixon->Draft Evader, since nothing on the graph indicates the attitude of non pacifists in relation to draft evasion, and indeed there exists draft evaders which are non pacifists. That case could nevertheless characterize a more subtle kind of decoupling since here the only way to arrive to the conclusion that Nixon is a draft evader would be considering that he is a pacifist (there is no direct negative link between Nixon and draft evader, or between republican and draft evader).

However, should we have Nixon-/->Pacifist and Nixon->Draft Evader as best answers, still the credibility of the second statement would be inferior to that of the first one, for whatever value υ might assume.

6 Conclusion

We presented here a formalism to treat imperfect information in semantic nets by the use of the possibilistic framework. We have also given a definition of extensions of possibilistic semantic nets and compared it to other models in the literature. We have showed that the formalism is able to cope with some of the problems faced by semantic nets, like multiple inheritance and the treatment of exceptions.

The formalism presented here gives semantic nets a greater expressibility power than the ones found in the literature. We have illustrated this feature in relation to the systems presented in [6] and [8], but not in relation to another important formal approach to the mathematical treatment of defeasible reasoning, given in [9]. Perhaps the most attractive feature of this latter approach in relation to the other two is that it separates defeasible from non defeasible reasoning. A preliminary analysis of the results our approach produces to the examples given in [9], considering a valuation $\alpha = 1$ to be non defeasible, and $\alpha < 1$ to be defeasible, shows however that our approach is capable of giving intuitively correct results when the system in [9] fails to give any answer. This analysis will be detailed in future work.

The main drawback of this formalism is the lack of a truly conclusive intuitive semantics for the qualified links. When only one interpretation for the links is used in one application this does not represent a problem, since the semantics of the valuations of the results is the same as those of the original links. The difficulty arises when more than one type of interpretation on the links is used in a single application. However, it is important to note that when possibility theory is employed in a knowledge-based system, what really matters is the ordering the valuations produce on its knowledge base. Let us imagine that the user of our semantic net has not given the valuations, but only ordered his pieces of knowledge in classes, such that the elements of the first class are more strongly supported than the second, and so on. Even if the system gives arbitrary values for the valuations on the classes, respecting however the ordering, the use of operator min is such that the final results produced will be coherent with the original ordering. This means that even if interpretations are mixed for the original valuations but the user consider the ordering they induce as reasonable, the valuations on the final results will be coherent with that ordering.

Last but not least, the formalism shown here is of a very simple conception, what makes it attractive both for knowledge engineers, experts and final users.

References

1. Quillian, M.R.: Semantic Memory. In: M.L. Minsky (ed): Semantic Information Processing. Cambridge, Ma: M.I.T. Press, pp. .216-270 (1968).

2. Bittencourt G., Marengoni M., Sandri S.: The use of Possibilistic Logic PL1 in a Customizable Tool for the Generation of Production-Rule Based Systems. Proc. of ECSQRAU' 93. Granada, Spain (1993).

3. Fahlman, S. E.: NETL: A System for Representing and Using Real-World Knowledge. Cambridge, Ma: M.I.T. Press (1979).

4. Froidevaux C., Kayser D.: Inheritance in Semantic Networks and Default Logic. In: P. Smets, E.H. Mandani, D. Dubois, H. Prade (eds): Non-Standart Logics for Automated Reasoning. Academic Press (1988).

5. Etherington D.W.: Formalizing Nonmonotonic Reasoning Systems. Artificial Intelligence, vol 31, n°1, pp. 41-85 (1987).

6. Touretzky D.S., Horty J.F., Thomason R.H.: A Clash of Intuitions: The Current State of Non-monotonic Multiple Inheritance Systems. Proc. 10th IJCAI, pp. 476-482 (1987).

7. Rossazza J-P.: Utilization de hierarchies de classes floues pour la representation de connaissances imprecises et sujettes a exceptions: Le systeme "SORCIER". Phd Thesis. Toulouse, France: Uni. Paul Sabatier (1990).

8. Sandewall E.: Nonmonotonic Inference Rules for Multiple Inheritance with Exceptions. Proc. of the IEEE, vol 74, n° 10, pp. 1345 - 1353 (1986).

9. Simari G.R, Loui R.P.: A mathematical treatment of defeasible reasoning and its implementation. Artificial Intelligence, vol 53, pp.125-157 (1992).

10. Dubois D., Prade H. (with the collaboration of H. Farreny, R. Martin-Clouaire, C. Testemale): Possibility Theory - An Approach to the Computerized Processing of Uncertainty. Plenum Press (1988).

11. Dubois D., Lang J., Prade H.: Possibilistic Logic. Report IRIT/91-98/R. Toulouse, France: IRIT (1991).

6. LOGICS

An Essay on Name and Extension of Rule-Given Properties

ENRIC TRILLAS

Dept. Inteligencia Artificial. U.P.M. 28600 Boadilla del Monte. Madrid, Spain.

SUSANA CUBILLO

Dept. Matemática Aplicada. U.P.M. 28600 Boadilla del Monte. Madrid, Spain.

A. R. DE SOTO

Dept. Dirección y Economía de la Empresa. U.L.E. 24071 León, Spain.

Abstract

In the crisp case, any non-empty set A of a universe X is a preordering's class under the material conditional $\rightarrow_A$. Then, the elements of A are to be considered as those of X satisfying the property of being the "followers" of any $a \in A$; we could see A as a "segment of the structure $(X, \rightarrow)$.

Based on this idea, it is studied the frequent case in which a vague or crisp predicate P is known through both a set of rules (representable by a fuzzy or crisp preorder) and by a crisp prototype. The paper shows how, on this hypothesis, the extension of P or the membership function of the (fuzzy or crisp, respectively) set labelled P is obtained.

1. Introduction

This essay is a first approach to an essential problem concerning Fuzzy Logic. By the moment, we will be placed on the philosophic side rather than on the mathematical one. We will talk from the philosophy, understanding it "à la Russell" as that which is not still established science but that may become science.

In some way, ours is a double problem. Firstly, the problem name/set, and secondly, the one of establishing general methods to set out the generalized compatibility function of a predicate P, or membership function of a fuzzy set labelled P. Both are subproblems of the problem of "knowing a predicate", considering the predicate as the name or linguistic label of a property that admits a quantification: of a property of the elements of a given set which is a matter of degree. This problem is also relevant if, simply, the property holds or not holds, but before Fuzzy Logic it was a partially invisible problem.

Let's begin seeing how any classic set can be "named" through a mathematically defined property. That is, if A is a non-empty subset of a given set X, there will be a predicate P that enunciates a mathematical property p of the elements of A such that $A = \{x \in X \ ; \ x \text{ is } P\}$; equivalently, A is the extension of P in X, or A is formed by the $x \in X$ which are P.

Of course, we will can say that both "x is P" and "x is into the extension of P" are equivalent propositions; we search the way to pass from simply *being* to *being* in a determinated place.

To a such an end, given A let it be $\rightarrow_A = A \times A \cup A^c \times X$ the preorder which, in Logic, is called the A-material conditional, playing A the role of a set of "true elements", in the sense that holds the Modus Ponens: if $a \in A$ and $a \rightarrow_A b$, then $b \in A$. Obviously, the classes of preordering module $\rightarrow_A$:

$$[x, \rightarrow_A) = \{y \in X \ ; \ x \rightarrow_A y\}.$$

are non-empty; effectively, it is $x \in [x, \rightarrow_A)$ because of reflexivity. Furthermore, for each $x \in A$ it is $[x, \rightarrow_A) = A$.

Theorem. If A and B are non-empty subsets of X, it is:

$$A = B \text{ iff } \rightarrow_A = \rightarrow_B .$$

Proof. Obviously, the necessary condition holds. Reciprocally, if $A = X$, from $X \times X = \rightarrow_B$ it trivially follows $B = X = A$.
Let's suppose $A \neq X$ and $x \in A$. Taking $y \notin A$ it is $(x, y) \notin \rightarrow_A$. Then $(x, y) \notin \rightarrow_B$ or $x \in B$ and $y \notin B$. That means $A \subset B$. Starting with $B \neq X$ and $x \in B$ it is analogously reached $B \subset A$. Then $A = B$, provided $A \neq X$ and $B \neq X$. □

Consequently, a non-empty subset is identified by its material conditional.

Now, if we suppose that given $A \subset X$, $A \neq \emptyset$, at least an element $a \in A$ is "known", then

$$A = [a, \rightarrow_A) = \{x \in X \ ; \ a \rightarrow_A x\}.$$

Thus, it is enough to consider the property p_a defined by:

x verifies p_a iff $a \rightarrow_A x$ (which can be enunciated as "x follows a"),

to accept that:

x verifies p_a iff x is P_a,

if P_a is the name of p_a.

Obviously, P_a is a crisp predicate on X verifying

$$A = \{x \in X \; ; \; x \text{ is } P_a\},$$

and then A is the extension of P_a in X: a name of A is P_a, because A is the set of the elements of X that are P_a. In a logic language:

$$A = \{x \in X \; ; \; \text{"}x \text{ is } P_a\text{" is a true proposition } \}$$

Taking, by definition, $P_a =$ "following a", *A is the set of the x in X which follows* a. As the extension of a crisp predicate P on X is usually written $\underline{P} = \{x \in X \; ; \; x \text{ is } P\}$, it is $A = \underline{P_a}$.

Obviously, being $A = \underline{P_x} = \underline{P_a}$ for each $x \in A$, both names *following a* and *following x* are synonyms.

Notes 1) *Giving name* is something conventional, but we have proposed a systematic method for giving name to any non-empty subset of X; of course, A is named as the "class of a module $\to_A$".
2) As A and $\to_A$ are equivalent, giving name to A is just to choose a and, without any other information, it simply becomes a problem of election. However, in the applications, it is needed more information; usually, it should be chosen an element being a prototype of the class. The subject, in general, is that a class can come from different contexts, that can lead or not to a prototype-element. Thus, $A = \{0, 1\}$ meaning the set of zeros of the polynomial $x(x-1)$ not easily has a prototype, but meaning the set of true values it admits 1 as the prototype.

With fuzzy predicates, of course, the situation is not essentially different but technically a little bit more complicate.

2. A Theoretical Method to Obtain the Extension of a Crisp Predicate Given by a Prototype and a Family of Rules

1. As we have seen, for all $a \in A$ it is $A = [a, \to_A)$. This way, choosing a "distinguished" element $u \in A$, the elements of A are those following u in the preorder $\to_A$.

So, if a predicate P on X has the extension $\underline{P} = \{x \in X \; ; \; \text{"}x \text{ is } P\text{" is a true proposition}\}$, both "$P$" and "following u according $\to_{\underline{P}}$" are synonymous expressions, given $u \in \underline{P}$: they have the same extension, or extensional meaning, in X. The predicate P is the name of a property p of the elements of X.

As long as the element u is concerned, it should be distinguished by means of some criterion relatively to the property p enunciated by P; i.e., affirming the truth of "u is P".

2. Now, let's see how, in a special case, it is possible to determine the extension in X on a predicate P such that it is known by its use translated by a family of rules like "If 'x is P', then 'y is P' ", and by an element u in E being a crisp prototype in the sense that "u is P" is a true proposition.

Let's call Px the proposition "x is P", and PX the set $\{Px \ ; \ x \in X\}$. We will write $Px \Rightarrow Py$ instead of "If Px, then Py" and we will suppose that the use of P in X is known through a family of rules $Px \Rightarrow Py$, for some couples $(x, y) \in X \times X$ and by the truth of Pu.

Taking advantage of the usual logical meaning of "If..., then..." we will amplify the given family of rules adding $Px \Rightarrow Pu$, for all $Px \in PX$. Clearly, Pu is a maximal element for the relational structure $(PX, \Rightarrow)$.

Furthermore, let's set the hypothesis that $\Rightarrow$ is reflexive and transitive; that is, that $\Rightarrow$ is a preorder. Now, because of the Representation Theorem for Preorders [3], it is:

$$\Rightarrow = \bigcap_{B \in L} \rightarrow_B ,$$

where $L = L(X, \Rightarrow)$ is the set of logical states for the preorder $\Rightarrow$.

Being $Px \rightarrow_B Pu$ for all $B \in L$ and all $Px \in PX$, it is enough to take each time a Px in B so that $Pu \in B$; so, $Pu \in \bigcap_{B \in L} B = B^*$, and as it is not empty, B^* is a logical state: $B^* \in L$.

Theorem. $B^* = \{Px \in PX \ ; \ Pu \Rightarrow Px\} = [Pu, \Rightarrow)$.

Proof. If $Pu \Rightarrow Px$, for all $B \in L$ it is $(Pu, Px) \in B \times B + B' \times X$ and, because of $Pu \in B$, it will be $Px \in B$ for any $B \in L$ too, and $Px \in B^*$. So, $[Pu, \Rightarrow) \subset B^*$. Reciprocally, if $Px \in B^*$, it is $Px \in B$ for all $B \in L$ and $(Pu, Px) \in B \times B \subset \rightarrow_B$ for any $B \in L$, or $(Pu, Px) \in \Rightarrow$. So, $B^* \subset [Pu, \Rightarrow)$. □

Clearly, if $Px \in B^*$, it is $Px \Rightarrow Pu \Rightarrow Px$.

Definition. Let it be $V : PX \rightarrow \{0, 1\}$ the function defined by

$$V(Px) = \varphi_{B^*}(x) = \varphi_{\cap_{B \in L} B}(Px) = \inf_{B \in L} \varphi_B(Px).$$

It is immediate to prove that function V verifies:

1) $V(Px) \leq \varphi_B(Px)$, for each $Px \in PX$ and each $B \in L$. Thus, it is $V(Px) = 1$ if and only if $\varphi_B(Px) = 1$ for all $B \in L$; in particular, it is $V(Pu) = 1$.

2) $V(Po) = 0$ if and only if $\inf_{B \in L} \varphi_B(Po) = 0$. Then, it exists some $B \in L$ such that $\varphi_B(Po) = 0$.

Theorem. It is $V^{-1}(1) = \{Px \in PX \ ; \ Px \in B^*\} = B^*$ and $V^{-1}(0) = \{Px \in PX \ ; \ Px \notin B^*\} = B^{*\prime}$., the complement of B^* in X.
Proof. It is

$$V^{-1}(1) = \{Px \in PX \ ; \ V(Px) = \inf_{B \in L} \varphi_B(Px) = 1\} = \{Px \in PX \ ; \ Px \in \bigcap_{B \in L} B = B^*\}$$

With all we have just set, and according with the assertion that Pu is a true proposition, it will be enough to define:

$$\text{"}Px \text{ is true} \quad \text{iff} \ V(Px) = 1 \ \text{iff} \ Px \in B^* \text{"}$$

to obtain $\underline{P} = \{x \in X \ ; \ Px \text{ is true}\} = B_0$, being $B_0 = \{x \in X \ ; \ Px \in B^*\}$.

Example 1. Let it be $X = \mathbb{N} - \{0\}$, P="even", and the rules:

$$Px \Rightarrow Py \quad \text{iff} \quad y = x + 2n \ (n \in Z).$$

Choosing $u = 2$ as prototype and because of $\Rightarrow$ is a preorder, it is

$$[P2, \Rightarrow) = \{Px \in PX \ ; \ P2 \Rightarrow Px\} = \{Px \in PX; x = 2{+}2n, \ n \in \mathbb{N}\} = \{P2, P4, P6, P8, ...\}.$$

Thus, $\underline{P} = \{2, 4, 6, 8, ...\}$, which of course, it is the set of even numbers. Any $o \in \{1, 3, 5, ...\}$ verifies $V(Po) = 0$.

3. The Case of a Rule-Given Vague Predicate

Let's generalize all that to the case of P a vague predicate known by some graduate, inexact or approximate rules. That is, provided it exists an inexact (or fuzzy) T-preorder[3],[6]:

$$I : PX \times PX \rightarrow [0, 1] \ , \quad I(Py/Px) \in [0, 1],$$

translating the approximate rules:

$$\text{"If } Px \text{, then } Py\text{" to the degree } I(Py/Px).$$

So, function I verifies the properties:
- Reflexive: $I(Px/Px) = 1$
- T-transitive: $T(I(Py/Px), I(Pz, Py)) \leq I(Pz/Px)$,

for all Px, Py, Pz in PX and any t-norm T.

As before, both the function I modelling the rules and the t-norm T such that I is T-transitive (the greatest t-norm which gets I T-transitive) they should come from a knowledge-base $K(X, P)$ relative to the subject problem. For example, if, as often holds, the graduate rules can be written as

$$I(Py/Px) = \begin{cases} 1, & \text{if } y = x \\ Min(f(Px), f(Py)), & \text{if } y \neq x, \end{cases}$$

for some function $f : PX \rightarrow [0,1]$, then it would be valid any t-norm, and $T = Min$ would be the greatest.

Because of the T-preorder's Representation Theorem [3],[6]:

$$I(Py/Px) = \inf_{f \in \mathcal{E}} I_f^T(Py/Px),$$

being $\mathcal{E}$ the family of logical T-states of I (that is, the family of the functions $f : PX \rightarrow [0,1]$ such that $T(f(Px), I(Py/Px)) \leq f(Py)$, for all Px, Py of PX, where I_f^T is the elemental T-preorder $I_f^T(Py/Px) = \sup\{z \in [0,1] \; ; \; T(z, f(Px)) \leq f(Py)\}$) which are not constant. The former inequation of the logical T-states is equivalent, because of the reflexivity of I, to the equation [3]

$$\sup_{Px \in PX} T(f(Px), I(Py/Px)) = f(Py).$$

Each logical state is a fuzzy possible world of true elements. Consequently,

a) If $I_f^T(Py/Px) = 0$, the only z such that $T((z, f(Px)) \leq f(Py)$ is $z = 0$. Then for all $z \in (0,1]$ it must be $f(Py) < T(z, f(Px)) \leq f(Px)$ and $f(Py) < f(Px)$. This way, if for a $f \in \mathcal{E}$, it is $I_f^T(Py/Px) = 0$, it will be $f(Py) \in [0, f(Px))$.

b) If $I(Py/Px) = 1$, it is $T(f(Px), 1) = f(Px) \leq f(Py)$ for all $f \in \mathcal{E}$.

Now, we will demand the additional hypothesis that there is $u \in X$, with $f(Pu) = 1$ for all $f \in \mathcal{E}$. It means that Pu is a true proposition in all "possible worlds of truth". In such case, as $I_f^T(Px/Pu) = f(Px)$ for every $f \in \mathcal{E}$, it is $I(Px/Pu) = \inf_{f \in \mathcal{E}} f(Px)$.

Defining

$$V_u : PX \rightarrow [0,1], \;\; V_u(Px) = I(Px/Pu) = \inf_{f \in \mathcal{E}} f(Px),$$

we obtain:

- $V_u(Px) \leq f(Px)$, for all $Px \in PX$ and for all $f \in \mathcal{E}$. Then, $V_u(Px) = 1$ iff $f(Px) = 1$ for all $f \in \mathcal{E}$. Particularly, $V_u(Pu) = 1$.

- $V_u(Px) = 0$ iff $\inf_{f \in \mathcal{E}} f(Px) = 0$; so, if $\mathcal{E}$ is finite, there is a $f \in \mathcal{E}$ for which $f(Px) = 0$.

- If there is a $o \in X$ such that $f(Po) = 0$ for all $f \in \mathcal{E}$, because of $I_f^T(Px/Po) = 1$ for every f, it follows that $I(Px/Po) = 1$ and $V_u(Po) = \inf_{f \in \mathcal{E}} f(Po) = 0$.

The relation $Px \equiv Py$ if and only if $V_u(Px) = V_u(Py)$, is an equivalence with classes $[Px]_u = \{Py \in PX \; ; \; V_u(Py) = V_u(Px)\} = V_u^{-1}(V_u(Px))$. Obviously,

$$[Pu]_u = \{Py \in PX \; ; \; V_u(Py) = 1\} = V_u^{-1}(1),$$

and the relation

$$[Px]_u \leq_u [Py]_u \quad \text{iff} \quad V_u(Px) \leq V_u(Py),$$

being independent of the representative element, is a total order.

Consequently, the function $\overline{V}_u : PX/\equiv \longrightarrow V_u(PX) \subset [0,1]$, defined by

$$\overline{V}_u([Px]_u) = V_u(Px) \in V_u(PX),$$

is one-to-one.

If there is an assertion $Po \in PX$ such that $V_u(Po) = 0$ (for they would be enough to be $f(Po) = 0$ for some $f \in \mathcal{E}$), then it is $\overline{V}_u([Po]_u) = 0$ and $\{0,1\} \subset V_u(PX) \subset [0,1]$. Clearly, for that f, it would be $I_f^T(Px/Po) = 1$ too.

With all that, and if it is the case that $V_u(PX) = [0,1]$, we can define:

$$\varphi_{\underline{P}}(x) = V_u(Px) = \overline{V}_u([Px]_u) \in [0,1],$$

in such a way that $\varphi_{\underline{P}}(u) = 1$ and $\varphi_{\underline{P}}(o) = 0$.

Note. In the former conditions, in $[0,1]$ there is the T-preorder I_P given by $I_P(r/s) = I_P(V_u(Py)/V_u(Px)) = I(Py/Px)$. The inexact preordered structure $([0,1], I_P)$ admits the Multiple-Valued Logic given by the following definitions:

- $r \rightarrow s = I_P(s/r)$
- $r' = r \rightarrow 0 = I_P(0/r)$
- $r + s = (r \rightarrow 0) \rightarrow s$
- $r \cdot s = (r' + s')'$.

Such definitions can be translated into $PX/\equiv$ by:

- For each $[Px]_u$, $[Px]_u'$ is the class $[Py]_u$ such that $V_u(Py) = V_u(Px) \rightarrow 0$.
- $[Px]_u + [Py]_u$ is the class $[Pz]_u$ such that $V_u(Pz) = (V_u(Px) \rightarrow 0) \rightarrow V_u(Py)$.
- etc.

Then in $PX/\equiv$ we obtain a logical structure naturally associated to the preorder I which, in fact, is given by the predicate P. The predicate is logically managed through the logical structure in $[0,1]$ obtained from I_P.

4. Some Examples with the Fuzzy Predicate B="Big" on X=[0,1]

As it is well known, for the fuzzy subset of [0,1] labelled BIG it is usually taken the membership function $\varphi_{BIG}(x) = x$ [7].

1. If the family of rules could be represented by a generic T-preorder I, then

$$I(By/Bx) = \inf_{f \in \mathcal{E}} I_f^T(By/Bx).$$

As "Bu is true" means $f(Bu) = 1$ for all $f \in \mathcal{E}$, it is:

$$I(Bx/Bu) = \inf_{f \in \mathcal{E}} sup\{z \in [0,1]\ ;\ T(z,1) \leq f(Bx)\} = \inf_{f \in \mathcal{E}} f(Bx).$$

Then, $Vu(Bx) = \varphi_{\underline{B}}(x) = \inf_{f \in \mathcal{E}} f(Bx)$ and, of course, $Vu(Bu) = 1$.

2. Let's consider the case in which $I = I_f^T$ with $f(Bx) = x$. Then

$$\varphi_{\underline{B}}(x) = I_f^T(Bx/Bu) = \sup\{z \in [0,1]\ ;\ T(z,u) \leq x\}.$$

As f is a logical state of I_f^T, it should be $f(Bu) = u = 1$. Then,

$$\varphi_{\underline{B}}(x) = x,$$

and $\varphi_{\underline{B}}(o) = 0$ implies $o = 0$.

It should be pointed out that a sufficient condition for the T-preorder $I_{f,P}^T(y/x) = sup\{z\ ;\ T(z,f(x)) \leq f(y)\}$ contains the total order $\leq$ of $[0,1]$, is that $f : [0,1] \to [0,1]$ be an increasing function; indeed, if $x \leq y$, because of $f(x) \leq f(y)$, it follows $T(z,f(x)) \leq T(z,f(y)) \leq f(y)$ for all z in $[0,1]$, and therefore $[0,1] \subset \{z\ ;\ T(z,f(x)) \leq f(y)\}$, which sup is $1 = I_{f,P}^T(y/x)$. In the chosen case, being $f = id$, $\leq$ is contained in I.

3. In particular, if $T = W$ (the Lukasiewicz' t-norm) and $f(Bx) = x$, the preorder is

$$I(By/Bx) = Min(1, 1 - x + y)$$

and, of course, $\varphi_{\underline{B}}(x) = x$, with $u = 1$ and $o = 0$.

Next cases correspond to T-preorders that are not elemental.

4. If

$$I(By/Bx) = \begin{cases} 1, & \text{if } x = y \\ Min(x,y), & \text{if } x \neq y, \end{cases}$$

with $T = Min$, it will be

$$V_u(Bx) = \begin{cases} 1, & \text{if } x = u \\ Min(x,u), & \text{if } x \neq u. \end{cases}$$

So, let it be $V_u(Bo) = Min(o, u) = 0$ (setting $u \neq o$); if $u < o$ it is $u = 0$, but it is an absurd that $\varphi_{\underline{B}}(u) = 0$; consequently, it must be $o < u$, and so on, $o = 0$.

It is easy to obtain

$$V_u(B1) = \begin{cases} 1, & \text{if } 1 = u \\ u, & \text{if } u < 1 \end{cases}$$

On the other hand, it is obvious that V_u verifies "$x \leq y$ implies $V_u(Bx) \leq V_u(By)$". Then, because of $u \leq 1$, it follows $1 = V_u(B1)$ and $u = 1$.

Finally,

$$V_u(Bx) = \left\{ \begin{array}{ll} 1, & \text{if } x = 1 \\ Min(1, x), & \text{if } x < 1 \end{array} \right\} = x,$$

and $\varphi_{\underline{B}}(x) = x$. Obviously, as before, $[Bx]_1 = \{Bx\}$.

5. If

$$I_f(By/Bx) = \begin{cases} 1, & \text{if } x = y \\ Max(1 - f(x), f(y)), & \text{if } x \neq y, \end{cases}$$

for a given $f : [0,1] \to [0,1]$ and $T = W$, it will be

$$V_u(Bx) = I_f(Bx/Bu) = \begin{cases} 1, & \text{if } x = u \\ Max(1 - f(u), f(x)), & \text{if } x \neq u, \end{cases}$$

that obviously verifies $V_u(Bu) = 1$. In order to $x \leq y$ implying $V_u(Bx) \leq V_u(By)$ it is a sufficient condition that f be an increasing function; so we will suppose it.

If we want $V_u(Bo) = 0$ it must be $0 \neq u$ and $Max(1 - f(u), f(o)) = 0$, i.e., $f(u) = 1$ and $f(o) = 0$. Thus, $V_u(Bx) = f(x)$ for all x in $[0,1]$, with $u = f^{-1}(1)$ and $o = f^{-1}(0)$.

The logic associated to B leads too $x' = I_{f,B}(0/x) = 1 - f(x)$. and if we want $x'' = x$ for all $x \in [0,1]$ it should be $x = x'' = (x')' = (1 - f(x))' = f(x)$; that is, $f = id$.

Finally, as before, $\varphi_{\underline{B}}(x) = x$, with $u = 1$ and $o = 0$.

Again, the preorder I_B does not contain the total order in $[0,1]$: it is $0.4 < 0.6$ but $I_B(0.6/0.4) = Max(0.6, 0.6) = 0.6 \neq 1$

References

[1] RUSSELL, B., *The Principles of Mathematics*, Routledge and Kegan Paul, London (1903).

[2] TRILLAS, E., *On Fuzzy Conditionals Generalizing the Material Conditional*, IPMU'92; Advanced Methods in Artificial Intelligence, Eds. B. Bouchon-Meunier, L. Valverde and R.- Yager, 85-100. Lecture Notes in Computer Science, Springer-Verlag (1993).

[3] TRILLAS, E., *On Logic and Fuzzy Logic*, to be shortly published in Int. Jour. of Uncertainty, Fuzziness and Knowledge-Based Systems.

[4] TRILLAS, E., ALSINA, C. *Logic: going farther from Tarski?*, Fuzzy Sets and Systems 53, 1-13, (1993).

[5] TRILLAS, E., ALSINA, C. *Some Remarks on Approximate Entailment*, International Journal of Approximate Reasoning (6), 525-533, (1992).

[6] VALVERDE, L., *On the structure of F-indistinguishability operators*, Fuzzy Sets and Systems 17, 313-328 (1985).

[7] ZADEH, L.A., *Fuzzy Sets*, Information and Control 8, 338-353, (1965).

On Non-Alethic Logic

Celina A. A. P. ABAR*, M. YAMASHITA*

* Catholic University of São Paulo
Department of Mathematics
Rua Marquês de Paranaguá, 111
01303-050 - São Paulo - Brazil
E_Mail: PUCSPDI@BRFAPESP.BITNET

Abstract. Non-alethic logic was introduced in da Costa [3]. In this kind of logic the principles of tertium non datur and of contradiction are not valid; furthermore, non-alethic logic constitutes a generalization of both paraconsistent and paracomplete logics. Nowadays, paraconsistent and paracomplete logics constitutes an important subject among non-classical logics, being studied in many countries, especially in Brazil, Australia, Italy and the U.S.A. In this note we present one propositional system of non-alethic logic N_1 and its corresponding first-order predicate system $N_1^=$.

1. Introduction

A theory T is said to be inconsistent if it has theorems a formula A and its negation ~ A ; and it is said to be trivial if every formula of this language is a theorem of the theory. Paraconsistent logic serves as a basis for inconsistent but nontrivial theories. Loosely speaking, a paraconsistent logic is a logic in which a proposition and its negation can be both true.

A logic is called paracomplete if, according to it, a proposition A and its negation ~A, can be both false. Intuitionistic logic and several systems of many-valued logic are paracomplete in this sense. In general a paracomplete logic can be conceived as the underlying logic of an incomplete theory in the strong sense, i.e. of a theory according to which a proposition and its negation are both false.

In this note we study one propositional system of non-alethic logic N_1 and its corresponding first-order predicate system $N_1^=$.

This logic was introduced in da Costa [3] and the author describes a new hierarchy of logics which constitute a generalization of both paraconsistent and paracomplete logics; that is, the principles of contradiction and excluded middle are not valid. In [7], N. Grana also studied a system of a mininal non-alethic predicate logic, the quantificational counterpart of a certain non-alethic propositional logic described in [8].

We can easily construct a hierarchy N_i $(0 \le i \le \omega)$ of non-alethic logics taking into account the calculi C_i $(0 \le i \le \omega)$ of da Costa [4] and P_i $(0 \le i \le \omega)$ of da Costa and Marconi [5].

The aim is to construct a logic such that when the principle of excluded middle holds, we obtain the paraconsistent logic; when the law of contradiction holds we have the paracomplete logic and when both principles hold we obtain the classical propositional logic.

Non-alethic logics are important for various philosophical problems (for instance, Hegel's logic) and in connection with constructivity, vagueness, logic programing, etc... The research about some of these relations is just starting. One of the recent applications of paraconsistent logic as well as paracomplete logic was its use in computer science. Reasoning in the presence of inconsistency is a field of growing importance in logic programming and knowledge representation. Details of these applications will appear elsewhere.
The symbols and terminology are those of Kleene [9], with obvious adaptations and extensions.

2. The Propositional Logics C_i $(1 \leq i \leq \omega)$ and P_i $(1 \leq i \leq \omega)$

2.1. The propositional calculus C_1

In a series of papers da Costa has presented a hierarchy of paraconsistent logic (see [6] for example). To make it easy for readers we transcribe, to begin, the underlying language to C_1 and its postulates. The language for C_1 contains the following primitive symbols: Propositional variables; Logical symbols: $\rightarrow$ (implication), $\wedge$ (conjunction), $\vee$ (disjunction), and $\sim$ (negation); Parentheses. Equivalence $\leftrightarrow$ is defined as usual.
$^\circ$ is defined $A^\circ \overset{def}{=} \sim(A \wedge \sim A)$.

The postulates of C_1 are:

1. $A \rightarrow (B \rightarrow A)$
2. $(A \rightarrow B) \rightarrow ((A \rightarrow (B \rightarrow C)) \rightarrow (A \rightarrow C))$
3. $A, A \rightarrow B / B$ (Modus Ponens)
4. $A \wedge B \rightarrow A$
5. $A \wedge B \rightarrow B$
6. $A \rightarrow (B \rightarrow A \wedge B)$
7. $A \rightarrow A \vee B$
8. $B \rightarrow A \vee B$
9. $(A \rightarrow C) \rightarrow ((B \rightarrow C) \rightarrow (A \vee B \rightarrow C))$
10. $\sim\sim A \rightarrow A$
11. $A \vee \sim A$
12. $B^\circ \rightarrow ((A \rightarrow B) \rightarrow ((A \rightarrow \sim B) \rightarrow \sim A))$
13. $A^\circ \wedge B \rightarrow (A \Delta B)^\circ$ where Δ is $\rightarrow, \wedge, \vee$.

The concepts of proof, deduction, etc., are defined as in Kleene [9].

Theorem 2.1.1. Adjoining to C_1 the principles of contradictions $\sim(A \wedge \sim A)$ we obtain the classical propositional calculus.

Theorem 2.1.2. If we set $\approx A \overset{def}{=} \sim A \wedge A^\circ$ we have in C_1:

$\vdash (A \rightarrow B) \rightarrow ((A \rightarrow \approx B) \rightarrow \approx A))$

$\vdash \approx\approx A \rightarrow A$

Theorem 2.1. 3. The following schemes are not valid in C_1:

$$\sim A \to (A \to B); \qquad (A \to B) \to (\sim B \to \sim A)$$
$$(A \to B) \to ((A \to \sim B) \to \sim A))$$
$$(\sim A \vee B) \to (A \to B)$$
$$\sim A \vee \sim B \to \sim (A \wedge B); \quad \sim A \wedge \sim B \to \sim (A \vee B)$$

Theorem 2.1.4. The following schemes are valid in C_1:

$$(A \to B) \to (\sim A \vee B); \quad (\sim A \to A) \to A$$
$$\sim (A \wedge B) \to \sim A \vee \sim B; \quad ((A \to B) \to A) \to A$$

Theorem 2.1.5. $\Gamma \vdash A$ iff $\Gamma \models A$ in C_1 (soundness and completeness theorems for C_1) ($\Gamma \models A$ means A is semantical consequence of the set of formulas Γ).

Problems of decidability related to the logics C_n, $1 \leq n \leq \omega$, and similar calculi, were investigated by da Costa, Alves, Loparic and Marconi [10], [11] and [12].

From the definition of valuation in C_n, $1 \leq n \leq \omega$, it is possible to define the quasi-matrices with which we prove that C_n, $1 \leq n \leq \omega$ is decidable.

2.2. The propositional calculus P_1

A hierarchy of paracomplete logic was introduced in da Costa and Marconi [5]. A logic is called paracomplete if, according to it, a proposition and its negation can be both false. We describe a propositional logic P_1 . The primitive symbols of P_1 are the following: Propositional variables; Connectives: $\to$ (implication), $\wedge$ (conjunction), $\vee$ (disjunction), and $\neg$ (negation); equivalence $\leftrightarrow$ is defined as usual. Parentheses.

We now formulate the postulates of P_1 where $A^* \overset{def}{=} A \vee \neg A$

1. $A \to (B \to A)$
2. $(A \to B) \to ((A \to (B \to C)) \to (A \to C))$
3. $A, A \to B / B$ (Modus Ponens)
4. $A \wedge B \to A$
5. $A \wedge B \to B$
6. $A \to (B \to A \wedge B)$
7. $A \to A \vee B$
8. $B \to A \vee B$
9. $(A \to C) \to ((B \to C) \to (A \vee B \to C))$
10. $((A \to B) \to A) \to A$

11. $A \to \neg\neg A$

12. $A \to (\neg A \to B)$

13. $\neg(A \wedge \neg A)$

14. $A^{*} \to ((A \to B) \to ((A \to \neg B) \to \neg A))$

15. $A^{*} \wedge B^{*} \to (A \,\Delta\, B)^{*}$ where Δ is $\to, \wedge, \vee$.

The concepts of proof, deduction, etc. are defined as in Kleene [9].

Theorem 2.2.1. Adjoining to P_1 the principle of excluded middle $A \vee \neg A$ we obtain the classical propositional calculus.

Theorem 2.2.2. If we set $\neg^{*} A \overset{def}{=} A \to (p \wedge \neg p)$ (p is a fixed propositional variable) we have in P_1:

$$\vdash (A \to B) \to ((A \to \neg^{*} B) \to \neg^{*} A))$$

$$\vdash \neg^{*}\neg^{*} A \to A$$

Theorem 2.2.3. The following schemes are not valid in P_1:

A^{*}

$(\neg A)^{*}$

$(A \to B) \to (\neg A \vee B)$

$\neg(A \vee B) \to \neg A \wedge \neg B$

$\neg(A \wedge B) \to \neg A \vee \neg B$

$(\neg A \to \neg B) \to (B \to A)$

$(A \to \neg A) \to A$

$\neg_{2k} A \to A$ ($\neg_{2k} A$ abreviates $\neg\neg \ldots \neg\neg A$, the symbol $\neg$ appears $2k$ times).

Theorem 2.2.4. The following schemes are valid in P_1:

$A \wedge \neg A \to B$

$\neg A \vee B \to (A \to B)$

$A \wedge B \to \neg(A \to \neg B)$

$A \wedge B \to \neg(\neg A \vee \neg B)$

Theorem 2.25. P_1 is not decidable by finite logical matrices.

A semantics of valuation (cf. Loparic and da Costa [10]) can be developed for P_1

Theorem 2.2.6. P_1 is decidable (by the method of valuations) (see Loparic and da Costa [10]).

3. The Propositional System N_1

Now, we describe a propositional logic N_1. To begin we present a description of N_1 followed by a semantic for that calculus and the proof of the completeness theorem.

We omit proofs when they are formally analogous to the proofs of the corresponding classical theorems.

The underlying language to N_1 contains the following primitive symbols: propositional variables; Connectives: $\rightarrow$ (implication), $\wedge$ (conjunction), $\vee$ (disjunction), and (negation); equivalence $\leftrightarrow$ is introduced as usual. Parentheses.

Definition: $A^\circ \overset{def}{=} \neg(A \wedge \neg A)$ and $A^* \overset{def}{=} A \vee \neg A$.

The postulates of N_1 are:

N_1 $A \rightarrow (B \rightarrow A)$

N_2 $(A \rightarrow B) \rightarrow ((A \rightarrow (B \rightarrow C)) \rightarrow (A \rightarrow C))$

N_3 $A, A \rightarrow B / B$ (Modus Ponens)

N_4 $A \wedge B \rightarrow A$

N_5 $A \wedge B \rightarrow B$

N_6 $A \rightarrow (B \rightarrow A \wedge B)$

N_7 $A \rightarrow A \vee B$

N_8 $B \rightarrow A \vee B$

N_9 $(A \rightarrow C) \rightarrow ((B \rightarrow C) \rightarrow (A \vee B \rightarrow C))$

N_{10} $((A \rightarrow B) \rightarrow A) \rightarrow A$

N_{11} $A^* \wedge B^\circ \rightarrow ((A \rightarrow B) \rightarrow ((A \rightarrow \neg B) \rightarrow \neg A))$

N_{12} $A^\circ \wedge B^\circ \rightarrow ((A \Delta B)^\circ \wedge (\neg A)^\circ)$ where Δ is $\rightarrow, \wedge, \vee$

N_{13} $A^* B^* \rightarrow ((A \Delta B)^* \wedge (\neg A)^*)$ where Δ is $\rightarrow, \wedge, \vee$.

N_{14} $A^\circ \rightarrow ((A \rightarrow \neg\neg A) \wedge (A \rightarrow (\neg A \rightarrow B)))$

N_{15} $A^* \rightarrow (\neg\neg A \rightarrow A)$

N_{16} $A^\circ \vee A^*$

The concepts of proof, deduction, etc. are defined as in Kleene [9].

Theorem 3.1. If we adjoin to N_1 the principle of contradiction we obtain the calculus P_1 and adjoining the principle of excluded middle we get C_1.

Theorem 3.2 . If we add both, principles of contradiction and excluded middle to N_1 we obtain the classical propositional calculus.

Theorem 3.3 . The axiom N_{16} is independent.

To prove the independence of axiom schema N_{16}, consider the following tables when 1 is the designated value:

	1 2 3	$\wedge$	1 2 3	$\vee$	1 2 3		
1	1 2 3	1	1 2 3	1	1 1 1	1	2
2	1 2 1	2	2 2 3	2	1 2 2	2	1
3	1 1 1	3	3 3 3	3	1 2 3	3	3

Theorem 3.4 . If we set $-A \overset{def}{=} A \to \neg A \wedge A^{\circ}$ we have in N_1:

$$\vdash (A \to B) \to ((A \to -B) \to -A)$$

$$\vdash --A \to A$$

Theorem 3.5 . N_1 contains the classical positive propositional logic.

Theorem 3.6 . We have in N_1:

1) $A \vdash A^{*}$	2) $\neg A \vdash A^{*}$
3) $A^{*}\ \neg\neg A \vdash A$	4) $A^{\circ}, A \vdash \neg\neg A$
5) $A^{\circ}, A, \neg A \vdash B$	6) $A^{*} \to B, A^{\circ} \to B \vdash B$
7) $A \vdash (\neg A)^{*}$	8) $A^{\circ} \vdash (\neg A)^{\circ}$
9) $\vdash (A \wedge \neg A)^{*}$	10) $\vdash A \vee A^{\circ}$
1) $\vdash (A^{\circ})^{*}$	12) $\vdash (A \wedge \neg A)^{\circ}$
3) $\vdash A^{\circ\circ}$	14) $\vdash \neg A^{\circ} \to A^{*}$
5) $\vdash -A^{\circ} \leftrightarrow A \wedge \neg A$	16) $\vdash \neg A^{\circ} \leftrightarrow -A^{\circ}$

Theorem 3.7. The following schemes are not valid in N_1:

$$A^{\circ} \wedge A^{*}; A \to \neg A^{*}; \neg A \wedge A^{\circ}; A^{*} \vee \neg A^{*}; (A^{*})^{\circ} \wedge (A^{*})^{*}.$$

4. A semantic for N_1.

We establish the decidability of N_1 by the method of valuation.

Definition. Let $e : F \to \{0,1\}$ be a function whose domain is F. e is said to be an *evaluation* if we have:

If A is axiom then $e\ (A) =\ $;
If $e\ (A) = e\ (A \to B) = 1$ then $e(B) =\ $;
There is a formula B such that $e(B) = 0$.

Definition. Let Γ be a set of formulas and F a formula. We say that Γ is F-satured if $\Gamma \nvdash F$ and for every B such that $B \notin \Gamma, \Gamma \cup \{B\} \vdash F$

Definition. A *valuation* is a evaluation that is the characteristic function of a F-satured set Γ.

Definition. A *valuation* in N_1 is the function $v : \mathrm{F} \to \{0,1\}$ such that:

v_1. $v(A \wedge B) = 1 \Leftrightarrow v(A) = 1$ and $v(B) = 1$
v_2. $v(A \vee B) = 1 \Leftrightarrow v(A) = 1$ or $v(B) = 1$
v_3. $v(A \to B) = 1 \Leftrightarrow v(A) = 0$ or $v(B) = 1$
v_4. $v(A) \neq v(\neg A) \Rightarrow v(\neg\neg A) \neq v(\neg A)$
v_5. $v(\neg(A \wedge \neg A) \neq v(A \wedge \neg A)$
v_6. $v(A^*) = v(B^*) = 1 \Rightarrow v(A \,\Delta\, B)^* = 1$,
where $\Delta \in \{\wedge, \vee, \to\}$
v_7. $v(A^\circ) = v(B^\circ) = 1 \Rightarrow v(A \,\Delta\, B)^\circ = 1$,
where $\Delta \in \{\wedge, \vee, \to\}$.

Definition. A valuation v is a model of a set Γ of formulas, if $v(A) = 1$ for every A in Γ. We say that A is a semantical consequence of the set of formulas Γ, and write $\Gamma \models A$, if every model of Γ is also model of $\{A\}$; in particular, $\models A$ means that A is valid in N_1.

Theorem 4.1. (Soundness theorem). If $\Gamma \cup \{A$ is a set of formulas of N_1, then $\Gamma \vdash A \Rightarrow \Gamma \models A$.

Theorem 4.2.(of Lindenbaum). For every set Γ and every formula F, if $\Gamma \not\vdash F$ then there is as set Δ that $\Gamma \subset \Delta$ and Δ is a set F-satured.

Theorem 4.3. Every set of formulas F-satured has a model.

As above and in Loparic and da Costa [10], with the help of the concept of valuation, we can define the notions of model of a set of formulas of semantic consequence, and the symbol $\models$. We then have

Theorem 4.4. $\Delta \vdash A$ iff $\Delta \models A$ in N_1. (soundness and completeness theorem for N_1.)

5. Non-Alethic Predicate Logic

Starting with the propositional calculus N_1, it is not difficult to construct the corresponding first-order predicate calculus $N_1^=$.
The underlying language of $N_1^=$ is a first-order language with equality in wich $\to$ (implication), $\wedge$ (conjunction), $\vee$ (disjunction), $\neg$ (negation), $\forall$ (the universal quantifier) and $\exists$ (the existencial quantifier) are primitive (as in N_1, the equivalence symbol, $\leftrightarrow$, is definid as usual). The postulates of $N_1^=$ are those of N_1, plus the following (with the restrictions as usual):

$B \to A / B \to \forall x A$

$\forall x A \to A_x[t]$

$A_x[t] \to \exists x A$

$A \to B / \exists x A \to B$

$\forall x A^{\circ} \to ((\forall x A)^{\circ} \wedge (\exists x A)^{\circ})$

$\forall x A^{*} \to ((\forall x A)^{*} \wedge (\exists x A)^{\circ})$

If A and B are two congruent formulas (in the sense of Kleene) or one of them is obtained from the other by the suppression of vacuous quantifiers, then $A \leftrightarrow B$ is an axiom.

$x = x$

$x = y \to (A \to A_x[y])$

Theorem 5.1. The semantics of a valuation of N_1 can be extended to $N_1^{=}$.

A semantic for $N_1^{=}$ is based on a generalization of the concept of structure in the sense of Shoenfield[13]. We define a structure for the language of $N_1^{=}$ and construct valuations to proof the soundness and completeness for $N_1^{=}$.

Theorem 5.2. $N_1^{=}$ is a conservative extension of N_1.

Starting with the system $N_1^{=}$, it is not difficult to construct a family of corresponding set theories as well as higher-order logics. Details will appear elsewhere.

References

1. Abar, C.A.A.P. and M. Yamashita, Remarks on Variable Binding Term Operators. *Polish Academy of Sciences, Bulletin of the Sect. of Logic*, v.15,n.4,pp.145-151,1986.
2. Abar, C.A.A.P., On a Modal Paracomplete Logic. *Proceedings of IPMU'92*, Universitat de Les Illes Balears, Spain,pp. 333-336, 1992.
3. da Costa, N.C.A., Paraconsistent and Paracomplete Logic. *Rendiconti dell'Accademia Nazionale dei Lincei*, 1989.
4. da Costa, N.C.A., Calcule propositionells pour les systèmes formels inconsistants. *Comptes Rendus de l'Academie des Sciences de Paris*, pp.3790-3792, 1963.
5. da Costa, N.C.A. and Diego Marconi, A Note on Paracomplete Logic. *Rendiconti dell'Accademia Nazionale dei Lincei*, pp.504-509. 1986.
6. da Costa, N.C.A., On the theory of inconsistent formal systems. *Notre Dame Journal of Formal Logic*, pp.497-510, 1974.
7. Grana, N., On a Minimal Non-Alethic Predicate Logic. *Boletim da Sociedade Paran. de Matemática*, v.11, n.1, 1990.
8. Grana, N., On Mininal Non-Alethic Logic. *Bulletin of the Sect. of Logic*, Polish Acad. Science, 1990.
9. Kleene, S.C., Introduction to Meta-mathematics. North-Holland, 1952.

10.Loparic, A. and Newton C.A. da Costa, Paraconsistency, Paracompleteness, and Valuations. *Logique et Analyse*, pp.119-131, 1984.

11.Loparic, A. and E.H.Alves, The semantics of the systems C_n of da Costa. *Proceedings of the Third Brazilian Conference on Mathematical Logic*, A. I. Arruda, N.C.A. da Costa, and A.M. Sette, editors, pp.161-172, 1980.

12.Marconi, D., A Decision-method for the calculus C_1. *Proceedings of the Third Brazilian Conference on Mathematical Logic.* A.I. Arruda, N.C.A. da Costa, and A.M. Sette, editors, pp.211-223, 1980.

13.Shoenfield, J.R., Mathematical Logic,.Addison-Wesley, 1967.

Marginal Problem in Different Calculi of AI[1]

Milan Studený

Academy of Sciences of Czech Republic, Inst. of Inform. Theory and Autom.
Pod vodárenskou věží 4, 18208 Prague 8, Czech Republic
E-mail: studeny@utia.cas.cz

Abstract. By the marginal problem we understand the problem of the existence of a global (full-dimensional) knowledge representation which has prescribed less-dimensional representations as marginals. The paper deals with this problem in several calculi of AI: probabilistic reasoning, theory of relational databases, possibility theory, Dempster-Shafer's theory of belief functions, Spohn's theory of ordinal conditional functions. The following result, already known in probabilistic framework and in the framework of relational databases, is shown also for the other calculi: the running intersection property is the necessary and sufficient condition for pairwise compatibility of prescribed less-dimensional knowledge representations being equivalent to the existence of a global representation. Moreover, a simple method of solving the marginal problem in the possibilistic framework and its subframeworks is given.

1 Introduction

Dealing with integration of knowledge in probabilistic expert systems one encounters the problem of consistency well-known as the *marginal problem* [8]: having prescribed a collection of less-dimensional probability measures (which represent pieces of knowledge given by experts – see [12]) one should recognize whether there exists a joint multidimensional probability measure having the prescribed less-dimensional measures as marginals (such a joint measure then could represent global knowledge kept by an expert system).

Of course, an analogous problem can be expected when one tries to model expert knowledge within another calculus for uncertainty management. Concretely, this paper is concerned with the marginal problem in the following branches of AI:

- probabilistic reasoning
- theory of relational databases
- theory of ordinal conditional functions
- possibility theory
- theory of belief functions.

[1]Supported by the grant n. 201/94/0471 "Marginal problem and its application" of the Grant Agency of Czech Republic.

As concerns the probabilistic framework[2] no direct method of solving the marginal problem is known but there exists an asymptotic method. Using the collection of prescribed less-dimensional measures one can define by means of the so-called *iterative proportional fitting procedure* [3] a sequence of multidimensional probability measures which is proved in [2] to converge iff there exists a joint measure having the prescribed measures as marginals. The limit measure then has the prescribed marginals and minimizes I-divergence within the class of such joint measures.

Nevertheless, one can sometimes evade this iterative procedure as the global consistency is under a certain structural condition put on the collection of underlying attribute sets[3] equivalent to the condition of pairwise compatibility which is easy to verify or disprove. Kellerer [9] showed that the global consistency is equivalent with the pairwise compatibility iff the collection $\mathcal{S}$ of underlying sets satisfies the *running intersection property* (see also [10], [8]):

$$(*)\left\{\begin{array}{l}\text{there exists an ordering } S_1,\ldots,S_n \text{ of } \mathcal{S} \text{ such that}\\ \forall j\geq 2 \quad \exists i \quad 1\leq i<j \quad S_j\cap(\bigcup_{k<j}S_k)\subset S_i\,.\end{array}\right.$$

This condition has a meaning of acyclicity of the hypergraph $\mathcal{S}$ (this terminology was accepted in [1]). For example, the chain in the following figure satisfies the running intersection property, but the cycle does not.

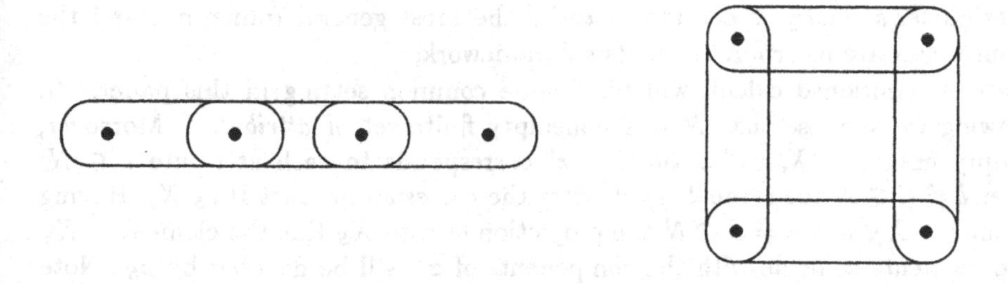

Fig. 1. Examples of hypergraphs.

Similar results were later shown in the theory of relational databases. A simple direct method of solving the marginal problem in this framework is known – see [10]. It consists in verifying whether a certain multidimensional database relation, made of the prescribed less-dimensional database relations by a concrete procedure, has the prescribed relations as marginals. Moreover, the running intersection property was also shown to be a necessary and sufficient condition for pairwise compatibility being equivalent to global consistency in this framework [1].

The aim of this paper is to show the results concerning the running intersection property for the other mentioned calculi as well. Note that this result for the calculus

[2]In this paper we restrict our attention to probability measures on finite sets.

[3]By an attribute we understand an elementary symptom or variable in consideration of an expert system. The corresponding knowledge is represented differently in different calculi. In probabilistic reasoning it is represented by a one-dimensional probability measure. Every expert can give evidence concerning certain area, i.e. his statements refer to a small set of attributes. Thus, the piece of knowledge given by an expert is represented by a less-dimensional probability measure embracing exactly attributes from the mentioned set (= the underlying attribute set).

of ordinal conditional functions was recently proved in [16]. The procedure from [16] can also be used in all the other calculi – one only has to give several basic constructions (specific for each calculus). This is done in this paper in order to make a comprehensive survey. Moreover, the above mentioned simple direct method of solving the marginal problem for relational databases is extended to the possibilistic calculus in this paper.

The next section recalls how knowledge is represented in all the calculi. The third section then describes the mentioned method of solving the marginal problem in the possibilistic calculus and gives an example showing that the global consistency of a collection of possibility measures is indeed strictly stronger than its global consistency in sense of Dempster-Shafer's theory. In the fourth section the main constructions allowing us to show the main results are given.

2 Basic Definitions

This section recalls how knowledge is represented in all the mentioned calculi of AI. The most of these calculi are constructed to be as general as possible and some readers can object that the definitions below are restrictive. But they express the essence of these calculi and make possible certain unifying point of view: the calculus of Dempster-Shafer's theory is considered to be the most general framework and the other approaches are incorporated as its subframeworks.

All above mentioned calculi will have some common setting in this paper. In the following we suppose that N is a nonempty finite set of *attributes*. Moreover, a nonempty finite set X_i called the *frame*[4] corresponds to each attribute $i \in N$. Whenever $\emptyset \neq S \subset N$ the symbol X_S denotes the cartesian product $\Pi_{i \in S} X_i$. Having an element $x \in X_N$ and $\emptyset \neq S \subset N$ the projection of x to X_S (i.e. the element of X_S whose components coincide with the components of x) will be denoted by x_S. Note that a marginal knowledge representation on a set of attributes will be always denoted by the symbol of the original knowledge representation having moreover as the upper index the symbol of the attribute set. The power set of a set Y will be denoted by $\exp Y$.

We start our definition survey with the probabilistic calculus which is probably the most developed approach for dealing uncertainty in AI.

Definition 1. (probability measure)
A *probability distribution* over N is a nonnegative real function $P : X_N \to [0, \infty)$ satisfying $\sum\{P(x); x \in X_N\} = 1$. The formula $P(A) = \sum\{P(x); x \in A\}$ (for $A \subset X_N$) then defines a set function (on $\exp X_N$) called a *probability measure* over N. Whenever $\emptyset \neq S \subset N$ and P is a probability measure over N, then its *marginal* on S is a probability measure P^S over S defined as follows (of course $P^N \equiv P$):
$P^S(A) = P(A \times X_{N \setminus S})$ for $A \subset X_S$ whenever $\emptyset \neq S \neq N$.

Another framework where the marginal problem has already been studied is the theory of relational databases.

[4] The frame is the set of "possible" values for the considered attribute.

Definition 2. (database relation)
A *database relation* over N is a nonempty subset of X_N. Whenever $\emptyset \neq S \subset N$ and R is a database relation over N, then its *marginal* on S is a database relation R^S over S defined as follows ($\mathsf{R}^N \equiv \mathsf{R}$):
$u \in \mathsf{R}^S \Leftrightarrow [\,(u,v) \in \mathsf{R}$ for some $v \in X_{N\setminus S}]$ whenever $u \in X_S$.

A further theory in our focus is Spohn's theory of ordinal conditional functions [15]. This theory gives a tool for mathematical description of dynamic handling of deterministic epistemilogy and in this sense it constitutes a counterpart of the probabilistic approach. Researchers in AI paid attention especially to a special class of natural conditional functions [7], [14]:

Definition 3. (natural conditional function)
Having a nonnegative integer function $\kappa : X_N \to \{0,1,\ldots\}$ satisfying $\min\{\kappa(x);\, x \in X_N\} = 0$, the formula $\kappa(A) = \min\{\kappa(x);\, x \in A\}$ (for $\emptyset \neq A \subset X_N$) defines a set function on $(\exp X_N) \setminus \{\emptyset\}$ called a *natural conditional function* (NCF) over N. Whenever $\emptyset \neq S \subset N$ and κ is an NCF over N, then its *marginal* on S is an NCF κ^S over S defined as follows ($\kappa^N \equiv \kappa$):
$\kappa^S(A) = \kappa(A \times X_{N\setminus S})$ for $\emptyset \neq A \subset X_S$ whenever $\emptyset \neq S \neq N$.

The next calculus is possibility theory which was proposed by Zadeh [17] as a model for quantification of judgements on the basis of fuzzy theory and later developed by Dubois and Prade [5].

Definition 4. (possibility measure)
A *possibility distribution* over N is a real function $\pi : X_N \to [0,1]$ satisfying $\max\{\pi(x);\, x \in X_N\} = 1$. The formula $\pi(A) = \max\{\pi(x);\, x \in A\}$ (for $\emptyset \neq A \subset X_N$) then defines a set function on $(\exp X_N) \setminus \{\emptyset\}$ called a *possibility measure* over N. Whenever $\emptyset \neq S \subset N$ and π is a possibility measure over N, then its *marginal* on S is a possibility measure π^S over S defined as follows ($\pi^N \equiv \pi$):
$\pi^S(A) = \pi(A \times X_{N\setminus S})$ for $\emptyset \neq A \subset X_S$ whenever $\emptyset \neq S \neq N$.

One of the most popular approaches for dealing with uncertainty in AI is Dempster-Shafer's theory [4], [13]. Knowledge can be described here in several equivalent ways (belief function or commonality function or plausibility function), we chose the concept of basic probability assignment.

Definition 5. (basic probability assignment)
A *basic probability assignment* (BPA) over N is a real function $\mathbf{m} : \exp X_N \to [0,\infty)$ satisfying $\sum\{\mathbf{m}(A);\, A \subset X_N\} = 1$ and $\mathbf{m}(\emptyset) = 0$. Whenever $\emptyset \neq S \subset N$ and $\mathbf{m}$ is a BPA over N, then its *marginal* on S is a BPA $\mathbf{m}^S$ over S defined as follows (of course $\mathbf{m}^N \equiv \mathbf{m}$):
$\mathbf{m}^S(A) = \sum\{\mathbf{m}(\mathsf{R});\, \mathsf{R} \subset X_N\ \ \mathsf{R}^S = A\}$ for $A \subset X_S$ (R^S from Definition 2).
Focal elements are the sets $A \subset X_N$ with $\mathbf{m}(A) > 0$.

The calculi above can be compared each other. For example one can assign a posssibility measure π_κ to every NCF κ by means of the formula:
$\pi_\kappa(A) = e^{-\kappa(A)}$ for $\emptyset \neq A \subset X_N$.

The mapping $\kappa \longrightarrow \pi_\kappa$ is injective and respects marginals i.e.
$(\pi_\kappa)^S = \pi_{(\kappa^S)}$ for every NCF κ over N, $\emptyset \neq S \subset N$.
If there exists an injective mapping respecting marginals from a calculus to another calculus we shall say that the former calculus is a *subframework* of the latter one. Thus, database relations are a subframework of possibility measures as a possibility measure π_{R} is assigned to every database relation R:

$$\pi_{\mathsf{R}}(A) = \begin{cases} 1 & \text{if } A \cap \mathsf{R} \neq \emptyset \\ 0 & \text{otherwise} \end{cases} \qquad (\text{for } \emptyset \neq A \subset X_N),$$

probability measures are a subframework of BPAs as a BPA m_P is assigned to every probability measure P:

$$\mathrm{m}_P(A) = \begin{cases} P(x) & \text{whenever } A = \{x\} \text{ for } x \in X_N \\ 0 & \text{otherwise} \end{cases} \qquad (\text{for } A \subset X_N),$$

and possibility measures are a subframework of BPAs as every possibility measure π can be identified with a BPA m_π whose collection of focal elements is a nest[5] and satisfies the relation $\pi(x) = \sum\{\mathrm{m}_\pi(B);\ x \in B \subset X_N\}$ for $x \in X_N$. We left to the reader to verify that m_π is determined uniquely by these two conditions and that all the mappings are injective and respect marginals.

Thus, the situation can be illustrated by the following picture.

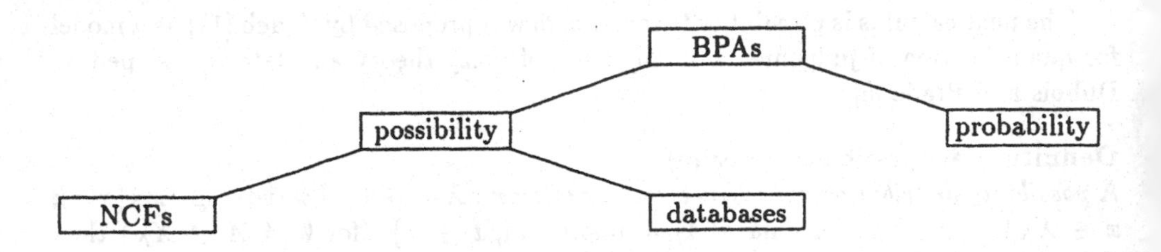

Fig. 2. Comparison of uncertainty calculi.

Remark The reader may think that database relations are a subframework of probability measures since one can assign a probability distribution P_{R} to every database relation R:

$$P_{\mathsf{R}}(x) = \begin{cases} (\text{card}\,\mathsf{R})^{-1} & \text{if } x \in \mathsf{R} \\ 0 & \text{otherwise.} \end{cases}$$

Nevertheless, this mapping does not respect marginals and therefore it is not interesting from our marginal problem point view. Indeed, one can take $N = \{1,2\}$, $X_1 = X_2 = \{0,1\}$, $\mathsf{R} = \{(0,0),(0,1),(1,0)\}$ and have $(P_{\mathsf{R}})^{\{1\}} \neq P_{(\mathsf{R}^{\{1\}})}$.

3 Marginal Problem in Possibilistic Framework

This section gives a simple direct method of solving the marginal problem in the possibilistic framework. Moreover, an example shows that a collection of possibility

[5] i.e. $A \subset B$ or $B \subset A$ for every two focal elements A, B

measures may be globally consistent within the BPA-framework but not within the possibilistic framework.

Firstly, we give exact definitions of concepts connected with the marginal problem. They are shared by all the calculi we deal with in this paper.

Definition 6. (pairwise compatibility, global consistency)
Let us have in mind any of the calculi mentioned in section 2. Suppose that $\{k_S; S \in \mathcal{S}\}$ is a collection of knowledge representations within that calculus (where the lower index S in k_S denotes the nonempty set of underlying attributes). The collection $\{k_S; S \in \mathcal{S}\}$ is called *pairwise compatible* iff $\forall S, T \in \mathcal{S}$ with $S \cap T \neq \emptyset$ $(k_S)^{S\cap T} = (k_T)^{S\cap T}$. Moreover, $\{k_S;\ S \in \mathcal{S}\}$ is called *globally consistent* iff there exists a global knowledge representation k (having the set of underlying attributes N) such that $\forall S \in \mathcal{S}\ \ k^S = k_S$.

It is evident that global consistency implies pairwise compatibility but the converse is not true. The following example shows it for all the mentioned calculi.

Example 1. (pairwise compatibility $\not\Rightarrow$ global consistency)
Consider the global attribute set $N = \{1, \ldots, n\}$ where $n \geq 3$, the frames $X_i = \{0, 1\}$ for $i \in N$ and the collection of attribute sets $\mathcal{S} = \{\, \{1, 2\}, \{2, 3\}, \ldots, \{n-1, n\}, \{n, 1\} \,\}$. Further details depend on calculi:

$\boxed{a}$ probability measures
Define probability distributions $\{P_S; S \in \mathcal{S}\}$ as follows:
$P_{\{n,1\}}(00) = P_{\{n,1\}}(11) = 0 \qquad P_{\{n,1\}}(01) = P_{\{n,1\}}(10) = 0.5$
$P_S(00) = P_S(11) = 0.5 \qquad P_S(01) = P_S(10) = 0 \quad$ for remaining $S \in \mathcal{S}$.
This collection is not globally consistent: having a probability distribution P over N with these prescribed marginals, the inequality $P(x) \leq P^S(x_S)$ implies $0 \leq P(x) \leq \min_{S \in \mathcal{S}} P_S(x_S) = 0$ for all $x \in X_N$ and this contradicts $\sum \{P(x);\ x \in X_N\} = 1$.

$\boxed{b}$ database relations
Put $\mathsf{R}_{\{n,1\}} = \{(0, 1), (1, 0)\}$ and $\mathsf{R}_S = \{(0, 0), (1, 1)\}$ for remaining $S \in \mathcal{S}$. This collection is not globally consistent as there is no database relation R over N having these prescribed marginals. Indeed, no $x \in X_N$ satisfies $\forall S \in \mathcal{S}\ \ x_S \in \mathsf{R}_S$ and this implies a contradictory conclusion $\mathsf{R} = \emptyset$.

$\boxed{c}$ NCFs
Let us define NCFs as point functions on X_S, $S \in \mathcal{S}$:
$\kappa_{\{n,1\}}(00) = \kappa_{\{n,1\}}(11) = 1 \qquad \kappa_{\{n,1\}}(01) = \kappa_{\{n,1\}}(10) = 0$
$\kappa_S(00) = \kappa_S(11) = 0 \qquad \kappa_S(01) = \kappa_S(10) = 1 \quad$ for remaining $S \in \mathcal{S}$.
Supposing an NCF κ over N has these marginals, the inequality $\kappa(x) \geq \kappa^S(x_S)$ implies $\kappa(x) \geq \max_{S \in \mathcal{S}} \kappa_S(x_S) = 1$ for $x \in X_N$ and this contradicts $\min\{\kappa(x); x \in X_N\} = 0$.

$\boxed{d}$ possibility measures
Take $\pi_S \equiv \pi_{\mathsf{R}_S}$, $S \in \mathcal{S}$ where $\{\mathsf{R}_S; S \in \mathcal{S}\}$ are database relations from $\boxed{b}$. Supposing π is a possibility distribution having these prescribed marginals, the inequality $\pi(x) \leq \pi^S(x_S)$ implies $\pi(x) \leq \min_{S \in \mathcal{S}} \pi_S(x_S) = 0$ for all $x \in X_N$ and this contradicts $\max\{\pi(x); x \in X_N\} = 1$.

$\boxed{e}$ BPAs
Take $\mathbf{m}_S \equiv \mathbf{m}_{\pi_S}$, $\ S \in \mathcal{S}\ $ where $\{\pi_S; S \in \mathcal{S}\}$ are possibility measures from $\boxed{d}$. To disprove global consistency realize that every $\mathbf{m}_S$ has only one focal element (with

assigned value 1): $m_{\{n,1\}}$ has $\{(01),(10)\}$ as its focal element, any other m_S has $\{(00),(11)\}$. If m is a BPA over N having $\{m_S; S \in \mathcal{S}\}$ as marginals and R is one of its focal elements, then its marginal R^S has to be a focal element of m_S (for every $S \in \mathcal{S}$). Hence by the procedure from b derive $R = \emptyset$ and this contradicts the definition of BPA.

As mentioned in Introduction, there is no direct method of testing global consistency within the probabilistic framework. Nevertheless, there exists such a method in the possibilistic framework.

Proposition 1. Suppose that $\{\pi_S; S \in \mathcal{S}\}$ is a collection of possibility distributions (when the lower index denotes the set of underlying attributes). Then $\{\pi_S; S \in \mathcal{S}\}$ is globally consistent iff the formula
$\pi_*(x) = \min_{S \in \mathcal{S}} \pi_S(x_S) \quad \text{for } x \in X_N$
defines a possibility distribution whose marginals are $\{\pi_S; S \in \mathcal{S}\}$.

Proof: The sufficiency is evident. For necessity suppose that π is a posibility measure having $\{\pi_S; S \in \mathcal{S}\}$ as marginals. Then
1. $\pi \leq \pi_*$
For each $x \in X_N$ and $S \in \mathcal{S}$ write $\pi(x) \leq \pi^S(x_S) = \pi_S(x_S)$ and then use the definition of π_*.
2. π_* is a possibility distribution
Evidently $0 \leq \pi_* \leq 1$; having $x_0 \in X_N$ with $\pi(x_0) = 1$ the preceding step gives $\pi_*(x_0) = 1$.
3. $\forall S \in \mathcal{S} \quad (\pi_*)^S = \pi_S$
Having fixed $S \in \mathcal{S}$ and $z \in X_S$ by the definition of marginal find $y \in X_N$ with $z = y_S$ and $(\pi_*)^S(z) = \pi_*(y)$. Using the definition of π_* write $\pi_*(y) \leq \pi_S(y_S) = \pi_S(z)$ i.e. $(\pi_*)^S(z) \leq \pi_S(z) = \pi^S(z)$. On the other hand $\pi^S(z) \leq (\pi_*)^S(z)$ follows from 1. and therefore $(\pi_*)^S(z) = \pi_S(z) = \pi^S(z)$. □

The result above gives already published criteria for subframeworks[6]. A collection of database relations $\{R_S; S \in \mathcal{S}\}$ is globally consistent iff the set $R_* = \bigcap_{S \in \mathcal{S}} R_S \times X_{N \setminus S}$ is a database relation having $\{R_S; S \in \mathcal{S}\}$ as marginals [10]. A collection of NCFs $\{\kappa_S; S \in \mathcal{S}\}$ is globally consistent iff the function $\kappa_*(x) = \max_{S \in \mathcal{S}} \kappa_S(x_S)$ determines an NCF over N having $\{\kappa_S; S \in \mathcal{S}\}$ as marginals [16]. These criteria can be derived from Proposition 1 owing to the following principle: supposing that possibility measures $\{\pi_S; S \in \mathcal{S}\}$ correspond to database relations (resp. NCFs) π_* gives a possibility measure corresponding to a database relation (resp. an NCF).

Therefore every collection of database relations (resp. NCFs) is globally consistent iff it is consistent within the possibilistic framework. Similarly, one can show that a collection of probability measures is globally consistent iff it is consistent within the BPA-framework[7]. Nevertheless, a collection of possibility measures need not be

[6] As concerns the test in the possibilistic framework I only found in [6] p. 6–7 a procedure where π_* was defined as one step of a procedure of approximate reasoning [18]. But the mentioned procedure computes one-dimensional marginals of π_* and the authors of [6,18] are not interested in the connection to the starting possibility measures which are not supposed to be pairwise compatible.

[7] Hint: Without loss of generality suppose $N = \bigcup \mathcal{S}$. Projections (=database marginals) of every focal element of a global BPA m having the prescribed marginals must be focal elements of marginals i.e. singletons. Hence, every focal element of m is a singleton.

globally consistent although it is globally consistent within the BPA-framework as the following example shows.

Example 2. (possibilistic consistency $\neq$ consistency within the BPA-framework)
Put $N = \{1,2,3\}$, $X_i = \{0,1\}$ for $i \in N$ and $\mathcal{S} = \{\{1,2\},\{1,3\},\{2,3\}\}$. Consider a collection of possibility distributions $\{\pi_S; S \in \mathcal{S}\}$ defined as follows:
$\pi_S(00) = \pi_S(11) = \frac{2}{3}$ $\pi(01) = \pi(10) = 1$ for $S \in \mathcal{S}$.
This collection is not globally consistent as the function $\pi_*(x) = \min_{S \in \mathcal{S}} \pi_S(x_S) = \frac{2}{3}$ is not a possibility distribution. But, the collection of BPAs $\{\mathrm{m}_{\pi_S}; S \in \mathcal{S}\}$ is globally consistent as one can consider the BPA m with three focal elements:
$\{(010),(011),(100),(101)\}$, $\{(001),(011),(100),(110)\}$, $\{(001),(101),(010),(110)\}$ with assigned values $\frac{1}{3}$.

4 Solvable Collections

In this section the collections of attribute sets for which pairwise compatibility is equivalent to global consistency are studied. We show that these collections are characterized within all mentioned calculi by means of the running intersection property.

We start with some definitions which are, of course, shared by all the studied calculi.

Definition 7. (solvable collection, reduced collection)
Having in mind any of the calculi mentioned in section 2 a collection of nonempty attribute sets $\mathcal{S}$ will be called *solvable* within that calculus iff every pairwise compatible collection of knowledge representations, whose collection of attribute sets is $\mathcal{S}$, is globally consistent. $\mathcal{S}$ will be called *reduced* iff $\forall A, B \in \mathcal{S}$ neither $A \subset B$ nor $B \subset A$. If $\emptyset \neq T \subset N$, then the *contraction* of $\mathcal{S}$ to T denoted by $\mathcal{S} \wedge T$ is defined as the collection of maximal sets[8] of $\{S \cap T; S \in \mathcal{S}\ S \cap T \neq \emptyset\}$.

The method used in [16] to show the necessity of the running intersection property for solvable collections within the NCF-framework in fact does not depend on a particular calculus (see Lemmas 8,9 and Theorem 2 in [16]). One only needs to show that a collection of attribute sets $\mathcal{S}$ is not solvable in two following cases:

$$[A] \quad \begin{cases} \mathcal{S} \text{ contains a sequence } S_1,\ldots,S_n\ (n \geq 3) \text{ such that} \\ \forall i = 1,\ldots,n \quad S_i \cap S_{i+1} \setminus \bigcup(\mathcal{S} \setminus \{S_i, S_{i+1}\}) \neq \emptyset \qquad (\text{where } S_{n+1} \equiv S_1). \end{cases}$$

$$[B] \quad \begin{cases} \mathcal{S} \text{ is a reduced collection with } \operatorname{card} \mathcal{S} \geq 2 \text{ satisfying} \\ \forall i,j \in \bigcup \mathcal{S} \quad \exists S \in \mathcal{S} \quad \text{with} \quad i,j \in S \end{cases}$$

Also the proof of sufficiency requires only to show that a collection $\mathcal{S}$ with $\operatorname{card} \mathcal{S} = 2$ is solvable – see [16]. Thus, in the sequel we only verify these facts for all the mentioned calculi.

Lemma 1. Supposing $\mathcal{S}$ is a solvable collection of attribute sets and $\emptyset \neq T \subset N$ the contraction $\mathcal{S} \wedge T$ is also solvable.

Proof: Supposing $\{k_L;\ L \in \mathcal{S} \wedge T\}$ is a pairwise compatible collection of knowledge representations we are to show that it is globally consistent. To this end we construct

[8] $A \in \mathcal{T}$ is maximal in $\mathcal{T}$ iff $[B \in \mathcal{T},\ A \subset B] \Rightarrow B = A$.

(and this step depends on a calculus) a pairwise compatible collection of knowledge representations $\{k'_S;\ S \in \mathcal{S}\}$ such that $\forall L \in \mathcal{S} \wedge T \quad [L = S \cap T \text{ for some } S \in \mathcal{S}]$ implies $(k'_S)^L = k_L$.
In the sequel we give the corresponding constructions for all studied calculi.
[a] probability measures
Having a pairwise compatible collection of probability distributions $\{P_L;\ L \in \mathcal{S} \wedge T\}$ select for each $i \in N \setminus T$ a probability distribution Q_i on X_i. Then put
$P'_S(x) = (P_L)^{S\cap T}(x_{S\cap T}) \cdot \prod_{i \in S\setminus T} Q_i(x_i) \qquad \text{for } x \in X_S,\ S \in \mathcal{S}$
where $L \in \mathcal{S} \wedge T$ with $S \cap T \subset L$ is arbitrarily chosen $((P_L)^{\emptyset}(-) \equiv 1 \equiv \prod_{i\in\emptyset} Q_i(-)\,)$.
[b] database relations
Having pairwise compatible database relations $\{\mathsf{R}_L;\ L \in \mathcal{S} \wedge T\}$, for each $S \in \mathcal{S}$ with $S \cap T \neq \emptyset$ find $L \in \mathcal{S} \wedge T$ with $S \cap T \subset L$ and put $\mathsf{R}'_S = (\mathsf{R}_L)^{S\cap T} \times X_{S\setminus T}$ ($\mathsf{R}'_S \equiv X_S$ in case $S \cap T = \emptyset$).
[c] NCFs
Having pairwise compatible NCFs $\{\kappa_L;\ L \in \mathcal{S} \wedge T\}$, for each $S \in \mathcal{S}$ with $S \cap T \neq \emptyset$ find $L \in \mathcal{S} \wedge T$ with $S \cap T \subset L$ and put $\kappa'_S(x) = (\kappa_L)^{S\cap T}(x_{S\cap T}) \quad \text{for } x \in X_S,\ S \in \mathcal{S}$ (where $(\kappa_L)^{\emptyset}(-) \equiv 0$).
[d] possibility measures
Having pairwise compatible possibility distributions $\{\,\pi_L;\ L \in \mathcal{S} \wedge T\,\}$, put:
$\pi'_S(x) = (\pi_L)^{S\cap T}(x_{S\cap T}) \qquad \text{for } x \in X_S,\ S \in \mathcal{S}$
(L has the same meaning as in [c], $(\pi_L)^{\emptyset}(-) \equiv 1$).
[e] BPAs
Supposing $\{\mathbf{m}_L,\ L \in \mathcal{S} \wedge T\}$ are pairwise compatible BPAs define for each $S \in \mathcal{S}$:

$$\mathbf{m}_S{}'(F) = \begin{cases} (\mathbf{m}_L)^{S\cap T}(E) & \text{whenever } F = E \times X_{S\setminus T} \text{ with } E \subset X_{S\cap T} \\ 0 & \text{for remaining } F \subset X_S \end{cases}$$

(L has the same meaning as in preceding steps and $(\mathbf{m}_L)^{\emptyset}(-) \equiv 1$). □

Consequence 1. Supposing [A] a collection of attribute sets $\mathcal{S}$ is not solvable.

Proof: Put $T = \{z_i;\ i = 1, \ldots, n\}$ where we chose $z_i \in S_i \cap S_{i+1} \setminus \bigcup(\mathcal{S} \setminus \{S_i, S_{i+1}\})$. Then use Lemma 1 and Example 1 in section 3 to get the desired conclusion. □

Lemma 2. Supposing [B] a collection of attribute sets $\mathcal{S}$ is not solvable.

Proof: Without loss of generality suppose $\bigcap \mathcal{S} = \emptyset$, otherwise put $T = N \setminus \bigcap \mathcal{S}$ and consider $\mathcal{S} \wedge T$ instead of $\mathcal{S}$. In all constructions below we put $X_i = \{0, 1\}$.
[a] probability measures
Denote $m = \text{card} \bigcup \mathcal{S} - 1$ and define for each $S \in \mathcal{S}$ a probability distribution P_S:

$$P_S(x) = \begin{cases} (m - \text{card}\, S) \cdot m^{-1} & \text{if } \sum_{i\in S} x_i = 0 \\ m^{-1} & \text{if } \sum_{i\in S} x_i = 1 \\ 0 & \text{if } \sum_{i\in S} x_i \geq 2 \end{cases} \qquad \text{for } x \in X_S\,.$$

It is no problem to verify pairwise compatibility of $\{P_S; S \in \mathcal{S}\}$. Now, suppose by contradiction that P is a probability distribution over N having $\{P_S; S \in \mathcal{S}\}$ as marginals.
(a) $P^{\cup S}(x) = 0 \qquad \text{whenever} \quad x \in X_{\cup S},\ \sum_{i\in\cup S} x_i \geq 2$.
Indeed: for fixed $x \in X_{\cup S}$ find $i, j \in \bigcup \mathcal{S}$ with $x_i = x_j = 1$ and then $S \in \mathcal{S}$ with $i, j \in S$. Hence $0 \leq P^{\cup S}(x) \leq P^S(x_S) = 0$.
(b) $P^{\cup S}(x) = m^{-1} \qquad \text{whenever} \quad x \in X_{\cup S}, \quad \sum_{i\in\cup S} x_i = 1$.
Indeed: for fixed $x \in X_{\cup S}$ take the only $i \in \bigcup \mathcal{S}$ with $x_i = 1$ and consider $S \in \mathcal{S}$ with $i \in S$.

Then $m^{-1} = P^S(x_S) = P^{\cup S}(x) + \sum\{ P^{\cup S}(x_S, y);\ y \in X_{\cup S \setminus S}\ \sum_{i \in \cup S \setminus S} y_i \geq 1 \}$, where the latter sum is zero by (a).
Evidently (b) gives a contradictory fact $\sum\{ P^{\cup S}(x); x \in X_{\cup S}\} \geq (m+1) \cdot m^{-1} > 1$.
[b] database relations
Put $\mathsf{R}_S = \{x \in X_S;\ \sum_{i \in S} x_i = 1\}$ for $S \in \mathcal{S}$. As $(\mathsf{R}_S)^V = \{x \in X_V;\ \sum_{i \in V} x_i \leq 1\}$ for every proper subset $V \subset S$, $\{\mathsf{R}_S;\ S \in \mathcal{S}\}$ is pairwise compatible. Nevertheless, no database relation R over N has $\{\mathsf{R}_S; S \in \mathcal{S}\}$ as marginals:
(a) Whenever $x \in X_N$ with $\sum_{i \in \cup S} x_i \geq 2$ then $x \notin \mathsf{R}$.
Indeed: choose $i, j \in \bigcup \mathcal{S}$ with $x_i = x_j = 1$ and $S \in \mathcal{S}$ with $i, j \in S$. Then $x_S \notin \mathsf{R}_S$ implies $x \notin \mathsf{R}$.
(b) Whenever $x \in X_N$ with $\sum_{i \in \cup S} x_i \leq 1$ then $x \notin \mathsf{R}$.
Indeed: find $S \in \mathcal{S}$ with $\sum_{j \in S} x_j = 0$ (fix contingent $i \in \bigcup \mathcal{S}$ with $x_i = 1$ and by $\bigcap \mathcal{S} = \emptyset$ find $S \in \mathcal{S}$ with $i \notin S$). Evidently $x_S \notin \mathsf{R}_S$ implies $x \notin \mathsf{R}$.
[c] NCFs
Define an NCF κ_S over N for each $S \in \mathcal{S}$ as follows:

$$\kappa_S(x) = \begin{cases} 0 & \text{if } \sum_{i \in S} x_i = 1 \\ 1 & \text{otherwise} \end{cases} \qquad \text{for } x \in X_S.$$

The collection $\{\kappa_S;\ S \in \mathcal{S}\}$ is pairwise compatible as for $\emptyset \neq V \subset S \in \mathcal{S}$, $V \neq S$ it holds:

$$(\kappa_S)^V(y) = \begin{cases} 0 & \text{if } \sum_{i \in V} y_i \leq 1 \\ 1 & \text{otherwise} \end{cases} \qquad \text{for } y \in X_V.$$

To disprove global consistency use the criterion mentioned below Proposition 1 and compute $\kappa^*(x) = \max_{S \in \mathcal{S}} \kappa_S(x_S)$ (for $x \in X_N$). Supposing $\kappa^*(x) = 0$ we get $\forall S \in \mathcal{S}\ \kappa_S(x_S) = 0$ i.e. $x_S \in \mathsf{R}_S$ where R_S is from [b] - but it was shown there that no $x \in X_N$ satisfies this requirement. Therefore $\kappa^* \equiv 1$.
[d] possibility measures
One can use for example the collection $\{\pi_{\mathsf{R}_S};\ S \in \mathcal{S}\}$ where $\{\mathsf{R}_S;\ S \in \mathcal{S}\}$ are database relations from [b] (see the reasoning before Example 2).
[e] BPAs
One can use $\{\mathbf{m}_{P_S};\ S \in \mathcal{S}\}$ where probability measures $\{P_S;\ S \in \mathcal{S}\}$ are from [a]. □

Lemma 3. A collection $\{I, J\}$ where $I, J \subset N$ is solvable within all mentioned calculi.

Proof: The constructions depend on calculi.
[a] probability measures
Having $\{P_I, P_J\}$ a compatible collection of probability distributions we put:

$$P(x) = \begin{cases} 0 & \text{if } (P_I)^{I \cap J}(x_{I \cap J}) = 0 \\ P_I(x_I) \cdot P_J(x_J) \cdot [(P_I)^{I \cap J}(x_{I \cap J})]^{-1} \cdot \prod_{i \in N \setminus I \cup J} Q_i(x_i) & \text{otherwise} \end{cases}$$

where $(P_I)^{\emptyset}(-) \equiv 1$ and Q_i are arbitrarily chosen one-dimensional prob. measures.
[b] database relations
Having $\{\mathsf{R}_I, \mathsf{R}_J\}$ a compatible collection of database relations we put:
$\mathsf{R} = (\mathsf{R}_I \times X_{N \setminus I}) \cap (\mathsf{R}_J \times X_{N \setminus J})$.
[c] NCFs
Having a couple of compatible NCFs $\{\kappa_I, \kappa_J\}$ we put:
$\kappa(x) = \max\{\kappa_I(x_I), \kappa_J(x_J)\}$ for $x \in X_N$.
[d] possibility measures
Having compatible possibility distributions $\{\pi_I, \pi_J\}$ one can use the formula from

Proposition 1:
$\pi_*(x) = \min\{\pi_I(x_I), \pi_J(x_J)\}$ for $x \in X_N$.

e BPAs

Having a compatible collection of BPAs $\{\mathbf{m}_I, \mathbf{m}_J\}$ we can define a BPA $\mathbf{m}$ over N having them as marginals as follows: focal elements of $\mathbf{m}$ will have the form $\mathsf{G} = (\mathsf{E} \times X_{N\setminus I}) \cap (\mathsf{F} \times X_{N\setminus J})$ where $\mathsf{E} \subset X_I$ is a focal element of $\mathbf{m}_I$, $\mathsf{F} \subset X_J$ is a focal element of $\mathbf{m}_J$ and $\mathsf{E}^{I\cap J} = \mathsf{F}^{I\cap J}$ (in case $I \cap J = \emptyset$ suppose $\mathsf{E}^{\emptyset} = \mathsf{F}^{\emptyset}$). Put $\mathbf{m}(\mathsf{G}) = \mathbf{m}_I(\mathsf{E}) \cdot \mathbf{m}_J(\mathsf{F}) \cdot [(\mathbf{m}_I)^{I\cap J}(\mathsf{E}^{I\cap J})]^{-1}$. □

Hence, one can conclude using [16]:

Proposition 2. A collection of (nonempty) attribute sets $\mathcal{S}$ is solvable within any of the mentioned calculi iff it satisfies the running intersection property:

$$(*)\begin{cases} \text{there exist an ordering } S_1, \ldots, S_n \text{ of } \mathcal{S} \text{ such that} \\ \forall j \geq 2 \ \exists i \ 1 \leq i < j \qquad S_j \cap (\bigcup_{k<j} S_k) \subset S_i . \end{cases}$$

5 Conclusion

The results proved in this paper have theoretical meaning first of all. The study of the marginal problem was so far limited to probability measures and dabase relations (resp. to NCFs). How, the horizons in this respect were broadened also to possibility measures and to the calculus of Dempster-Shafer's theory.

The reader can object that the study of "ideal" consistency of input knowledge may be unrealistic, but I think it is useful to be aware of these results. For example, the results concerning the running intersection property highlight the significance of *decomposable models* [11] which correspond uniquely to collections satisfying this condition. If one is interested in the "internal coherence" of his (her) procedures (i.e. whether the "input" pieces of knowledge and the "output" piece of knowledge are coherent) one should take advantage of these models no matter which calculus one decided on to represent knowledge.

I hope that the method of testing global consistency for possibility measures is of some benefit, as well.

References

1. Beeri, C., Fagin, R., Maier, D., Yannakis, M.: On the desirability of acyclic database schemes. J. Assoc. Comp. Mach. **30** (1983) 479–513.
2. Csiszár, I.: *I*-divergence geometry of probability distributions and minimization problems. Ann. Prob. **3** (1975) 146–158.
3. Deming, W.E., Stephan, F.F.: On a least square adjustment of a sampled frequency table when expected marginal totals are known. Ann. Math. Statis. **11** (1940) 427–444.
4. Dempster, A.P.: Upper and lower probabilities induced by a multivalued mapping. Ann. Math. Statis. **38** (1967) 325–339.
5. Dubois, D., Prade, H.: Possibility Theory: An Approach to Computerized Processing of Uncertainty. Plenum Press, New York - London (1988).

6. Dubois, D., Lang, J., Prade, H.: Possibilistic logic. Report IRIT/97-98/R, Institut de Recherche en Informatique de Toulouse (1991).

7. Hunter, D.: Graphoids and natural conditional functions. Int. J. Approx. Reasoning **5** (1991) 485-504.

8. Jiroušek, R.: Solution of the marginal problem and decomposable distributions. Kybernetika **27** (1991) 403-412.

9. Kellerer, H.G.: Verteilungsfunktionen mit gegebenem Marginalverteilungen (in German). Z. Wahrsch. Verw. Gebiete **3** (1964) 247-270.

10. Malvestuto, F.M.: Existence of extensions and product extensions for discrete probability distributions. Discr. Math. **69** (1988) 61-77.

11. Pearl, J.: Probabilistic Reasoning in Intelligent Systems. Morgan Kaufman, San Mateo CA (1988).

12. Perez, A., Jiroušek, R.: Constructing an intensional expert system INES. In: J.H. von Bemmel, F. Grémy, J. Zvárová (eds.): Medical Decision Making: Diagnostic Strategies and Expert Systems. North-Holland, Amsterdam (1985), 307-315.

13. Shafer, G.: A Mathematical Theory of Evidence. Princeton University Press, Princeton - London (1976).

14. Shenoy, P.P.: On Spohn's rule for revision of beliefs. Int. J. Approx. Reasoning **5** (1991) 149-181.

15. Spohn, W.: Ordinal conditional functions: a dynamic theory of epistemic states. In: W.L. Harper, B. Skyrms (eds.): Causation in Decision, Belief Change, and Statistics vol. II. Kluwer, Dordrecht (1988), 105-134.

16. Studený, M.: Conditional independence and natural conditional functions. Int. J. Approx. Reasoning **12** (1995) 43-68.

17. Zadeh, L.A.: Fuzzy sets as a basis for a theory of possibility. Fuzzy Set and Systems **1** (1978) 3-28.

18. Zadeh, L.A.: A theory of approximate reasoning. In: J.E. Hayes, D. Michie, L.I. Mikulich (eds.): Machine Intelligence 9. Ellis Horwood, Chichester (1979), 149-194.

Implication connectives for logics with right weakening

Philippe Besnard **Yves Moinard**
IRISA, Campus de Beaulieu
35042 RENNES Cedex, FRANCE
{besnard,moinard}@irisa.fr

Abstract. We study non-monotonic inference relations, focusing on the properties of the underlying monotonic consequence relation. This sheds new light on the general properties that may or may not be satisfied by a non-monotonic relation. We let the underlying consequence relation to be any monotonic logic, indicating under what precise conditions it gives rise to a non-monotonic inference relation that obeys major principles such as right weakening and the rule of detachment. We begin to explore the border line that non-monotonic relations cannot cross if they are to enjoy these properties. This work may be considered either as a constructive opposition, or as a complement, to the previous work about the general properties of non-monotonic inference relations.

1 Introduction

Incomplete information cannot be dealt with adequately using standard logics — that are characterized as inference relations satisfying a few properties including the so-called monotony. In fact, inference relations failing monotony (i.e., non-monotonic systems) have emerged as a solution to the formalization of some common sense reasoning problems.

Only recently have the properties of non-monotonic inference relations been studied carefully [Gab85, Mak88, Bes88, KLM90, FLM91]. Most of these papers focus on weakening the requirement for monotony and investigate various ways to do it. Here, we focus on the properties of a non-monotonic inference relation $\mid\!\sim$ relative to an underlying monotonic inference relation $\vdash$ (that is, a standard logic). Such a combination is indeed the case in almost all the non-monotonic formalisms up to date. We aim at identifying how the properties of the "basic" relation $\vdash$ influence the properties of the relation $\mid\!\sim$.

This paper may be considered as a sequel to [Bes88] which defined several notions such as CMP (*compound modus ponens*), later known as *"right weakening"* when $\vdash$ is classical logic:

$$\frac{\mathcal{S} \mid\!\sim A \qquad \mathcal{S} \vdash A \rightarrow B}{\mathcal{S} \mid\!\sim B}.$$

We keep the former name CMP to emphazise that we do not demand $\vdash$ to be classical logic but we only require $\vdash$ to obey *reflexivity:*

$$\mathcal{S} \vdash A \quad \text{whenever} \quad A \in \mathcal{S}$$

and *cut:*

$$\frac{\mathcal{T} \vdash B \qquad \mathcal{S}, B \vdash A}{\mathcal{S}, \mathcal{T} \vdash A}.$$

Observe that $\vdash$ satisfying both reflexivity and cut implies that $\vdash$ also satisfies *monotony*:

$$\frac{\mathcal{S} \vdash A}{\mathcal{S}, \mathcal{T} \vdash A}.$$

Also, an inference relation obeying cut satisfies a particular case of cut called ***cumulative transitivity*** [Mak88]:

$$\frac{\mathcal{T} \vdash B \qquad \mathcal{T}, B \vdash A}{\mathcal{T} \vdash A}.$$

Conversely, an inference relation satisfying monotony and cumulative transitivity satisfies cut. As usual, an inference relation satisfying both reflexivity and cut is called a *consequence relation* [Tar30, Gen33].

A word about notation is in order: we write $\mathcal{S}, \mathcal{T}$ and $\mathcal{S}, A$ as a shorthand for $\mathcal{S} \cup \mathcal{T}$ and $\mathcal{S} \cup \{A\}$ respectively, where $\mathcal{S}$ and $\mathcal{T}$ denote sets of formulas, and letters $A, B, C, \cdots$ denote formulas (or subformulas as in $A \rightarrow B$ below).

In addition to CMP, we focus our attention on the *rule of detachment* (RD),

$$\frac{\mathcal{T} \mid\!\sim A \rightarrow B}{\mathcal{T}, A \mid\!\sim B}.$$

which is also known as the *"easy half of deduction theorem"*. [KLM90] describes the class of inference relations C, in which the rule of detachment is equivalent to monotony. We claim that the equivalence established in [KLM90] relies on imposing the underlying logic $\vdash$ to be a very specific one, namely classical logic. We mean to examine the matter in a more general way, only requiring $\vdash$ to be a standard logic. Thus, we are in a position to determine under what conditions the rule of detachment, a rather desirable inference schema, may fail or hold in non-monotonic systems. We indeed show how the status of RD evolves when we weaken the underlying logic $\vdash$.

In the next section, we give the general definitions used in the paper. Also, we define an underlying consequence relation $\vdash$ from any given non-monotonic inference relation $\mid\!\sim$. We give two forms of non-monotony which are satisfied by virtually all the non-monotonic formalisms proposed in the literature. In section 3, we give three results which follow the general schema: if $\mid\!\sim$ is non-monotonic in some way, then either $\mid\!\sim$ must fail for at least one rule among {CMP, RD}, or the underlying monotonic consequence relation $\vdash$ must falsify some well-identified axiom. A fourth result shows that compactness and non-monotony are not incompatible. Also we provide a new notion of compactness, which is more useful than the classical one when it comes to non-monotonic inference relations.

2 Non-monotonic inference relations

As explained in the introduction, we make the following assumption: any non-monotonic inference relation $\mid\!\sim$ contains an underlying "classical" inference relation $\vdash$. This we call ***conservativeness*** [Bes88] of the non-monotonic inference relation

$$\mathcal{S} \mid\!\sim A \qquad \text{whenever} \qquad \mathcal{S} \vdash A,$$

later known as ***"supraclassicality"*** when $\vdash$ is classical logic.

Be careful that $\mid\!\sim$ denotes the "non-classical" inference relation, and $\vdash$ denotes the underlying consequence relation. Of course, we will be interested in $\mid\!\sim$ being ***non-monotonic***, that is, for some sets of formulas $\mathcal{S}, \mathcal{T}$ and some formula A,

$$\mathcal{S} \mid\!\sim A \qquad \text{and} \qquad \mathcal{S}, \mathcal{T} \not\mid\!\sim A.$$

We now justify our assumption by the fact that in any "natural" non-monotonic $\mid\!\sim$, at least one consequence relation $\vdash$ is included. The problem is to define what a "natural" $\mid\!\sim$ is. We suppose that it must at least be reflexive. Then we can define a sub-relation $\vdash_0$ by

$$\mathcal{T} \vdash_0 A \qquad \text{if and only if} \qquad A \in \mathcal{T}.$$

As $\vdash_0$ clearly satisfies the cut rule, this defines the smallest (but not the most interesting one...) consequence relation underlying a reflexive non-monotonic inference relation $\mid\!\sim$. Another easy way to define a monotonic sub-relation of a given $\mid\!\sim$ is

$$\mathcal{T} \vdash_1 A \quad \text{if and only if,} \quad \text{for any } \mathcal{T}\text{' we have} \quad \textit{if} \quad \mathcal{T} \subseteq \mathcal{T}' \quad \textit{then} \quad \mathcal{T}' \mid\!\sim A.$$

It is routine to verify that $\vdash_1$, when indeed a consequence relation, is the largest consequence relation underlying the inference relation $\mid\!\sim$. Clearly, if $\mid\!\sim$ is reflexive, so is $\vdash_1$. Moreover, if $\mid\!\sim$ satisfies cumulative transitivity, then $\vdash_1$ satisfies cut. If these two conditions are satisfied by $\mid\!\sim$, then we can regard $\vdash_1$ as the ***canonical consequence relation attached to*** $\mid\!\sim$. However, it would be nice to find some weaker conditions for $\mid\!\sim$ which suffice to insure that $\vdash_1$ is truly a consequence relation. Referring to well-known examples, we find that $\vdash_1$, as the canonical consequence relation attached to circumscription, or to normal default reasoning, is classical logic (at least under some reasonable conditions about the formulas involved). Circumscription obeys the property of cumulative transitivity, but normal default reasoning does not [Mak88], which shows that this condition is not necessary to ensure that $\vdash_1$ is actually a consequence relation.

In order to be as general as possible, we mention only a few symbols, namely the connectives $\rightarrow$ and $\neg$, and specify as few conditions as possible about these connectives. This contrasts to the work in [KLM90] where all the connectives $\neg, \rightarrow, \wedge, \vee, \leftrightarrow$ are fixed and conform to classical logic. On the other hand, various authors have argued that a more appropriate basis for non-monotonic reasoning is intuitionistic logic [Gab82], many-valued logic [Tur84], conditional logic [Del88], or paraconsistant logic [PB91]. This variety motivates our present work.

A specific form of non-monotony of interest to us is one where for at least one case, non-monotony is "finitely generated" (so to say): there exists a set of formulas $\mathcal{S}$ and some formulas $\{B_1, \ldots, B_n\}$ such that $\mathcal{S} \mid\!\sim A$ but $\mathcal{S}, B_1, \ldots, B_n \not\mid\!\sim A$ for some n. Let us call this property ***"finitistic non-monotony"***.

If a negation connective $\neg$ is available in the language, we call *"literal non-monotony"* the situation where there exists some A and T such that $T \mathrel{|\!\sim} A$ while $T, \neg A \not\mathrel{|\!\sim} A$.

Note that for all the non-monotonic formalisms proposed up to date, both the finitistic and literal forms of non-monotony hold. Indeed, non-monotonic formalisms have been introduced in order to formalize in an appropriate manner reasoning that involves some common sense notions such as rules with exceptions. Let us take the archetypal example about birds. Let T be a theory expressing that "birds fly" and that "Tweety is a bird". Denoting the two pieces of information "Tweety is a penguin" and "penguins never fly" as $info_1$ and $info_2$ respectively, we expect that $T \mathrel{|\!\sim}$ "Tweety flies" while $T, info_1, info_2 \not\mathrel{|\!\sim}$ "Tweety flies". Using the terminology previously introduced, this is a case for finitistic non-monotony. If we add the definite information that "Tweety does not fly" (meaning that there is no way Tweety can fly but we do not know why), then we expect that $T, info_3 \not\mathrel{|\!\sim}$ "Tweety flies", where $info_3$ is the negation of "Tweety flies". Using the above terminology, this is a case for literal non-monotony.

So, it makes sense to investigate the properties of non-monotonic systems in terms of the above two notions: finitistic non-monotony and literal non-monotony. In fact, non-monotonic systems falling under neither of these two categories would be rather odd in aiming at common sense reasoning.

We now end the presentation of the general notions and definitions by a short discussion about negation and the absurdity symbol.

It sometimes happens that we have an absurdity symbol $\perp$ in the language (in which case, negation often comes from interpreting $A \rightarrow \perp$ as $\neg A$). Such an absurdity symbol is convenient to express *relative consistency*

$$\frac{S \mathrel{|\!\sim} \perp}{S \vdash \perp},$$

which is the most significant feature for non-monotonic inference relations. In order to understand why, consider what makes non-monotony to arise in the first place: we want certain specific conclusions to be drawn in the absence of definite information. Clearly, the conclusions we want to draw are only palliative: we are not willing to let them contradict blatant evidence, we rather commit ourselves to withdraw them in such a case (i.e. to withdraw them if inconsistency threatens). This is exactly what is stated by the rule of relative consistency. As well-known examples, the normal fragment of default logic, and the well-founded fragment of circumscription satisfy relative consistency with respect to classical logic.

Some interesting axioms and rules can also be expressed by means of $\perp$ such as the *law of non-contradiction*:

$$\vdash A \rightarrow ((A \rightarrow \perp) \rightarrow \perp),$$

and the *elimination of double negation*

$$\frac{S, A \rightarrow \perp \vdash \perp}{S \vdash A},$$

which is formulated here as the principle of *reductio ad absurdum*.

3 Monotony from the rules CMP and RD

Central to our work are the rules CMP

$$\frac{\mathcal{S} \mathrel{|\!\sim} A \qquad \mathcal{S} \vdash A \rightarrow B}{\mathcal{S} \mathrel{|\!\sim} B}$$

and RD

$$\frac{\mathcal{T} \mathrel{|\!\sim} A \rightarrow B}{\mathcal{T}, A \mathrel{|\!\sim} B}.$$

On the one hand, CMP is the basis for any non-monotonic system: one can hardly imagine a "natural" non-monotonic system $\mathrel{|\!\sim}$ denying CMP with respect to *every* consequence relation $\vdash$ when an implication connective $\rightarrow$ is present in the language (we disregard the trivial case of the premisses membership relation). On the other hand, RD is especially meaningful because it relates implication to hypothetical reasoning in the intuitive way: you conclude "if A then B" in the situation $\mathcal{S}$ when the situation $\mathcal{S} \cup \{A\}$ makes you to conclude B.

"Unfortunately", CMP and RD are rather close to monotony. Hence we mean to explore the border line that non-monotonic systems cannot cross to meet CMP and RD. Precisely, we want to establish several incompatibility results of the general form: if $\mathrel{|\!\sim}$ fails monotony of type X and $\vdash$ satisfies an axiom of type Y then either CMP or RD does not hold for $\mathrel{|\!\sim}$.

• First, let us suppose that in the underlying logic, the connective $\rightarrow$ obeys the axiom schema called *Lewis paradox*:

$$\vdash A \rightarrow (B \rightarrow A).$$

Now, if $\mathcal{S} \mathrel{|\!\sim} A$, as we have $\mathcal{S} \vdash A \rightarrow (B_1 \rightarrow A)$ by monotony of $\vdash$, we get $\mathcal{S} \mathrel{|\!\sim} B_1 \rightarrow A$ by CMP. Now, if RD holds, $\mathcal{S}, B_1 \mathrel{|\!\sim} A$, using $\mathcal{S}, B_1 \vdash A \rightarrow (B_2 \rightarrow A)$, we get $\mathcal{S}, B_1, B_2 \mathrel{|\!\sim} A$ (by CMP and RD). Repeating the obvious pattern n times, we end up with $\mathcal{S}, B_1, \ldots, B_n \mathrel{|\!\sim} A$ for any finite sequence of formulas $B_1, \ldots, B_n$. Then, a contradiction with the finististic non-monotony assumption arises.

Thus, if $\mathrel{|\!\sim}$ is to be a finistic non-monotonic logic conforming to RD and CMP, then $\mathrel{|\!\sim}$ cannot be constructed out of a logic $\vdash$ satisfying Lewis paradox. This tells us that, in such a case, we should rather consider a relevant logic as the underlying $\vdash$.

This result extends [KLM90] in the sense that their equivalent statement concerning monotony and the rule of detachment is based on the validity of

$$\frac{\mathcal{S}, A, B \vdash C}{\mathcal{S}, A \wedge B \vdash C}$$

which excludes a number of logics (see for instance [RB80]).

Indeed, we always deal with axioms involving implication and sometimes negation (or the absurdity symbol) but nothing about other connectives. This contrasts with [KLM90], notably because we need not to enforce any constraint on conjunction for instance, we even need not have any conjunction connective. Does it matter? Yes, it does because the logics that Kraus, Lehmann and Magidor consider do not admit competing conclusions. Technically, it all boils down to the rule AND

$$\frac{\mathcal{S}, A \mid\!\sim B \qquad \mathcal{S}, A \mid\!\sim C}{\mathcal{S}, A \mid\!\sim B \wedge C}$$

which Kraus, Lehmann and Magidor apply when proving that monotony in the form

$$\frac{\vdash A \rightarrow B \qquad B \mid\!\sim C}{A \mid\!\sim C}$$

and RD are equivalent. The proof that monotony entails RD is as follows. Assume $A \mid\!\sim B \rightarrow C$. According to classical logic, $\vdash A \wedge B \rightarrow A$. Applying monotony yields $A \wedge B \mid\!\sim B \rightarrow C$. Other easy steps involving reflexivity (for $\mid\!\sim$) and right weakening lead to $A \wedge B \mid\!\sim B$. Using AND, we then obtain $A \wedge B \mid\!\sim B \wedge (B \rightarrow C)$. In view of $\vdash B \wedge (B \rightarrow C) \rightarrow C$, right weakening gives us $A \wedge B \mid\!\sim C$ as desired.

Let us return to the reason why the approach by Kraus, Lehmann and Magidor is too restrictive. By enforcing the AND rule, their approach does not account for the "multiple extensions" phenomenon of default logic for instance. (Remember that, in default logic, we can sometimes have $\mathcal{S} \mid\!\sim A$ and $\mathcal{S} \mid\!\sim \neg A$ but $\mathcal{S} \not\mid\!\sim A \wedge \neg A$.)

•• Second, let us suppose that a negation connective $\neg$ is available, that satisfies an axiom schema related to Lewis paradox, the *ex falso quodlibet*:

$$\vdash A \rightarrow (\neg A \rightarrow B).$$

If $\mathcal{S} \mid\!\sim A$, we get $\mathcal{S} \mid\!\sim \neg A \rightarrow B$ using CMP, then RD yields $\mathcal{S}, \neg A \mid\!\sim B$ for every formula B. Choosing B as A, this is a contradiction with the assumption of literal non-monotony.

Again, there exist logics which are weaker and more appropriate than classical logic: paraconsistent logics are good candidates. An illustration is the logic defined by [PB91].

Note that in fact we even did not use the full ex falso quodlibet axiom, requiring only the weaker axiom schema

$$\vdash A \rightarrow (\neg A \rightarrow A),$$

which is an instance of both Lewis paradox and ex falso quodlibet. It is definitely a less controversial axiom because it even holds in relevant logics with mingle, for instance the logic RM [AB75].

••• Third, we drop the negation connective, but consider an absurdity symbol $\perp$. Now, suppose $\mathcal{S} \mid\!\sim A$. By the law of non-contradiction, $\mathcal{S} \vdash A \rightarrow ((A \rightarrow \perp) \rightarrow \perp)$. By CMP, $\mathcal{S} \mid\!\sim (A \rightarrow \perp) \rightarrow \perp$. Applying RD yields $\mathcal{S}, A \rightarrow \perp \mid\!\sim \perp$. Due to relative consistency, $\mathcal{S}, A \rightarrow \perp \vdash \perp$. Then, the elimination of double negation gives $\mathcal{S} \vdash A$. Thus, we have proved that if $\mathcal{S} \mid\!\sim A$ then $\mathcal{S} \vdash A$. Conservativeness is the converse. So, $\mid\!\sim$ collapses into $\vdash$. As a consequence, $\mid\!\sim$ cannot be non-monotonic. Thus, CMP, RD and the law of non-contradiction together with the elimination of double negation for $\vdash$ and with relative consistency of $\mid\!\sim$ with respect to $\vdash$ make that any tentatively non-monotonic $\mid\!\sim$ collapses into its monotonic mate $\vdash$. Note that the law of non-contradiction as expressed here can be considered as a particular instance of the axiom

$$\vdash A \rightarrow ((A \rightarrow B) \rightarrow B),$$

which is satisfied by a large number of the logics containing an implication connective $\rightarrow$.

4 About compactness for non-monotonic logics

Our last result is of a slightly different kind, but it is also a result showing that non-monotonic inference relations may satisfy more properties than is sometimes argued.

It has sometimes been argued that "a non-monotonic relation cannot be compact". We do not agree with this statement. We explain why by analyzing compactness, trying to retain most of that property while dropping monotony. Clearly, if $\mid\!\sim$ is non-monotonic, it is not expected to satisfy the following "upward" part of compactness:

$$\text{if } \mathcal{T} \subseteq_f \mathcal{S} \quad \text{and} \quad \mathcal{T} \mid\!\sim A \quad \text{then} \quad \mathcal{S} \mid\!\sim A$$

where $\subseteq_f$ means "is a finite subset of". (Notice that fairly artificial non-monotonic inference relations can be defined that satisfy upward compactness after all.) For a non-monotonic relation, we should rather aim at keeping the "downward" part of the compactness:

$$\mathcal{S} \mid\!\sim A \text{ only if for some } \mathcal{T} \subseteq_f \mathcal{S} \text{ we have } \mathcal{T} \mid\!\sim A.$$

However, it appears that this definition is generally too weak, for a non-monotonic relation, to be interesting. The appropriate definition for a non-monotonic relation seems to be:

$$\mathcal{S}, \mathcal{T} \mid\!\sim A \text{ only if} \quad \text{for some} \quad \mathcal{T}' \subseteq_f \mathcal{T} \quad \text{we have} \quad \mathcal{S}, \mathcal{T}' \mid\!\sim A.$$

We call *"bicompactness"* this property of an inference relation (clearly, bicompactness implies compactness).

We justify our claim that monotony and compactness are not closely related:
(1) There exist non-monotonic relations which are (bi)compact.
(2) There exist non-monotonic relations which are not (bi)compact.
(3) There exist monotonic relations which are (bi)compact.
(4) There exist monotonic relations which are not (bi)compact.
As case (1) is probably the only controversial one, we give an example:

The language $\mathcal{L}$ consists of $\{A_i\}_{i \in I} \cup \{\top\}$. We have, as a definition for $\mid\!\sim$:
$\{A_i\} \mid\!\sim \top$ (for any $i \in I$),
$\{A_j\}_{j \in J} \mid\!\sim A_i$ (for any $i \in J$, any $J \subseteq I$, $J \neq \emptyset$),
$\{A_j\}_{j \in J} \cup \{\top\} \mid\!\sim \top$ and $\{A_j\}_{j \in J} \cup \{\top\} \mid\!\sim A_i$ (for any $J \subseteq I$, any $i \in J$).

Clearly, $\mid\!\sim$ is non-monotonic: $A_1 \mid\!\sim \top$, $\{A_1, A_2\} \not\mid\!\sim \top$ and it is clearly (bi)compact.

It remains to examine under which conditions for the underlying $\vdash$ (if any), bicompactness and non-monotony are really exclusive properties.

We make now a comparison between our notion of bicompactness and other notions found in the literature. Recently a few other authors have examined the problem of compactness in a non-monotonic framework. The most advanced studies we have found are [Fre90] and [Her94]. However, these two texts examine only what happens when $\mid\!\sim$ is cumulative, which is a rather severe restriction. To help the reader understanding our comparison, it is easier to introduce here the following classical notation: $C_{\mid\sim}(\mathcal{S}) = \{A \in \mathcal{L}, \mathcal{S} \mid\!\sim A\}$. Remind that an inference relation $\mid\!\sim$ is *cumulative* iff $\mathcal{T} \subseteq \mathcal{S} \subseteq C_{\mid\sim}(\mathcal{T})$ implies $C_{\mid\sim}(\mathcal{T}) = C_{\mid\sim}(\mathcal{S})$. [Fre90] writes that a cumulative $\mid\!\sim$ is *supracompact* iff, for any formula $A \in \mathcal{L}$ and for any set $\mathcal{T}$ of formulas in $\mathcal{L}$, we have $\mathcal{T} \mid\!\sim A$ iff there exists some $\mathcal{T}' \subseteq_f \mathcal{T}$ such that for any set $\mathcal{S} \subseteq C_{\mid\sim}(\mathcal{T})$, we have $\mathcal{T}', \mathcal{S} \mid\!\sim A$.

The last "iff" could appear unexpected for a definition dealing with non-monotonic inference relations. However, the "if" part is always true here: if there exists some $\mathcal{T}' \subseteq_f \mathcal{T}$ such that for any set $\mathcal{S} \subseteq C_{\mid\sim}(\mathcal{T})$, we have $\mathcal{T}', \mathcal{S} \mid\!\sim A$, then, taking $\mathcal{S} = \mathcal{T}$, we get $\mathcal{T} \mid\!\sim A$. So, this definition of supracompactness is equivalent to: $\mid\!\sim$ is supracompact iff $A \in C_{\mid\sim}(\mathcal{T})$ implies that there exists some $\mathcal{T}' \subseteq_f \mathcal{T}$ such that for any set $\mathcal{S} \subseteq C_{\mid\sim}(\mathcal{T})$, we have $A \in C_{\mid\sim}(\mathcal{T}', \mathcal{S})$, i.e.

$\mid\!\sim$ is supracompact iff

$$C_{\mid\sim}(\mathcal{T}) \subseteq \bigcup_{\mathcal{T}' \subseteq_f \mathcal{T}} \bigcap_{\mathcal{S} \subseteq C_{\mid\sim}(\mathcal{T})} C_{\mid\sim}(\mathcal{T}', \mathcal{S}).$$

We may also write our own definition in these terms:

$\mid\!\sim$ is bicompact iff $C_{\mid\sim}(\mathcal{T}) \subseteq \bigcap_{\mathcal{S} \subseteq \mathcal{T}} \bigcup_{\mathcal{T}' \subseteq_f \mathcal{T} - \mathcal{S}} C_{\mid\sim}(\mathcal{T}', \mathcal{S})$.

Now, if $\mathcal{T}' \subseteq_f \mathcal{T}$, we have $\mathcal{T}' = \mathcal{T}_1 \cup \mathcal{T}_2$ with $\mathcal{T}_1 \subseteq_f \mathcal{T} - \mathcal{S}, \mathcal{T}_2 \subseteq \mathcal{S}$, and $\mathcal{T}', \mathcal{S} = \mathcal{T}_1, \mathcal{S}$, thus $\bigcap_{\mathcal{S} \subseteq \mathcal{T}} \bigcup_{\mathcal{T}' \subseteq_f \mathcal{T} - \mathcal{S}} C_{\mid\sim}(\mathcal{T}', \mathcal{S}) = \bigcap_{\mathcal{S} \subseteq \mathcal{T}} \bigcup_{\mathcal{T}' \subseteq_f \mathcal{T}} C_{\mid\sim}(\mathcal{T}', \mathcal{S})$.

Thus, $\mid\!\sim$ is bicompact iff

$$C_{\mid\sim}(\mathcal{T}) \subseteq \bigcap_{\mathcal{S} \subseteq \mathcal{T}} \bigcup_{\mathcal{T}' \subseteq_f \mathcal{T}} C_{\mid\sim}(\mathcal{T}', \mathcal{S}).$$

Remind now that $\mathcal{T} \subseteq C_{\mid\sim}(\mathcal{T})$, thus: $\bigcup_{\mathcal{T}' \subseteq_f \mathcal{T}} \bigcap_{\mathcal{S} \subseteq C_{\mid\sim}(\mathcal{T})} C_{\mid\sim}(\mathcal{T}', \mathcal{S}) \subseteq \bigcup_{\mathcal{T}' \subseteq_f \mathcal{T}} \bigcap_{\mathcal{S} \subseteq \mathcal{T}} C_{\mid\sim}(\mathcal{T}', \mathcal{S})$. We know that we have: $\bigcup_{\mathcal{T}' \subseteq_f \mathcal{T}} \bigcap_{\mathcal{S} \subseteq \mathcal{T}} C_{\mid\sim}(\mathcal{T}', \mathcal{S}) \subseteq \bigcap_{\mathcal{S} \subseteq \mathcal{T}} \bigcup_{\mathcal{T}' \subseteq_f \mathcal{T}} C_{\mid\sim}(\mathcal{T}', \mathcal{S})$ (this is true for any sets $C_{\mid\sim}(\mathcal{T}', \mathcal{S})$). We have shown:

$$\bigcup_{\mathcal{T}' \subseteq_f \mathcal{T}} \bigcap_{\mathcal{S} \subseteq C_{\mid\sim}(\mathcal{T})} C_{\mid\sim}(\mathcal{T}', \mathcal{S}) \subseteq \bigcap_{\mathcal{S} \subseteq \mathcal{T}} \bigcup_{\mathcal{T}' \subseteq_f \mathcal{T}} C_{\mid\sim}(\mathcal{T}', \mathcal{S}).$$

This establishes that if $\mid\!\sim$ is supracompact, then it is a fortiori bicompact. In fact, the notion of supracompactness is far stronger than the notion of bicompactness. From the sudies of [Fre90, Her94], it seems that, in the case of cumulative inference relations, this strong notion of supracompactness is sometimes needed. Note that in the above example of a relation being bicompact without being monotonic, the relation is supracompact, too. [Her94] also introduces the following definition: An inference relation $\mid\!\sim$ is ***weakly compact*** iff

$$\mathcal{S} \mid\!\sim A \text{ only if for some } \mathcal{T} \subseteq_f C_{\vdash}(\mathcal{S}) \text{ we have } \mathcal{T} \mid\!\sim A.$$

This notion of weak compactness is a slight variant (an even weaker one) of the downward part of compactness as introduced above, thus it is much weaker than bicompactness. Not unexpectedly, Herre shows that this notion is not very powerful.

Note that we could have weakened a bit our definition of bicompacity, requiring only :

$$C_{\mid\sim}(\mathcal{T}) \subseteq \bigcap_{\mathcal{S} \subseteq_f \mathcal{T}} \bigcup_{\mathcal{T}' \subseteq_f \mathcal{T}} C_{\mid\sim}(\mathcal{T}', \mathcal{S}).$$

This notion is slightly weaker than the bicompacity defined above, because we have obviously: $\bigcap_{\mathcal{S} \subseteq \mathcal{T}} \bigcup_{\mathcal{T}' \subseteq_f \mathcal{T}} C_{\mid\sim}(\mathcal{T}', \mathcal{S}) \subseteq \bigcap_{\mathcal{S} \subseteq_f \mathcal{T}} \bigcup_{\mathcal{T}' \subseteq_f \mathcal{T}} C_{\mid\sim}(\mathcal{T}', \mathcal{S})$.

Here is a slight variant of this last notion:

$\mid\!\sim$ is ***weakly supracompact*** iff, for any theory $\mathcal{T}$, we have:

$$C_{\mid\sim}(\mathcal{T}) \subseteq \bigcap_{\mathcal{S} \subseteq_f C_{\vdash}(\mathcal{T})} \bigcup_{\mathcal{T}' \subseteq_f C_{\vdash}(\mathcal{T})} C_{\mid\sim}(\mathcal{T}', \mathcal{S}).$$

This notion is introduced and studied in [Her94], where an example of a kind of preferential entailment satisfying weak supracompactness is given.

More results should be obtained for these various notions related to compactness of non-monotonic inference relations, in order to determine which definition is the most appropriate in a given context.

5 Conclusion and future work

These preliminary investigations counter some widely expressed claims about non-monotonic reasoning. We wanted to show that those particular claims are not true in general, that they are true only under special conditions, more or less implicit in the literature, and which we indertook to make more precise. Clearly, a lot of work has to be done from these first steps to clarify the matter thoroughly. For example, there remains to see if some condition can actually turn non-monotony and (bi)compactness into incompatible properties. Also, there remains to characterize the "interesting" properties that a "natural" non-monotonic inference relation should enjoy. But we have given some hints, together with a few formal results which help to understand the relationships between a non-monotonic inference relation and its underlying monotonic consequence relation. Finally, there remains to examine in detail what non-monotonic inference relations look like when the underlying monotonic consequence relation is not classical logic but is a rival of it, e.g. intuitionistic logic or some relevance logic.

References

[AB75] A.R. Anderson and N.D. Jr. Belnap. *Entailment: The Logic of Relevance and Necessity.* Princeton University Press, 1975.

[Bes88] Philippe Besnard. Axiomatizations in the metatheory of nonmonotonic inference systems. In *7th Conf. of the Canadian Society for Computational Studies of Intelligence (CSCSI)*, pages 117–124, Edmonton, Morgan-Kaufmann, June 1988.

[Del88] James P. Delgrande. An approach to default reasoning based on a first-order conditional logic: Revised report. *Artificial Intelligence*, 36:63–90, 1988.

[FLM91] Michael Freund, Daniel Lehmann, and Paul Morris. Rationality, Transitivity, and Contraposition. *Artificial Intelligence*, 52:191–203, 1991.

[Fre90] Michael Freund. Supracompact inference operations. In *Nonmonotonic Inductive Logic, in LNAI 543*, pages 59–73, Karlsruhe, Springer-Verlag, December 1990.

[Gab82] Dov M. Gabbay. Intuitionistic bases for nonmonotonic logic. In D.W. Loveland, editor, *LNCS 138*, pages 260–273. Springer-Verlag, 1982.

[Gab85] D.M. Gabbay. Theoretical foundations for nonmonotonic reasoning in expert systems. In K.R. Apt, editor, *Proc. Logics and Models of Concurrent Systems, NATO ASI Series F*, volume 13, page xxx. Springer Verlag, 1985.

[Gen33] G. Gentzen. Untersuchungen über das logische Schliessen. *Mathematische Zeitschrift*, 39:176–210, 405–431, 1933. *Trad. française publiée aux Presses Universitaires de France, Paris, 1955*
English translation in Collected papers of Gerhard Gentzen *(Szabo ed.), North Holland, Amsterdam, 1969.*

[Her94] Heinrich Herre. Compactness Properties of Nonmonotonic Inference Operations. Proc. of JELIA'94. In *Logic in Artificial Intelligence, LNAI 838*, pages 19–33, York, Springer-Verlag, September 1994.

[KLM90] Sarit Kraus, Daniel Lehmann, and Menachem Magidor. Nonmonotonic reasoning, preferential models and cumulative logics. *Artificial Intelligence*, 44:167–207, 1990.

[Mak88] David Makinson. General theory of cumulative inference. In *Non-Monotonic Reasoning, in LNAI-346*, pages 1–18. Springer-Verlag, June 1988.

[PB91] T. Pequeno and A. Buschbaum. The logic of epistemic inconsistency. In *Proc.* 2nd *Conf. on Principles of Knowledge Representation and Reasoning*, pages 435–460, Toronto, 1991.

[RB80] N. Rescher and R. Brandom. *The Logic of Inconsistency.* Blackwell, 1980.

[Tar30] A. Tarski. Über einige fundamentale Begriffe der Metamathematik. *C.R. Soc. Sciences et Lettres de Varsovie*, 23, cl. III:22–29, 1930. *Trad. française parue dans* Logique, sémantique, métamathématique, *tome 1, Tarski, A., Armand Collin publ., Paris 1972.*
English translation in Logic, Semantics, Metamathematics. Papers from 1923–1938 *(Woodger ed.), Clarendon Press, Oxford, 1956.*

[Tur84] Raymond Turner. *Logics for artificial intelligence.* Hellis Horwood Ltd, 1984.

Stochastic Logic

Silviu Guiasu
Department of Mathematics and Statistics, York University
North York, Ontario, M3J 1P3, Canada

Abstract. Stochastic logic is a probabilistic logic that essentially takes interdependence and global connection into account. It generalizes classical and fuzzy logics. The paper contains the basic properties of this kind of logic and the relationship between its concepts and the standard ones.

1 Introduction

Stochastic logic is a nonstandard logic. It is not a rival of classical logic but rather a generalization of both classical and fuzzy logics. Its main objective is to take interdependence and global connection into account. It is both probabilistic and global. Stochastic logic is a supplementary logic in the sense that it is compatible both to classical logic, which is nonprobabilistic, or strictly deterministic, and to fuzzy logic, which may be viewed globally as being a probabilistic logic with independent components.

The probabilistic feature of the proposed logic is motivated by the necessity of coping with uncertainty in dealing with real life problems. Such uncertainty could have objective or subjective causes. Sometimes, the qualifications for being something are imprecise. Sometimes, the qualifications for being something else are precise, but there is real difficulty in determining whether or not certain subjects satisfy them. Stochastic logic, however, should not be identified to the standard probabilistic logic where the logical concept of probability replaces the two truth values (true, false) from the classical logic. Its aim is mainly to take interdependence and global connection into account. Interdependence is universal and essentially influences the values of truth. When there is interdependence among entities, the degree of truth with which one of these entities satisfies a certain property depends in general on the degrees of truth with which the other entities satisfy that property or some other properties. Sometimes this interdependence is relatively weak and may be ignored, but when it is strong, such a simplification would distort reality.

The aim of this paper is to present basic properties of stochastic logic and the relationship between its concepts and the standard ones. The different probability distributions involved may be objective, i.e. based either on stable relative frequencies determined by repeating probabilistic experiments or by solving optimization problems with constraints (like the principle of maximum entropy, for instance), and subjective, in which case they are rather called credibility distributions.

2 Stochastic Sets

Stochastic sets play for stochastic logic the same role as that played by Cantor's sets for classical logic. In order to simplify the presentation we are going to deal mainly with two-valued stochastic sets referring to a finite number of entities. This means that the properties or predicates considered here take on only two possible values, namely 1, meaning yes (or satisfied), and 0, meaning no (or not satisfied). Let us emphasize, however, that although a property p can have only two possible values, the degree of truth of the statement 'the entity x satisfies (has) property p' may be an arbitrary number from the numerical interval $[0, 1]$. Generalizations of stochastic sets to cases involving multi-valued properties or predicates and an infinite set of entities will be discussed at the end of the paper.

Let $X = \{x_1, \ldots, x_m\}$ be a crisp (Cantor) set of entities called universe. Denote by $\{0,1\}^X$ the class of all 2^m elementary binary m-dimensional cylinders $[k_1, \ldots, k_m; x_1, \ldots, x_m]$, where $k_i \in \{0,1\}, (i = 1, \ldots, m)$. Such a cylinder specifies the function which associates the binary 0-1 symbols $k_1, \ldots, k_m$ to the elements $x_1, \ldots, x_m$ of X, respectively. Denote by $\{0,1\}^m$ the set of all binary 0-1 vectors with m components and by $\mathcal{P}(X)$ the class of all crisp subsets of X. Obviously, $\{0,1\}^X$ is equivalent both to $\{0,1\}^m$ and to $\mathcal{P}(X)$. These equivalences are based on the obvious correspondence between the m-dimensional binary cylinder $[k_1, \ldots, k_m; x_1, \ldots, x_m]$, the m-dimensional binary vector $(k_1, \ldots, k_m)$, and the subset $S(k_1, \ldots, k_m) = \{x_i; k_i = 1\} \subseteq X$.

The *stochastic set* A corresponding to the property or predicate p_A is a probability distribution $\varphi_A : \{0,1\}^X \longrightarrow [0,1]$, called membership probability distribution corresponding to p_A. The number $\varphi_A([k_1, \ldots, k_m; x_1, \ldots, x_m])$ is the probability (credibility) that x_1 has (if $k_1 = 1$) or has not (if $k_1 = 0$) property p_A, and,..., and x_m has (if $k_m = 1$) or has not (if $k_m = 0$) property p_A. We have

$$\sum_{k_1, \ldots, k_m \in \{0,1\}} \varphi_A([k_1, \ldots, k_m; x_1, \ldots, x_m]) = 1. \tag{1}$$

Two stochastic sets A, B on X are *equal* if their membership probability distributions are equal, i.e. $\varphi_A = \varphi_B$. Throughout this paper, a stochastic set A is identified to its membership probability (credibility) distribution φ_A on $\{0,1\}^X$.

The *restriction* of a stochastic set φ_A to a crisp subset X^* of the universe X has as membership probability distribution the marginal probability distribution of φ_A relative to the elements of X^*. Thus, for instance,

$$\varphi_A([k_i; x_i]) = \sum_{k_1, \ldots, k_{i-1}, k_{i+1}, \ldots, k_m \in \{0,1\}} \varphi_A([k_1, \ldots, k_m; x_1, \ldots, x_m]) \tag{2}$$

represents the probability (credibility) that x_i has (if $k_i = 1$) or has not (if $k_i = 0$) property p_A regardless of what happens with the other entities of X. It defines the membership probability distribution of the restriction of φ_A to the subset $X^* = \{x_i\} \subset X$.

The *complement* with respect to X of the restriction of φ_A to the crisp subset $X^* \subset X$ is the restriction of φ_A to the crisp subset $X - X^*$. Thus, for instance, the complement of (2) with respect to X is

$$\varphi_A([k_1,\ldots,k_{i-1},k_{i+1},\ldots,k_m;x_1,\ldots,x_{i-1},x_{i+1},\ldots,x_m]) = \sum_{k_i\in\{0,1\}} \varphi_A([k_1,\ldots,k_m;x_1,\ldots,x_m]).$$

3 Particular Cases

(a) If $f_A : X \longrightarrow [0,1]$ is a *fuzzy set* ([1],[2]) and the elements of X are independent entities, then it generates the particular stochastic set

$$\varphi_A([k_1,\ldots,k_m;x_1,\ldots,x_m]) = \varphi_A([k_1;x_1])\ldots\varphi_A([k_m;x_m]), \tag{3}$$

where $\varphi_A([1;x_i]) = f_A(x_i)$, and $\varphi_A([0;x_i]) = 1 - f_A(x_i), (i = 1,\ldots,m)$.

Conversely, if A is a stochastic set with the membership probability distribution φ_A, then it induces a fuzzy set on X having the membership function $f_A(x_i) = \varphi_A([1;x_i]), (i=1,\ldots,m)$. Thus, we can state the following proposition:

Proposition 1: Every stochastic set on X relative to a property (predicate) p_A induces a fuzzy set on X relative to p_A but the converse is true only if the elements of the universe X are independent with respect to p_A.

(b) A *crisp (Cantor) set* A is a particular stochastic set whose membership probability distribution φ_A is degenerate, i.e. there is only one m–dimensional cylinder $[k_1,\ldots,k_m;x_1,\ldots,x_m]$ such that $\varphi_A([k_1,\ldots,k_m;x_1,\ldots,x_m]) = 1$. The corresponding crisp subset of X is $A = \{x_i; k_i = 1\}$. Conversely, to any crisp subset $A = \{x_{i(1)},\ldots,x_{i(r)}\} \subseteq X$ it corresponds the stochastic set φ_A for which $\varphi_A([k_1,\ldots,k_m;x_1,\ldots,x_m])$ is equal to

$$\begin{cases} 1 & \text{if } k_{i(1)} = \ldots = k_{i(r)} = 1, k_i = 0, (i \neq i(1),\ldots,i(r)); \\ 0 & \text{for all the other vectors } (k_1,\ldots,k_m). \end{cases}$$

4 Joint Stochastic Sets

Let p_A and p_B be two two-valued (i.e. yes-no) properties (predicates) refering to the same universe $X = \{x_1,\ldots,x_m\}$. The joint stochastic set corresponding to these two properties is defined by the membership probability (credibility) distribution:

$$\varphi_{A,B} : \{0,1\}^X \times \{0,1\}^X \longrightarrow [0,1],$$

where $\varphi_{A,B}([k_1,\ldots,k_m;x_1,\ldots,x_m],[l_1,\ldots,l_m;x_1,\ldots,x_m])$ is the probability (credibility) that x_1 has (if $k_1 = 1$) or has not (if $k_1 = 0$) property p_A,and,...,x_m has (if $k_m = 1$) or has not (if $k_m = 0$) property p_A, *and* x_1 has (if $l_1 = 1$) or has not (if $l_1 = 0$) property p_B,and,...,x_m has (if $l_m = 1$) or has not (if $l_m = 0$) property p_B.

For instance, $\varphi_{A,B}([1,0,1,0;x_1,x_2,x_3,x_4],[0,1,1,0;x_1,x_2,x_3,x_4])$ is the probability (credibility) that x_1 has (or satisfies) property p_A but not p_B, x_2 has (or satisfies) p_B but not p_A, x_3 has (or satisfies) both p_A and p_B, while x_4 has (or satisfies) neither p_A nor p_B.

Let A and B be two stochastic sets on X. The stochastic set A is *included* in B, and we write $A \subset B$, if

$$\varphi_{A,B}([k_1,\ldots,k_m;x_1,\ldots,x_m],[l_1,\ldots,l_m;x_1,\ldots,x_m])=0$$

whenever there is at least a pair $(k_i,l_i),(1\le i\le m)$, such that $k_i=1$ and $l_i=0$.
Two stochastic sets A,B on X are *compatible* if $A\subset B$ and $B\subset A$.

5 Conditional Stochastic Sets

Given two properties or predicates p_A and p_B, the class of *conditional stochastic sets* corresponding to p_A given p_B, denoted by $A \mid B$, is characterized by 2^m membership probability (credibility) distributions on $\{0,1\}^X$, denoted by $\varphi_{A|B}$, whose general component

$$\varphi_{A|B}([k_1,\ldots,k_m;x_1,\ldots,x_m]\mid[l_1,\ldots,l_m;x_1,\ldots,x_m])$$

signifies the probability (credibility) that x_1 has (if $k_1=1$) or has not (if $k_1=0$) property (predicate) p_A, and,..., and x_m has (if $k_m=1$) or has not (if $k_m=0$) property (predicate) p_A *given* that x_1 has (if $l_1=1$) or has not (if $l_1=0$) property (predicate) p_B, and,..., and x_m has (if $l_m=1$) or has not (if $l_m=0$) property (predicate) p_B. A similar definition may be given for the class of conditional stochastic sets $\varphi_{B|A}$. In what follows, let us denote by

$$K=[k_1,\ldots,k_m;x_1,\ldots,x_m],\text{ and } L=[l_1,\ldots,l_m;x_1,\ldots,x_m] \tag{4}$$

two arbitrary m-dimensional cylinders referring to the properties (predicates) p_A and p_B, respectively. If φ_A and φ_B are two stochastic sets, then we have

$$\varphi_{A,B}(K,L)=\varphi_A(K)\,\varphi_{B|A}(L\mid K)=\varphi_{A|B}(K\mid L)\,\varphi_B(L).$$

Two properties (predicates) p_A,p_B are *independent* with respect to the universe X if $\varphi_{A|B}=\varphi_A$ and $\varphi_{B|A}=\varphi_B$, in which case $\varphi_{A,B}(K,L)=\varphi_A(K)\,\varphi_B(L)$.

6 Logical Operations

(a) *Intersection.* Two properties (predicates) p_A,p_B referring to the universe X being given, the *intersection* (or *conjunction*) $A\cap B$ is the stochastic set corresponding to the compound property (predicate) 'p_A *and* p_B', whose corresponding membership probability (credibility) distribution $\varphi_{A\cap B}:\{0,1\}^X\longrightarrow[0,1]$ is defined by

$$\begin{aligned}&\varphi_{A\cap B}([j_1,\ldots,j_m;x_1,\ldots,x_m])\\&=\textstyle\sum^*\varphi_{A,B}([k_1,\ldots,k_m;x_1,\ldots,x_m],[l_1,\ldots,l_m;x_1,\ldots,x_m]),\end{aligned} \tag{5}$$

where $\sum^*$ is taken in the following way: if $j_i=1$, then the sum contains all the terms for which $k_i=1$ and $l_i=1$; if $j_i=0$, then the sum contains all the terms for which *either* $k_i=1$ and $l_i=0$ *or* $k_i=0$ and $l_i=1$, *or* $k_i=0$ and $l_i=0$. Obviously, if the two properties p_A and p_B are independent, then, in the above sum, $\varphi_{A,B}$ is replaced by the product $\varphi_A \varphi_B$.

Remark: If p_A and p_B are independent properties (predicates) and the elements of the universe X are independent with respect to these two properties (predicates), then, according to (5), the intersection of the fuzzy sets f_A and f_B induced by the stochastic sets φ_A and φ_B, respectively, is the fuzzy set

$$\begin{aligned} f_{A\cap B}(x_j) &= \varphi_{A\cap B}([1;x_j]) = \varphi_{A,B}([1;x_j],[1;x_j]) \\ &= \varphi_A([1;x_j])\,\varphi_B([1;x_j]) = f_A(x_j)\,f_B(x_j), \end{aligned}$$

which, is a generalization for fuzzy sets of the usual intersection between crisp (Cantor) sets.

(b) *Union.* Two properties (predicates) p_A, p_B referring to the universe X being given, the ***union*** (or ***disjunction***) $A \cup B$ is the stochastic set corresponding to the compound property (predicate) 'p_A *or* p_B', whose corresponding membership probability (credibility) distribution $\varphi_{A\cup B} : \{0,1\}^X \longrightarrow [0,1]$ is defined by

$$\begin{aligned} &\varphi_{A\cup B}([j_1,\ldots,j_m;x_1,\ldots,x_m]) \\ &= \textstyle\sum^{**} \varphi_{A,B}([k_1,\ldots,k_m;x_1,\ldots,x_m],[l_1,\ldots,l_m;x_1,\ldots,x_m]), \end{aligned} \tag{6}$$

where $\sum^{**}$ is taken in the following way: if $j_i = 0$, then the sum contains all the terms for which $k_i = 0$ and $l_i = 0$; if $j_i = 1$, then the sum contains all the terms for which *either* $k_i = 1$ and $l_i = 0$ *or* $k_i = 0$ and $l_i = 1$, *or* $k_i = 1$ and $l_i = 1$. Obviously, if the two properties p_A and p_B are independent, then, in the above sum, $\varphi_{A,B}$ is replaced by the product $\varphi_A \varphi_B$.

Remark: If p_A and p_B are independent properties (predicates) and the elements of the universe X are independent with respect to these two properties (predicates), then, according to (6), the union of the fuzzy sets f_A and f_B induced by the stochastic sets φ_A and φ_B, respectively, is the fuzzy set

$$\begin{aligned} f_{A\cup B}(x_j) &= \varphi_{A\cup B}([1;x_j]) = \varphi_A([1;x_j])\,\varphi_B([0;x_j]) \\ &+\varphi_A([0;x_j])\,\varphi_B([1;x_j]) + \varphi_A([1;x_j])\,\varphi_B([1;x_j]) \\ &= f_A(x_j)\,[1-f_B(x_j)] + [1-f_A(x_j)]\,f_B(x_j) + f_A(x_j)\,f_B(x_j) \\ &= f_A(x_j) + f_B(x_j) - f_A(x_j)\,f_B(x_j), \end{aligned}$$

which, is a generalization for fuzzy sets of the usual union between crisp (Cantor) sets.

(c) *Complement.* If A is a stochastic set corresponding to property (predicate) p_A having the membership probability (credibility) distribution φ_A, then its ***complement*** (or ***negation***) with respect to the property (predicate) p_A is the stochastic set $\overline{A}$ or $\mathbf{C}A$ corresponding to the property (predicate) '*non* p_A', whose membership probability (credibility) distribution is

$$\varphi_{\bar{A}}([k_1,\ldots,k_m;x_1,\ldots,x_m]) = \varphi_A([\bar{k}_1,\ldots,\bar{k}_m;x_1,\ldots,x_m]), \tag{7}$$

where $\bar{0} = 1$ and $\bar{1} = 0$.

Remark: If $f_A : X \longrightarrow [0,1]$ is a fuzzy set on X, then the membership function of the complementary fuzzy set is $f_{\bar{A}} = 1 - f_A$. If the elements of X are independent, then, according to (3), f_A and $f_{\bar{A}}$ completely determine the membership probability distributions φ_A and $\varphi_{\bar{A}}$ of the stochastic sets A and $\overline{A}$, respectively. As $\varphi_A([k;x_i])$ is equal to $f_A(x_i)$ if $k=1$ and to $1-f_A(x_i)$ if $k=0$, and $\varphi_{\bar{A}}([k;x_i])$ is equal to $f_{\bar{A}}(x_i)=1-f_A(x_i)$ if $k=1$ and to $1-f_{\bar{A}}(x_i)=f_A(x_i)$ if $k=0$, we have $\varphi_{\bar{A}}([k;x_i])=\varphi_A([\bar{k};x_i])$ and (3) implies (7).

7 De Morgan's Relations

If p_A and p_B are two two-valued properties (predicates) referring to the elements of the same universe X, then we have:

Proposition 2: For any two stochastic sets A and B corresponding to the properties (predicates) p_A and p_B, respectively, we have $\overline{A\cup B}=\overline{A}\cap\overline{B}$.

Proof: Using (6) and (5), we have

$$\begin{aligned}
&\varphi_{\overline{A\cup B}}([j_1,\ldots,j_m;x_1,\ldots,x_m])=\varphi_{A\cup B}([\bar{j}_1,\ldots,\bar{j}_m;x_1,\ldots,x_m])\\
&=\sum{}^{**}\varphi_{A,B}([k_1,\ldots,k_m;x_1,\ldots,x_m],[l_1,\ldots,l_m;x_1,\ldots,x_m])\\
&=\sum{}^{*}\varphi_{A,B}([\bar{k}_1,\ldots,\bar{k}_m;x_1,\ldots,x_m],[\bar{l}_1,\ldots,\bar{l}_m;x_1,\ldots,x_m])\\
&=\sum{}^{*}\varphi_{\bar{A},\bar{B}}([k_1,\ldots,k_m;x_1,\ldots,x_m],[l_1,\ldots,l_m;x_1,\ldots,x_m])\\
&\qquad=\varphi_{\bar{A}\cap\bar{B}}([j_1,\ldots,j_m;x_1,\ldots,x_m]).
\end{aligned}$$

Proposition 3: For any two stochastic sets A and B corresponding to the properties (predicates) p_A and p_B, respectively, we have $\overline{A\cap B}=\overline{A}\cup\overline{B}$.

Proof: Using (5) and (6), we have

$$\begin{aligned}
&\varphi_{\overline{A\cap B}}([j_1,\ldots,j_m;x_1,\ldots,x_m])=\varphi_{A\cap B}([\bar{j}_1,\ldots,\bar{j}_m;x_1,\ldots,x_m])\\
&=\sum{}^{*}\varphi_{A,B}([k_1,\ldots,k_m;x_1,\ldots,x_m],[l_1,\ldots,l_m;x_1,\ldots,x_m])\\
&=\sum{}^{**}\varphi_{A,B}([\bar{k}_1,\ldots,\bar{k}_m;x_1,\ldots,x_m],[\bar{l}_1,\ldots,\bar{l}_m;x_1,\ldots,x_m])\\
&=\sum{}^{**}\varphi_{\bar{A},\bar{B}}([k_1,\ldots,k_m;x_1,\ldots,x_m],[l_1,\ldots,l_m;x_1,\ldots,x_m])\\
&\qquad=\varphi_{\bar{A}\cup\bar{B}}([j_1,\ldots,j_m;x_1,\ldots,x_m]).
\end{aligned}$$

8 The Law of Double Negation

If p_A is a two-valued property (predicate) referring to the elements of the universe X, then we have:

Proposition 4: For any stochastic set A corresponding to the property (predicate) p_A, we have $\overline{\overline{A}}=A$.

Proof: Applying (7) twice, we get

$$\begin{aligned}
&\varphi_{\overline{\overline{A}}}([k_1,\ldots,k_m;x_1,\ldots,x_m])=\varphi_{\bar{A}}([\bar{k}_1,\ldots,\bar{k}_m;x_1,\ldots,x_m])\\
&\varphi_A([\bar{\bar{k}}_1,\ldots,\bar{\bar{k}}_m;x_1,\ldots,x_m])=\varphi_A([k_1,\ldots,k_m;x_1,\ldots,x_m]).
\end{aligned}$$

9 Product Stochastic Sets

Let $X=\{x_1,\ldots,x_m\}$ and $Y=\{y_1,\ldots,y_n\}$ be two universes and p_A, p_B two properties (predicates) referring to the elements of X and Y, respectively. The ***product*** stochastic set $A\times B$ corresponding to p_A relative to X ***and*** p_B relative to Y is defined by the membership probability (credibility) distribution $\varphi_{A\times B}$: $\{0,1\}^X \times \{0,1\}^Y \longrightarrow [0,1]$, such that

$$\varphi_{A\times B}([k_1,\ldots,k_m;x_1,\ldots,x_m],[l_1,\ldots,l_n;y_1,\ldots,y_n]) =$$
$$\varphi_A([k_1,\ldots,k_m;x_1,\ldots,x_m])\,\varphi_{B|A}([l_1,\ldots,l_n;y_1,\ldots,y_n] \mid [k_1,\ldots,k_m;x_1,\ldots,x_m]),$$

where $\varphi_{B|A}$ is the class of conditional membership probability (credibility) distributions of B relative to Y given A relative to X. The membership probability (credibility) distribution $\varphi_{A\times B}$ describes a relationship between A and B, while the class of membership probability (credibility) distributions $\varphi_{B|A}$ defines a correspondence from the source A to the destination B.

Remark: A joint stochastic set (A, B) relative to the universe X is a product stochastic set $A \times B$ relative to X and Y when $Y = X$.

10 Syllogisms

(a) *Modus ponens:* Given $\varphi_{B|A}$, if φ_A then φ_B, where

$$\varphi_B([l_1,\ldots,l_n;y_1,\ldots,y_n]) =$$
$$\sum \varphi_{B|A}([l_1,\ldots,l_n;y_1,\ldots,y_n] \mid [k_1,\ldots,k_m;x_1,\ldots,x_m])\varphi_A([k_1,\ldots,k_m;x_1,\ldots,x_m])$$

the sum being taken with respect to all $k_1,\ldots,k_m \in \{0,1\}$. The above equality may be written, in an abbreviated form, as $\varphi_B = \varphi_{B|A} \circ \varphi_A$. Let us notice also that if $\varphi_A([k_1,\ldots,k_m;x_1,\ldots,x_m]) \geq \alpha$ and

$$\varphi_{B|A}([l_1,\ldots,l_n;y_1,\ldots,y_n] \mid [k_1,\ldots,k_m;x_1,\ldots,x_m]) \geq \beta,$$

then $\varphi_B([l_1,\ldots,l_n;y_1,\ldots,y_n]) \geq \alpha\,\beta$.

(b) *Modus tollens:* Given $\varphi_{B|A}$, if φ_B then φ_A, provided that the system of equations $\varphi_{B|A} \circ \varphi_A = \varphi_B$ may be solved with respect to the unknown probability (credibility) distribution φ_A.

11 Belief and Plausibility Induced by a Stochastic Set

Let A be a stochastic set with the membership probability (credibility) distribution $\varphi_A : \{0,1\}^X \longrightarrow [0,1]$. As noticed in Section 2, $\{0,1\}^X$ is equivalent to $\mathcal{P}(X)$; therefore φ_A induces a probability (credibility) distribution on $\mathcal{P}(X)$, i.e.

$$\varphi_A(E) \geq 0, \quad E \subseteq X, \quad \sum_{E\subseteq X} \varphi_A(E) = 1.$$

Let $\mathcal{F}_A$ be the class of focal subsets of X with respect to φ_A, i.e.

$$\mathcal{F}_A = \{E; E \subseteq X, \varphi_A(E) > 0\} \subset \mathcal{P}(X).$$

The *belief* and *plausibility* induced by the stochastic set φ_A on $\mathcal{P}(X)$ are defined by

$$Bel_A(E) = \sum_{F\subseteq E, F\neq\emptyset} \varphi_A(F), \quad Pl_A(E) = \sum_{F\cap E\neq\emptyset} \varphi_A(F),$$

for every nonempty $E \in \mathcal{P}(X)$. The standard definitions of the belief and plausibility functions ([3],[2]) assume that the basic probability assignment on $\mathcal{P}(X)$ has to be equal to zero for the empty subset of X. This is not the case here, as $\varphi_A(\emptyset)$, which is $\varphi_A([0,\ldots,0;x_1,\ldots,x_m])$, is not necessarily equal to zero. In our case, for each nonempty $E \subset X$ we have:

$$\varphi_A(\emptyset) + Bel_A(E) + Pl_A(X - E) = 1.$$

Remark: As in the standard case, if the elements of $\mathcal{F}_A$ are nested, i.e. $E_1 \supseteq E_2 \supseteq \ldots$, then the belief and plausibility induced by φ_A on $\mathcal{P}(X)$ are called ***necessity*** and ***possibility***, respectively.

12 Weighted Stochastic Sets

If φ_A is the membership probability (credibility) distribution of the stochastic set A, then the weighted stochastic set $w \circ A$ is defined by

$$\varphi_{w \circ A}(F) = \sum_E w(F \mid E)\, \varphi_A(E),$$

for all $F \in \mathcal{P}(X)$, where the sum is taken with respect to all $E \in \mathcal{F}_A$. The weights $w(\cdot \mid \cdot)$ are arbitrary nonnegative numbers satisfying the only constraint

$$\sum_F \sum_E w(F \mid E)\, \varphi_A(E) = 1.$$

The number $w(F \mid E)$ may be interpreted as being the credibility of $F \subseteq X$ given the focal subset $E \in \mathcal{F}_A$. As shown in [4], in dealing with joint or product stochastic sets, special weights $w(\cdot \mid \cdot)$ may be chosen in order to get the classical decision rules proposed by Hooper, Dempster, Bayes, and Jeffrey.

13 Measures of Uncertainty

As the stochastic logic deals with membership probability (credibility) distributions, the classical measures of uncertainty may be used. Thus, with the notation (4), we have:

(a) ***Shannon's entropy:*** The amounts of uncertainty contained by the stochastic sets φ_A and $\varphi_{A,B}$, respectively, are measured by:

$$H(\varphi_A) = -\textstyle\sum_K \varphi_A(K) \ln \varphi_A(K);$$
$$H(\varphi_{A,B}) = -\textstyle\sum_{K,L} \varphi_{A,B}(K,L) \ln \varphi_{A,B}(K,L).$$

(b) ***Kullback-Leibler's indicator:*** If φ_A and φ_B are two stochastic sets corresponding to the properties (predicates) p_A and p_B, respectively, both referring to the same universe X, the following indicator shows how much φ_A differs from the reference distribution φ_B:

$$I(\varphi_A : \varphi_B) = \sum_K \varphi_A(K) \ln[\varphi_A(K)/\varphi_B(K)],$$

provided that $\varphi_B(K) = 0$ implies $\varphi_B(K) = 0$.

(c) ***Kullback-Leibler's divergence:*** If φ_A and φ_B are two stochastic sets corresponding to the properties (predicates) p_A and p_B, respectively, both referring to the same universe X, a measure of how different these stochastic sets are is:

$$J(\varphi_A, \varphi_B) = I(\varphi_A : \varphi_B) + I(\varphi_B : \varphi_A),$$

provided that $\varphi_A(K) = 0$ if and only if $\varphi_B(K) = 0$.

(d) ***Watanabe's measure of interdependence:*** If φ_A and φ_B are two stochastic sets corresponding to the properties (predicates) p_A and p_B, respectively, both referring to the same universe X, and $\varphi_{A,B}$ is the joint stochastic set corresponding to the compund property (predicate) 'p_A *and* p_B', then the amount of interdependence between p_A and p_B is measured by:

$$W(\varphi_{A,B}; \varphi_A, \varphi_B) = H(\varphi_A) + H(\varphi_B) - H(\varphi_{A,B}).$$

Also, if φ_A is the membership probability (credibility) distribution corresponding to the property (predicate) p_A relative to the universe $X = \{x_1, \ldots, x_m\}$ and $\varphi_{A;i}$ is the restriction of φ_A to the subset $\{x_i\} \subset X$, then the amount of interdependence among the elements of the universe X with respect to p_A is:

$$W(\varphi_A; \varphi_{A;1}, \ldots, \varphi_{A;m}) = \sum_{i=1}^{m} H(\varphi_{A;i}) - H(\varphi_A),$$

where $H(\varphi_{A;i}) = -\varphi_A([0; x_i]) \ln \varphi_A([0; x_i]) - \varphi_A([1; x_i]) \ln \varphi_A([1; x_i])$. A similar formula may be used for measuring the amount of interdependence among the components $X_1, \ldots, X_r$ of a partition (dissection) of X with respect to p_A, by replacing the sum of $H(\varphi_{A;i})$, for $i = 1, \ldots, m$, with the sum of the entropies of the restrictions of φ_A to X_j, for $j = 1, \ldots, r$.

(e) ***The entropic distance:*** If φ_A and φ_B are two stochastic sets corresponding to the properties (predicates) p_A and p_B, respectively, both referring to the same universe X, and $\varphi_{A,B}$ is the joint stochastic set corresponding to the compund property (predicate) 'p_A *and* p_B', then the distance between p_A and p_B is measured by:

$$d(p_A, p_B) = H(\varphi_{A,B}) - W(\varphi_{A,B}; \varphi_A, \varphi_B).$$

When dealing with two different universes X and Y, the joint probability (credibility) distribution $\varphi_{A,B}$ has to be replaced in the above measures (a), (d), and (e) by the product probability (credibility) distribution $\varphi_{A \times B}$. For other measures of uncertainty induced by plausibility, belief, and fuzziness and their relationship with the classical measures defined above, see [5].

14 Paradoxes

It is well-known that self-referential paradoxes appear in classical logic when the same set proves to have both values (yes, no) of a two-valued property (predicate) p_A. There is no such paradox in stochastic logic because, as shown in section 2, each stochastic set is itself self- contradictory. In particular, the extreme events 'all the elements of the universe X have the property p_A' and 'no element of X has the property p_A' may be possible if

$$\varphi_A([1,\ldots,1;x_1,\ldots,x_m]) > 0, \text{ and } \varphi_A([0,\ldots,0;x_1,\ldots,x_m]) > 0.$$

15 Generalizations

The above considerations may be generalized in a straightforward way to the case of a multivalued stochastic logic. In such a case, the membership probability (credibility) distribution (1) should be replaced by

$$\varphi_A : \{0_A, 1_A, 2_A, \ldots, M_A\}^X \longrightarrow [0,1],$$

where $0_A, 1_A, 2_A, \ldots, M_A$ are the possible values of property (predicate) p_A. Now,

$$\varphi_A([k_1,\ldots,k_m;x_1,\ldots,x_m]), \quad k_i \in \{0_A, 1_A, 2_A, \ldots, M_A\}$$

is the probability that the element x_1 has the value k_1 of property (predicate) p_A, and,..., and the element x_m has the value k_m of property (predicate) p_A. In such a multivalued case,

$$\bar{k}_i = \{0_A, 1_A, 2_A, \ldots, M_A\} - \{k_i\}.$$

The above formalism may be extend to the case when the universe X is an infinite (countable or continuous) crisp set by using the standard way of extending the cylinder sets (see [6]). For applications of the stochastic logic to classification problems and decision making see [7] and [8], respectively.

References

1. L.A. Zadeh: Fuzzy sets. Information and Control 1, 338-353 (1965)
2. B. Bouchon-Meunier, La logique floue. Paris: Presses Univ. de France 1993
3. G. Shafer: A mathematical theory of evidence. Princeton: Princeton Univ. Press 1976
4. S. Guiasu: Weighting independent bodies of evidence. In: M. Clarke, R. Kruse, S. Moral (eds.): Symbolic and quantitative approaches to reasoning and uncertainty. Lecture Notes in Computer Science 747. Berlin-Heidelberg: Springer 1993, pp.168-173
5. S. Guiasu: A unitary treatment of several known measures of uncertainty induced by probability, possibility, fuzziness, plausibility, and belief. In: B. Bouchon-Meunier, L. Valverde, R.R. Yager (eds.): Uncertainty in intelligent systems. Amsterdam: North-Holland 1993, pp.355- 365
6. J. Yeh: Stochastic processes and the Wiener integral. New York: Marcel Dekker 1973, pp.3-21
7. S. Guiasu: Fuzzy sets with interdependence. Information Sciences 79, 315-338 (1994)
8. S. Guiasu: Reaching a verdict by weighting evidence. In: P.P. Wang (ed.): Advances in fuzzy theory and technology, vol.2. Raleigh, North Carolina: Bookwrights 1994, pp.167-180

An Approach to Handle Partially Sound Rules of Inference

Siegfried Gottwald

Institut für Logik und Wissenschaftstheorie, Universität Leipzig, D-04109 Leipzig, Germany

Abstract. One of the common features in fuzzy sets, many-valued logic, and fuzzy logic is that all three areas in some sense are related to partiality, to kinds of degrees. In fuzzy logic this partiality appears in the degrees to which formulas are given as premises as well as in the degrees of truth of such formulas. Calculi, yet, work with formulas and inference rules. But there is an asymmetry here: formulas have truth degrees, inference rules have to be (absolutely) sound. We discuss the problem how to generalise this soundness to partial soundness.

1 Background

Fuzzy sets describe properties which suitable objects, i.e. the elements of the universe of discourse, may have only *partially*, i.e. only to some degree (membership degree).

Formulas of many-valued propositional logic describe propositions which may be only *partially* true (or partially false), i.e. which have a truth degree. The same is the case for closed formulas of many-valued first order logic. Additionally, there, formulas which contain free individual variables describe properties which hold for the objects, of which they are properties, only to some (truth) degree. Hence, having membership degrees and truth degrees identical, formulas of first order many-valued logic correspond to fuzzy sets.

Finally, fuzzy logic combines the idea of a logical calculus as a formal system for inferring formulas from given (sets of) premises with the idea of fuzzy sets in the way that fuzzy logic allows for only *partially* "given" premises, i.e. for fuzzy sets of premises.

As systems of logic, both many-valued as well as fuzzy logic have a syntactic and a semantic aspect. The semantic considerations for many-valued logic are based (i) in the propositional case on valuations, i.e. mappings $v : V \to \mathcal{D}$ from the set V of propositional variables into the set $\mathcal{D}$ of truth degrees and completed by a consequence operation $\mathrm{Cn}_{\mathcal{S}} : \mathbb{P}(\mathcal{FOR}) \to \mathbb{P}(\mathcal{FOR})$ over subsets of the class $\mathcal{FOR}$ of all (well formed) formulas, and (ii) in the first order case on interpretations A which map individual constants into a given universe of discourse $|A|$ and predicate letters (of arity n) to (n-ary) functions from $|A|$ into the set $\mathcal{D}$ of truth degrees and again completed by a consequence operation $\mathrm{Cn}_{\mathcal{S}} : \mathbb{P}(\mathcal{FOR}) \to \mathbb{P}(\mathcal{FOR})$.

The semantic considerations for fuzzy logic are based (i) on the valuations $v : V \to \mathcal{D}$ which map the set of propositional variables V into the set $\mathcal{D}$ of membership degrees and completed by a consequence operation $\mathrm{Cn}^{\star}_{\mathcal{S}} : \mathbb{F}(\mathcal{FOR}) \to \mathbb{F}(\mathcal{FOR})$ over the fuzzy subsets (with membership degrees in $\mathcal{D}$) of the class $\mathcal{FOR}$ of formulas, and (ii) in the first order case on interpretations A which map individual constants into the universe of discourse $|A|$ and predicate letters (of arity n) to (n-ary) fuzzy relations in $|A|$ and completed again by a consequence operation $\mathrm{Cn}^{\star}_{\mathcal{S}} : \mathbb{F}(\mathcal{FOR}) \to \mathbb{F}(\mathcal{FOR})$ over the fuzzy subsets (with membership degrees in $\mathcal{D}$) of the class $\mathcal{FOR}$ of formulas.

The syntactic considerations for many-valued as well as fuzzy logic are based on suitable calculi fixed by their respective (crisp) sets of axioms and sets of inference rules. (Inferences in any case are finite sequences.) For fuzzy logic each inference rule R with k premises splits into two parts $\mathsf{R} = (\mathsf{R}^1; \mathsf{R}^2)$ such that R^1 is a (partial) k-ary mapping from $\mathcal{FOR}$ into $\mathcal{FOR}$ and R^2 is a k-ary mapping from $\mathcal{D}$ into $\mathcal{D}$. The idea is that R^2 associates with the degrees to which actual premises of R are given a degree to which the actual conclusion of R is given.

In any case, one of the main goals of those calculi is the axiomatization of the set of logically true formulas or of the logical part of elementary theories.

2 The problem

Let us first look at many-valued logic. Of course, not every inference schema is accepted as an inference rule: a necessary restriction is that to *sound* inference schemata, i.e. to inference schemata which led "from true premises only to a true conclusion". For many-valued logic this actually shall mean either that in case all premises have truth degree 1, which we suppose to be the only designated one, also the conclusion has truth degree 1 or that the conjunction of all premises has a truth degree not greater then the truth degree of the conclusion.

For fuzzy logic a comparable soundness condition is assumed, called "infallibility" in (Bolc and Borowik 1992) instead of "soundness" as e.g. in (Novak 1989, 1992) or "Korrektheit" as in (Gottwald 1989).

Now, looking back at the approaches we mentioned, one recognises that many-valued logic deals with partially true sentences, fuzzy logic deals with even partially given (sets of) premises – but in both cases soundness of the rules of inference is taken in some absolute sense.

What about partial soundness for inference schemata?

Is it possible to consistently discuss such a notion of partial soundness? And if possible: how such a notion can be integrated into many-valued as well as into fuzzy logic?

Regarding applications of partially sound rules of inference, it is quite obvious that an analysis of the heap paradox or the bold man paradox or other types of paradoxes may, instead of referring to the use of implications with a truth degree a little bit smaller then 1, become based upon the use of inference schemata which are "not completely sound".

3 A first approach

As already in the preceding remarks we will restrict the considerations here to Lukasiewicz type many-valued systems: truth degrees a subset of the real interval $[0,1]$, bigger truth degrees as the "better" ones, and 1 the only (positively) designated truth degree.

As the inference schema under discussion let us consider the schema

$$(\mathsf{R}) : \quad \frac{H_1, \ldots, H_n}{H}$$

with n premises $H_1, \ldots, H_n$ and the conclusion H. Furthermore, we use $[\![G]\!]$ to denote the truth degree of the formula G. The dependence of $[\![G]\!]$ from the supposed interpretation and a valuation of the (free) individual variables can be supposed to be clear from the context.

The usual soundness condition for (R) we shall consider is the inequality

$$[\![H_1 \& \ldots \& H_n]\!] \leq [\![H]\!] \tag{1}$$

with & for a suitable conjunction operation, e.g. some t-norm. We prefer (1) because it is more general then demanding only

$$\text{IF} \quad [\![H_1]\!] = \ldots = [\![H_n]\!] = 1 \quad \text{THEN} \quad [\![H]\!] = 1 \ . \tag{2}$$

A partially sound schema (R) of inference, which is not sound in the usual sense, has to deviate from (1) to some degree. And this degree of deviance we intend to use to "measure" the deviance of schema (R) from soundness, i.e. to define the "degree of soundness" of schema (R).

Referring to a measure of deviance of schema (R) from soundness condition (1) it is quite natural to look at the value $\delta(\mathsf{R})$ defined as

$$\delta(\mathsf{R}) = \sup\left(\max\{0, [\![H_1 \& \ldots \& H_n]\!] - [\![H]\!]\}\right) \tag{3}$$

where the sup has to be taken with respect to all interpretations and all valuations of the individual variables. But other approaches as (3) can be discussed too.

Now, with respect to a suitable implication connective $\rightarrow$ one should have

$$[\![G_1]\!] \leq [\![G_2]\!] \quad \text{iff} \quad [\![G_1 \rightarrow G_2]\!] = 1 \tag{4}$$

for any formulas G_1, G_2 and hence

$$[\![H_1 \& \ldots \& H_n \rightarrow H]\!] = 1 \tag{5}$$

as an equivalent soundness condition instead of (1). Using the symbol $\models$ for the consequence or satisfaction relation one has thus usually

$$\begin{aligned} (\mathsf{R}) \text{ sound} &\Leftrightarrow \text{always } [\![H_1 \& \ldots \& H_n]\!] \leq [\![H]\!] \\ &\Leftrightarrow \text{always } [\![H_1 \& \ldots \& H_n \rightarrow H]\!] = 1 \\ &\Leftrightarrow\ \models (H_1 \& \ldots \& H_n \rightarrow H) \ . \end{aligned}$$

These equivalences give another way to approach partial soundness instead of (3).

Definition 1 *For a schema of inference* (R) *its degree of soundness shall be*

$$\kappa(\mathsf{R}) =_{\text{def}} \inf([\![H_1 \& \ldots \& H_n \rightarrow H]\!]) \tag{6}$$

with the infimum taken over all interpretations and all valuations of the individual variables.

Corollary 1 *With* $\rightarrow$ *the Lukasiewicz implication, characterised by the truth degree function* $\text{seq}_L(u,v) = \min\{1, 1-u+v\}$, *one has*

$$\kappa(\mathsf{R}) = 1 - \delta(\mathsf{R})$$

and thus (6) as a suitable generalisation of the idea which led to (3).

For $\kappa(\mathsf{R})$ the degree of soundness of inference rule (R) one thus always has

$$\kappa(\mathsf{R}) \leq [\![H_1 \& \ldots \& H_n \rightarrow H]\!]$$

or even, accepting $\underline{\kappa(\mathsf{R})}$ as a constant of the language to denote the degree $\kappa(\mathsf{R})$ and having (4) satisfied,

$$[\![\underline{\kappa(\mathsf{R})} \rightarrow (H_1 \& \ldots \& H_n \rightarrow H)]\!] = 1$$

which via importation and exportation for the implication and commutativity for the conjunction operations is equivalent to

$$[\![H_1 \& \ldots \& H_n \rightarrow (\underline{\kappa(\mathsf{R})} \rightarrow H)]\!] = 1 \ .$$

This is a first way to "code" partially sound rules: it presupposes that one has to have each degree of soundness as a truth degree constant available within the language.

4 Partially sound rules in many-valued logic

The use of sound rules of inference within the process of inference of new propositions from given premises can be seen as a transfer of confidence in the premises to a confidence in the conclusion.

For classical logic, this type of interpretation looks completely unproblematic: "confidence" can be understood as the assumption of truth. A sound rule of inference does then rationally transfer this confidence from the premises to the conclusion. For many-valued logic, one way to interpret "confidence" in a formula or proposition H is to translate this into the statement $[\![H]\!] = 1$, i.e. again into the assumption of the (complete) truth of H. But it also seems natural to give "confidence in H" another reading meaning $[\![H]\!] \geq u$ for some truth degree u. In this second sense confidence itself is graded in some sense, and this is in very interesting coincidence with the basic idea of many-valued logic, i.e. with the graduation of truth.

Soundness condition (1) now can be read as allowing only such inference schemata as sound ones which transfer the "common confidence" in the premises, i.e. the confidence in the conjunction of the premises into a suitable confidence in the conclusion.

This reading of soundness shows that and why soundness condition (1) seems to be preferable over soundness condition (2). And this reading is based on a kind of identification of truth degrees with degrees of confidence (in the sense that degree u of confidence in H means $[\![H]\!] \geq u$) – and this does not seem completely unreasonable. Yet, here this interpretation of truth degrees as (lower bounds of) confidence degrees either refers to an identification of truth degrees with degrees of confidence – or it can simpler be seen as an addition of confidence degrees to classical logic.

Up to now the discussion was related simply to soundness. What about partially sound rules of inference? Of course, depending on their degree of soundness the confidence in the conclusion, given the confidences in the premises, should be smaller than in case of a completely sound rule of inference (with the same premises) - and becoming as smaller as the degree of soundness is becoming smaller.

Looking again at the schema (R) and first assuming full confidence in the premises, i.e. assuming $[\![H_i]\!] = 1$ for $1 \leq i \leq n$ or $\models H_i$ for $1 \leq i \leq n$, it seems reasonable to assume

$$\kappa(\mathsf{R}) \leq \text{confidence in } H \ ,$$

i.e. to assume

$$\text{IF} \quad [\![H_1]\!] = \ldots = [\![H_n]\!] = 1 \quad \text{THEN} \quad \kappa(\mathsf{R}) \leq [\![H]\!] \ . \tag{7}$$

This last condition in the present case also can be written as

$$\kappa(\mathsf{R}) \leq [\![H_1 \& \ldots \& H_n \rightarrow H]\!] \leq [\![H]\!] \tag{8}$$

simply assuming additionally that $[\![1 \rightarrow H]\!] = [\![H]\!]$ holds always true, which is only a mild and natural restriction concerning the implication connective $\rightarrow$.

Thus, from the intuitive point of view an application of a partially sound rule of inference can be understood via the additional idea of confidence degrees.

But, what about repeated applications of partially sound rules of inference? And is it really a convincing idea to identify degrees of confidence with (lower bounds of) the truth degrees?

At least for the second one of these questions a more cautious approach may help to separate confidence and truth degrees: a change from formulas to ordered pairs consisting of a formula and a degree of confidence. The consequence of this idea for inference schemata like (R) now is that premises and conclusions have to become ordered pairs, i.e. (R) changes into the modified schema

$$(\hat{\mathsf{R}}) : \quad \frac{(H_1, \alpha_1), \ldots, (H_n, \alpha_n)}{(H, \beta)} \ .$$

Choosing here confidence degrees, like membership degrees and truth degrees, from the real unit interval $[0,1]$ means that in $(\hat{\mathsf{R}})$ usually $\alpha_i > 0$ for $1 \leq i \leq n$ as well as $\beta > 0$; furthermore $\beta = \beta(\boldsymbol{\alpha})$ has to be a function of $\boldsymbol{\alpha} = (\alpha_1, \ldots, \alpha_n)$.

A modified soundness condition for the inference schema $(\hat{\mathsf{R}})$ now is easily at hand.

Definition 1. A modified inference schema of the form $(\hat{\mathsf{R}})$ is sound* iff it always holds true that in case one has $[\![H_i]\!] \geq \alpha_i$ for all $1 \leq i \leq n$, then one also has $[\![H]\!] \geq \beta$.

The weaker form (2) of the usual soundness condition for (R) often is formulated in model theoretic terms as the condition that each model of $\{H_1, \ldots, H_n\}$ also is a model of H. In its stronger form (1) this model theoretic version in many-valued logic simply reads: each interpretation is a model of $H_1 \& \ldots \& H_n \rightarrow H$.

Slightly modifying the notion of model, cf. (Gottwald 1989), and calling for $\alpha \in [0,1]$ an interpretation A an α-model of a formula H iff $[\![H]\!] \geq \alpha$ in A (for all valuations of the individual variables) enables a model theoretic reformulation also of soundness*.

This formulation becomes even simpler if one calls A an $(\alpha_1, \ldots, \alpha_n)$-model of a sequence $(H_1, \ldots, H_n)$ of formulas iff A is an α_i-model of H_i for all $1 \leq i \leq n$.

Corollary 2 *A modified inference schema* $(\hat{\mathsf{R}})$ *is sound* iff every* $(\alpha_1, \ldots, \alpha_n)$-*model of* $(H_1, \ldots, H_n)$ *is a* β-*model of* H.

Obviously, Definition 1 subsumes the traditional soundness condition (2), but also (1) and even condition (3) – simply depending on a suitable choice of β as a function of $\boldsymbol{\alpha} = (\alpha_1, \ldots, \alpha_n)$.

Regarding condition (1) and assuming that (R) fulfils condition (1) one may take in $(\hat{\mathsf{R}})$ the degree $\beta = \mathrm{et}^{(n)}(\boldsymbol{\alpha})$ with $\mathrm{et}(u,v)$ the truth degree function characterising the conjunction & and $\mathrm{et}^{(n)}$ the n-ary generalisation of et, then of course this schema $(\hat{\mathsf{R}})$ is sound*. On the other hand, having the inference schema $(\hat{\mathsf{R}})$ sound* and using $\alpha_i = [\![H_i]\!]$ for all $1 \leq i \leq n$, so in case $\beta(\boldsymbol{\alpha}) = \mathrm{et}^{(n)}(\boldsymbol{\alpha})$ the corresponding "reduced" schema (R) is simply sound.

But if $(\hat{\mathsf{R}})$ is sound* and one does not always have $\beta(\boldsymbol{\alpha}) \geq \mathrm{et}^{(n)}(\boldsymbol{\alpha})$, then for $\delta^*(\mathsf{R}) = \inf_{\alpha_1,\ldots,\alpha_n}(\beta(\boldsymbol{\alpha}) - \mathrm{et}^{(n)}(\boldsymbol{\alpha}))$ one has always

$$1 + \delta^*(\mathsf{R}) \leq 1 - \mathrm{et}^{(n)}(\boldsymbol{\alpha}) + \beta(\boldsymbol{\alpha}) \,. \tag{9}$$

With $\kappa^*(\mathsf{R}) = 1 + \delta^*(\mathsf{R})$ this can be written as

$$\kappa^*(\mathsf{R}) \leq \mathrm{seq}_L(\mathrm{et}^{(n)}(\boldsymbol{\alpha}), \beta(\boldsymbol{\alpha})) \tag{10}$$

and obviously corresponds to (the first inequality of) condition (8).

Therefore, inference schemata of type $(\hat{\mathsf{R}})$ cover the usual sound inference rules as well as the partially sound rules of inference, and the soundness* for $(\hat{\mathsf{R}})$ covers usual soundness together with partial soundness.

Having a closer look at schema $(\hat{\mathsf{R}})$ one recognises that this schema can be split into two parts, viz. the old schema (R) together with a mapping R_{sem} : $(\boldsymbol{\alpha}) \mapsto \beta(\boldsymbol{\alpha})$. Taking $(\hat{\mathsf{R}}) = (\mathsf{R}_{syn}; \mathsf{R}_{sem})$ with now (R_{syn}) for (R), because (R) describes the "syntactic part" of $(\hat{\mathsf{R}})$, one has that $(\hat{\mathsf{R}})$ corresponds to

$$\left(\frac{H_1, \ldots, H_n}{H} \,;\quad \frac{\alpha_1, \ldots, \alpha_n}{\beta(\alpha_1, \ldots, \alpha_n)} \right) . \tag{11}$$

Thus $(\hat{\mathsf{R}})$ splits in the same way as inference rules in fuzzy logic are composed of a syntactic and a semantic part.

Moreover, starting a deduction from a set Σ of "premises" using inference rules of type $(\hat{\mathsf{R}})$ needs premises which have the form of ordered pairs (formula; degree), i.e. one in such a case has to assume

$$\Sigma = \{(H_1, \alpha_1), (H_2, \alpha_2), \ldots\} \ . \tag{12}$$

But this exactly means to consider

$$\Sigma : \mathcal{FOR} \rightarrow \mathcal{D} \tag{13}$$

if one "completes" the set Σ of (12) with all the ordered pairs $(H, 0)$ for which $H \notin \{H_1, H_2, \ldots\}$.

With $\mathcal{D} = [0,1]$ or $\mathcal{D} \subseteq [0,1]$ as was always supposed up to now, (13) means that Σ is a *fuzzy subset* of the set $\mathcal{FOR}$ of all well-formed formulas. That means, looking at partial soundness of inference rules via their formulation in the form $(\hat{\mathsf{R}})$ leads immediately into the field of fuzzy logic: the sets of premises to start a deduction from become fuzzy subsets of the set of all formulas, and the inference schemata become in a natural way the form those schemata have in fuzzy logic.

5 Partially sound rules in fuzzy logic

The rules of inference in fuzzy logic usually, as e.g. in (Novak 1989), are given in the form of an ordered pair and written down as

$$(\tilde{\mathsf{R}}) : \quad \frac{H_1, \ldots, H_n}{r^{syn}(H_1, \ldots, H_n)} \quad \left[\frac{\alpha_1, \ldots, \alpha_n}{r^{sem}(\alpha)}\right]$$

with $r^{syn}(H_1, \ldots, H_n) \in \mathcal{FOR}$ and $r^{sem}(\alpha) \in \mathcal{D}$ a truth and membership degree.

For such rules soundness preferably is defined via model theoretic terms.

To do this one first needs the semantic consequence operation Cn^*_S of fuzzy logic. We will consider only the first order case, i.e. refer to interpretations A which map individual constants of the language to objects of the universe $|A|$ and the (n-ary) predicate symbols of the language to (n-ary) fuzzy relations in $|A|$. The truth degree $\mathrm{val}_A(H)$ of a formula H with respect to A is defined in the standard manner. (For the moment we prefer the notation $\mathrm{val}_A(H)$ over $[\![H]\!]$.) Furthermore, for each fuzzy set $\Sigma \in \mathbb{F}(\mathcal{FOR})$ of formulas and each interpretation A one defines

$$A \quad \text{model of} \quad \Sigma =_{\text{def}} [\![H \,\varepsilon\, \Sigma]\!] \leq \mathrm{val}_A(H) \quad \text{for all } H \in \mathcal{FOR} \tag{14}$$

using the notation "$[\![H \,\varepsilon\, \Sigma]\!]$" instead of $\mu_\Sigma(H)$ as in (Gottwald 1989) because of the natural interpretation of the membership degrees as truth degrees of a many-valued membership predicate ε.

Then one defines the fuzzy set $\mathrm{Cn}^{\star}_{S}(\Sigma) \in \mathbb{F}(\mathcal{FOR})$ by the condition

$$[\![H \,\varepsilon\, \mathrm{Cn}^{\star}_{S}(\Sigma)]\!] = \inf\{\mathrm{val}_{A}(H) \mid A \text{ model of } \Sigma\} \ . \tag{15}$$

Now soundness of $(\tilde{\mathsf{R}})$ is the condition that for each fuzzy set

$$\Sigma_R = \alpha_1/H_1 + \alpha_2/H_2 + \ldots + \alpha_n/H_n$$

with $(H_1, \ldots, H_n) \in \mathrm{dom}\,(r^{syn})$ one has for any interpretation A

$$r^{sem}(\mathrm{val}_A(H_1), \ldots, \mathrm{val}_A(H_n)) \ \leq\ \mathrm{val}_A(r^{syn}(H_1, \ldots, H_n)) \ , \tag{16}$$

i.e. one has for the fuzzy singleton

$$\mathrm{Cn}^{\star}_{R}(\Sigma_R) = \langle\!\langle r^{syn}(H_1, \ldots, H_n)\rangle\!\rangle_{r^{sem}(\alpha)}$$

with membership degree $r^{sem}(\alpha)$ the model theoretic soundness condition

$$\text{IF} \quad A \text{ model of } \Sigma_R \quad \text{THEN} \quad A \text{ model of } \mathrm{Cn}^{\star}_{R}(\Sigma_R) \ . \tag{17}$$

What now about partial soundness for rules of type $(\tilde{\mathsf{R}})$ in fuzzy logic?

Most natural it seems to modify condition (17) in such a way that given $\Sigma_R \in \mathbb{F}(\mathcal{FOR})$ and A as a model of Σ_R one only has that A is a model of $\mathrm{Cn}^{\star}_{R}(\Sigma_R)$ *to some degree.*

A precise meaning for this last phrase now shall be for any $\Sigma \in \mathbb{F}(\mathcal{FOR})$ and any $\alpha \in \mathcal{D}$

$$\mathrm{mod}_{\alpha}(A, \Sigma) \ =_{\mathrm{def}} [\![H \,\varepsilon\, \Sigma]\!] \leq \mathrm{val}_A(H) + (1 - \alpha) \quad \text{for all } H \in \mathcal{FOR} \tag{18}$$

which is equivalent with

$$\mathrm{mod}_{\alpha}(A, \Sigma) \ \Leftrightarrow \alpha \leq \sup_{H \in \mathcal{FOR}} \ (1 - [\![H \,\varepsilon\, \Sigma]\!] + \mathrm{val}_A(H)) \tag{19}$$

and which generalises (14) in the sense that

$$A \ \text{model of} \ \Sigma \ \Leftrightarrow \ \mathrm{mod}_1(A, \Sigma) \ .$$

Partial soundness of $(\tilde{\mathsf{R}})$ now can be read as having instead of (17)

$$\text{IF} \quad A \text{ model of } \Sigma_R \quad \text{THEN} \quad \mathrm{mod}_{\alpha}(A, \mathrm{Cn}^{\star}_{R}(\Sigma_R)) \tag{20}$$

for some $\alpha \in \mathcal{D}$. Quite naturally the ***soundness degree*** $\kappa(\tilde{\mathsf{R}})$ of $(\tilde{\mathsf{R}})$ can be taken as the supremum of all α such that $\mathrm{mod}_{\alpha}(A, \mathrm{Cn}^{\star}_{R}(\Sigma_R))$ holds true for all $(H_1, \ldots, H_n) \in \mathrm{dom}\,(r^{syn})$, all $\alpha_1, \ldots, \alpha_n \in \mathcal{D}$ and all models A of Σ_R:

$$\begin{aligned}\kappa(\tilde{\mathsf{R}}) =_{\mathrm{def}} \sup\{\alpha \mid \mathrm{mod}_{\alpha}(A, \mathrm{Cn}^{\star}_{R}(\Sigma_R)) \text{ for all } (H_1, \ldots, H_n) \in \mathrm{dom}\,(r^{syn}),\\ \text{all } \alpha_1, \ldots, \alpha_n \in \mathcal{D} \text{ and all models } A \text{ of } \Sigma_R\}.\end{aligned}$$

As in the case of many-valued logic using a partially sound rule of inference $(\tilde{\mathsf{R}})$ in a deduction inside fuzzy logic should mean to lower the "degree" of the rules conclusion depending on its soundness degree $\kappa(\tilde{\mathsf{R}})$. A reasonable candidate for this lowered degree instead of $r^{sem}(\alpha)$ then is

$$r_*^{sem}(\alpha)) = \max\{0, r^{sem}(\alpha) - (1 - \kappa(\tilde{\mathsf{R}}))\}$$
$$= \mathrm{et}_L(r^{sem}(\alpha)), \kappa(\tilde{\mathsf{R}}))$$

with $\mathrm{et}_L(u,v) = \max\{0, u+v-1\}$ the Lukasiewicz conjunction operation of many-valued logic.

But the result of this approach is exactly the same as if one replaces a partially sound schema of inference $(\tilde{\mathsf{R}})$ with soundness degree $\kappa(\tilde{\mathsf{R}})$ by the new schema

$$(\tilde{\mathsf{R}})^* \; : \quad \frac{H_1, \ldots, H_n}{r^{syn}(H_1, \ldots, H_n)} \quad \left[\frac{\alpha_1, \ldots, \alpha_n}{\mathrm{et}_L(r^{sem}(\alpha)), \kappa(\tilde{\mathsf{R}}))}\right].$$

Therefore, in fuzzy logic the use of partially sound rules of inference can equivalently be managed by the use of suitable sound rules of inference.

6 Conclusions

The generalisation of the soundness of rules of inference to partial soundness proves to be manageable. Adding partially sound rules to many-valued logic leads in a natural way to fuzzy logic. But adding partially sound rules to fuzzy logic can equivalently be understood as starting from specific fuzzy sets of premises. Hence, in fuzzy logic it is possible to infer even with partially sound inference rules.

References

BOLC, L. BOROWIK, P., *Many-Valued Logics, 1.* Springer: Berlin, 1992.

GOTTWALD, S., *Mehrwertige Logik.* Akademie-Verlag: Berlin, 1989.

NOVAK, V., *Fuzzy Sets and Their Applications.* A. Hilger: Bristol, 1989.

NOVAK, V., *The Alternative Mathematical Model of Linguistic Semantics and Pragmatics.* Plenum Press: New York, 1992.

Using Classical Theorem-Proving Techniques for Approximate Reasoning: Revised Report

Stefan Brüning[1] and Torsten Schaub[2]*

[1] Intellektik, TH Darmstadt, Alexanderstraße 10, D-64283 Darmstadt
[2] Theoretische Informatik, TH Darmstadt, Alexanderstraße 10, D-64283 Darmstadt

Abstract. We propose an approach to approximate classical reasoning via well-known theorem-proving techniques. Unlike other approaches, our approach takes into account the interplay of knowledge bases and queries and thus allows for query-sensitive approximate reasoning. We demonstrate that our approach deals extremely well with the examples found in the literature. This reveals that conventional theorem-proving techniques can account for approximate reasoning.

1 Introduction

A distinguishing mark of human reasoning is its *speed*. Usually, we answer questions in a spontaneous way. In rare cases, however, our answers are wrong or we do not even arrive at answers at all. In computer science, reasoning processes are usually formalized in logical systems like first-order logic. But even though this logical approach offers several indispensable benefits, problem solving in logical systems is usually intractable. For instance, query-answering in propositional logic is co-NP-complete [5].

Recently, a couple of approaches to approximate reasoning in logical systems have been proposed in order to overcome this gap. In [13] a method for ***knowledge compilation*** is proposed. The idea is to compile a knowledge base (KB) represented as a propositional formula into a logically stronger and a logically weaker Horn-formula for which query-answering is polynomial [4]. [2] extend the idea of *limited inference* found in [8] by proposing a stronger and a weaker version of propositional entailment. The idea is to restrict classical satisfiability to a certain subset of the language. This leads to an extended notion of interpretations in which some propositions and their negation may be both true or both false. Both approaches offer a *complete* and a *sound* approximation of propositional entailment. However, both methods cannot be directly mapped onto existing automated theorem provers. In the first case, one needs special-purpose algorithms for compilation. Moreover, this approach might be to rigid whenever we are faced with rapidly changing KBs. In the second case, one has to deal with non-classical entailment relations requiring non-standard proof procedures. In addition, it is still unclear how one has to partition the language for determining the non-classical entailment relations.

In what follows, we introduce an approach to approximate reasoning that relies on existing theorem-proving techniques. The idea is to take an existing calculus for propositional (or even first-order) logic which can be decomposed

* On leave from IRISA, Campus de Beaulieu, 35042 Rennes Cedex, France

into a "fast" and a "slower" part; thereby the "fast" (and possibly incomplete) submethod is enriched by the "slower" one such that the entire method is complete. This enrichment should allow for a stepwise approximation that finally meets the entire method. A prime candidate for this purpose is the combination of *unit-resulting resolution* (URR) [14] and *reasoning by cases* (CASE). Together, both methods provide a sound and complete calculus for propositional and first-order logic. The "fast" part is taken up by URR. Roughly speaking, URR is a restriction of hyper-resolutions to those yielding unit clauses only [12]. Notably, URR allows for polynomial query-answering beyond the class of Horn-formulas. Moreover, URR constitutes a special purpose but high-speed reasoning mode in high-performance theorem provers like OTTER [10]. The "slower" part is played by CASE. As regards approximate reasoning, our intended purpose of CASE is (roughly) to enable URR whenever the latter is inapplicable. In such a case, CASE splits clauses into several ones that then allow for URR. In our approach, the control of such CASE-steps is accomplished by so-called *reachability relations*, which also constitute a well-known concept in automated theorem proving. These relations provide means for accessing the part of a KB relevant to a given query. This approach takes into account the interplay of KBs and queries and thus allows for query-sensitive approximate reasoning.

2 From Classical Towards Approximate Reasoning

We consider propositional formulas in conjunctive normal form. In this case, a KB can be represented as a set of clauses, where a clause is a disjunction of literals. We verify whether a query q is entailed by a KB Σ by checking whether the set of clauses representing $\Sigma \wedge \neg q$ is unsatisfiable. In this case we also say that $\neg q$ is refuted by KB.

URR restricts derivable clauses to *unit clauses* (ie. clauses with a single literal) and the *empty clause* $\Box$. Let L^d denote the negation of the literal L.

Definition 2.1 *Let $C = \{L_1, \ldots, L_n\}$ and $C_1 = \{L_1^d\}, \ldots, C_m = \{L_m^d\}$ be clauses. Define*

$$\mathsf{URR}(C, C_1, \ldots, C_m) = \begin{cases} \{L_m\} & \text{if } n = m+1 \\ \Box & \text{if } n = m \end{cases}$$

For a set of clauses S, define $\mathsf{URR}^0(S) = S$ and $\mathsf{URR}^{i+1}(S) = \mathsf{URR}^i(S) \cup \{C \mid C = \mathsf{URR}(C_1, \ldots, C_n),\ C_i \in \mathsf{URR}^i(S) \text{ for } 1 \le i \le n\}$. Then, a clause C is derivable from S by URR iff $C \in \bigcup_i \mathsf{URR}^i(S)$.

A query q is entailed by a KB Σ if $\Box$ is derivable by URR from the clauses representing $\Sigma \wedge \neg q$. Consider the KB consisting of clauses (1) to (4) along with the query D resulting in clause (5):

$$(1)\ \{\neg A, B\},\ \ (2)\ \{A, B, \neg C\},\ \ (3)\ \{\neg B, \neg C, D\},\ \ (4)\ \{C\},\ \ (5)\ \{\neg D\}$$

The sole possibility to start with is to perform URR with clauses (3), (4), and (5), resulting in the new unit clause (6) $\{\neg B\}$. Further URR involving clause (1) and (6) yields (7) $\{\neg A\}$. Now, the empty clause is derivable via (2), (4), (6), and

(7). Hence, clauses (1) to (5) are unsatisfiable and so D is entailed by clauses (1) to (4).

A set of clauses can be shown to be unsatisfiable by URR iff it contains an unsatisfiable subset of clauses which is *renamable-Horn* (ie. switching signs of literals in a suitable way yields a Horn-clause set [9]). In the propositional case, the satisfiability of such a set S can be decided in $O(n^2)$ time and $O(n)$ space, where n is the number of literals occurring in S.

But URR relies on the existence of unit-clauses. So, we might sometimes be enforced to reason by cases before applying URR. In fact, we obtain a complete inference mechanism by adding the concept of case-analysis (CASE) [3]:

Definition 2.2 *Let $S = \{C_1, \ldots, C_m\}$ be a clause set and let $\langle S_i \rangle_{i \in I}$ be a family of clause sets such that $\bigcup_{i \in I} S_i \models S$ and $S_i = \{C_{i_1}, \ldots, C_{i_m}\}$ with $C_{i_j} \subseteq C_j$ for $1 \leq j \leq n$ and $i \in I$. Then, S is satisfiable if S_i is satisfiable for some $i \in I$.*

Observe that each case S_i is strictly smaller (in the number of occurring literals) than S. In fact, this definition is a very general one and thus admits various instantiations (see Section 3 and 4).

Consider the clause set obtained by replacing clause (4) by (4') $\{C, E\}$ in the above example. Now, URR is not applicable. However, we may generate two cases by splitting clause (4'). In the first case, we delete literal E in (4'), which allows for deriving the empty clause by URR, as shown above. The second case, obtained by deleting C in (4'), however yields a satisfiable set of clauses which is perceivable after further CASE-steps. This implies that the original set of clauses is satisfiable. Hence D is not entailed by the KB consisting of (1), (2), (3), and (4').

The task of approximation can also be seen as identifying the relevant information for (dis)proving queries. The notion of relevance relies on the interplay between KBs and queries. In our approach, the concept of relevance is addressed by means of a *reachability relation*[3] that allows for measuring the inferential distance between queries and clauses in the KB.

Definition 2.3 (Reachability) *Let S be a clause set and let L be a literal occurring in a clause C of S. The set of literals reachable from L is defined inductively as follows:*[4]

1. *Any complementary literal L^d in a clause $C' \in S$ is reachable from L if $C \neq C'$.*
2. *If $E = \{K, L_1, \ldots, L_n\}$ is a clause of S and K is reachable from L, then any literal L_i^d contained in a clause $E' \in S$ is reachable from L if $E \neq E'$, where $1 \leq i \leq n$. We call K an* ancestor *of each L_i^d.*

Consider the query D in the previous example; it is represented by clause (5) $\{\neg D\}$ in the corresponding clause set. Following Definition 2.3, for instance, the literal D in clause (3), the literal B in clause (1) and (2), the literal C in clause (4), etc. are reachable from $\neg D$. Roughly, a literal K is reachable from a

[3] Reachability was introduced in [11] in a modified way.

[4] Recall that L^d denotes the negation of the literal L.

literal L if the derivation of the clause $\{K\}$ might contribute to a refutation of L. For instance, B and C are reachable from $\neg D$. In fact, the derivation[5] of the unit clauses $\{B\}$ and $\{C\}$ allows for refuting $\neg D$ by URR.

For approximate reasoning, we proceed by defining two concepts for measuring the relevance of literals occurring in a KB to putative queries. That is, we define the notions of *distance* and *weight* of a literal K to another literal L. Both measures are incrementally computed for each literal while parsing a KB. In this way, we can easily compute both measures for each query in turn.

Definition 2.4 *Let S be a clause set and let L and K be distinct literals occurring in S such that K is reachable from L.*

The distance of K to L, $d(L, K)$, is defined as the minimal length of a sequence $K_1, \ldots, K_n$ of literals such that $K_1^d = L$, $K_n = K$ and K_i is an ancestor of K_{i+1} for $1 \leq i < n$. We set $d(L, L) = 0$ and $d(L, K) = \infty$ if K is not reachable from L.

The weight of K to L, $w(L, K)$, is defined as the number of sequences of different[6] literals $K_1, \ldots, K_n$ such that $K_1^d = L$, $K_n = K$ and K_i is an ancestor of K_{i+1} for $1 \leq i < n$. We set $w(L, L) = \infty$ and $w(L, K) = 0$ if K is not reachable from L.

In the above example, we obtain $d(\neg D, C) = 2$ from the minimal sequence D, C. Accordingly, we know that in order to refute $\neg D$, we might have to use clause (4) after 2 inference steps. Thus, if the distance of a literal K to a literal L is small, K is likely to constitute an important part of a refutation of L. We have $w(\neg D, C) = 3$ in the above example. This tells us that there are three different derivations starting with the goal $\neg D$ using clause (4). Thus, if the weight of K to L is high, the chance of using K in a refutation of L is also high.

Given two clauses c_1 and c_2, the weight of c_2 to c_1 is defined as the *maximal weight* of a literal in c_2 to a literal in c_1. Similarly, the distance of c_2 to c_1 is defined as the *minimal distance* of a literal in c_2 to a literal in c_1.

Of course, the concept of reachability allows for a variety of different measures, among which we have chosen those of *distance* and *weight* as exemplars. In practice, such measures are combined and then lead to a selection function among clauses. A thorough treatment of such selection functions is however beyond the scope of this paper.

3 Correct (but Incomplete) Approximate Reasoning

We introduce a correct but incomplete method for approximate reasoning. That is, if a query can be answered approximately, then it is also entailed by the given KB but not vice versa. The idea is to process a set of clauses representing the KB and the negated query by URR as long as possible. If the resulting set of clauses contains the empty clause, it is unsatisfiable, and we are done. If not, we perform reasoning by cases via CASE and then return to URR. The (possibly

[5] Observe that we consider *putative* and not actual derivations.

[6] Note that we do not have to consider sequences containing multiple occurrences of a literal. A derivation corresponding to such a sequence can always be pruned.

exponential) number of cases generated by CASE-inferences is bound and thus serves as a parameter for an approximation. This leads to an overall complexity of $O(n^2 \times p)$, where n is the number of literals in the KB and p is the number of permissible cases generated by CASE-inferences. Since p is fixed, our method in polynomial in n.

Clearly, we have to make precise how cases are generated. For the purpose of correct (but incomplete) approximate reasoning, it is sufficient to select a clause $C = \{L_1, \ldots, L_n\}$ from the actual clause set S and to generate the cases $S_i = (S \backslash C) \cup \{L_i\}$. The unsatisfiability of S then follows from the unsatisfiability of the S_i.

While testing this approach on the examples in the literature, the most striking observation is that they can be solved without reasoning by cases. This substantiates the use of URR as the fundamental inference mechanism. Its power helps to minimize the need for reasoning by cases for many practical problems. As a first example, consider the KB Σ represented by clauses (1) to (7). In [6], Σ is used to demonstrate that the compiled Horn-KB for correct approximate reasoning may be of exponential size, which in turn leads to exponential query-answering relative to the original KB. In our approach, however, any query q to Σ is solvable in polynomial time provided the clausal representation of $\Sigma \wedge \neg q$ is renamable-Horn (note that Σ itself is renamable-Horn). In this case, we can answer any query by pure URR without any case-analysis. Consider how the query ($CompSci \wedge ReadsDennet \wedge ReadsKosslyn \rightarrow CogSci$) represented by clause (5) to (8) (labeled by Q) is answered in our approach by means of the URR-steps (9) to (12):

(1) KB $\{\neg CompSci, \neg Phil, \neg Psych, CogSci\}$
(2) KB $\{\neg ReadsMcCarthy, CompSci, CogSci\}$
(3) KB $\{\neg ReadsDennet, Phil, CogSci\}$
(4) KB $\{\neg ReadsKosslyn, Psych, CogSci\}$
(5) Q $\{CompSci\}$
(6) Q $\{ReadsDennet\}$
(7) Q $\{ReadsKosslyn\}$
(8) Q $\{\neg CogSci\}$
(9) URR(4, 7, 8) $\{Psych\}$
(10) URR(3, 6, 8) $\{Phil\}$
(11) URR(1, 8, 9, 10) $\{\neg CompSci\}$
(12) URR(5, 11) $\square$

The next KB given by clause (1) to (8) along with the query ($cow \rightarrow molar\text{-}teeth$) given in (9) and (10) are taken from [2]. Again, this clause set can be decided employing URR only, even though it is non-Horn:

(1) KB $\{\neg cow, grass\text{-}eater\}$
(2) KB $\{\neg dog, carnivore\}$
(3) KB $\{\neg grass\text{-}eater, \neg canine\text{-}teeth\}$
(4) KB $\{\neg grass\text{-}eater, mammal\}$
(5) KB $\{\neg carnivore, mammal\}$
(6) KB $\{\neg mammal, canine\text{-}teeth, molar\text{-}teeth\}$
(7) KB $\{\neg mammal, vertebrate\}$
(8) KB $\{\neg vertebrate, animal\}$
(9) Q $\{cow\}$
(10) Q $\{\neg molar\text{-}teeth\}$

(11) URR(1,9) {*grass-eater*}
(12) URR(3,11) {¬*canine-teeth*}
(13) URR(4,11) {*mammal*}
(14) URR(6,12,13) {*molar-teeth*}
(15) URR(10,14) □

In [2], a certain subset of the language, namely {*cow*, *grass-eater*, *mammal*, *canine-teeth*, *molar-teeth*}, has to be determined (in a non-obvious manner) in order to yield the same result.

In general, the selection of a clause for reasoning by cases is the point where we have to look for relevant information. In fact, we pick clauses for case-analysis in a "query-oriented manner". That is, we select clauses which seem to be relevant for solving a query by choosing clauses that probably contribute to a successful derivation. This is accomplished by means of distance and weight: A clause is selected for case-analysis if its distance to a clause of the query is low and its corresponding weight is high.

Consider for instance the following KB containing train connections. An implication $A \longrightarrow B$ means that there is a train connection from A to B.

(1) KB {¬*Rome*, *Pisa*}
(2) KB {¬*Paris*, *Rome*}
(3) KB {¬*Berlin*, *Paris*}
(4) KB {¬*Warsaw*, *Berlin*}
(5) KB {¬*Warsaw*, *Moskow*}

Suppose that we have the query q asking whether there is a train connection from *Paris* or *Berlin* to *Rome* and *Pisa*. This query is derivable from the first three clauses whereas the two last clauses are of no importance. Taking into account the clausal representation of $\neg q$, namely

$$\{\{Paris, Berlin\}, \{\neg Rome, \neg Pisa\}\},$$

we recognize that, initially, URR is not applicable and therefore case-analysis has to be performed. Now we may split any of the above clauses. But splitting clause (5) is totally useless since it does not contribute to a successful derivation. This is recognized by the concept of reachability. No literal in clause (5) is reachable. Thus the distance of clause (5) to a clause of the query is ∞. Correspondingly the weight of clause (5) to a query-clause is 0. Since the clauses most relevant to a query are the clauses encoding the query (the distance of such clauses is equal to 0, and the weight equals ∞), the clause {*Paris*, *Berlin*} is (besides {¬*Rome*, ¬*Pisa*}) the best choice for case-analysis. This yields two clause sets, one containing the clause {*Paris*} and another containing {*Berlin*}. Both sets can be shown to be unsatisfiable by URR. Thus, the original clause set is unsatisfiable and the query is provable by means of a single case-analysis.

4 Complete (but Incorrect) Approximate Reasoning

We describe a complete but incorrect method for approximate reasoning. That is, if a query cannot be answered approximately, then it is neither entailed by the given KB. Here, we are thus interested in disproving queries in an approximate way. The basic idea is as follows. First, case-analysis is performed on a set of clauses representing the KB and the negated query. This yields a set of cases. Afterwards the satisfiability of one (or several) such cases is tested. If one of these cases is satisfiable, the original clause set is satisfiable and the query is disproved. Again, our method is parameterized by the number p of cases generated by CASE-inferences, which in turn leads to a complexity of $O(n^2 \times p)$ being polynomial in the number of literals in the KB, n.

First of all, let us make precise how such cases are generated. For complete approximate reasoning, it suffices to turn sets of clauses into certain sets of Horn-clauses. This offers several advantages. First, Horn-clauses are a prime candidate for fast URR. Second, we can decide the satisfiability of a set of Horn-clauses purely by URR, without any case-analysis. We generate Horn-cases by deleting all but one positive literal from each non-Horn-clause. In this way, a Horn-case consists of maximal (wrt set inclusion) non-empty Horn-clauses. Since the disjunction of all such Horn-cases is equivalent[7] to the original set of clauses, our approach meets propositional entailment whenever we consider the entire set of Horn-cases.

The possibly exponential number of Horn-cases in the worst case is controlled by the parameter p limiting the number of cases generated by CASE-inferences. The major problem is to identify the Horn-cases most relevant for disproving the given query. We address this problem by means of the notion of reachability and its induced measures. Recall that we are looking for satisfiable cases. We thus try to turn the KB into a case *not* entailing the query. Candidates for such cases are generated by removing positive literals from non-Horn-clauses which are (possibly) relevant for deriving the query.[8] In terms of our measures, we select literals with either a minimal distance or a maximal weight.

Consider the following KB taken from [2]:

(1) KB $\{\neg person, child, youngster, adult, senior\}$
(2) KB $\{\neg youngster, student, worker\}$
(3) KB $\{\neg adult, student, worker, unemployed\}$
(4) KB $\{\neg senior, pensioner, worker\}$
(5) KB $\{\neg student, child, youngster, adult\}$
(6) KB $\{\neg pensioner, senior\}$
(7) KB $\{\neg pensioner, \neg student\}$
(8) KB $\{\neg pensioner, \neg worker\}$

The goal is to show that the query ($child \rightarrow pensioner$) is not derivable from this set of clauses. Accordingly, we have to show that the clause set S containing (1) to (8) and $\{child\}$ and $\{\neg pensioner\}$ is satisfiable. To this end, we have to generate a set of Horn-clauses by deleting positive literals relevant to the query. We proceed again in a "query-oriented manner". First, consider the query-clause

[7] This is a corollary to Theorem 3 in [6].
[8] Without loss of generality, we assume that a query is a set of unit clauses (so that we can refrain from modifying the query).

{*child*}. Since no literal in S is reachable from *child*, no literal in S is relevant to this part of the query. Therefore, {*child*} does not influence the deletion of literals in S. This is different for the second query-clause, {¬*pensioner*}. The literal most relevant to this part of the query is obviously *pensioner* in (4); its distance to ¬*pensioner* is 1 rendering it a prime candidate for deletion. Then, turning (4) into a Horn-clause by removing the literal *pensioner* yields {¬*senior*, *worker*}. Now, there is no way left for refuting ¬*pensioner*. Since no clause apart from (4) allows for refuting ¬*pensioner*, we thus may remove positive literals (to a single one) from the remaining clauses in an arbitrary way. In any case, the resulting set of Horn-clauses will be satisfiable and so the query is disproved.

5 Discussion

In [13], a KB Σ is translated into two Horn-KBs Σ_{glb} and Σ_{lub} guaranteeing correct but incomplete and complete but incorrect inferences wrt Σ. Σ_{glb} is referred to as the *greatest lower bound* (GLB) of Σ since it is required to be a Horn-formula with a maximum set of models $\mathcal{M}(\Sigma_{glb}) \subseteq \mathcal{M}(\Sigma)$. Accordingly, the *lowest upper bound* (LUB) of Σ, Σ_{lub}, is a Horn-formula with a minimum set of models $\mathcal{M}(\Sigma_{lub}) \supseteq \mathcal{M}(\Sigma)$. The computation of such Horn-approximations is NP-hard. Hence the idea is to compute them "off-line" in order to allow for polynomial "on-line" approximations wrt Σ_{glb} and Σ_{lub}. In general, there are many GLBs whereas there is only one LUB of Σ. Also, in general, the size of Σ_{glb} is linear whereas the size of Σ_{lub} is exponential in the size of Σ. Nonetheless, the approach is very appealing in the case of application domains not involving modifications of the KB, as recently demonstrated in [7].

For a comparison with our approach, let us first deal with the correct (but incomplete) case. The exponential size of LUBs is a serious drawback: First, it leads to exponential query-answering relative to the original KB. Second, it may even be impossible to store the resulting LUB. This difficulty applies even to genuine examples like the one given in Section 3 on a person's reading habits. In this example, the LUB is of exponential size, as shown in [6]. In contrast, we have seen that our methods behaves polynomially in this example if the clausal representation of the KB and the negation of the query is renamable-Horn. In general, we control the exponential factor by limiting the number of cases generated by CASE-inferences. So in the unlimited case, our method corresponds to propositional entailment, which in turn may require the investigation of exponentially many cases. Finally, finding a LUB is a problem that cannot be parallelized [1] while case-analysis is (in general) perfectly suited for parallelization.

In the complete (but incorrect) case, our approach resembles the one taken in [6]. In both approaches, the original KB is approximated by regarding some of its Horn-variants. That is, a case of the KB corresponds to one of its lower bounds. So the difference rests on how these Horn-variants are chosen. In [6], all approximations are compared in an anytime manner. Initially, an arbitrary lower bound LB_1 is selected. Then, it is tested whether another lower bound LB_2 is entailed by LB_1. If $\mathrm{LB}_1 \models \mathrm{LB}_2$, then LB_1 is replaced by LB_2. This continues as long as no weaker lower bound can be found. Apart from the fact that comparing two lower bounds is NP-hard, this approach suffers from the problem that many lower bounds are not even comparable, ie. neither $LB_1 \models LB_2$ nor $LB_2 \models LB_1$ holds. Recall that this results in multiple GLBs. Moreover, it is then extremely difficult to select the right GLB being appropriately "weak" for each query in turn. Since we pursue an "on-line" approach, we cannot afford the aforementioned difficulties. Therefore, we leave it to our reachability measures to select the Horn-cases most relevant to the given query. The price we pay for avoiding computationally expensive comparisons of approximations is that the number of Horn-cases may exceed the number of GLBs. However, our approach is (in general) extremely flexible in adjusting the selection of approximations wrt to the given query.

Now, let us turn to the approach proposed in [2] which relies on two extended notions of interpretations (on literals[9]), *S-1-* and *S-3-interpretations*. For a subset S of the underlying alphabet L, an *S-1*-interpretation maps every letter in $L \setminus S$ and its negation into *false*; and an *S-3*-interpretation does not map both a letter in $L \setminus S$ and its negation into *false*. The letters in S are treated in the standard way. The two notions of entailment, $\models^1_S$ and $\models^3_S$, are defined in the usual way. It is instructive to verify that $\models^1_S$ is complete but incorrect and that $\models^3_S$ is correct but incomplete. Both entailment relations are parameterized by S and coincide with standard entailment in the case of $S = L$. For a KB Σ, this approach has in both cases an overall complexity of $O(|\Sigma| \times |S| \times 2^{|S|})$ [2].

As in our approach, the exponential factor in the approximation is controlled by the parameter of the approximation. In both approaches, this allows for a stepwise convergence to standard entailment. Apart from its appealing duality, we note that the approach of [2] is semantically well-founded. The extremely problematic point is the choice of S, the letters treated in the standard way. In fact, there are no means for determining an S appropriate for answering a particular query. In our approach, we solve a similar problem by reachability relations providing means for accessing the part of a KB relevant for answering a given query. This technique should also be applicable for choosing S. But even without reachability relations, we can identify classes of formulas for which our method is polynomial, which is not the case in [2]. This is due to the fact that our basic inference mechanism, URR is much more powerful than the ones corresponding to $\models^1_S$ and $\models^3_S$, in the case of $S = \emptyset$ or other insufficient subsets of L (cf. Section 3).

6 Conclusion

We have proposed an approach to approximate reasoning which relies on the combination of unit-resulting resolution and reasoning by cases; thereby taking advantage of well-known theorem-proving techniques. This has revealed that conventional theorem-proving techniques can indeed account for approximate reasoning. We have shown how approximate reasoning can be controlled by means of reachability relations — another well-known theorem-proving technique —

[9] That is, a standard interpretation maps each letter and its negation into opposite truth-values.

and their induced measures. This approach takes into account the interplay of KBs and queries and thus allows for query-sensitive approximate reasoning. Our method is orthogonal to the ones found in the literature inasmuch it is neither compiling KBs as [13] nor restricting satisfiability to certain subsets of the language as [2]. In particular, we have demonstrated that our method performs very well on the examples found in the literature. However, it remains future work to test ours as well as other approaches on large-scaled KBs.

Acknowledgments The first author is supported by DFG grant no. Bi 228/6-2. The second author is partially supported by CEC grant no. ERB4001GT922433.

References

1. M. Cadoli. Semantical and computational aspects of horn approximations. In *Proc. of IJCAI*, pages 39–45, 1993.
2. M. Cadoli and M. Schaerf. Approximate Entailment. In *Proc. of the Conf. of the Ital. Ass. for AI*, pages 68–77. Springer, 1991.
3. C. Chang. The decomposition principle for theorem proving systems. In *Allerton Conf. on Circuit and System Theory*, pages 20–28, U. of Illinois, 1972.
4. W. P. Dowling and J. P. Gallier. Linear-time algorithms for testing the satisfiability of propositional horn formulae. *Jour. of Logic Programming*, 1:267–284, 1984.
5. M. R. Garey and D. S. Johnson. *Computers and Intractability: A Guide to the Theory of NP-Completeness.* Freeman, 1979.
6. H. Kautz and B. Selman. Forming Concepts for Fast Inference. In *Proc. of AAAI*, 1992.
7. H. Kautz and B. Selman. An empirical evaluation of knowledge compilation. In *Proc. of AAAI*, 1994.
8. H. Levesque. Logic and the complexity of reasoning. *Jour. of Phil. Logic*, 17:355–389, 1988.
9. H. R. Lewis. Renaming a Set of Clauses as a Horn Set. *Jour. of the ACM*, 25(1):134–135, 1978.
10. W. McCune. Otter users' guide. Argonne Nat'l Laboratories, Argonne, 1988.
11. G. Neugebauer. From Horn Clauses to First Order Logic: A Graceful Ascent. Technical Report AIDA-92-21, TH Darmstadt, 1992.
12. J. A. Robinson. Automatic deduction with hyper-resolution. *Jour. of Computer Math.*, 1:227–234, 1965.
13. B. Selman and H. Kautz. Knowlede Compilation Using Horn Approximations. In *Proc. of AAAI*, 1991.
14. L. Wos, R. Overbeek, E. Lusk, and J. Boyle. *Automated Reasoning, Introduction and Applications.* Prentice-Hall, 1984.

Translation-Based Deduction Methods for Modal Logics

Olivier Gasquet, Andreas Herzig

IRIT - UPS
118 route de Narbonne
F-31062 Toulouse Cedex, France

Abstract. The aims of this paper are twofold: First, we review the automated deduction method for normal multi-modal logics which has shown to be the most general and fruitful, namely translation into first-order theories, and more precisely, the functional translation into equational theories with ordered sorts Second, to show how this method can be extended to monotonic modal logics through a translation from the latter into normal modal logics.

1 Introduction

Modal logics are useful tools to formalize a large number of natural reasonings. According to the desired use the formula $\Box p$ will be read "It is necessarily the case that p", "p will hold forever", "p is believed", "p is known", etc. Traditionnally, they allow to take into account only one concept at a time: One can only consider time (and not knowledge), or the beliefs of only one agent (and not of several ones), etc. In a more general manner multi-modal logics allow to reason about several theories in the same language: knowledge of several agents, knowledge changing through time or actions, etc. This is enabled by the use of formulas involving several modal operators. Moreover, in the multi-modal framework it is possible to state *interaction principles* between modal operators. For example: $[you][tomorrow]p \rightarrow [tomorrow][you]p$ can be read "what you know today about tomorrow, you will know it tomorrow". The expressive power of these logics make them important tools to formalize multi-agent universes where time, actions, intentions, knowledge and belief naturally interact. There are numerous examples of the use of these logics. First of all, dynamic logics [17] are multi-modal logics. [16], [15] and [23] have studied knowledge and belief (and time in the latter). Intentions are taken into account in [7]. In [21, 22] norms are formalized.

For the sake of clarity, we have restricted ourselves to propositional logics. The methods can be extended easily to the predicate case.

The paper is organized as follows: We introduce the translation method for mono-modal normal logics (serial and non-serial ones) in section 2, and for multi-modal normal logics in section 3. We show how these methods can be modified in order to enlarge their range of application: to non-normal logics without necessitation rules in section 4, and to non-normal logics without the K axiom (i.e. with a neighbourhood semantics) in section 5.

2 Mono-modal Normal Systems

A modal propositional language is defined from a classical language (built from a set of propositions p, q, r ... together with the connectives $\wedge, \vee, \neg, \rightarrow$) and a set M of *modal operators*. In the mono-modal case M is reduced to $\Box$. Then $\Diamond p$ is defined as an abreviation of $\neg\Box\neg p$. Thus $\Diamond(p \wedge \Box q)$ is a mono-modal propositional formula. Formulas are denoted by $A, B, \ldots$. In the multi-modal case, each of the modal operators is indexed by a parameter.

2.1 Syntax

Mono-modal normal systems have been intensively studied. The translation methods have been designed initially for them. Axiomatically, these systems are obtained by adding axioms to the smallest of them, namely the system K. This system contains all theorems and inference rules of classical propositional logic and:

- The axiom K: $\Box(p \rightarrow q) \rightarrow \Box p \rightarrow \Box q$
  This axiom is not always desirable in any interpretation of $\Box$. E.g. in an epistemic interpretation it means that knowledge is closed under material implication. The rejection of this principle leads to weaker systems (see section 5);
- The inference rule (called necessitation rule): from A infer $\Box A$.
  Again this principle can be rejected under some interpretations (section 4).

The standard normal systems are based on the system K plus some of the following axioms:

D: $\Box p \rightarrow \Diamond p$, what is obligatory is permitted (deontic interpretation), what is believed is envisageable (doxastic interpretation).
T: $\Box p \rightarrow p$, what is known is true (epistemic reading);
4: $\Box p \rightarrow \Box\Box p$, what is known is consciously known, or if I know something, I know that I know it (epistemic reading). It is called positive introspection axiom.

There are also 5: ($\Diamond p \rightarrow \Box\Diamond p$), B: ($p \rightarrow \Box\Diamond p$), De: ($\Diamond p \rightarrow \Diamond\Diamond p$), 2: ($\Diamond\Box p \rightarrow \Box\Diamond p$), G: ($\Box(\Box p \rightarrow p) \rightarrow \Box p$) that is used in the logic of provability, F: ($\Diamond p \rightarrow \Box p$) that can be read: "if p can be true then p will be true in any case" and allows to express determinism (e.g. of actions). The standard notation to denote a normal system that includes the axioms A1, ..., An is KA1..An.

2.2 Semantics

The standard semantics of these logics is possible worlds semantics. A *frame* is a pair (W, R) where W is a set of possible worlds, and R is a binary relation over W also called *accessibility relation*. In the sequel, we will sometimes use $(u, v) \in R$ or $v \in R(u)$ instead of uRv. A *model* is of the form $M = (W, R, m)$ where (W, R) is a frame and m is an *interpretation function* that associates with each proposition the set of all the possible worlds where it holds. Such a model is said to be based on the frame (W, R). An *interpretation* is made of a model and of a particular world w, the actual world, in which the formula to be interpreted is evaluated. The *satisfaction relation* is defined inductively by:

$M, w \models p$ iff $w \in m(p)$
$M, w \models \Box A$ iff for all $u \in R(w)$: $M, u \models A$
$M, w \models \Diamond A$ iff there exists $u \in R(w)$ such that $M, u \models A$

and is as usual for the classical connectives. In the rest of the paper will only give the truth condition concerning the modal operators.

2.3 Completeness

In order to develop efficient deduction methods for these logical systems, one must find semantical characterizations that allowing to identify theoremhood and validity. The best situation is when, for a given system L, there exists a class of frames that characterize it and this class can be defined in addition by a first-order formula (thus the system KT is characterized by the class of reflexive frames). When this is the case deduction reduces to deduction in a first-order theory for which everything that has been designed for the first-order framework can be used (theorem provers, strategies ....). There are the following characterizations (see [20]):

- K by the class of frames,
- KT by the class of reflexive frames,
- KD by the class of serial frames (each world has at least one successor),
- KDe by the class of dense frames ($\forall u, v(uRv \rightarrow \exists w(uRw$ and $wRv))$)

In fact, there are more general results: Properties can be associated with each of $D, T, 4, 5, B, 2, F, De$ (resp. seriality, reflexivity, transitivity, euclideanity, symmetry, confluence [1], functionality, density) such that if $L = KA1 \ldots An$, with $A1, \ldots, An \in \{D, T, \ldots\}$ then L is characterized by the class of frames satisfying the properties associated with each of the Ai's. The most general results are due to [28].

2.4 Translation

We are first going to restrict ourselves to the case of those systems that can be translated easily. Then we will extend it to all mono-modal systems.

We are going to use first-order sorted theories. Variables are indexed with a sort symbols that specify on which domain they are interpreted. Thus, $x\colon s$ denotes that x is of sort s. A function symbol having a *profile* $f\colon s_1, \ldots, s_n \rightarrow s$ is interpreted by a function $Ds_1 \times \ldots \times Ds_n \rightarrow Ds$, where each of the Ds's is the domain associated to the sort s. For more details we refer to [29].

Systems with Axiom D In these lines we are going to restrict ourselves to serial systems which contain the axiom of seriality D and whose semantical characterizations involve serial accessibility relations. The central idea of the functional translation is that any serial relation R can be represented by a set F of *total functions* in the sense that $(u, v) \in R$ iff $\exists f \in F\colon f(u) = v$. These functions are called *access functions*. Thus

---
[1] A relation R is confluent iff $\forall t, u, v\colon tRu$ and $tRv \rightarrow \exists w\colon uRw$ and vRw.

with the property of reflexivity there is associated the fact that the identity function ($f(x) = x$) is an access function. This idea has been first explored in [1], [2, 3], [10, 11], [18], [24, 25].

The first idea is to look for immediate correspondances between relational and functional properties. Thus reflexivity is expressed by $\forall u \exists f \in F: f(u) = u$. The other properties can be expressed in the same way. Then one can define a translation of modal formulas (we suppose that formulas are in negative normal form, the negation symbol occuring only at the atom level) :

For literals: $tr(L, u) = L(u)$ where L is a literal.

The intuition is that u represents the world where L is to be interpreted. Formally, with each proposition p we associate a unary predicate p. As we will see below, u is obtained by successive applications of access functions to the initial world.

$tr(\Box A, u) = \forall f: F\, tr(A, f(u))$ where $f: F$ denotes that f is of sort F;

$tr(\Diamond A, u) = \exists f: F\, tr(A, f(u))$.

tr is homomorphic with respect to the other connectives. (This will always be the case in the sequel).

The translation of a formula is given by:

$TR(A) = \exists w: W tr(A, w)$

Note that we need several sort symbols: W denoting possible worlds, F denoting access functions (and moreovec D denoting individuals in the predicate case). The first-order theories that are the target of the translation additionally contain axioms expressing properties of functions, namely associativity of composition and existence of a neutral element.

For example the formula $p \rightarrow \Diamond p$ is a theorem of the system KT iff, after translation and skolemization, the set:

$S_1 = \{p(w_0), \neg p(g(w_0))\}$ is inconsistent in the first-order theory that contains (among others): $\forall u: W \exists f: F f(u) = u$. According to the the theory there exists a function f such that $f(w_0) = w_0$. If we instantiate g by f we get $S_2 = \{p(w_0), \neg p(f(w_0))\}$, i.e. $\{p(w_0), \neg p(w_0)\}$. Therefore S_1 is inconsistent[2].

It turns out the sort W is not necessary here, and that it is sufficient to take into account the path (in terms of function composition) between the initial world and the actual one ([2, 3], [10, 11], [14], [25]). Thus the formula $p \wedge \Box\neg p$ is translated into $p(Id) \wedge \forall g: F \neg p(g \bullet Id)$, where Id denotes the identity function and $\bullet$ denotes inverse function composition (i.e. $f \circ g = g \bullet f$). The equational axioms are modified in the same way, for example that corresponding to reflexivity is: $\exists f: F\, f = Id$: the identity function *is* an access function. This was not guaranteed in the theory $\forall u: W\, \exists f \in F f(u) = u$. In [1], [9], [18], [26, 24]) there have been designed optimized skolemization procedures that take advantage of the fact that translated formulas have a particular structure. We will not investigate this point here.

Systems Without the Axiom D The above method only applies to logics having the axiom $\Box A \rightarrow \Diamond A$, which corresponds to the property of seriality of the accessibility relation. This is not the case if there is a world w without successors ($R(w) = \emptyset$).

[2] This refutation could have been obtained using resolution and paramodulation.

Then no access function is defined at w. But it is difficult to handle partial functions in classical logic (which is the target logic).

Several solutions have been given. In [18], [10, 11] translations are defined into an logic that lays between classical logic and modal logic, and a resolution principle gas been given for it. The translation given in [24] consists in adding in the translation process a unary predicate *end* that handles the fact that the actual world is a world without successors. The main clause of the translation is

$$tr(\Box A, w) = \forall f{:}F\ end(f(w)) \vee tr(A, f(w))$$

This translation adds one occurrence of the predicate *end* at each resolution step.
We present below a new translation for these logics. Each logic $KA1\ldots An$ is translated into a logics $KDA1\ldots An$. If it is followed by the serial functional translation, we obtain that of [24]. Such a translation is more elegant and simpler from the point of view of soundness and completeness proofs. It is defined by

- $TR(p) = p$
- $TR(\Box A) = end \vee \Box TR(A)$
- $TR(\Diamond A) = \neg end \wedge \Diamond TR(A)$.

where *end* is a propositional variable that is added to the initial language. For example, the formula $\Box p \wedge \Box\neg p$ (which is K-consistent), is translated into $TR(\Box p \wedge \Box\neg p) = (end \vee \Box p) \wedge (end \vee \Box\neg p)$ which is KD-consistent. (It is equivalent to the formula *end*.)

3 Multi-Modal Normal Systems

3.1 Syntax

Just as mono-modal systems, multi-modal systems are obtained by adding axioms to some basic system which is K_M (where M is the set of modal operators of the language). In the sequel, we will omit the subscript M. In addition to its classical basis, this system contains one K axiom for each modal operator $[a]$ (denoted by $K([a])$) and the necessitation rule: from A infer $[a]A$.

Each of the axioms from section 2 can be added, e.g. $D(a) : [a]p \rightarrow \langle a\rangle p$, and so on for the other axioms.

The supplementary feature of the multi-modal framework is that there are axioms involving *several* modal operators such as $[b]p \rightarrow [a]p$. Other interesting axioms are:

- Inclusion(a,b): $[a]p \rightarrow [b]p$: "b knows what a knows".
- Persistence: $[a][tomorrow]p \rightarrow [tomorrow][a]p$: "If a knows something about tomorrow, then tomorrow a will know (remember) it.
- Knowledge of the law: $[obligatory]p \rightarrow [a][obligatory]p$: "If p is obligatory, a knows it is so", or "Everybody is supposed to know the law"

3.2 Semantics

The semantics of these systems is a generalization of the usual one. Frames are pairs (W, R) where R is no longer a relation but a set R_a, R_b, ... of relations: one for each modal operator. The satisfiability relation is modified as follows:

$M, w \models [a]A$ iff for all $u \in R_a(w)$: $M, u \models A$;

$M, w \models \langle a \rangle A$ iff there exists $u \in R_a(w)$: $M, u \models A$.

3.3 Completeness

For the propositional case, the largest overview has been given in [4, 5]. In particular, axioms of the form $[a_1] \ldots [a_n]p \rightarrow [b_1] \ldots [b_m]p$ correspond to the property: $R_{a_1} \bullet \ldots \bullet R_{a_n} \subseteq R_{b_1} \bullet \ldots \bullet R_{b_m}$. (The operation $\bullet$ is the relation product defined by $(u, v) \in R \bullet S$ iff $\exists w$: uRw and wSv). By convention, if $i = 0$ then $[a_1] \ldots [a_n]p$ reduces to p, and $R_{a_1} \bullet \ldots \bullet R_{a_n} = I = \{(u, u): u \in W\}$.

3.4 Translation

The translation method is very close to the one for the mono-modal case, except that several access functions sets are used instead of just one. Consequently the target (first-order) logic must have one sort symbol for each modal operator, plus the sort W.

Unification algorithms have been given in [8], [13] for several multi-modal systems characterized by a first-order frame semantics. In these papers, as well as in [12] and [14], inclusion axioms of the form $[a]p \rightarrow [b]p$, are handled in the target logic by a subsort declaration ($b \leq a$). This means that a variable of sort a can be instantiated by a variable of sort b.

Example: The formula $[a]p \rightarrow \langle b \rangle p$ is first negated: $[a]p \wedge [b]\neg p$ and then translated into: $\exists w : W(\forall f : a(p(f(w))) \wedge \forall g : b(\neg p(g(w))))$. After a skolemization step, we obtain two clauses $p(f(w_0))$ and $\neg p(g(w_0))$, where f and g are variables of sort a and b respectively. In a system like $KD(a).D(b).Inclusion(a, b)$, b is a subsort of the sort a. Hence, f and g can be unified and the resolution rule applies.
The optimizations that have been mentioned in 2.4 can be used in the multi-modal case as well. For the sake of clarity we will now restrict ourselves to the mono-modal case.

4 Non-Normal Systems Without Necessitation

All systems considered so far contained the necessitation rule: from A infer $\Box A$. This principle is natural in many reading of the $\Box$ operator (such as the temporal one). It has been criticized in the case of a deontic [21, 22] and epistemic reading [16] because it leads to omniscience [3].

4.1 Semantics

The semantics must guarantee that $\Box\top$ is not valid.
Formally, we have the same models as in the case of normal systems, except that for

[3] Another aspect of the omniscience principle is that knowledge is closed under material implication and is expressed by the axiom K. Systems without it will be treated in the next section.

some $w \in W$, the set of successors of w: $R(w)$ can be undefined (which is different from being empty). In other words, viewing a relation as a function from W to 2^W, such functions are partial here (while they are total in the case of normal systems). The satisfiability relation becomes:

$M, w \models \Box A$ iff $R(w)$ is defined and for all $u \in R(w)$: $M, u \models A$

$M, w \models \Diamond A$ iff $R(w)$ is undefined or there exists $u \in R(w)$: $M, u \models A$.

4.2 Translation

We translate each non-normal system in a normal one. The translation is defined as follows:

- $TR(\Box A) = normal \wedge \Box TR(A)$
- $TR(\Diamond A) = \neg normal \vee \Diamond TR(A)$

This translation preserves satisfiability and validity.

The most famous non-normal systems are Lewis' $S2$ and $S3$ [19]. Their semantics must satisfy the following supplementary requirements:

- R is semi-reflexive: forall w, if $R(w)$ is defined then wRw;
- R is transitive in the case of the system $S3$;
- Validity of formulas is checked only in normal worlds.

The first condition means that both systems satisfy the axiom $\Box p \rightarrow p$, and the second condition means that $S3$ validates the axiom $\Box p \rightarrow \Box\Box p$. The last condition means that both validate the rule: from A infer $\Box A$ *if A is classical*. This last rule can be handled by an extension of the above translation:

$TR'(A) = normal \wedge TR(A)$

that preserves satisfiability and allows to translate Lewis' $S2$ and $S3$ systems into KT and $KT4$ respectively. Thus the formula $\Diamond(p \vee \neg p)$ which is unsatisfiable w. r. t. $S2$ and $S3$ is translated into: $normal \wedge (\neg normal \vee (\Diamond(p \wedge \neg p))$ which is inconsistent w. r. t. KT (resp. $KT4$). (The latter can be proved using the serial translation). Hence $\neg\Diamond(p \wedge \neg p)$ is a theorem both $S2$ and $S3$.

5 Non-Normal Systems Without the Axiom K

We are going to treat modal logics that do not contain the axiom K: $(\Box(p \rightarrow q) \wedge \Box p) \rightarrow \Box q$. In particular, the corresponding semantics neither validates $\Box(p \wedge q) \rightarrow \Box p \wedge \Box q$ nor the converse. These formulas have been criticized by several philosophers in the case of a deontic [21, 22], epistemic or doxastic [6] reading. In the logics of 'local reasoning' of [16], $\Box p$ is read 'p is explicitely believed'. In such an interpretation one does not want to derive $\Box p \wedge \Box q \rightarrow \Box(p \wedge q)$. This corresponds to the suggestion that the knowledge of an agent is organized as a set of 'frames of mind' which can even contain contradictory knowledge. In [27] it has been shown that the multi-modal framework can be used for representation of uncertainty, and in particular that it provides an approximation of probabilities: $[a]A$ can be read "probability of A is at least a". Clearly, in this case we do not want to derive $[a]A \wedge [a]B \rightarrow [a](A \wedge B)$. On the other hand, the converse $[a](A \wedge B) \rightarrow [a]A \wedge [a]B$ should be valid.

5.1 Semantics

For these non-normal logics without K axiom there have been designed 'neighbourhood semantics'. With [6] we distinguish two types of it.

Classical modal logics have semantics in terms of minimal models: a *minimal frame* is a pair (W, N) where W is a set of possible worlds and N is an mapping from W to 2^{2^W} associating worlds with families of sets of worlds. A *minimal model* is a triple (W, N, m) where (W, N) is a minimal frame and m is a valuation function as ususal. For a given world w, $N(w)$ is called the neighbourhood of w.

The satisfiability relation is defined as:

$M, w \models \Box A$ iff $\exists U \in N(w)$ such that $\forall v$: $(v \in U \Leftrightarrow M, u \models A)$.

The basic classical modal logic is denoted by E. Extensions of E can be obtained BY restricting N. For example, if w is in each of the sets of $N(w)$ then the axiom $T = \Box p \rightarrow p$ is validated. The resulting logic is called ET.

Monotonic modal logics have the same semantics except that equivalence is replaced by implication:

$M, w \models \Box A$ iff $\exists U \in N(w)$ such that $\forall v$: $(v \in U \Rightarrow M, u \models A)$.

The basic monotonic modal logic is called EM. Monotonic modal logics have been used to model implicit and explicit belief [16]. They validate the monotonicity axiom $M = \Box(p \wedge q) \rightarrow \Box p \wedge \Box q$. Note that the converse of M is not valid.

5.2 Translation for Monotonic Modal Logics

We are going to give a translation from monotonic modal logics to normal multi-modal logics. The target logic is a bi-modal one with a set of modal operators $M = \{[1], [2]\}$. The translation is defined by:

$TR(\Box A) = \langle 1 \rangle [2] TR(A)$

$TR(\Diamond A) = [1] \langle 2 \rangle TR(A)$

The intuition behind this translation is the following: To each neighbourhood of the minimal frame we associate an new world that can be accessed via a relation R_1, and from this world each of the worlds of this neighbourhood are accessible via R_2. Thus in the truth condition for $\Box$, "there exists $U \in N(w)$" corresponds to "there exists $U \in R_1(w)$", and $(v \in U \Rightarrow M, u \models A)$ corresponds to $v \in R_2(U) \Rightarrow M, u \models A$.

6 Conclusion

We have presented several translation methods that have recently been designed for multi-modal logics. We have extended them to several non-normal logics which are important in the domain of reasoning on knowledge, belief and obligations.

References

1. Y. Auffray. *Résolution modale et logique des chemins*. PhD thesis, Université de Caen, France, 1989.

2. Y. Auffray and P. Enjalbert. Modal theorem proving: An equational viewpoint. In *Int. Joint Conf. on AI*, 1989.
3. Y Auffray and P. Enjalbert. Modal theorem proving: An equational viewpoint. *Journal of Logic and Computation*, 1990.
4. L. Catach. Normal multi-modal logics. In *Proc. Nat. Conf. on AI (AAAI'88)*, pages 491–495, 1988.
5. L. Catach. *Les logiques multi-modales*. PhD thesis, Université Paris VI, France, 1989.
6. B. Chellas. *Modal logic: an introduction*. Cambridge University Press, 1980.
7. Ph. Cohen and H. Levesque. Intention = choice + commitment. In *Proc. of AAAI*, Seattle, 1987.
8. F. Debart, P. Enjalbert, and M. Lescot. Multi-modal deduction using equational and order-sorted logic. In M. Okada and S. Kaplan, editors, *2nd Conf. on Conditional Rewriting Systems*, LNCS. Springer Verlag, 1990.
9. L. Fariñas del Cerro and A. Herzig. Linear modal deductions. In *9th Int. Conf. on Automated Deduction*, volume 310 of *LNCS*, pages 500—516. Springer Verlag, 1988.
10. L. Fariñas del Cerro and A. Herzig. *Machine Learning, Meta Reasoning and Logics*, chapter Deterministic Modal Logics for Automated Deduction. Kluwer Academic Publishers, 1989.
11. L. Fariñas del Cerro and A. Herzig. Deterministic modal logic for automated deduction. In *9th European Conference on Artificial Intelligence*, 1990.
12. L. Fariñas del Cerro and A. Herzig. *Handbook of Logic in Artificial Intelligence*, chapter Automated Deduction for Epistemic and Temporal Logics. Oxford University Press, 1995.
13. O. Gasquet. A unification algorithm for multimodal logics with persistence axiom. In *5th International Workshop on Unification (UNIF'91)*, Barbizon, France, 1991.
14. O. Gasquet. *Applied Logic: How, What and Why*, chapter imization of Deduction for Multi-Modal Logics. Kluwer Academic Publishers, 1995.
15. J. Halpern. Reasoning about knowledge: an overview. In J. Halpern, editor, *Conf. on Theoretical Aspects of Reasoning about Knowledge*, pages 1–17. Morgan Kauffmann, 1986.
16. J. Halpern and Y. Moses. A guide to the modal logic of knowledge and belief. In *Proc. of IJCAI*, 1985.
17. D. Harel. *Handbook of Philosophical Logic III*, chapter Dynamic Logic". Reidel Publishing Company, 1986.
18. A. Herzig. *Déduction automatique en logique modale et algorithmes d'unification*. PhD thesis, Université Paul Sabatier, Toulouse, France, 1989.
19. G. Hughes and M. J. Cresswell. *An Introduction to Modal Logic*. Methuen & Co. Ltd, 1968.
20. G. Hughes and M. J. Cresswell. *A Companion to Modal Logic*. Methuen & Co. Ltd, 1984.
21. A. J. I. Jones and I. Pörn. Ideality, sub-ideality and deontic logic. *Synthese*, 65, 1985.
22. A. J. I. Jones and I. Pörn. 'ought and must'. *Synthese*, 66, 1986.
23. S. Kraus and D. Lehmann. *Automata, Languages and Programing*, volume 226 of *Lecture Notes in Computer Science*, chapter Knowledge, belief and time, pages 186–195. Springer-Verlag, 1986.
24. H. J. Ohlbach. A resolution calculus for modal logics. In *9th Int. Conf. on Automated Deduction*, volume 310 of *LNCS*. Springer-Verlag, 1988.
25. H. J. Ohlbach. Semantics-based translation method for modal logics. *Journal of Logic and Computation*, 1(5):691–746, 1991.
26. H. J. Ohlbach. imized translation of multi modal logic into predicate logic. In *Proc. of International Conference on Logic Programming and Automated Reasoning (LPAR'93)*, LNAI. Springer-Verlag, 1993.
27. H. J. Ohlbach and A. Herzig. Parameter structures for parametrized modal operators. In *Proc. of Int. Joint Conf. on Artificial Intelligence*, 1991.

28. H. Sahlqvist. Completeness and correspondence in the first and second order semantics for modal logics. In S. Kanger, editor, *Proc. 3rd Scandinavian Logic Symposium 1973*, volume 82 of *Studies in logic*. North-Holland, 1975.
29. M. Schmidt-Schauß. *Computational Aspects of an Order-Sorted Logic with Term Declarations*. LNAI. Springer-Verlag, 1989.

A System of Knowledge Representation Based on Formulae of Predicate Calculus Whose Variables Are Annotated By Expressions of A "Fuzzy" Terminological Logic

Rita Maria da Silva[1], Antônio Eduardo C.Pereira[2], Márcio Andrade Netto[3]

[1]IRIT- Université Paul Sabatier- Toulouse France
e-mail: rita@irit.fr
UFU- Universidade Federal de Uberlândia
Dept. Informática - UFU
Av. Universitária S/N- Campus Santa Mônica
38400-902- Uberlândia, M.G., Brazil
[2]UFU Dept.Engenharia Elétrica
Av. Universitária S/N- Campus Santa Mônica
38400-902- Uberlândia, M.G., Brazil
[3]Dept. Engenharia Elétrica - UNICAMP
C.P. 6101 - 13081-000- Campinas, S.P., Brazil

Abstract. In this paper one presents a system for introducing assertions in a knowledge base (kb). These assertions are represented as formulae of Predicate Calculus (PC) whose variables are annotated by concepts of Terminological Logic (TL). The system answers questions by using methods of inference both from PC and TL. The terminological language used is characterized by a fuzzy treatment of concepts. The main contribuitions are: a unification algorithm which closes the semantic gap between PC and TL; introduction of uncertainty in subsumption; use of subsumption to simplify the tracing of the proof; use of Partial Evaluation to link assertional and terminological reasoning.

1. Introduction

This paper originates from the LESD project (financed by CNRS) which aims to build a system to store assertions about software specification in a kb. The assertions and queries are expressed in a restricted form of natural language related to Space Sciences [19]. A syntatic and semantic parser translates every assertion into a form of PC. Among the problems related to the storage of assertions and to information retrieval, one can emphasize:

a) *Expressivity of the answer*. This problem is related to pragmatics. The answer must satisfy some expectation of the user. One kind of failure of expressivity is giving too little information in the answer. The user will also be very insatisfied if the system produces a long tracing of the proof.

b) *Subsumption of concepts.* A method to achieve expressivity is to focus the information provided to the user on a set of concepts describing objects and properties of objects. If a litteral is in the set, this litteral may be part of the tracing. Later on the authors address the problem of composing the tracing. For now, it is enough to know that they have used TL to simplify the result of composition. For example, if the theorem prover discover that X is *human & man & ¬woman*, it may just answer that X is a *man*. To better understand this kind of simplification, the reader may consult texts about TL [4,14, 16]. If one uses TL to simplify concepts, a problem arises when there is neither the complete subsumption as exists between the concepts *human* and *man*, neither a complete disjunction as exists between *human & man* and *human & woman*. Therefore, should one consider the simple automatic device which controls the altitude of the spacecraft a computer? One proposes a subsumption algorithm which treats the primitive concepts by means of a method of dealing with uncertainty.

c) *Composing the answer by Explanation Based Learning (EBL) (or by Partial Evaluation* [11]). The answer given to the user is a partial evaluation [7] of the query in relation to the kb. A knowledge representation system (krs) is formed by a set Fs of formulae and by a set of inference rules. There are two recursive subsets of Fs, D and P, such that Fs = D ∪ P. All formulae belonging to P are axioms. The formulae of D are considered data. The set of inference rules is called semantics of the krs. Among the inference rules, there is one which will specify how and when the others will be used. This special rule is called computational rule. Let's suppose that D and P are described by some set of syntatical rules (this is always possible because D and P are recursive). Let Dp and Pp be the sets of productions describing D and P, respectively. A formula belongs only to D if it is described by Dp - (Dp ∩ Pp). A formula has components only in P if one can describe it by using Pp - (Dp ∩ Pp). Let's consider the knowledge representation system krs(D, P, semantics). If an inference f has components only in P, it is partial. We call inference any deduction from an input data called query. The set chosen to express the answer is recognized by a subset of Pp - (Dp ∩ Pp), i.e., the answer is a partial evaluation. The subset of Pp - (Dp ∩ Pp) was chosen in such a way as to define a terminological language. This means that the user will be able to use methods of TL to analyse the answer (subsumption and classification).

d) *Insertion in the knowledge base.* Let's suppose that we want to insert the formula:

(i) ∀(X, computer(X) & on-board(X), controls(ioi-gs, X)).

If someone else has already stored the formula:

(ii) ∀(X, on-board(X), controls(ioi-gs, X)),

there is no need to insert the formula (i) because formula (ii) subsumes it.

The algorithms to verify subsumption are very complex and it is impracticable to use them with a general well formulated formula. For this reason, one restricts their use to those parts of the formulae belonging to a very restricted terminological language.

2. Representation of Knowledge

The domain we are working with is software specification for spacecraft design. There are situations when the designers find it best to work with short sentences which they call *requirements*. The problem is to transform these requiremets into a form of knowledge representation suitable to storage and retrieval [19]. Next one shows an example of requirement:

E1: "The ioi-gs shall control the computers on-board the space vehicle".

This analysis is made by a British parser called Alvey-GDE [10]. This parser translates the requirement given in *E1* into the following representation:

$$
\begin{aligned}
F1\text{:}\quad & \tau(X1, \text{ioi-gs}(X1) \\
& \quad \tau(x2, \tau(X3, \text{space-vehicle}(X3) \\
& \qquad \text{computer}(X2)\ \& \\
& \qquad \text{on-board}(X2, X3)) \\
& \quad \text{control}(X1,X2)))
\end{aligned}
$$

Sentences like *F1* will be asserted in a kb. In order to do this, they will be skolemized [1]. Analougously, a query posed by the user is first transformed into an expression similar to *F1* that has also to be skolemized. One implemented the retrieval information system in such a way that the answer corresponds to a partial evaluation of the query with respect to the second argument of the quantifiers. This means that the answer belongs to a subset of $\{Dp - (Dp \cap Pp)\}$. Therefore, the second argument of the quantifiers must also belong to a subset of $\{Dp - (Dp \cap Pp)\}$.

The subset of $\{Dp - (Dp \cap Pp)\}$ used to express the answer is chosen in such a way that it can be easily expressed as formulae of TL. Therefore the user is able to use the various resources of TL [14] to deal with that answer,e.g., subsumption and classification. It is usual in TL to work with two distinct languages, that is, a terminological language and an assertional language [4, 14]. One uses Partial Evaluation to establish a link between assertional and terminological reasoning. The terminological language used is characterized by a fuzzy treatment of the primitive concepts. In classical TL, a primitive concept c1 subsumes a primitive concept c2 if c1 = c2. The subsumption algorithm used in the project is such that c1 subsumes c2 if the extension of c1 includes the extension of c2. The notion of extension is defined in texts about TL. The extension used in the project is such that $c1 \subseteq c2$ iff $\forall X\ (\mu_{c1(X)} \leq \mu_{c2(X)})$, where $\mu_{c(X)}$ is the membership degree of $X \in c$ [6].

The authors believe that the combination of TL and Fuzzy Logic has helped them to present a solution to the problem of Hilbertian reference in Natural language. This solution is acceptable in the chosen domain. The Hilbertian reference problem is the following: a definite article may be used to point to an object about which one has mentioned before. Given an object X, introduced by the Hilbertian quantifier $\tau(X,\_,\_)$, the system must find its reference in the kb. It will, then, compare X with the other objects in the base. This comparison is made by using the notions of subsumption and classification of TL. The subsumption becomes considerably more appopriate to Natural Language processing due to the introduction of ideas from Fuzzy Logic. Besides, Explanation Based techniques [2, 11, 13] allow us to perform the classification without having to compare X with every object in the kb. As one will see later, it will be necessary to compare X only with objects which unify with it during the partial evaluation.

The skolemization process is performed in two steps: in the first step, the expressions corresponding to the second arguments of the quantifiers are transformed into TL expressions (free of quantifiers expressions composed of concepts and roles), as shown in *F2* expression below:

F2: τ(x1, ioi-gs(x1), τ(x2, computer(x2), τ(x3, space-vehicle(x3), on-board(x2,x3) & control(x1, x2)))).

In the second and final step of skolemization, the second argument of the quantifiers (TL expressions) are written as annotations to the variable introduced in the first argument of the quantifier, as one can see in the following *F3* expression:

F3: control(X/ τ(ioi-gs), X3/ τ(computer & on-board(space-vehicle))).

Every argument of a litteral has the form *X/exp*, where *X* is a variable or constant annotated by *exp* and *exp* is an expression of TL which specifies the set to which *X* belongs (Nebel,[14]). Besides the semantics given by Nebel, the expressions of TL can also be considered as predicates which the variables must satisfy. In this case, they will receive the following semantics (Vilain, [20]):

Expression	*Semantics*
x	$\lceil x \rceil$
c1 & c2	$\lambda X(\lceil c1 \rceil X) \,\&\, \lceil c2 \rceil X)$
r(c)	$\lambda X\, \forall Y(\lceil r \rceil X\, Y) \Rightarrow (\lceil c \rceil Y)$
τ(c)	$\lambda X\, \Vert\{X \mid (\lceil c \rceil X)\}\Vert = 1$

To simplify the discussion, the concept τ(c) will not be considered from now on. This concept τ was introduced to allow the skolemization of the Hilbertian quantifier and of other quantifiers related to cardinality of sets (at least, at most, the majority etc). Nebel shows how to treat *at-least* and *at-most*. One should deal with τ(c) in a similar way.

3. Semantic Unification

The annotations of the variables are expressions of TL which can be considered as predicates. These predicates should be true for the variable under consideration. If this is the case, the result of proving formula *F1* is the same as proving formula *F3*. After all, verifying *r(X1/c1, X2/c2, X3/c3, ...) & ...* making sure that *(⌈c1⌉ X1), (⌈c2 ⌉ x2), (⌈ c3 ⌉ X3...) & ...* are all true is the same thing as verifying *r(X1, X2, X3, ...) & c1(X) & c2(X2) & c3(X3) &....*

The first approach, however, can be performed with greater efficiency since it will not use all the resources of a PC theorem prover. Besides, one is going to collect the concepts related to every variable. The collected concepts will restrict the variable of the query and this restriction, by itself, may be the desired answer. For example, if one queries:

?- control(X/ioi-gs, Y/C)

and get the answer *Y/computer & on-board(space-vehicle),* one knows that the ioi-gs controls all computers on-board the space-vehicle. Although one can not get a listing of the computers which the ioi-gs controls, one is able to know that they must satisfy the predicates: *λ(X)computer(X)* and *λ(Y)on-board(Y, space-vehicle).* The collection of concepts is made during unification. The idea is that if a variable X/cx unifies with a variable Y/cy, the unification between X and Y will be performed like in classical unification. This means that all results of classical unification will be true of the semantic algorithm (complexity, correction etc). The process of storing the concepts *cx* and *cy* will be concurrent with the unification. This process will be carried out in such a way that one can discover, at the end of proof, whether the concept of the variable X will be cx & cy. In other words, X, after the unification, will suffer the restrictions of Y as well as its own restrictions. We can easily prove it. In fact, *r1(..., X/cx, ...)* is equivalent to *r1(..., X, ...) & (⌈cx ⌉ X)* and *r2(..., Y/cy, ...)* is equivalent to *r2(..., Y, ...) & (⌈cy⌉ Y).* After the resolution and unification, we have *X=Y & (⌈cx ⌉ X) & (⌈cy ⌉ Y).* Therefore, by the semantics of composition of two concepts, X will suffer the restriction *cx & cy.* Of course one must simplify the resulting formula, e.g., *cx & cy.* In the simplification process, if *cx = c1x & c2x & c3x* and *c2x* subsumes *cy*, then one can use *c1x & c3x & cy* instead of *c1x & c2x & c3x & cy .* In presenting the unification algorithm, the authors follow Robinson very closely. Definitions:

(D1) Entity is a data structure denoted by X/c with the following selectors:
identifier(X/c) = X
concept(X/c) = c
where identifiers may be variables or constants.

(D2) Binding is a data structure denoted by *X/cx = Y/cy*, where X is a variable. The selectors are: *car(X/cx = Y/cy),* that produces *X/cx* and *cdr(X/cx = Y/cy),* that produces *Y/cy.*

(D3) Environment is a set of bindings whose variables of the car are different from each other. Given a variable X and an environment E, the function *(assoc X E)* will give a binding whose variable of the car is X.

(D4) Immediate value of X in E is defined by: iv ≡ λX cdr(assoc(X, E)).

(D5) Composition of concepts c1 and c2 is a simplification of *c1 & c2*.

(D6) If *c1 ≡ cl1 & cm1 & cr1* and *cm1* subsumes *c2*, then *cl1 & cr1 & c2* is a simplification of *c1 & c2*. Another simplification of *c1 & c2* is *c1 & c2* itself.

(D7) Extension of c is the set {X | ⌈c⌉ (X)}. The extension of c is written as c^e.

(D8) c subsumes d if $c^e \supseteq d^e$. With this definition of subsumption, one can prove that a simplification of *c1 & c2* is equivalent to *c1 & c2* itself.

(D9) The following subsumption algorithm is correct but not complete:

```
subsumes ≡ ( Y λ sub λ C1 λ C2
  (a) if C1 = X & Z then
        sub(X, C2) ∧ sub(Z, C2)
      else
        (b) if C2 = U & W then
              sub(C1, U) ∨ sub(C1, W)
            else
              (c) if primitive(C1) ∧ primitive(C2) then
                    ∀X μC2(X) ≤ μC1(X)
                  else
                    (d) if C1 = R(D1) ∧ C2 = R(D2) then
                          sub(D1, D2)
                        else
                          (e) if C1 = R(D1) ∧ C2 = U & W then
                                sub(C1, U) ∨ sub(C1, W)
                              else false)
```

One should remember that Y is the least fixed point operator. It is easy to prove that the algorithm is correct but not complete. The proof that it stops follows from induction over the length of a flattened TL expression. On the other hand, one can easily verify that clauses (a), (b) (c) (d) and (e) satisfy the definition of subsumption.

(D10) Final value of an entity X is defined by:

```
fvalue ≡ (Y λ fv λ X λ E
  let V/CV = assoc(identifier(X), E) in
  if V/CV = ⊥ then X
  else
    fv(V/combine(concept(X), CV), E),
```

where "⊥" indicates "bottom".

This algorithm should produce VX/CX as final value of X, where VX is the final value as proposed by Robinson and CX is a combination of the concepts of all variables directly or indirectly bound to X.

(D11) Let G and H be two flat litterals (one makes this restriction as one only needs this kind of litterals; however, there is no difficulty in extending the ideas presented in this paper to litterals having functions as arguments). Unification of G and H in the environment E is finding the most general extension E' so that G'= H', where G' and H' are the dereferencing of G and H in E. To dereference a flat litteral, one needs only to substitute the final value for every argument. One wants a unification algorithm that stores in the environment all concepts that annotate the variables of the litterals one tries to unify. The concepts must be organized in the environment in such a way as to allow the final value algorithm of a classical theorem prover to collect them. By doing so, one is able to assure that our final theorem prover will be a partial evaluator of an incomplete hypothesis (or query) with respect to the kb. The partial evaluator answers questions by collecting in the environment (built from the kb and from the query) all concepts that correspond to appropriate constraints to the variable of the hypothesis (i.e., the variable of the query). As one has already mentioned, these concepts belong to the set {Dp - (Dp ∩ Pp)}. In other words, answering questions in this paper corresponds to "specialize" as much as possible the incomplete hypothesis (query) from information stored in the kb. Next, one shows the unification algorithm:

```
unify ≡ λ G λ H λ E
  if E = ⊥ then E
  else
     if  G = ¬Gp ∧ H = ¬Hp ∧ functor(Gp) =
        functor(Hp) ∧ arity(Gp) = arity(Hp) then
         unifyargs(Gp, Hp, 1, E)
     else
         if functor(G) = functor(H) ∧ arity(G) =
            arity(H) then
              unifyargs(G, H, 1, E)
         else ⊥


unifyargs ≡ (Y λ uniargs λ G λ H λ I λ E
  if I > last_argument_of(G) then
   E
  else
   if E = ⊥ then
     ⊥
   else
     uniargs(G,H, I+1, unifyarg(fvalue(argument I of G, E), fvalue(argument I of H,
            E),E)))
```

```
unifyarg ≡ λX λZ λE
 if X = ⊥ or Z = ⊥ then ⊥
 else
  let VX/CX = X and W/CW= Z in
  if occurs(X,Z) then
  ⊥
  else
   if variablep(VX) then
    [X = Z|E]
   else
    if variablep(W) then
     [Z = X | E]
    else ...
```

As one can see, except for the annotations, this algorithm is not different from a traditional unification algorithm, which was proved to be correct. Therefore, one has only to prove that the annotations are correctly collected. One can do it by examining the semantic tree that is built during partial evaluation. Sweeping down the tree and examining the unification of the variables, one notices that the unifications have built a chain of bindings which the final value algorithm is able to follow.

As one has already mentioned, the hypothesis (as well as the requirements) is transformed into a conjunction of litterals [P, r(X1/C1, X2/C2, ...), Q,...] that is equivalent to [P, r(X1, X2, ...), C1(X), C2(X2), ..., Q]. The sequence of resolution performed during the proof will erase all litterals of the hypothesis but not the annotations. These annotations which were not erased will constitute the partial evaluation result.

4. Conclusion

The authors implemented a prototype of the system described in this article. The results are satisfactory for our domain of requirements whose principal problem is hilbertian referencing. We think that the use of the subsumption inference method of the TL is an interesting option for kb consulting, specially when kb stores information presented in natural language. The reason for this is that hilbertian reference, i.e., reference introduced by the definite article, can be easily resolved by terminological classification. Extending TL with resources from Fuzzy Logic eliminates the main difficulty of using it in the chosen domain. The authors believe, however, that there are other ways of dealing with uncertainty that also produce good results. The use of partial evaluation to separate the assertional part of the knowledge base from the definitional part greatly simplified the task of the theorem prover. Most of the sentences used in testing the assertions were reduced to a single litteral or to a disjunction of two litterals by the partial evaluator.

References

1. Chang,C. and Lee, R., Symbolic Logic and Mechanical Theorem Proving, Academic Press, New York San Francisco London (1973)

2. DeJong, G. F. and Gratch, J., Steve Minton, Learning Search Control knowledge: An Explanation Based Approach, in: Artificial Intelligence 50 (1991) 117 - 127
3. Donini,F.M., Lenzerini, M. and Nardi, D., The Complexity of Existential Quantification in Concept Languages, in: Artificial Intelligence 53 (1992) 309-327
4. Doyle, J. and Patil, R.S., Two theses of knowledge representation: language restrictions, taxonomic classification, and the utility of representation services, in: Artificial Intelligence 48 (1991) 261 - 297
5. Dowty, D. Wall, R. and Peters, S., Introduction to Montague Semantics, D. Reidel Publishing Company (1985)
6. Dubois, D.and Prade, H., Théorie des Possibilités, MASSON (1985)
7. Ershov, A. P., Mixed Computation: Potential Applications and Problems for Study in: Theoretical Computer Science 18(1982) 41-67
8. Gazdar, F., Klein, E., Pullum, G. and Sag.I, Generalized Phrase Structure Grammar, in: Harvard University Press, Cambridge, 1985
9. Gazdar, G. and Mellish, Natural Language Processing in LISP
10. Grover,C., Briscoe, E., Carrolis, G. and Boguraev, The Alvey Natural Language Tools Grammar, in: University of Cambridge, Computer Lab.
11. Harmelen, F. and Bundy, A., Explanation-Based Generalisation = Partial Evaluation, in: Artificial Intelligence 36 (1988) 401-412
12. Jones,S.L.P., The Implementation of Functional Programming Languages
13. Minton, S., Carbonell, J. G., Knoblock, C.A., Kuokka, R., Etizioni, O., and Gil, Explanation Based Learning: A Problem Solving Perspective, in: Artificial Intelligence 40 (1989) 63-118
14. Nebel, B., Reasoning and Revision in Hybrid Representation- Lectures Notes in Artificial Intelligence n.422
15. Nillson, N. (1980) Principles of Artificial Intelligence, Tioga Publishing Co; Palo Alto, California.
16. Patel-Schneider, P.F., Undecidability of Subsumption in NIKL, in: Artificial Intelligence 39 (1989) 263-272
17. Pereira, F.C.N., Incremental Interpretation, in: Artificial Intelligence 50 (1991) 37-82
18. Shesov, S.D. and Myasnikov, A.G., Logical-Semantic Structure of a Terminology and its Formal Properties, in: Nauchno-Teckhnicheskaya Informatsiya, Seriya 2, Vol. 21, N.3, 1987
19. Toussain,Y., Methodes Informatiques et Linguistiques Pour L' aide à la Specification de Logiciel, Phd thesis (1992), Université Paul Sabatier Toulouse, France
20. Vilain, M., The Restricted Language Architecture of a Hybrid Representation System, Proceedings IJCAI-85, Los Angeles, CA (1985)
21. Westerstâl,D., Quantifiers in Formal and Natural Languages, in: Handbook of Philosoph.. Logic, Gabbay, D. et Guenthner, F.,Eds.D.Reidel Publ. Co., 1989, chap.1, pp 1-131

Using Preference Based Heuristics To Control Abductive Reasoning

John Bigham

Department of Electronic Engineering,
Queen Mary & Westfield College, University of London,
Mile End Road, London E1 4NS

Abstract. Explanations for symptoms can be selected using a variety of criteria. When numerical information is scarce, approaches which depend on partial orders could be of value. The problem of generating explanations for symptoms when only preference relationships between possible causes is considered. The lack of information leads to large number of possible solutions and control of the reasoning process is very important. A computational approach based on focusing the reasoning using a cost bounded ATMS is described.

1 Introduction

In diagnosis an explanation can be considered as a set of assumptions which if true either predict the symptoms or are at least consistent with the symptoms. A preference based ordering of assumptions can be used to induce an ordering over sets of assumptions. An abductive explanation for a set of symptoms is a set of assumptions which the model of the system can use to deduce the symptoms. In the context of explanations for symptoms, if preferences between assumptions are represented by explicit preferability relations, then an ordering over the explanations can be deduced. More specifically, suppose A is a set of assumptions with an associated binary preferability relation $\leq$ induced by some preference criterion on A, then $a \leq b$ iff assumption $a \in A$ is less preferable to assumption $b \in A$. There are many ways of defining a preference relation $\leq\leq$ on the power set of A, i.e. on the explanations.

2 Using heuristics to generate preferred explanations: a simple example

In this section a simple example is presented which will be used to motivate and illustrate the subsequent sections. Figure 1 shows a functional model, built out of functional entities called units. Each unit corresponds to a specific functionality of the modelled system (telecommunication network, or medical system, or space craft.) A unit is for example a multiplexing functionality or the functionality of transmitting data from one unit to another unit, or the behaviour of a test to produce test results. Units are connected to other units by using ports. A port defines how one unit should connect to other units, and has an associated domain which defines the nature of the interaction. A port connects to the port of the other unit. An important point is that no logic is represented in the connections between ports. A line between two ports just establishes the equivalence of the ports. An input port can have only one input connection from another unit, though an output port can be connected to any number of input ports of other units. This is not a limiting restriction, as it simply to ensure that the behaviour is be explicitly written in the behavioural description of the unit and not implied by the connectivity.

Units also have states which indicate possible internal states a unit can have, and also observed (operational) states which the unit is known to have. Each unit has an associated behaviour which is represented by behavioural rules. The rules dictate the relationship between the input ports and the output ports and states of the unit. The behavioural rules can also be used to express uncertain propagation, but this is not illustrated here. The overall model behaviour is characterised by interaction of individual unit behaviours via the port connections between units. The language for modelling functional entities using units and ports was developed in the area of the management of telecommunication networks and is described more fully in [1] and [2]. Since the language is applicable to many domains where model based reasoning is appropriate, a simple form is used here for illustration.

Suppose a system has been modelled as connected functional entities. The behaviour of each of the functional entities A, B, C, E, and F depends on the state of an internal variable, called self with domain {working, not-working}, and the status of an input port called upstream-port with domain {anomalous, not-anomalous}. The output port is called downstream-port and has domain {anomalous, not-anomalous}. The ruleset is:
if upstream-port = anomalous then downstream-port = anomalous
if self = not-working then downstream-port = anomalous
if upstream-port = not-anomalous & self = working then downstream-port = not-anomalous

D is slightly different having ruleset:
if upstream-port1 = anomalous then downstream-port = anomalous
if upstream-port2 = anomalous then downstream-port = anomalous
if self = not-working then downstream-port = anomalous
if upstream-port1 = not-anomalous & upstream-port2 = not-anomalous & self = working then downstream-port = not- anomalous

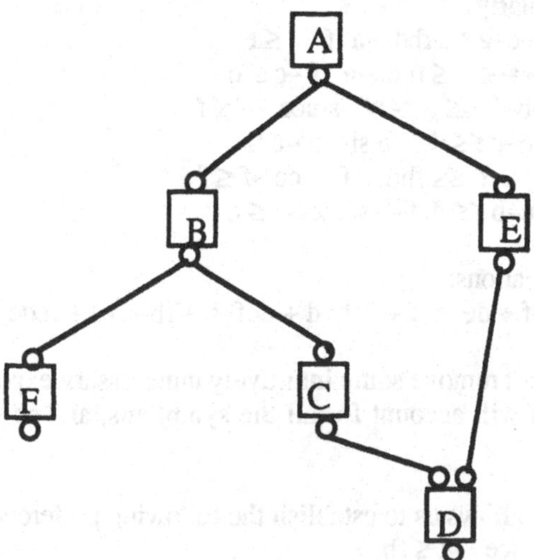

Figure 1 : A functional model

Suppose that anomalous outputs are observed at the outputs of D and F. Call these symptoms s1 and s2. s1 at the downstream port of D results in the following explanations by backward chaining from the symptoms:-

$$\sim d + d\sim c + dc\sim b + dcb\sim a + d\sim e + de\sim a$$

Note that one would normally only generate $\sim d + \sim c + \sim b + \sim a + \sim e$ as these are sufficient. Working assumptions are necessarily introduced when a symptom corresponding to a normal state is observed. However the working assumptions are kept here to illustrate the points.

s2 at the downstream port of F results in the following explanations:-

$$\sim f + f\sim b + fb\sim a$$

Combining both explanations removing supersets and inconsistent explanations gives:-
$\sim f\sim d + d\sim c\sim f + dc\sim b\sim f + dcb\sim a\sim f + d\sim e\sim f + de\sim a\sim f + f\sim b\sim d + df\sim b\sim c + dcf\sim b + df\sim e\sim b + def\sim a\sim b + fb\sim a\sim d + fbd\sim a\sim c + fbdc\sim a + fbd\sim e\sim a + fbde\sim a$

Certain explanations whilst valid in a general case are clearly unnecessarily cautious and complex in most situations. Typical heuristics we may be prepared to make are:-

A1 Any fault assumption is less likely (a priori) than any working assumption, i.e. $\forall x \forall y (\sim x \leq y)$
This includes the case $\sim x \leq x$. Whilst this is not true for all propositions it is valid for simple working, not working assertions.

A2 Any fault assumption is less likely than any conjunction of working assumptions, i.e. $\forall x \forall y \ldots \forall z (\sim x \leq\leq y \ldots z)$ This is a generalisation of A1.

A3 Some working assumptions are preferred to other working assumptions. (Or equivalently, some not working assumptions are preferred to other not working assumptions.) For this example only one such relationship will be used, namely, $c \leq e$.

Using A1 allows us to say that explanation def~a~b will be less preferred to fbde~a provided $\sim b \leq b$. Similarly

$$fbd\sim e\sim a \leq\leq fbde\sim a \text{ if } \sim e \leq e$$
$$fbd\sim a\sim c \leq\leq fbdc\sim a \text{ if } \sim c \leq c$$
$$dc\sim b\sim f \leq\leq dcf\sim b \text{ since } \sim f \leq f$$
$$df\sim b\sim c \leq\leq dcf\sim b \text{ since } \sim c \leq c$$
$$dcb\sim a\sim f \leq\leq fbdc\sim a \text{ since } \sim f \leq f$$
$$df\sim e\sim b \leq\leq dcf\sim b \text{ since } \sim e \leq c$$

This leaves the explanations:
$\sim f\sim d + d\sim c\sim f + d\sim e\sim f + de\sim a\sim f + f\sim b\sim d + dcf\sim b + fb\sim a\sim d + fbdc\sim a + fbde\sim a$

On its own A1 does not remove some intuitively unnecessary explanations. An example is de~a~f. Clearly ~a will account for all the symptoms, and so ~f as well seems less likely.

Using assumption A2 allows us to establish the following preferences:
$de\sim a\sim f \leq\leq fbde\sim a$ since $\sim f \leq fb$
$f\sim b\sim d \leq\leq dcf\sim b$ since $\sim d \leq dc$

fb~a~d ≤≤ fbde~a since ~d ≤≤ de or fb~a~d ≤≤ fbdc~a since ~d ≤≤ dc

This leaves the explanations :
~f~d + d~c~f + d~e~f + dcf~b + fbdc~a + fbde~a

Using assumption A3: fbdc~a ≤≤ fbde~a since c ≤ e leaving
~f~d + d~c~f + d~e~f + dcf~b + fbde~a

Additionally d~e~f ≤≤ d~c~f since c ≤ e tells us that ~e ≤ ~c in the binary working/ not working case. The explanations left are therefore
~f~d + d~c~f + dcf~b + fbde~a

Retaining both working and not working assumptions is expensive. A cheaper way is to ignore working assumptions and just retain not working assumptions. Doing this in the above example gives ~f~d + ~f~c + ~f~e + ~b + ~a

If partial order information is available on the not working assumptions (in the running example we have ~e ≤ ~c) then this could be utilised. For example ~f~e ≤ ~f~c gives ~f~d + ~f~c + ~b + ~a as the resultant explanations. In the binary case the approach of only considering not working assumptions will give a same number of solutions as using assumption A2 and enough assumptions of type A3 to discriminate between different paths to the same non working set of hypotheses. Not using A3 will lead to more solutions as the alternatives such as fbdc~a and fbde~a are in the same equivalence class.

In assumption A2 we are in fact saying that any single fault assumption is more likely than a double fault containing that single fault, any double fault less likely than a triple fault containing the double fault. (e.g. ~b~d ≤≤ ~b in fact follows from A2; f~b~d ≤≤ dcf~b since ~d ≤≤ dc) This means that the reduction for A2 can be done by simple subsumption checks on the *fault assumptions.* i.e.~f~d + d~c~f + d~e~f + dcf~b+ fbdc~a + fbde~a can be generated by subset checks on the fault assumptions ignoring the working assumptions. In the non binary case similar assumptions can be made, e.g. that any fault mode is less likely than any conjunction of working assumptions, and simplifications can be made.

The disadvantage of only retaining not working assumptions is that since the complete explanation is not retained the belief in a particular explanation cannot be readily computed. However in a practical problem retaining only non working assumptions is an attractive option, especially when we have scarce information to rank the paths.

In this example explanations have been systematically eliminated. In practice elimination is not always wanted and the preference criteria are used to provide an ordering. The problem is that a lot of explanations are generated an a procedure for controlling the generation of unnecessary explanations is required. The following sections look at this problem in more detail.

3 Democratism and Elitism

An approach which can be used to relate the preferences between explanations to orderings in the underlying assumptions has been developed by Cayrol , Royer & Saurel

[3, 4]. Here a preference based ordering is used to induce an ordering over sets of beliefs. In the context of explanations, if preferences between assumptions in an explanation are represented by explicit preferability relations, then an ordering over the explanations can be deduced. They define two selection modes for choosing preferred sets of beliefs, which they call democratism and elitism. Elitism and democratism are two ways of defining a preference relation $\leq\leq$ on the power set of A.

If F and G are two non-empty subsets of A, then F is elitistically preferred to G, written $G \leq\leq^e F$, iff for any element f in F\G there is an element g in G\F such that $g \leq f$ and not $f \leq g$.(everything kept must be better than something removed)

If F and G are two non-empty subsets of A, then F is democratively preferred to G, written $G \leq\leq^d F$, iff for any element g in G\F there is an element f in F\G such that $g \leq f$ and not $f \leq g$. (everything removed is replaced by something better)

Examples:
G= {c, d, e} F = { c, d} $G \leq\leq^e F$ is true
G= {c, d, e} F = {a, b, c, d} and $e \leq a$ and $e \leq b$ then $G \leq\leq^e F$ and $G \leq\leq^d F$ is true
G= {c, d, e} F = {a, b, c, d} and $e \leq a$ then $G \leq\leq^e F$ is not true and $G \leq\leq^d F$ is true
G= {c, d, e, f} F = {a, b, c, d} and $e \leq a$ and $e \leq b$ then $G \leq\leq^e F$ is true and $G \leq\leq^d F$ is not true.

Cayrol et al show that if $\leq$ is finitely chained partial ordering on A (i.e. no infinite strictly increasing chain) then $\leq\leq$ is also such a partial order on the power set of A.

3.1 Relationship between A2 and Elitism

Returning to the definitions of elitism and democtratism in the context of G and F representing explanations. It may seem odd to always prefer what could be a long conjunction of assumptions to a shorter conjunction. For example if G= {c, d, e} and F = {c, d, a, b, f, g, h} and $e \leq a$, $e \leq b$, $e \leq f$, $e \leq g$, $e \leq h$, then $G \leq\leq^e F$ and $G \leq\leq^d F$. However in the context of the application, e could be a not working assumption and a, b, f, g, h working assumptions and *then* $\leq\leq^e$ and $\leq\leq^d$ corresponds to A2. This form of elitism, which will be called *weak elitism*, will be used to *eliminate* solutions. Weak elitism will be defined as follows:
If F and G are two non-empty subsets of A, then F is w-elitistically preferred to G, written $G \leq\leq^{we} F$, iff for any element f in F\G which corresponds to a working assumption there is an element g corresponding to a non working assumption in G\F such that $g \leq f$ and not $f \leq g$

The decision to eliminate of explanations using weak elitism means that we should ignore ranking amongst explanations based on rankings of working assumptions. This is because any explanation can mapped to an explanation involving all the the not working assumptions in the explanantion conjoined with all the other working assumptions. For example, if α~x is an explanation and β~x is an explanation, where α and β are conjunctions of working assumptions, then
α~x + β~x = $\alpha\beta$~x + α~β~x + $\alpha\beta$~x + ~$\alpha\beta$~x which can be reduced to $\alpha\beta$~x after elimination using weak elitism. Additionally by the same argument $\alpha\beta$~x can be written $\alpha\beta\gamma$~x where γ are the working assumptions for assumptions not explicitly mentioned in

$\alpha\beta$~x. Therefore *elimination* using weak elitism is equivalent to only using not working assumptions.

Intra working or intra not working comparisons are more complex and application dependant. A modification to elitism, called strong elitism, to require preference to different elements for each comparison is intuitively more acceptable when comparing, for example, all non working solutions - though consequently is less likely to be applicable in the general case.

Strong elitism can be defined as: If F and G are two non-empty subsets of A, then F is s-elitistically preferred to G, written G $\leq\leq^{se}$ F, iff for any element f in F\G there is an element g in G\F such that g $\leq$ f and not f$\leq$g and there is no other element of F\G already selected because g $\leq$ f . This means that if G= {c, d, e} and F = {c, d, a, b, f, g, h} and e $\leq$ a, e $\leq$ b, e $\leq$ f, e $\leq$ g, e $\leq$ h, then it is not true that G $\leq\leq^{se}$ F. If G= {c, d, e, a} and F = {c, d, b, f} and e $\leq$ f, e $\leq$ b, a $\leq$ f, then G $\leq\leq^{se}$ F because e $\leq$ b, a $\leq$ f. This comparison is potentially computationally expensive as the check for strong elitism requires that all possible combinations of preferences be checked.

Because strong elitism is not frequently applicable, in what follows weak elitism has been used to *eliminate* candidate solutions and elitism is used as a criterion to *prioritise* the generation of solutions.

4 Use of a cost bounded ATMS

This section assumes the reader is familiar with ATMS terminology, as described in [5], and [6] (for example). However it is hoped that the rest of the paper can be understood without this section.

One computational approach to abduction is to construct, by backward chaining from each symptom, a directed graph of propositions which we will treat as a ATMS network. Explanations for all the symptoms are generated by augmenting the network by justifications which combine the individual networks. In figure 2 the ATMS network for "Explanations for F" and "Explanations for D" have been combined to create a node the label for which contains the explanations for both symptoms. Environments correspond to explanations for symptoms. We will assume that when a justification is used to create a new environment then any new environment is a superset of any environment in the antecedent of the justification (hyper resolution is not required or the network has been simplified). In the network shown elimination using weak elitism has been asssumed, so only not-working assumptions are present. This is not necessary, but is taken as an example. Justifications are created from the rules. For example the rule of unit F
if self = not-working then downstream-port = anomalous
is used to create the justification ~f -> S2, and the rule
if upstream-port = anomalous then downstream-port = anomalous
of F, and the rule
if self = not-working then downstream-port = anomalous
of B, are used to create the justification ~b -> S2.

Following Ngair & Provan [7] each environment has a cost computed using a cost function which preserves the partial ordering on sets of assumptions induced by set inclusion, i.e. the cost function has to satisfy $\rho(e_1) \leq \rho(e_2)$ iff $e_1 \subseteq e_2$. Here we define ρ as a preference function such that $\rho(e_1) \leq \rho(e_2)$ iff $e_1 \leq\leq^e e_2$. Since $e_1 \subseteq e_2 \Rightarrow e_2 \leq\leq^e e_1$ we can also use simple superset removal. To reduce the number of environments supersets are removed.

Ngair & Proven's algorithm is as follows:

Set the cost bound to be the lowest possible, i.e. the cost associated with the empty environment and introduce all assumptions with cost lower or equal to the current bound. Run the basic ATMS algorithm with the current cost bound
Here if l_1 and l_2 are labels, then rather than $l_1 \wedge l_2$ being computed as the smallest subsets in $\{ e_1 \cup e_2 \mid e_1 \in l_1, \text{and } e_2 \in l_2\}$, it is computed as:
the smallest subsets $\{e = e_1 \cup e_2 \mid e_1 \in l_1, \text{and } e_2 \in l_2, \text{and } \rho(e) \leq \text{current cost bound}\}$

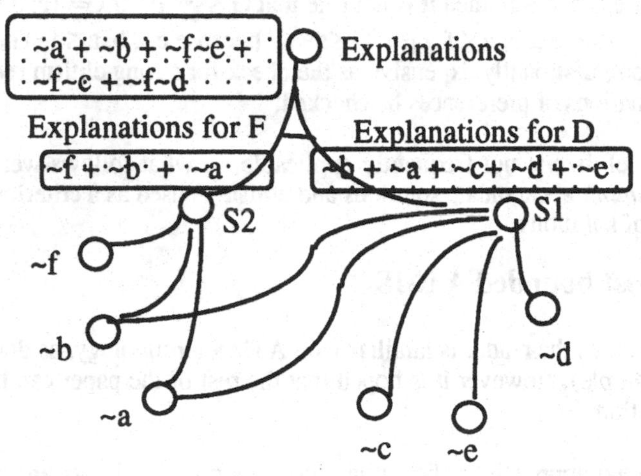

Figure 2 ATMS network which generates explanations of symptoms at D and F

If $\rho(e) >$ current cost bound then the environment is blocked and is not allowed to propagate further.
As before if l_1 and l_2 are labels then $l_1 \vee l_2$ is computed as the smallest subsets in $\{l_1 \cup l_2\}$.

The CBATMS has a special node in which the environments are considered as solutions. In our application, this node corresponds to the "ExplanationsNode". If an environment appears in this label a result has been found, otherwise the cost bound is increased to the next higher cost and introduce more assumptions. The CBATMS ensures that the cost of explanations increase monotonically as assumptions are added. The subset condition above ensures that any environment which is created as the result of combining existing environments has a cost greater than any of the environments used in the combination. A computational approach which allows incremental updating of the labels has been implemented.

5 Use of Heuristics

The difficulty with a general approach is that the lack of information inevitably makes it difficult to order solutions and hence keep control over the number of environments created. A solution is to add additional preference conditions to ensure that solutions are generated in a reasonable order. These preference conditions are added after all the symptoms have been observed.

An example is the heuristic that in a particular context the fault is a clocking fault (rather than say a line fault). In a numerical probability type approach, these heuristics can sometimes be accommodated by adjusting probabilities appropriately or modifying the cost function in the cost bounded approach described. Probabilities of non clocking faults can be reduced, or the cost of non clocking assumptions in an explanation can be increased. When only preference structures are used such situations can be modelled by creating preferences for clocking faults over other faults. $\sim e \leq \sim c$ is an example of this where C may lie in a clocking causal path. Another common heuristics is that fault reports are more likely to appear near the faulty unit. This is usually because the modelling of the functional entities did not include any predictions on the reliability of a unit reporting its own fault. (This could of course be rectified by adding additional dependencies and probabilities in the causal model, but again the information on the prior probabilities can be hard to collect.) In the example the heuristic suggested can be used to augment the preference relationships to that shown in figure 3 below.

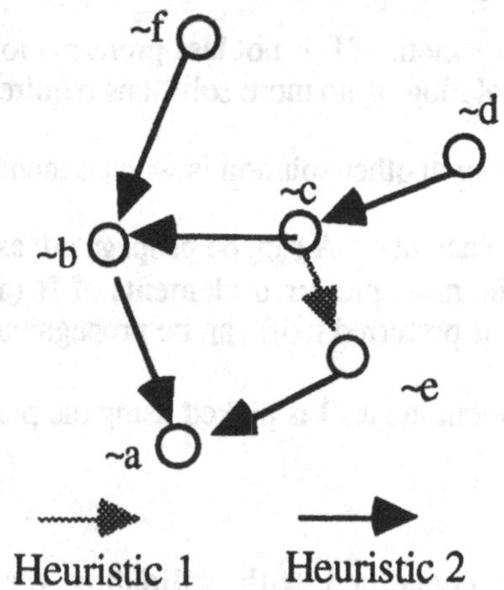

Figure 3 Incorporating a preference partial order

6 Algorithm Details

Generally when assumptions are added by preference then by assumption A2 all the working assumptions will be added first of all. This will usually not allow any propagation, except in the case where some of the observed symptoms are consistent with working hypotheses. If environments less preferred under $\leq\leq^{we}$ are to be eliminated then we do not include the working assumptions at all. The computational approach is:-

1. Let A' be the set of all possible assumptions. A is the set of assumptions which have yet to be selected for propagation. Initially A is A'. At any stage those assumptions in A' more preferred to those in A, have already been used for propagation in the network. H is the set of preference relations. Initially these are provided by the heuristics, though the use of elitism will also induce relationships which are added to H. E is the set of explanations for all the symptoms based on the assumptions in A' \A, i.e. the label of the node corresponding to the explanations in the ATMS network. B is the set of blocked environments. The threshold partial order T is defined as the partial order involving elements in $B \cup A$. T is augmented as new environments are created. The best elements of T are those which are not provably less preferred to any other element. At any time the set T* contains the best elements of T. This means that if a propagation generates an environment for a node, and this environment is less preferred to an element of T* then propagation with this environment is stopped at this point, and the blocked environment is added to B and T updated. B and the explanations set (E) are initially empty.

$E=\emptyset$; $B=\emptyset$, $A=A'$, and so $T= \langle A', H \rangle$

2. Remove one of the best assumptions in A and adjust T* to contain the new best assumptions. Propagate in the ATMS using the best assumption. Block propagation of all environments less preferred to any element of T*, adding any blocked environments to B, any new environments in the explanations node to E, new preferences to H, and update T.

3. If $E \neq \emptyset$ check if an element of E is not less preferred to an element of T* using $\underline{\leq\leq}^{e}$. If so that element is a solution. If no more solutions required then stop.

4. If there is no solution or another solution is wanted then:

> if $A \neq \emptyset$, then an element of A <u>can</u> be propagated as in 2.
> if $B \neq \emptyset$, one of the most preferred elements of B (an element of B such that no other element of B is preferred to it) <u>can</u> be propagated as in 2.

The most preferred element in $A \cup B$ is picked using the preferences in T. Go to 3

7 Future Work

Future work will be concerned with validating the approach, considering its computational complexity, and incorporating the use of temporal information into the control of the generation of explanations. Other algorithms to focus the search for the best explanations will also be considered.

Acknowledgements

This work is was undertaken as part of the Esprit DRUMS II and UNITE projects. Thanks to the referee for helpful remarks.

References

1 Bigham, J., Pang D., Kehl, W., Newstead, M., Neuman, C. (1992) "A generic maintenance system for telecommunications", in "The Management of Telecommunications Networks", eds. R. Smith, E.H. Mamdani, & J. Callaghan, Ellis Horwood. (Also 6th TMN conference, Madeira September 92).

2 Bigham, J., Pang, D., (1993) "A model based reasoning system for the maintenance of telecommunications networks", International Journal of Engineering Intelligent Systems for Electrical Engineering and Communications", No 1, Vol. 1

3 Cayrol, C., Royer, V., & Saurel, C. (1993) "Methodological Approach to Preference Based Diagnosis" ToolDiag'93 April 93 Available from Centre d'Etudes et de Recherches de Toulouse, 2 Avenue Edouard Belin, B.P. 4025, 31055 Toulouse Cedex.

4 Cayrol, C., Royer, V., & Saurel, C. (1992) "Management of Preferences in Assumption Based Reasoning", IPMU 92 Springer Verlag

5 de Kleer, J. (1986) "Problem solving with the ATMS", Artificial Intelligence, Vol. **28**, pp197-224, North Holland.

6 de Kleer, J. & Williams, B.C. (1987) "Diagnosing multiple faults" Artificial Intelligence, **32**, pp 97-130.

7 Ngair, T. H. and Provan G. (1993) "A Lattice-Theoretic Analysis of Assumption-based Problem Solving", Institute of System Science, National University of Singapore, Kent Ridge, S(0511), Republic of Singapore

Efficient Interpretation of Propositional Multiple-valued Logic Programs *

Gonzalo ESCALADA-IMAZ[1] and Felip MANYÀ[1,2]

[1] Institut d'Investigació en Intel.ligència Artificial (IIIA)
Spanish Council for Scientific Research(CSIC)
Campus Universitat Autònoma de Barcelona
08193 Bellaterra, Barcelona, Spain

[2] Departament d'Informàtica i Enginyeria Industrial
Universitat de Lleida (UdL)
Ap. Correus 471, 25080 Lleida, Spain
e-mail:{gonzalo,felip}@iiia.csic.es

Abstract. Logic programming languages such as Prolog are widely used. A clear shortcoming of these languages is that every predicate can take only two truth values. A natural development is to consider that predicates could have many possible values. Thus, the main goal of this paper is to present an interpreter for infinitely-valued propositional logic programming. Some issues concerning the efficiency of the interpreter are discussed, and the negation as failure and the cut operator are also defined and integrated in the present multiple-valued context. The properties of the interpreter algorithm are carefully analyzed. Some of the areas of application of this work are expert systems and logic programming.

1 Introduction

Interest in logic programming languages is growing as new applications are emerging in areas such as Deductive Databases [14] and Artificial Intelligence. Nevertheless, one of their drawbacks is the lack of natural mechanisms to deal with uncertainty or fuzzy knowledge. This is due to the fact that classical programming languages are based on two-valued logic. Some solutions to this problem have been proposed in the literature [1, 2, 5, 13, 15, 16, 18, 20], mainly by basing the languages on uncertainty logics (possibilistic, probabilistic, evidential), as well as on fuzzy or multiple-valued logics. In this paper we describe an efficient interpreter for multiple-valued propositional logic programs. In contrast to other approaches, the interpreter is able to deal with a family of infinitely-valued logics, since we can modify the interpretation function for the conjunction and implication connectives. The interpretation function for the conjunction connective, noted T, has to be a T-norm (i.e. a binary continuous operation such that for all a, b, c in $[0, 1]$ satisfies the following properties:

* Research partially supported by the project P.B91-0334-C03-03 funded by the DGICYT and the project 94-13 funded by the Universitat de Lleida.

$T(a,b) = T(b,a)$, $T(a, T(b,c)) = T(T(a,b),c)$, $T(0,a) = 0$, $T(1,a) = a$ and if $a \leq b$ then $T(a,c) \leq T(b,c)$ for all c). The interpretation function for the implication connective, noted I_T, is defined by residuation wrt the T-norm (i.e. $I_T(a,b) = Sup\{c \in [0,1] | T(a,c) \leq b\}$). See [7] for a more detailed explanation. Moreover, the interpreter includes the negation as failure rule and the control operator cut adapted to our multiple-valued context.

Succinctly, the basis of the family of multiple-valued logics considered in this work is as follows. A multiple-valued sentence is an ordered pair $(S;\alpha)$, where S is a classical-like sentence and $\alpha \in [0,1]$ denotes a truth value. In our context, a program P is a set of multiple-valued sentences $(S;\alpha)$, where S is a classical-like fact or rule. Interpretations are mappings from the first components of sentences to the set of truth values $[0,1]$. An interpretation $\mathcal{I}$ satisfies a multiple-valued sentence (S,α) iff the assigned truth value to the sentence $\mathcal{I}(S)$ is greater than or equal as α. An interpretation is a model of a program P iff it satisfies all the facts and rules of P.

Given a program P and a goal q, the objective of the interpreter is to determine the greatest value $\alpha \in [0,1]$, called maximum degree of logical consequence of q wrt P, such that $P \models (q,\alpha)$. The algorithmic aspects of the interpreter are carefully studied in such a way that the interpreter finds the maximum degree of logical consequence of a goal q wrt a program P in linear time. This linearity is obviously of crucial importance in practical applications. Besides, the interpreter apply some pruning strategies.

A direct application of the interpreter is in the knowledge-based systems area [17] where the interpreter acts as a backward chaining inference engine. For a detailed survey of multiple-valued theorem proving you can consult [11].

This paper is organized as follows. In the next section we define the syntax, semantics and logical inference of the family of infinitely-valued logics the interpreter can deal with. Section 3 introduces the interpreter, beginning with a simple version. Section 4 introduces the negation as failure rule and Section 5 describes the cut adapted to our multiple-valued context. Section 6 outlines the implementation of a programming environment based on the interpreter. Finally, some conclusions and future work are given in the last section. The proofs of the theorems can be found in [8].

2 Multiple-valued Horn logics

As our aim is to design an interpreter for multiple-valued logic programs, first of all we present first the syntax, semantics and logical inference restricted to the Horn case.

Syntax: A program P in this language is formed by a finite set of multiple-valued facts (BF) and rules (BR) defined as follows:

- **fact:** it is an ordered pair $(p;\alpha)$, where p is a classical-like propositional atom and α is a value in $[0,1]$.

- **rule:** it is an ordered pair $(q \leftarrow p_1, \ldots, p_k; \alpha)$, where $q \leftarrow p_1, \ldots, p_k$ is a classical-like propositional rule and α is a value in $[0,1]$.

Semantics: Given a program P and a goal q, the objective of a classical propositional interpreter is to verify that for each model of P, q is also satisfied. If so, we say that q is a logical consequence of P. In the multiple-valued case, each model of P may assign a different truth value to q. This leads us to define the notion of maximum degree of logical consequence, i.e. the greatest α such that $P \models (q, \alpha)$. The aim of the interpreter will be to find this degree. Let us give some formal semantics definitions:

Definition 1 interpretation. An interpretation $\mathcal{I}$ is a mapping from the first components of the facts and rules to the infinite set of truth values $[0,1]$ verifying the following properties:

- $\mathcal{I}(p) \in [0,1]$,
- $\mathcal{I}(q \leftarrow p) = I_T(\mathcal{I}(p), \mathcal{I}(q))$ and
- $\mathcal{I}(q \leftarrow p_1, \ldots, p_n) = I_T(T(\mathcal{I}(p_1), T(\mathcal{I}(p_2)), \ldots T(\mathcal{I}(p_{n-1}), \mathcal{I}(p_n)), \ldots)), \mathcal{I}(q))$.

It should be noticed that the interpretation of a rule can be computed once the interpretation of the atomic components and the functions I_T and T are given.

Definition 2 satisfiability relation. An interpretation $\mathcal{I}$ is a model of a sentence (S, α), where S is a classical-like fact or rule, iff $\mathcal{I}$ satisfies the sentence, noted $\mathcal{I} \models (S, \alpha)$. The satisfiability relation is defined as follows:

$$\mathcal{I} \models (S; \alpha) \text{ iff } \mathcal{I}(S) \geq \alpha.$$

Definition 3 logical consequence. Let P be a program and (p, α) a fact. (p, α) is a logical consequence of P iff all the models of P are also models of (p, α).

Definition 4 maximum degree of logical consequence. Let P be a program and q a goal. The maximum degree of logical consequence of q wrt P is the truth value $\alpha_q = Sup\{\alpha | P \models (q; \alpha)\}$.

Proposition 5. *Let P be a program, q a goal and $\alpha \in [0,1]$ a truth value. Then,* $Sup\{\alpha | P \models (q; \alpha)\} = Inf\{\mathcal{I}(q) | \mathcal{I} \models P\}$.

Logical inference: The deduction relation $\vdash$ is defined by the following inference rule and axiom:

- *Multiple-valued Modus Ponens* [19, 10]:

$$\frac{(p_1; \alpha_1), \ldots, (p_n; \alpha_n) \quad (q \leftarrow p_1, \ldots, p_n; \alpha_{R_1})}{(q; T(\alpha_R, T(\alpha_1, \ldots, T(\alpha_{n-1}, \alpha_n)) \ldots))}$$

- *Axiom:* $\vdash (p; 0)$.

We define the syntactic counterpart of maximum degree of logical consequence:

Definition 6 maximum degree of deduction. Let P be a program and q a goal. The maximum degree of deduction of q wrt P is the value

$$\alpha_q = Sup\{\alpha | P \vdash (q;\alpha)\}.$$

Theorem 7 soundness and completeness. *Let P be a program and q a goal. $P \models (q;\alpha)$, where α is the maximum degree of logical consequence of q wrt P, iff $P \vdash (q;\alpha)$, where α is the maximum degree of deduction of q wrt P.*

Restriction: In the programs considered here we suppose that we do not have the "recursivity problem". This means that there are no rules of the kind $(q \leftarrow p;\alpha_1)$ and $(p \leftarrow q;\alpha_2)$, i.e. there are no cases such that the value of a proposition p depends on itself.

Justification: Graphically, the program P can be represented as classically by an AND-OR graph:

- For each propositional atom p we associate a node labeled p.
- For each rule $(q \leftarrow p_1,\ldots,p_k;\alpha)$ we associate a connector $(q,(p_1,..,p_k))$ formed by k arcs (q,p_i), $1 \leq i \leq k$.

An AND-OR graph representing a program P without indirect recursivity implies that the graph has no cycles. On the one hand, it has been shown that the algorithms for binary propositional logic dealing with cycles are fairly difficult [4], [9] (we consider this problem as future work in our multiple-valued logic context). On the other hand, in contrast to classical first order logic, it can be verified that, in the propositional case, cycles are not included in any solution but they complicate the search for the possible existing ones.

Remark: It should be noticed that the couple $(p;\alpha)$ (resp $(q \leftarrow p_1,\ldots,p_k;\alpha)$) does not denote that the truth value of p (resp $q \leftarrow p_1,\ldots,p_k$) is α, but is comprised in the interval $[\alpha,1]$. The main reason for working with these intervals is of practical kind. Thus, lower bounds are used in areas as in expert systems [17] where a fact $(p;\alpha)$ means that p is true at leats with degree α. Another technical reason is to ensure that the base composed by the facts and rules is satisfiable. Indeed, if we consider that $(p;\alpha)$ indicates that the interpretation of p is only the value α (and not the interval $[\alpha,1]$), we could deduce by two different deduction paths (applying the multiple-valued Modus Ponens inference rule) that p has two different concrete values and so P would be unsatisfiable. Similarly, if intervals $[\alpha,\alpha']$ are envisaged, we could deduce that p is in two disjoint intervals $[\alpha_1,\alpha_2]$ and $[\alpha_3,\alpha_4]$. Thus, taking this kind of intervals ensures that the two different deduction paths lead to intervals $[\alpha_1,1]$ and $[\alpha_2,1]$ whose intersection is obviously not empty, and therefore there exists at least one interpretation $\mathcal{I}$ such that $\mathcal{I}(p) \geq max(\alpha_1,\alpha_2)$ for both deduction paths.

3 The interpreter

The aim of the interpreter is to automate the application of the Multiple-valued Modus Ponens inference rule to determine the maximum degree of deduction of a goal wrt a program. For this purpose, it applies the Modus Ponens rule in a reverse way as in classical backward inference systems. That means, rather than applying the rule to $p, q \rightarrow r$ when we have determined p and q for deducing r, it searchs first for whether r can be deduced which in turn leads to check whether p and q can be deduced.

The whole interpreter presents some difficulties in carrying out the evaluation of a multiple-valued propositional logic. In order to facilitate its comprehension, we present first a simple version. On this basis, we will then complete the algorithm in a second stage.

3.1 A simple version

Principle of the algorithm: We have seen that the problem of finding the maximum degree of deduction of a goal q wrt a program P can be represented by an AND-OR graph. The principle of the interpreter algorithm to find the maximum degree of deduction relies on a depth-first search for this graph.

The algorithm has two main recursive functions called OR-EXTENSION and AND-EXTENSION. The former scans all the rules $R : (p \leftarrow p_1, \ldots, p_n; \alpha_R)$ whose consequent is a given proposition p. For this purpose, OR-EXTENSION calls AND-EXTENSION that visits recursively, calling OR-EXTENSION, the successor nodes $p_1, \ldots, p_n$ associated to the antecedents of each rule R. Given the value α_R of rule R and after obtaining the maximum degree of deduction α_i of each successor node p_i, AND-EXTENSION calculates the deduction value α of the consequent p by using rule R. This computation applies the Modus Ponens rule; i.e. $\alpha = T(\alpha_R, T(\alpha_1, \ldots, T(\alpha_{n-1}, \alpha_n)) \ldots))$. The scanning of all the rules by OR-EXTENSION enables us to determine the maximum degree of deduction for p given by the different deduction paths defined by the different possible applications of the rules. This degree is, in turn, given back to the corresponding AND-EXTENSION function performing thus a bottom-up propagation of the maximum degree of deduction from the facts up to the query.

In a brute force search in the AND-OR graph, redundant computations can appear by examining certain parts of the AND-OR graph more than once, since a node p can be a succesor node of several connectors. To overcome this problem we will use a marking technique for nodes when scanning the graph. Thus, when a proposition node is visited the first time all the possible deduction paths concluding this proposition are explored. The maximum degree of deduction obtained by this exploration will be assigned to the proposition node. Further visits to an already visited node will retrieve the previous computed maximum degree of deduction, thus avoiding redundant expansions of the same node.

Data Structures: For each node, we use the following data structure:

•$val(p) = \alpha$ represents the maximum degree of deduction of p in a given stage of the algorithm. Initially, $val(p) = \alpha$ if there exists a fact $(p; \alpha)$ in BF and $val(p) = 0$, otherwise.

•$visit(p) \in \{t, f\}$ where $visit(p) = t$ iff p has been expanded by OR-EXTENSION. Initially, $visit(p) = f$ for all the propositions.

•rules-conseq$(p) = \{R : p$ is the consequent of $R\}$, the set of rules concluding p.

For each rule, we use the following data structure:

•val$(R) = \alpha_R$ represents the value attached to R.

•antecedents$(R) = \{p : p$ is an antecedent of $R\}$, the set of antecedent propositions of rule R.

Algorithm

Input: BF, BR and a query q.
Output: the maximum degree of deduction of q wrt $P = BF + BR$.

```
INTERPRETER(BF, BR, q)
for all p : val(p) ← α if (p; α) ∈ BF
        val(p) ← 0 if (p; α) ∉ BF
        visit(p) ← f
for all R = (p ← p1, ..., pn; α) do:
        val(R) ← α
        antecedents(R) ← {p1, ..., pn}
        rules − conseq(R) ← rules − conseq(p) ∪ R
return(OR-EXTENSION(q))

OR-EXTENSION(p)
   if visit(p) then return(val(p))
   visit(p) ← t
   for all R ∈ rules − conseq(p) do:
        and ← AND-EXTENSION)(R)
        if and > val(p) then val(p) ← and
   return(val(p))

AND-EXTENSION(R)
   value ← val(R)
   for all p ∈ antecedents(R) do:
        value ← T(value, OR-EXTENSION(p))
   return(value)
```

Theorem 8 correctness. INTERPRETER(BF, BR, q) *returns* α *iff* α *is the maximum degree of logical deduction of q wrt P* $(P = BF + BR)$.

Theorem 9 linearity. INTERPRETER(BF, BR, q) *is in* $O(N)$, *where N is the total number of occurrences of propositions in P* $(P = BF + BR)$.

3.2 An improved version

In the previous simple version of the algorithm all the outgoing arcs from a particular node p are crossed, but some techniques for pruning the search in the whole graph can be applied. In fact, this pruning is based on the following property of the function T:

$$T(x_1, x_2) \geq T(T(x_1, x_2), x_3) \geq \ldots \geq T(\ldots T(T(x_1, x_2), x_3), \ldots, x_n).$$

Thus, let us suppose that the scanning of the first connector produces the value α. Then, we begin to explore the subsequent connectors. The value of a proposition node applying a certain connector is given by the application of the Modus Ponens employing the rule represented by the connector. Thus, if we have a connector $(p, (p1, .., pk); \alpha_R)$ and the computed value of p_i is α_i, then the value for the father proposition node of the connector is given by $T(\ldots T(T(\alpha_R, \alpha_1), \alpha_2), \ldots, \alpha_k)$. Using the previous property, we have that

$$T(\alpha_R, \alpha_1) \geq \ldots \geq T(\ldots T(T(\alpha_R, \alpha_1), \alpha_2), \ldots, \alpha_i)$$
$$\geq \ldots \geq T(\ldots T(T(\alpha_R, \alpha_1), \alpha_2), \ldots, \alpha_k), i < k.$$

Therefore, as this function is incrementally calculated, as soon as

$$T(\ldots T(T(\alpha_R, \alpha_1), \alpha_2), \ldots, \alpha_i) \leq \alpha, i < k,$$

we can prune the remaining search space defined by the subgraphs rooted by the nodes $p_{i+1}, .., p_k$.

Algorithm: the function INTERPRETER(BF, BR, q) which is in charge of the initialization of the data structures is the same as that of the previous algorithm.

```
OR-EXTENSION(p)
   if visit(p) then return(val(p))
   visit(p) ← t
   for all R ∈ rules − conseq(p) do:
         and ← AND-EXTENSION(R, val(p))
         if and > val(p) then val(p) ← and
   return(val(p))

AND-EXTENSION(R, v)
   if val(R) ≤ v then return(val(R))
   value ← val(R)
   for all p ∈ antecedents(R) do:
         value ← T(value, OR-EXTENSION(p))
         if value ≤ v then return(value)
   return(value)
```

Theorem 10 correctness. INTERPRETER(BF, BR, q) *returns* α *iff* α *is the maximum degree of deduction of q wrt P* $(P = BF + BR)$.

Theorem 11 linearity. INTERPRETER(BF, BR, q) *is in* $O(N)$, *where* N *is the total number of occurrences of propositions in P* $(P = BF + BR)$.

Then, it is clear that if t is the running time of INTERPRETER(BF, BR, q) of the former version and t' is the running time of the later version then $t \geq t'$.
Remark: The prunning strategy used is general for all T-norms. However, some other strategies could be defined for particular T-norms (e.g. when the T-norm is the *min* function).

4 Negation

One of the most commonly used rules in logic programming to deduce negative information is the Negation as Failure Rule (NFR). Before presenting the NFR for multiple-valued propositional logic we will first overview the NFR in the classical ***propositional*** case.

In the classical propositional case we can have rules where some of the antecedent propositions are negated, i.e. $(q \leftarrow p_1, \ldots, \neg p_i, \ldots, p_k; \alpha)$. The NFR is as follows: if $P \not\vdash p_i$ then $P \vdash \neg p_i$ (note that in the propositional case NFR coincides with the Closed World Assumption). From a semantic point of view if $P \not\models p_i$ implies that p_i is false (or 0) in some models of P and it is possibly true (or 1) in others. In this case, the NFR establishes that we restrict the models of P to those models $\mathcal{I}$ of P such that $\mathcal{I}(p_i) = 0$ $(\mathcal{I}(\neg p_i) = 1)$.

An extension of the NFR to the multiple-valued case is as follows. If p_i is deduced with maximum degree of deduction 0 then we apply the NFR considering only the interpretations $\mathcal{I}(p_i) = 0$ and so, $\mathcal{I}(\neg p_i) = 1$.

In a more general case, if $P \models (p_i; \alpha)$, where α is the maximum degree of logical consequence, then we consider only the models $\mathcal{I}$ of P such that $\mathcal{I}(p_i) = \alpha$. This means that the NFR for the multiple-valued case considers only those models of P where $\mathcal{I}(\neg p_i) \geq 1 - \alpha$, namely in the interval $[1 - \alpha, 1]$.

Example 1. Let us consider the following simple example where the query is q.

$$(q \leftarrow p_1, \neg p_2, p_3; \alpha)$$
$$(p_1; \alpha_1)$$
$$(p_2; \alpha_2)$$
$$(p_3; \alpha_3)$$

We can see that the maximum degree of deduction for p_2 is α_2, hence the NFR assumes $1 - \alpha_2$ as the maximum degree of deduction for $\neg p_2$. Hence, q is deduced with the maximum degree of deduction $T(T(T(\alpha, \alpha_1), 1 - \alpha_2), \alpha_3)$.

Remark: As is well known [14], when we allow negative literals to occur in the antecedent of a rule some inconsistencies can be detected. For instance, let us consider the following simple case: $(p \leftarrow \neg p; 1), (p; 0)$ we can see that p and $\neg p$ are deduced with maximum degree of deduction 1. It should be noted that this also happens in classical propositional logic. Some conditions under which the consistency of a program is warranted could be studied [6], but they are beyond the scope of this paper.

Algorithm: It is like the previous version changing the AND-EXTENSION function by the following one:

```
AND-EXTENSION(R, v)
  if val(R) ≤ v then return(val(R))
  value ← val(R)
  for all L ∈ antecedents(R) do:
        if L = p then
            value ← T(value, OR-EXTENSION(p))
        else
            value ← T(value, 1 − OR-EXTENSION(p))
      if value ≤ v then return(value)
  return(value)
```

Theorem 12 correctness. INTERPRETER(BF, BR, q) *returns* α *iff* $BF, BR \vdash (q; \alpha)$, *where* α *is the maximum degree of deduction of* q *using the deduction relation and the negation as failure rule.*

Theorem 13 linearity. INTERPRETER(BF, BR, q) *is in* $O(N)$, *where* N *is the total number of occurrences of propositions in* $P = BF + BR$.

5 Cut

In classical logic programming, the cut "/" is a control facility to be used by the programmer for pruning some portions of the search space. For example, assume that we have a rule $q \leftarrow p_1, p_2, /, p_3$ (we do not have included the first order arguments). Once the cut is executed, it prunes the possible alternatives for all the literals to its left in the body of the rule as well as the alternatives to satisfy the consequent (in our rule, the alternatives for q, p_1 and p_2 are not considered provided the cut is executed).

If we are dealing with propositional logic, the cut influences only the search space of the consequent of the rule, since there are no variables to be instantiated.

In our multiple-valued context, we could define a multiple-valued cut as follows:

Definition 14 multiple-valued cut $(/, \alpha)$. If the interpreter finds a rule of the form $q \leftarrow p_1, p_2, \ldots, p_i, (/, \alpha), p_{i+1}, \ldots, p_k$ then if the maximum degree of deduction α_{1i} of the conjunction $p_1 \wedge \ldots \wedge p_i$ verifies that $\alpha > \alpha_{1i}$ (resp. $\leq$) then the remaining alternatives for q are (resp. are not) considered.

Note that in our multiple-valued context it makes no sense to consider the alternatives of satisfying the literals on the left of the cut in the body since, before arriving at the cut, the maximum degrees of deduction for those literals have been computed and so no search space involving those literals have been left without scanning.

Example 2. Let us study the following portion of a program:

$$(q \leftarrow p_1, p_2, (/, 0, 6), p_3, p_4; 1)$$
$$(q \leftarrow p_4, p_5; 1)$$
$$(q \leftarrow p_5, p_6, p_7; 1)$$
$$\ldots\ldots\ldots\ldots\ldots$$
$$(p_1; 0.8)$$
$$(p_2; 0.8)$$

Supposing that $T(x, y) = x * y$ then the deduction value for $p_1 \wedge p_2$ is 0.64 and hence, in this example, only the first rule would be considered among those concluding q. The maximum degree of deduction of q would be given as previously by the one associated to the rule and those found for p_1, p_2, p_3 and p_4.

If we have the same context but with the cut (/, 0.8) then the maximum degree of deduction of q would be obtained taking into account the other deduction paths by using the remaining rules concluding q.

Algorithm: The complete algorithm embodying the NFR and "/" is defined below. The INTERPRETER function is as the previous one. OR-EXTENSION scans the rules concluding a proposition p but now the scanning is stopped if a rule $R : p \leftarrow p1, \ldots, p_i, (/, \alpha)p_{i+1}, \ldots, p_n$ is found and $T(\mathcal{I}(p_1, T(\mathcal{I}(p_2, \ldots T(\mathcal{I}(p_{i-1}, \mathcal{I}(p_i) \ldots) \geq \alpha$. For that purpose, AND-EXTENSION returns a couple (value, boolean) where "value" is defined as before and "boolean" is a boolean variable such that "boolean"=t iff a rule $R : p \leftarrow p1, \ldots, p_i, (/\alpha)p_{i+1}, \ldots, p_n$ is found and $T(\mathcal{I}(p_1, T(\mathcal{I}(p_2, \ldots T(\mathcal{I}(p_{i-1}, \mathcal{I}(p_i) \ldots) \geq \alpha$.

```
OR-EXTENSION(p)
  if visit(p) then return(val(p))
  visit(p) ← t;cut ← f
  for all R ∈ rules − conseq(p) and while cut = f do:
      (and, cut) ← AND-EXTENSION(R, val(p))
      if and > val(p) then val(p) ← and
  return(val(p))

AND-EXTENSION(R, v)
  if val(R) ≤ v then return(val(R))
  value ← val(R);cut ← f
  for all L ∈ antecedents(R) do:
      if L = (/, α) then: if value ≥ α then cut ← t
      else do:
          if L = p then OR ← OR-EXTENSION(p)
          else (L = ¬p) OR ← 1 − OR-EXTENSION(p)
          value ← T(value, OR)
          if value ≤ v then return(value, cut)
  return(value, cut)
```

Theorem 15 correctness. INTERPRETER(BF, BR, q) *returns* α *iff* $BF, BR \vdash (q; \alpha)$, *where* α *is the maximum degree of deduction of* q *using the deduction relation, the negation as failure rule and the cut operator.*

Theorem 16 linearity. INTERPRETER(BF, BR, q) *is in $O(N)$, where N is the total number of occurrences of propositions in$P$ $(P = BF + BR)$.*

6 Implementation

A programming environment has been implemented in C++ to execute multiple-valued propositional logic programs [3]. The main features of this environment are the following:

- It has a compiler that from the source program creates the data structures for facts and rules. This compiler is formed by a lexical analyzer and a syntactical analyzer which have been implemented with Lex and Yacc, respectively. It also has facilities for handling lexical and syntactical errors.
- The T-norm used by the program can be modified. Some T-norms are pre-defined, but the user can define its own T-norms.
- Programs can be executed in debug mode. The user can know the state of the execution when entering and leaving the functions AND-EXTENSION and OR-EXTENSION. He can also select a particular list of atoms to be executed in debug mode.
- A friendly user interface has been implemented using Object Windows Library 2.0 from Borland for IBM PC-like computers.

7 Concluding remarks

One of the restrictions of logic programming languages is that each predicate can take only two values. In this paper, we have presented a way of overcoming this drawback by giving an interpreter for propositional multiple-valued logic programming.

We have studied the properties of the interpreter algorithm. In particular, we have shown that the proposed interpreter has linear behavior in the worst case. This property is of particular interest in practical applications. Aiming at these practical goals, we have provided the language with the negation as failure rule and the cut operator.

In addition to the cut control facility employed by the programmer, the interpreter algorithm has been designed in such a way that some pruning of the search space are performed whenever possible. Thus, our interpreter improves significantly a "brute force" search algorithm. As future work we plan to move the proposed interpreter to the first order case.

References

1. Atanassov, K., and Georgiev, C. Intuitionistic fuzzy prolog. *Fuzzy Sets and Systems 53* (1993), 121–129.
2. Baldwin, J. Evidential support logic programming. *Fuzzy Sets and Systems 24* (1987), 1–26.

3. Béjar, R. *Implementación de un intérprete proposicional y de un intérprete de primer orden para programación lógica multivaluada.* EUP-Universitat de Lleida, 1993. (graduating project).
4. Dowling, W. F., and Gallier, J. H. Linear-time algorithms for testing the satisfiability of propositional horn formulae. *Journal of Logic Programming 3* (1984), 267–284.
5. Dubois, D., Lang, J., and Prade, H. Poslog, an inference system based on possibilistic logic. *Proceedings North American Fuzzy Information Processing Society Congress* (1990), 177–180.
6. Escalada-Imaz, G. *Optimisation d'Algorithmes d'Inference Monotone en Logique des Propositions et du Premier Ordre.* Université Paul Sabatier, Toulouse, 1989. (PhD Thesis).
7. Escalada-Imaz, G., and Manyà, F. A linear interpreter for logic programming in multiple-valued propositional logic. In *Proceedings of IPMU'94* (Paris, 1994), pp. 943–949.
8. Escalada-Imaz, G., and Manyà, F. *Efficient Interpretation of Propositional Multiple-valued Logic Programs.* IIIA Research Report 95–03, 1995.
9. Ghallab, M., and Escalada-Imaz, G. A linear control algorithm for a class of rule-based systems. *Journal of Logic Programming 11* (1991), 117–132.
10. Godo, L. *Contribució a l'Estudi de Models d'inferència en els Sistemes Possibilístics.* FIB-UPC, Barcelona, 1990. (PhD Thesis).
11. Hähnle, R. *Automated Deduction in Multiple-Valued Logics.* Oxford University Press, 1993.
12. Lee, R. C. T. Fuzzy logic and the resolution principle. *Journal of the Association for Computing Machinery 19*, 1 (1972), 109–119.
13. Li, D., and Liu, G. *A Fuzzy Prolog Database System.* Research Studies Press and John Wiley and Sons, 1990.
14. Lloyd, J. W. *Foundations of Logic Programming.* Springer-Verlag, 1987.
15. Martin, T., Baldwin, J. F., and Pilsworth, B. W. The implementation of fpprolog: A fuzzy prolog interpreter. *Fuzzy Sets and Systems 23* (1987), 119–129.
16. Mukaidono, M. Fundamentals of fuzzy prolog. *International Journal of Aproximate Reasoning 3* (1989), 179–193.
17. Puyol, J., Godo, L., and Sierra, C. A specialisation calculus to improve expert systems communication. In *ECAI'92* (Vienna, 1992 (extended version:IIIA Research Report 92/8), pp. 144–148.
18. Tamburrini, G., and Termini, S. Towards a resolution in a fuzzy logic with lukasiewicz implication. In *Proceedings of IPMU'92* (Paris, 1992), pp. 271–277.
19. Trillas, E., and Valverde, L. On mode and implication in approximate reasoning. In *Approximate Reasoning in Expert Systems*, M. M. Gupta et al., Ed. North Holland, 1985.
20. Weigert, T. J., Tsai, J., and Liu, X. Fuzzy operator logic and fuzzy resolution. *Journal of Automated Reasoning 10* (1993), 59–78.

Many-valued Epistemic States. An Application to a Reflective Architecture: Milord-II*

Lluís GODO[1‡], Wiebe van der HOEK[2], John-Jules Ch. MEYER[2], Carles SIERRA[1‡]

[1]Institut d'Investigació en Intel·ligència Artificial, (IIIA)
Spanish Council for Scientific Research, (CSIC)
Campus Universitat Autònoma de Barcelona, 08193 Bellaterra, Spain
e-mails: {godo,sierra}@iiia.csic.es

[2]Department of Computer Science, Utrecht University
P.O. BOX 80.089, 3508 TB Utrecht, The Netherlands
e-mails: {jj,wiebe}@cs.ruu.nl

Abstract. Halpern and Moses' theory on epistemic states and minimizing knowledge is a formalism with which one can infer what is known and, more importantly, what is unknown by an agent. This formalism has been used up to now in a classical two-valued framework. In this paper we formulate an extension of it when the underlying logic is many-valued, in order to deal with knowledge possibly pervaded with fuzziness. Then we apply this extension to the meta-level architecture *MILORD II*. The object level is an approximate reasoning component based on many-valued logics. The meta-level component makes use of some special meta-predicates to reason about the different states of knowledge of the object level . Our generalization of Halpern and Moses' theory allows us to interpret meta-level reasoning in terms of many-valued epistemic states, providing in turn a modal interpretation of *MILORD II* meta-predicates.

1 Introduction

The knowledge states of a rational and introspective agent are usually modelled as *stable sets* of epistemic formulas. Epistemic formulas are formulas in a language with a standard pair of epistemic (modal) operators standing for knowledge and possibility. In [Halpern & Moses, 84], Halpern and Moses define and characterize what a *minimal epistemic state* associated to a set of premises is, using the notions of stable set and S5-Kripke models. Based on such epistemic states, Halpern and Moses define an entailment relation with which one can infer what is known and, more importantly, what is unknown by an agent. This entailment relation is obviously non-monotonic, and provides the link of this epistemic theory to logical *meta-level architectures*. Namely, the entailment relation can be used to derive *meta-knowledge* about what is known and what is not from object-level formulas represented as non-modal formulas. Moreover, in [Meyer & van der Hoek, 93a, 93b], a default logic based on epistemic notions is introduced where the above mentioned meta-knowledge is used as input to derive default beliefs.

* Research supported by the ESPRIT III Basic Research Action n° 6156 DRUMS II
‡ Research also supported by the Spanish CICYT project ARREL TIC92-0579-c02-01

This formalism has been used up to now in a classical two-valued framework. However, many times we want to model agent states coping with *fuzzy knowledge*, in the sense that an agent's knowledge can incorporate propositions which can be assigned partial degrees of truth, and thus one is led to a many-valued calculus. To do this, an extension of the formalism is necessary. First of all, in section 2 and 3 we extend the epistemic framework to be defined on top of a many-valued logic. Then we apply in section 4 this extension to the meta-level architecture *MILORD II*. This is an architecture for Knowledge Based Systems (KBS) that combines reflection and modularization techniques, together with an approximate reasoning component based on many-valued logics. In this way, the system is able to deal with complex reasoning patterns in the large. In this paper we investigate the logical foundation of a core fragment of *MILORD II*, focusing in particular on giving a modal interpretation of the *MILORD II* meta-predicate *WK* through the notion of Halpern & Moses' epistemic states, extended to the many-valued case. This is done in section 5. The key idea behind the modal interpretation we propose is to view the meta-level component as a system reasoning about many-valued epistemic states representing different states of knowledge of the object level component. An example of the use of this modal interpretation of meta-predicates is provided in section 6. Finally, section 7 contains some concluding remarks.

2 The Modal Many-Valued Logic *MVEL*

In this section we introduce a particular modal logic *MVEL*, standing for *Many-valued Epistemic Logic*, based on a particular many-valued semantics, in the sense that propositional variables are interpreted in an arbitrary (finite) set A_n of n truth-values. The set of truth-values is taken as a scale of partial degrees of truth, ranging from *False* to *True*. However, as we will see, a special kind of unary connectives, known as *indicators* in the literature of many-valued logics, will make possible to always evaluate any formula to either *True* or *False*. This modal logic will serve us as the logical framework where to put into relation *MILORD II* reasoning system and Halpern and Moses' epistemic states.

From now on, we will take a generic chain of n elements

$$0 = a_1 < a_2 < \dots < a_n = 1$$

where 0 and 1 are the booleans *False* and *True* respectively as the set A_n of truth-values on top of which formulas of the *MVEL language* will be valued. We begin with a description of the syntax and semantics of the *MVEL* language.

Formulas: Given a finite set of atomic symbols At, formulas of the language are built upon the set $\{(z)p \mid z \in A_n, p \in At\}$, whose elements are called *quasi-atoms,* in the usual way with connectives $\rightarrow$ and $\neg$, the modal operator $\Box$, denoting "it is known that", and its dual operator $\Diamond$:

- every quasi-atom is a formula
- if φ is a formula, so is $\neg\varphi$
- if φ and ψ are formulas so is $\varphi \rightarrow \psi$
- if φ is a formula, so are $\Box\varphi$ and $\Diamond\varphi$

• nothing else is a formula

Other connectives are definable: $\varphi \wedge \psi$ is $\neg(\varphi \rightarrow \neg\psi)$ and $\varphi \vee \psi$ is $\neg\varphi \rightarrow \psi$. Formulas built in this way will be referred as *MVEL-formulas*. Non-modal formulas will be also called *objective*.

Semantics: A *MVEL* Kripke model is a structure $K = \langle W, \models, R \rangle$, where W is a set of possible worlds, $\models : At \times W \rightarrow A_n$ is, for each world, a valuation mapping of atoms into the set of truth-values A_n, and $R = W \times W$ is the universal accessibility relation. We will write $w(A) = z$ for $\models (A, w) = z$. Within such a model K, we will use the same symbol $\models$ to denote its induced interpretation of *MVEL*-formulas, i.e. the mapping

$$\models : MVEL\text{-}Formulas \times W \rightarrow \{0, 1\}$$

defined as follows (we use the expression $w \models \varphi$ to denote $\models(\varphi, w) = 1$):

- $w \models (i)A$ iff $w(A) = i$
- $w \models \neg P$ iff $w \not\models P$
- $w \models P \rightarrow Q$ iff $w \not\models P$ or $w \models Q$
- $w \models \Box P$ iff $w' \models P$, for all $w' \in W$
- $w \models \Diamond P$ iff $w' \models P$, for some $w' \in W$

where A is an atomic symbol and P and Q are arbitrary *MVEL*-formulas.

Notice that quasi-atoms are not atoms in the classical sense, since they cannot be valued independently. Indeed, they depend on a previous valuation of their propositional variables. However, once quasi-atoms are interpreted, the rest of formulas are interpreted according to the classical two-valued semantics. So, in some sense we have classical interpretations on top of many-valued semantics. As usual, we can also define the standard notions of satisfaction relation and of logical consequence

Definition 2.1 Given a model $K = \langle W, \models, R \rangle$ and a *MVEL*-formula φ, φ is satisfied in K, written $K \models \varphi$, if $w \models \varphi$ for all $w \in W$. Analogously, we write $MVEL \models \varphi$, if $K \models \varphi$ for all *MVEL* models K. Finally, φ is a logical consequence of a set of *MVEL*-formulas Γ, written $\Gamma \models \varphi$, if for all *MVEL* model K, $K \models \psi$ for all $\psi \in \Gamma$ implies $K \models \varphi$.

The particular semantics we have described for *MVEL* leads us to propose for obvious reasons the following axiomatic system, the same as for the classical propositional S5 modal logic, except for the explicit mention to the characteristic relationships between quasi-atoms.

MVEL **Axioms**: the axioms of *MVEL* are the following ones

(a) classical tautologies ,

(1) $\bigvee_{i \in A_n} (i)B$, for $B \in At$

(2) $\bigwedge_{i \neq j} \neg((i)B \wedge (j)B)$, for $B \in At$ and $i,j \in A_n$

(b) classical modal axioms for *S5*

MVEL **Deduction rules**: modus ponens and necessitation

Axioms (1) and (2) state that every propositional variable is given a truth-value and only one.

Definition 2.2 *MVEL* $\vdash \varphi$ if, and only if, φ follows from the set of axioms of propositional (two-valued) calculus (a), many-valued axioms (1) and (2), modal axioms (b), and by applying the Modus Ponens and Necessitation deduction rules.

Proposition 2.3. Let D be the set of all instantiations of axioms (1) and (2). Then it holds that *MVEL* $\vdash \varphi$ iff $S5 \cup D \vdash \varphi$.

Theorem 2.4 (Completeness) *MVEL* $\vdash \varphi$ iff φ has the value 1 in all worlds of all *MVEL* many-valued Kripke models, i.e. iff *MVEL* $\models \varphi$

3 *MVEL* Epistemic States

We now want to apply the approach of Halpern and Moses onto the epistemic language *MVEL*. To do so we treat formulas of the form *(z)A* as 2-valued propositional atoms. In this way we need not really extend the original approach to many-valued logic, since formulas of the above form are 2-valued: they are always true or false. Of course, we should keep in mind that these atoms are not completely independent, but are sometimes logically related, such as given by the axioms (1) and (2). As in the original approach we also consider stable sets of *MVEL*-formulas.

Definition 3.1 A set Φ of *MVEL*-formulas is *stable* if it satisfies the following:

1) closed under *MVEL*-propositional tautologies, that is, Φ includes all classical two-valued propositional tautologies together with the axioms (1) and (2)
2) if $A \in \Phi$ and $A \rightarrow B \in \Phi$ then $B \in \Phi$
3) $A \in \Phi$ iff $\Box A \in \Phi$
4) $A \notin \Phi$ iff $\neg \Box A \in \Phi$
5) Φ is many-valued propositional consistent

Stable sets enjoy the next property that will be used to define minimal epistemic states.

Proposition 3.2 A stable set of epistemic formulas is uniquely determined by the objective formulas it contains.

In the following, Prop(Φ) will denote the subset of Φ that exactly contains all objective formulas of Φ, and $Th_{MVEL}(\Phi)$ the set of formulas derivable in MVEL from Φ.

Proposition 3.3 Let Φ be *MVEL*-stable. Then there is an *MVEL* Kripke-model K_Φ such that its theory (the formulas true in it) is Φ. In particular $K_\Phi \models$ axiom(1), axiom(2).

Sketch of proof. Take $K_\Phi = (W, \models, R)$ such that for all $w \in W : w \models \text{Prop}(\Phi)$. Then the propositional part of the theory of K_Φ is Prop(Φ). Since both the theory of K_Φ and Φ are stable sets, and stable sets are determined uniquely by their non-modal part, we get that the theory of K_Φ is Φ.

Proposition 3.4 A *MVEL*-stable set Φ is closed under S5-consequence.

Proof. Suppose $\Phi \models_{S5} \varphi$. (To prove that $\varphi \in \Phi$) This entailment means that any model satisfying Φ satisfies φ. Since the model $K_\Phi \models \Phi$, we obtain that $K_\Phi \models \varphi$. That is to say that φ is contained in the theory of K_Φ, which is equal to Φ.

Now suppose that φ is a (objective) formula that describes all the facts that have been learnt by the agent. We are interested in defining the epistemic state of the agent if he "only knows φ". This state must be minimal in some sense. Halpern & Moses take this epistemic state to be that stable set containing φ for which the objective part is the least (with respect to set inclusion). However such a least stable set does not exist for every formula φ. A formula is called honest if this least stable set does exist. Formally, this is stated by the following definition.

Definition 3.5 A formula φ is *honest* if there exists a stable set Σ^φ that contains φ and such that for all stable sets Σ containing φ it holds that $Prop(\Sigma^\varphi) \subseteq Prop(\Sigma)$.

The intention is that Σ^φ denotes the stable set representing the state of knowledge of the agent who "knows only φ" (if φ is honest). Every objective formula can be proved to be honest, so that we can indeed speak about the epistemic state associated with knowing only some objective formula.

Proposition 3.6 Every objective formula is honest.

Next proposition contains some further interesting results.

Proposition 3.7 Let φ be objective and D as defined in Proposition 2.3. Then:
(i) $Prop(Th_{MVEL}(K_\varphi)) = Prop(\Sigma^\varphi)$, where K_φ stands for the 'greatest' MVEL model of φ, i.e. $K_\varphi = \cup\{ K \mid K \models \varphi\}$.
(ii) $Prop(\Sigma^\varphi) = Prop(Th_{MVEL}(\varphi))$, or equivalently, $Prop(\Sigma^\varphi) = Th_{PropCalc}(\{\varphi\} \cup D)$,
(iii) If ψ is objective then: $\psi \in \Sigma^\varphi \Leftrightarrow D \models \varphi \rightarrow \psi$

Proof.
(i) Straightforward adaptation of the proof of Theorem 3.1.11(ii) of Meyer & van der Hoek (1995).
(ii) Since φ is objective, $\varphi \in \Sigma^\varphi$, and, since stable sets are by definition closed under MVEL-consequences, $Prop(Th_{MVEL}(\varphi)) \subseteq Prop(\Sigma^\varphi)$. For the converse, suppose that $\psi \notin Prop(\Sigma^\varphi)$. By (i), $K_\varphi \not\models \psi$; hence there is a model K for φ with a state w such that $w \not\models \psi$. Since K is a model for the objective formula φ we have $w \models \varphi$, thus $\varphi \not\models_{MVEL} \psi$.
(ii) $\psi \in \Sigma^\varphi \Leftrightarrow \psi \in Prop(\Sigma^\varphi) = Th_{PropCalc}(\{\varphi\} \cup D) \Leftrightarrow D \models_{PropCalc} \varphi \rightarrow \psi$.

The notion of an epistemic state determined by an honest formula naturally leads to a non-monotonic entailment relation which, given an honest formula φ, yields all things that are known in the epistemic state associated with φ.

Definition 3.8 *(non-monotonic entailment)* For an honest φ, we define $\varphi \mid\!\sim_{MVEL} \psi$ if, and only if, $\psi \in \Sigma^\varphi$.

4 Linking *MILORD II* to *MVEL*

A Knowledge Base (KB) in *MILORD II* (see Fig. 1) consists of a set of hierarchically interconnected modules. A module can be understood as a functional abstraction, by fixing both the set of components it needs as input and the type of results it can produce. Each module contains an Object Level Theory (*OLT*) and a Meta-Level Theory (*MLT*) interacting through a reflective mechanism. This meta-level approach, based on reflection techniques and equipped with a declarative backtracking mechanism, is used by the system to deal with non-monotonicity

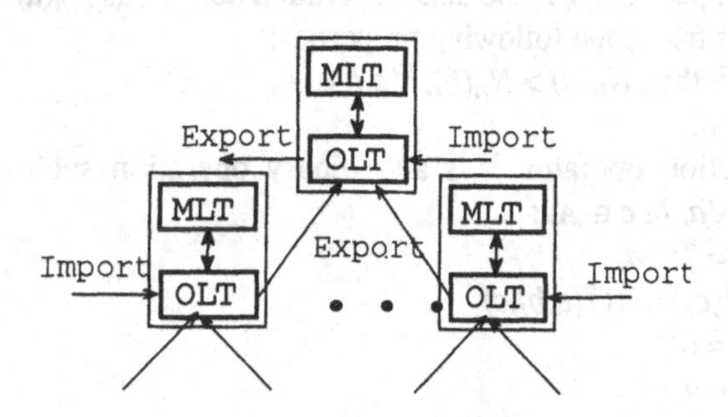

Figure 1. Milord II KB structure.

Reflection mechanisms may be understood, in this context, as a clear separation between domain and control knowledge. Besides, the system provides at the object level a family of representation languages based on multiple-valued logics to deal with approximate reasoning, where the sets of truth-values stand for different scales of linguistic terms representing degrees of truth. Because of the scope of the paper, we will focus on the logical description of different reasoning components of a *MILORD II* module, i.e. the object- level, meta-level and reflection components, as well as on the reasoning dynamics of module as a whole.

4.1 Object and meta level *MILORD II* languages

The propositional language $OL_n = (A_n, \Sigma_O, C, OS_n)$ of the *object level* is defined by:

- An ordered **set of truth-values** A_n
- A **Signature** Σ_O, composed of a set of atomic symbols plus *true* and *false*
- A set of **Connectives** $C = \{\neg, \wedge, \rightarrow\}$
- A set of **Sentences** OS_n = *Mv-Literals* $\cup$ *Mv-Rules*

Sentences are pairs of classical-like propositional sentences and intervals of truth-values. The classical-like propositional sentences are built from a set of atomic symbols and the above set of connectives, but restricted to literals and rules. Thus, the sentences of the language are of the following types:

- *Mv-Atoms*: $\{ (p,V) \mid p \in \Sigma_O$ and V is an interval of truth-values of $A_n\}$
- *Mv-Literals*: $\{ (p,V) \mid (p, V) \in$ *Mv-atoms* or $p = \neg q$ and $(q, V) \in$ *Mv-atoms*$\}$
- *Mv-Rules*: $\{ (p_1 \wedge p_2 \wedge ... \wedge p_n \rightarrow q, V^*) \mid p_1, p_2, ..., p_n$ and q are literals, $V = [a, 1]$ is an upper interval of truth-values of A_n, with $a > 0$, and $p_i \neq p_j, p_i \neq \neg p_j, q \neq p_j, q \neq \neg p_j$ for all i and j $\}$

The semantics is basically determined by the *conjunction* and *implication* connective operators, i.e. by the truth-value algebra. Such operators enjoy most of the properties of usual connectives in Fuzzy Logic since they are the finite counterpart of the well-known *t-norms* and related operators.

Definition 4.1 A *MILORD II algebra of truth-values* is a finite and ordered algebra $A_{n,T} = <A_n, 0, 1, T, I_T, N>$ such that:

1) A_n is a chain of n elements $0 = a_1 < a_2 < ... < a_n = 1$, denoting an ordered set of truth-values, where *0* and *1* are the booleans *False* and *True* respectively.

2) The negation operator N_n is the unary operation defined as $N_n(a_i) = a_{n-i+1}$, the only one that fulfils the following properties:

N1: if $a < b$ then $N_n(a) > N_n(b), \forall a,b \in A_n$

N2: $N_n^2 = \mathrm{Id}$.

3) The conjunction operator T is any binary operation such that the following properties hold $\forall a, b, c \in A_n$:

T1: $T(a,b) = T(b,a)$

T2: $T(a,T(b,c)) = T(T(a,b),c)$

T3: $T(0,a) = 0$

T4: $T(1,a) = a$

T5: if $a \leq b$ then $T(a,c) \leq T(b,c)$ for all c

4) The implication operator I_T is defined by residuation with respect to T, i.e.

$I_T(a,b) = Max\{c \in A_n \mid T(a,c) \leq b\}$

From now on, we will take a generic truth-value algebra $A_{n,T} = <A_n, 0, 1, T, I_T, N>$ as the truth-value algebra on top of which formulas will be interpreted. Notice that having truth-values in the sentences enables us to define a classical satisfaction relation in spite of the models being multiple-valued assignments.

- **Models** M_ρ are defined by valuations ρ from Σ to A_n such that $\rho(true) = 1$ and $\rho(false) = 0$, and they extend to other first components of sentences as follows:

  $\rho(\neg p) = N_n(\rho(p))$

  $\rho(p_1 \wedge p_2 \wedge ... \wedge p_n \rightarrow q) = I_T(T(\rho(p_1), ..., \rho(p_n)), \rho(q))$
- The **Satisfaction Relation** between models and sentences is defined by:

  $M_\rho \models (p, V)$ *iff* $\rho(p) \in V$
- The **Semantical entailment** between sets of sentences Γ and sentences A is defined as usual: $\Gamma \models A$ iff for any model M_ρ, $M_\rho \models \Gamma$ implies $M_\rho \models A$

The object level deduction system $(OL_n, \vdash_O)$ is based on the following axiom scheme:

(AS) $(\varphi, [0, 1])$

on the following axioms

(A-1) *(true, 1)* (A-2) *(false, 0)*

and on the following inference rules:

(RI-1) *weakening*: $(\varphi, V_1) \vdash_O (\varphi V_2)$, where $V_1 \subseteq V_2$,

(RI-2) *not-introd.*: $(p,V) \vdash_O (\neg p, N^*_n(V))$, $(\neg p,V) \vdash_O (p, N^*_n(V))$, for p atom

(RI-3) *composition*: $\{(\varphi,V_1), (\varphi,V_2)\} \vdash_O (\varphi,V_1 \cap V_2)$

(RI-4) *specialization*: for any literals $p_1 \dots p_n$ and q

$$\{(p_i,[a_i, b_i]), (p_1 \wedge p_2 \wedge \dots \wedge p_n \rightarrow q,[a_r, 1])\} \vdash_O (p_1 \wedge p_2 \wedge \dots \wedge p_{i-1} \wedge p_{i+1} \wedge \dots \wedge p_n \rightarrow q,[T(a_i,a_r), 1]).$$

It is easy to check that this deductive system is sound. It is also complete considering some restrictions in the structure of theories [Puyol-Gruart et al, 92].

The *meta-level language* is based on a first order logic with special meta-predicates *WK* and *ASS*. We will concentrate only on the formalization of meta-predicate *WK*, leaving meta-predicate *ASS* out of the scope of this paper. Given a current Object Level Theory (*OLT*), $WK(p,V)$ will be true in the corresponding meta-level theory *MLT* when (p, V) belongs to *OLT*, i.e. $OLT \vdash_O (p, V)$. The connection between object and meta languages is done through the next reification and reflection rules:

$$\frac{(p, V) \in OLT}{\vdash_M WK(p,V)} \qquad \frac{(p, V) \notin OLT}{\vdash_M \neg WK(p,V)} \qquad \frac{\vdash_M ASS(p, V)}{\vdash_O (p, V)}$$

Besides meta-predicates *WK* and *ASS*, there also exists meta-predicate *K*. The semantics of $K(p, V)$ is that V is the minimal interval such that the proposition (p, V) belongs to the *OLT*. This meta-predicate *K* is definable from meta-predicate *WK* as follows:

$$K(p, [a_i, a_j]) \equiv WK(p, [a_i, a_j]) \wedge \neg WK(p, [a_{i+1}, a_j]) \wedge \neg WK(p, [a_i, a_{j-1}])$$

4.2 Embedding of the *MILORD II* object level into *MVEL*

MILORD II object-level formulas will be interpreted as objective *MVEL* formulas in a first step. The second step will be to interpret the Meta-level Theory in a particular reasoning stage as the epistemic state determined by the formulas of the current Object-level Theory formulas. We will use the following abbreviations:

- $(\geq a_i)A$ for $(a_i)A \vee (a_{i+1})A \vee \dots \vee (1)A = \bigvee_{j \geq i} (a_j)A$,
- $(\leq a_i)A$ for $(0)A \vee (a_1)A \vee \dots \vee (a_i)A = \bigvee_{j \leq i} (a_j)A$
- $(a_i{:}a_j)A$ for $(a_i)A \vee (a_{i+1})A \vee \dots \vee (a_j)A = \bigvee_{i \leq k \leq j} (a_k)A$

and specially, the following abbreviations are important to understand the intuition behind the embedding of *MILORD II* into *MVEL*:

- $(z)(\neg A)$ for $(N(z))A$
- $(z)(A \wedge B)$ for $\bigvee_{T(x,y)=z} (x)A \wedge (y)B$
- $(z)(A \vee B)$ for $\bigvee_{S(x,y)=z} (x)A \wedge (y)B$
- $(z)(A \rightarrow B)$ for $\bigvee_{I(x,y)=z} (x)A \wedge (y)B$

where $S(x,y) = N(T(N(x),N(y)))$. Notice that introducing these abbreviations into the fragment of non-modal *MVEL* formulas amounts to consider a propositional language $L(At)$ built in the usual way from the set of propositional symbols *At*, unary connectives such as $\neg$, interval indicators (V), being V any interval of truth-values, and from the binary connectives $\wedge$, $\vee$, and $\rightarrow$. Due to the many-valued nature of both

MILORD and *MVEL*, it is easy to establish a one-to-one correspondence between *MVEL* possible worlds and *MILORD II* object-level models when the corresponding languages are built over the same set of atomic symbols, denoted Σ_O in *MILORD II* and *At* in *MVEL* (see section 2). The above abbreviations give the hint of how to translate *MILORD II* object-level formulas into *MVEL* formulas. We will denote by φ^* the translation of a *MILORD II* formula φ following the next table schema.

MILORD II ⇒	*MVEL*
mv-literals: $(p, [a_i, a_j])$	$(a_i{:}a_j)p$
mv-rules: $(p_1 \wedge ... \wedge p_n \rightarrow q, [a_i, 1])$	$(\geq a_i)\,(p_1 \wedge ... \wedge p_n \rightarrow q)$

In order to make explicit the semantical equivalence of the pairs of formulas in the above table, let $\Sigma_O = At$, and denote by w both a *MVEL* possible world and the corresponding *MILORD II* model. We will write $\models_{MILORD}$ and $\models_{MVEL}$ to differentiate between the *MILORD* and *MVEL* entailment relations.

Lemma 4.2 For any $A, B \in L(At)$ and for any possible world w, the following properties hold:

- $w \models_{MVEL} (z)(\neg A)$ iff $z = N(w(A))$
- $w \models_{MVEL} (z)(A \wedge B$ iff $z = T(w(A) \wedge w(B))$
- $w \models_{MVEL} (z)(A \vee B)$ iff $z = S(w(A) \wedge w(B))$
- $w \models_{MVEL} (z)(A \rightarrow B)$ iff $z = I(w(A) \wedge w(B))$

where w is supposed to be extended to non-atomic sentences of $L(At)$ using operations N, T, S and I to interpret negation, conjunction, disjunction and implication respectively.

As a consequence of this lemma, next proposition proves that the embedding of the *MILORD II* object language in *MVEL* is faithful.

Proposition 4.3 For all $\varphi \in$ *Mv-literals* $\cup$ *Mv-rules*, we have: $w \models_{MILORD} \varphi$ if, and only if, $w \models_{MVEL} \varphi^*$, where φ^* is the translation of φ according to the above table.

Proof. Let $\varphi \in$ *Mv-Literals*, i.e. $\varphi = (p, [a_i, a_j])$. By definition, $w \models_{\text{MILORD}} \varphi$ iff $w(p) \in [a_i, a_j]$, that is, iff there exists $l \in [i, j]$ such that $w(p) = a_l$, and hence iff $w \models_{\text{MVEL}} \bigvee_{i \leq k \leq j} (a_k)p$. Now let $\varphi = (p_1 \wedge p_2 \wedge ... \wedge p_n \rightarrow q,\ [a_i, 1]) \in$ *Mv-rules*. Again by definition $w \models_{\text{MILORD}} \varphi$ iff $w(\varphi) \geq a_i$, i.e. iff $r \geq i$, being $a_r = I(T(w(p_1), w(p_2), ..., w(p_n)), w(q))$, that is $a_r = I(w(p_1 \wedge p_2 \wedge ... \wedge p_n), w(q))$, and thus, $w \models_{\text{MVEL}} (a_r)(p_1 \wedge p_2 \wedge ... \wedge p_n \rightarrow q)$, and consequently $w \models_{\text{MVEL}} \bigvee_{j \leq k} (a_k)(p_1 \wedge p_2 \wedge ... \wedge p_n \rightarrow q)$. Conversely, if $w \models_{\text{MVEL}} \bigvee_{j \leq k} (a_k)(p_1 \wedge p_2 \wedge ... \wedge p_n \rightarrow q)$ there exists only one index r not smaller than i such that $w \models_{\text{MVEL}} (a_r)(p_1 \wedge p_2 \wedge ... \wedge p_n \rightarrow q)$ and moreover it must be the case that $w(p_1 \wedge p_2 \wedge ... \wedge p_n \rightarrow q) = a_r \geq a_i$. Therefore, $w \models_{\text{MILORD}} (p_1 \wedge ... \wedge p_n \rightarrow q,\ [a_i, 1])$ holds.

As a consequence of this theorem, we can consider the *MILORD II* object-level language as a fragment of *MVEL*, consisting of formulas of type $(\geq z)A$, or more generally, of formulas of type $(x{:}y)A$, where A is an objective implication with a conjunction (possibly empty) of literals as premise and with a literal as conclusion. Next section is devoted to extend the embedding when the *MILORD II* meta-level language is also considered.

5 Modal Interpretation of *K* and *WK* Meta-Predicates

Our aim in this paper is to prove that meta-predicates can be given a modal interpretation when the meta-level architecture of *MILORD II* is used to model non-monotonic epistemic reasoning in the sense that the meta-level component system is to reason about the state of knowledge of the object-level component. A MILORD-II reasoning flow inside a module can be modelled as a sequence of object and meta level theories as depicted in Fig. 2.

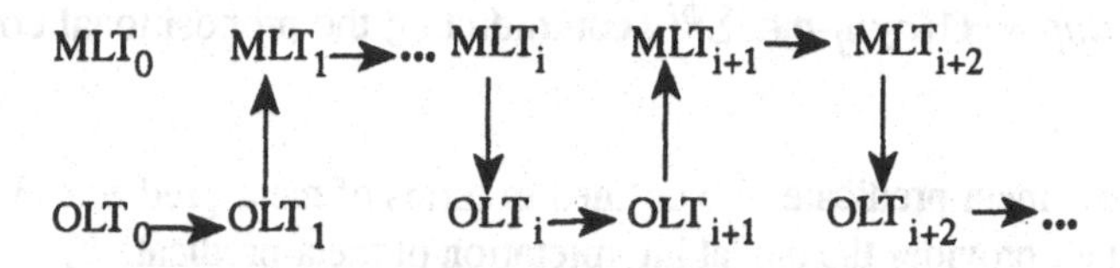

Figure 2. Dynamics of the reasoning process in MILORD II.
The symbol → stands for deductive closure (w.r.t the corresponding languages), ↑ for the upwards reflection and ↓ for the downwards reflection.

In this setting, the object-level theory at stage i OLT_i is considered to be the closure, w.r.t. the object-level deduction system (see section 4), of OLT_{i-1}. On the other hand, it is also quite natural to represent the state of knowledge of the object-level component at a given step i by the set of mv-literals OLT_i contains. This would be very close to the many-valued counterpart of the concept of *partial model* used in the *BMS* system ([Tan & Treur, 91], [Tan, 92]). More concretely, if we denote by $Lit(OLT_i)$ the set of mv-literals belonging to OLT_i, the corresponding meta-theory MLT_i is built as follows:

$$MLT_i = MLT_o \cup \{WK(A) \mid A \in Lit(OLT_i)\} \cup \{\neg\, WK(A) \mid A \notin Lit(OLT_i)\}$$

where MLT_o is the set of initial meta-rules of the Knowledge Base. Notice that, because of being OLT_i deductively closed, and in particular under the *weakening* rule of inference RI-1, if $(p, V) \in Lit(OLT_i)$, then $(p, U) \in Lit(OLT_i)$ for all $U \supseteq V$. The idea to modally interpret the meta-predicate *WK* (the *K* predicate can be defined in terms of *WK*) is to consider the proposition

$$\varphi_i = \bigwedge \{ A^* \mid A \in Lit(OLT_i) \}$$

describing the state of knowledge of the object level system at the step i, where A^* denotes the translation of A to its corresponding *MVEL* formula (see previous section). Inside *MVEL*, φ_i is a finite conjunction of objective formulas and thus it is honest. Therefore, it makes sense to consider the epistemic state Σ^{φ_i}. In this context, the modal interpretation of meta-predicate *WK* is given by the following theorem.

Theorem 5.1. For any literal p it holds:

$$WK(p, [a_i, a_j]) \in MLT_i \text{ iff } \Box(a_i{:}a_j)p \in \Sigma^{\varphi_i}$$

or equivalently, in terms of the non-monotonic entailment:

$$WK(p, [a_i, a_j]) \in MLT_i \text{ iff } \varphi_i \mid\!\sim_{MVEL} \Box(a_i{:}a_j)p$$

Proof: By definition of the MILORD II upward reflection rules, $WK(p, [a_i, a_j]) \in \mathrm{MLT_i}$ if, and only if, there exists $V \in Int(A_n)$ such that $(p, V) \in$ OLT$_i$ and $V \subseteq [a_i, a_j]$, but since OLT$_i$ is deductively closed, this amounts to say that $WK(p, [a_i, a_j]) \in$ MLT$_i$ iff $(p, [a_i, a_j]) \in Lit(\mathrm{OLT}_i)$, and consequently, by definition of φ_i, this is the necessary and sufficient condition for $(a_i{:}a_j)p$ to appear in φ_i. Therefore, on the one hand, if $(a_i{:}a_j)p$ appears in φ_i then it must belong to the epistemic state Σ^{φ_i} too, as well as $\Box(a_i{:}a_j)p$. On the other hand, suppose $\Box(a_i{:}a_j)p$ belongs to the epistemic state Σ^{φ_i} and therefore so does $(a_i{:}a_j)p = \bigvee_{i \le k \le j}(a_k)$. To prove that $(a_i{:}a_j)p$ is in φ_i, suppose not: then $D \not\models \varphi_i \rightarrow (a_i{:}a_j)p$. So, by Proposition 3.7, $(a_i{:}a_j)p \notin \Sigma^{\varphi_i}$, and hence, as Σ^{φ_i} is MVEL-stable $\neg\Box(a_i{:}a_j)p \in \Sigma^{\varphi_i}$ and hence $\Box(a_i{:}a_j)p \wedge \neg\Box(a_i{:}a_j)p \in \Sigma^{\varphi_i}$, contradicting the propositional consistency of a stable set.

According to how meta-predicate K is defined in terms of meta-prediacte WK (see section 4.1), next corollary provides the modal interpretation of meta-predicate K.

Corollary 5.2. For any literal p it holds that:

$$K(p, [a_i, a_j]) \in MLT_i \quad \text{iff} \quad \Box(a_i{:}a_j)p \wedge \neg\Box(a_{i+1}{:}a_j)p \wedge \neg\Box(a_i{:}a_{j-1})p \in \Sigma^{\varphi_i}$$

The above results show that it actually makes sense to provide a modal interpretation for the meta-predicates, shown in the next table, completing the one of section 4.2.

MILORD II $\Rightarrow$	*MVEL*
mv-literals: $(p, [a_i, a_j])$	$(a_i{:}a_j)p$
mv-rules: $(p_1 \wedge \ldots \wedge p_n \rightarrow q, [a_i, 1])$	$(\geq a_i)(p_1 \wedge \ldots \wedge p_n \rightarrow q)$
mv-meta-literals: $WK(p, [a_i, a_j])$, $K(p, [a_i, a_j])$	$\Box(a_i{:}a_j)p$, $\Box(a_i{:}a_j)p \wedge \neg\Box(a_{i+1}{:}a_j)p \wedge \neg\Box(a_i{:}a_{j-1})p$
mv-meta-rules[1]: $WK(p,[a,1]) \wedge \neg WK(q,[b,1]) \rightarrow ASS(r, [c,1])$	$\Box(\geq a)p \wedge \neg\Box(\geq b)q \rightarrow \Delta(\geq c)r$

As a summary, we can consider the *MILORD II* Object and Meta Level Languages as fragments of *MVEL*, namely:

- The Object Level Language corresponds to formulas of type $(\geq z)A$, or more generally $(x{:}y)A$, where A is an objective implication with a conjunction (possibly empty) of literals as premise and with a literal as conclusion.
- The Meta Level Language corresponds to epistemic implicative formulas of the general form $\Box(\geq x)p \wedge \neg\Box(\geq y)q \rightarrow \Delta(\geq z)r$.

6 Example

Consider the next example of a *MILORD II* KB:

Truth-values {0 = false < moderately-possible (mp) < possible (p) <
< very-possible (vp) < certain = 1}

Connectives: T(x,y) = min(x,y)

[1] mv-meta-rules, the behavior of ASS and its corresponding Δ modal operator and their role in the reflection downward process are not dealt within this paper (see [Meyer & van der Hoek, 93b], for a possible treatment of a reflection downward process).

$I(x,y) = 1$, if $x \le y$; $I(x,y) = y$, otherwise

Object level formulas (mv-literals, mv-rules):

(Low_WBC, [vp, 1])

(Low_WBC → Compromised_Host, [vp, 1)

(Gram_neg_inf → Est_Coli, [p, 1])

(I_Dont_know → Gram_Pos_inf, [vp, 1])

Meta Level formulas (mv-meta-rules):

WK(Compromised_Host, [p, 1]) ∧ ¬ WK(Gram_Pos_inf, [modp, 1]) →
ASS(Gram_neg_inf, [vp, 1])

This KB is rewritten in *MVEL* formalism as:

Object level (Objective formulas):

(≥vp) low_WBC

(≥vp) (low_WBC → Compromised_Host)

(≥p) (Gram_neg_inf → Est_Coli)

(≥vp) (I-Dont_know → Gram_Pos_inf)

Meta level (Modal Formulas):

$[\,\square\,(\geq p)\text{Compromised\_host} \wedge \neg\square\,(\geq modp)\text{Gram\_Pos\_inf}\,] \rightarrow \Delta\,(\geq vp)\text{Gram\_Neg\_inf}$

The dynamics of the reasoning process is represented in the next figure and described below.

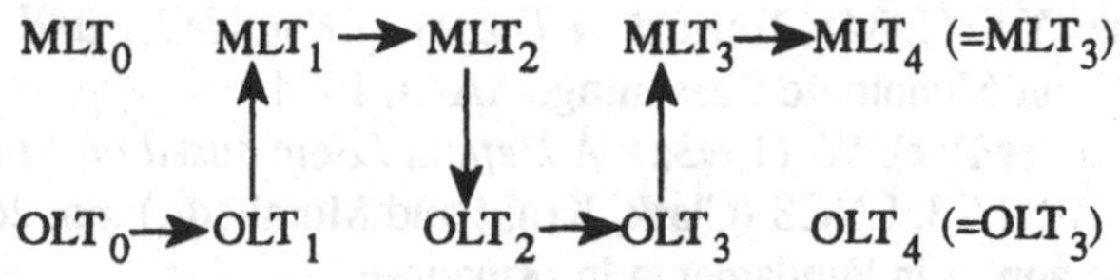

Figure 3. Dynamics of the reasoning process in the example considered.

Begin

1. $OLT_0 = KB_{object}$, $MLT_0 = KB_{meta}$
2. $OLT_1 = OLT_0 + \{(\geq vp)\text{Compromised\_Host}\}$
3. $MLT_1 = MLT_0 + \{\square\,(\geq vp)\text{Compromised\_host}, \square(\geq vp)\text{low\_WBC}\} +$
   $\{\neg\square\,(a_i{:}a_j)\text{Gram\_Pos\_inf}, \neg\square\,(a_i{:}a_j)\text{Gram\_neg\_inf}, \neg\square\,(a_i{:}a_j)\text{Est\_Coli},$
   $\neg\square\,(a_i{:}a_j)\text{I-Dont\_know} \mid (a_i{:}a_j) \neq (0{:}1)\}$
4. $MLT_2 = MLT_1 + \{\Delta\,(\geq vp)\,\text{Gram\_Neg\_inf}\}$
5. $OLT_2 = OLT_1 + \{(\geq vp)\,\text{Gram\_Neg\_inf}\}$
6. $OLT_3 = OLT_2 + \{(\geq p)\,\text{Est\_Coli}\}$
7. $MLT_3 = MLT_0 + \{\square\,(\geq vp)\text{Compromised\_host}, \square(\geq vp)\text{low\_WBC}\} +$
   $+ \{\square\,(\geq vp)\,\text{Gram\_Neg\_inf}, \square\,(\geq p)\,\text{Est\_Coli}\} +$
   $+ \{\neg\square\,(a_i{:}a_j)\text{Gram\_Pos\_inf}, \neg\square\,(a_i{:}a_j)\text{I-Dont\_know} \mid (a_i{:}a_j) \neq (0{:}1)\}$
8. $MLT_4 = MLT_3$
9. $OLT_4 = OLT_3$

Stop

In terms of epistemic states, this reasoning flow is equivalently expressed as follows:

a. $KB_{object} \vdash (\geq vp)\text{Compromised\_host}$

b. $\varphi_1 = (\geq vp)\ \text{low\_WBC}\ (\geq vp) \wedge (\geq vp)\text{Compromised\_host}$

c. $\varphi_1 \vdash \Box$ $(\geq vp)$Compromised_host (or equivalently, $\Box(\geq vp)$Compromised_host $\in \Sigma^{\varphi 1}$)
$\varphi_1 \vdash \neg \Box$ $(\geq modp)$Gram_Pos_inf (or equivalently, $\neg\Box(\geq modp)$Gram_Pos_inf $\in \Sigma^{\varphi 1}$)
d. $\varphi_1 \cup KB_{meta} \vdash \Delta$ $(\geq vp)$ Gram_Neg_inf
e. $KB_{object} \cup \{(\geq vp)$Gram_Neg_inf$\} \vdash (\geq p)$ Est_Coli
f. $\varphi_2 = \varphi_1 \wedge (\geq vp)$Gram_Neg_inf $\wedge (\geq p)$ Est_Coli
g. $\varphi_2 \vdash \Box$ $(\geq p)$Est_Coli

7 Conclusions

In this paper, a many-valued extension of the concept of epistemic states is presented. Its application to the meta-level architecture *MILORD II* is developed by giving a modal epistemic interpretation of some meta-predicates that allow reasoning about the object level knowledge state. In this sense, it is mainly a contribution in studying the relationship between special meta-predicates in meta-level architectures for non-monotonic reasoning, such as *MILORD II* or *BMS* , and modal operators in non-monotonic epistemic logics. Although it needs further research, this work has also a clear relation with the formalization of default reasoning via epistemic logics as in [Meyer & v.d. Hoek, 93a-b].

References

Halpern J.Y., Moses Y.O.(1984): *Towards a Theory of Knowledge and Ignorance*, Proc. Workshop on Non-Monotonic Reasoning, AAAI, 1984

Meyer J.-J. Ch., van der Hoek W. (1993a): *A Default Logic based on Epistemic States*, in Proc. ECSQUARU '93, LNCS (Clark, Kruse and Moral eds.), pp. 265-273, Springer Verlag. Also to appear in Fundamenta Informaticae.

Meyer J.-J. Ch., van der Hoek W. (1993b): *An Epistemic Logic for Defeasible Reasoning Using a Meta-Level Architecture Metaphor*, Technical Report IR-329, Free University Amsterdam.

Meyer J.-J. Ch., van der Hoek W. (1995): *Epistemic Logic for AI and Computer Science*, forthcoming.

Puyol-Gruart J., Godo Ll., Sierra C. (1992): *A Specialization Calculus to improve Expert Systems Communication*, IIIA Research Report 92-08, May. Short version in Proc. of ECAI'92, (Bernd Neumann Editor), Vienna, pp. 144-148.

Sierra C., Godo L. (1992): *Modularity, Uncertainty and Reflection in MILORD II*", Proc of the 1992 IEEE International Conference on Systems, Man and Cybernetics. Chicago, October, Vol. 1, pp. 144-148.

Sierra C., Godo L. (1993): *Specifiying Simple Schedulling Tasks in a Reflective and Modular Architecture*, in: Treur J. and Th. Wetter (eds.) Specification of Complex Reasoning Systems, Ellis Horwood pp. 199-232.

Tan Y.H. (1992): *BMS - A Meta-level Approach to Non-Monotonic Reasoning*, in van der Hoek et al. (eds.) Non-Monotonic Reasoning and Partial Semantics, Ellis Horwood.

Tan Y. H., Treur J. (1991): *A Bi-modular Approach to Non-Monotonic Reasoning*, in Proc. First World Congress on the fundamentals of AI, (M. De Glas & D. Gabbay, eds.), WOCFAI-91, pp. 461-475, Paris.

7. CHAOS

Acoustical Chaotic Fractal Images for Medical Imaging

Woon S Gan

Acoustical Services Pte Ltd
29 Telok Ayer Street
Singapore 0104
Republic of Singapore

Abstract. In this paper, we incorporate in diffraction and wave nature for the ultrasound propagation in a fractal structure. Ultrasound propagation in a fractal and inhomogeneous medium causes chaotic scattering. This gives rise to chaotic fractal images. We call this chaotic imaging. This has application in medical imaging as human heart and human brain are known to have fractal structure. The mathematical model of Diffusion Limited Aggregation (DLA) is used ti describe the fractal structure. To describe the wave nature, the KZK equation with parabolic approximaton is used. The statistical properties of the fluctuations of wave field is characterized by using correlation functions. These correlation functions are modified for the DLA model. The method of iteration is used in the inverse part of the problem. It is found that nonlinear inversion displays chaotic behaviours. The iteration produces a sequence of velocity estimates. This forms the chaotic velocity images.

1 Introduction

In recent years chaos theory has found application in many fields such as the earth science [1,2], turbulence [3], laser science [4], etc. However, not much work has been done which relate the chaos theory with inverse problems [5,6,7], maybe because the inversion is purely mathematic and physically non-realistic. In this paper we apply chaos theory to the inhomogeneous medium, taking account of wave nature and diffraction. A theory of diffraction tomography is then formulated which yields chaotic velocity images.

2 Forward Problem

The purpose here is to determine the scattered field. the nonlinear wave equation is used here. Various approaches to the solution of the equation have been attempted. One dimensional solutions, such as those of Blackstock [8] can provide some useful information for medical ultrasound systems but are unable to reproduce the fine detail and phase variations seen in "real" pressure fields. Highly diffracted and focused pressure fields require more rogorous treatment. Smith and Beyer [9] commented on the "lack of appropriate theoretical analysis" when they published nonlinear measurements on a focussed acoustic source operated of 2.3MHz. One of the most significant theoretical advances came in 1969 when Zabolotskaya and Khokhlov [10] published a solution of the nonlinear wave equation for a confined sound beam in which it was assured that the shape of the wave varies, slowly both along the beam and transversely to it. In 1971 Kuznetsov [11] extended their treatment to include absorption and the resulting equation is now widely known as the KZK equation. He also obtained solution which is known as the parabolic approximation to the nonlinear wave equation and is equivalent to the paraxial approximation used in optics. The KZK equation accounts for diffraction, absorption and nonlinearity and is valid for circular apertures that are many wavelengths in diameter and will accept arbitrary source conditions.

2.1 General Wave Equation

The KZK equation is a nonlinear wave equation for a scalar potential Φ, in consideration of the dynamics of a viscous heat conducting fluid. It is correct to the second order with terms for diffraction, absorption and nonlinearity.

$$\frac{\partial^2\Phi}{\partial t^2} - c^2\nabla^2\Phi = \frac{\partial}{\partial t}[2\alpha c k^2\nabla^2\Phi + (\nabla\Phi)^2 + \frac{B}{2A}\frac{1}{c^2}(\frac{\partial\Phi}{\partial t})^2] \quad (1)$$

where c =speed of sound, α =absorption coefficient, k =wave number.

The left-hand side of eq(1) is the three dimensional linear Helmholtz wave equation. Of the three terms on the right-hand side, the first term is a linear term and accounts for absorption, the second term is due to convective nonlinearity in the equation of state.

2.2 Parabolic Approximation

Kuznetsov [11] also showed that eq (1) could be simplified by approximation, in the case of a quasi-plane wave field and the Laplacian (∇^2) can be replaced by the transverse Laplacian (∇_1^2).

A circular aperture that is many wavelengths in diameter (ie, ka is large) falls in this category since most of the energy is confined to a beam in the axial direction. This is known as the parabolic (or paraxial) approximation and is equivalent to the Fresnel approximation that is sometimes used in the diffraction integral for near-field calculations. Kuznetsov(11) 's parabolic approximation can be expressed in a normalised form:

$$[4\frac{\partial^2}{\partial\tau\partial\sigma} - \nabla_1^2 - 4\alpha R_0\frac{\partial^3}{\partial\tau^3}]\overline{P} = 2\frac{R_0}{l_D}\frac{\partial^2}{\partial\tau^2}\overline{P}^2 \quad (2)$$

whre $\overline{P} = (\underline{P}/\underline{P_0})$ is the acoustic pressure normalised by the source pressure and

$\tau = (\omega t - kz)$ is retarded time, ie includes a phase term for a plane wave travelling in the z direction, R_0=Rayleigh distance=$\frac{ka^2}{2}$, l_D=shock distance and a=aperture radius.

In this equation σ is the Rayleigh distance and ζ is the radial coordinate normalised by the aperture radius, i.e. $\sigma = \frac{2z}{ka^2}$ and $\zeta = r/a$.

A trial solution is then assumed in the form of a Fourier series (for the time wave form) with amplitude and phase that are functions of the spatial coordinates, ie

$$P(\sigma,\zeta,\tau) = \sum_{n=1}^{\infty} q_n(\sigma,\zeta,\tau)\sin[n\tau + \Psi_n(\sigma,\zeta,\tau)]$$

or

$$P(\sigma,\zeta,\tau) = \sum_{n=1}^{\infty} g_n(\sigma,\zeta,\tau)\sin(n\tau) + h_n(\sigma,\zeta,\tau)\cos(n\tau) \qquad (3)$$

where $g_n = q_n \cos\varphi_n$, $h_n = q_n \sin\varphi_n$ and n is the harmonic number with n=1 representing the fundamental frequency, q=Fourier solution amplitude and ψ=Fourier solution phase. Substituting the trial solution (3) in eq (2) and collecting terms in $\sin(n\tau)$ and $\cos(n\tau)$ gives a set of coupled differential equations for g_n and h_n:

$$\frac{\partial g_n}{\partial \sigma} = -n^2 \alpha R_0 g_n + \frac{1}{4n}\nabla_1^2 h_n + \frac{nR_0}{2l_D}[\frac{1}{2}\sum_{k=1}^{n-1}(g_k g_{n-k} - h_k h_{n-k}) - \sum_{p=n+1}^{\infty}(g_{p-n} g_p + h_{p-n} h_p)] \qquad (4)$$

$$\frac{\partial h_n}{\partial \sigma} = -n^2 \alpha R_0 h_n + \frac{1}{4n}\nabla_1^2 g_n + \frac{nR_0}{2l_D}[\frac{1}{2}\sum_{k=1}^{n-1}(h_k g_{n-k} - g_k h_{n-k}) - \sum_{p=n+1}^{\infty}(h_{p-n} g_p + g_{p-n} h_p)] \qquad (5)$$

Eqs(4) and (5) form the basis of the numeriacal solution which can be implemented in a FORTRAN program.

3 Fractal Structure as a Diffraction Medium

Sound propagation is related to the elastic properties of the medium. Most solids have tensorial elasticity and can support transverse as well as longitudinal sound waves. Hence, it is necessary to investigate the nature of their elasticity first. To attack this

problem, fractals is formulated as a growth problem and we use the Diffusion Limited Aggregration (DLA) model [12], a growth model. For DLA, one needs to calculate the growth probabilities. Here one focuses on multi-fractality's relation to transport properties of the fractal medium.
Let $P(r,t)=$ probability of finding the random walker on sites at a fixed distance r from the starting point. The probability $P(r,t)$ to find the walker at l at time t is a Gaussian,

$$P(l,t)=P(o,t)\exp(-l^2/4Dt) \tag{6}$$

where D=fractal dimension. The moments of the probability density $<P^q(r,t)>$ can be written as a convolution integral:

$$<P^q(r,t)>=\int_0^\infty Q(r/l)P^q(l,t)dr \tag{7}$$

where $Q(r/l)$=probability of finding the sites separated by a chemical distance l and Euclidean distance r. The chemical distance is the shortest path between two sites on the cluster. In the general case, the qth moment $<P^q(r,t)>$ can be written as

$$<P^q(r,t)>=\frac{1}{N_r}\sum_{i=1}^{N_r}P_i^q(r,t) \tag{8}$$

where the sum is over all N_r sites located a distance r from the origin (N_r may include many configurations or a single configuration with a very large number of cluster sites). The sum equation (8) can be separated into sums over different l values (N_m values of l_m):

$$\begin{aligned}<P^q(r,t)>&=\frac{1}{N_r}\{\sum_{i=1}^{N_1}P_i^q(l_1,t)+\sum_{i=1}^{N_2}P_i^q(l_2,t)+\cdots\cdots\}\\&=\frac{1}{N_r}\sum_m N_m\times\frac{1}{N_m}\sum_{i=1}^{N_m}P_i^q(l_m,t)\\&=\frac{1}{N_r}\sum N_m<P^q(l_m,t)>\end{aligned} \tag{9}$$

This covers all the scattering points within the fractal medium. In this problem it is assumed that the random walker starts at the origin D and after t time steps can be found at r[x] with very different probabilities at differenc sites.
For the scattering of sound by a fractal medium one needs to treat all sites of the fractal as starting points and the various parameters like sound velocity, attenuation coefficients etc have to be modified for fractals. Fractal media are characterized by not having a very characteristic length scale and they have a very inhomogeneous density distribution. One can therefore expect to find very different physical properties in

materials with fractal structuure compared to the ordinary solids. Furthermore, real fractals are disordered and highly irregular. In some sense they can be regarded as ideally disordered materials. In conventional diffraction tomography theory, one considers only scattering by one point by ignoring the object size. This is known as Born approximation. Here the object size is taken into account as consisting of several scattering points and all sites of the fractal are considered as scattering points. We call this type of diffraction "fractal diffraction".

3.1 Wave Scattering Modified by the Fractal Medium

The expression for the scattered acoustic pressure wavefield is modified by the correlation coefficient which contains the fractal dimension of the medium. We have obtained scattered wavefield amplitude fluctuation as

$$p(\sigma,\zeta,\omega t-kz)=\sum_{n=1}^{\infty}q_n(\sigma,\zeta,\omega t-kz)\sin[n(\omega t-kz)\bullet$$
$$+\Psi_n(\sigma,\zeta,\omega t-kz)] \tag{10}$$

For diffraction of sound wave by a fractal medium, one needs to consider all sites of the fractal as scattering points. For this reason, the correlation coefficient is chosen as eq (9). By modifying eq (10) by eq (9), then the auto-correlation function for the amplitude fluctuation is given by the following formula:

$$\overline{P_1(t)P_2(t)}=\int_0^{R_1}\int_0^{R_2}\iint\iint_{-\infty}^{+\infty}P_1(\sigma_1,\zeta,\omega_1 t-kz_1)P_2(\sigma_2,\zeta,\omega_2 t-kz_2)\bullet$$
$$<P^q(r,t)>d\sigma_1 d\sigma_2 dz_1 dz_2 d\omega_1 d\omega_2 \tag{11}$$

where the coordinates of the receivers are $(R_1,0,0)$ and $(R_2,0,0)$.

The power spectral density (PSD) of the scattered field = Fourier transform of autocorrelation function= $\int_{-\infty}^{\infty}\overline{P_1(t)P_2(t)}e^{-j2\pi ft}dt$ (12)

where f=frequency. The overall amplitude of the acoustic pressure of the scattered field is proportional to the square root of the PSD.

4. Inverse Problem

The purpose here is to obtain sound velocity field in the medium from the scattered sound pressure field. The method of nonlinear iteration will be used. The aim is to obtain velocity images under diffraction tomography format. Our purpose is to apply to medical imaging. The nonlinearity at tomographic inversion here is related to heterogenerity of the human tissue. For instance, the problem of inverse scattering in a homogeneous

background is linear because straight rays are involved. The inverse scattering with small disturbances belongs to quasi-linear as raypaths are smooth curves of small curvature. The vital difficulties in the inverse scattering problems are the typical nonlinearity caused by the strong disturbances which cannot be solved by direct employment of Born or Rytov approximations. In order to study the characteristics of nonlinear inversion, one needs instructions from the theory of nonlinear systems. In nonlinear dynamics, chaos means a state of disorder in a nonlinear system. Usually chaotic solutions of nonlinear system are considered only in forward problem such as nonlinear oscillation etc. In this paper, one is dealing with chaos in the inverse problem of nonlinear iteration instead of occuring in the solution of differential equation.
First of all, the scattered wavefield (acoustic pressure) during the forward problem will be needed in the inverse problem. This will be the square root of the P.S.D given by eqs (12) and (11)

4.1 Reconstruction Algorithm

At a start, the following inhomogeneous planar wave equation is used:

$$[\nabla-\frac{1}{c^2(x)}\frac{\partial^2}{\partial t^2}]u(x,t)=\alpha' \tag{13}$$

The following iteration formulae are introduced:

$$c_k^{-2}(x)=c_{k-1}^{-2}(x)+\gamma_k(x) \tag{14}$$

$$\text{and } u_K(X,T)=u_{K-1}(X,T)+v_K(X,T) \tag{15}$$

$$\text{with } \lim_{k\to\infty} c_k(x)=c(x) \tag{16}$$

where u=scattered wavefield, v and γ are disturbances and γ is proportional to α. Putting eqs.(14) and (15) into (13) yields

$$[\nabla-\frac{1}{c_{k-1}^2(x)}\frac{\partial^2}{\partial t^2}]\tilde{u}(x,t)=\gamma_k(x)\frac{\partial^2 u(\tilde{x},t)}{\partial t^2} \tag{17}$$

The solution of (17) becomes

$$u_k(x,t)=u_{k-1}(x,t)+\iint G_{k-1}(x,x',t,t')u(x',t')\gamma_k(x')dx'dt' \tag{18}$$

where the Green's function satisfies

$$[\nabla-\frac{1}{c_{k-1}^2(x)}\frac{\partial^2}{\partial t^2}]G_{k-1}(x,x',t,t')=-\frac{\partial^2}{\partial t^2}\delta(t-t')\delta(x-x') \tag{19}$$

Now, one puts $v_k = \mu v_{k-1}$ for slow iterations, then

$$u_k(x,t) - u_{k-1}(x,t) = \iint G_{k-1}(x,x',t,t')[u_{k-1}(x,t') + \mu v_{k-1}(x,t')]\gamma_k(x)dxdt' \tag{20}$$

where μ is a small number, $0 < \mu < 1$. Equations (20) and (14) can be used for successive iterations as follows. The initial scattered wavefield (acoustic pressure) can be obtained form the square root of the P.S.D given by eqs (12) and (11). Then γ_1, can be found from (20) by setting $v_k = 0$. Following iteration is to calculate u_k, G_k and v_k, then to solve (20) for γ_k. The iteration produces a sequence of velocity estimates $c_k(x)$, $k = 1, 2, \cdots$.

4.2 Chaotic Solutions

The iteration formulae (14) and (15) are the so-called Poincare maps. In fact they are a type of Standard map. The characteristics of the nonlinear iteration depend upon the Poincare maps together with the iteration parameters. Complicated Poincare maps or nonlinear variation of the iteration parameters can cause chaos iteration and disorder output sequences. The inner entropy for a system given by eq (14) corresponding to inversion errors increases with k. In other words, the output sequence $c_k(x)$ would become disorder when k as well as the inner entropy become larger. When k>5 the output suddenly goes to disorder and irregular, giving rise to chaos. The irregularity is caused by the nonlinearity of the Poincare map due to small errors existing in the data. To plot the Poincare map given by $c_k^{-2}(x)$ versus $c_{k-1}^{-2}(x)$, one needs to find $\gamma_k(x)$ and $\gamma_1(x)$ can be found from (20) by setting $v_k = 0$. For numerical computation of the Poincare map, the following parameters have to be known and in this paper, for human tissue: $\alpha, l, D, R_0, a, l_D$.

This paper is a preliminary report on our work on chaotic imaging. Works are presently being carried out (a) on the computation of the Poincare map and to prove numerically the existence of chaos for certain limit of the values of parameters, (b) the computation of the reconstructed velocity images and this will be the chaotic images.

5. Conclusions

Chaotic images do exist in acoustical imaging especially when the medium is highly inhomogeneous and fractal. The most likely candidates of human tissue for the observation of chaotic images are the human heart and the human brain which have fractal structure [13,14]. The advantage of chaotic images are their high sensitivity to the change in initial parameter and this makes it useful for the detection of early stage cancerous tissue. It would be more sensitive than the B/A nonlinear parameter diffraction tomography [15] as this is limited only to the quadratic term.

References

1. Shida Liu: Earth system modelling and chaotic time series. Chinese Journal of Geophysics 33, 155-165 (1990)

2. Yong Chen: Fractal and fractal dimensions. Academic Journal Publishing Co. Beijing 1988.

3. D. Ruelle and F.Takens: On the nature of turbulence. Chaos II. World Scientific 1990, pp120-145

4. P.W.Milonni, M.L.Shih, J.R. Ackerhalt: Chaos in laser-matter interactions. World Scientific Lecture Notes in Physics Vol. 6, 1987

5. W.S.Gan: Application of chaos to sound propagation in random media. Acoustical Imaging Vol 19. Plenum Press 1992, pp 99-102

6. W.S.Gan, C.K.Gan: Acoustical fractal images applied to medical imaging. Acoustical Imaging Vol 20. Plenum Press 1993, pp 413-416

7. W.Yang, J.Du: Approaches to solve nonlinear problems of the seismic tomography. Acoustical Imaging Vol 20. Plenum Press 1993, pp591-604

8. D.T.Blackstock: Generalised Burgers equation for plane waves. J Acoust Soc Am 77, 2050-2053(1985)

9. C.W.Smith, R.T.Beyer: Ultrasonic radiation field of a focusing spherical source at finite amplitudes. J Acoust Soc Am 46, 806-813 (1969)

10. E.A.Zabolotskaya, R.V. Khokhlov: Quasi-plane waves in the nonlinear acoustics of confined beams. Sov Phys Acoust 15,35 (1969)

11. Y.P.Kuznetsov: Equations of nonlinear acoustics. Sov Phys Acoust 16, 467 (1971)

12. H.E.Stanley: Fractals and multifractals: the interplay of physics and chemistry. fractals and Disordered Systems. Springer-Verlag 1991, pp1-50

13. B J West, A.L Goldberger. American Scientist 75, 354 (1987)

14. C A Skarda, W.I. Freeman: Behav. Brain Sci 10, 161 (1987)

15. A.Cai, Y.Nakagawa, G.Wade, M Yoneyama: Imaging the acoustic non-linear parameter with diffraction tomography. Acoustical Imaging Vol 17. Plenum press 1989, pp273-283

Chaos Causes Perspective Reversals for Ambiguious Patterns

Kazuo Sakai*, Tsuyoshi Katayama**, Satoshi Wada*** and Kotaro Oiwa****

*Meiji University, Izumi Campus, 1-9-1 Eifuku, Suginami-ku, Tokyo 168, Japan
**Japan Automobile Research Institute, Inc., Yatabe, Tsukuba, Ibaraki 305, Japan
***Aomori Chuo Junior College, 12 Yokouchi Aza Kanda, Aomori 030-12, Japan
****Faculty of Education, Oita University, 700 Dan-no-haru, Oita, Oita 870-11, Japan

Abstract. A new chaotic mechanism is found to cause perspective reversal for ambiguious patterns. And the new senario to cause perspective reversal is proposed with use of chaos, by reexamining the psychological parameters of schema. The psychological parameters are incorporated into a PDP schema model proposed by Rumelhart et al., and a simple one-dimensional map is derived in a mean-field approximation. The numerical experiments for a one-dimensional logistic map are performed, and the results are compared with new psychological experiments of perspective reversals for a Necker cube by 96 students. It is shown that these are qualitatively good agreement with each other. This gives a concrete example of the role of chaos in cognitive science.

1 Introduction

The study of ambiguous patterns such as a Necker cube[1] (Fig. 1) has intrigued psychologists for a long time. Peculiar is oscillatory behaviors between possible two percepts (Fig. 2) in the Necker cube. The best known hypothesis thus far to explain these findings is that of saturation or fatigue proposed by Köhler[2]. Upon this basis, a variety of theories has been published[3, 4, 5]. But there are two cricial discrepancies between experimental facts and the fatigue assumption discussed below.

It has been known by experiments[7] that there are two types of observers, *i.e.*, fast and slow observers. Fast observer has larger frequency of perspective reversals than slow observer. We here point out that the slow observers become faster as experiments are repeated. In fact, the observers with no experiences are almost all slow observers at the first time. This is supported in our experiments, since only 5 persons are fast in all the 96 observers who was made experiment at the first time. Listed below are the experimental facts which conflict with fatigue effect:

Fact 1. The persistent times (durations) staying one of the two percepts are of the order of one second or at most ten seconds.

Fact 2. The persistent times become shorter and shorter as experiments are repeated.

The *Fact 1* indicates that the fatigue effect appeares within several seconds or more. If we intend to interprete this fact by the fatigue effect, the following question becomes arose:

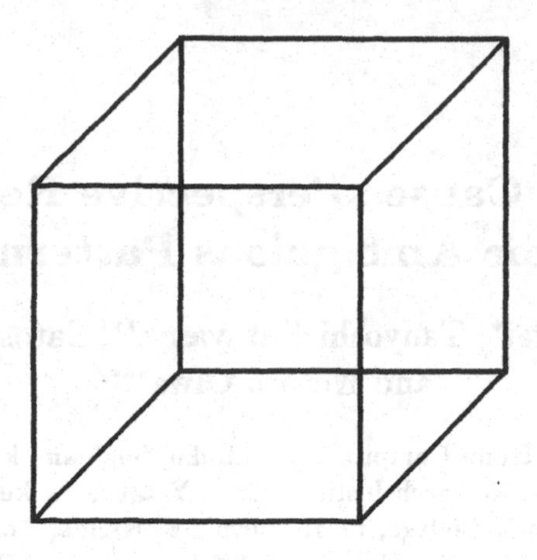

Fig. 1. Necker cube.

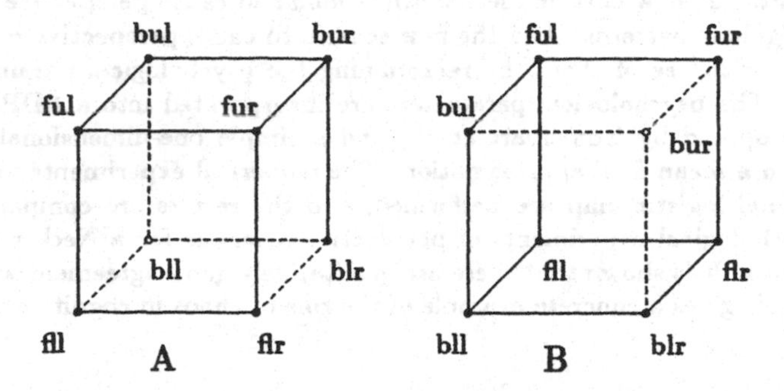

Fig. 2. Two perspects of a Necker cube, and 16 hypotheses of the vertices which we assign to network units.Here 'ful' is meant by 'forward-upper-left', 'blr' by 'backward-lower-right', and so on.

Question 1. ***Does the fatigue effect works so quick?***

On the other hand, *Fact 2* means that the observer familiar with the Necker cube become easy to be fatigued. And the next question is taken place:

Question 2. ***Does the observer used to seeing become easy to be fatigued?***

If say yes, this conflicts with the fact that the observer used to seeing recover from fatigue quickly.

In Sec. 2, we reexamine the psychological schema parameters so as to give consistent interpretations with the above, and show a new experiment to clarify psychological state. In Sec. 3, a network model is introduced, and a one-dimensional map is derived in a mean-field approximation, finally the associative schema model is proposed. Sec. 4 is devoted to discussions and conclusions. The preliminary reports are presented in Refs. [10], [12] – [14].

2 Perspective Reversal

We here discuss psychological parameters of a schema model, and clarify the mechanism responsible for perspective reversals by estimating characteristic time of change in these parameters. And we propose a new senario of perspecitve reversals.

2.1 Psychological Schema Parameters

We propose two parameters of schama as follows.

(a)*Strength of schema formation* which denotes tendency to have a simpler interpretation of line figure of Necker cube as cubic figure.

This is weak for an observer at a first experience, and strong for one used to seeing the Necker cube. Thus the characteristic time of change is at most of the order of number of experimental trials. In a experimental sequence of one trial (10 min.), there are many cases in which the frequency of oscillation becomes large. These give a minimum estimation of the characteristic time of formation strength to be at least several minites.
On the other hand, the characteristic time of perspective reversals is of the order of second, and the reversal mechanism is not governed by this strength of schema formation. The strength of schema formation is reflected to the frequency of perspective reversals as discuss in Sec. 4.
As the another parameter of schema, we propose

(b)*Strength of schema conviction* which denotes tendency to be convinced of one of schemata.

The conviction strength is thought to be weak immediately after an occurrence of the perspective reversal, since new schema just occurred is unexpected one. But soon later, the new shcema seems to be plausible for an observer, and the conviction strength becomes larger. The characteristic time of conviction strength is thought to be of the order of second, since changes in conviction strength are accompanied with perspective reversals. We may now conclude that the perspective reversals is closely related to a dynamical process of conviction strength.
Although the characteristic time of conviction strength coincides with that of perspective reversals, the larger conviction strength never bring reversal process by itself. In order to show oscillatory behaviors in perspective reversals, some kinds of switching mechanism are required which has been thus far due to a fatigue effect. We propose in Sec. 3.3 such a switching mechanism as a new chaotic switching making use of nonlinear effect.

2.2 Uncertain States of Percepts

Our senario of perspective reversals is as follows. Immediately after an occurrence of the perspective reversal, the conviction strength is weak. But soon after, it becomes large. Further increments of the conviction strength make its state unstable, and chaotic responses occur due to a nonlinear effect. Through this chaotic wandering, a new state corresponding to another percept is found and stabilized. The above processes are repeated, and perspective reversals are realized.
In order to justify this senario, we have to find the chaotic wandering states in a sequence of psychological experiment. These states are considered to be uncertain, and neither one of percepts nor another is certain. Since the published experiments

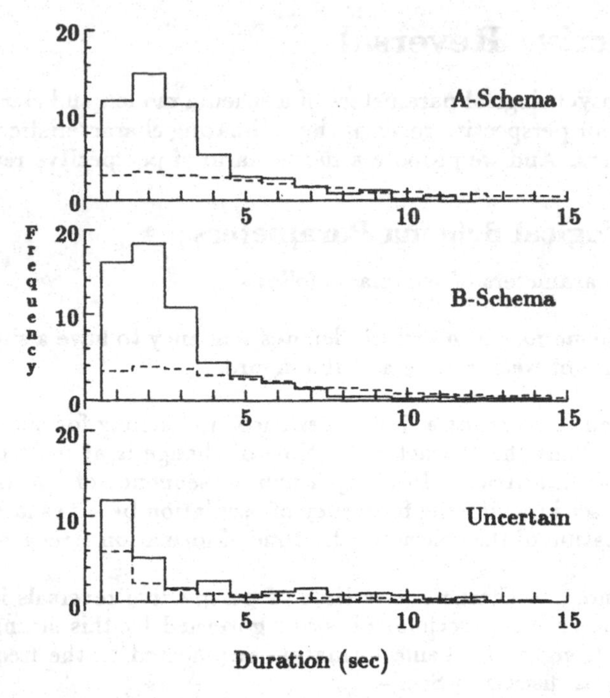

Fig. 3. Frequencies in experiments by observers.The solid line denotes averaged frequencies of five fast observers with averaged durations less than 3 sec. The broken line denotes those of ninety-one slow observers greater than 3 sec.

of perspective reversal[7, 8] are not concerned to the existence of uncertain state, we have made experiments with a new method to extract the uncertain state of percepts.

The basic concept of our experiment is to take three kinds of times, *i.e.*, (1)times persisting A-schema, (2)B-shcema and (3)uncertain. The frequencies of these times are plotted in Fig 3. The observers are 96 students aged 18 to 22 years, who did not know the aims of the experiments. They are required not to move their eyes from the center of the Necker cube figure. The experimental run lasted at 10 min.

As shown in Fig. 3, the distribution of the persistent time in uncertain state is quite different from those of A- and B-schema. The uncertain distribution is a monotonically decreasing, although the others have a hump. This type of monotonically decreasing features is known to appear when chaotic burst[9] takes place. Therefore this could be seen as a candidate for an experimental evidence of chaotic behaviors.

3 Associative Schema Model

We now construct a theoretical model describing the psychological phenomenon of perspective reversals. Starting from a connectionist model proposed by Rumelhart *et al.*[6] of PDP (Parallel Distributed Processing) group, we relate the psychological parameters of formation and conviction strength to their schema model. Their original model does not show perspective reversals, but does moderate dynamics finding each one of perspects. Therefore we will add new dynamics corresponding to that of the conviction strength, and realize oscillatory behaviors of the perspective reversals.

3.1 PDP Schema Model

The PDP schema model[6] is a simple constraint satisfaction model of a Necker cube (Fig. 1). We assume that 16 units represent hypotheses about the correct interpretations of the vertices as shown in Fig. 2. In the PDP schema model, these hypotheses are taken to be units whose activation denotes certainty for the hypothesis.
The activation rule of a single unit is given by[6]

$$a_i(t+1) = \{1 - |net_i(t)|\}\, a_i(t) + net_i(t)\, \theta(net_i(t)) \,, \tag{1}$$

where $net_i(t)$ indicates a net-input from other units, defined by

$$net_i(t) = \sum_j w_{ij}\, a_j(t) + bias_i \,. \tag{2}$$

This model is called an *interactive activation model*[11]. Here w_{ij}'s denote inter-unit weights between i-th and j-th units, $bias_i$ a bias of i-th unit. These weights have been chosen so as to guarantee the balance condition (sum rule [10]) for each i,

$$\sum_{j=\{A\}} w_{ij} + \sum_{k=\{B\}} w_{ik} = 0 \,, \tag{3}$$

between A- and B-schemata. Furthermore, the symmetricity is also assumed at each unit i and k in A- and B-schemata, respectively,

$$\sum_{j=\{A\}}^{(i\in\{A\})} w_{ij} = \sum_{l=\{B\}}^{(k\in\{B\})} w_{kl} \,. \tag{4}$$

There are two kinds of parameters in PDP schema model, *i.e.*, the interconnection weight w_{ij} and bias $bias_i$. The weight parameter denotes firmness of the schema formation, and we may regard it as the strength of schema formation. On the other hand, the unit becomes easy to activate as the bias grows, and it is natural to assign the bias to the strength of schema conviction.

3.2 Derivation of One-Dimensional Map

We here indicate that the interactive activation model implies chaos, because a one-dimensional logistic map $X(t+1) = AX(t)(1-X(t))$ can be extracted from the above activation rule (1) with (2) in a mean-field approximation as followings. For simplicity, we first assume the simplest case of $bias_i = 0$ and $w_{ij} = w > 0$. This is single schema case in which all units are equivalent and cooperative with others. The net-input (2) thus becomes positive definite, *i.e.*, $net_i(t) = w\sum_j a_j(t) = Nw\langle a(t)\rangle > 0$. Here $N = \sum_j 1$ indicates total number of units, and $\langle a(t)\rangle$ is meant by taking an average of activations $\sum_j a_j(t)/\sum_j 1$. By taking an average of the activation rule (1), we obtain $\langle a(t+1)\rangle = (1+Nw)\langle a(t)\rangle - Nw\langle a(t)\rangle^2$. This becomes a well-known one-dimensional logistic map $X(t+1) = AX(t)(1-X(t))$, where $X(t) = (Nw/(1+Nw))\langle a(t)\rangle$ and $A = 1+Nw$.
From this discussion, we can conclude in a mean-field approximation that *PDP schema model implies chaos*, since the logistic map is a representative one to cause chaos. It should be noted that the unit itself has no origin to cause chaos .

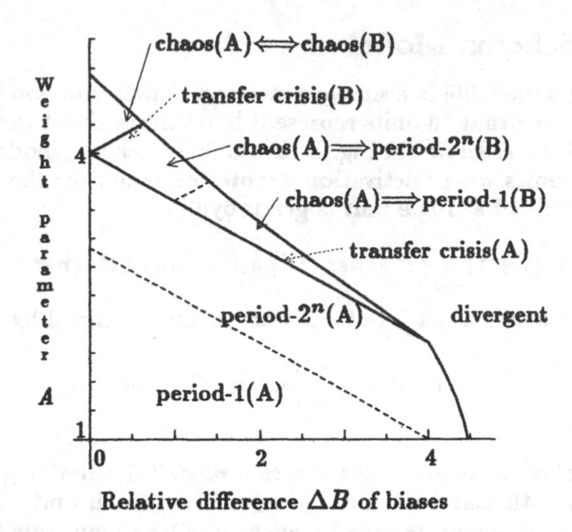

Fig. 4. Phase diagram calculated by a one-dimensional logistic map (7). The initial value of $X(0)$ is set to be 0.001.'period-2^n' denotes a period-doubling route to chaos.

We extend this discussion to two schemata case such as in the Necker cube[10, 12, 13]. Assuming a bias in A-schema to be $+\Delta B/2$ and in B-schema $-\Delta B/2$, the net-inputs (2) can be read as

$$\langle net_i(t)\rangle_A = -\langle net_i(t)\rangle_B = 3c\,\Delta a(t) + \frac{\Delta B}{2}\,, \tag{5}$$

where

$$\Delta a(t) = \langle a(t)\rangle_A - \langle a(t)\rangle_B = \frac{\sum_{i=\{A\}} a_i(t)}{\sum_{i=\{A\}} 1} - \frac{\sum_{i=\{B\}} a_i(t)}{\sum_{i=\{B\}} 1}\,. \tag{6}$$

We have assumed no self-connections of units ($w_{ii} = 0$), and the sum-rules[10] for inter-unit weights have been used. The parameter c characterizes a weight parameter between units, and is related to an average of weights within same schema.
We now introduce scaled variables $X(t) = \{(A-1)/A\}\,\Delta a(t)$and $Z = \Delta B/(2A)$. The map functions $X(t+1) = F(X(t))$ is obtained with

$$F(X) = AX\{1 - |X + Z|\} + (A-1)\,Z\,. \tag{7}$$

Here the definition $A \equiv 1+3c$ has been used. It is apparent that Eq. (7) is reduced to the well-known one-dimensional logistic map with vanishing Z for positive X, and it is confirmed that the PDP schema model surely shows chaotic responses in a mean-field manner.
Period-1 fixed points of Eq. (7) is determined by $X^* = F(X^*)$ which gives three fixed points, $X^* = X_1^*, -X_1^*$ and $-Z$ where we have introduced an period-1 fixed point $X_1^* = (A-1)/A$ in a one-dimensional logistic map $f(X) = AX(1-X)$. The theoretical phase diagram calculated by Eq. (7) is plotted in Fig. 4, from which the relative difference ($\Delta B = 2AZ$) of biases between schemata is also appeared to be a bifurcation parameter besides the weight parameter $A = 1+3c$. It is noted that Eq. (7) is invariant under interchanges $X \to -X$ and $Z \to -Z$. This means that

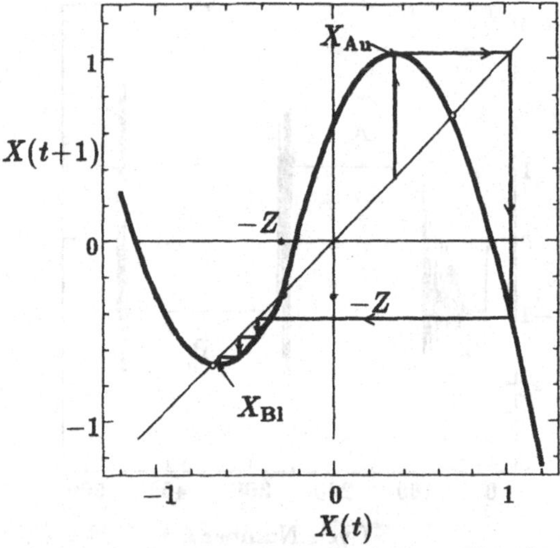

Fig. 5. A typical situation to occur a transfer crisis. The chaotic orbit has a chance transferring to B-schema, through the narrow channel $[1, F(X_{Au})]$. It is noted that, at this parameter set, the orbit falling in B-schema becomes stable at a period-1 fixed point near the lowermost point X_{Bl}.

an interchange of A- and B-schemata holds good if Z is also interchanged into $-Z$. Therefore, we hereafter concentrate ourselves only on a case of $Z = \Delta B/(2A) > 0$. Characteristics of the phase diagram Fig. 4 are the occurrence of *transfer crises*[12, 15]. The typical situation to take place the transfer crisis is shown in Fig. 5. This is a kind of 'local' boundary crisis against the chaotic attractor of A-schema. The *transfer crisis* is found and named by Yamaguchi and Sakai[15] in 1983 for a forced one-demensional logistic map by an alternating disturbance. This phenomenon has been found in physical system of a Josephson junction[16, 17] by numerical calculations. Eq. (7) is now found to be the simplest map showing the *transfer crisis* between coexisting two attractors.
Peculiar is a hysteresis[15] associated with the transfer crisis, which is seen in Fig. 4 along the line at constant weight parameter A. Starting from vanishing bias $\Delta B = 0$, the orbit is confined in A-schema. When the ΔB is increasing along the constant A line, the orbit becomes unstable and chaotic. Further increase takes an occurrence of the *transfer crisis*, and the orbit is falled into B-schema, when ΔB exceeds the critical line of 'transfer crisis(A)'. But, the reverse process never hold because of the hysteresis effect, even if ΔB is switched back to decrease.
The hysteresis is arose from breaking of the symmetry in Eq. (7). This symmetry breaking is taken from the existence of the relative difference ΔB of biases between A- and B-schemata. At the next subsection, we make use this peculiar feature of a hysteresis, and the new interpretation is given for the oscillatory behaviors in psychological phenomena observed such as in a Necker cube.

3.3 Dynamics of Perspective Reversal

We are now at the stage to relate PDP schema model and psychological schema parameters of formation and conviction strength.
In order to realize the oscillatory behaviors of perspective reversals, we should incor-

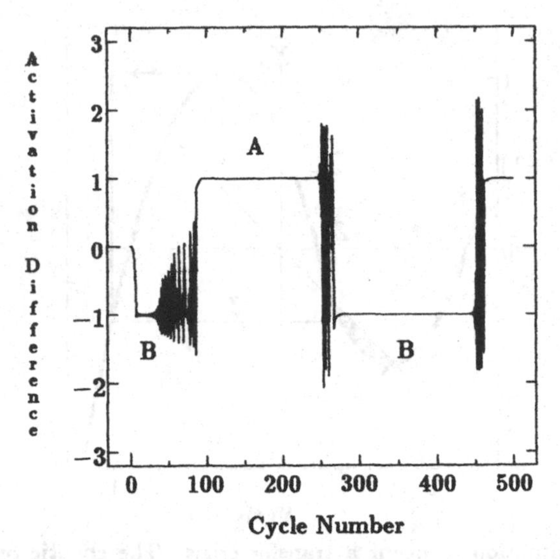

Fig. 6. Time-series profile of an activation difference $\Delta a(t)$ by maps (7) and (9). $\Delta a = 1$ and -1 denotes A- and B-schema, respectively. The adopted initial parameters are $c = 0.6 (A = 2.8), \epsilon = 0.01, \Delta B(t=0) = Z = 0, X(t=0) = -0.001$. Note that plotted are not $X(t)$ but $\Delta a(t) = X(t)/X_1^*$ which is chosen because the period-1 fixed points are not affected by the bifurcation parameter A.

porate some dynamics into the conviction strength (bias). We adopt the following update rule for biases[13],

$$bias_i(t+1) = (1-\epsilon)\, bias_i(t) + \epsilon\, net_i(t)\,, \tag{8}$$

where ϵ is a positive smallness parameter to govern the velocity of changes.
The cognitive meanings of Eq. (8) is as followings. When the particular hypothesis (say, i-th unit) is feasible ($net_i > 0$), enhanced is the tendency ($bias_i$) to believe that the hypothesis holds good.
Eq. (8) can be rewritten in a mean-field approximation as

$$Z(t+1) = Z(t) + \epsilon X(t)\,, \tag{9}$$

where $Z(t) = \Delta B(t)/(2A)$. An introduction of the enhancement mechanism of Eq. (9) leads to movement along a line of constant weight parameter A on the phase diagram Fig. 4. If the initial state is located at period-1(A), it moves to the right and finally falls into the window of B-schema. In Fig. 6, we plot the time-series profile calculated by maps Eqs. (7) and (9). The oscillatory behaviors are obtained as required in realizing perspective reversals. An alternation or switching phenomena seen in Fig. 6 is caused by a tranfer crisis[12, 15]. It is seen as an abrupt change of orbits from chaos confined in one attractor to a stable fixed point apart from the chaotic attractor.
The dynamical behaviors in Fig. 6 might be interpreted in a cognitive sense as followings. Once a certain schema is confirmed, the confirmation becomes greater (corresponding to a growth of $|\Delta B|$ according to (8) or (9)). Further confirmation gives rise to a loss of confidence corresponding to chaotic state, and leads to a sudden discovery of an another perspect via chaotic transition due to transfer crisis.

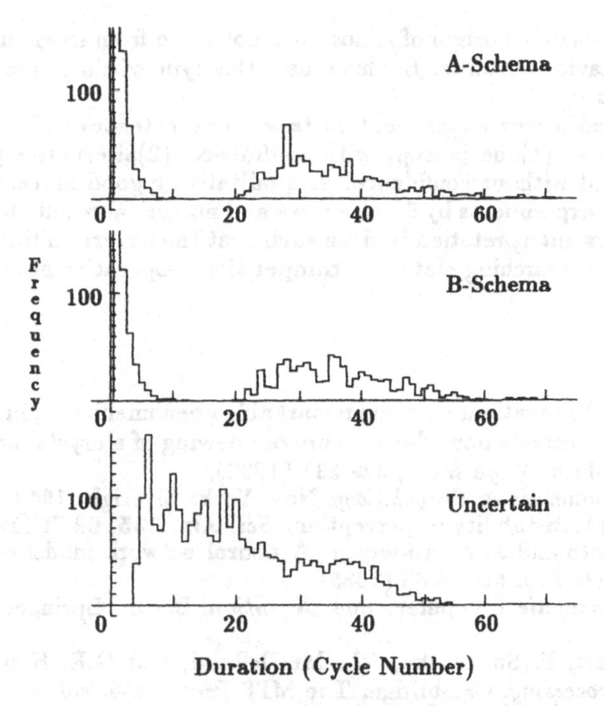

Fig. 7. Frequency distributions by maps (7) and (9). Performing 100,000 iterations, the resulting time-series are classified into three categorizes: A-schema if $\Delta a = 1 \pm 0.1$, B-schema if $\Delta a = -1 \pm 0.1$, otherwise, 'Uncertain'.

The characteristic features in Fig. 6 are the existence of uncertain region in the time-series, *i.e.*, the chaotic wandering region. The numerical calculations are performed up to 100,000 iterations with use of Eqs. (7) and (9), and are plotted in Fig. 7. Comparing with Fig. 3, the qualitative agreement is obtained, if we neglect the lower duration below 20 cycles in Fig. 7. These neglections are rather meaningful, because such a shorter duration should not really recognized by an observer.

4 Discussions

We here discuss the effect of formation strength. As indicated in *Fact 2* in Sec. 1, the frequency of perspective reversals are larger for an observer used to seeing the Necker cube. This can be explained by our model as follows. The observer used to seeing has a larger strength of schema formation which is corresponding to the bifurcation parameter $A = 1 + 3c$ in the phase diagram of Fig. 4. Since the horizontal width of ΔB at constant A is propotional to the average period of chaotic switching, the larger A causes lager frequency of oscillatory behaviors in perspective reversals. In summary, we propose a new idea to interpret perspective reversals of ambiguous patters such as a Necker cube. Based on a PDP schema model, we derive a one-dimensional logistic map in a mean-field approximation, the obtained map shows a varieties of dynamics including chaos and *transfer crises*[12, 15] with hysteresis. These are based on the following findings: (1) PDP schema model implies chaos. (2) The chaos can be specified by a one-dimensional map in a mean-field approximation. (3) A transfer of an orbit can be taken place from the chaotic attractor of one schema into another attractor corresponding to different schema.

It should be noted that the origin of chaos does not come from single unit itself but from averaged behaviors of units. In this sense, this type of chaos should be called as a *mean-field chaos*.

We have introduced a new experiment to take three categories of times for perspective reversal, *i.e.*, (1)one percept with confidence, (2)alternative percept, and (3)uncertain percept without confidence. A qualitatively good agreements are obtained between the experiments by 96 observers and numerical results by the derived logistic maps. A new interpretation is given such that the uncertain times could be a chaotic wandering or searching state in a competitive-cooperative network system.

References

1. L. A. Necker: Observations on some remarkable phenomena seen in Switzerland; and an optical phenomenon which occurs on viewing of a crystal or geometrical solid. *Philosophical Magazine* **3**, 329-337 (1832).
2. W. Köhler: *Dynamics in Phsychology*. New York: Liveright 1940
3. F. Attneave: Multistability in perception. *Sci. Am.* **225**, 62-71 (1971)
4. A. H. Kawamoto and J. A. Anderson: A neural network model of multistable perception. *Acta Psy.* **59**, 35-65 (1985)
5. H. Haken: *Synergetic Computers and Cognition*. Berlin: Springer 1991, Chap. 13
6. D.E. Rumelhart, P. Smolensky, J.L. McClelland, and G.E. Hinton: *Parallel Distributed Processing*. Cambridge: The MIT Press 1986, vol. 2
7. A. Borsellino, *et al.*: Effects of visual angle on perspective reversal for ambiguous patterns. *Perception* **11**, 263-273 (1982)
8. J. R. Price: Perspective duration of a plane reversible figure. *Psychon. Sci.* **9(12)**, 623-624 (1967)
9. Y. Aizawa: Global aspects of the dissipative dynamical systems I. *Prog. Theor. Phys.* **68**, 64-84 (1982)
10. K. Sakai, T. Katayama, K. Oiwa and S. Wada: Theory of chaotic dynamics with transfer crises in a mean-field PDP schema model (Discovery of transfer crises and its cognitive meanings). *The Bul. of Arts and Sci.* Meiji Univ. **249**, 105-150 (1992)
11. J.L. McClelland and D.E. Rumelhart: An interactive activation model of context effects in letter perception: Part 1. An account of basic findings. *Psy. Rev.* **88**, 375–407 (1981)
12. K. Sakai, T. Katayama, K. Oiwa and S. Wada: New mechanism to transfer schemata caused by transfer crises. *Proc. IIZUKA92*. Iizuka, Japan, 149-152 (1992)
13. K. Sakai, T. Katayama, S. Wada and K. Oiwa: Perspective Reversal Caused by Chaotic Switching in PDP Schema Model. *Proc. of the 1993 IEEE Int. Conf. on Neural Networks*. San Francisco, CA, **3**, 1938-1943 (1993)
14. K. Sakai, T. Katayama, S. Wada and K. Oiwa: Chaos Causes Perspective Reversals for Ambiguious Patterns. *Proc. of the Fifth Int. Conf. on Information Processing and Management of Uncertainty in Knowledge-Based Systems*. Paris, France, (1994)
15. Y. Yamaguchi and K. Sakai: New type of 'crisis' showing hysteresis. *Phys. Rev.* **A27**, 2755-2758 (1983)
16. K. Sakai and Y. Yamaguchi: Nonlinear dynamics of a Josephson oscillator with a $\cos\phi$ term driven by dc- and ac-current sources. *Phys. Rev.* **B30**, 1219-1230 (1984)
17. M. Marek and I. Schreiber: *Chaotic Behaviour of Deterministic Dissipative Systems*, Cambridge: Cambridge Univ. Press 1991

Implications of a Continuous Approach to Chaotic Modeling

Maurice E. COHEN*,**, Donna L. HUDSON*,
Malcolm F. ANDERSON*,***, Prakash C. DEEDWANIA*,***

*University of California, San Francisco, 2615 E. Clinton Avenue, Fresno, CA 93703 USA
**California State University, Fresno, California 93740 USA
***Veterans Affairs Medical Center, Fresno, California 93703 USA

Abstract. Application of chaos theory to medicine has yielded intriguing and controversial results. In this paper, a continuous conjecture for the solution of the Poincaré equation suggests that chaos provides a framework for the measurement of disorder within a system. However, as demonstrated by the behavior of the continuous versus the discrete approach to chaos, the general use of the term chaos must be used with care. The theoretical results are illustrated in two medical applications dealing with hemodynamic flow and congestive heart failure.

1 Introduction

The basic common thread in chaos theory is the recursive evaluation of seemingly simple functions which produce unexpectedly complex results. An iterative function does not suddenly become chaotic, but rather goes from the stage of convergence to a single value to a bifurcation, or convergence to two values. Additional bifurcations occur, and finally chaos results [1].

For the Poincaré equation

$$a_n = A\, a_{n-1}(1 - a_{n-1}) \qquad (1)$$

where A is a constant whose value changes the behavior of the function. The recursion is dependent on the selection of a_0, which must be chosen between 0 and 1. For increasing values of A, the equation progresses from single value convergence to chaos. Within the chaotic area, regions of stability unexpectedly appear [2]. For integer values of n, this function exhibits chaotic properties for $A > 3.57$. These properties include apparent lack of periodicity and sensitivity to initial conditions. It will be shown in this paper that the regions of stability are only a matter of perception when discrete values of n are considered. If continuous values are taken, these apparent dichotomies disappear [3].

Many processes in medicine appear to exhibit chaotic properties [4]. Chaotic analyses have become increasingly useful, especially in cardiology [5,6]. Chaotic analysis is still in its infancy, with new techniques needed to determine the degree to which a data set appears to be chaotic. Such an approach is described below.

In the traditional approach to chaotic modeling, a recurrence relation is established which is evaluated at integer values, generally with each integer corresponding to a fixed time interval. These values are then plotted and connected with straight lines. However, actual non-integer values of these recurrence relations are unknown. We have developed an approach which permits an approximate solution of the Poincaré equation not only for integer values but for all real values of n and for any value of A in the important range $2 \leq A \leq 4$ [7].

These results show that in fact the chaotic behavior of this function is not apparent when viewed as a continuous, and not as a discrete, model except in the narrow mathematical definition. In fact, this work emphasizes the danger of approximating any continuous process by a discrete model when the underlying principles are not understood [8].

2 Conjecture for Continuous Solution of Poincaré Equation

2.1 Conjecture

The Poincaré equation (1) has an exact solution for A=2 and A=4. The exact solution for A=4 is

$$a_n = [1 - T_{2^n}(1-2a_0)]/2 \tag{2}$$

where $T_n(x)$ is the Chebyshev function [9].

The solution given in equation (2) is valid only for A = 4, a point in the region of chaos. As no exact solution is available for other values within the region of chaos, we constructed a method for approximating solutions for any value of A, $2 \leq A \leq 4$ [7].

Assume a solution of the type

$$a_n = \sum_{k=0}^{\ell} \alpha_k T_k(2^n x) \tag{3}$$

where $T_k(x)$ is the Chebyshev function of the first kind. We assume ℓ to be the number of points in the interval $0 \leq n \leq 1$. Thus

$$a_n^2 = \sum_{k=0}^{\ell} \alpha_k^2 T_k^2(2^n x) + 2 \sum_{\substack{j>i \\ i=0,1,\dots,\ell-1 \\ j=1,2,\dots,\ell}} \alpha_i \alpha_j T_i(2^n x) T_j(2^n x) \tag{4}$$

The conjecture adopted assumes that progressing from one point to the next implies the addition of a Chebyshev polynomial. Hence

$$a_{n+1}^2 = \sum_{k=0}^{2\ell} \beta_k T_k(2^n x) \tag{5}$$

where n is assumed to be a real number.

Feeding (3), (4), and (5) in the Poincaré equation (1), simplifying and comparing coefficients gives nonlinear equations involving the unknowns α_i's and β_j's and the arguments x of the Chebyshev polynomials. We have solved 300 equations involving 300 variables and have chosen the appropriate α_i's and β_j's and the argument of the Chebyshev to satisfy a_n to be strictly monotonic increasing in the interval $0 \le n \le 1$.

It is important to note that the conjecture adopted by the authors permit the extension of unique solutions to be valid not only for integers but for all real values of n, $2 \le A \le 4$.

2.2 Computer Solution

The approximation is implemented on a SUN SPARCserver 470 computer as well as a CRAY super computer to ensure accuracy. Five programs are used, each of which is briefly described below. All results are computed to at least 10 places of accuracy.

Program 1 solves nonlinear simultaneous equations to determine the α_i's from equation (3). The results will become more accurate as the number of α_i's increases. The algorithm is based on Newton's method for solution of nonlinear equations [10]. It requires the computation of the Jacobian, which is defined by J(x):

$$\begin{bmatrix} \frac{\delta f_1(\mathbf{x})}{\delta x_1} & \frac{\delta f_1(\mathbf{x})}{\delta x_2} & \cdots & \frac{\delta f_1(\mathbf{x})}{\delta x_n} \\ \frac{\delta f_2(\mathbf{x})}{\delta x_1} & \frac{\delta f_2(\mathbf{x})}{\delta x_2} & \cdots & \frac{\delta f_2(\mathbf{x})}{\delta x_n} \\ & \vdots & & \\ \frac{\delta f_n(\mathbf{x})}{\delta x_1} & \frac{\delta f_n(\mathbf{x})}{\delta x_2} & \cdots & \frac{\delta f_n(\mathbf{x})}{\delta x_n} \end{bmatrix} \tag{6}$$

A general pattern has been established for the generation of the n equations for $f_i(\mathbf{x})$, which compare coefficients of the Chebyshev functions, as well as a general pattern for the partial derivatives of these equation. The computer can thus generate the simultaneous nonlinear equations for any value of n. The maximum number we have computed to this point is n=300. The algorithm finds roots in an iterative manner. The general algorithm has been modified to permit only positive roots for the α_i's.

Program 2 uses the α_i's from program 1 to determine the β_j's according to equations (4) and (5). Again, these equations are automatically generated for any value of n. This is a straightforward computational program.

Using the α_i's from program 1, program 3 finds the appropriate root for the Chebyshev polynomial which will result in a monotonic increasing function. This algorithm is based on the Newton-Raphson method.

Program 4 computes the additional points using the root from program 4 and the β_j's from program 2, and equations (4) and (5). At the completion of this algorithm, values for a_n have been established for all values, $0 \leq n \leq 1$. The number of values is determined by the number of simultaneous equations which were solved, currently 300 points for n between 0 and 1.

In program 5, the Poincaré equation (1) is applied to the non-integer values from program 4, which produced a continuous plot for $0 \leq n \leq 1$, to produce a continuous plot for larger values. The accuracy of this plot as n increases will be dependent on the number of simultaneous solutions in the first program.

The above method is illustrated for two values of A. Figure 1 shows the approximation for A = 4 compared to the exact solution which is known for this value of A and is given in equation (2). Note that the results are virtually identical. Figure 2 shows the approximation for A = 3.5, an area where bifurcation is observed for the integer-valued Poincaré equation. Note that this conjecture accounts for the bifurcation, and for higher values of A, the further bifurcations observed for integer values. There is no known solution at this value, so no comparison is possible.

3 Discrete and Continuous Difference Plots

3.1 Poincaré Plots

Chaotic equations are sometimes used to generate graphs which are known as Poincaré plots. For example, using the Poincaré equation (1), a Poincaré plot is obtained by plotting a_{n+1} vs. a_n. The resulting plot is a measure of the degree of chaos in the system. We produced similar graphs using the continuous solution. Figure 3 shows a discrete Poincaré plot at A = 4.0, while Figure 4 shows the continuous Poincaré plot for the same value. Note that the character of these plots is totally different. Nonetheless, the Poincaré plot remains a good measure of the degree of variability in the system.

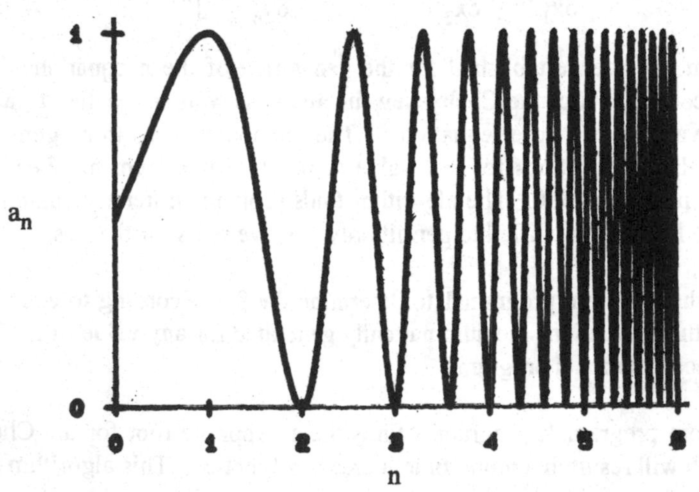

Fig. 1. Approximation (Solid Line) and Exact Solution (Dotted Line): A = 4.0

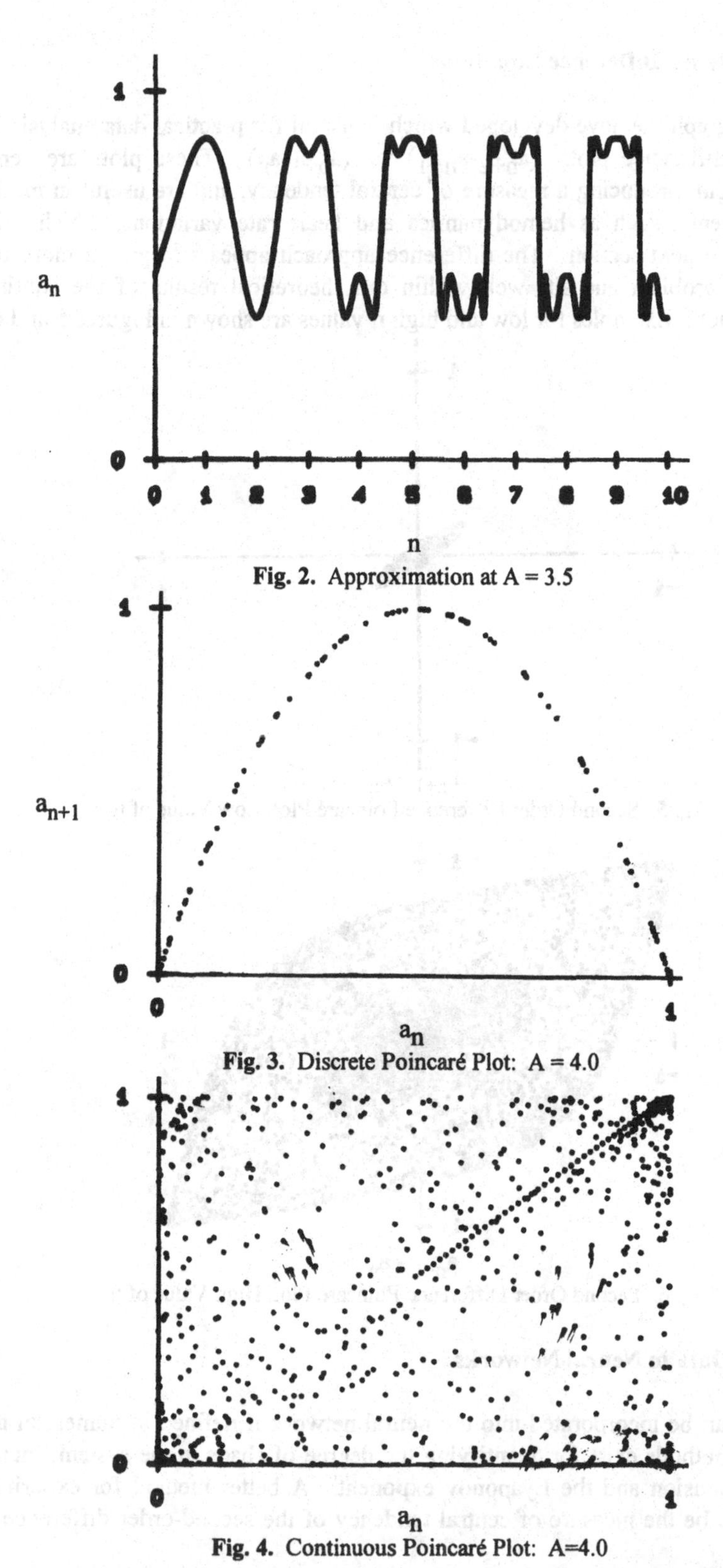

Fig. 2. Approximation at A = 3.5

Fig. 3. Discrete Poincaré Plot: A = 4.0

Fig. 4. Continuous Poincaré Plot: A=4.0

3.2 Second Order Difference Equations

Another graph we have developed which is useful for practical data analysis is the second order difference plot: $(a_{n+2}-a_{n+1})$ vs. $(a_{n+1}-a_n)$. These plots are centered around the origin, producing a measure of central tendency, and are useful in modeling biological systems, such as hemodynamics and heart rate variations, which will be illustrated in the next section. The difference approach appears to give a more robust picture of the problem and fits well within our theoretical results of the continuous Poincaré equation. Examples for low and high n values are shown in Figures 5 and 6.

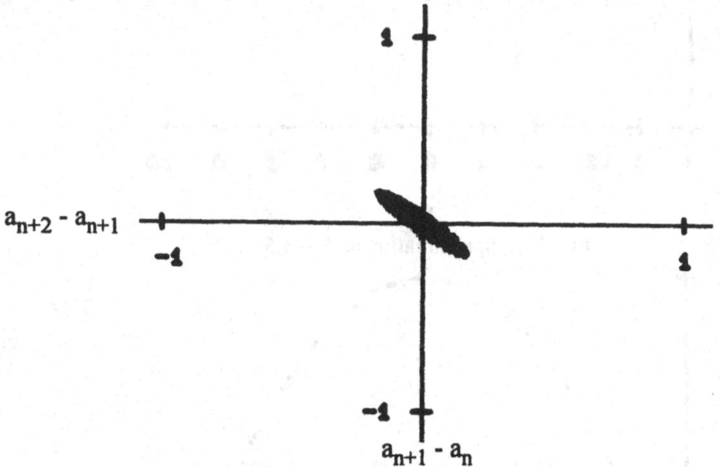

Fig. 5. Second Order Difference Poincaré Plot, Low Value of n

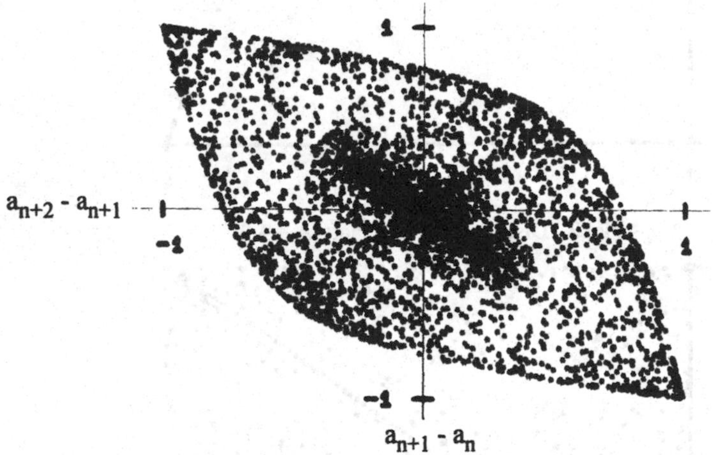

Fig. 6. Second Order Difference Poincaré Plot, High Value of n

3.3 Chaotic Data in Neural Networks

Chaotic data can be incorporated into the neural network if defined in numerical terms. A number of methods exist for quantifying the degree of chaos in the system, including the fractal dimension and the Lyaponov exponent. A better method for experimental data appears to be the measure of central tendency of the second-order difference plots discussed above. For data analysis, the degree of central tendency is computed by selecting a circular region around the origin of radius r, counting the number of points which fall within the radius, and dividing by the total number of points. Let t = total number of points, and r = radius of central area. Then

$$n = \left[\sum_{i=1}^{t-2} \delta(d_i)\right]/(t-2) \qquad (7)$$

where

$$\delta(d_i) = \begin{cases} 1 & \text{if } [(a_{i+2}-a_{i+1})^2+(a_{i+1}-a_i)^2]^{.5} < r \\ 0 & \text{otherwise} \end{cases}$$

The node value n will thus be a number between 0 and 1, inclusive. This information could be used in a neural network in conjunction with other patient data, with the value added to the network like any other node.

4 Use in Medical Decision Making

4.1 Blood Flow

Hepatic blood flow data were generated through an implanted pulsed Doppler flow meter using conscious dogs [11]. With the use of an analog-digital converter, the data were stored on an IBM PC and were subsequently transferred to a SUN SPARCserver 470 computer for analysis. Second-order difference plots for the hepatic blood flow data under different conditions are shown below. Each point v_n represents the volume of blood flow at time n. Blood flow in the conscious dog with no drugs (Figure 7) appears the most chaotic.

The introduction of the vasoactive drugs significantly alters the picture. The plot in Figure 8 shows the blood flow after the injection of nicotine which appears to significantly reduce the chaos. This is emphasized by comparison with Figure 5 and Figure 6, which show the theoretical non-chaotic and chaotic plots [12].

4.2 Congestive Heart Failure

Data were obtained from twenty-four hour Holter monitoring tapes which recorded the R-R intervals, which are the times between heart beats. Studies were done on normal individuals and on patients with congestive heart failure (CHF). The average number of points per 24 hour study is around 100,000.

Figure 9 shows a second-order difference graph for a normal patient. By comparison, Figure 10 shows a patient with congestive heart failure. In both figures, h_n indicates the R-R interval at time n. Note that the normal patient exhibits much less variation than the patient with CHF. This pattern was found consistently in normal patients. In the CHF patients, a wider range of patterns were detected.

In order to quantify the obvious graphical differences illustrated in Figures 9 and 10, equation (7) is used to compute central tendencies. Examples for the two graphs are:

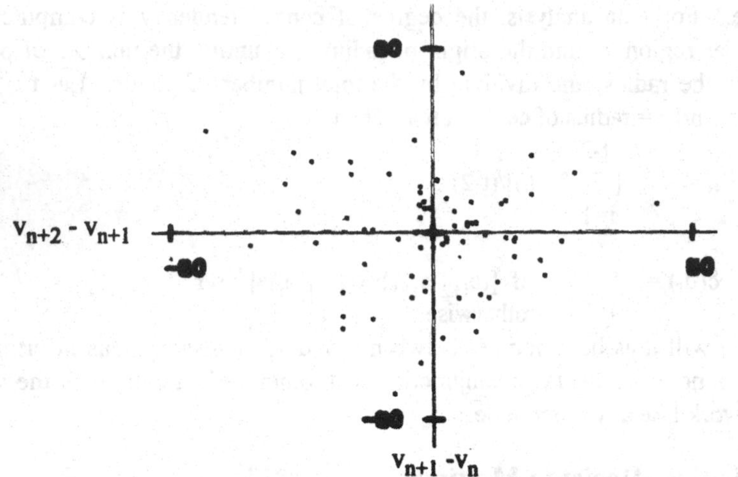

Fig. 7. Second Order Difference Plot, Hepatic Blood Flow: Control

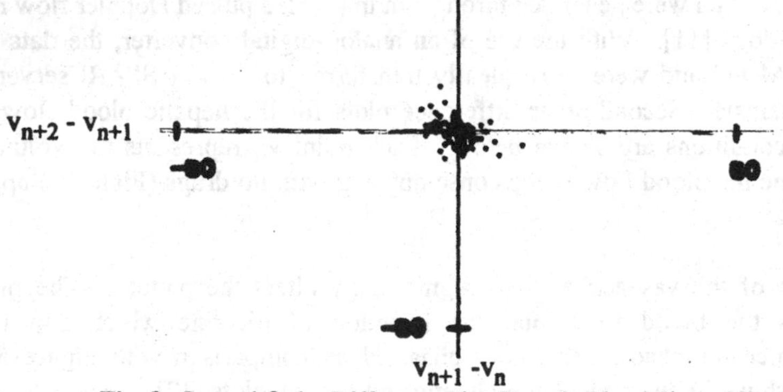

Fig. 8. Second Order Difference Plot, Hepatic Blood Flow: Nicotine

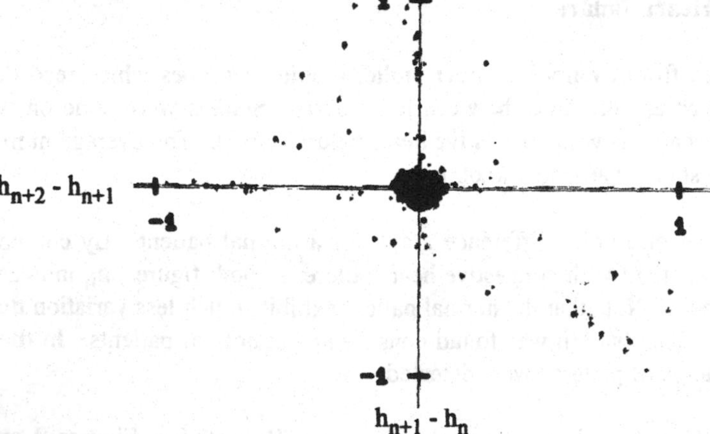

Fig. 9. Second Order Difference Plot, Individual with Normal Heart Function

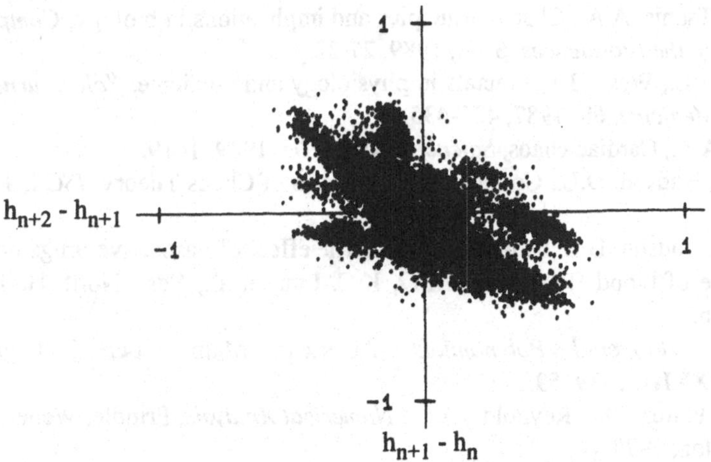

Fig. 10. Second Order Difference Plot, Patient with Congestive Heart Failure

$t_{normal} = 104443$ $\qquad t_{chf} = 109374$

$n_{normal} = 0.986$ $\qquad n_{chf} = 0.232$

Note that a rule could also be established in which a subjective measure of the degree of chaos is used. These numbers can also be incorporated into a neural network model, as mentioned above.

5 Conclusions

The use of chaotic modeling in nonlinear systems which previously produced intractable solutions has provided useful new insights to many applications, including medicine. Viewing chaotic functions as continuous rather than discrete systems, as we have done in this paper, raises new questions about the precise definition of chaos. However, many practical tools result from this approach, including second-order difference plots of the continuous Poincaré equation and measures of central tendencies. The first can be used to provide a graphical frame of reference in the determination of variability in a system, while the second provides a numerical value which can be used in conjunction with other parameters in neural network models.

The two medical examples given here illustrate that chaotic modeling can provide insight into practical problems for which previous modeling techniques failed. It offers both a visual and computational approach to interpretation of medical data.

References

1. Poincaré. H., Memoire sur les courbes definies par les equations differentielles, *J. Math. Pures et Appliq.*, **4th Ser.**, **1**, 1885, 167-244.
2. Gleick, J., *Chaos, Making a New Science*, Penguin Books, New York, 1987.
3. Cohen, M.E., Hudson, D.L., Anderson, M.F., A conjectured continuous approach to chaotic modeling, *Nonlinear Theory and Its Applications*, 1993, 783-786.

4. Tsonis, P.A. Tsonis, A.A., Chaos: principles and implications in biology, *Computer Applications in the Biosciences*, **5 (1)**, 1989, 27-32.
5, Goldberger, A.L., West, B.J., Fractals in physiology and medicine, *Yale Journal of Biology and Medicine*, **60**, 1987, 421-435.
6. Goldberger, A.L., Cardiac chaos, *Science*, **243 (2987)**, 1989, 1419.
7. Cohen, M.E., Hudson, D.L., Computational Aspects of Chaos Theory, *ISCA*, 1994, 89-93.
8. Cohen, M.E., Hudson, D.L., Anderson, M.F., The effect of vasoactive drugs on the chaotic nature of blood flow, *MEDINFO*, K.C. Lun, et al., Eds., North Holland, 1992, 931-936.
9. Szego, G., *Orthogonal Polynomials*, American Math. Soc. Colloquium Publications, **XXIII**, 1939, 59.
10. Burton, R.L., Faires, J.D., Reynolds, A.C., *Numerical Analysis*, Prindle, Weber, and Schmidt, Boston, 1978.
11. Cohen, M.E., Hudson, D.L., Moazamipour, H., Anderson, M.F., Chaotic blood flow analysis in an animal model, *Computer Applications in Medical Care*, 1990, **14**, 323-327.
12. Cohen, M.E., Hudson, D.L., Anderson, M.F., Vazquez, C., Blood flow data exhibit chaotic properties, *Int. J. of Microcomputer Applications*, **12 (3)**, 1993, 82-87.

This work was supported in part by a grant from University of California Valley Medical Education Foundation.

"Homeostasic" Control of Imbalanced StrangeAttractors With or Without Asking to a Change in Chaotic Dynamics

Elie. Bernard-Weil

Fondation Alphonse de Rothschild, 25-29 Rue Manin,
F-75940 Paris Cedex 19)

Abstract. Previously, we showed how the problem of the strange attractors (SA) did not only concern the nature of their dynamics and how the aim of the control did not depend only on the fact that such a dynamics was considered as favorable or wrong for the functioning of the system modeled by these SA. The topology in the phase space of a SA is as much important, because the ergodic theory allows us to take inot account the mean values of the variable trajectories in this SA. As far as biomathematical models are concerned, these mean values may correspond either to physiological values or to pathological values. So, control should in certain cases not try to make disappear (or reestablish) a chaotic dynamics (CD), but to displace the SA in order that the mean values would return to a physiological state. This paper reports a theoretical study of such type of SA control with a model so-called "model for the regulation of agonistic antagonistic (AA) couples" or "AA networks", which include equations giving the values of control variables. Then it would be possible to reestablish the balance between the variables with or without making disappear the CD of the attractor. In the last case, the control variables have themselves a CD.

This paper mainly concerns chaotic dynamics (CD) and the problems elicited by the control of corresponding systems. However, given that we will use a general model (or a model of function in the sense proposed by Rosen), i.e. the "model for the regulation of agonistic antagonistic couples" (MRAAC), it has seemed firstly necessary to recall the structure of this model.

1 The Model for the Regulation of Agonistic Antagonistic Couples and its Use in Control of Imbalanced Limit-cycles

1.1. In fact, our biomathematical direction was due to the existence of a problem often encountered in medical therapy, but which is, so to speak, *repressed*, with a few exceptions, because its solving is not consistent with the reductionist kind of thought, still prevailing to-day. Schematically, the usual therapeutical policies are infered from investigations *in vivo* and experiments *in vitro* which have pointed out a defect or an excess of a link in a metabolic chain (release of a ligand, receptor or genome disturbances, or at the level of "cascade" cytoplasmic reactions, including protein kinase, cAMP, Ca^{++} release....): if there is an agent, agonist or antagonist for this effect, or another type of strategy capable of stimulating the deficient link or inhibiting the prevailing element, this

agent or strategy *should* be able to provoke an improvement of the disease where these anomalies have been observed.
Now we have to distinghish some cases where this type of reasoning give rise to favourable and especially lasting effects (it is the case of high blood pressure treated by angiotensin conversion inhibitor, or of gastric ulcer treated by anti-H2 agents...), and, conversely, other cases where therapeutical trials was proving delusive, either owing to the lack of favourable effects, or owing to a rapid escape from the desired effects, or yet on account of effects opposed to the waited effects.
In order to explain such events, we supposed and partially verified a mechanism called "pathological homeostasis", which may occur under two forms:
Pathological homeostasis of I-type. Pathological cell control is going to react against a well-intentioned measure (such as administering the deficient factor) by increasing the factor already in excess. When we studied the hormonal couple vasopressin (VP)/ cortisone thirty years ago, of which imbalances were considered as giving rise to various pathologies , it seemed obvious to us that the escape from usual cortisone actions was due to a phenomenon we called later a "neuro-postpituitary response to a corticosteroid load": this phenomenon consists in further increase of an already oversecretion of VP .
Pathological homeostasis of II-type. It seems possible that the escape from the effects of an agent is not necessary coupled with an increase of the agent generally associated with the first one and forming together a couple of agents with opposite actions. If this increase is cancelled by some appropriate measures for instance, another imbalance factor belonging to another couple would take the place of the agent prevented from increasing
We did not make reference to the well-known phenomena of *desensitization* and *down-regulation* which may intervene in some steps of the mechanisms of pathological homeostasis but which cannot resume these steps.
1.2 How is it allowed to elaborate a control strategy under these conditions? Either there is no possibility of control due to a counteraction of body pathological regulations which will always have "the last word", or we will succeed in finding the "vulnerable spot" of those adverse regulations. The first idea consists in the addition to the "good" agent of a strategy ending to inhibit the agent with opposite effects. So we might perhaps succeed in doing this task, but, as we saw *supra*, pathological homeostasis of II-type risks substituting another counteracting agent.

We argued otherwise and we proposed a *bipolar control* [4] which associated stimulating and inhibiting agents, i.e. the same fundamental mechanisms that Nature uses in control of health and diseases - the result differences corresponding to balance or imbalance between the opposite actions. However, *interactions between bipolar strategies of Nature and bipolar strategies of the physicians* should be well understood.

1.3. The model for the regulation of agonistic antagonistic couples (MRAAC) is supposedly known [1, 2, 3, 5]. Most of our researches on the MRAAC is concerned with the behaviour of one couple of agents, thus forbidding the observation of SA. However a super-MRAAC (now called an *AA network*) has already been built and a study of SA produced by this model has been carried out. At present, we will state only some characteristics of the elementary MRAAC.

1.3.1. It is necessary to understand the epistemological background, the principles of the model functioning, its meanings and the intended objectives on which the MRAAC must appear. Let us say here only that it is a *general model*, common to various systems, or the *model of a function* [the function of balance and growth (or decrease)]. Mathematically, it is formed by non-linear différential equations, with phenomenological and not physical parameters. When functioning is *physiological*, it enables the antagonistic balance ($x = y$ for instance) and/or the agonistic balance ($x + y = m$ for instance) of both variables to be restored after a perturbation. But, in case of a *pathological* functioning, another different critical point becomes stable ($x \neq y$, $x + y \neq m$) and we have to propose a *control* (cf. *infra*).
1.3.2. The mathematical formalization of the elementary MRAAC is the following:

$$
\begin{aligned}
\dot{x} &= k_1(u+r)+k_2(u+r)^2+k_3(u+r)^3+c_1(v+s)+c_2(v+s)^2+c_3(v+s)^3 \\
\dot{y} &= k_1'(u+r)+k_2'(u+r)^2+k_3'(u+r)^3+c_1'(v+s)+c_2'(v+s)^2+c_3'(v+s)^3 \\
\dot{X} &= k_5(u+r)+k_6(u+r)^2+k_7(u+r)^3+c_5(v+s)+c_6(v+s)^2+c_7(v+s)^3 \\
\dot{Y} &= k_5'(u+r)+k_6'(u+r)^2+k_7'(u+r)^3+c_5'(v+s)+c_6'(v+s)^2+c_7'(v+s)^3
\end{aligned}
\quad (1)
$$

$u(t) = x(t) - y(t) + n$; $r(t) = X(t) - Y(t)$; $v(t) = x(t) + y(t) - m + q(t)$; $s(t) = X(t) + Y(t)$; x and y = state variables; X(t) and Y(t) = control variables, of the same nature as x(t) and y(t); k_i, c_j, m, n = constant parameters (usually $n = 0$) or sometimes variable in relation to time; q(t) represents a synchronizer such as: $q = A \sin(\omega t + \phi)$. In the simulation of circadian rhythms in biomedicine, the synchronizer simulates the day-night alternance.
1.3.3. Principles of AA control. Since 70, we have suggested controlling such a model in the cases in which it simulates an imbalance between x and y (antagonistic and/or agonistic) *by adding control equations (X, Y) similar in shape(i.e. merely differing in parameter values) to state equations*. The trajectories of the composed variables (x + X) and (y+Y) may thus be again balanced. Therefore, *an imbalance between two endogenous AA variables has to be controlled by the administration of the same but exogenous variables* (bilateral strategies or bipolar therapies). Details about control techniques, particularly for identification of control variable parameters may be found in [1]).
We have also pointed out that the administration of the *deficient* variable was *ineffective*; on the contrary, *the administration of the sole variable already in excess may reestablish the balance.*
Another approach consists in leaving the control equations of (1) and using instead of them new equations in X(t) and Y(t) according to pharmacokinetic dynamics (after an identification of the corresponding parameters by minimization of a performance index).
1.3.4. In order to aim at a better understanding of these principles, we stated that by "superposing" the state- and the control-phaseportraits (both imbalanced), it becomes possible to reconstruct the physiological phase-portrait (balanced). Or we may state that this type of control consists in a a kind of *remodeling the epigenetic landscape*. Such a control corresponds also to the definition given by P. Delattre of the ideal control process in general, a combination of the controlled system and the controlling system resulting in an *auto-control* [7].

2 The Problem of the Strange Attractors in the Light of Agonistic Antagonistic Networks

2.1. The strange attractors (SA) seem to correspond to a universal dynamical behaviour in biology.
SA therefore would take the place of limit-cycles which had previously replaced the asymptotical critical points in biological modelling. It is doubtful that such a generalization would be pertinent in all cases (due to the peak of a circadian rhythm for instance, although a SA seems to be able to simulate ultra-, infra- and circadian rhythms together), however the proposed view elicits some new problems. Two types of facts have to be considered:
a) We should not be puzzled by the term "chaotic" [chaotic determinism or dynamics (CD)]: on the contrary, SA establish an order from chaos, and this order may be defined, among other properties, by the *topology* (or the place of the SA in the phase space), which becomes a kind of "invariant". Indeed *Birkhoff ergodic theory applies to chaotic determinism: the temporal mean of a variable on a "sufficient" long trajectory is constant.*
b) Although, in regards to aperiodic systems, people usually take an interest *in the transitions between aperiodic and periodic functioning,* by trying to connect such a type of functioning to the physiological behaviour of a system, or by pointing out how passing from one to another, such transitions could correspond to some physiological functions (and *a contrario* by determining what could be a pathological functioning of the system under these conditions). But, given that our interest in modelling originated in the possibility of defining imbalances (inside the so-called agonistic antagonistic couples such as cortisone/vasopressin couple, or stimulating/inhibiting agents couple, or insulin/glucagon couple, or phosphorylation/dephosphorylation couples in normal or cancerous cell ...), it became necessary to continue with this type of modelling as far as SA were concerned. *The notion of imbalanced SA has resulted from these facts.*
2.2. A review of the biomedical research about CD [5] has allowed us to conclude that deterministic chaos has mainly been considered up to now *from the point of view of temporal sequences and fractal dimension,* and not in the manner evoked *supra*.
2.3. Diverse opinions have been expressed as to the meaning of SA. In general SA seem to be a beneficial property of biomedical systems (cf. [6]): given that chaotic motion includes an infinite number of unstable periodic motions, this enables a system to preadapt to the variability of the environment. Goldberger et al. [9] considered deterministic chaos as the "wisdom of the body", by opposing it to the concept of homeostasis: an opposition which does not seem to us fully pertinent when this concept applies to the notion of balanced and imbalanced SA. Moreover, when we speak about the rapid loss of information leading to unpredictable events, we also should take into account the concept of ergodicity (cf. *supra*), which allows us justly to predict the general behaviour of a system (the mean value of a variable for instance) (cf. also SA considered as an invariant set by Stewart [15]).
2.4.. Attempts to use control in SA generally supposed that to suppress a chaotic motion would be desirable. Ott et al. [12], Garfinkel et al. [8], Petrov et al. [13], Shinbrot et al [14}postulated that it should possible to stabilize a system around one of periodic motions by using the definitive feature of chaos, i.e. the extreme sensitivity to initial conditions. A more easier type of control consists of forcing a system (chaotic oscillations of cAMP in Amoebas) by small-amplitude periodic input [of cAMP in this example (Li et al.)] [10].

2.5.. Then was approached *the problem of SA production and control.* Contrary to research carried out on the elementary MRAAC - which was used for several years in a practical field of biomedicine (parameter identification, therapeutical inferences) - this part of the paper is primarily of theoretical meaning.

2.5.1. A possible formalization of agonistic antagonistic networks is the following one (cf. more general equations in [2]):

$$\dot{x}_i = \left[\sum_j k_{ij}(u_i + r_i)^j + \sum_j c_{ij}(v_i + s_i)^j\right] + \left[\sum_j \bar{k}_{ij}(\sum_i u_i + r_p)^j + \sum_j \bar{c}_{ij}(\sum_i v_i + s_p)^j\right]$$

$$\dot{y}_i = \left[\sum_j k'_{ij}(u_i + r_i)^j + \sum_j c'_{ij}(v_i + s_i)^j\right] + \left[\sum_j \bar{k}'_{ij}(\sum_i u_i + r_p)^j + \sum_j \bar{c}'_{ij}(\sum_i v_i + s_p)^j\right] \quad (2)$$

$$\dot{X}_p = \left[\sum_j \hat{k}_{pj}(u_p + r_p)^j + \sum_j \hat{c}_{pj}(v_p + s_p)^j\right] + \left[\sum_j \tilde{k}_{pj}(\sum_i u_i + r_p)^j + \sum_j \tilde{c}_{pj}(\sum_i v_i + s_p)^j\right] + \Lambda_1$$

$$\dot{Y}_p = \left[\sum_j \hat{k}'_{pj}(u_p + r_p)^j + \sum_j \hat{c}'_{pj}(v_p + s_p)^j\right] + \left[\sum_j \tilde{k}'_{pj}(\sum_i u_i + r_p)^j + \sum_j \tilde{c}'_{pj}(\sum_i v_i + s_p)^j\right] + \Lambda_2$$

with $\Lambda_1 = \sum_j \lambda_{pj}(X_p - \overline{X}_p)^j$ and $\Lambda_2 = \sum_j \lambda'_{pj}(Y_p - \overline{Y}_p)^j$; i = 1,2...,m; j = 1,2...,n;.

$u_i = \sum_i (x_i - y_i); v_i = \sum (x_i + y_i - m_i); r_i = X_p - Y_p,\ s_i = X_p + Y_p$ for $i = p$.

This is just a simplified version of AA-networks due to the fact that only *one single pair of variables* (X_1, Y_1) was used for the entire system in an attempt to reestablish global balance; Λ_1, Λ_2 represents an anti-drift device ($\overline{X}$, $\overline{Y}$ and λ_{pj} are considered not to vary with time and are also used in the elementary MRAAC in order to avoid the drift of the limit-cycle dimension four formed by x, X, y and Y). Moreover, the (2m)-dimensional vector $(\dot{x}_i, \dot{y}_i)$ and the vector $(\dot{X}_p, \dot{Y}_p)$ are defined to be composed of: (a) a sum (see first bracket) directly corresponding to its counterpart in (1) and (b) an interconnection modulus (see second bracket) combining all the balances (or imbalances). Synchronizers q_j (cf. *supra*) were also added in v_j, their use is important for the appareance of SA's.

2.5.2. In order to produce a SA, first we identified the parametric values of (2) so that the curves x_1, y_1 fit the experimental values (periodic) of the circadian rhythms of cortisol and vasopressin. As far as curves x_2, y_2 were considered, for instance variations of other anti-growth factor and growth factor, there was a lack of data, and it was enough to obtain periodic values of these assumed variables (fig.1). Period of the synchronizers was 24 hours for both equations. Then, by modifying the parametric values and by adding a constraint of aperiodicity, the new identification ended to observe the SA of figure 2 [when an oscillating network has a dominant free period different from the period of a stable oscillating input, SA's may appear for some precise values of the free period (cf. a chart in [5]); so a system could "choose" between different types of dynamics according to a physiological goal]. Finally, these procedures were repeated for the pathological values of x_1, y_1: we give only in fig. 3 the chaotic aspect of the curves in question.

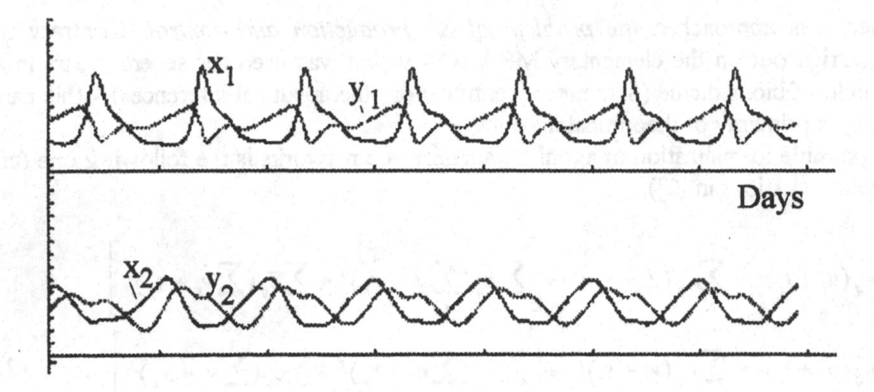

Figure 1 - Physiological rhythms of cortisol (x_1) and vasopressin (y_1)

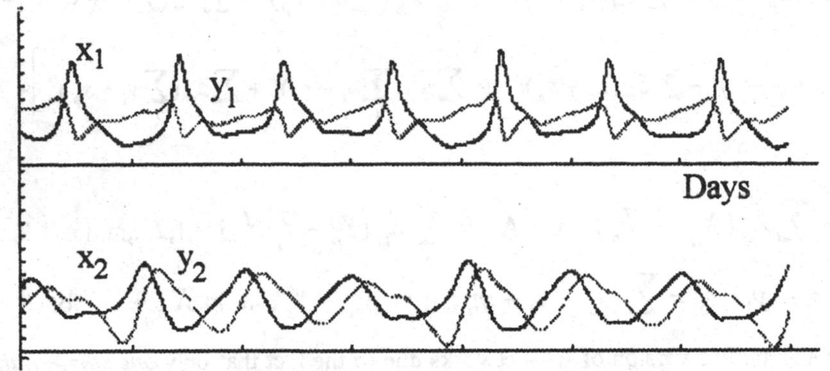

Figure 2 - Idem fig. 1, but under the form of a strange attractor

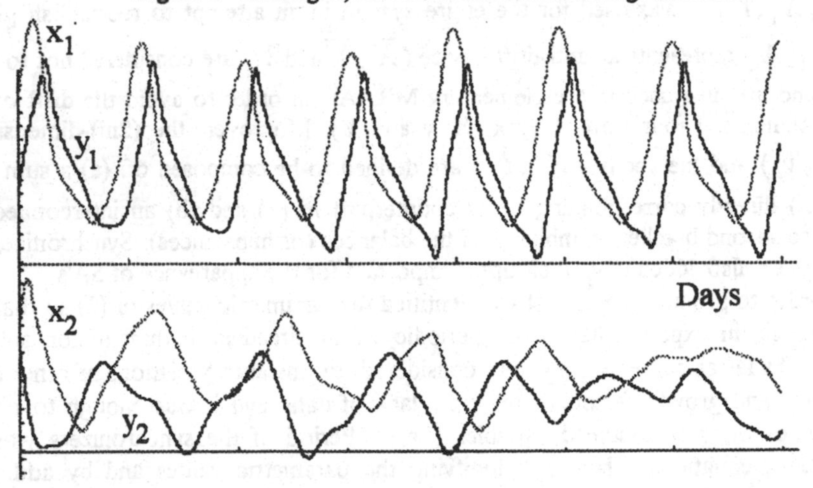

Figure 3 - Imbalanced strange attractor: mean values of (x_1-y_1)=-.2029; (x_1+y_1-m)=.4341; (x_2-y_2)=-.1907; (x_2+y_2-m)=.1958.

2.5.3. The same control principles can be applied to AA networks, however we only attempted to control the network by a number of control equations lower than that of state

equations (as in equs (2)). Then, in 1981 [11], we proofed that *a global imbalance in an AA network may be controlled by a couple of agents similar to a subset of this network.*
2.5.4. What is the goal of a control in relation to this type of imbalance? *to restore the mean normal values of the differences x_i, y_i which we set equal to zero, or at least to approach these values, without aiming at the elimination of its chaotic nature.* As it has been already said, *only one control couple should be used, and the couple X_1, Y_1 was chosen*; by controlling the less imbalanced network couple, it was hoped that the x_2, y_2 couple would also benefit from this control. To reach this goal, minimization of the following performance index was carried out:

$$J_{\hat{k}_{jp},\hat{c}_{jp},\overline{X},\overline{Y},\lambda_i} = \sum_n ((x_{1n}+X_{1n})-(y_{1n}+Y_{1n}))^2 + \sum_n (x_{2n}-y_{2n})^2 \quad (3)$$

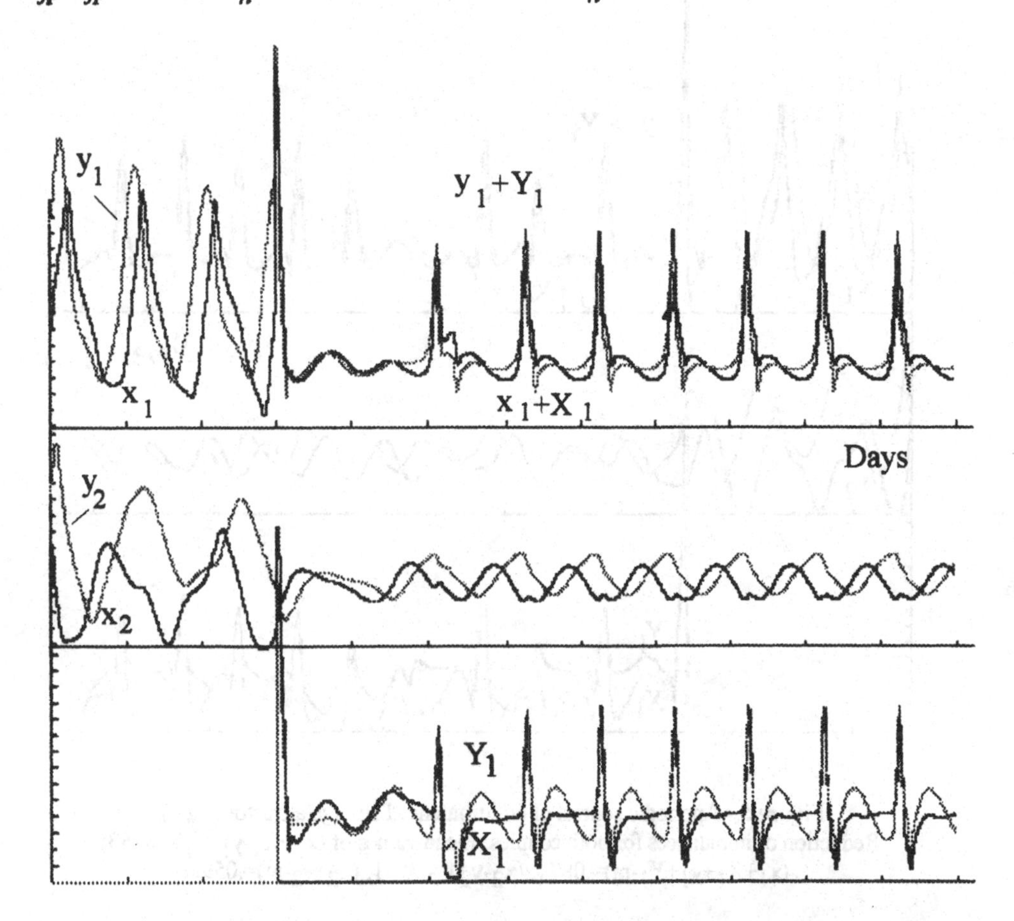

Figure 4 - Periodic control of an imbalanced strange attractor (fig. 3). Reduction of imbalances for both couples: mean values of $(x_1+X_1-y_1-Y_1)$=.0087; $(x_1+X_1+y_1+Y_1-m)$=.0808);(x_2-y_2)=-.0561; (x_2+y_2-m)=.0666 (curves of x_2 and y_2 not represented in this figure and the following ones)

In our preceding papers where other initial curves were used, this method elicited a decrease in agonistic and antagonistic imbalance (for both curves). Nevertheless although we were not trying to make CD disappear, it was replaced by a quasi-periodic dynamics. In the present paper, two solutions are proposed: the first control was also a periodic control, and the other shows that CD is going on despite of the improvement of the imbalances (fig. 4 and 5). *As it were, control of imbalanced SA has been performed by the addition of another SA.*

We ought to emphasize the fact that, *when once the optimization has established the values of control parameters, it is still possible to obtain an efficient control by changing the starting time of the control*: the control parameter field becomes itself an invariant.

2.5.5 Now it may be asked whether this method could be applied in the case of real systems. When the imbalances in periodic oscillations were considered (cf. 1.3.3), the dynamics of the control variables, themselves periodic, can be used for an

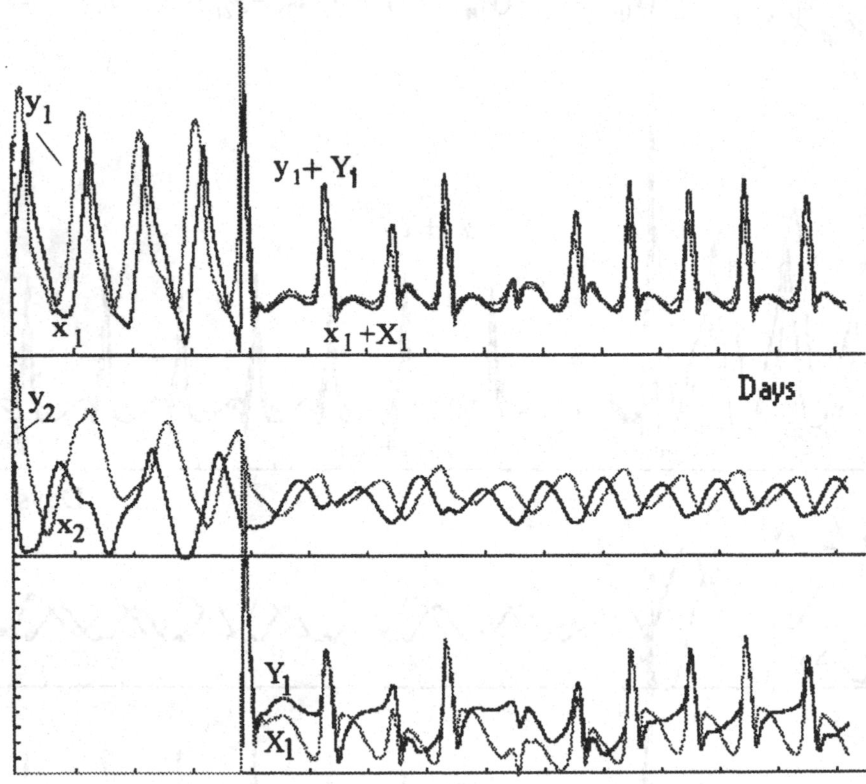

Figure 5 - Aperiodic control of an imbalanced strange attractor (fig.3).
Reduction of imbalances for both couples: mean values of $(x_1+X_1-y_1-Y_1)$=.0183; $(x_1+X_1+y_1+Y_1-m)$=.0874; (x_2-y_2)=-.0581; (x_2+y_2-m)=.0598

effective control (it was only necessary to choose convenient initial conditions). In the case of chaotic control, conditions similar to those observed at the beginning of the control will never again be observed: for instance, it was not possible to reinject the X and Y curves of the figures 4 and 5 at any moment in SA dynamics. However, we may imagine the following steps: a): identify the parameters of SA corresponding to a real system (this task some problems!); b) identify the control parametric fiedl by the proposedd method; c) use in parallel the administration of agents corresponding to X and Y and deduce, in real time, their successive values by injecting again the new values of the controlled variables into the mathematical model - in a kind of feed back control which yet has to be defined.

2.5.6. In this model, a SA can easily be changed into a periodic attractor. It is sufficient to use *Fourier series control* instead of equs (2) control equations and to choose a

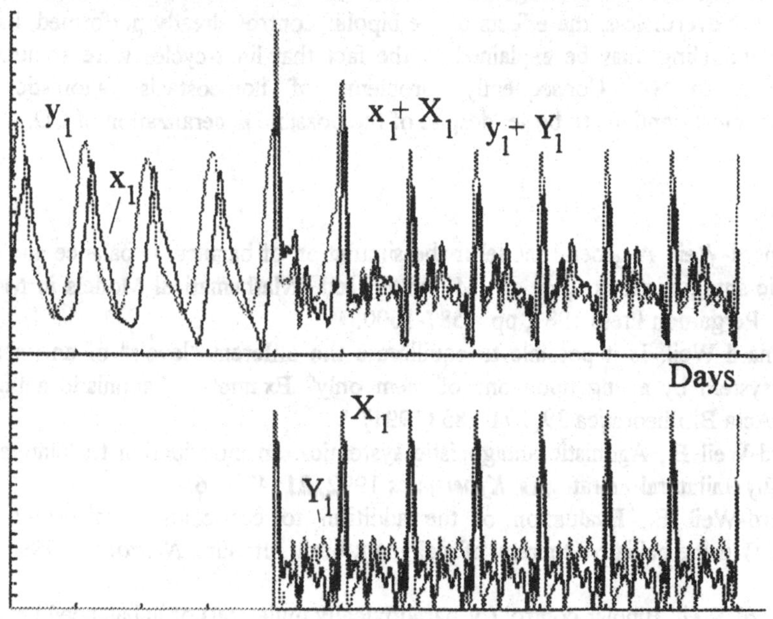

Figure 6 - Periodic control of an imbalanced strange attractor (fig. 3) by means of periodic inputs (Fourier series). Reduction of imbalances for both couples: $(x_1+X_1-y_1-Y)=.0293$; $(x_1+X_1+y_1+Y_1-m)=.259743$; $(x_2-y_2)=-.0869$; $(x_2+y_2-m)=.1402$

periodicity constraint as a performance index in order to identify Fourier series coefficients. By using forced periodic inputs (Fourier series of order 7) and adding the performance index (3) which now includes periodicity constraint, a decrease in the imbalance was obtained, simultaneously to a disappearance of the CD (fig. 6). It is worth noting that such a type of inputs may be used with other initial values (starting at different times). These results would simplify the problem of actually applying this research. Nevertheless, simulations showed that the control of the imbalances gave different results according to the new time of starting, even if the CD continues to disappear.

2.5.7. Finally, *power spectrum*, *greater non-null Lyapounov's exponent*, measure of *fractal dimension* by the method of Procaccia and Grassberger have been performed for all the reported simulations. They demonstrated the CD of the figures 2, 3 and 5 (value of the fractal dimension nearly equal to 3.5).

3 Conclusions

It seems pertinent to distinguish two modalities of control in case of SA: one of them concerns the CD itself; the other concerns the mean values of the trajectories, according to the ergodic theory. In the biomedical field, but perhaps also in other fields, we ought to decide what type of control (or both) is (are) recommended. The proposed model enlightens on this subject, although its practical use could elicit some difficulties at the present time. Nevertheless, the effects of the bipolar control already performed, based on limit-cycles modeling, may be explained by the fact that limit-cycles were sometimes an approximation to SA. Consequently, problems of homeostasis (agonistic and/or antagonistic) must continue to be set despite of the possible generalization of CD.

References

1. E. Bernard-Weil: A general model fr the simulation of balance, imbalance and control by agonistic antagonistic couples. In: M. Witten (ed.):Mathematical Models in Medicine. New York: Pergamon Press 1987, pp. 1587-1600.

2. E. Bernard-Weil: Is it possible to equilibrate the different "levels" of an imbalanced biological system by acting upon one of them only? Example of agonistic antagonistic networks. Acta Biotheoretica 39, 271-285 (1991).

3. Bernard-Weil E., Agonistic antagonistic systemics: an introduction to bilateral - and paradoxically unilateral - strategies; *Kybernetes* 1992; **21**: 47 - 66.

4. Bernard-Weil E., Evaluation of the addition to corticoids of a growth factor (vasopressin) in the palliative therapy of malignant brain tumours, *Neurol Res* 1991; **13**: 94 - 109.

5. E. Bernard-Weil: Bipolar control (or paradoxically unilateral) of imbalanced limit-cycles and strange attractors. Chaos and homeostasis. J Biol Systems 1, 311 - 334 (1992)..

6. M.E. Cohen, D.L. Hudson, M.F. Anderson and C. Vazquez: Blood flow data exhibit chaotic properties. Int. J. Microcomput. 12, 82-7 (1993).

7. Delattre P.: Direct and inverse regulation control in transformation systems. Math Biosc 34, 303-324 (1981).

8. A. Garfinkel, M.L. Spano, W.L. Ditto and J.N. Weiss: Controlling cardiac chaos. Science 257, 1230-35 (1992).

9. A.L. Goldberger and B.J. West: Fractals in biology and medicine. Yale J Biol Med 60, 421-435 (1987).

10. Y.X. Li, J. Halloy, J.L. Martiel, B. Wurster and A. Goldbeter: Suppression of chaos by periodic oscillations in a model for cyclic AMP signalling in Dictyostelium cells. Experientia 48, 603-606 (1991).

11. P. Nelson and E. Bernard-Weil: Justification d'un modèle ago-antagoniste. Int. J. Biomed. Comput. 11, 145-62 (1980).

12. E. Ott, G. Grebogi and J. Yorke: Controlling chaos. Phys Rev Lett 64 1196 (1990).

13. V. Petrov, B. Peng and K. Showalter: A map-based algorithm for controlling low-dimensional chaos. J Chem Phys 96 ,7506-13 (1992).

14. T. Shinbrot, C. Grebogi, E. Ott and J.A. Yorke: Using small perturbations to control chaos. Nature 363, 411-417 (1993)

15. I. Stewart: Does God Play Dice? Oxford: Basic Blackwell, 1989.

8. REUSABILITY

A Feature Based Reuse Library

Sooyong Park and James D. Palmer

Center for Software Systems Engineering
School of Information Technology and Engineering
George Mason University
Fairfax, VA 22030-4444

ABSTRACT. Software reuse is widely recognized as a solution to improving software productivity, quality and reliability. While there have been high expectations for the benefit of software reuse, the actual gain has been relatively small. One of the problems is that code components are typically developed with an application-orientation without consideration of reuse. As a result, they are generally difficult to identify for reuse even though they are quite reusable. One approach to this problem is to search for code component based on software system features where features are a composite set of predefined facet attributes. To support feature-based reuse, a feature-based reuse library (FRL) is proposed.

1. Introduction

Software reuse is widely recognized as a key solution to improving software productivity, quality, and reliability. The idea is to reuse work accomplished in prior software developments to reduce the effort for subsequent tasks that have common features.

While there have been high expectations for the benefit of software reuse, the actual gain has been relatively small. One of the problems is that code components are typically developed in an application-orientation without consideration of reuse. As a result, they are generally not suited for reuse even though they have common features in both the same and other problem domains. The reason is that when software engineers develop a software, for a same feature of software, their design and code can be different. Therefore when a software engineer reuse the codes, he/she might try to find different specification of code from the reuse library. In this situation, the reuse is failed since he/she could not find a reusable code even though the codes are available for the feature.

One approach of solution to this problem is to search for a code component objects based on software system features, as they may be more "alike" than code components. To support feature-based reuse, a feature-based reuse library (FRL) is proposed that is composed of four models. The first model supports faceted classification of feature objects. The second model is a domain-specific feature hierarchy. The third model is a relation table that maps the feature objects to component objects. The fourth model is a domain-specific component object hierarchy.

2. Feature-based retrieval

Kang et al.[KCH+ 90] defined feature as a prominent or distinctive user-visible aspect, quality, or characteristic of software system. In this paper, the definition of feature is limited definition of Kang et al -i.e. a prominent or distinctive user-visible aspect. In this section, a high-level overview of the steps involved in retrieval of software components by software features is presented.

Assume we are developing an aircraft cruise control system that one need is to develop a software component for speed calculation. By using a feature-based reuse library (FRL), we can retrieve reusable software in the requirement analysis phase, since features can be identified during that activity.

As shown in table 1, the feature "speed calculation", he/she identifies facets and their attributes of this feature is defined by 4 attributes of facets: designation, domain, objects, and operations.

Facet	*Attribute*
Designation	Speed Calculation
Domain	Aircraft, Cruise Control
Objects	Speed Sensor, kph
Operations	Calculation, Convert

Table 1. A facet information of speed calculation

With this feature definition, the potential matches are sought with existing features in the FRL. In this example, the matched features are shown in table 2, speed calculation in automobile cruise control domain and position calculation in navigation system domain. These two are selected as possible matches because of the matched attributes for objects and operation facets.

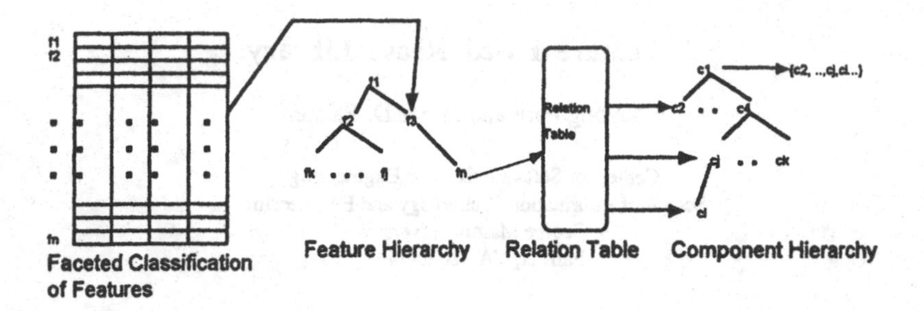

Figure 1 Feature Based Reuse Process

Facet	*Attribute*
Designation	Speed Calculation
Domain	Automobile Cruise Control
Objects	Speed Sensor, mph
Operations	Calculation

Facet	*Attribute*
Designation	Position Calculation
Domain	Navigation System
Objects	Speed Sensor, Satellite
Operations	Calculation

Table 2. Retrieved features for calculate speed

Based on selected features, the user select most possible feature. If he/she selected feature "Speed Calculation" in automobile cruise control(ACC) domain, the FRL search related or necessary features by feature hierarchy model in the domain of automobile cruise control. With the retrieved feature, the FRL search the code components which developed for the retrieved features by using feature/component relation. After code component(s) are retrieved, the FRL search the related code components by using component hierarchy model in the domain. Followings can be selected from the FRL.

Package: Get-RPM	Package:RPM-to_MPH	Package: Send-MPH
Object:RPM	Object: RPM, MPH	Object:MPH
Function:Get current RPM	Function:Convert rpm to mph	Function:Send current MPH
Medium: Ada	Medium: Ada	Medium: Ada

In summary, the FRL operation involves five steps.
STEP 1: Identify feature by specifying the facet attributes
STEP 2: Find potentially matching features in the FRL
STEP 3: Find related or necessary features from the feature hierarchy
STEP 4: Find code components for retrieved feature(s) using feature to component relational table
STEP 5: Retrieve related or necessary code component(s) for retrieved code component(s) from the component hierarchy.
These steps are described in figure 1. A key advantage of the feature-based reuse library approach is that it can be extended to generate broader set of reusable software components.

3. Related Work

The reuse library problem has been one of the classical problems in reuse. Different classification approaches have been proposed. Ostertag et al.[OHDB92] categorized the approaches as three types: free-text keywords, faceted index, and semantic-net based.

Frakes and Nejmeh[FN87a] used a free text indexing approach for C software components. In this system, index terms were automatically extracted from the descriptive headers on C modules and functions were stored in an inverted file in an information retrieval system. Retrieval from this system was done using Boolean search techniques.

The faceted index approach proposed by Prieto-Diaz[PDF87] relies on a defined set of keywords extracted by experts from program descriptions and documentation. These keywords are arranged by facets into a classification scheme and used as standard descriptors for software components. A thesaurus is derived for each facet to provide vocabulary control and to add a semantic component to retrieval. Keywords can only be used within the context of the facet they belong to and ambiguities are resolved through the thesaurus.

A major advantage of a faceted classification over a hierarchical scheme is the freedom it gives the indexes to synthesize terms to express complex concepts. All concepts do not have to be predetermined at the time of creating the classification system. Rather, the indexes can synthesize terms from facets to create concepts as needed. A faceted scheme is also easier to update and modify since you can change one facet without affecting any others.

Semantic-net based approaches have been proposed by Wood and Sommerville[WS88] and by Devanbu et al.[DBS90]. Devanbu et al have used frames to represent software components from system 75. Their reuse system, called Lassie, attempts to support multiple views of system 75: a functional view which describes what components are doing, architectural view of the hardware and software configuration, a feature view that relates basic system functions to features such as call forwarding, and a code view which captures how code components relate to each other.

Wood and Sommerville have used case frames for indexing reusable components. A case frame is a frame whose slots are fundamental nominals, actions, and modifiers from some domain. Nominals are simple things that can be conceptualized without the need to relate them to other objects. Actions are simple verbs, and modifiers specify attributes of nominals and actions. Nominals, actions, and modifiers are related via semantic-net.

Ostertag et al.[OHDB92] proposed the AIRS system that is a hybrid of the faceted index and semantic network approaches. The domain information inherent in the facets is used largely to reduce the rigidity and the laborious creation of a semantic structure. A hierarchical frame system is used to maintain information about which of the objects in the reuse libraries have which features, how these objects are grouped, and how the features are related.

The proposed approach is similar to that of Ostertag et al. in that faceted index is used to reduce the large searching space, but the meaning of a feature is different. Their usage of feature is to differentiate the code component i.e. feature of code component. However, the proposed approach focus on user's requirement as discussed in section 2.

4. The Feature-based Reuse Library(FRL) model

The FRL model for classification and retrieval is based on the features and the code components of a problem domain. The code component is called component object in the FRL. The reusable component objects are retrieved by feature. A feature is treated as an object so it is called feature object.

The FRL is composed of four models: A faceted classification model of feature object, a feature object hierarchy model, a relation model, and a component object hierarchy model. In this section, the details of these models is discussed.

4.1 Faceted classification model of feature object

The classification of feature objects is based on facets similar to those defined by Prieto-Diaz[PDF87]. Prieto-Diaz was mostly concerned with functional components and based his classification on imperative statements with the triplet: function, objects, medium/agent

This is not very applicable in classifying feature objects. First of all, feature objects are brief description of a system. Thus the best way to describe a feature object would be to find a designation that fits it. This is called a designation facet. Also, feature objects are domain oriented. Thus another way to describing the feature is to find a domain that fits. This is called the domain facet. However, this facet is not used when a user wants to search for a feature object in multiple domains. As noted, sometimes there is no feature in a domain, or there is more interest in the functionality of the feature. For the functionality of a feature object operation facet is used. Also then might be interest in what objects are involved in the feature. For the involved objects facet is added. The operation and object facets are similar to Prieto-Diaz's function and object facet. In summary, the proposed four facets are:
Designation, Operation, Object, Domain

4.2 Feature object hierarchy model

In addition to faceted classification of feature objects, the FRL has a feature object hierarchy model for each domain. The feature object hierarchy model is based on a generalization/aggregation hierarchy[1]. Figure 2

[1] e.g. a printer is a generalization of a dot matrix printer and a laser printer. A dot matrix printer is a aggregation of ribbons and pins

provides an example of a feature object hierarchy. In this hierarchy, we can interpret that f2 is an aggregation of f5, f6, and f7 and f4 is a generalization of f8 and f9.

Once a feature object is selected, the FRL retrieves related feature objects by using the associated feature object hierarchy for domain. In retrieving related feature objects, following rules apply:

RULE1: For any selected feature, if its descendant feature objects are aggregated feature objects then retrieve all the descendants
RULE2: For any selected feature, if its descendant feature objects are generalized feature objects then retrieve at least one feature
RULE3: For any selected feature, if its predecessor feature object is generalized feature, then retrieve its descendant feature

Figure 2 explains the usage of the rules in the context the example.
If feature object f2 was selected, feature objects f5, f6, and f7 would be retrieved by rule 1, and f1 would be retrieved by rule 3. If feature object f4 was selected, the FRL would retrieve f1 by rule 3, and by rule 2

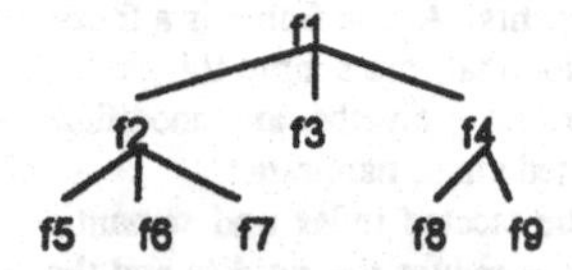

Figure 2 A feature object hierarchy

one feature object between f8 and f9. By using the feature hierarchy, we can retrieve feature objects that are either related to or necessary for the selected feature objects.

4.3 Relation table model

The FRL includes domain-specific relations between feature objects and components within domains. This model is an important element of the FRL as it is the bridge between feature objects and component objects.

A relation object R can be represented as a three place tupel, R <F, C, D >, where F is a set of feature objects, C is a set of component objects, and D is a domain. In building the relations between feature objects and component objects, there can be four cases.

CASE 1: In domain d, for a feature object fi, if multiple component objects are involved, the relation can be represented as: R < fi, AND{Cj, . .,.Cn}, d >
CASE 2: In domain d, for a feature fi, if multiple component objects can perform the fi, it can be represented as: R < fi, OR{Cj, . .,.Cn}, d >
CASE 3: In domain d, for a component ci, if multiple feature objects are involved, it can be represented as: R < AND{fj, . .,.fn}, ci, d >
CASE 4: In domain d, for a component ci, if multiple feature objects can be performed, it can be represented as: R < OR{fj, . .,.fn}, ci, d >

Among the four cases, since we are focusing on the path from feature objects to component objects, the CASE 1 and 2 can be used. We will discuss the relation model by example relations based on the feature and component objects in figure 2. Following are example relations:
R1 < f1, c1, d1 >
R2 < f2, c2, d1 >
R3 < f5, c4, d1 >
R4 < f6, OR{c6, c8}, d1 >
R5 < f7, AND{c5, c7}, d1 >

The description of the feature hierarchy model showed that, with the feature object f1, we could get related feature objects f1, f2, f5, f6, f7. If we apply these feature objects to example relations, we can get the following result.
f1 $\rightarrow$ c1 by R1
f2 $\rightarrow$ c2 by R2
f5 $\rightarrow$ c4 by R3
f6 $\rightarrow$ c6 by R4
f7 $\rightarrow$ {c5, c7} by R5
With these retrieved component objects, if we use component object hierarchy in figure 3, the actual retrieved component objects are {c1, c2, c4, c5, c6, c7}.

4.4 Component object hierarchy model

The component object hierarchy model is also based on a generalization/aggregation hierarchy. The same rules discussed in feature hierarchy object model are applied to this model. Figure 3 shows an example component hierarchy.

As discussed in previous sections, the retrieval on feature object f1 could get component objects {c1, c2, c4, c5, c6, c7} using the feature object hierarchy model and the relation table model. With those retrieved component objects, the FRL supports a search on related or necessary component objects by using the component hierarchy in figure 3 with the following results.

c1 → c1 by exact matching
c2 → {c4, c5, c6} by Rule 1
c4 → {c9, c10} by Rule 1
c5 → c5 by exact matching
c6 → c11 by Rule 2
c7 → c7 by exact matching

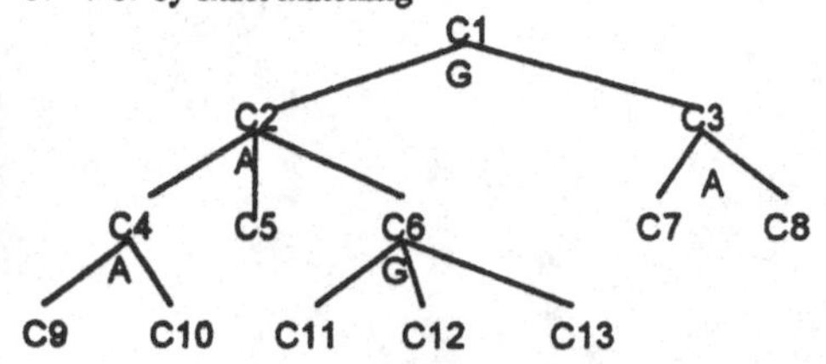

Figure 3 Component Object Hierarchy

As retrieved component objects, we get {c1, c2, c4, c5, c6, c7, c9, c10, c11}. This example shows that searching for a feature results in 9 potential reuse component objects.

5 Conclusion

Using feature-based component retrieval helps maximize reuse since multiple component objects are retrieved. The idea of retrieving multiple component objects is similar in concept to the mega programming method based on the reuse of mega modules suggested by Wiederhold et al.[WWC92]. While the mega programming approach proposes reuse of a unit of a mega module, the concept of the FRL is to supports reuse multiple moderate size modules with the actual reuse effect being essentially the same.

This paper has focused on the reuse library problem. The methods to build feature object hierarchies, relation tables, component object hierarchies, and component object interconnections are not addressed in detail as they require further research

References

[AS87] S. Arnold and S. Stepoway. The reuse system: Cataloging and retrieval of reusable software. Proceedings of COMPCON'87, pages 376-379, 1987.

[BR87] T. Biggerstaff and C. Ritcher. Reusability frame work, assessment, and directions. IEEE Software, pages 41-49, March 1987.

[CB91] G. Caldiera and V. Basili. Identifying and qualifying reusable software components. IEEE Computer, pages 61-70, February 1991.

[DBS90] P. Devanbu, R. Brachman, and P. Selfridge. Lassie: A classification-based software information system. Proceedings of the 12th Intl. Conf. on Software Engineering, March 1990.

[FG89] W. Frakes and P. Gandel. Representation method for software reuse. Proceedings of TRI-Ada '89, pages 302-314, 1989.

[FN87a] W. Frakes and B. Nejmeh. An information system for software reuse. Proceedings of the 10th Minnowbrook Workshop on Software Re-use, pages 142-151, 1987.

[FN87b] W. Frakes and B. Nejmeh. An information system for software reuse. Proceedings of the Tenth Minnowbrook Workshop on Software Reuse, 1987.

[KCH+ 90] K. Kang, S. Cohen, J. Hess, W. Novak, and A. Peterson. Feature-oriented domain analysis(foda) feasibility study. Technical Report CMU/SEI-90-TR-21, Software Engineering Institute, Pittsburgh, PA, November 1990.

[NC90] S. Navathe and A. Cornelio. Modeling physical system by complex structural object and complex functional objects. International Conference on Extending Database Technology, March 1990.

[OHDB92] E. Ostertag, J. Hendler, R.P. Diaz, and C. Braun. Computing similarity in a reuse library system: An ai-based approach. acm Transactions on Software Engineering and Methodology, pages 205-228, July 1992.
[Onu87] E. Onuegbe. Software classification as an aid to reuse: Initial use as part of a rapid prototyping system. Proceedings of the Twentieth Annual Hawaii International Conference on Systems Sciences, 1987.
[PDF87] R. Prieto-Diaz and P. Freeman. Classifying software for reusability. IEEE Software, 4(1):6-16, January 1987.
[RW91] H. Reubenstein and R. Waters. The requirements apprentice: Automated assistance for requirements acquisition. IEEE Transactions on Software Engineering, pages 226-240, March 1991.
[WS88] M. Wood and I. Somerville. An information system for software components. SIGIR Forum, 22(3):11-25, 1988.
[WWC92] Gio Wiederhold, Peter Wegner, and Stefano Ceri. Toward mega programming. Communications of the ACM, pages 89-99, Nov 1992.

Managing Uncertainty in an Object Recovery Process

Harald Gall and René Klösch
Vienna University of Technology
Institute of Information Systems, Department of Distributed Systems
Argentinierstrasse 8, A-1040 Vienna, Austria, Europe.

Abstract

Object-oriented concepts facilitate the reusability as well as the maintainability of existing software. Due to the great amount of existing procedural software, object identification in procedural programs is an important approach. The object recovery process for this identification of objects within procedural programs presents several uncertainties and ambiguities, which have to be resolved by acquisition of additional knowledge from the application domain and a human expert. In this paper we show the basic concepts of the object recovery process and describe those uncertainties and ambiguities, as well as our way of managing them in order to identify objects in procedural programs.

1 Introduction

Based on the increasing importance of reusability and maintainability in software development research in this field has uncovered many problems. Especially the reuse and maintenance of procedural software is a difficult problem (e.g. the components to be reused are difficult to define, maintaining existing programs is involved, and the composition of new systems reusing existing components is subject of intensive research).

The object-oriented paradigm offers characteristics (e.g. abstraction, encapsulation, information hiding, combination of data and behaviour, inter-object communication through message passing) that support the resolution of many of these problems: Objects, as the basic components of the object-oriented paradigm, are the elements to be reused. The concepts of encapsulation (information hiding) and combination of data and behaviour simplify the extraction of reusable objects from existing systems and facilitate the maintainability of systems.

Furthermore, the process of generating new systems using existing objects (system composition) is also supported by the concept of combining data and behaviour. The system composition process using objects is inherent in the object-oriented systems engineering process. Therefore, problems of module interconnection can be reduced in the object-oriented paradigm.

Motivated by these advantages of the object-oriented paradigm and the existence of the large amount of procedural software to be maintained and reused, we have developed a program transformation process that transforms procedural programs into an object-oriented architecture. This transformation process is used to identify objects within originally procedural programs (object recovery).

Within this object recovery process a few problems concerning the acquisition of additional knowledge from the application domain and the resolution of uncertainties have to be managed. This paper concentrates on problems of this knowledge acquisition and resolution of uncertainties during our object recovery process.

2 Related Work

Object recovery methods for identifying objects in procedural programs have been mainly developed at the source-code level. Objects are extracted from non-object-oriented source code by identifying the abstract data types of the program. These abstract data types form the objects that are formally expressed, e.g. using Z++ [11].

The design recovery method Desire [3, 4] uses informal information that is incorporated in the source code. This information is matched to domain information represented in a domain model in order to derive design information from the input program.

In the knowledge-based approach [6] for maintaining and modifying source code, a set of rules provides a reasoning framework to help detect several classes of bugs.

PAT [10] uses an object-oriented framework to represent programming concepts and a heuristic-based concept-recognition mechanism to derive high-level concepts from the source code using program plans.

Recognizer [15] automatically finds occurrences of a given set of clichés (commonly used programming structures) in Common Lisp programs and builds a hierarchical description of the program in terms of the clichés it identifies.

These automated methods usually lack of commercial applicability since the search for all possible derivations of a given structure is far too expensive and complex.

Our approach differs from the above mentioned approaches in the following way: 1) Object-oriented domain engineering methods are used and the domain information is represented in an object-oriented application model; 2) We use intermediate representations of the procedural input program at different levels of abstraction (e.g., structure charts, dataflow diagrams, entity-relationship diagrams, object-oriented application models); 3) A human expert is integrated into the object recovery process in order to introduce the necessary domain information; and 4) Application-semantic decisions are heuristic based, and supported by the human expert.

At present the method works with Pascal programs and identifies objects within the procedural source code. The resulting object structure can be transformed at the implementation level into an Object-Pascal program. The basic concepts are also applicable to C programs.

3 The Object Recovery Process

This section gives a brief overview of our object recovery process which tries to identify potential objects in a procedural program and is based on a program transformation process that transforms a procedural input program into a functional equivalent object-oriented architecture [8, 9].

The input of the object recovery process is a program, implemented in a conventional procedural programming language, e.g. Pascal or C. It cannot be assumed that either the data of the program are normalized (the third normal form would be desirable) or the analysis and design has been done by using conventional structured analysis methods. We do not assume that any development documentation exists. The only assumption for this object recovery process is the chance to perform a kind of requirements analysis of the input program.

The object recovery process consists of three main steps:

1. *Design Recovery*

   The goal of the design recovery step is the generation of design documents at different levels of abstraction:

   - Structure charts (SCs) and dataflow diagrams (DFDs) are reconstructed automatically using the techniques of [1, 2].
   - These structure charts and dataflow diagrams are used for the reverse generation of an entity-relationship diagram (ERD). The object recovery process works on a flat file struture, so we had to develop our own technique. For underlying relational or hierarchical schemas some approaches for reverse generating ERDs exist ([7, 14, 12]).
   - Based on the ERD an object-oriented application model of the examined program can be constructed (RooAM). This RooAM consists of objects with attributes and structural relations, e.g. is-a, part-of, (mainly derived from the ERD), services, instance and message connections (derived from the source code).

   The *reverse* generated object-oriented application model (RooAM) represents the procedural input program in an object-oriented form and is used as the first input of the following object mapping step.

2. *Application Modeling*

   During the design recovery step different uncertainties may arise (e.g. procedures/functions that become services belong to more than one object). To resolve these ambiguities it is necessary to add higher-level information from the application domain in the object recovery process.
   An object-oriented application model is, therefore, developed independently of the implementation of the input program, but based on the same requirements analysis. This model is called the *forward* generated object-oriented application model (FooAM) and is used as the second input for the following object-mapping step.

   Different object-oriented analysis methods can be used for modeling the application. For our considerations we use the terminology of Coad and Yourdon's OOA [5].

3. *Object-Mapping*

   The goal of the object-mapping process is to find a mapping of similarities between the elements of the FooAM and the RooAM. The FooAM thereby works as a pattern for the object recovery process.

   During this object-mapping process different problems, e.g. varying naming conventions, structural differences, not identifiable parts, etc., have to be considered.

   The result of the object-mapping process is a target application model that is defined as the synthesis of the FooAM and the RooAM and incoporates all objects that could have been identified in the source code of the procedural input program.

During the above mentioned steps of the object recovery process there exist some uncertainties and ambiguities that have to be resolved through acquisition of additional knowledge.

The next sections point out these uncertainties as well as their management in more detail.

4 Uncertainty and Domain Knowledge Acquisition in Application Modeling

Modeling an application is based on the requirements analysis of the desired system. Furthermore different kinds of information from the application domain are also necessary during the modeling process. It is the task of a human engineer to perform the application modeling process, which can only be supported by modeling tools, but not fully automated. An optimal integration of these modeling tools within the modeling process will result in well defined application models.

During the application modeling process the human engineer has to choose among a great variety of development possibilities, e.g., whether an entity should become an object on its own, or if it is part of another object, or whether to use a master/slave approach or a client/server approach (see [13]), etc. The client/server approach is often claimed to be the "real" object-oriented approach for this modeling process, but nevertheless some analysts generate object-oriented application models based on the master/slave approach.

The variety of development possibilities during the application modeling process causes uncertainty, which directly influences the results of the object-mapping process during our object recovery process.

5 Uncertainty in Design Recovery

The design recovery step covers the generation of various low-level design documents from procedural source code. No additional information is necessary for this generation process.

In addition to low-level design documents, an ERD is generated, but this generation cannot be fully automated. Additional information is necessary to define the cardinalities for each relation, or to identify the key attributes for the definition of inter-entity data dependencies.

The cardinality of a relation is determined by underlying constraints. For the definition of the cardinality it is therefore necessary to find those statements in the source code which represent the examined constraints.

This can be easy for the cardinilaties of structural relations (e.g. is-a, part-of), because these cardinalities are possibly represented in the type declarations, e.g. `item` is part-of `delivery_note` with a range of [1..4] (at least one, maximum four `items` per `delivery_note`) shown in the following example:[1]

```
TYPE delivery_note = RECORD
     dn#:INTEGER;
     ...
     item_list: ARRAY[1..4] OF item;
                         END;
```

The definition of the cardinality of general relations has to be made by a human engineer, who has to identify the relevant statements within the source code (e.g. for/while/loop-statements, if-conditions, etc.), but it is possible and useful to restrict the area of relevant statements by a tool. Especially dynamic constraints that have to be met within some program executions (e.g., a `delivery_note` has to be part of an invoice, sometime after the `delivery_note` was generated) are difficult to identify, even for a human engineer.

Based on the ERD the RooAM is reverse generated by adding dynamic aspects (services and service relationships) to the static aspects derived from the ERD. This process of adding dynamic aspects cannot be fully automated, because it may not be determined to which object (entity) a particular service candidate has to be encapsulated. The domain information for this decision is represented in the FooAM and used during the object mapping step.

The result of the design recovery is a RooAM that exhibits objects, structural relationships (is-a and part-of relations derived from the ERD), and dynamic aspects. These dynamic aspects are represented through services and corresponding service relationships. For many service candidates it was possibly not determinable to which object they belong; those service candidates are called ambiguous service candidates and, as a consequence, assigned to each potential object. They are marked with an asterisk for the resolution of this ambiguity during the following object-mapping process.

During design recovery it is necessary to deal with uncertainty on different occasions and not to make premature decisions. Such decisions would be extremely dangerous for the introduction of failures during the object recovery process. Our approach postpones such uncertain decisions until additional domain knowledge for a reliable resolution of those ambiguities is available. To postpone the decision is not a weakness, but a serious and even necessary task.

6 Managing Uncertainty in the Object-Mapping Process

The two application models (RooAM and FooAM) are matched in order to form a resulting target ooAM. The similarities between the FooAM and the RooAM are the basis for this matching process. The FooAM serves as a pattern in which additional domain information is represented: it exhibits a target object structure, an unambiguous service definition for each object, and the corresponding message connections.

[1] Consider that if a dynamic data structure (dynamic list) is used instead of an ARRAY, it is not possible to define the cardinality of a part-of relation.

The goal of the object mapping process is to resolve the ambiguities regarding the service candidates and to define the necessary message connections. This can only be performed for objects that can be mapped between the two application models.

The mapping process is based on various typical similarities between the two application models: structural similarities concerning objects and their interrelationships (part-of, is-a relations), attributes and instance connections, and services and message connections.

Object interrelationships can vary, especially in the RooAM. It is, therefore, often difficult to match these relations between the two application models. As an example we will discuss below the is-a relation and its possible variations within the RooAM in more detail.

6.1 Variations of the is-a Relation in the RooAM

The pattern for the mapping of an is-a relation is a generalization/specialization (gen/spec) structure of the FooAM. Figure 1 shows the structure of a `delivery_note` and its specializations in a commercial invoice system.

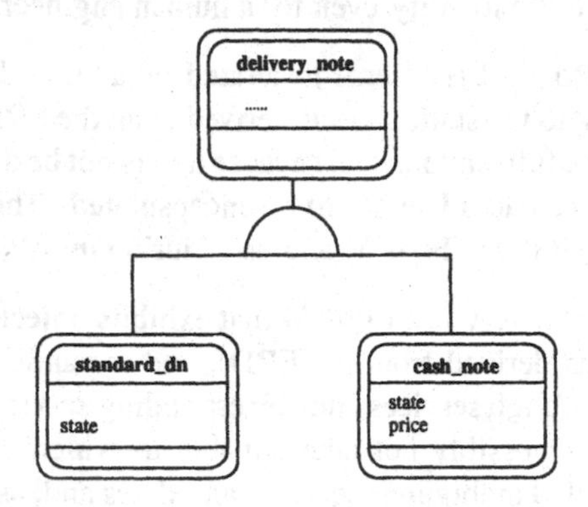

Figure 1: Gen/spec structure of the FooAM

For the identification of this is-a relation within the RooAM different possibilities of implementation have to be considered:

1. *varying subtypes*:

   A gen/spec structure can be implemented using a different subtype structure. Figure 2 shows the above mentioned gen/spec-structure with `delivery_note` as supertype and only one subtype `cash_note`. Thereby the attribute `state` of the supertype `delivery_note` (with the range of values `(invoicing_done, invoicing_not_done)`) is redefined in the subtype `cash_note` (with the range of values `(outstanding, paid)`), and the attribute `price` is added to the subtype `cash_note`.

   Since the gen/spec structure is partially implemented this uncertainty problem can be (quite easily) managed by an experienced human engineer. The person has to match the two gen/spec structures against each other to find out the similarities and, as a consequence, the structure of the RooAM, which is a variation of the pattern structure of the FooAM.

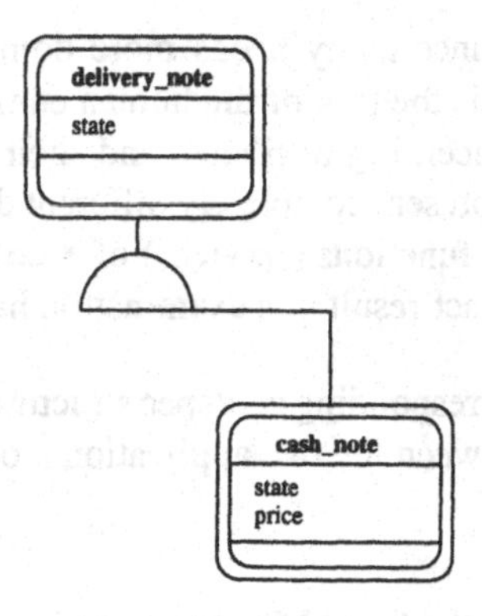

Figure 2: Varying subtypes of an is-a relation

2. *no gen/spec structure within the RooAM*:

   A `delivery_note` can also be implemented by using only a single type. The range of values of the attribute `state` is enhanced to represent the range of values of both subtypes within one type (`state = (invoicing_done, invoicing_not_done, outstanding, paid)`). The attribute `price` is defined within the type, but only used in `cash_note`.

   This uncertainty can be managed by examining the usage of the attribute `price` in different program paths: One path works with `price == 0` to represent the `delivery_note` and the other path works with `price > 0` for the `cash_note`. A similar examination can be performed for the attribute `state` to discover in which program paths a special range of its values (`(invoicing_done, invoicing_not_done)` or `(outstanding, paid)`) is used.

3. *different types*:

   The gen/spec pattern structure of the FooAM can be realized through separate types in the RooAM (none of them incorporating a gen/spec structure). In the example shown in Figure 3 the implementation of the gen/spec structure is realized through two separate types `cash_invoice` and `delivery_note`.

   This is often the case when large old systems are maintained over and over again so that the maintenance staff simply adds a new (very similar) type in order not to produce any side effects (because of a lack of program understanding).

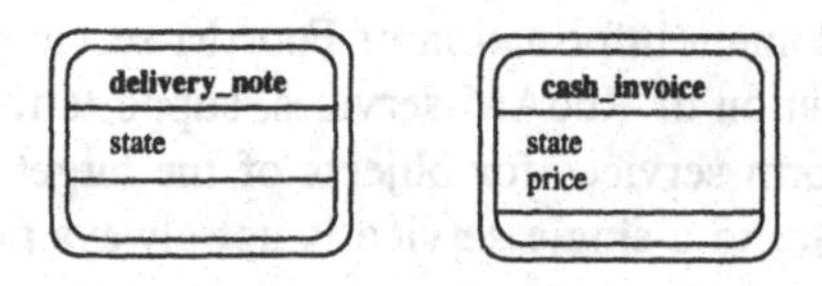

Figure 3: Different types form an is-a relation

Managing this kind of uncertainty needs more domain knowledge than the above mentioned variations. It is the task of the human engineer to find out the similarities of the defined types concerning attributes and their range of values, and services although they seem to represent completely different domain artifacts. The signatures of the procedures and/or functions (services) of each type can help to find out some similarities, but for an exact result this examination has to be done in detail.

Based on these similarities a corresponding gen/spec structure may be defined in the RooAM. After that step the mapping between the two application models is possible.

6.2 Attribute Similarities

The objects of the FooAM and the RooAM exhibit similar attributes and attribute relationships (instance connections). The main problem is the different naming of the attributes: Attributes of FooAM-objects have more abstract names and usually no type declarations, RooAM-objects have detailed implementation specific names and concrete types.

The human engineer can resolve these differences only by using his domain knowledge. RooAM-objects usually exhibit an application-semantic name, as it was shown by Biggerstaff [3]. This makes the mapping process easier, because the decision for the mapping is based upon semantic (domain) information.

Moreover the specific data types (e.g., integer, char, etc.) of the RooAM-objects provide additional information about the domain concepts they represent.

Both the application-semantic names, and the specific data types of each attribute of an RooAM object enable the human engineer to map the two application models in order to identify objects within the procedural input program.

6.3 Service Similarities

Services and service relationships (message connections) also offer information for the mapping of FooAM-objects to RooAM-objects.

Services of the FooAM are usually more abstract and exhibit a general interface description that may be specified in pseudo-code. Although the specification is general it offers some domain information (e.g., kind of operation on a specific object, message connection to other FooAM-objects, etc.) that is useful for the object mapping process.

Services of the RooAM are detailed and contain specific information about the data types of the parameters in the interface of a service (that is a procedure or function), the required input, and the output provided by the service. Furthermore there are message (procedure) calls to other object-services that provide information for the mapping process.

Using both, the abstract domain information of FooAM-services, and the detailed implementation specific information of RooAM-services, supports the human engineer during the mapping process to form services for objects of the target ooAM. It seems obvious that mapping one procedure to a single service is not always possible and a grouping of procedures/functions is necessary.

The resulting target ooAM contains only unambiguous services (for those service candidates that could be mapped to FooAM-services) and the corresponding message connections. The remaining ambiguities have to be resolved in a further step, described in the next section.

6.4 Ambiguous Service Candidates

One uncertainty in this object mapping process is that not all of the target objects (of the FooAM) can be mapped to recovered object candidates (of the RooAM). The more difficult uncertainty concerns the resolution of the ambiguous service candidates: In some cases this resolution cannot be performed without grouping a few service candidates (i.e. procedures/functions) of the RooAM to a single service of the FooAM.

Some service candidates could not be assigned to a particular object and are, therefore, called ambiguous service candidates. A service candidate becomes an ambiguous service candidate, because of the following reasons:

- The service candidate manipulates[2] components/attributes of more than one entity/object.
- The service candidate uses components/attributes of more than one entity/object and no other service candidate manipulates components/attributes of the same entity/object.

The uncertainty to which object a service candidate should be assigned can be resolved in the object mapping step, since the necessary application domain knowledge (represented through the FooAM) is available. Ambiguous service candidates of the RooAM are mapped to services of the FooAM. If an exact matching cannot be achieved, the human engineer has to consider that the ambiguous service candidate can represent only a partial functionality of an FooAM-service. In this case the mapping process becomes more complex, since some service candidates have to be conceptually combined in order to allow a mapping to a particular FooAM-service.

Managing both, the complexity of combining service candidates of the RooAM to form a particular FooAM-service, and resolving the ambiguities of service candidates, can be done by using the given application domain knowledge (represented in the FooAM) and comparing the specifications of RooAM-service candidates (e.g., signatures and the performed function) and FooAM-services (e.g., pseudo-code specification of functions and the manipulated/used attributes of objects).

7 Results

The problem of managing uncertainty in our object recovery process can only be solved by introducing application domain knowledge. This domain knowledge is represented in an object-oriented application model (FooAM), which is developed independently of the examined implementation, but based on the same requirements analysis. During the development of the FooAM, domain knowledge is introduced by the human engineer who performs the application modeling task.

The object mapping process can only be performed in connection with a human engineer, since some uncertainties exist:

[2] A service candidate manipulates an entity/object means that the value of a component/attribute of this entity/object is defined or changed by this service candidate.

- Some service candidates of the RooAM have to be conceptually combined to form a particular FooAM-service.
- Ambiguous service candidates have to be resolved to form unique services for the resulting target implementation, using the domain information represented in the FooAM.

These uncertainties prevent the matching process from being fully automated. On the contrary, in the current state of our approach a human engineer is necessary to make the application-semantic decisions.

For future research there seem to be some possibilities for a further automation of our approach, especially in the area of domain knowledge representation. This means an expansion of our method to use domain analysis instead of application analysis.

8 Conclusion

In this paper, we introduced an object recovery process that integrates application domain knowledge. We briefly indicated the basic concepts of our object recovery process and the resulting uncertainties and ambiguities. We described how we restrict these areas of uncertainty and how we lead a human engineer to resolve those uncertainties and ambiguities. In the current state of our approach we rely on a human engineer and a concrete application model to introduce the additional required knowledge, but we try to integrate domain analysis concepts for a further automation of our method.

References

[1] P. Benedusi, A. Cimitile, and U. De Carlini. A reverse engineering methodology to reconstruct hierarchical data flow diagrams for software maintenance. *IEEE Conference on Software Maintenance*, pages 180–189, October 1989.

[2] P. Benedusi, A. Cimitile, and U. De Carlini. Reverse engineering processes, design document production, and structure charts. *The Journal of Systems and Software*, 19(3):225–245, November 1992.

[3] T.J. Biggerstaff. Design recovery for maintenance and reuse. *IEEE Computer*, 22(7):36–49, July 1989.

[4] T.J. Biggerstaff, B.G. Mitbander, and D.E. Webster. Program understanding and the concept assignment problem. *Communications of the ACM*, 37(5):72–83, May 1994.

[5] P. Coad and E. Yourdon. *Object Oriented Analysis*. Yourdon Press Computing Series. Prentice Hall, Inc., Englewood Cliffs, 1991.

[6] B.K. Das. A knowledge-based approach to the analysis of code and program design language (pdl). *IEEE Conference on Software Maintenance*, pages 290–296, October 1989.

[7] S.R. Dumpala and S.K. Arora. *Entity-Relationship Approach to Information Modeling and Analysis*, chapter Schema Translation Using the Entity-Relationship Approach. North-Holland, 1983.

[8] H. Gall and R. Klösch. Capsule oriented reverse engineering for software reuse. *4th European Software Engineering Conference, ESEC '93, Garmisch-Partenkirchen, Germany*, pages 418–433, September 1993.

[9] H. Gall and R. Klösch. Program transformation to enhance the reuse potential of procedural software. In *ACM Symposium on Applied Computing, SAC '94*, pages 99–104. Phoenix, Arizona, ACM Press, March 1994.

[10] M.T. Harandi and J.Q. Ning. Knowledge-based program analysis. *IEEE Software*, 7(1):74–81, January 1990.

[11] H.P. Haughton and K. Lano. Objects revisited. *IEEE Conf. on Software Maintenance, Sorrento, Italy*, pages 152–161, October 1991.

[12] P. Johannesson and K. Kalman. A method for translating relational schemas into conceptual schemas. In C. Batini, editor, *Proceedings of the 7th International Conference on Entity-Relationship Approach*, pages 279–294. North-Holland, 1988.

[13] J.D. McGregor and D.A. Sykes. *Object-Oriented Software Development: Engineering Software for Reuse*. Van Nostrand Reinhold, New York, 1992.

[14] S.B. Navathe and A.M. Awong. Abstracting relational and hierarchical data with a semantic data model. In S. March, editor, *Proceedings of the Sixth International Conference on Entity-Relationship Approach*. North-Holland, 1987.

[15] Ch. Rich and L.M. Wills. Recognizing a program's design: A graph-parsing approach. *IEEE Software*, 7(1):82–89, January 1990.

Juggling in Free Fall: Uncertainty Management Aspects of Domain Analysis Methods

Mark A. Simos

Organon Motives - Software Reuse Consulting Services
36 Warwick Road, Watertown, MA 02172, U.S.

Abstract. Although software reuse research has borrowed extensively from artificial intelligence techniques and methods, there has been little explicit discussion in reuse research of uncertainty management, an area of critical importance in many AI applications. Yet several fundamental reuse issues, particularly in domain analysis methods and processes, can be usefully framed as problems of uncertainty. This paper characterizes ad hoc reuse, design for reuse, domain-specific reuse and domain analysis from an uncertainty-based perspective, and presents and motivates key aspects of a specific DA method, Organization Domain Modeling (ODM) as examples of uncertainty management strategies in domain analysis methods and processes.

1 Introduction

Software engineers routinely apply uncertain techniques to uncertain problems, specified by incomplete, inconsistent, ambiguous, untestable, or inaccurate requirements [9]. Reuse of trusted, validated software components can, in principle, reduce engineering uncertainty. But as researchers and practitioners have ruefully discovered, reuse processes are themselves rife with uncertainty. This paper proposes an interpretation of uncertainty as an organizing principle for distinguishing a spectrum of approaches to reuse, ranging from ad hoc reuse to systematic methods for engineering domain-specific reusable assets.

Section 2 surveys areas of overlap between artificial intelligence (AI) and software reuse, and suggests factors in the lack of emphasis to date on uncertainty issues in reuse. Section 3 characterizes ad hoc reuse, design for reuse, domain-specific reuse and domain analysis in terms of the uncertainty issues addressed and raised by each approach. Section 4 describes key aspects of the Organization Domain Modeling (ODM) domain analysis method as examples of uncertainty management strategies at the method and process level. The conclusion suggest possibilities for richer cross-fertilization of AI uncertainty management techniques and reuse-based software engineering.

2 Background: AI and Reuse

Software reuse research has borrowed extensively from AI representations, techniques and methods. In the reuse context, *domain analysis* (DA) connotes a discipline of eliciting and formalizing developers' expertise in building well-defined families or classes of software applications [1]. DA has been characterized as knowledge engineering for software-intensive domains [10]. Semantically rich knowledge representation formalisms, such as structured inheritance (i.e., semantic) networks have been usefully applied to facilitation of program understanding for large systems [7], construction of domain models for reuse [16], and constraint-driven configuration and composition of software systems [19]. Case-based or analogical reasoning has also been applied to DA [2], and has influenced work in programming, design, and architectural patterns or "idioms" [8].

In light of this conceptual and technical cross-fertilization, it is curious that uncertainty management has received relatively little direct attention in the software reuse context, despite its crucial role in many AI applications. One possible explanation is the level of inference required. AI systems that address uncertainty management typically do so in the context of applying sophisticated inference techniques; few reuse environments to date have employed this level of inference support.

Where uncertainty has been dealt with in inference-based techniques applied in reuse, it has arisen primarily in the *utilization* problem. This can be seen in both major approaches to engineering of reusable assets, *compositional* and *generative* reuse. Rule-based systems have been applied to the problem of partial or fuzzy matching of requirements to reusable component features, to guide utilizers as they search for, retrieve, and adapt software components for reuse [11]. Inference techniques have been applied in automatic programming using domain-specific generators [3]. But generally, domains chosen for generative approaches tend to be those in which uncertainty management issues are less obvious. In generative reuse, decisions are usually encapsulated in the form of automated code generators or component configuration agents. Domain models may include representations of routinized decision support systems, generating instances of programs within a well-defined space of possible implementations [18].

Refined uncertainty management techniques in reuse could support domains amenable to inference-based support, yet not deterministic enough for a uniformly generative solution, thus widening the scope in which reuse technology can be effectively applied. But an area that has received much less attention is the upstream process of creating reusable assets (whether component libraries or generators), including the processes of domain selection, scoping and modeling as well as asset implementation.

3 Uncertainty Aspects in Reuse

This section articulates an uncertainty-based perspective on software reuse, focusing in turn on ad hoc software reuse, design for reuse, domain-specific reuse and domain analysis.

3.1 Ad Hoc Reuse

To distinguish an essential software reuse process from related concepts such as portability, modularity, etc., we propose an informal working definition of *ad hoc reuse* as adaptation and use of a software workproduct in a context for which it was not originally developed.* As illustrated in Figure 1, uncertainty issues inherent in this definition can be characterized by these key questions:

---

* This definition sidesteps some persistent definitional problems in reuse. A "new context of use" could be within the same system as the original context. Conversely, a "context for which originally developed" could be a multiple-system context, e.g., a software product developed for a marketplace of user contexts. The workproduct's original context of use, or *intended scope of applicability*, is the defining aspect.

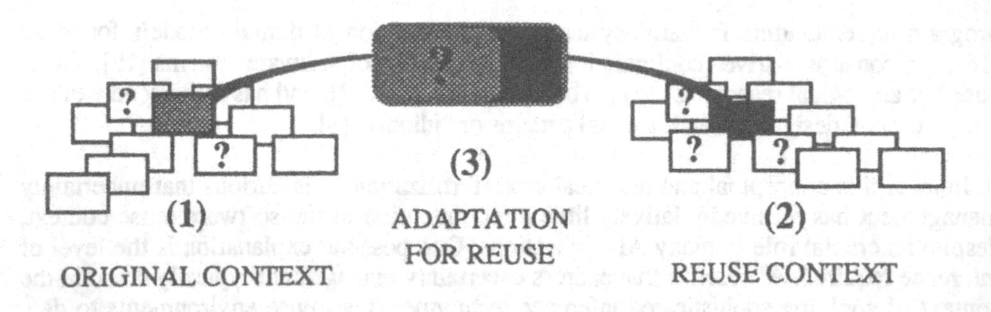

Fig. 1. Potential sources of uncertainty problems in ad hoc reuse

1) *How does a given workproduct really behave in its original context?* Workproducts drawn from a single-system context may be ill- or under-specified. For example, the workproduct may not have satisfied its requirements, or may not have been adequately tested; some requirements may have been satisfied by the external system context.

2) *How will a reused workproduct behave in a new context?* Even components completely compliant with documented requirements may behave in unpredictable ways in new contexts. Requirements characterize behavior in terms of contextual assumptions, made during initial system development, that are often implicit, undocumented, and *emergent*: that is, each new context considered reveals unnoticed features of the original context. Each context is thus in reality a synergy of multiple contexts and/or constraints; some aspects of the workproduct may prove reusable in a shifted context, while other aspects may clash with new contextual constraints. Like exhaustive testing, determining a component's behavior in potential contexts of use is a theoretically infinite undertaking.

3) *How will adaptations affect behavior of the workproduct?* Adaptation may be necessary to incorporate workproducts into a new context. Adaptations are often performed on the basis of only partial understanding of workproducts, (rarely designed for ease in "lifting out of context"). Workproducts can be designed to remain robust and predictable under certain pre-determined kinds of adaptations. Changes beyond such designed adaptations compromise any certainty about a workproduct's behavior.

The uncertainty associated with these questions tends to frustrate trade-off analysis to determine whether "to reuse or not to reuse," or which of several candidate workproducts to reuse. Reuse and adaptation of a component may prove more rather than less labor-intensive, or may save labor but decrease quality. Hence most reuse researchers discourage technology or management emphasis on ad hoc reuse. What may be less apparent is that the effectiveness of ad hoc reuse is compromised largely by uncertainty issues.**

** In fact, successive ad hoc adaptations of a workproduct may increase rather than reduce uncertainty. This resembles "architecture erosion" in system maintenance and reengineering, where a system design's original organizing principle decays over time. Reuse is closely related to system evolution, for *differences* between two contexts of use can be analogous to *changes* in one context over time.

3.2 Design for Reuse

Acknowledgment of problems with ad hoc reuse has led most reuse researchers to the position that software must be *designed for reuse*. As illustrated in Figure 2, two complementary strategies can be applied to reduce the uncertainty involved in uncontrolled adaptation of reused components: better documentation of the original rationale and context of the reused component; and controlled adaptation techniques.

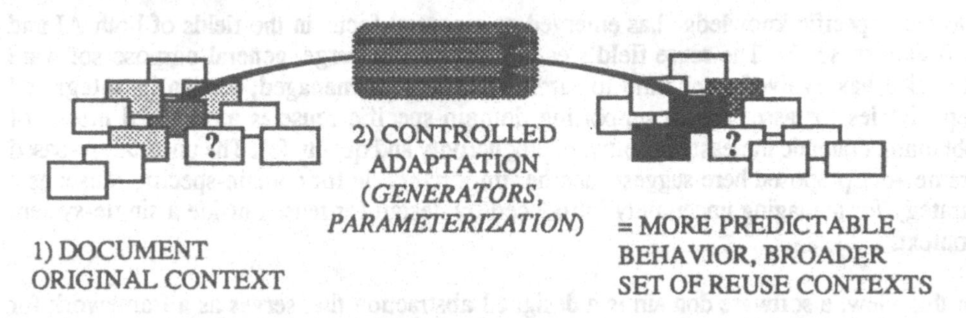

Fig. 2. Design-for-reuse uncertainty management strategies

Contextual Documentation. Some reuse researchers emphasize thorough documentation of the original contextual assumptions of a given workproduct [2]. If we cannot anticipate all future uses, or test components for these uses, we can aid future reusers by recording rationale for key decisions and contextual assumptions (e.g., system dependencies, calling structure, assumed pre- and post-conditions). This aspect of reuse has overlaps with techniques for reverse engineering and re-engineering, and can be viewed as an uncertainty management strategy addressing questions 1) and 2) above.

Controlled Adaptation. A number of adaptation techniques can be considered part of the repertoire of reuse-based software engineering: extended parameterization, compile-time macros, templates, application generators, Ada generics, and object-oriented techniques such as information hiding and code inheritance. Properly employed, these "design for reuse" techniques can maintain clean boundaries between the component and utilizer-written adaptations (e.g., hand-written code), thus extending conventional envelopes of controlled tailoring and adaptation to maintain "black-box reuse," with attendant benefits for configuration management and predictability. As uncertainty management strategies, they address questions 2) and 3) above.

Uncertainty in Design for Reuse. Design for reuse techniques do address the primary uncertainty issues in ad hoc reuse. In addition, they seem indisputably worthwhile from a general software engineering standpoint. (It is hard to imagine better documentation *not* improving at least the maintainability of software!) Yet a new dimension of uncertainty is introduced because these techniques are no longer grounded in a specific application context. Though many software engineers equate reusability to "generic" software solutions (not unlike the AI field's early attempts at "general problem solvers"), reusable software must be designed, either implicitly or explicitly, with an intended scope of applicability—perhaps not defined in relation to a single system, but also not infinite.

Neither better documentation standards nor expanded adaptation techniques alone help to determine the best scope for a component. Each adaptation technique involves design tradeoffs excluding some contexts of use, encompassing others. Thus, defining an intended multi-context scope of applicability for a reusable component emerges as a central uncertainty issue in design for reuse.

3.3 Domain-Specific Reuse and DA

Domain-specific knowledge has emerged as a central focus in the fields of both AI and software reuse [3]. The reuse field's early emphasis on large, general-purpose software libraries has evolved over time towards a concept of managed, vertically integrated repositories (or asset bases) supporting domain-specific reuse as a principal means of obtaining dramatic increases in software productivity and quality [5]. The uncertainty-based framework proposed here suggests another important role for domain-specific reuse: as a strategy for managing uncertainty introduced by design for reuse outside a single-system context.

In this view, a software domain is a designed abstraction that serves as a framework for developing collections of reusable components (or *assets*) likely to be reused in concert within applications. A domain model provides a basis for specifying an intended range of applicability for each asset and for the asset base as a whole. Design of individual assets may be influenced by their anticipated placement within the asset base.

By integrating design for reuse techniques within a strategic domain-specific framework, DA addresses many of unresolved uncertainty issues discussed above. Re-engineering for reuse renders component behavior more predictable; re-engineering cost is amortized across multiple potential contexts of use. Asset implementation can draw on re-engineered legacy workproducts in combination with the full repertoire of flexible adaptation techniques to facilitate "black-box" reuse of components. DA is best suited for precedented domains with relatively mature and stable technology, where context and rationale for legacy systems can be rediscovered through analysis of legacy artifacts and interviews with domain informants.

Uncertainty in DA. As with ad hoc reuse and design for reuse, DA resolves some uncertainty issues, while introducing new kinds of uncertainty in its own right. The domain analyst faces a simultaneously independent, yet mutually interacting set of variables, with neither the luxury of a fixed problem statement nor a fixed set of solutions or capabilities as a guide. Who are the targeted customers for an asset base, and the other stakeholders in the domain? What is the domain boundary, the proper set of systems to study? What defines a feature of the domain? When is domain modeling complete?

This suggests the image of this paper's title: in free fall outside the gravity of specific system requirements, many competent engineers have tried to juggle the complex tradeoffs involved, only to watch the pins float from their hands. The next section describes one DA method's approach to "juggling in free fall."

4 Uncertainty Management in ODM

This section describes some aspects of the Organization Domain Modeling (ODM) method as example strategies for managing uncertainty issues discussed in previous sections. Space constraints preclude a general overview of the entire ODM method, which has been described in detail elsewhere (see ODM Background below). The intent here is to highlight uncertainty management rationale for key ODM features.

Surprisingly, uncertainty provides a remarkably coherent unifying view of the overall structure of ODM, comparable to the centrality of risk management and risk mitigation in the Spiral Development Model [4]. Understanding the uncertainty management strategies supported at the level of the ODM process model and methods themselves should provide a basis for exploring appropriate supporting technology. While the discussion will highlight uncertainty issues that are confronted by other DA methods and techniques, detailed comparison of ODM to other approaches is beyond the scope of this paper.

4.1 ODM Background

ODM provides a systematic basis for engineering assets for reuse, based on explicit documentation of their intended scope of applicability; and for maximizing use of legacy artifacts as a basis for both domain knowledge elicitation and re-engineering. ODM includes a structured set of workproducts, a tailorable process model and a set of modeling techniques and guidelines, that span the context-setting activities of defining and scoping the domain; gathering and modeling domain information; and specifying requirements for both individual assets to be developed and the structure of the asset base as a whole.

ODM was first prototyped as part of the design process for the Reuse Library Framework (RLF), as reported in [12, 17]. The method was further developed through the author's work as an independent consultant on software reuse [13, 14]. In addition, considerable support and collaboration in refining the ODM method has come from Hewlett-Packard Corporate Engineering's Reuse Program and Unisys Corporation, as a prime contractor for ARPA's STARS program. The method has been applied on a small scale within Hewlett-Packard, and is currently being applied on a larger scale as part of the STARS Army/Unisys Demonstration Project. A guidebook describes the conceptual foundations of ODM, and outlines the processes and workproducts defined by the method [15]. Other papers detail the experiences to date with application of ODM, within HP [6] and in the context of the multi-year STARS/Army Demonstration Project [20].

The following sub-sections focus on those aspects of ODM with direct relevance to uncertainty management.

4.2 Defining Organizational Context

Managing organizational uncertainty is a central concern in DA. There are multiple stakeholders in any system development effort. But domain knowledge is typically distributed, not only across multiple systems, but across organizational, departmental, project and product line boundaries. Stakeholders may include both domain informants and potential customers for reusable assets. DA modeling decisions can therefore be viewed as boundary definition negotiations between diverse stakeholder interests. This creates complex tradeoffs, where requirements may differ in priority or even be inconsistent, and even basic domain terminology may be interpreted in diverse ways. (Although what constitutes a single system context will vary from organization to organization, a domain will, almost by definition, represent multiple contexts *in the organization's own terms.*)

As the name Organization Domain Modeling suggests, ODM integrates explicit models of organizational context into domain modeling. The method deals, not with domains in the abstract, but with *organization domains*. This organizational context grounds the entire modeling life cycle, including articulations of domain analysis objectives and domain selection criteria, domain identification, definition and scoping, and detailed modeling. Articulating organizational context and stakeholder expectations establishes entry and exit criteria for modeling activities otherwise quite difficult to bound.

4.3 Iterative Scoping/Context-Setting

Establishing clear domain boundaries is of critical importance for domain-specific reuse. These boundaries establish an explicit scope of intended reusability for assets; poorly defined boundaries will result in users not looking for assets that are there, or looking for assets that are not. As illustrated in Figure 3, ODM structures the domain modeling life cycle as a series of incremental scoping and context-setting decisions. Each step builds on previous decisions and offers opportunities for further refinement and focusing. These scoping decisions represent potential "roll-back" points; for example, new exemplar systems can be added incrementally if deemed necessary after an initial modeling cycle.

ODM domains are initially defined and scoped with rules of inclusion and exclusion that determine whether a given system should be considered a member of the domain. Definition principles are empirically grounded in a set of example systems, or *exemplars*, a distinct "population" of applications that share certain defining characteristics. Rationale for new or shifted boundary decisions is documented as issues surface during modeling. Emphasis on explicitly documenting scoping decisions and maintaining traceability across the steps is motivated by the need to reduce risk and maintain a bounded DA process. Iterative and explicit scoping serve as a combined risk mitigation strategy, constraining and managing the overall ODM process.

A subset (or *representative set*) of exemplar systems is selected to provide detailed data for the domain model. The set is chosen as a strategic sampling of the range of diversity within the domain. Historical relations among the exemplar systems are documented explicitly to ensure that interpretation of common and variant features within the representative set is not skewed by sensitivity to the distinction between independently developed and historically related systems.

In successive scoping steps: domain-relevant portions of each representative system are identified; a sample set of artifacts from various system life cycle phases is selected for detailed analysis; and key *domain informants* are identified as sources of domain information (including not only domain experts but end-users and other stakeholders as well). Traceability to domain information sources acknowledges the inherent uncertainty of modeling domain knowledge that may vary from system to system, and from expert to expert. Traceability information also helps provide a scoping mechanism for the models, by ensuring that there are data points for at least all individual features in the descriptive models (thus providing a kind of feasibility analysis).

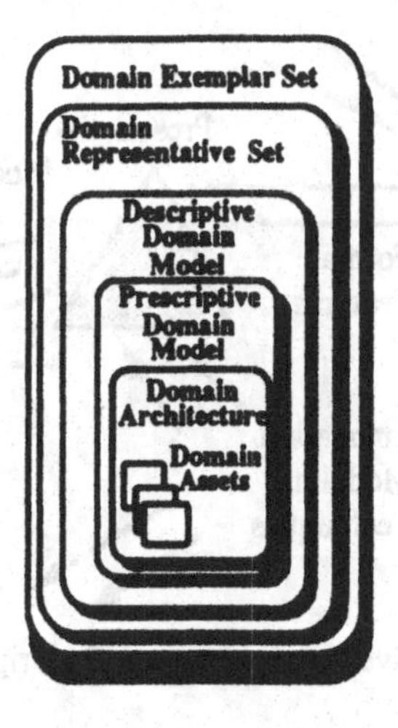

Fig. 3. Iterative Scoping Steps in ODM

4.4 Descriptive and Prescriptive Modeling

The domain analyst must "juggle" two ever-present risks. The first, often voiced by skeptics of reuse, is over-reliance on precedented features and legacy implementations, which may bias new development towards older, possibly inferior system designs and stifle valuable innovation. The second risk is less often discussed: freed from accountability to individual system requirements, domain modeling may become merely unconstrained system modeling, representing modelers' "pet designs," however sound or unsound they may be, and however reflective these designs may be of modelers' domain expertise (or lack thereof).

ODM addresses this creative tension between precedent and innovation by partitioning the modeling process into distinct *descriptive* and *prescriptive* phases. In the descriptive modeling phase, each element of the domain model is grounded in a *precedent* observed in one of the representative exemplars or through interviews with domain informants and general information sources. After descriptive models are validated and refined, a new rescoping is performed to identify the range of functionality to be supported by reusable assets. In this *prescriptive* phase, modelers make binding decisions, select a specific set of targeted *customer contexts*, and prioritize features relative to these contexts. Not all features are committed to in the prescriptive models, and not all exemplar systems are targeted as potential customer contexts.

Between these phases is a transitional *innovative modeling* activity (illustrated in Figure 4) that allows exploration of feature interdependencies and novel feature combinations. The intent in the innovation phase is to create a logical and orthogonal closure of the domain feature space, using a repertoire of model transformation techniques. Innovative features may suggest new potential customer contexts for reusable assets. Innovation activities can be dynamically tuned to align with organization resources and need for conservative vs. novel engineering solutions. This descriptive-innovative-prescriptive process model, which distinguishes ODM from many other DA methods, is an important cognitive technique as well as a plan for producing specific artifacts in a specific sequence.

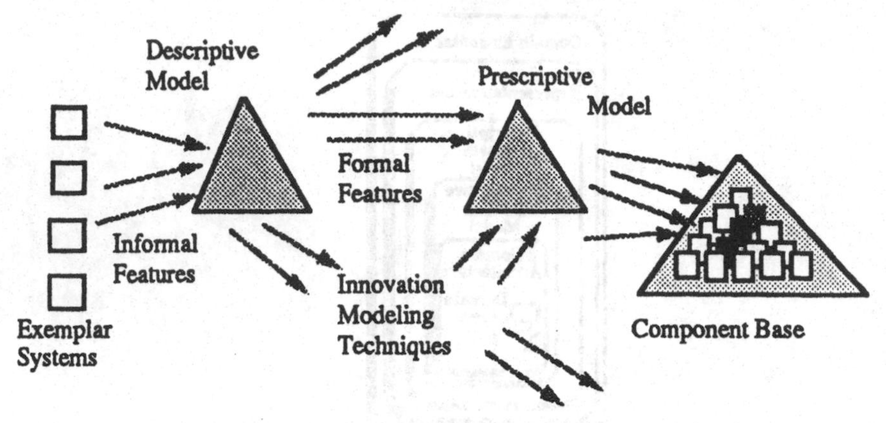

Fig. 4. Descriptive, Innovative and Prescriptive Modeling

4.5 Domain Features

Domain features are the connecting link between descriptions of exemplar artifacts and prescriptive asset specifications. Arbitrariness in feature extraction is a significant source of uncertainty in DA. ODM combines a fine-grained modeling discipline, supported by formal representation techniques, with empirical emphasis on features of interest to various stakeholders. Empirical results, based on stakeholder perspectives are used to cross-check and validate formal models.

Prioritized features are mapped to specifications for assets in the context of an overall asset base architecture. Sets of features likely to be used in concert are clustered into feature utilization patterns; these in turn suggest "asset ensemble" specifications which contain and localize the impact of technology evolution in asset implementation (e.g., migrating from a family of components to a generator). Asset implementers may develop only a subset of the assets specified in the architecture context; this represents yet another successive scoping stage.

5 Conclusion

The ODM method has been presented, not as an example of direct application of uncertainty management techniques from AI, but to illustrate a set of uncertainty issues that any systematic DA method must address, whether explicitly or implicitly. Uncertainty management offers a useful unifying perspective on reuse, particularly in terms of DA methods. By framing the processes of DA in terms of uncertainty management, it should be possible to apply specific uncertainty management techniques in such areas as defining domain stakeholders in organizations, fuzzy definition rules in bounding domains, lexicon formation and tradeoff analysis in feature modeling. Conversely, strategies for managing uncertainty in reuse methods and processes may serve to enrich the AI field's repertoire of approaches to these issues.

References

1. Arango, G., R. Prieto-Diaz: Domain analysis concepts and research directions. Domain Analysis and Software Systems Modeling, ed. R. Prieto-Diaz and G. Arango, IEEE Computer Society Press, 1991.
2. Bailin, S., KAPTUR: Knowledge Acquisition for Preservation of Tradeoffs and Rationale. CTA, Rockville Maryland, 1992.
3. Barstow, D., "Domain-Specific Automatic Programming," IEEE Transactions on Software Engineering, Vol. SE-11, Nov, 1985.
4. Boehm, B., "Spiral Model of Software Development Enhancement," IEEE Computer Surveys: 61-72, May 1988.
5. Boehm, B., W. Scherlis. "Megaprogramming." Proceedings, DARPA Software Technology Conference, Arlington VA, Apr 1992.
6. Collins, P., "Toward a Reusable Domain Analysis," Proceedings, 4th Annual Workshop on Software Reuse (WISR), Reston Virginia, Nov 1991.
7. Devanbhu, P., R. Brachman, P. Selfridge, B. Ballard, "LaSSIE: A Knowledge-Based Software Information System," in Domain Analysis and Software Systems Modeling, ed. R. Prieto-Diaz and G. Arango, IEEE Computer Society Press, 1991.
8. Garlan, D., M. Shaw, "An Introduction to Software Architecture", Advances in Software Engineering and Knowledge Engineering, Vol. I, World Scientific Publishing Company, 1993.
9. Giddings, R., "Accommodating Uncertainty in Software Design," CACM, v. 17, n. 5, May, 1984.
10. Prieto-Diaz, R., "Domain Analysis for Reusability," Proc. COMPSAC '87, 1987.
11. ibid., Reuse Library Process Model. TR AD-B157091, STARS IBM 03041-002, 1991.
12. Simos, M., "The Growing of an *Organon*: A Hybrid Knowledge-Based Technology and Methodology for Software Reuse"; in Domain Analysis and Software Systems Modeling, ed. R. Prieto-Diaz and G. Arango, IEEE Computer Society Press, 1991.
13. ibid. "Software Reuse and Organizational Development," Proceedings, 1st Int'l Workshop on Software Reuse, Dortmund, July 1991.
14. ibid. "Navigating Through Soundspace: Modeling the Sound Domain At Real World," Proceedings, 5th Workshop on Software Reuse, Herndon, VA., Nov, 1991.
15. ibid., Organization Domain Modeling (ODM) Guidebook, Version 1.0, Unisys STARS Technical Report STARS-VC-A023/011/00, Advanced Projects Research Agency, STARS Technology Center, 801 N. Randolph St., Suite 400, Arlington VA, 22203, March 1995.
16. Solderitsch, J., K. Wallnau, J. Thalhamer., "Constructing Domain-Specific Ada Reuse Libraries," Proceedings, 7th Annual National Conference on Ada Technology, March 1989.
17. Reusability Library Framework AdaKNET/AdaTAU Design Report, NRL, System Development Group, Unisys Defense Systems, PAO D4705-CV-880601-1, 1988.
18. Virginia Center of Excellence for Software Reuse andTechnology Transfer (VCOE), Domain Engineering Guidebook, Software Productivity Consortium, Herndon, VA, SPC-92019-CMC, Version 01.00.03, Dec 92.
19. Wallnau, K., "Toward an Extended View of Reuse Libraries," Proceedings, 5th Annual Workshop on Software Reuse (WISR), Palo Alto, CA., Nov 1992.
20. Wickman, G., J. Solderitsch, M. Simos, "A Systematic Software Reuse Program Based on an Architecture-Centric Domain Analysis," Proceedings, 6th Annual Software Technology Conference, Salt Lake City, Utah, April 1994.

Hybrid Fuzzy Metrics for Software Reusability

Bradley J. Balentine[1], Mansour K. Zand[1] and Mansur H. Samadzadeh[2]

[1] University of Nebraska-Omaha, Omaha NE 68182, USA
[2] Oklahoma State University, Stillwater OK 74078, USA

Abstract. Software reuse is an important new technology with the potential to increase software productivity dramatically. While software reuse is possible at each stage of the software life- cycle, code reuse could offer benefits in terms of reduced development time. This paper examines the nature of reusable code, focusing on measures of understandability of code, the assumption that code which is easy to understand is more likely to be reused. Software metrics for measuring understandability are examined. A selection of promising measures are combined using fuzzy set theory to propose a method of assessing the degree of reusability of structured source code written in conventional, imperative programming languages.

1 INTRODUCTION

The cost of both commercial and government software is rising rapidly. There are several reasons for rising software costs, including increasing complexity of systems, shortage in qualified software professionals, and lack of improvement in development technologies [3]. Because we cannot control either system complexity or the number of software professionals, we must search for ways to increase software productivity. One technology which could increase productivity by an order of magnitude or more is software reuse [11].

Prieto-Diaz and Freeman define the term reuse as "the use of previously acquired concepts and objects in a new situation" [11]. They suggest two major aspects of software reuse: (1) reuse of ideas and knowledge, and (2) reuse of specific software artifacts. Thus, software reuse can be accomplished at any stage of the software development life cycle. Reuse at the early stages of the life cycle is easier than at the later stages, because the material is of a more abstract nature. However, the benefit of reuse at later stages is that the reuser has skipped the preceding stages. For instance, reuse of code means the reuser has skipped the design phase. In addition, the benefits are even greater if the code has been tested thoroughly by the developer, and perhaps reused before. This allows the new reuser to reduce testing time, yet be assured the code is relatively error-free.

To be beneficial, code reuse must save time over developing similar code from scratch [11]. With a reuse library available, the steps for code reuse are: 1) access the code, 2) understand the code, and 3) adapt the code, if necessary. Understanding the code, is one of the main factors for successful code reuse. Selecting the component(s) which can potentially be more easily understood completes the code-access step and reduces the time required in the adaptation step. Obviously, within a specific domain, code that is easily understood has greater potential for reuse than complex code.

Understandability hinges on several factors including: programmer ability, program form, and program structure [2]. Programmer ability includes level of fluency in the code's programming language and the programmer's familiarity with the problem domain. We are interested in the measuring intrinsic understandability of source code, so we will consider programmer ability a constant. Program form includes indentation, paragraphing, capitalization rules, etc. Gordon states that studies of program form have shown no statistically significant effect on comprehension for conventional programming languages [2]. Program structure includes number of statements, complexity of control flow, and depth of statement nesting, among other things. Because of its great impact on understandability, we will focus on program structure.

This paper proposes a method of measuring the feasibility of reusing structured source code by concentrating on factors affecting the understandability of the code. The proposed method follows the example of the evaluation mechanism described by Prieto-Diaz and Freeman [11]. We'll begin by examining various metrics used to quantify the level of understandability of source code, and select appropriate measures for our purpose. Then, we will apply fuzzy set theory to each metric to derive measures of the code's potential for reuse. By weighting these measures and again applying fuzzy set theory, we will derive an overall measure of the feasibility of reusing the code based on its level of understandability.

2 SOFTWARE METRICS

Software metrics fall into two major categories: static and dynamic [8]. Since we are interested in a programmer examining unfamiliar code, we are interested in static metrics, which are based on static analysis of source code. Typical static metrics include program size, data structures and data flow, and control flow metrics [6]. In addition, several hybrid metrics have been developed which combine two or more elements of the other metric categories [1, 6]. Various categories of metrics are discussed in the following subsections.

2.1 Size Metrics

Intuitively, large programs should be more difficult to understand simply because of the greater amount of information to be assimilated. The question is: how do we measure program size? Different ways to count lines of code causes problems when comparing metrics data of various research projects. The best line counts seem to be the number of lines without comments and the number of statements, according to [8].

An alternative to counting the lines of code was developed by Halstead [4]. His method is based primarily on the number of operators, *n1*, and operands, *n2*, in a program. The other two of his four basic code counts are *N1*, the total number of occurrences of operators, and *N2*, the total number of occurrences of operands. Based on the four basic code counts, Halstead defines a number of measures including the following:

1) program vocabulary $n = n1 + n2$,

2) program length $N = N1 + N2$,

3) program volume $V = N log_2 n$,

4) potential volume $V^* = N log_2 n^*$, where n^* is the size of the potential vocabulary required if the program were a built-in function,

5) program level $L = V^*/V$.

Halstead suggests that program volume and level can be used to measure the effort required to generate a piece of software using the formula $E = V/L$ for effort. According to Li and Cheung, Halstead's effort formula is an acceptable measure for program size [4]. Size metrics have been very successful [6]. Experiments have shown that larger programs cost more to maintain than smaller ones. However, where size differences are minimal, size metrics fail to distinguish among programs. In these cases, other characteristics including data structures, data flow, and control flow should also be considered for differentiating among programs.

2.2 Data Structure and Data Flow Metrics

Data structure and data flow metrics are based on the configuration and use of data within programs [6]. This metric is especially useful for programs with a large number of modules, but there is little empirical evidence of its validity. Chapin provides another data metric called the Q measure [6]. In this technique, data is categorized differently according to how it is used in a program: a) input data needed to produce output, b) data that is changed or created, c) control data, and d) tramp data. Chapin contends that each category affects complexity differently, and he weights each category differently. The Q measure is calculated for each module as follows:

$$Q = \sqrt{R + W'}$$

where R is the repetition factor and W' is the sum of the weighted counts. The Q measure complexity of a program is the arithmetic mean of the module complexities. Like the other data structure and flow metrics, Q measure is not comprehensive and lacks supporting empirical evidence.

2.3 Control Flow Metrics

Program control flow has been the object of major research on software metrics [6]. Program control flow is essentially the sequence of possible executions of the statements. Typically, control flow is represented by a directed graph, which has a single source node and a single sink node. The nodes represent blocks of sequential code and the edges are possible paths of execution among these sequential blocks. Fig. 1 presents a typical control flow diagram. The first control flow metric was proposed by McCabe [10]. His goal was to develop a technique to provide a quantitative basis for modularization, so that testing time and maintenance costs could be reduced. His complexity measure counts the number of basic paths through a program. McCabe's technique revolves around the definition of the cyclomatic number $V(G)$. Cyclomatic number of a graph G, with n vertices, e edges, and p connected components, is

$$V(G) = e - n + 2p.$$

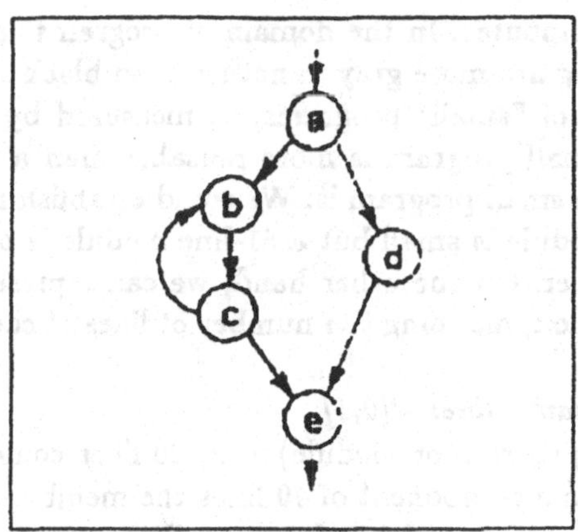

Fig. 1 - A control flow diagram.

Applying McCabe's technique to the control flow diagram in Fig. 1, we get the cyclomatic number of 3.

McCabe's technique can be used operationally to limit the cyclomatic complexity to 10, which McCabe indicates is a reasonable, although not a magical, number. Cyclomatic complexity correlates well with other metrics, including Halstead's effort and the number of statements metrics [8].

Harrison and Magel developed a complexity metric based on McCabe's metrics [5]. They noted that McCabe's calculation of cyclomatic complexity is an effective complexity measure, but since it only looks at basic paths, it doesn't account for the nesting levels of the control structure. They defined the SCOPE metric, which they believe is a more complete control complexity metric. The SCOPE metric more closely matches our concept of complexity by reflecting an increase in complexity due to control structure nesting.

The research presented in this paper uses hybrid metrics which take into account most of the above-mentioned metrics. Harrison and Magel's SCOPE metric is used as a measure of control flow complexity. We also use Halstead's volume metric to measure the complexity of sequential blocks of code. In this way, we will account for both program complexity and program size measurements. We will also account for the level of documentation in the code by measuring the ratio of source comments to lines of code, and the data used in a program by evaluating the data types used. Finally, we will attempt to establish a measure of reusability due to the number of between-module linkages in the source code. We will use fuzzy set theory to evaluate these metrics, and to combine the measures into an overall measure of understandability, which will in turn indicate the code's potential for reuse.

3 FUZZY SET THEORY

In conventional set theory membership in sets is binary: an object is either a member or a non-member. Membership is typically assigned the value 1, and non-membership is assigned the value 0. In fuzzy sets, on the other hand, membership is not binary. Fuzzy sets allow a gradual transition from membership to non-membership, with the degree of membership represented by a value between 0 and 1 [13]. By assigning a degree of membership to each object in the set, ideas such as gray, or off-white, can be modeled in the computer. In the domain of programs and software metrics, the measures we are using are more gray in nature than black or white. As an example, consider the domain of "small" programs, as measured by lines of code. We want to indicate that a small program is more reusable than a large one, but we must first establish what a small program is. We could establish hard and fast limits and say that a 50-line module is small but a 51-line module is not small. This does not seem realistic, however. On the other hand, we can represent program size by the following fuzzy function, mapping the number of lines of code to a value between 0 and 1:

smallprogram : lines - [0,1]

A component (a program or module) with 20 lines could be assigned the membership value 0.7, and a component of 40 lines the membership value 0.4. [11]. This function allows a gradual transition from small to non-small, which more closely models our perception of small than establishing a rigid size limit. To implement fuzzy functions on a computer, we need a mathematical formula which models the membership function. We will use Zadeh's standard function with variable parameters [13]:

$$m_s(u) = 0 \quad for u \leq \alpha$$
$$m_s(u) = 2[\tfrac{u-\alpha}{\gamma-\alpha}]^2 \quad for \alpha \leq u \leq \beta$$
$$m_s(u) = 1 - 2[\tfrac{u-\gamma}{\gamma-\alpha}]^2 \quad for \beta \leq u \leq \gamma$$
$$m_s(u) = 1 \quad for u \geq \gamma$$

Where α, β, and γ are minimum, median, and maximum values on the membership curve, respectively, and u is the value of data that its membership is being computed.

Measured values, when evaluated using this standard function, result in an S-shaped curve, as illustrated in Fig. 2.

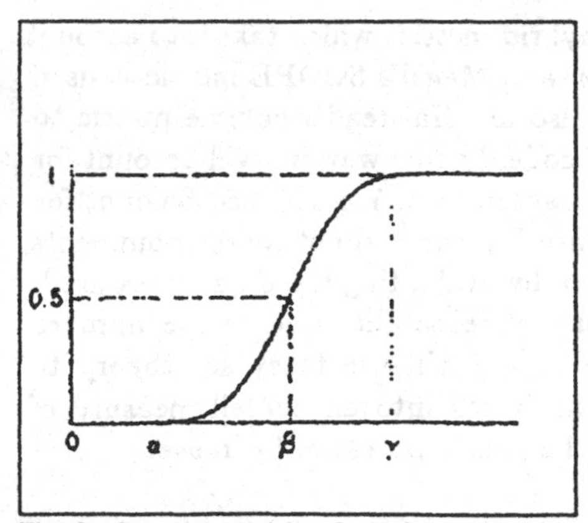

Fig. 2 - S-curve resulting from fuzzy function evaluation.

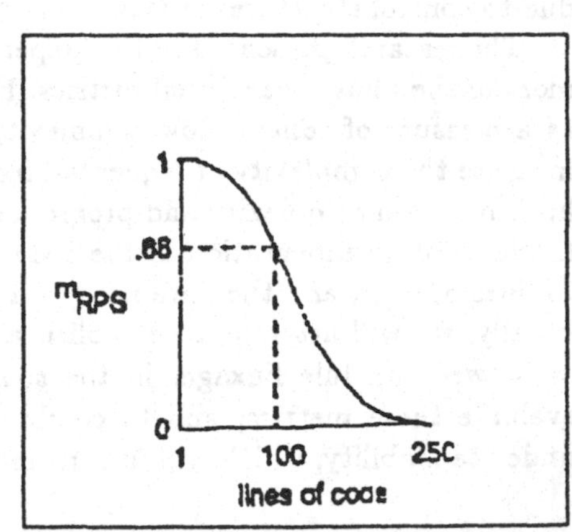

Fig. 3 - 100-line program membership in class of reusable program sizes.

Fuzzy functions can be modified by fuzzy modifiers. These are operations that change the membership functions of a fuzzy set by spreading the transition between membership and non-membership, sharpening the transition, or moving the transition zone [13]. Prieto-Diaz and Freeman used "reuser experience" as a fuzzy modifier for each of their metrics. For this paper, the goal is to rate the ease of understanding of a piece of code without reference to the reuser, so we will treat reuser experience as constant. Intuitively, however, it appears that each of the metrics should be modified by the size of the program under consideration. For instance, the same cyclomatic complexity value for one small program and one large program would indicate that the small program was easier to understand than the large program. This result is similar to Gill and Kemerer's complexity density metric, where the cyclomatic complexity is divided by program size [1].

Because each of the metrics considered depend on program size, we will use program size in terms of non-comment source lines of code as a fuzzy modifier for each function.

3.1 Fuzzy Modifier - Size

In their paper describing a study of error density in Ada source code, Withrow and her colleagues stated that program size is believed to be related to error density [12]. As program size increases, programmers in general have more trouble managing all of the details, which can lead to errors. From this, one can surmise that error density, complexity, and understandability are interrelated. Withrow and her colleagues performed a study of error density in Ada software, and concluded that optimal size of a program is somewhere between 200 and 250 lines. We will make use of these values for our size modifier by choosing 250 as the maximum reasonable number of non-comment source lines of code in a reusable software component. Therefore, the class of Reusable Program Sizes, *RPS*, includes all programs with size 250 lines and less. Any program size can be evaluated for membership in the class of reusable program sizes by using the standard function defined previously. For example, a program of size 100 lines of code, is evaluated as follows:

$$m_{RPS}(100) = 1 - 2(\frac{u-\alpha}{\gamma-\alpha})^2$$

$$m_{RPS}(100) = 1 - 2(\frac{100-l}{250-1})^2 = .68$$

Thus, a program of size 100 has a 0.68 membership value in the class of reusable program sizes. This evaluation is shown graphically in Fig. 3. Note that the curve in Fig. 3 is the 'negation' of the curve in Fig. 1. Negation is performed by subtracting the standard functions from 1, thereby reversing the curve.

3.2 Fuzzy Complexity Metric

As stated previously we are using Harrison and Magel's nesting level metric, the SCOPE number, as our complexity measure. Therefore, we need to establish a reasonable maximum value for SCOPE. We examined several control flow diagrams in order to find the diagram with the highest SCOPE measure and a cyclomatic complexity of 10. Because non-structured control flow negatively affects understandability, we limited our research to diagrams using only structured constructs (i.e., sequential, conditional, and iterative). Of course, the SCOPE number of a control flow diagram may be increased, by adding non-branching nodes, while the cyclomatic complexity remains constant. In order to obtain a reasonable SCOPE limit, we examined only the "branching skeleton" of our diagrams by combining all consecutive sequential nodes into a single node.

The control flow diagrams investigated included diagrams consisting entirely of conditional structures, entirely of iterative structures, and of alternated conditional and iterative structures. The highest value for SCOPE number in the control flow diagrams considered occurred in the control flow diagram consisting entirely of nested "if" structures. The branching skeleton of this graph is given in Fig. 4, which shows the control flow graph with the adjusted complexity for each node.

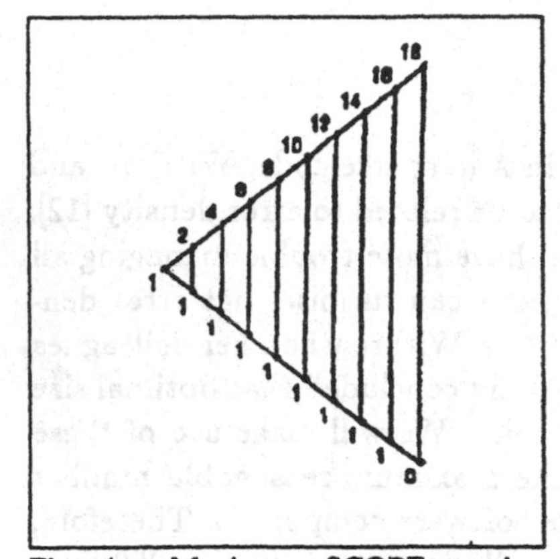

Fig. 4 - Maximum SCOPE number when cyclomatic complexity = 10.

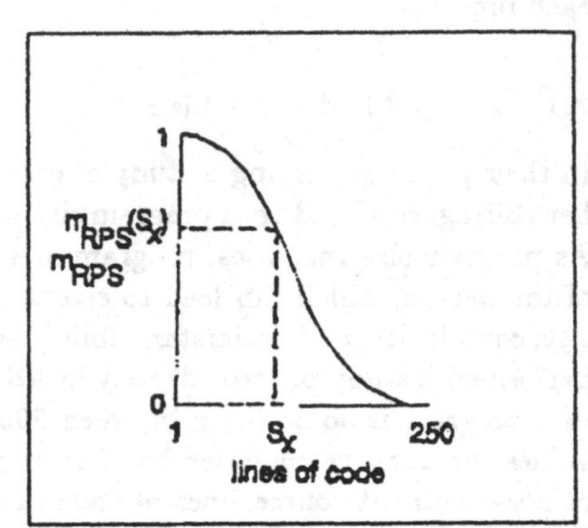

Fig. 5 - Membership of program size S_x in class of reusable program sizes.

The SCOPE number for Fig. 4 is calculated by adding the adjusted complexities:

$$SCOPE = 18 + 16 + ... + 4 + 2 + 9(1) = 99$$

This value becomes the upper limit for our complexity function, or the point at which complexity evaluates to 0 membership in the class of reusable program complexities. This upper limit for complexity can be considered the maximum reasonable complexity for a program of average size. That is, a program whose size evaluates to 0.5 membership in the class of reusable program sizes, will have its complexity evaluated on an S- curve with 99 as the point where membership becomes 0. For programs of other than average size, this fuzzy complexity function will be modified by moving the point where membership becomes 0. We will do this by first calculating the value of a modifier to the complexity maximum value based on a specific program's size. This modifier will be used to shift the maximum complexity value of 99 either right or left to produce the fuzzy complexity function for the given program. An example of this process follows.

Given a program X, we begin by evaluating its size, S_X, using our standard functions and Fig. 2. The result is the membership of x in the class of reusable program sizes,

$$m_{RPS}(S_x) = 1 - 2(\frac{S_x - l}{250 - 1})^2$$

For programs of larger-than-average size this membership value will be less than 0.5, and smaller-than-average programs will evaluate to more than 0.5. This evaluation is shown graphically in Fig. 5.

As stated previously, the membership of an average size program in the class of reusable program sizes, $m_{RPS}(S)$, will be 0.5. The difference between 0.5 and the actual membership value of a specific program X, $m_{RPS}(S_X)$, is the value of the size modifier, mod_X, for program X:

$$mod_X = m_{RPS}(S_X) - 0.5$$

Since the membership value $m - RPS(S - X)$ of larger-than-average programs will be less than 0.5, the modifier mod_X for larger-than-average programs will be a negative number. Conversely, the modifier for smaller-than-average programs will be a positive number. Once calculated, the modifier mod_X is applied to the maximum reasonable complexity of average-sized programs, 99, to determine the maximum reasonable complexity value $CMAX_X$ for a program of size S_X.

$$CMAX_X = 99 + (99 * mod_X)$$

This value becomes the point on the complexity graph where membership evaluates to 0. This modification of the fuzzy complexity function is illustrated graphically in Fig. 6.

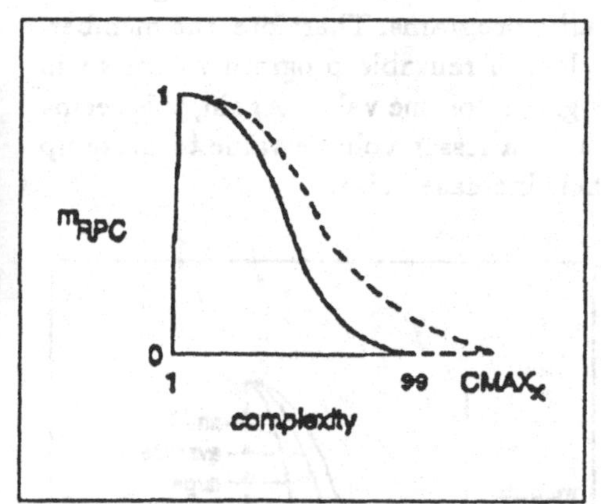

Fig. 6 - Modification of complexity fuzzy function due to program size modifier.

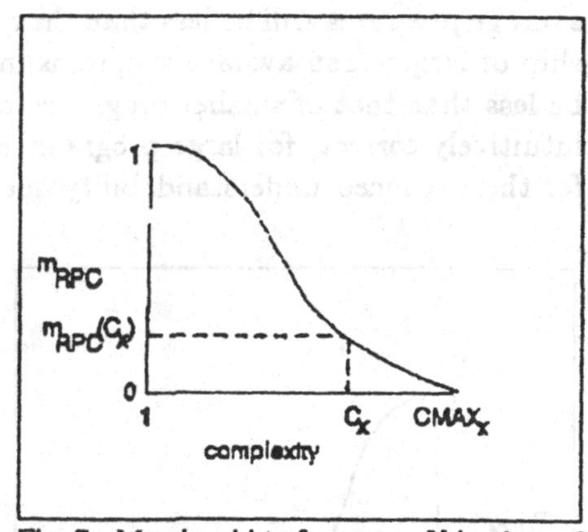

Fig. 7 - Membership of program X in class of reusable program complexities.

Since the modifier mod_X for larger-than-average programs will be a negative number, the value of $CMAX_X$ will be smaller than 99. Conversely, the value of $CMAX_X$ for smaller-than-average programs will be larger than 99. The final step in calculating the membership of program X in the class of reusable program complexities, $m_{RPC}(X)$, is to evaluate the complexity C_X of program X against the modified complexity function:

$$m_{RPS}(C_X) = 1 - 2(\frac{C_X - l}{CMAX_X - 1})^2$$

The process of calculating the membership of program X in the class of reusable program complexities is illustrated in Fig. 7.

Because the maximum reasonable complexity value $CMAX_X$ for larger-than-average programs will be less than the maximum reasonable complexity value for average-sized programs, 99, the membership of larger-than-average programs in the class of reusable program complexities will be less than that of smaller programs for the same complexity value. This result seems intuitively correct, for a larger program is more difficult to understand due simply to its size. It therefore must have less control-flow complexity to be equally as understandable as a shorter program.

3.3 Fuzzy Volume Metric

As stated earlier, Halstead's formula for program volume metrics is $V = N\log_2 n$, where n is the total number of operators and operands, and N is the total occurrences of operators and operands [4]. To implement this metric, we need to derive a maximum reasonable volume value for an average-sized program. Preliminary investigation on a limited set of programs produced the value of 1500 for maximum volume. This value will be used until further investigation can be accomplished. This fuzzy metric was implemented and modified by program size as was the fuzzy complexity metric. Fig. 8 presents the volume fuzzy function for programs of average size.

As in the complexity fuzzy metric, the maximum volume value for larger-than-average programs will be less than that of smaller programs. Therefore, the membership of larger-than-average programs in the class of reusable program volumes will be less than that of smaller programs for any given volume value. Again, this seems intuitively correct, for large programs must have a lesser volume value to make up for their reduced understandability due to their increased size.

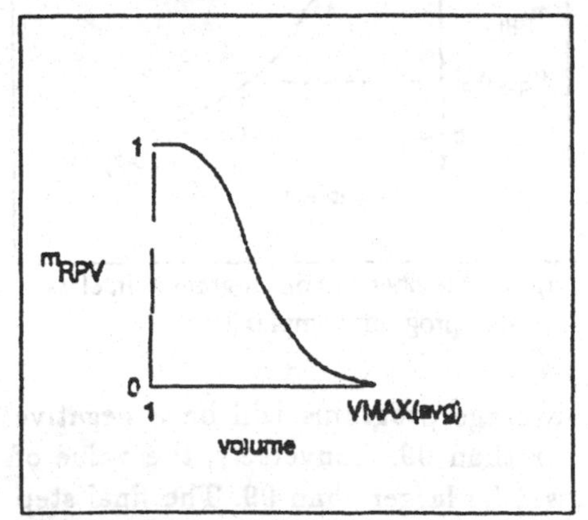

Fig. 8 - Volume fuzzy function for average size programs.

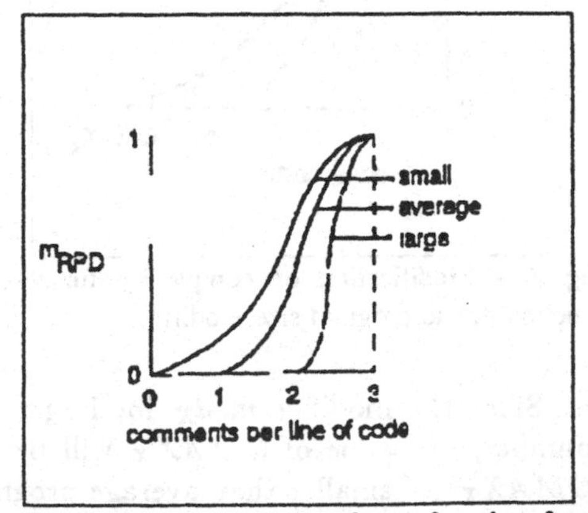

Fig. 9 - Documentation fuzzy function for small, average, and large size programs.

3.4 Fuzzy Documentation Metric

The significance of source code documentation to program understanding is debatable. Intuitively, it seems that the more quality comments included in a program, the easier it would be to understand the program. But, experienced programmers probably pay more attention to the code than to the comments, as indicated by Love [9]. This is possibly a result of two factors: the code is less verbose, and the programmer may question whether the comments are up-to-date.

Without studying the source comments themselves, there is not much we can do to assess the quality of the comments in the code. To approximate the quality of the documentation quantitatively, we simply concentrate on the amount of documentation. Similar to our complexity and volume metrics, the documentation metric should be modified by program size, a smaller program requiring less documentation to be understandable. However, it will be difficult to determine the number of comments required for an average-size program to be reusable, as was done for the previous metrics. Instead, we will implement a measure based on the ratio between the number of comments in the code and the number of lines of code. The size modification will be implemented by subjective establishment of minimum values for each program size curve, instead of by applying the process described in Fig. 2. Fig. 9 shows our fuzzy implementation of this metric, using a subjective value for the required ratio of comments to lines of code for each category of program.

We also have applied the fuzzy set theory to structure and linkage metrics in a similar approach [14]. The next section discusses the overall reuse feasibility measure.

4 OVERALL FEASIBILITY

By applying fuzzy set theory to each of the selected metrics, we produce independent measures of a program's potential for reuse. Next, we want to combine these values into an overall measure of the feasibility of reusing a program. After gathering the required statistics and evaluating each of the individual metrics, our next step is to weigh each of the measures appropriately. The documentation, linkage, and structure metrics are very subjective. Their weight will be unity to preclude them from overpowering the more objective measures. The remaining metrics, i.e., control complexity and program volume, seem equally significant and require only a value of 2 for their combined weight to dominate the overall reusability assessment. The final step is to apply fuzzy set theory once more. Once the individual metrics are evaluated, we multiply each value by its weight. Then, summing the results, we will obtain a number between 0 and 7. This interval can be plotted on an S- curve, and the membership value can be calculated. This value is the overall reuse feasibility measure, m_{ORF}, for a given program. Figure 10 illustrates this process.

5 CONCLUSION AND FUTURE WORK

This paper proposes a method for evaluating programs for their reuse potential. The method focuses on control flow, program size, and other metrics which affect code understandability. Fuzzy set theory allows the independent results of various metrics

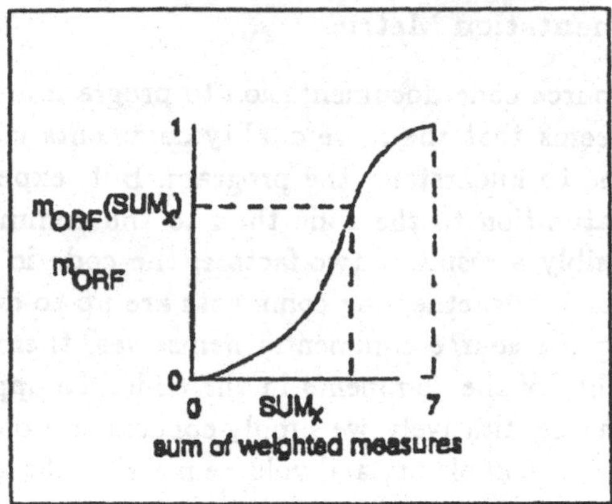

Fig. 10 - Overall reusability indicator fuzzy function.

to be combined into an overall, comprehensive measure of the feasibility of reusing software. This measure could be used in numerous ways: to select between similar components in a reuse library, to determine whether to place a new component in the reuse library, and to operationally limit the complexity of newly developed code. Validation of this method would require its implementation in an evaluation tool capable of processing source code, extracting the required metric information, and applying the functions to evaluate programs for potential reusability. Automated results of the tool would be compared to the subjective, manual ranking of the order of reusability for a suite of candidate programs for reuse. If the rankings correlated well, this method could be deemed a significant indicator of the type of software which has a high likelihood of reuse.

6 REFERENCES

1. Gill, G.K., and Kemerer, C.F. "Cyclomatic complexity density and software maintenance productivity." *IEEE Transactions on Software Engineering*, SE-17 (12) : 1284-1288, December 1991.

2. Gordon, R.D., "Measuring improvements in program clarity" *IEEE Transactions on Software Engineering*, SE-5 (2,) : 79-90, March 1979.

3. Griss, M., "Software reuse: from library to factory," *IBM Systems Journal*, 32(4):1-23, November 1993.

4. Halstead, M.H.,"Elements of software science," New York: *Elsevier North-Holland*, Inc., 1977.

5. Harrison, W.A., and Magel, K.I.,"A complexity measure based on nesting level," *ACM SIGPLAN Notices*, 16 (3) : 63-74, March 1981. 6. Harrison, W., et al., "Applying software complexity metrics to program maintenance,", *Computer* September 1982 : 65-79.

7. Horowitz, E., and Munson, J., "An expansive view of reusable software," *IEEE Transactions on Software Engineering*, SE-10 (5) : 477-482, September 1984.

8. Li, H.F., and Cheung, W.K., "An empirical study of software metrics," *IEEE Transactions on Software Engineering*, SE-13 (6) : 697-708, June 1987.

9. Love, T. ,"An experimental investigation of the effects of program structure on program understanding," ***ACM SIGPLAN Notices*** **12 : 105-113, March 1977.**

10. McCabe, T. ,"A complexity measure," ***IEEE Transactions on Software Engineering,*** **SE-2 (4) : 308-320, December 1976.**

11. Prieto-Diaz, R., and Freeman, P., "Classifying software for reusability," ***IEEE Software,*** **January 1987 : 6-16.**

12. Withrow, C., "Error density and size in Ada software," ***IEEE Software*** **January 1990: 26-30.**

13. Zadeh, L.A., "Making computers think like people," ***IEEE Spectrum*** **21 (8) : 26-32, August 1984.**

14. Balentine, B., and Zand, M., "Fuzzy Set theory and Reusability Metrics," ***Technical Report TR-CS-93-6,*** **UNO 1993.**

A Program Transformation System For Software Reuse

Mustapha BAKHOUCH

LAFORIA-IBP Université Paris VI - Case 169
4, place jussieu 75252 Paris cedex 05 France

Abstract. This paper describes a transformation system for numerical algorithms. This system takes as a specification, a simple and clear algorithm manipulating entities represented by a data structure, and generates another, equivalent algorithm that manipulates the same entities but represented differently by another data structure . The output algorithm is less clear and more complex than the input one, but it is more efficient because it uses data representation properties. The system allows software reuse: the final algorithm is not built from scratch, but it is generated from the initial one by transformation rules.

1. Introduction

Scientific softwares often manipulate data which are constrained to respect a fixed representation. When these data are represented differently, one must use an interface to make them acceptable by the program. This way allows the program to work safely, but it does not use data properties to improve the program. Completly rewriting the program for a better use of the new data representation is expensive and ensuring its correctness with respect to the first program is difficult.

In this paper we propose a transformational method that makes it possible to generate a program both efficient and correct with respect to its specification (the initial program). Efficiency is due to good use of data representations, and correctness is guaranteed by applying correctness-preserving transformation rules. Our system deals also with the fuzzification of algorithms manipulating real numbers. The final algorithm is obtained from the fuzzification of usual operations such that *min*, *max*, *addition* or *multiplication*.

Our approach is object-oriented [11,15]: both the programs and entities (mathematical beings) they manipulate are represented by object nets. The modelling of mathematical beings uses both object-oriented representation and data abstraction techniques [10,16]. This project subsequently intends to complete a tool box called "Tweety", which aims to help users develop scientific programs [14].

This paper is organized as follows: section 2 provides state of the art of program transformation systems. In section 3, we present the problem to be solved. In section 4 we give an object-oriented process of modelling algorithms (ODA objects) and data (MB objects and DS objects). In section 5, we introduce our transformation system and the transformation rules it uses for solving the problem presented in section 3, and we discuss a strategy for applying them. In section 6 we give an example of its application. Finally, in section 7, we discuss system perspectives and future work.

2. Program transformation systems: state of the art

Transformational programming is a method of program construction by successive applications of transformation rules. Usually this process starts with a (formal) specification, that is, a formal statement of a problem or its solution, and ends with an executable program. The individual transitions between the various versions of a program are obtained by applying correctness-preserving transformation rules. A program transformation system is an implemented system for supporting transformational programming (see [6, 12] for a complete survey). They are generally used for performing code modification or generating actual programs from formal specifications. An important characteristic of a program transformation system is the rule selection mechanism which determines the degree of automation of the system: user-selection, semi-automatic-selection, fully-automatic-selection.

General transformation systems support various aspects of the transformational approach to software development, special purpose systems use the transformational concept to achieve a specific goal. They may be characterised by the particular aspect of the transformational activity (e.g., optimization, synthesis, verification or adaptation) on which they focus and the particular class of programs that they manipulate.

Very few program transformation systems deal with scientific programs. Among them we present:

SPECIALIST [9] deals with the optimisation of executable programs in the domain of matrix problems. The input of the system is an Algol-like program, bearing local predicate constraints on data structures, such as "the matrix used is a diagonal matrix". The target is a simplified Algol-like program.

TAMPR (Transformational-Assisted Multiple Program Realization) is a special-purpose transformation system, since its primary goal is to use transformations to adapt numerical algorithms to particular hardware, software, and problem environments [3]. TAMPR is designed to abstract from the details of several numerical subroutine packages for slightly different machines or languages and to express their commonality in the form of a prototype program, a program from which a set of variant, systematically related mathematical subroutines can be automatically derived.

In the field of automatic programming, COGITO [7] is a general-purpose transformation system. It uses an expert system for generating algorithms from a declarative specification.

In [13], H. Partsh uses abstract data types as a framework to describe transformation rules but only for transforming recursive programs which has been studied early in [3, 4].

In [4, 8] a mechanical transformation of data types is described. The basis of data type transformations developed in the paper is a "commutative square" of functions comprising horizontally a user-defined function and its corresponding concrete version, and vertically the abstraction functions between the domain and the range types. Then concrete versions are obtained as a composition involving abstraction functions, their inverses and user-defined functions.

3. The problem to be solved

First let us introduce useful notations. We assume that we have the following tools:

♦ A source language (SL) and a target language (TL): an imperative language to express numerical algorithms. This language uses a poor syntax but it is sufficiently powerful to express an important number of algorithmic constructions. Further, we will give an overview of STL (Source and Target Language: we use the same language for the source and the target language) which is particularly simple. Although one may choose another source/target language like, for example , Pascal, Fortran or Ada, we choose to define a specific language (STL) for reasons of convenience.

♦ A translator from the source language constructs a net of objects called ODA (Object-oriented Algorithm Description). Each SL construction is represented by an ODA object. This representation by objects is intended to be the input of a transformation system which aims at improving algorithms.

♦ A translator from ODA description to a target language.

♦ A transformation system manipulating an objects net. The main goal of our paper is to study this system.

Our problem is as follows:

Let P1 be a program written in SL, DS1 the data structure manipulated by P1 and DS2 another data structure representation equivalent to DS1 (it means that the same data is represented differently by DS1 and DS2).

We wish to write another program P2 that performs the same task as P1 but manipulating DS2. To achieve this, we first convert P1, DS1 and DS2 into an ODA description. The P1-ODA obtained is then transformed by applying transformation rules until obtaining a P2-ODA version satisfying a certain criteria. This version is, then, translated to SL to obtain P2.(**Fig. 1.**). Using an ODA representation of the program makes the system completely independent from the source and target language.

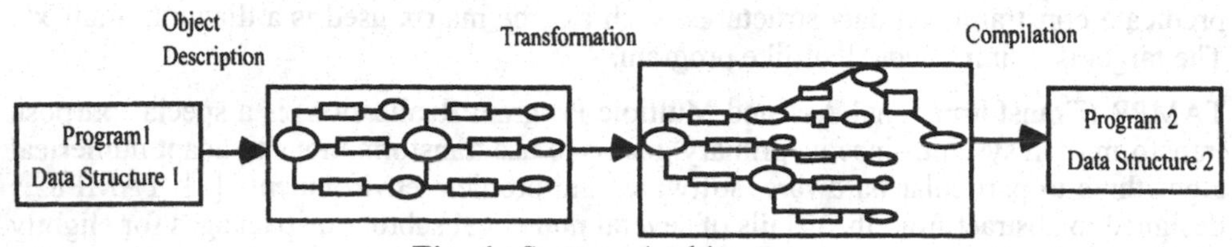

Fig. 1. System Architecture

The practical motivation of this investigation is program improvment by taking advantage of data structure properties. So, P1 is assumed to be easy to understand but not efficient. P2 is expected to be more efficient, correct with respect to P1, but probably less transparent.

Examples:

♦ transforming a program version of a matrix-vector product to a version in which the matrix is sparse. This example illustrates one of our system goals: matrix involved with the modelling of physical phenomena are usually sparse, and an implementation taking advantage of their sparsity is required to improve computations.

♦ transforming a program version of a matrix-vector product using matrix rows to a version using matrix columns. This example illustrates another goal of our system: the final version runs efficiently on some kinds of machine and can be better parallelised.

♦ transforming a program implementing a mathematical function on real numbers to a program implementing the extension of this function to fuzzy numbers.

4. Object-oriented modelling

4.1. A language for expressing source and target algorithms (STL)

We use the same language both for source and target language (STL). STL is an imperative language that makes it possible to express the essential algorithmic constructions in numerical analysis. A STL program is a function that takes input data and performs instructions to give output data. Here we give an example of STL program (for a complete syntax see [1]).

Example: Inner Product of two vectors

```
(Function Inner-Product
; input
 ((Declare U Vector) (Declare V Vector))
; output
 (Declare P Real)
; Instructions
 ((Declare I Integer)
  (P --> <- 0.0)
  (For I From 1 To (V --> Dimension) Do
       (P --> <- (P --> +((U --> Component I) --> * (V --> Component I))))
  EndFor)))
```

4.2. Object-oriented Description of Algorithms (ODA)

ODA is a hierarchical description by objects of STL constructions. A STL program may be converted into an ODA object net. Here is the graph representing the hierarchy of ODA objects named prototypes **(Fig. 2.)**.

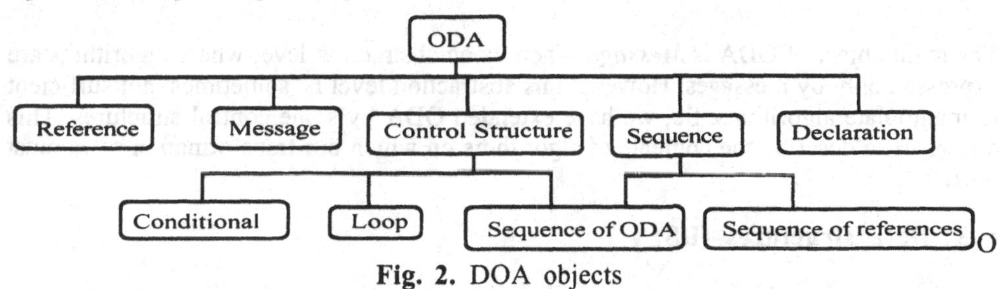

Fig. 2. DOA objects

An ODA is an object representing an algorithmic construction. It knows methods making it convertible to many programming languages. Although we are interested by the translation to STL, it would be easy to convert ODA objects to other target languages like Fortran, Pascal or Ada. We prefer STL, because its conversion into ODA objects is less complicated. We use CRL (Carnegie Representation Language) to represent our ODA objects [5]. Prototypes of ODA are: Reference, Message, declaration, sequence of references and control structures.

Reference. ODA is intended to describe general algorithms that manipulate any real entities represented by objects. These entities are manipulated by means of references. A reference is an object with a slot *associated-being*; the value of its slot is the entity associated with the reference.

Message. Message is an ODA prototype representing a STL Function-Call. It is described by its properties: (i) *receiver*: the receiver of the message. It should be a reference, a message or a sequence of objects that are references or messages. (ii) *selector*: a name of an operation. (iii) *arguments*: like receiver. (iv) *invoked method*: a method representing the operation of naming the value of *selector*.

Declaration. tells that the value of the slot *Asociated-Being* of a reference is of certain type. A ODA Declaration is an object with two slots: *Declared-Being* and *Declared-Reference.*

Control Structure. Conditional, Loop, Sequence.

Example: The STL instruction

```
(For i From 1 To (V --> Dimension) Do
        ((V --> Component i) --> <- (U --> Component i))
EndFor)
```

is represented by the following ODA objects net: **(Fig. 3.)**

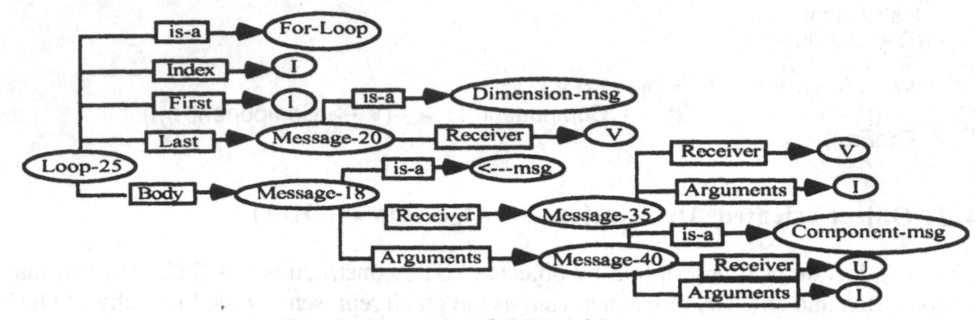

Fig. 3. Example of ODA objects net

The main object of ODA is *Message*. There is an abstraction level where algorithms are expressed only by messages. However this abstraction level is, sometimes, not sufficient to manipulate algorithms. So, we have extended ODA by some control structures. This allows us to describe the contents of algorithms on which our transformation rules must operate.

4.3. Data structures (DS)

Computer entities where data are stocked are represented by objects called data structures (DS). Lists and arrays are elementary data structures. Other data structures which may be constructed upon the elementary ones are used to represent more complex data (for example, a *storage* of a sparse matrix may be a DS object with slots: *Coefficients*, *Row-Index*, *Column-index* which are valued DS).

4.4. Mathematical Beings (MB)

As we saw below, ODA objects are used to describe more general algorithms that manipulate any object-oriented represented entities. In this section we will decribe a modelling process for mathematical beings. These mathematical beings will be abstract entities manipulated by our algorithms. Therefore ODA objects make it possible, by means of references to an associated mathematical being, to express numerical algorithms. So algorithms formally defined below will now take a mathematical semantic.

A mathematical being (MB) is characterised by a set of properties. We may distinguish three types of properties:

♦ Properties that allow a MB to be related to computer entities (data structures); for example a vector has a *Storage* property that relates it to a DS Array.

♦ Properties that allow a MB to be related to other MB (a mathematical description of MB). For example the MB *Matrix* is related to MB *Vector* by the property *Row*. Such properties are also described by objects and are instances of a prototype called *Method*. A method is characterised by: (i) *Name*: the name of the operation implemented by the method. (ii) *Input, Output*: references or sequence of references associated to the mathemathical beings manipulated by the algorithm associated with the method. (iii) *Associated-Algorithm*: a ODA objects net.

♦ *Axioms:* when the MB object is viewed as an abstract data type, axioms define the semantic of the MB by the way of relations between its properties.

5. A Transformation system for numerical algorithms

5.1. Transformation rules

A transformation rule is a partial mapping from program to program. It can be represented in two ways: by an algorithm taking a program and producing a new one (procedural rule), or by a pair of program schemes (schematic rule or pattern replacement rule).

For our problem, in order to transform an algorithm A_1 manipulating objets in a set E_1 to an algorithm A_2 manipulating objets in a set E_2, we need three kinds of transformation rules:

♦ ODA transformation rules, which are syntactic rules expressing relationships between ODA objects (for example: introducing new references, suppressing references, or loop merging).

♦ Representation change rules express constraints on E1 and E2 objects, meaning that E1 and E2 are representations of the same data. These rules allow replacement of a program manipulating one representation by a program manipulating another representation. They are expressed by means of the relationships between data structures used in the two representations.

♦ Each object manipulated in the program is an instance of a mathematical being (MB) or data structure (DS) prototype. This prototype encapsulates methods that implement operations, and axioms that express their properties (like abstract data type, but operations are implemented). This representation allows us to have more information on the manipulated object and to complete our transformations by folding and unfolding rules that are included in the methods implementing operations. Other rules (rewriting rules) can be deduced from the axioms representing operations properties.

5.1.1. ODA rules

In section 4 we saw how to represent LS statements by object nets. So, even the syntactic equivalence between LS statements can be represented by object net equivalence. This equivalence makes it possible to define ODA rules. These rules, which represent some programming tactics, (for example: introducing new references, suppressing references, loop merging or specializing loops), are used to perform some optimisations at the end of the transformation processus, or to make other transformations rules applicable. Here are some examples of ODA rules.

Loop decomposition rule. It deals with the decomposition of an iterating loop on a set E_1 in two loops on two sets E_2, E_3 such that $E_2 \cup E_3 = E_1$ and $E_2 \cap E_3 = \emptyset$: this rule deals with loops whose *bodies* are *conditionnals* and their *conditions* slots are predicates on the loop *index*. Let : $E_1 = [1 \ldots N]$, $E_2 = \{i \in E_1 : p(i)\}$, $E_3 = E_1 - E_2$; p is a logic predicate and B a loop in the following form:

(For i ∈ E1 Do (If p(i) <instruction-1(i)> <instruction-2(i)> EndIf) EndFor)

then under some conditions the loop B can be transformed into two loops B1 and B2 with the following form:

B_1: (For i ∈ E_2 Do <instruction-1(i)> EndFor)
B_2: (For i ∈ E_3 Do <instruction-2(i)> EndFor)

Introducing references. This transformation replaces a *message* by a *reference,* by introducing an assignment message before the ODA in which the replaced message occurs.

Suppressing references. This rule consists, for a reference, to see if there is an assignment message taking this reference as receiver, to delete it, and replace all its occurrences by the argument of the deleted assignment message.

Conditional property. A message with a *Conditional* as *Receiver*

((If <condition> <if-true> <if-false> EndIf) --> <selector> <args>)

can be transformed to a conditional:

(If <condition> (<if-true> --> <selector> <args>)
(<if-false> --> <selector> <args>) EndIf)

Simplification rules. like, for example,

♦ if m is a message with selector * then ,
- if arguments (m) = 0, then replace m by 0
- if receiver (m) = 0, then replace m by 0
- if receiver (m) = Sequence and if 0 is a member of this sequence, then replace m by 0
- if arguments (m) = Sequence and if 0 is member of this sequence, then replace m by 0

♦ if m is a message with selector +, then
- if receiver(m) = 0 then replace m by arguments(m)
- if arguments(m) = 0 then replace m by receiver(m)

♦ if m is a assignment message and if receiver(m) = arguments(m), then delete (m)

♦ if B is a loop and Body(B) = empty then delete(B)

5.1.2. Data representation changes

Two algorithms are equivalent if they perform the same task: their running on all possible input data returns equivalent data as output (equivalence by execution), or an algorithm can be obtained from the other by correctness-preserving transformation rules (equivalence by rewriting) (see [1]). Here we are not interested in the execution, so we assume some elementary equivalences to be proven by execution; these elementary equivalences plus other transformations allow us to check more complex equivalences. The elementary equivalences are represented by a pair of ODA objects They are horizontal transformation rules: the left hand side of the rule manipulates data structures involved in the first representation and the right hand side manipulates data structure involved in the other representation.

Example: one may represent a squared matrix in two ways:

♦ in representation 1 (by M_1) coefficients are stored in a two-dimension array T1

♦ in representation 2 (by M2) coefficients are stored by rows in a one-dimension array T2 ;

an elementary equivalence is:

$T_1[i , j] \rightarrow T_2[n^*(i - 1) + j]$

5.1.3. Transformations about mathematical beings

Following the modelling of mathematical beings and data structure, two types of transformation rules appear:

♦ folding and unfolding rules are involved by the implementations of type operations;

♦ rewriting rules are obtained by directing axioms involved in the data abstract representation.

Contrarily to the previous transformation rules relating the elementary equivalence between algorithms, transformation rules related to mathematical beings are vertical: data structure, manipulated before and after the transformation, belongs to the same representation.

Unfolding rules. Unfolding is the operation that replaces a function call by the body of the function correctly instanciated. This transformation is well known in program transformation synthesis [4]. Following our process of modelling, unfolding will be replacing an objects net representing a message by the the objects net representing the body of the method invoked by this message.

Example: if we transform the message (written in STL): (V_1 --> Inner-Product V2), we obtain by unfolding an ODA which can be translated in STL as:

```
(p --> <- 0.0)
(For I From 1 To (V1 --> Dimension) Do
    (p --> (p --> + (((V1 --> Component I) (V2 --> Component I)) --> *)))
EndFor))
```

Folding rules. When an ODA is identified as the body of a method defined in a MB or DS object, we can replace it by a message. It is a means to bring algorithms to high abstraction level in order to transform them at this level. This transformation is called a folding rule.

Folding is the inverse operation of unfolding, but identifying an ODA as a body of a method is a complex task. Ther are two possibilities:

♦ an ODA (usually a sequence of ODA) is an instance of the body of a method defined in a prototype of a mathematical being or a data structure (sometimes other transformations are needed to achieve a form that can be matched to the method body). In this case, we replace the current ODA by a message which has as its selector the name of the method, as receiver an instance of the MB or DS in which the method is defined, and as its arguments a part of the parameters of the method.

♦ the ODA cannot be identified as a body of a method; in this case one must define a new primitive.

A search algorithm is required to define which prototype the method is associated with.

Abstract data types. Here, we do not introduce abstract data type theory: for more details see [10], [16]. In our system, abstract data types are a source of expertise which are used to complete our transformation rules base. Axioms involved in the abstract data type definition are directed to rewrite rules.

example: one dimension real array (array-1D)

operations profils:

Length: Array-1D → Integer
At: Array-1D x Integer → Real
Find-Index: Array x Real → Integer
Contains-Value: Array x Real → Boolean

axioms

Axiom-1. ∀ T∈ Array-1D, ∀ Val ∈ Real
If (T --> Contains-Value Val) Then (T --> At (T --> Find-Index Val)) = Val

Axiom-2. ∀ T∈ Array-1D, ∀ I ∈ Integer
If I ∈ [1 ... (T --> Length)] Then (T --> Find-Index (T --> At I)) = I

5.2. About strategy

When we have identified all the transformation rules needed to solve our problem, we have no idea of how to use them in order to transform an initial algorithm (taking a specification role) into an algorithm satisfying specific criteria. Firstly, there is no algorithm capable of performing this task without user interaction. Secondly, rewriting systems are not terminating.

It is utopian to conceive of a system which uses all the rules we introduced and is capable of solving the problem without user interaction. This is due to the fact that some rules can sometimes be applied from left to right, others from right to left.
To be realistic, we do not claim to build a fully automatic system, but we think that an interactive system is feasible. While some steps of the system will be fully automatic, others will require user interaction. Research in this area must focus on reducing to a minimum the work imposed on the user.

At present, the system is nearly completely formalized, ODA objects, DS objects and MB-objects have been implemented, and most transformation rules that are manually selected have been identified.

A user rule selection mechanism is being used to test the system in order to produce a heuristical expertise to make the system automatic in most cases.

6. Example

In this example, we will see how to transform the inner product algorithm of two vectors. Let V1 ∈ *Non-Sparse-Vector* and V2 ∈ *Sparse-Vector* be two instances of *Vector* representing the same vector V: in V1 all coefficients are stored, but in V2 only non-zero coefficients are stored. And let U ∈ *Non-Sparse-Vector* be a representation of another vector. V1, V2 and U have the same dimensions. Let A1 be the algorithm computing the inner product of V and U when V is represented by V1 (i.e. V is seen as *Non-Sparse-Vector*). All V1 components are stored in the array TV1, only non-zero coefficients of V2 are stored in the array TV2, the coefficients index are stored in the array IV2

We transform A_1 by applaying transformation rules in order to obtain a version A2 of A1 in which V1 is replaced by V2. In the transformation process we use axioms related to the abstract data type array-1D (5.1.3) and identity constraints (data representation changes) between V1 in *Non-Sparse-Vector*and V2 in *Sparse-Vector*:

Constraint-V1V2. (TV_1 --> at I) =(If ((IV_2 --> Find-Index I) --> ≠ False)
(TV_2 --> at (IV_2 --> Find-Index I)) 0.0 EndIf)

Constraint-V2V1. (TV_2 --> at J) = (TV_1 --> at (IV_2 --> at J))

Transformations:

Initial algorithm A1

```
(Declare P Real)
(P --> <- 0.0)
(For I From 1 To N Do (P --> <- ( P --> + ((TV1 --> at I) --> * (TU --> at I))) EndFor)
```

applying Constraint-V1V2 and the conditional property rule we obtain:

```
(Declare P Real)
(P --> <- 0.0)
(For I From 1 To N
  (If ((IV2 --> Contains-Value I) (P --> <-(P --> +
               ((TV2 --> at (IV2 --> Find-Index I)) --> * (TU --> at I))))
                        (P --> <- (P --> + ( 0.0 --> * (TU --> at I)))) EndIf)
EndFor)
```

applying the loop decomposition rule, with: p(I) = ((IV2 --> Contains I), E2 = {I ∈ [1... N]: p(I)} and E3 = {I ∈ [1... N]: Not p(I)}. we obtain:

```
(Declare P Real)
(P --> <- 0.0)
(For I ∈ E2
  (P --> <- (P --> + (TV2 --> at (IV2 --> Find-Index I)) --> * (TU --> at I))))
 EndFor)

(For I ∈ E3 (P --> <- (P --> + ( 0.0 --> * (TU --> at I)))) EndFor)
```

The second loop can be reduced to an empty instruction by simplification rules. In the first loop, the mapping , J ∈ [1...M] ---> (IV2 --> at J) is a bijection, we can then replace in this loop I by (IV2 --> at J) and E2 by [1...M]. We obtain:

```
(Declare P Real)
(P --> <- 0.0)
(For J ∈ [1...M]
  (P --> <- (P --> + ((TV2 --> at (IV2 --> Find-Index (IV2 --> at J)))
                          --> * (TU --> at (IV2 --> at J)))))
EndFor)
```

Finally, applying Axiom-2 provided by the abstract data type array-1D, we can replace: (IV2 --> Find-Index (IV2 --> at J)) by J; we obtain A2:

```
(Declare P Real)
(P --> <- 0.0)
(For J ∈ [1...M]
    (P --> <- (P --> + ((TV2 --> at J) --> * (TU --> at (IV2 --> at J)))))
EndFor)
```

7. Conclusion and perspectives

We show by this example that three kinds of transformation rules are needed to solve our problem. These transformation rules are representation changes that provide a starting point and guide the transformation process to obtain a final algorithm satisfying specific criteria, syntactic transformations (ODA rules) and transformation rules related to mathematical beings or data structures.

The experimentation currently in progress will make it possible to develop a strategy for the application of transformation rules. Work will continue to enrich the modelling process by the addition of new concepts dealing with the modelling by objects of transformation rules and heuristics for applying them. A model of our problem in terms of constraints on initial and final algorithms is being developed.

References

1. M. Bakhouch, *Manipulation d'algorithmes numériques, approche transformationnelle et modélisation par objets.* Thèse de doctorat de l'université Paris VI, LAFORIA TH94/03, 11 juillet 1994.
2. J. M. Boyle, *Software adaptability and program transformation. In Software Engineering,* H. Freemann and P. M. Lewis, Eds. Academic Press, New York, pp. 75-93.
3. M. Burstall, J. Darlington, *A system which automatically improves programs.* Acta Inf. 6. PP 41, 60. (1976).
4. M. Burstall, J. Darlington, *A transformation system for developing recursive programs.* J; ACM 24, 1. Janvier. PP 44, 67. (1977).
5. *Carnegie Group Incorporation.* Technical documentation for Knowledge Craft ,1990.
6. S. Feather, *A survey and classification of some program transformation approaches and techniques* ; In L. G. L. T. Meertens, editor, Program Specification and transformation, pp. 165-195, IFIP, Elsevier Science Publishers B. V. (North-Holland), 1987.
7. B. Ginoux, *Génération automatique d'algorithmes par systèmes expert, à partir de spécifications de haut niveau : Le système COGITO.* Thèse de doctorat de l'université de Paris IX, 1991.
8. P. G. Harrison and H. Khoshnevisan, *The mechanical transformation of data types.* The computer journal, vol 35, N° 2, 1992
9. Kibler, D, F *Power, efficiency, and correctness of transformation systems.* Ph.D. dissertation, Univ of California, Irvine. 1978
10. Liskov and J. Guttag, *Abstraction and specification in program developpement,* The MIT Press, 1986.
11. Masini/A. Napoli/D. Colnet/D. Léonard/K. Tombre, *Les langages à objets : langages de classe, langages de frames, langages d'acteurs.,* iia InterEditions, 1990.
12. Partsh et P. Pepper, *Program transformations systems.* Computing Surveys SI, Vol 15, n° 3, september 1983.
13. Partsh et P. Pepper, *Program transformations expressed by algebraic type manipulations.* TSI, Vol 5, n° 3, 1986.
14. Ryckbosch, *La boite à outils Tweety :,* note interne EDF HR34-2270, Février 1991.
15. Ryckbosch, *Le logiciel scientifique : Conception par objets,* ed, Lavoisier, 1992.
16. M. Wirsing, *Algebraic Specification* (chapter 13), in Handbook of theoretical computer science, vol B, Edited by J. van Leeuwen. Elsevier Sience Publishers B.V., 1990.

Abstract Visualization of Software
A Basis For a Complex Hash-Key?

Roland T. Mittermeir and Lydia Würfl
Institut für Informatik, Universität Klagenfurt
A-9022 Klagenfurt / Austria

Abstract. Software retrieval constitutes - in spite of several advances - still a practical problem for software reuse. This might be partly due to the fact that the fuzzy representation of a program sketch, as it emerges in the programmers mind, has little to do with the more abstract specifications needed by current retrieval tools.

This paper shows how to augment retrieval features of software repositories by facilities allowing to match the structure of a sketch of software against the structure of components in the repository. This approach yields not only an alternative access path to components, it also provides a rough handle for assessing the cost of change with proxy-matches.

1. Introduction

In spite of high-productivity aspirations, systematic software reuse is not yet a "natural" part of the development process with most professional software developers. As indicated in [17], this might be due to many non-technical factors. But technical issues remain and the software retrieval problem is one of them.

In this paper we present after a brief survey of existing approaches to the retrieval problem an approach based on procedural sketches. The merit of this approach should be that it allows programmers to express themselves in terms they are used to in their routine constructional work.

2. Approaches to Software Retrieval

As one can see from Krueger's survey [10], software retrieval still constitutes an important problem in all approaches towards software reuse. Here, we specifically address the issue of reusing components of source code.

Approaches to software retrieval can be categorized broadly into three classes: classificatory, specification based, and AI-based approaches.

The most traditional and most widely used among them are *classificatory approaches*. Within this group, keyword search [12] is the simplest and oldest method. The faceted approach proposed by Prieto-Diaz [16] can be considered as an improvement of the keyword approach. The original version of the software archive, which built on classification structures [14] can be considered as a variant of simple classificatory approaches. The retrieval structure described in [8] is based on similarity of functionality and thus leads already to the next category, specification based approaches.

Specification based approaches allow for more precise descriptions of the components sought. They either build just on the signature of components [6] or use even full specifications [5]. The reward gained for the additional effort in describing components by specifications lies in the preciseness thus gained and in the lattice properties of specifications [4]. They constitute the theoretical basis on which powerful retrieval tools based on theorem proving can be built [13]. Thus, the link to methodologies from *artificial intelligence* is made. Representatives to be named here are the PARIS system [9], which is based on relating specifications, or Maareks approach based on lexical affinities [11], and the AIRS Model [15].

Here, we introduce an approach based on *structural abstractions*, which is to complement the methods mentioned above. It builds on the premise, that software developers are highly trained in procedural thinking. Hence, they tend to understand problems quite often in procedural terms, even if this preliminary understanding might still be vague and mapping this vague procedural understanding to purely descriptive specifications might be hard for them. Hence, we propose in the sequel to use this vague procedural understanding, which tends to be related to the main data structures to be manipulated, as search key. It is to be mapped against the main structures of a set of components of the repository, which might have been pre-selected by some other (e.g. faceted) means. Obviously, such a mapping will not be straight one-to-one. It has to allow for differences in levels of abstraction as well as for fuzzyness. How one may obtain such a coarse "visualization" of core structures and how to fuzzily map them against each other is described in section 3 and 4 respectively.

3. Structural Abstraction

3.1. Control Structure driven Search

While search based on complete specifications has the advantage of preciseness, it has the disadvantage that a substantial section of today's programmers is not sufficiently trained in formal methods to use them easily. Further, it has the disadvantage that in cases, where only a partial match could be obtained, the degree of mismatch between the specification driving the search and the specification of the component(s) found is in no way indicative for the effort needed to change the software such that it will completely satisfy the needs of the new application. The often heard advice: "if you have to change it, don't reuse it" might be partly grounded on this fact.

Based on these considerations, we were looking for a representation which might be more appealing for programmers but still be suitable as search key. In this attempt, we built on the axiom that programmers will soon after being told about a problem start to reason about its solution in terms of high level (control-) structures of the language they are finally to use.

While this axiom still needs to be empirically substantiated, there is more than personal experience for its backing. This backing comes from the literature on quantitative software engineering. Software metrics came to age and notably for initial project assessment, one uses software metrics. Apparently, the specifics of these metrics applied in early phases of a project depend on various aspects. In case McCabe Metrics are used [7], one has direct evidence, that control structures are used already in very early phases. In case one uses e.g. function points or related measures for early effort estimation, one departs from rather static aspects. However, one has the data structures to be processed as parameters and from them again it would be relatively easy to determine the most fundamental structure of the control flow of the program to be developed.

Evidently, in all cases mentioned above, the "control flow" used for early assessment only contains the core structures of the algorithm to be developed. In the final program, these core structures, carrying most of the semantic of the program, are usually hidden in a wealth of "rococo", materialized in structures for taking care of special boundary conditions, missing values, proper checking of input, preparation of output, etc., etc.. Hence, if considered from the syntactic point of view of the programming language used, the control flow of the initial sketch and the control flow of the final program will have rather little correspondence. The problem is, that the syntactic perspective is blind with respect to the pragmatics of its structures. It neither can distinguish between small loops "plastered" on top or within the main semantic loops, nor is it possible to differentiate between core-alternatives of the algorithm and some added on conditions which are necessary to set the stage for the core algorithm.

To account for these problems, we propose to "weigh" control structures by the "work" they carry. This can be done by considering the data structures processed in a given portion of the control structure. If this can be achieved, one would have a mechanism to automatically see the coarse picture even in the fine-print code of the final program. Given this, one could perform some kind of pattern matching between the picture given by a program sketch and a readymade executable program stored in some repository. Apparently, this matching cannot aim for an exact match but has to allow for fuzzy correspondence. Next, we describe how the broad-brush picture can be extracted from (sketches of) source code.

3.2. Visualizing data structures

To allow for an adequate representation of the control flow as described above, one has to consider the interaction between control flow and data flow. An adequate notion to highlight the interaction between control flow and data flow is the concept of program slices [18], [2]. A program slice can be defined as the sequence of statements which directly or indirectly influences the data computed at a certain point of the program.

Since the final program constitutes an intertwining of all the possible program slices, exactly this intertwining has to be visualized. In order to do so, we will "color" the streams of data running through the program (resp. the slices) such that each data-item (named variable or referenced instance) is allocated a specific color and width.

While color will serve to distinguish variables and possibly indicate their relative importance with respect to the application semantic, width will indicate the relative weight of an item. Thus, skalar variables will be indicated by slim lines while complex records or multi-dimensional arrays will be indicated by broad lines. Width might also be indicative of the repetition factor inherent in the index-type of an array. We are reluctant though, to put too heavy emphasis on this aspect, since it might be more akin to a specific application than to the structure of the software at hand.

3.3. Visualizing weighed control patterns

Following the idea of slices, we need a way to visualize the (potential) flow of data through a program such that it visually highlights those portions of the program, where the most important (in terms of most complex or most heavy/voluminous) work will be done. This, of course, will apply not only for some specific slice, but for the union of all of them. Hence, it applies for the interaction between data- and control flow within the complete program.

Departing from the notion of a control graph, we isolate, in accordance to the primitives of structured programming, three basic notions of graphic primitives:

sequence: represented by a line,
alternative: represented by a line that forks into branches and recombines thereafter,
repetition: represented by a circular line (or circles).

However, since we are not interested in a mathematical abstraction - like in McCabe graphs, where sequences would be collapsed into arcs connecting only nodes which influence the flow of control - we might make full use of a two-dimensional (graphic) plane. Dimensions needed in excess could be abstracted either into color - as already indicated in connection with data representation - and texture - as used for the role of items. Nevertheless, we cannot refine the control graph to the statement-level. This might be premature, if this approach is to be used already early during design time and if the difference in cost between broad-brushing and fine painting is to be exploited.

Based on these considerations, the following picture will emerge:

Representation of usage: We distinguish between reading, writing and reflexive modification. In classical programming languages (with the assignment-operator separating left hand side (lhs) and right hand side (rhs) of an assignment operation) this corresponds to the exclusive rhs-usage, exclusive lhs-usage, and usage on both sides of the assignment operator. It is suggested to use light texture for rhs-usage, heavier for lhs-usage, and heaviest (full) for usage on both sides of the assignment.

Sequences: Sequences are represented by vertical structures, such that the items used within the sequence are indicated as vertical lines with the respective width and color of the item and in the texture corresponding to the usage pattern within this sequence.

Assuming that a surface of standardized size is available for this graph, the length of this vertical portion will be proportional to the number of different "ideas" tapped within this sequence with respect to the overall number of "ideas" expressed by this program. The notion of "idea" is apparently a fuzzy one. In case of actual code, this will be the number of statements. In case of pre-code (pseudo-code, Nassi-Shneiderman diagrams, ...) it will be the number of entries made in the respective formalism.

Alternatives: Alternatives, be they IF-THEN-ELSE- or CASE-constructs will be simply indicated by branching lines. This visualizes not so much the arrangement of conditional statement sequences in code. It is much more an abstraction of the way branching is handled by Nassi-Shneiderman diagrams.

Of course, there rests the problem of how to indicate IF-THEN constructs (having just one branch). We choose to consider even in these cases the control graph to have two branches. One of them figures in the data-flow-/slice-representation; the other one is empty. Hence, IF-THEN constructs are represented as bends.

The length of the individual branches will be given by the length of the longest branch, i. e. the branch who has internally the highest number of "ideas" to be processed. The width of the branching construct has no semantics. It is left to be defined by the necessity of displaying the internals (nesting) of the individual branches properly.

Loops: Loops are cyclic structures. Hence, it is more than fair to represent them as cycles. In order to be consistent with the argument about the correspondence between length and number of "ideas" and also to allow for proper nesting of other control structures, these cyclic lines will finally get an elliptical shape.

As of current, we propose that the repetition factor controlling a loop is not represented. This is analogous to the decision not to represent the number of elements in an array but to strictly stick to structural information. It will be subject to empirical studies, whether an indication of bulky loops (e.g. by increasing the diameter of the circle or the width of the item-lines) might be more suggestiv.

Nesting: Nesting of alternatives seems obvious. A branch will be further split into subbranches. Following the same logic though, we will indicate nested loops as circular structures residing on the loop within which they are executed. Hence, graphically, they will be piggy-backed by the graph of the structurally enclosing loop.

Procedure calls: Procedure calls leave a context and branch into another one. Hence, they are shown as tubes branching off from the picture of the (main) program (or from the procedure already called) like see-flowers from a reef. On some blank spot of the drawing surface, their flow of control will be indicated in the same way as indicated above for the main control structures (possibly drawn with some scale factor).

For stilistic reasons, we suggest that procedural nesting is indicated by drawing nested procedures to the right of their environment's body. If the same procedure is called from several spots, one will have a different "stemm" (passage of control) to the identical "blossom" (visualization of the procedure) for each call.

Recursion: Recursion has to do with both, repetition and procedure calls. Hence, we recommend to draw recursive procedures in such a way, that the image of the body is replicated on its right side in diminishing size. The reduction in size will be such, that finally a triangular shape will emerge.

3.4. Visual mapping

The "program-pictures" emerging from the visualizations indicated above apparently have no application semantic whatsoever. They are colorful abstract images, comparable to and inspired by graphical images of fractals. Their advantage is though, that it will be rather easy for a human to diagnose, where the center of activity might be. Further, the high capability of humans to recognize patterns might even allow programmers to identify similarity between programs on this level.

However, since our initial claim was to give the programmer a search key conformant to his/her everyday thinking, we would feel to have failed, if we would stop at this level. The argument is, that such "program-pictures" can be automatically drawn from the code of programs which are already stored in the repository and that the programmer can also write a sketch of code which could either be conformant to the syntax of the programming language, but be just of sub-prototyping quality (even this would incure already substantial savings [3]), or one could indicate the main control structure and the usage of items will be just given in terms of usage lists (e.g.: x, y, z := a, b, x, z) or in terms of low level design notations such as Nassi-Shneiderman diagrams.

The image thus generated will apparently not be fully conformant to any image generated from the repository. However, it will - if written with the same ideas in mind - be structurally conformant and hence lead to a similar picture on the broad brush level. If it is known in advance that such a correspondence will be very unlikely (perhaps because a very intricate algorithm has been used in the implementation of a stored component), it would also be possible to prepare a naive description yielding the same result as a "footprint" of the algorithm in order to improve the chances for a match (E.g.: In addition to its generated real "picture", a quicksort routine might have associated to it also the "footprint" of a simple insertion sort).

To relieve the programmer from the burden of comparing abstract images which are not understood, we propose in the following section a scheme to automate this task.

4. Automating Fuzzy Mapping - Hashing Images

4.1. The nature of the problem

The problem of mapping a search image against images of programs available in some repository is to some extent a pattern matching problem. On the other hand it has to do with the problem of reducing the information given in a lengthy search key to its essence. This is a general problem in information retrieval, related to the question how to optimally condense information. We try to approach this question from the side of hashing, knowing that other, broader approaches are still to be pursued.

Considering hashing as an alternative to relate these images, one has first to be aware, that the hashing problem is in general reverse to the problem we are investigating. With hashing, one aims for an exact match on a restricted address space - collisions are considered the price to be paid for gaining storage (and access) efficiency. In our case, we rather want to obtain collisions for similar search pattern. The approach to be pursued might be similar though. In both situations, one will prior to the application of the hashing algorithm perform a key analysis to identify those portions of the key which yield the best performance with respect to the success criteria sought.

This "key-analysis" of conventional hashing techniques will be replaced in our case by an image analysis. To support this, we will represent the images of programs described above not as pixel-information but as data structures representing the program's control graph. This data structure will contain structural information of the graph as well as "layout-information" such as color, texture, elongation etc..

4.2. Hashing images - weighing of substructures

Remembering that the "program-picture" obtained is in essence the collapsed set of program slices, we will use data-flow information as a first clue for identifying the primary nature of the algorithm.

In general, the highest weighed branch(es) will be close to the exteriour and close to the left side of the image. They should contribute substantially to the value of the "hash-key". The farther away from this main road a branch (syntactical construct, sub-image) is, the less its weight. This principle can be approximated by a left to right weighing strategy.

It should be noted, that this mapping and weighing scheme is just based on heuristics. It will be subject to empirical evaluation and further refinement.

5. Further applications

The "program-pictures" as described in section 3 present an abstract visualization of the control structure of a program and of the intertwining between control flow and data flow. Hence, the complexity of the picture is directly proportional to the complexity of the code.

Standard complexity measures could be used for estimating the effort needed for adaptations. However, with these metrics one can assess only those base costs of modifications involved with understanding of the piece of code to be modified. The extent of modifications required cannot be assessed by them, since they have no basis for assessing the difference between the component at hand and the component needed.

The "program-pictures" proposed here provide some hints in this direction. While they could also give a first indication of the overall complexity of the component, they would also allow to assess "on a glimpse" the difference between the component needed and the component(s) at hand. While this assessment will not be precise, it will still be a big step with respect to the current situation, where the programmer seeking some readymade component would have to take the decision to modify or to redo it almost blindfoldedly.

6. Summary

The paper presented a proposal to visualize software in terms of the intertwining of control flow and data flow. Departing from the hypothesis, that (some/most) programmers will think about solutions (at least after reflecting briefly about the problem) in terms of these characteristics, it was proposed, to use this information as auxiliary access-path to repositories of reusable software components.

As side benefits of this form of description, it would allow programmers to get at least a vague idea about the effort involved with component modification, in case such a need arises.

The project to investigate this approach is still in its initial phases. To carry it to its end, a lot of empirical work as well as technical work needs to be done.

7. References

[1] BARNES B. et al: "A Framework and Economic Foundation for Software Reuse"; in TRACZ W. (ed.): "Tutorial: Software Reuse: Emerging Technology", IEEE-CS Press, 1988, pp. 77-88.

[2] BERZINS V., LUQI, DAMPIER D.: Proceedings "Increasing the Practical Impact of Formal Methods for Computer-Aided Software Development: Software Slicing, Merging and Integration"; Naval Postgraduate School, Monterey, CA, 1993.

[3] BOEHM B.W.: "Software Engineering Economics"; Prentice Hall, Englewood Cliffs, N.Y., 1981.

[4] BOUDRIGA N., ELLOUMI F., MILI A.: "On the Lattice of Specifications. Applications to a Specification Methodolgy"; Formal Aspects of Computing, Vol. 4, 1992, pp. 544 - 571.

[5] BOUDRIGA N., MILI A., MITTERMEIR R.: "SemanticBased Software Retrieval to Support Rapid Prototyping"; Structured Programming, Vol. 13, 1992, pp. 109-127.

[6] CHEN P.S., HENNICKER R., JARKE M.: "On the Retrieval of Reusable Software Components"; Proc. "Advances in Software Reuse", Lucca, 1993; IEEE-CS Press, 1993, pp. 99-108.

[7] FUCHS N.: "Cost Management with Metrics of Specifications"; Proc. Conf. Software Reliability, Elsevier, 1990, pp. 176-184.

[8] FUGINI M.G., FAUSTLE S.: "Retrieval of Reusable Components in a Development Information System"; Proc. "Advances in Software Reuse", Lucca, 1993; IEEE-CS Press, 1993, pp. 89-98.

[9] KATZ S., RICHTER Ch.A., THE K.-S.: "PARIS - A System for Reusing Partially Interpreted Schemas"; in: Proc. 9th ICSE, Monterey, 1987 and BIGGERSTAFF T.J. and PERLIS A.J.: "Software Reusability", Vol. 1, acm & Addison Wesley, 1989, pp. 257-273.

[10] KRUEGER Ch. W.: "Software Reuse"; ACM Computing Surveys, Vol. 24/2, June 1992, pp. 131-183.

[11] MAAREK Y.S. and BERRY D.M.: "The Use of Lexical Affinities in Requirements Extraction"; Proc. 5th Intern. Workshop on Software Specification and Design; Pittsburgh, 1989, IEEE-CS Press, 1989, pp. 196-202.

[12] MATSUMOTO Y.: "A Software Factory: An Overall Approach to SW Production"; in: Proc. of the ITT Workshop on Reusability in Programming, Newport R. I., 1983 and in: FEEMAN P.: Tutorial "Software Reusability"; IEEE-CS Press, 1987, pp. 155-178.

[13] MILI A., MILI R., MITTERMEIR R.: "Storing and Retrieving Software Components: Preliminary Implementation"; Proc. 16 ICSE, Sorrento 1994, pp. 91-100.

[14] MITTERMEIR R. and ROSSAK W.: "Software Bases and Software-Archives - Alternatives to support Software Reuse"; Proc. of the FJCC '87, Dallas, TX, Oct. 1987, pp. 21-28.

[15] OSTERTAG E., HENDLER J., PRIETO-DIAZ R., BRAUN Ch.: "Computing Similarity in a Reuse Library System: An AI-Based Approach"; acm Trans. on Software Engieering and Methodology, Vol. 1/3, July 1992, pp. 205-228.

[16] PRIETO-DIAZ R. and FREEMAN P.: "Classifying Software for Reusability", IEEE Software, Vol. 4/1, Jan 1987, pp. 6-16.

[17] TRACZ W. "Software Reuse: Motivators and Inhibitors"; Proc. COMPCON '87 and TRACZ W. (ed.): "Tutorial: Software Reuse: Emerging Technology", IEEE-CS Press, 1988, pp. 62-67.

[18] WEISER M.: "Program slicing"; IEEE Trans on Software Engineering, Vol. SE-10/4, July 1984, pp. 352-357.

Appendix:

An example program (InsertionSort) and its "program-picture". (For the sake of reproduction, different colors are indicated by texture. Hence, the texture-parameter is not used).

```
PROGRAM Sort;
VAR n         : INTEGER;
    SortArray : [1..10] OF CHAR;

PROCEDURE InsertionSort(VAR A: ARRAY OF CHAR; len:INTEGER);
  VAR    i, j : INTEGER;
         c    : CHAR;

  BEGIN (* InsertionSort *)
    IF len = 1
     THEN Writeln('This list does not have to be sorted')
     ELSE BEGIN
       FOR i := 2 TO len DO BEGIN
         c := A[i];
         j := i;
         WHILE A[j-1] > c DO BEGIN
           A[j] := A[j-1];
           j    := j-1;
         END{WHILE};
         A[j] := c;
       END{FOR};
     END{ELSE};
  END (* InsertionSort *);

BEGIN (* Sort *)
  SortArray := 'An Example';
  n  := Length(SortArray);
  InsertionSort(SortArray, n);
  Write(SortArray);
END (* Sort *).
```

9. APPLICATIONS

Retrieval of Images Using Pixel based Object Models

Masayuki MUKUNOKI, Michihiko MINOH and Katsuo IKEDA

Department of Information Science, Kyoto University. Kyoto, 606-01, Japan.

Abstract. We apply our Pixel-based object labeling method to the problem of indexing images. Our method is a means to assign an object label to each pixel in out-door scenes. It is suitable for automatic object labeling and applicable to automatic indexing problems.
The recognition rate of our method is about 77%. To apply our method to the problem of indexing, we introduce a pictorial query method for retrieval. We conducted some retrieval experiments, and examined the usability of our method to the indexing problem.

1 Introduction

At present, the ability of computers to recognize objects is not as good as that of human beings, still some applications do exist, which can make use of the present object recognition results. One of such applications is to use the results as the index of images. We have conducted research on object recognition of out-door scenes [1], and introduce our object recognition method to image retrieval.

In traditional image databases, someone must set keys to images. It is boring routine work and takes a lot of time, particularly when there are many images in the database. Object recognition techniques will help to solve this problem. The ultimate goal of object recognition in this case is to describe the objects in images automatically. Our "Pixel-based object labeling method" for object recognition is suitable for this purpose.

There have been some research on object labeling of out-door scenes [2, 3, 4, 5]. They basically employ such strategies that generate initial regions with signal-level segmentation and then assign the object labels to the regions. But it is difficult to generate the initial regions for out-door scenes because there are various objects in them and the conditions for the camera and the light source cannot be fixed. As a result, it prevents the automatic object labeling. In our "Pixel-based object labeling method," we do not generate initial regions, hence we can avoid this problem.

Next, we use the result of object labeling of out-door scenes as the index of the images. We employ a pictorial query method for retrieval [6, 7]. Users draw an object sketch as the condition for the retrieval. The retrieval system compares this sketch with the index, and then presents images in the order of similarity. The indexes generated by object recognition do contain errors, but we can realize a fault tolerant retrieval system with the pictorial query method.

In the following sections, we explain our "Pixel-based object labeling method" and its application to image retrieval.

2 Pixel-Based vs. Region-Based Approach for Object Labeling

The goal of object labeling of an image is to segment it into regions each of which corresponds to one object and has a unique object label. Most of the studies so far adopted the "region-based" approach. They initially segment an image into regions and label them with object models. The object models are mappings between the features derived from the image and the object labels. However, in this strategy, there is a problem in generating the initial regions.

The initial regions are generated based on signal-level similarity of pixels: They are generated only from the statistical similarity or the spatial closeness of the pixels in an image, and the knowledge about the objects is difficult to be introduced. Because of the variety of objects, generation of the initial regions based on the signal-level similarity does not guarantee that a region actually corresponds to an object.

The results of the initial segmentation have a great impact on the object labeling. But the results are largely affected by the parameters chosen for the signal-level processing. Figure 1(a)(b) shows example out-door scenes, and Figure 1(c)(d) shows the results of signal-level segmentation with the same processing parameters. In Figure 1(c), a region almost corresponds to an object, while in Figure 1(d), regions are over-segmented and there are too many small regions each of which does not correspond to one object. It is difficult to select adequate processing parameters, particularly for out-door scenes. In most studies, these parameters are heuristically determined for each image. Consequently, it prevents automatic labeling.

On the other hand, since objects in out-door scenes are natural objects, they do not have any definite shape. The features of shape and size are not essential for the object labeling, even if we could get a region that corresponds to a whole object. Therefore, it is pointless to make the initial regions for object labeling.

Considering the above discussion, we have proposed a new approach to the problem of object labeling of out-door scenes [1]. It is called "Pixel-based object labeling method." It proceeds as follows (Figure 2):

1. Calculating feature values of every pixel.
2. Evaluating the object labels of each pixel using the Pixel-based object models, which are mappings between the features, derived from a pixel, and the object labels. The result is the evaluation value for each object label.
3. Assigning an object label to every pixel based on the evaluation values obtained in step 2.

Regions are generated by assembling the pixels that have the same object labels. Though pixels do not have the features of shape and size, they are not essential for the labeling of natural objects. Other features, such as color, texture and location, can be obtained from the pixels.

There are no heuristic thresholding parameters in the processing. The processing does not be interfered by such parameters and, as a consequent, automatic object labeling can be realized.

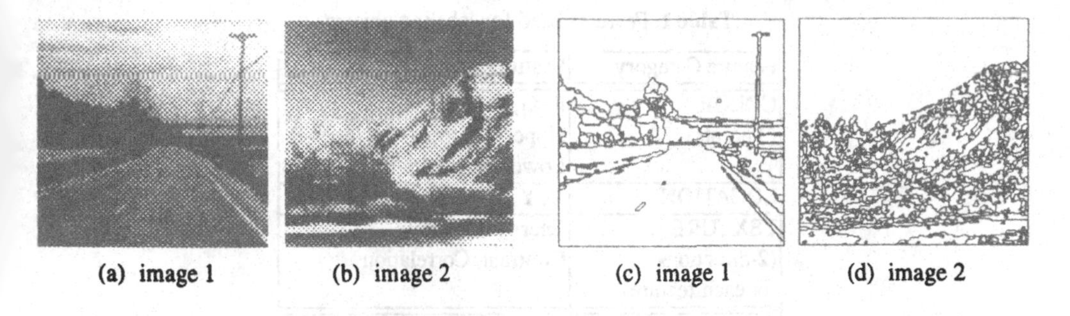

Fig. 1. Examples of the out-door scenes and their signal-level segmentation results

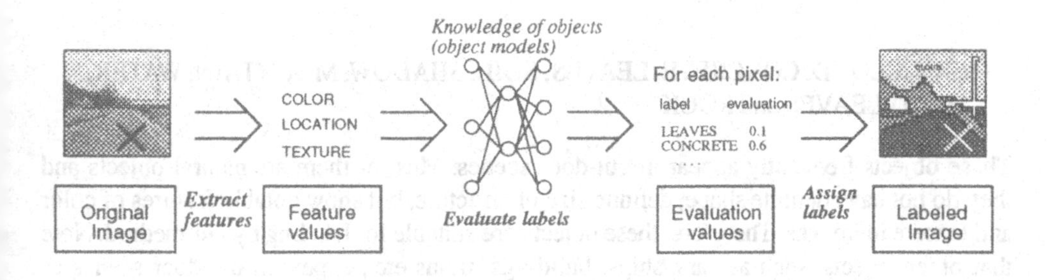

Fig. 2. Pixel-based object labeling method

3 Construction of Pixel-Based Object Models

3.1 Object Models with Neural-Network

We construct the object models with Neural-Network(NN). We use a 3-layered perceptron. There is one input node for each feature and one output node for each object label. Each feature value is normalized into [0.0, 1.0] and used as input to the corresponding node in the input layer.

We use a "back propagation algorithm" for training, which minimizes the mean square error between the outputs of NN and the teaching signals. The teaching signal is 1 for the correct output node, and 0 for the other output nodes. As the result of this training process, the network realizes the mapping between the local features of a pixel and its object label.

In our approach, it is essential to construct object models by training. Although we cannot describe the object models clearly in the pixel-based approach, we can give samples of the correspondence between a pixel and its object label, and we can calculate the local features of the pixel. In this way, we can easily construct the object models. And there are other benefits of constructing the object models with NN: We can get the result of the evaluation numerically; NN can simulate any mapping between the features and the object labels and can show high performance after the training.

3.2 Objects

We use the following 10 object labels:

Table 1. Features used for labeling objects

Feature Category	Features
COLOR (3 points)	R, G, B Upper-R, Upper-G, Upper-B Lower-R, Lower-G, Lower-B
LOCATION	X, Y
TEXTURE (2-directions for each feature)	Energy, Entropy, Contrast, Correlation
SIZE (region only)	Size, Width, Height
SHAPE (region only)	Ith, Iratio, ISratio

SKY, CLOUD, CONCRETE, LEAVES, SOIL, SHADOW, MOUNTAIN, WATER, DEAD-LEAVES and ROCK

These objects frequently appear in out-door scenes. Most of them are natural objects and they do not have definite shape, definite size or structure, but show notable features of color and texture in images. Therefore, these objects are suitable for labeling by our method. Note that other objects, such as cars, ships, buildings, trains etc., appear in out-door scenes as well. We do not deal with these artificial objects, because the features of color or texture are not the key for the object labeling in this case.

3.3 Features

Table 1 shows the features used for labeling the objects. Only COLOR, LOCATION and TEXTURE features are used in the "Pixel-based method." SIZE and SHAPE features are used in comparative experiments in Section 3.4.

In the "Pixel-based method," the COLOR and TEXTURE features are computed within 15×15 neighboring pixels of each pixel. This suppresses noise in the image and enables us to calculate TEXTURE features for each pixel.

R, G and B indicate the mean values of red, green and blue values within the neighboring pixels, respectively. The original images are color images and each pixel has red, green and blue values.

Upper-R, Upper-G, Upper-B, Lower-R, Lower-G and Lower-B are R, G and B features of the pixel at the position at 128 pixels above and below, respectively. If the position is beyond the size of the image, the features of the pixels at the upper-most or lower-most is used. We use these features because the color features are not stable, that is, easily affected by the camera condition or the light condition. For example, if we take pictures of the same place both in a fine day and in a cloudy day, the colors of objects in them are different. The color features of other pixel may supplement this information. The positions (128 pixel above or below) of the pixels are decided experimentally.

X and Y indicate the location of a pixel in an image. The objects near the top of the image are likely SKY, CLOUD, MOUNTAIN etc.. The objects near the bottom of the image

Table 2. Recognition rate

	Pixel	Region *with* SHAPE and SIZE	Region *without* SHAPE and SIZE
testing set (%)	77.3	69.5	68.2
training set (%)	94.4	92.1	91.5

are likely CONCRETE, WATER, LEAVES etc.. The features of X and Y can contribute to the categorization of these objects.

TEXTURE features are calculated from a co-occurrence matrix [8]. The co-occurrence matrix $P(i, j|d, \theta)$ is a set of probabilities that two pixels, each of which has pixel-value i and j respectively, appear at distance d in direction θ within the neighborhood of the pixel. Energy, Entropy, Contrast and Correlation stand for the uniformity of the texture, the complexity of the texture, the contrast of the texture, and the slope of the texture, respectively. These features are frequently used to distinguish textures. We calculate these features with a distance $d = 1$ and with two directions $\theta = 0, 90°$. As a result, we get 8 feature values for TEXTURE.

SIZE and SHAPE features are used in the experiments in Section 3.4. They are the features for regions. SIZE is the number of pixels in a region. Width and Height are the horizontal and vertical length of a region, respectively. SHAPE features are calculated from the moments. Ith, Iratio and ISratio stand for the direction of a region, the slenderness of a region and the roundness of a region, respectively.

3.4 Experiments and Results

For experiments, we used 66 out-door scenes. We selected 975 sample points from 33 scenes for training and 970 sample points from the remaining 33 scenes for testing. The training was terminated when the mean square error became less than 0.05. The number of nodes in the intermediate layer was 11, which was determined empirically. The recognition rate of trained NN is shown in Table 2.

For comparison, we constructed two region-based object models. One uses COLOR, LOCATION and TEXTURE features, just like the pixel-based object models. The other uses SHAPE and SIZE features, in addition to the COLOR, LOCATION and TEXTURE features. In both cases, the feature values are calculated within the region in which the sample point is included. Regions of the images are generated by signal-level segmentation. Parameters of the segmentation are determined heuristically. The recognition rates of the region-based object models are also shown in Table 2.

In Table 2, the two region-based object models show almost equal recognition rate. This result shows that the features of SHAPE and SIZE are not essential for object labeling. Pixels do not have these features but we can see that this is not a problem.

The recognition rate of the pixel-based object models is higher than those of the region-based ones, even when the SHAPE and SIZE features are added. This shows that the features derived from a pixel have as much information as features derived from a whole region for object labeling.

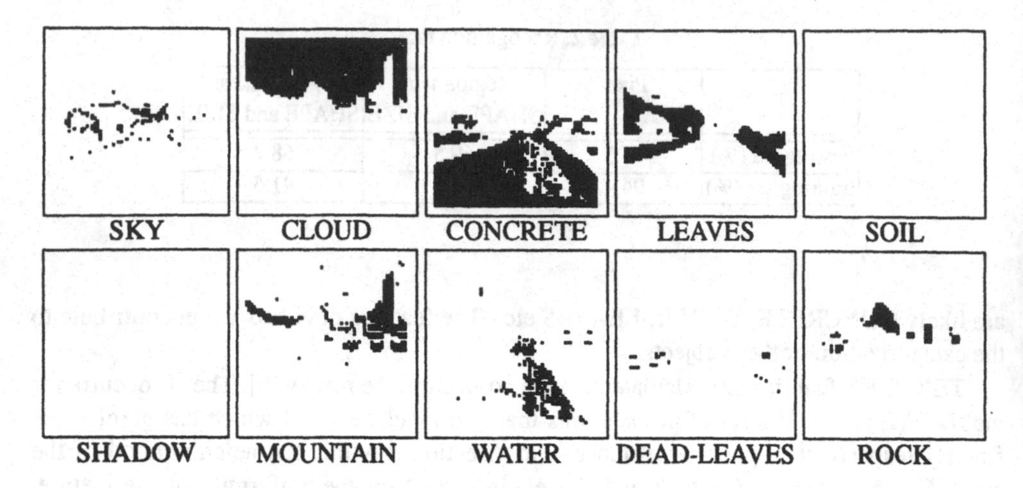

Fig. 3. Results of object labeling

Next, we apply these pixel-based object models to whole images. Figure 3 shows the result of object labeling of Figure 1(a). The output node which has the highest output value in the output layer is taken as the object label of the pixel. It is the simplest way to assign the object label to each pixel. There are some pixels which are labeled incorrectly. The object models are simple mappings between the feature values and the labels, and they do not consider the labels of other pixels. To improve the result, other knowledges, for example the spatial relation of objects, could be introduced to object labeling.

4 Retrieval with Object Sketch

4.1 Application of our Method to Image Retrieval

We apply our object labeling method to image retrieval of out-door scenes. General process of image retrieval is as follows: a user gives some keys; the retrieval system compares the keys with the indexes of images; and show some candidate images, the indexes of which match the keys, to the users.

One useful kind of indexes for image retrieval is the contents of the image, i.e. the objects within the image. For example, if a user wants to retrieve some images which contain "mountain," it is natural to use "the-mountain-is-contained" as the index.

The result of our object labeling is the image assigned an object label to each pixel. It includes the information of the contents of the image and it can be used as the index. But, as shown in the previous section, the recognition rate of our method is not 100% and there are errors in the result. This means that some objects which do not exist in the image may appear in the result, while some objects which exist in the image may not appear. As a result, if we use the list of object labels appearing in the result as the index of the image, the image retrieval may fail. In order to cope with such errors, we employ a pictorial query method for retrieval.

4.2 Pictorial Query Method for Retrieval

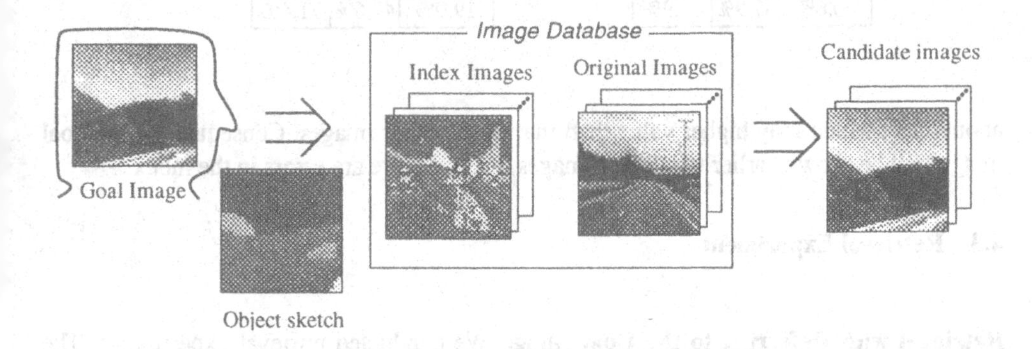

Fig. 4. Retrieval method

A pictorial query method is a method that uses an image to express the condition of retrieval. We use an object sketch of the image for retrieval. The object sketch is an image where each pixel value represents the object label of that pixel in the original image.

Our object labeling method assigns labels to all the pixels in the image. We use the labeled image as the index of the original image and call it "the index image." The size of the original images is 256 × 256 pixels. In order to reduce the size of the index images, we assign an object label to every 4 pixels. The size of the index images thus becomes 64 × 64. This is sufficient resolution for the retrieval of out-door scenes images. The indexing process is automatically performed by computer.

To retrieve images, a user draws an object sketch of a goal image that he wants to retrieve. User does not have to specify all labels of the pixels in the goal image. The pixels that users do not specify labels are labeled as UNKNOWN by the retrieval system. The size of the object sketch is 64 × 64.

To select "candidate images," the system calculates the similarity S between the object sketch and the index images. The similarity is calculated as $S = \frac{M}{N}$ where

M : number of pixels residing at the same position in the object sketch and the index image and having the same object label.

N : number of pixels in the object sketch which have labels specified by the user, i.e. 64 × 64 - (number of UNKNOWN pixels).

The system then shows the candidate images in the order of similarity.

An object sketch is one useful way to express the goal images. Since images have much information, it is difficult to express the sort of objects, the positions of them, the spatial relations between them etc. with linguistic description.

Furthermore, the pictorial query method is robust against errors in the index images. The recognition rate of our object labeling method is about 77%. We cannot know which pixels are correctly labeled and which pixels are incorrectly labeled only from the result of object labeling. But, at the retrieval moment, if a user were able to give the correct labels, the similarity between the index image of the goal image and the object sketch would be

Table 3. Recall ratio

(a) Referring to the goal image

The Best	Best 5	Best 20
60.6%	87.9%	93.9%

(b) Without referring to the goal image

The Best	Best 5	Best 20
19.0%	47.6%	71.4%

about 77%, which may higher value than that of the other images. Consequently, the goal image will be shown earlier than other images, even if there are errors in the index.

4.3 Retrieval Experiment

Retrieval with Referring to the Goal Image We conducted retrieval experiments. The number of the images in the database is 461 which include the images used in Section 3.4. First, we made the index images for all the images in the database using the Pixel-based object models constructed in Section 3.4. Second, we selected the 33 images from them which were not used to construct the object models. And third, we retrieved these images among the 461 images. We drew the object sketches referring to the goal images. It means that the goal image is displayed and we set the object labels on the goal image. As a result, the sketches drawn here are correct sketches of the goal images in a sense. It takes about 1.7 seconds for a retrieval on HP9000/735(124 MIPS).

The recall ratio of the retrieval is shown in Table 3(a). The recall ratio is the ratio that the goal image is within the top n-candidates. The Table shows that 60.6% of the images are presented as the best candidate. In this experiment, the object sketch had almost correct labels because it was drawn referring to the goal image. The recall ratio of the best candidate should be almost 100% in the ideal situation. But there are several images which are similar each other in the database. They decrease the recall ratio.

There were four images which were not within the best 5 candidates. In the worst case, the goal image was the 36th candidate. The similarities between the index image of the goal image and the object sketch were within $0.43 \sim 0.52$ for those four images. They are worse values than those of the other images. This shows that there are some images which cannot be labeled correctly. One reason of this problem is the light condition of the images. To improve the recall ratio, we should consider the method to get rid of such affection, as well as to improve the average recognition rate of object labeling.

Retrieval without Referring to the Goal Image In a real situation, a user retrieves an image without referring to the goal image. We conducted an experiment simulating such a situation.

We showed several images in our database to each retriever. Each retriever selected an image from them as the goal image, and drew the object sketch without referring to it. We tested for 18 retrievers, with a total of 42 retrievals. The results of retrievals are shown in Table 3(b). The goal image was retrieved within the best twenty candidates for 71.4% of the cases. In the worst case, the goal image was the 184th candidate.

Examples of the typical cases of retrieval are shown in Figure 5. Figure 5(a) shows the successful case. There are 3 typical cases of retrieval failure. Figure 5(b) shows a retrieval

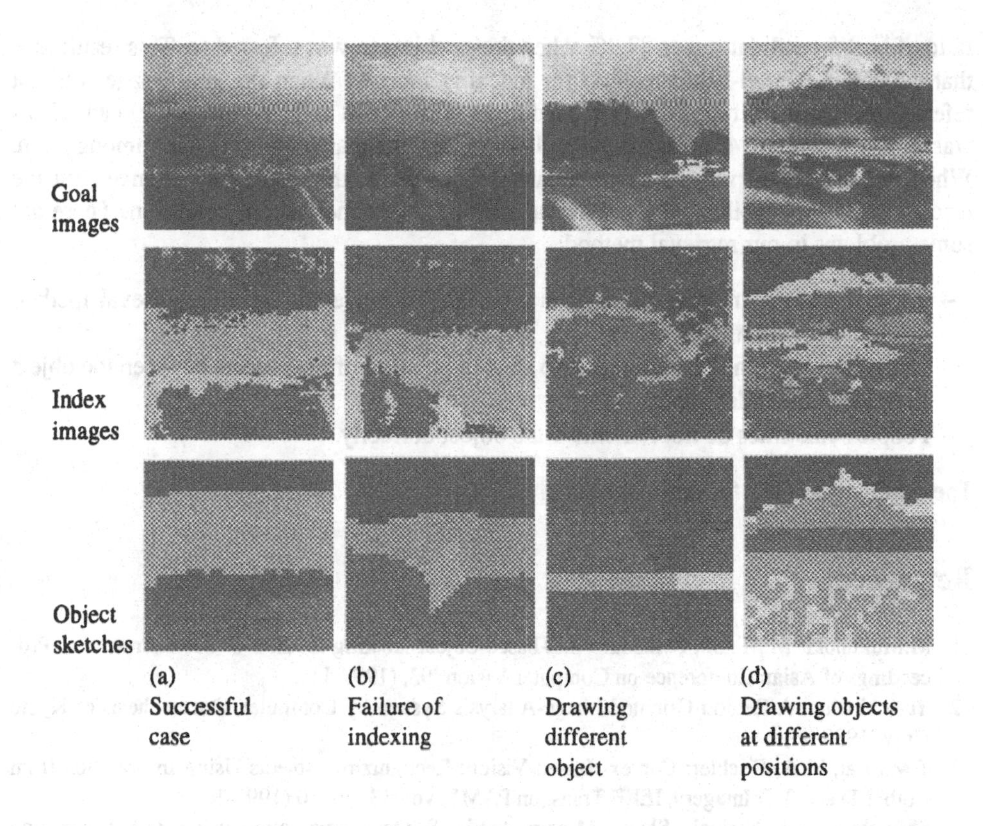

Fig. 5. Typical examples of retrieval

that failed because of a wrong index image. This is the problem of object labeling described in Section 4.3.

Figure 5(c) and (d) show failure caused by wrong object sketches. In Figure 5(c), the retriever drew the sky as a cloudy sky, but it was labeled as a blue sky in the index image. There were several cases where the retrievers mixed up blue sky and cloudy sky, especially when the sky was not clear blue. It seems that some retrievers do not remember whether a sky is clear or cloudy. To cope with this problem, we should introduce only one "sky" category which is compatible to these two categories. In Figure 5(d), the retriever drew the objects at wrong positions. The system compares the label of pixels at the same positions in the index image and the object sketch. This method is vulnerable to shift of position. We should also consider a solution to deal with this problem.

5 Conclusion

We described our Pixel-based object labeling method. The recognition rate of the Pixel-based method is higher than that of region-based methods. In addition, we need not use signal level segmentation in our method. As a consequence, we can assign object labels automatically.

Next we applied our Pixel-based method to the problem of indexing images. The recall ratio of best 5 candidates was 87.9% when the goal image was referred to. This result tells that we can use Pixel-based method for indexing images. When the goal image was not referred to, the recall ratio of best 5 candidates was 47.6% and that of best 20 candidates was 71.4%. There are 461 images in the database and there are similar images among them. When the number of images in the database is increase, the result may change. But the result shows the usability of our retrieval method under the current condition. There are some problems in our retrieval method:

- There are some images which cannot be labeled correctly and our retrieval method cannot cope with this.
- Our retrieval method is vulnerable to shift of position of the objects between the object sketch and the index image.
- People sometimes do not remember the object correctly.

The problems are left for future research.

References

1. M.Mukunoki, M.Minoh, K.Ikeda: Pixel-based object labeling method of out-door scenes, Proceedings of Asian Conference on Computer Vision '93, (1993-11).
2. Yu-ichi Ohta: A Region-Oriented Image-Analysis System by Computer, Doctor thesis of Kyoto Univ.(1980-03).
3. T.M.Strat, M.A. Fischler: Context-Based Vision: Recognizing Objects Using Information from Both 2-D and 3-D Imagery, IEEE Trans. on PAMI., vol. 13, no. 10 (1991-10).
4. Shinichi Hirata, Yoshiaki Shirai, Minoru Asada: Scene interpretation using 3-D information extracted from monocular color images (in Japanese), Trans. IEICE vol. J75-D-II, no. 11, pp. 1839-1847 (1992-11).
5. Hidehiro Ohki, Tsutomu Endo: Scene understanding system using contextual information – Hypothesis generation and conflict resolution – (in Japanese), Technical report of IEICE PRU92-43 (1992-10).
6. T.Kato, T.Kurita, N.Otsu, K.Hirata: A Sketch Retrieval Method for Full Color Image Database – Query by Visual Example –, 11th ICPR, pp. 530-533 (1992).
7. M.Kurokawa, J.K.Hong: An Approach to Similarity Retrieval of Images by Using Shape Features (in Japanese), Trans. IPS Japan, vol. 32, no. 6, pp. 721-729 (1991-06).
8. R.M.Haralick: Statistical and structural approaches to texture, Proc. of the 4th IJCPR, pp. 45-69 (1978).

Object Recognition Based on Pattern Features

Weijing ZHANG and Anca RALESCU[1, 2]

The Laboratory for International Fuzzy Engineering Research
Siber Hegner Bldg. 3F, 89-1 Yamashita-cho
Naka-ku, Yokohama 231, Japan
Telephone: (81) 45-212-8240 Fax: (81) 45-212-8255
zhang@fuzzy.or.jp, anca@fuzzy.or.jp

Abstract. We propose an approach to object recognition based on low level image pattern features. This approach is intended to avoid creating complex high level object model. Our goal is to achieve *"quick and (possibly) rough"* object recognition from images. An object is defined by a collection of patterns and relations among them. A pattern is a visual clue to an object, such as color, edges texture, etc. Operations like edge detection and region segmentation are performed in a predefined area, which is a small window over which the pattern is defined. Relations among patterns are used to provide constraints on the search space for patterns and for an intelligent pattern matching. Fuzzy variables are used to express patterns in order to build the system's tolerance to imprecision. Sample data is summarized into fuzzy sets. Pattern matching is realized by fuzzy reasoning. The overall evaluation of the matching can be provided by aggregations (i.e. fuzzy integral).

1. Introduction

Often in visual information processing quick and rough recognition of the image content is required. On the other hand daily experience shows that humans also recognize objects by a few visual clues. They respond to these clues very quickly and, depending on the goal of the recognition, they neglect other information. Sometimes the clue is a color, sometimes it is a shape; in general humans don't need much knowledge to recognize an object in a scene [1,2].

A recognition process can be very simple. For example, from "something white in something blue", we may understand that it is actually "a cloud in the sky". Instead of complicated and abstract knowledge representation paradigms (as often done in AI) more direct input-output connections can be very effective and efficient in many cases [3].

Zadeh [4] has pointed out that *precision and certainty carry a cost.* In fact, Zadeh formulated the guiding principle for soft computing as "Explore the tolerance for imprecision, uncertainty and partial truth to achieve tractability, robustness and low solution cost." Our goal is not precise recognition but rough recognition.

---

[1] On leave from the Computer Science Department, University of Cincinnati, Cincinnati Ohio, U. S. A.

[2] This work was partially supported by the NSF Grant INT91-08632.

Simplified object model representation also results in the reduction of processing time. The meaning of a pattern along with the relations between patterns contribute to a quick recognition by restricting the search space for patterns.

In the following sections, we describe a pattern features based approach to object recognition. After a general description, we discuss what a pattern is, how to define a pattern, what an object is from the pattern based point of view, and how object recognition is realized by intelligent pattern matching.

2. Pattern Features Based Approach

Fig. 1(a) shows the block diagram of a conventional image recognition system: features are extracted from the input image after edge detection and/or region segmentation; these features are matched with the object model in a knowledge base [5, 6].

Usually, the features are abstract and the object models are complicated, without a direct connection to the low level image data. For example, a desk has a `rectangular` surface as its top. Here, `shape = rectangle` is an abstract feature for the model of desk. In the literature this problem is referred to as the gap between image and concepts [7] and it is towards filling this gap that typical image understanding efforts are aimed at.

To narrow this gap and to make the recognition process less complicated, we consider the pattern feature based approach (Fig. 1(b)).

A pattern feature is defined at a low level, i.e., it is available directly from the input image data; moreover it provides immediate support for its associated object.

Informally, a pattern feature corresponds to a visual clue with which a human can recognize an object very easily. These clues provide the "it looks like __" information about an object.

Humans use visual clues for recognition even when such clues are not intrinsic to the object to be recognized: If some one's office has a beautiful calendar on the wall, it is not difficult to find the office even when we do not know the person. "Turn left onto the third street. There is a big office building on the right corner" provides another example to show how pattern features are used in our daily life.

It may be argued that since image patterns vary under different situations it is impossible to make a database to include all kinds of patterns associated to an object. This is only partially true because not every situation needs to be taken into account. We don't need to remember what an object looks like in all situations from all viewpoints (if we don't look at a desk from the back, the rear view based image patterns are not needed). We only need the information useful for the cases in which we need to use.

In addition, some assumptions may be made to reduce the amount of image patterns. For example, if we know the position and the angle of the camera, we can predict a lot of information from the image, such as how large the object should be in the image, how the object should look like [8], etc. Knowing these parameters contributes to the reduction of the amount of patterns to be remembered.

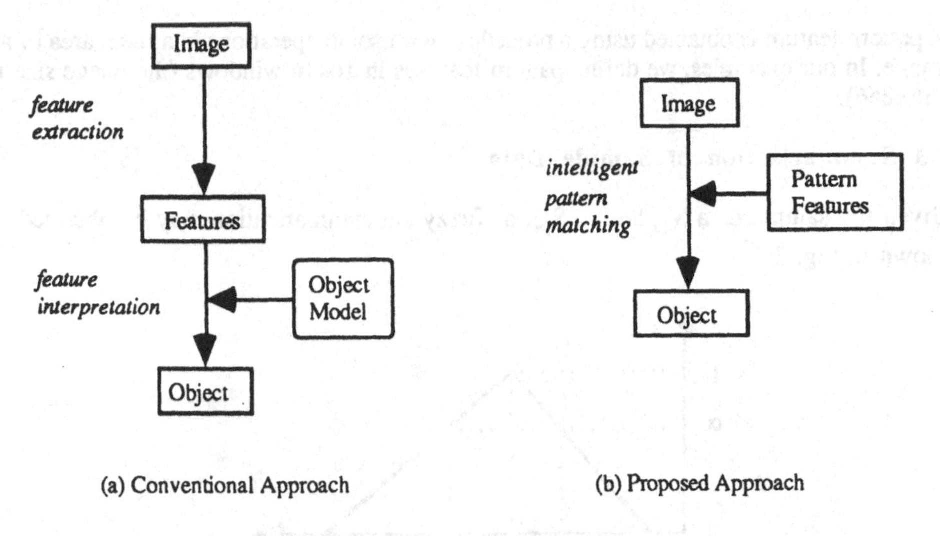

(a) Conventional Approach

(b) Proposed Approach

Figure 1 Image Recognition System

3. Patterns and Pattern Features

3.1 Pattern

We take a closer look at pattern features but the first question is: what is a pattern?

Definition: *A pattern is a visual clue to an object under a given situation which provides support for the existence of the object. A pattern can be obtained directly from the image data.*

A pattern is defined at a much lower level compared with the conventional concept of features. Therefore, a pattern can be a color pattern, edge pattern, texture pattern, etc.

3.2 Pattern Features

Pattern features are expected to satisfy the following requirements:

1. Pattern features are values of a pattern. They are associated with objects under various situations: viewpoints, lighting conditions, etc.

2. There should be more than one pattern feature or pattern groups for one object. Because pattern features are defined at low level, it is possible that pattern matching based on a particular feature fails. In this case other features could be used instead. One object may have more than one pattern based models.

3. Pattern features are associated with specific objects. However, as a visual clue to the target object, a pattern feature does not have to be intrinsic to the object, e.g. a pattern for a desk may be an image pattern of a calendar.

A pattern feature is obtained using a procedure with some operations in a fixed area of an image. In our examples, we define pattern features in 16x16 windows (the image size is 256x256).

3.3 Summarization of Sample Data

Given the sample data $S_1, S_2, ..., S_N$, a fuzzy set summarization may be obtained as shown in Fig. 2:

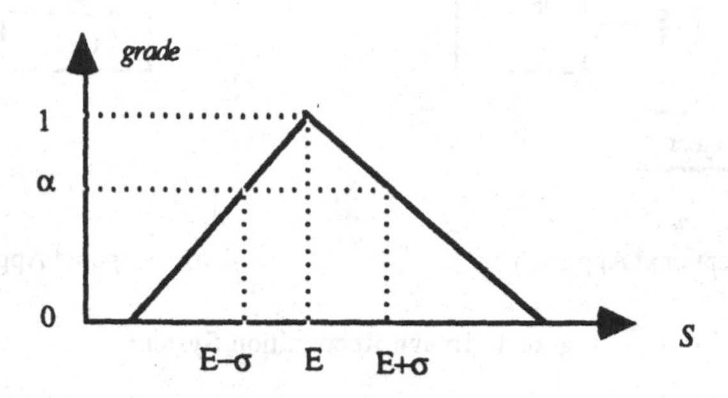

Figure 2 A fuzzy set from sample data

where $E = \frac{1}{N}\sum_{i=1}^{N} S_i$ is the sample average, $\sigma = \frac{1}{N}\sum_{i=1}^{N} \left|S_i - E\right|$ is the average absolute deviation and α is a specified value.

Pattern matching is evaluated by the overlapping of two fuzzy sets. We insist here neither on the summarization technique nor on the matching algorithm and only note that several alternatives are possible for each.

3.4 Basic Patterns

The patterns are based on the results of the image processing: color, edge, and texture information. Here we outline the use of the first two patterns.

3.4.1 Color Pattern. Color is a very important visual clue for object recognition and is often used in region segmentation [9]. In many cases, color plays the most important role in locating an object from an image and sometimes, color may be the only clue available for some objects.

We use the Hue, Saturation and Intensity (HSI) color space obtained from the RGB color information through the following transformation:

$$I = \frac{R+G+B}{3}$$

$$S = 1 - \frac{\min(R,G,B)}{I}$$

$$H = \arctan \frac{\sqrt{3}(G-B)}{(R-G)+(R-B)}$$

The HSI system is used because it has a straightforward interpretation in terms of human perception [9, 10] and hue, saturation and intensity are independent.

A color pattern is a fuzzy vector $C = (C_h, C_s, C_i)$, where C_h, C_s, C_i are fuzzy sets on H, S, I respectively in a 16x16 region of image. We calculate the average and the average absolute deviation of 16x16 HSI values and obtain $D = (D_h, D_s, D_i)$. The matching degree of the region with the color pattern is expressed by the fuzzy rule:

IF (H is C_h) and (S is C_s) and (I is C_i) THEN Color is C

where the degrees to which (H is C_h), (S is C_s) and (I is C_i) hold are obtained from the matching of D_h with C_h, D_s with C_s, and D_i with C_i respectively. For color variations under different situations like lighting, operations on the membership function may reflect the differences. This would be a simple and effective approach to manipulate complex situations.

3.4.2 Edge Pattern. Edge is another important visual clue for object recognition. Since edge detection has been one of the most active subjects for researchers in image processing and understanding [6, 7] several results can be used to define edge patterns.

Compared to color patterns edge patterns present much more variations in their features. Fig. 3 shows some examples of corner patterns for a desk.

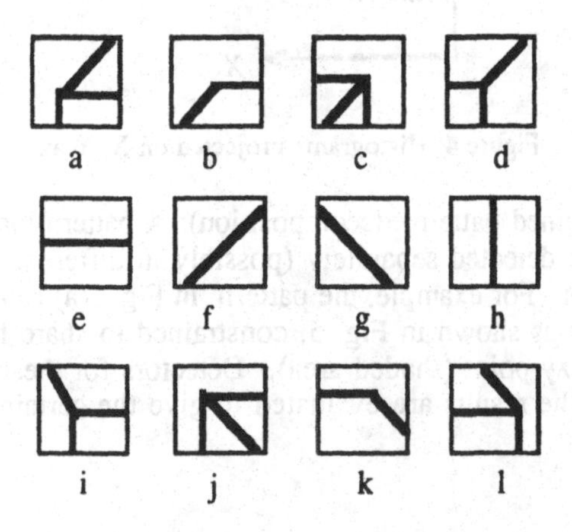

Figure 3 Edge patterns for desk

Using techniques developed for edge detection (e.g. the Sobel operator) in a 16x16 region of image P, we have a 16x16 matrix E whose elements represent the degrees that the pixel is on an edge.

$$\left(P_{ij}\right)_{16x16} \xrightarrow{\text{Sobel filter}} \left(E_{ij}\right)_{16x16}$$

The basic form of an edge pattern is a matrix E which can be thought of as a membership function for a two-dimensional fuzzy set.

To look for an edge pattern in an image, we use the same filter on the image and get the fuzzy set for the image data. The result is the matching degree between the fuzzy set obtained from data and the fuzzy set representing pattern.

Procedures need to be developed to detect pattern features. Every procedure works like an agent. Examples of such procedures are as follows:

Example 1 (histogram): A histogram of the edge matrix E projected on X or Y axis can be used to define some edge patterns. For example, the edge pattern (e) in Fig. 3 can be denoted by two fuzzy sets as shown in Fig. 4. Actually, to detect this pattern, only the changes in the Y direction need to be detected. Because, in some sense, pattern features are very similar to character patterns techniques developed for character recognition can be very helpful for the creation of pattern detection procedures.

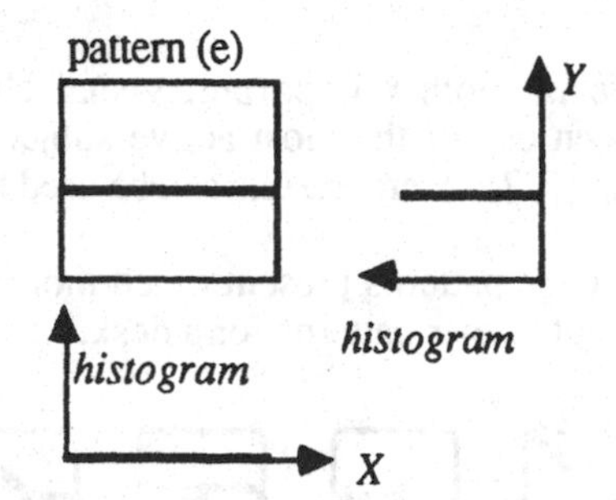

Figure 4 Histograms projected on X, Y axis

Example 2 (constrained pattern decomposition): A pattern may be decomposed into sub-patterns to be detected separately (possibly in different ways) but which must satisfy a constraint. For example, the pattern in Fig. 3(a) can be decomposed as the three sub-patterns as shown in Fig. 5. constrained to share the point O, which we represent as a fuzzy point (shaded area). Detectors for these sub-patterns are used respectively and the results are evaluated to give the certainty on the existence of pattern (a).

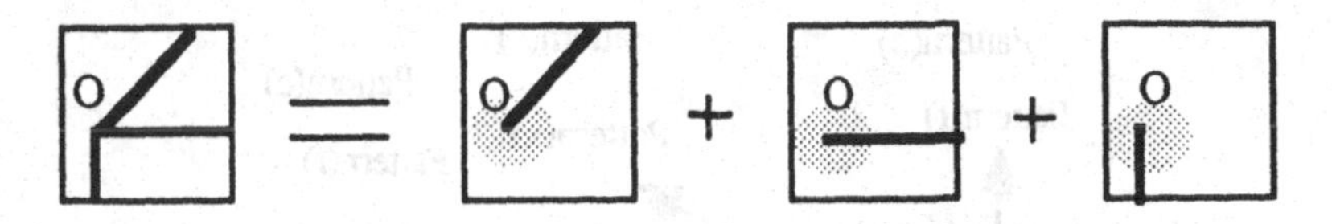

Figure 5 Constrained pattern decomposition

4. Object Recognition

The single most important requirement for model based object recognition is the compatibility between the object model and the information extracted from the image (i.e. the results of the image processing).

4.1 Pattern Based Object Model

Using pattern features, an object is defined as a group of patterns and relations among the patterns [11]. That is,

$$\text{Obj} = (\boldsymbol{P}, \boldsymbol{R})$$

where $\boldsymbol{P} = \{ P_k, k=1,...n \}$ is the collection of patterns and $\boldsymbol{R} = \{R(P_i, P_j) \mid P_i, P_j \in P\}$ is the collection of spatial relations between the patterns in $\boldsymbol{P}$.

Fig. 6 shows some pattern features of a desk, whose edge patterns have been shown in Fig. 3. Pattern (x) is a color pattern, not an edge pattern.

Fig. 7 gives part of a model of a desk top based on pattern features. A desk top is given by some patterns and also the relations among them. In the figure, for simplicity, only the relations starting from pattern(a) is shown (in any case not all the possible relations between patterns need to be specified).

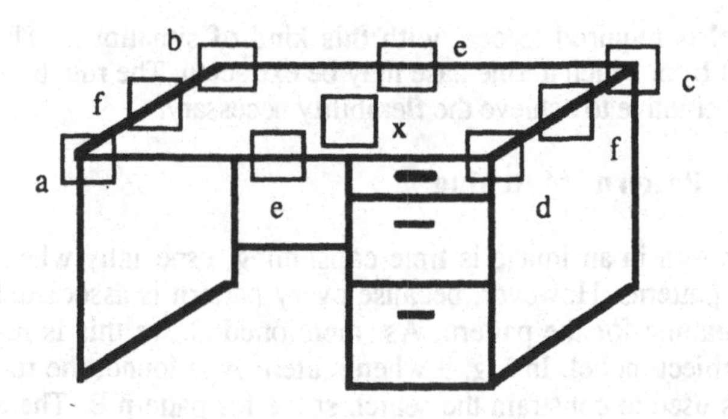

Figure 6 Pattern features of a desk

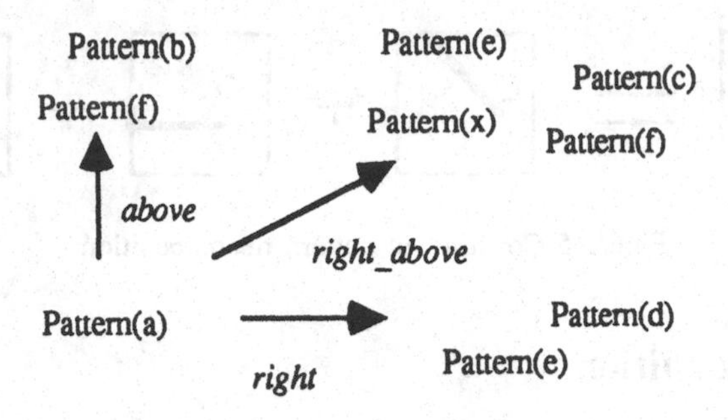

Figure 7 Object model based on patterns

Since the pattern features based model is not precise and due to the imprecision which may result from the operations on fuzzy sets the recognition result based on this approach will be rough. However, it will also be robust to noise. Sometimes, because of occlusions, not all patterns are available in a complex scene. For example, in Fig. 8, we will never find the pattern from Fig. 6(b).

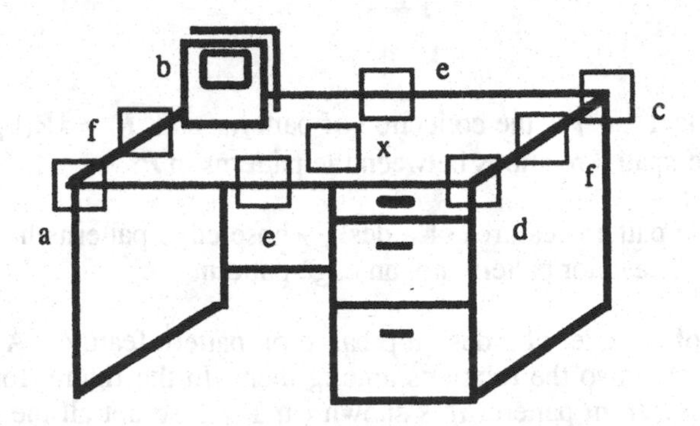

Figure 8 Unavailability of a pattern due to occlusion

The object model is required to deal with this kind of situations. The model may be derived as in [11] from which a rule base may be extracted. The rule based representation may be a good alternative to achieve the flexibility necessary.

4.2 Intelligent Pattern Matching

Looking for a pattern in an image is time consuming, especially when we have a very large number of patterns. However, because every pattern is associated with an object, there is some meaning for the pattern. As mentioned above this is represented by the relations in the object model. In Fig. 9 when pattern A is found, the relation between A and B, R(A,B) is used to constrain the search space for pattern B. The area specified by R(A,B) is a fuzzy area because the relation R(A,B) itself is a fuzzy relation.

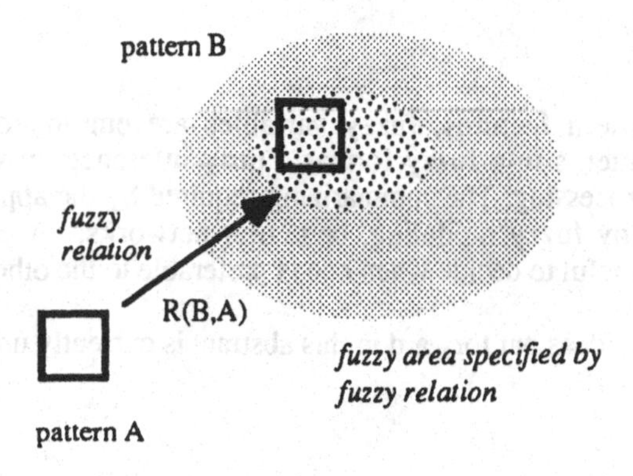

Figure 9 Spatial relation constraint on the search space

The search should be started from the area with highest grade satisfying the relation. In other words, we look for pattern B from the α–cut area with the biggest α value.

4.3 Rules for Object Recognition

Using a pattern features based object model, rules can be generated for object recognition. These rules can be combined according to an interpretation of the model (conjunctive, disjunctive, aggregation [12] etc.)

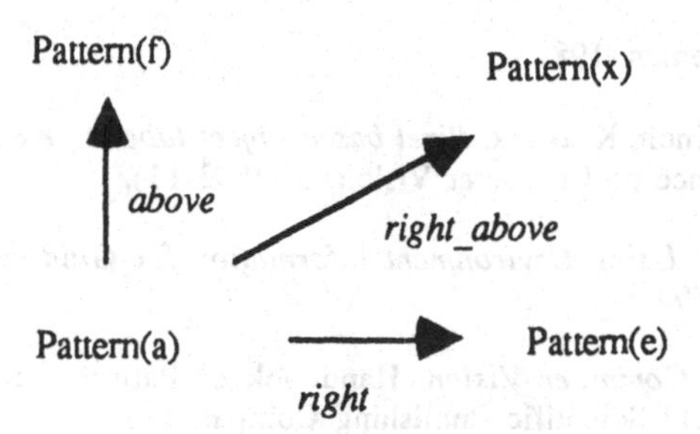

Figure 10 A simplified object model

For example, from the object model in Fig. 10, we have

R1: IF X is Pattern(a) and Y is Pattern(f) and R(Y, X) is *above* THEN Object is desk

R2: IF X is Pattern(a) and Y is Pattern(x) and R(Y, X) is *right_above* THEN Object is desk

R3: IF X is Pattern(a) and Y is Pattern(e) and R(Y, X) is *right* THEN Object is desk

5. Conclusion

To define image pattern features, fuzzy variables are employed. It is effective to manipulate image patterns in a fuzzy format. Fuzzy inference provides a platform for flexible knowledge processing. The summarizing required by the approach proposed here may be done either by fuzzy modeling, or neural networks. A comparison of these techniques would be useful to decide when one is preferable to the other.

A system based on the ideas put forward in this abstract is currently under development.

6. References

[1] M. Sugeno, W. Zhang: *An approach to scene understanding using fuzzy case based reasoning*, Proc. Sino-Japan Joint Meeting on Fuzzy Sets and Systems, 1990.

[2] W. Zhang, M. Sugeno: *A Fuzzy Approach to Scene Understanding*, Proc. FUZZ-IEEE'93, 564-569, San Francisco, Mar. 28-Apr. 1, 1993.

[3] R. A. Brooks: *Intelligence without Representation*, Artificial Intelligence 47, 139/159, 1991.

[4] L. A. Zadeh: *Fuzzy logic and soft computing*, LIFE lecture, February 1994.

[5] Y. Shirai: *Three-Dimensional Computer Vision*, Springer-Verlag, 1987.

[6] D. Marr: Vision, Freeman, 1982.

[7] M. Mukunoki, M. Minoh, K. Ikeda: *Pixel based object labeling method for out-door scenes*, Proc. Asian Conference on Computer Vision'93 (1993-11).

[8] J. Gasos, A. Ralescu: *Using Environment Information for Guiding Object Recognition*. to appear in Proc. IPMU'94.

[9] Q. Luong: *Color in Computer Vision*, Handbook of Pattern Recognition and Computer Vision, 311/368, World Scientific Publishing Company.

[10] F. Perez and C. Koch: *Toward Color Image Segmentation in Analog VLSI: Algorithm and Hardware*, Int'l J. Computer Vision, 12(1), 17/42 (1994).

[11] A. Ralescu and J. F. Baldwin: *Concept Learning from Examples with Applications to a Vision Learning System*, The Third Alvey Vision Conference, Cambridge England, September 15-17 1987.

[12] K. Miyajima, T. Norita, A. Ralescu: *Management of Uncertainty in Top-down, Fuzzy logic-based Image Understanding of Natural Objects*, LIFE TR-3E0011-E, 1992.

Key-picture Selection for the Analysis of Visual Speech with Fuzzy Methods

Hans H. Bothe and Nicolai von Bötticher

Department of Electronics, Technical University Berlin
Einsteinufer 17, D-10587 Berlin, Germany
hans@ife-lipps.ee.tu-berlin.de

Abstract. The goal of the described work is to model visual articulation movements of prototypic speakers with respect to custom-made text. A language-wide extension of the motion model leads to a visible speech synthesis and further more to an artificial computer trainer for speechreading. The developed model is based on a set of specific video key-pictures and the interpolation of interim pictures. The key-picture selection is realized by a fuzzy-c-means classification algorithm.

1 Introduction

The movements of the speech organs are structurally interrelated within the spoken context. The speech organs produce sound in the course of a fully overlapping phonal coarticulation. The projection of these movements on the speakers face may be seen as a visual speech which mostly contains sufficient information to enable hearing impaired persons to speechread a spoken text. The visual recognition is largely focussed on the speakers mouth region, especially on the lips. Since lip movements contain the larger part of the visually perceptible information, this paper proposes the modeling of face movements with the help of related lip shapes and the distance chin-nose only.

A realistic appraisal of the research effort leads to necessary limitations due to the high number of influencing factors (e.g. facial physiognomy, dialect, speed of delivery, sentence and word stress). On one hand, the word material on which the movement analysis is based has to be fixed on a representative subset of the existing phonetic sequences. Thus, the developed motion model is an extension of this subset. On the other hand, the investigation is limited to prototypic speakers.

The subject of modeling visual speech and coarticulation effects has been addressed by several authors [1-4] for different languages. There are two principle classes of models that transform a given text into corresponding facial movements: The movements are either related to an interpolation between selected key-pictures out of a specific codebook or directly controlled by certain visual features. In the first case, the key-pictures are usually determined within a video film by a classification process which is based on visual features. In the second case a set of equations or rules has to be fixed depicting the course of the trait pattern. The parameter driven models are usually related to Parke's 3D-face-model [5] and implemented on specific animation computers.

In this paper a method is described that allows an easy implementation of visible speech synthesis on PCs by using a codebook of key-pictures; only low-cost computers make possible to use the animation system in schools and rehabilitation centers for hearing-impaired people.

2 Feature Extraction

The speech signal and movement data were recorded on videotape. Those pictures which fit best with the subjective impressions of different experts in speechreading for good articulation of each sound were interactively indicated with the help of both the acoustic and visual material. The acoustic phone boundaries - determined with the help of oscillogram, sonagram and playback - indicate the scanning range of each wanted picture.

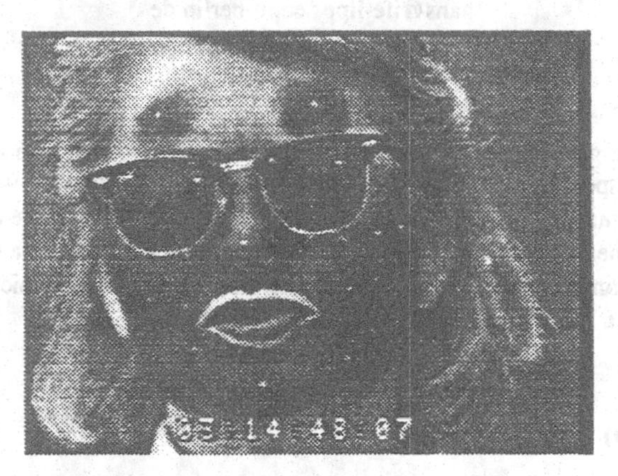

Fig. 1. Set points and contours in the speakers face

In order to allow a repeat examination resulting in the same data several points on nose and forehead, as well as the lip contours, were marked with a contrasting color. A typical picture is shown in figure 1. The set points on nose and forehead determine the face coordinate system. Together with the proposed contours they were localized with the help of an automatic contrast search program based on an active contour algorithms [6].

Experts in speechreading determined certain key-pictures which correlate each with the optimum articulation positions for each spoken phoneme [7]. These key-pictures were classified on base of a set of representative visual features. A typical example using inner and outer lip contour and the distance chin-nose is shown in figure 2. These primary features may be used for the composition of more complicated secondary features as described in [8].

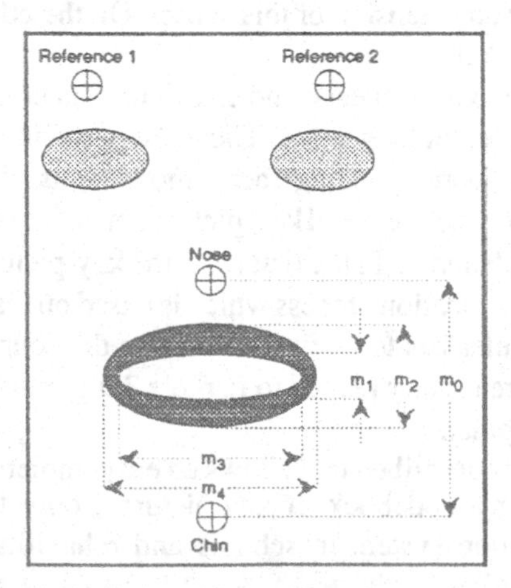

Fig. 2. Exemplary extraction of primary visual features in the speaker's face

3 Key-picture Selection

The visual feature vectors of the selected pictures create a distribution in the feature space. By clustering this space, representative key-pictures can be calculated with the help of a simple classification algorithm: the centroids of the clusters are taken as prototype representatives and point out the key-vectors by a nearest-neighbor method. The related grey-scale pictures are key-pictures. The clustering technique depends on the underlying animation model.

3.1 Diphone Related Key-picture Selection Based on Visemes

Visemes are the smallest speaker independent visual units of a language. For instance, the phonemes /b, p, m/ belong to the same viseme (B) since they are visually not distinguishable in fluent speech [9,10].

A first order approximation for modeling backward and forward coarticulation effects is gained by grouping the correlating key-pictures with respect to the viseme structure and takes into account the next neighboring phonemes. The phonematic text is split into overlapping diphones, and diphthongs are represented by two closely connected single phones, whereas the phoneme-related picture of the second are classified with respect to the first one. This process leads to a deterministic diphone related phoneme-to-key-picture mapping. In order to reduce the large amount of resulting pictures (41×41 according to the phonematic structure of the German language) , the phonemes were grouped with respect to the viseme structure. Assuming 14 visemes, this procedure leads to 13×13=169 key-pictures.

In the later animation those feature vectors closest to the resulting cluster centers are taken to depict the set of key-pictures of the investigated articulation movements.

3.2 Key-picture Selection by Fuzzy-c-means Algorithm

The developed diphone model had been implemented as a PC animation and evaluated with respect to the naturalness of the generated movies. This system produces artefacts with certain phonetic structures which may be related mainly to coarticulation effects extending beyond the immediate next neighboring sounds. Besides integrating syllable structures as proposed in [11], a further step is to classify the key-pictures without a previous grouping. In the later visual speech synthesis an artificial neural network is trained to select the pictures with respect to the phoneme sequence, since in this case there is no a priori correlation between phonemes and key-pictures. In order to generate the movie, interim pictures between the selected key-pictures are calculated by an interpolation algorithm.

The feature vectors have been classified by means of the fuzzy c-means algorithm as described in [12, 13]. The algorithm generates optimum location of the clusters automatically with respect to a given number of clusters. The representatives of the clusters are again taken as key-pictures.

The motion model consists of a multi-layer neural network; it selects the specific key-pictures with respect to the surrounding next 3+3 neighbor phonemes.

The fuzzy c-means clustering algorithm does iteratively calculate the cluster centers and with them the new membership grades of the objects. It may be processed mainly in the following four steps, with $\underline{m}$ being a feature vector (object) and N the number of objects:

Step 1: Choose

- the desired number of clusters n with $2 \leq n \leq N$,
- a distance measure as, for instance, the Euclidian distance,
- a suitable contrast parameter q,
- the initial condition of the membership matrix $\underline{U}^{(0)} = (u_{ij})^{(0)}$ of the objects $\underline{m}_i$ and the clusters Q_j with

$$\Sigma_{j=1,\ldots,n}\, u_{ij} = 1, \quad 0 < \Sigma_{j=1,\ldots,N}\, u_{ij} < N .$$

Step 2: Let s be the actual iteration step. Adjust the n cluster centers $\underline{v}_j^{(s)}$ with

$$\underline{v}_j^{(s)} = \frac{\Sigma_{i=1,\ldots,N} (u_{ij}^{(s)})^q * \underline{m}_i}{\Sigma_{i=1,\ldots,N} (u_{ij}^{(s)})^q} .$$

Step 3: Be $\underline{U}^{(s+1)} = (u_{ij}^{(s+1)})$. With

$$I_i = \{ j \in \{1, \ldots, n\} \mid d(\underline{m}_i, \underline{v}_j) = 0 \},$$

$$I_i^C = \{1, \ldots, n\} \setminus I_i ,$$

calculate the membership grades for the new cluster centers with case $I_i = 0$:

$$u_{ij}^{(s+1)} = \frac{1}{\Sigma_{k=1,\ldots,n}\, d(\underline{m}_i, \underline{v}_j)^{(s)} / d(\underline{m}_i, \underline{v}_k)^{(s)}}$$

case $I_i \neq 0$:

$$u_{ij}^{(s+1)} = 0 \quad \text{for all} \quad j \in I_i^C .$$

Normalize the membership grades with

$$\Sigma_{j \in I_i} (u_{ij}^{(s+1)}) = 1.$$

Step 4: Let $\|.\|$ be a matrix norm matching d_{ij} (e.g., Euclidian norm) and $\varepsilon > 0$. Stop iteration, if

$$\| \underline{U}^{(s+1)} - \underline{U}^{(s)} \| \leq \varepsilon .$$

The fuzzy c-means algorithm results in a set of n key-pictures for each individual speaker.

Given n =15, 28 and p = 2.0, the projection of the distribution of the cluster centers on the features height over width of the inner lip contour is shown in figure 3 by the centers of the circuits. The radius of which is a measure of the amount of objects in that cluster. Figure 3 shows the resulting clusters of one individual person.

The motion model has again been implemented on a PC. It is still in the state of experimentation. The necessary interim pictures are calculated by a morphing algorithm. For this purpose, a fixed amount of set points were interactively placed in each key-picture serving as a framework for the triangulation. These set points are related to specific physiological points in the speakers face. The interim pictures are placed in equidistant time intervals.

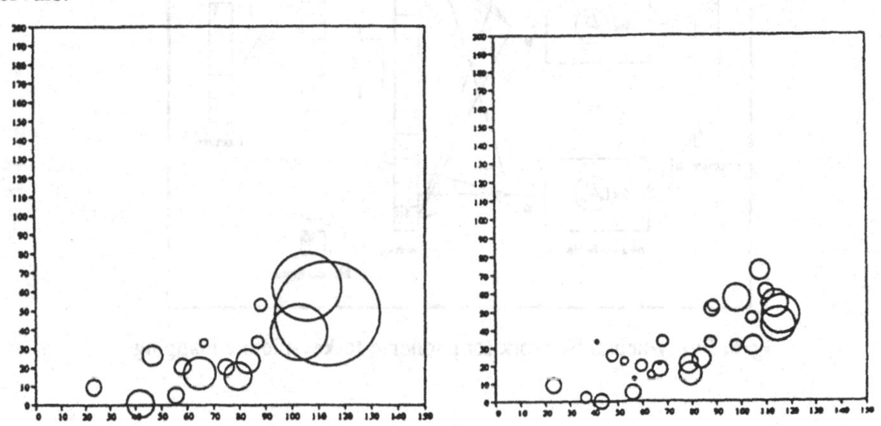

Fig. 3. Cluster distribution and object frequency for n=15, 28, p=2.0, height over width of the inner lip contour in screen pixel

4 Text-to-Key-picture Mapping

Several different mapping algorithms have been developed. In the case of diphone clustering, the key-picture selection for a given text is realized with the help of a 2D look-up table with pointers on the key-picture number as shown in figure 4.

	B	F	D	L	...	--> 1st viseme
B	kp_{BB}	kp_{FB}	kp_{DB}	...		
F	kp_{BF}	kp_{FF}	...			
D	kp_{BD}	...				
L	...					
...						
2nd viseme						

Fig. 4. Viseme related diphone text-to-key-picture mapping

In the case of automatic clustering, the selection algorithm is either based on a fuzzy neural network (FNN) [14] as shown in figure 5, or on a fuzzy inference engine (FIE) with IF...THEN...-selection rules. In both cases, coarticulation effects are considered by the use of a 3+1+3 phoneme selection frame for the mapping.

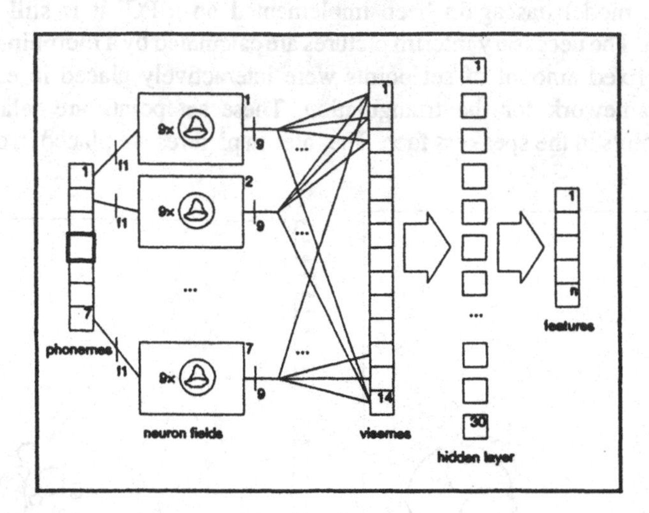

Fig. 5. Fuzzy-neural Network for phoneme-to-key-picture mapping

The input coding of the FNN can be realized with the help of Kohonen's self-organizing feature map The production rules of the FIE are generated with the help of a genetic algorithm (GA), using mutation and recombination processes. The GA does also optimize the membership functions related to the visual features.

5 Facial Animation System

The later facial animation with phonematic text input uses the depicted key-pictures as a framework for the film generation. The time control is based on averaged key-picture distances and phoneme-related standing times of the single frames of the film.

The developed articulatory models were presented in schools or centers for deaf or hard-of-hearing children as a PC animation system. Several investigations with different motion models show that, as an assistive aid, it can be of high value for improving the children's ability to lip-read.

References

1. M. Cohen and D.W. Massaro, Modeling Coarticulation in Synthetic Visual Speech. In: N.M. Thalmann and D. Thalmann (Eds.), Computer Animation '93, 1993.

2. H.H. Bothe, G. Lindner and F. Rieger; The development of a computer animation program for the teaching of lipreading, In: E. Ballabio, I. Placencia-Porrero and R. Puig de la Bellacasa (Eds.), Technology and Informatics 9, Rehabilitation Technology: Strategies for the European Union, Amsterdam, (1993), 45-49.

3. M. Saintourens, M.H. Tramus, H. Huitric, and M. Nahas: Creation of a synthetic face speaking in real time with a synthetic voice, Proceedings of the Workshop of Speech Synthesis, Autrance, (1990), 381-393.

4. D. Storey and M. Roberts: Reading the speech of digital lips: motives and methods for audio-visual speech synthesis, Visible Language 22 (1989), 112-127.

5. F.I. Parke: Parametrized Model of Facial Animation, IEEE Computer Graphics and Applications 2, 61-68.

6. M. Kass, A. Witkin and D. Terzopoulos: Snakes: Active contour models, Int. J. Comput. Vision 1 (1987) 4, 321-331.

7. H.H. Bothe and F. Rieger: Zusammenhang zwischen visuellen und akustischen Korrelaten lautlicher Artikulationsprozesse, Proceedings of DAGA 94, Dresden (1994).

8. H.H. Bothe and P. Gutjahr: Optimization of a fuzzy classification of closed curves by a Genetic Algorithm. Proceedings of the Sixth International Fuzzy Systems Association World Congress (IFSA), Sao Paulo, July 22-28, (1995).

9. G. Alich: Zur Erkennbarkeit von Sprachgestalten beim Ablesen vom Munde. (Dissertation) Bonn, (1961).

10. Owens, E. and B. Blazek: Visemes Observed by Hearing-impaired and Normal-hearing Adult Viewers. Journal of Speech and Hearing Research 28 (1985), 381-393.

11. H.H. Bothe, F. Rieger and R. Tackmann: Visual coarticulation effects in syllable environment, Proceedings of the EUROSPEECH, Berlin, (1993), 1741-1744.

12. J.C. Bezdek: Pattern recognition with objective function algorithms, London, 1981.

13. H.H. Bothe: Fuzzy Logic - Einführung in Theorie und Anwendungen, (2. Ed.), Berlin, 1995.

14. H.H. Bothe: Artificial Visual Speech, Synchronized with a Speech Synthesis System. In: W.L. Zagler, G. Busby and R.R. Wagner (Eds.): Lecture Notes in Computer Science, Vol. 860, Springer-Verlag, (1994).

Effective Scheduling of Tasks Under Weak Temporal Interval Constraints

Frank D. Anger Rita V. Rodriguez
Computer Science Department, University of W. Florida
Pensacola, FL 32514 USA {fdang,rrodrigu}@dcs106.dcsnod.uwf.edu

Abstract

Numerous AI planning applications and real-time system scheduling problems do not fit the traditional scenarios of the scheduling literature; instead, they are better expressed in terms of the temporal interval relations between the tasks. Given a set of tasks and a set of constraints expressed in terms of the atomic temporal interval relations, the problem of finding the shortest consistent schedule often arises. In the most general situation, the interval constraints leave some degree of uncertainty: the problem is under-specified. It is first shown herein that, in the completely specified case, the greatest lower bound of all schedule lengths can be calculated as the "*Size*" of a chain of intervals playing a role similar to that of a critical path in the familiar critical path analysis. Subsequently, a heuristic search algorithm is presented to reduce the general under-determined case to a completely specified one.

Key Words: *Constraint Propagation, Planning, Real-Time Systems, Scheduling, Temporal Interval Relations, Temporal Reasoning.*

1 Introduction

The performance of a collection of tasks of known duration often gives rise to the problem of determining the minimal schedule length for the complete set of tasks [7]. A variant problem surfaces as well: determining if all the tasks can be completed in a given time T. Although such situations have been considered in the literature when the constraints are in the form of limited resources, precedence relations, or both [14,13,2], the problems are not usually addressed for the case of general temporal relation constraints. If we know, for example, that task A is *before* task B, B is *during* C, A *overlaps* C, and C *overlaps* D and the lengths of the tasks are known, how can the minimum schedule length be determined? The succeeding sections answer the question, when the temporal relation is known between every pair of intervals, in terms of an easily calculated quantity termed the "*Size*" of a chain of intervals. (To our knowledge, the problem of optimal scheduling under interval constraints has not previously been addressed and the concept of critical chains is original.) An effective heuristic algorithm is then suggested for the case when there is uncertainty about some or all of the temporal relations. An unusual feature of the class of problems discussed stems from relations such as *overlaps* or *before* which do not allow for a *minimal* schedule since the greatest lower bound of feasible schedules is strictly less than any particular feasible schedule.

Although the completely specified problem can always be converted to a problem about a labeled, directed graph of (*start* and *finish*) endpoints of the given set of tasks, the attractive *longest-path algorithm* does not uncover the shortest schedule length. Moreover, to extend the results to the case of incomplete information, the problem must be discussed directly in terms of temporal interval relations. In the general case the network of such intervals and temporal relations is *non-pointizable*: not reducible to a network of temporal point relations [21].

The study of qualitative temporal relations between activities which take place over intervals of time has flourished in AI over the past decade [1,11,15,20]. Whereas most of the work considers time as *linear*, representable on a number line, other more exotic temporal models have also been investigated ([3,4,16,17]) for applications to planning and distributed systems. Constraint propagation methods advance the approaches to temporal representation enabling the automation of the reasoning process about possible ordering of events [1,21], with recent encouraging results [12,19] increasing the attractiveness of such reasoning methods. Methods for solving corresponding quantitative challenges include [8,2,9] but are less extensive. For more traditional scheduling problems, a host of classification and complexity of results have been published ([13,14,18]).

Section 2 sets forth the background for the rest of the paper, including the assumptions and definitions related to the temporal interval relations in linear time; then establishes a few basic properties of schedules of intervals under relational constraints and presents the two main results which establish an equivalence between minimal schedule length and maximal critical chain size. Section 3 presents and discusses a heuristic algorithm to find a "good" schedule in the under-determined case in which constraints are not always "atomic." Section 4 concludes with a summary and further suggestions.

2 Definitions and the Fully-Determined Case

Throughout the remainder of the paper certain assumptions and definitions will remain in effect leading up to the concept of a *critical chain* and its *Size*. The given assumptions lead to the major result proved in the paper.

Assumptions: Given a collection S of tasks together with the *length* $|I|$ of each task (or *interval*) I and, for any pair I, J in S, the exact (one of 13) interval relation between them. These interval relations, $I\ r\ J$, are denoted by (with $r^{\smile}$ denoting the *converse* of r)

before (<), *meets* (m), *overlaps* (o), *starts* (s), *during* (d), *finishes* (f), *equals* (=), *after* (>), *met-by* ($m^{\smile}$), *overlapped-by* ($o^{\smile}$), *started-by* ($s^{\smile}$), *contains* (cn), *finished-by* ($f^{\smile}$),

Result: The greatest lower bound (*glb*) of all feasible schedules for the tasks of S is given by the ***maximum Size of all critical chains in*** S, or, equivalently, by the *Size* ***of any complete critical chain in*** S.

Definitions:

1. Sched(S) = *glb* of the lengths of all feasible schedules for S.

In order to achieve as short a schedule as possible, the scheduler should attempt to start each task as early as possible, although only certain *critical* tasks actually affect the total time. Such a critical task I may be constrained by its relation to another critical task from starting at time 0: we say that the other task ***supports*** I. More precisely:

2. I ***ε-supports*** J in a particular schedule, written I *ε-supp* J, iff $0 \leq J_b - I_a < \epsilon$, where a and b are in {start,finish} (or $\{s, f\}$). (By obvious convention we confuse here I_a with the time at which I_a is scheduled in the selected schedule.)
3. Given S, define $S^* = S \cup \{PRE, POST\}$, where PRE and $POST$ are two new intervals of length 0. Add the temporal relations:
   - $PRE\ m\ I$ for all I in S which have no other intervals starting before I,
   - $PRE < I$ for all other intervals in S
   - $POST > I$ for all intervals in S.

   By convention, all schedules must schedule PRE at time 0. It is clear that Sched(S^*) = Sched(S).
4. I_a ***supports*** J_b in S^*, written I_a *supp* J_b, (or just I *supp* J) is defined inductively as follows:
   - If GLB(J) = 0, then PRE_a *supp* J_s for both $a = s$ and $a = f$ (or just "PRE *supp* J").
   - I_a *supp* J_b if there exists a (possibly empty) sequence of intervals $A_1, A_2, \ldots, A_{n-1}$ such that PRE *supp* A_1 *supp* A_2 ... *supp* A_{n-1} *supp* I, and, for all $\epsilon > 0$, any schedule which starts J within ϵ of GLB(J) must satisfy I_a *ε-supp* J_b. We usually simply write I *sup* J.
   - The only I *supp* J relations are those obtained by the preceding two rules.

   It is easily established that *any* schedule starting J within ϵ of GLB(J) starts all the A_i (and I) within ϵ of their respective GLB's. Furthermore, PRE *ε-supp* A_1 *ε-supp* A_2... *ε-supp* A_{n-1} *ε-supp* I in all such schedules. Intuitively, if I supports J, then scheduling I later forces J to be scheduled later than GLB(J); or I blocks J from being scheduled earlier than GLB(J).

The inductive definition of *supp* is not vacuous because there must always be some task or tasks with GLB(J) = 0.

Unfortunately, *supp* has some unpleasant aspects. It is not transitive, symmetric, or anti-symmetric; nonetheless, *supp* is reflexive, while I *supp* J *supp* K *supp* I is possible with distinct I, J, and K (for example, if $PRE\ m\ I\ m\ J\ f\ K\ si\ I$).

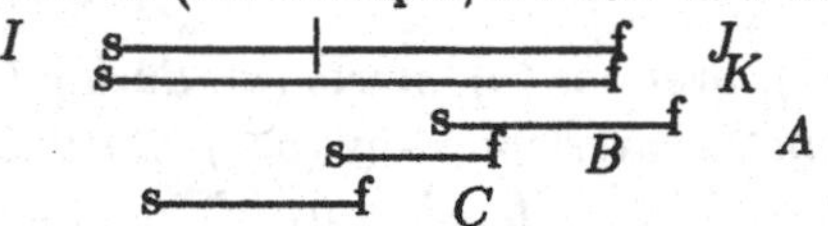

The picture also illustrates that there can be distinct chains supporting the same interval, A, and that the only chain supporting a task C may contain intervals that can be scheduled later than C.

To define the concept of a *critical chain*, in analogy with the *critical path* of PERT charts, it might at first appear sufficient to adopt chains of tasks, each of which supports the next; however, in the example pictured, with $|A| = 1$, $|B| = 2$, $|C| = 3$, and $|D| = 1$,

A B s f
C s f
s f s f D

even though A *supp* B *supp* C *supp* D, A and B are not critical to the schedule, as C can be started as close to 0 as desired. This kind of redundancy, as well as that of repeated intervals in the chain, is eliminated through the introduction of the *level* of a task.

5. For all I in S^*, define the *"level"* of I, *lev*(I), as:
   - *lev*$(PRE) = 0$
   - *lev*$(I) = n > 0$ iff there exists J of level $(n-1)$ such that J *supp* I, $J \neq I$, and I is not supported by any interval of level $< (n-1)$. (The condition $J \neq I$ is only needed to assure that PRE is not assigned level 1 as well as level 0.)
6. A ***critical chain*** in S is a sequence $A_0 r_1 A_1 r_2 A_2 \ldots r_n A_n$ of intervals and relations in S such that A_i *supp* A_{i+1} and *lev*$(A_{i+1}) =$ *lev*$(A_i) + 1$ for $i = 0, 1, \ldots, n-1$.
7. A critical chain is *complete* if $A_0 = PRE$ and $A_n = POST$. In this case, *lev*$(A_i) = i$ for all i.

At this point we are tempted to define the *length* of a critical chain by adding up the lengths of its tasks. Unfortunately, some links in the chain actually "go backwards," as illustrated below:

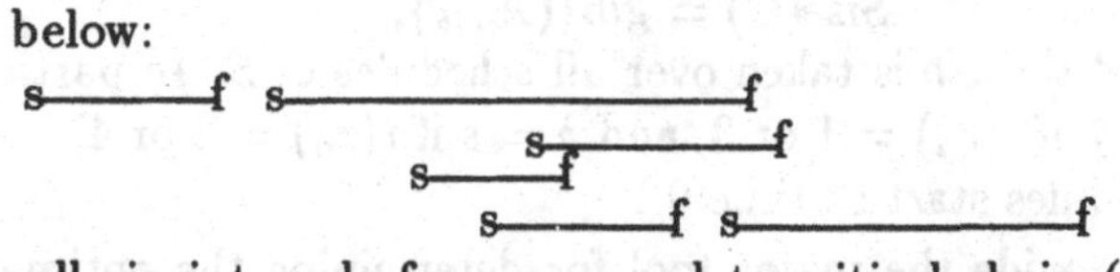

In the figure, all six intervals form a complete critical chain (removal of any one shortens GLB(S)), yet the third and fourth tasks' lengths are *subtracted* from the sum of the remaining lengths to obtain GLB(S). Whether the next task in the chain progresses forward or backward depends on which pair of endpoints inhibit this next task from being moved to an earlier start time, giving rise to five possibilities.

8. The interval temporal relations, $I\ r\ J$, are divided into 5 *types*, $t(r)$, according to the way in which I could support J. These types are given in the Table 1.

Any of the first four types could be present when I *supp* J, but none implies that I *supp* J. In the case of *equals* and *overlaps* with $|I| = |J|$, the type can be both 2 and 3 when I *supports* J at both endpoints; however, this type of support cannot appear in a critical chain, since the two intervals so related always have the same level.

9. The *Size* of a critical chain $C = A_0 r_1 A_1 r_2 A_2 \ldots r_n A_n$ with $A_0 = PRE$ is given by:

Table 1: **Possible Support Types**

TYPE	RELATIONS	POSSIBLE SUPP	EXAMPLE
1	*b, m*	$I_f supp J_s$	
2	*d, f, fi, =, o*	$I_f supp J_f$	
3	*di, s, si, =, o*	$I_s supp J_s$	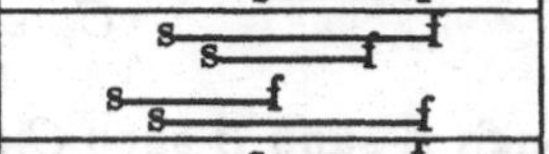
4	*oi, mi*	$I_s supp J_f$	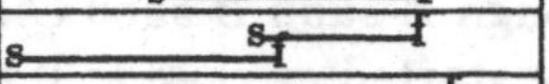
5	*bi*	*I cannot supp J*	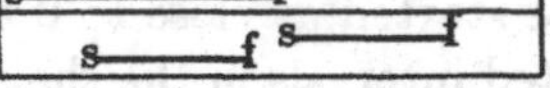

- $Size(C) = 0$ if $C = A_0$,
- $Size(C) = |A_1|$ if $n = 1$, and
- $Size(C) = Size(A_0 r_1 A_1 ... r_{n-1} A_{n-1}) + D * |A_n|$ if $n > 1$,

where
$$D = \begin{cases} +1 & \text{if } t(r_n) = 1 \text{ or } 3 \\ -1 & \text{if } t(r_n) = 2 \text{ or } 4. \end{cases}$$

Main Results: With the definition of a *critical chain* and its *Size* in hand we proceed in Proposition 1 to establish the relationship between the *Size* of a critical chain and length of a schedule of the tasks in the chain. The development culminates in Theorems 1 and 2 establishing complete critical chains as the determinants of Sched(S). Proofs of all results can be found in [5] or [6]. (The latter presents the case of completely determined relations, the proofs being quite involved. In this paper we extend, in Section 3, to the under-determined case.)

Proposition 1: If $C = A_0 r_1 A_1 r_2 ... r_n A_n$ is a critical chain in S with $A_0 = PRE$, then
$$Size(C) = glb\{(A_n)_b\},$$
where b in $\{s, f\}$ and the *glb* is taken over all schedules of S. In particular,
$$b = f \text{ if } t(r_n) = 1 \text{ or } 3, \text{ and } b = s \text{ if } t(r_n) = 2 \text{ or } 4.$$
(Recall that all schedules start at time 0.)

Theorems 1 and 2 provide the major tool for determining the optimal schedule in the presence of temporal interval constraints. The former reveals that if a *complete critical chain* can be found, its *Size* renders the greatest lower bound (*glb*) of schedules for a set of tasks, S, while the latter guarantees the existence of a complete critical chain. In Section 3, methods will be discussed for exploiting these results even if the constraints are incompletely specified.

Theorem 1: If C is any complete critical chain in S, then Sched(S) = $Size(C)$. (Recall that Sched(S) is the *glb* of all schedule lengths for S.)

Proof: A complete critical chain C starts with PRE and ends with $POST$, and in particular has $t(r_n) = 1$; hence, by Proposition 1, $Size(C) = glb\{POST_s\} = glb\{(A_{n-1})_f\} =$ Sched(S). (Or, $Size(C) = glb\{POST_f\} =$ Sched(S^*) = Sched(S). □

It remains to show that there always exists a complete critical chain, from which it follows that we can always obtain Sched(S) by computing *Size*(C) for all critical chains and choosing the maximum, or by finding a complete critical chain and computing its *Size*.

Proposition 2: Every interval I in S has a supporting chain

$$PRE\ supp\ A_1\ supp\ A_2\ supp \ldots supp\ A_m = I.$$

It follows from Proposition 2 that the level of I is defined for every I in S: all I have supporting chains, and the level is just the least number of tasks in any of these.

Theorem 2: For any collection S of tasks and relations, there exists a complete critical chain.

3 The Case of General Temporal Constraints

In the foregoing presentation the assumption was made that every pair of tasks was related by a known ***atomic*** interval relation. Often, however, a problem is ***under-determined***: some tasks are not related by any temporal constraint, or they are related by a ***non-atomic*** constraint such as "*disjoint*," "*starts before*," or "$< +m+o$." In such cases, even finding one ***feasible*** schedule is NP-hard, since it is equivalent to solving the ***satisfiability*** problem for the given set of constraints [21]. Nonetheless, if the set of constraints is consistent, one has a reasonable chance of finding a solution by arbitrarily selecting, for a pair of tasks I, J, a possible relation $I\ r\ J$ and applying constraint propagation to the resulting problem to deduce the implied constraints on other tasks. If this leads to no contradictions (no "empty" constraints), repeat the process; if a choice proves inconsistent, backtrack. Experimental results of Ladkin and Reinefeld [12] indicate that modifications of this approach tend to be successful in reasonable (cubic) time in the vast majority of cases.

In our situation, however, we would also like to obtain a near-optimal choice of relations—one that produces a completely determined problem which has a schedule near to the minimal schedule possible for the original under-determined problem. The basic heuristic that applies can be phrased: ***the more the tasks are executed in parallel*** (overlapping) ***the shorter the schedule***. Interval relations such as *before* and *meets* allow no overlapping, while *during* and *finishes* allow maximum overlapping. Secondarily, we would like to commit to relations which allow maximum choice of positioning: for example, I ***starts*** J ties I to J, whereas I ***during*** J permits I to be moved around inside of J to perhaps shorten the schedule. We therefore divide the 13 atomic relations into several categories, or ***priorities***. For the sake of simplicity we consider only relations $I\ r\ J$ with $|I| \leq |J|$, which excludes $s^{\smile}$, $f^{\smile}$, and *cn*.

Priority	*Relations*
1	$d, =$
2	s, f
3	o
4	$o^{\smile}$
5	$<, m, m^{\smile}, >$

The heuristic must order all the given constraints $I\ r\ J$ with $|I| \leq |J|$ in decreasing order by *importance* (to schedule length) of the choice of *atomic* relation between I and J. Writing each such r as $r = r_1 + r_2 + \cdots + r_k$, with each r_i atomic (the + indicating *or*), the "importance" of r can be defined as

$$\max_{i,j}\{|P(r_i) - P(r_j)|\},$$

where P indicates Priority as defined previously. For example, the *importance* of $< +m + o^{\smile}$ is $P(<) - P(o^{\smile}) = 5 - 4 = 1$, whereas that of $d + o$ is $3 - 1 = 2$. We would therefore want to rank $I\ (d + o)\ J$ before $I'\ (< +m + o^{\smile})\ J'$ and would choose $I\ d\ J$. Note that the *importance* of any relation, all of whose atomic disjuncts have the same priority, is 0, indicating that there is no advantage to considering the relation until all other choices have been made.

Examples: Assume a network, $\mathcal{N}$, consisting of a collection S of tasks T_i and non-atomic relations $r_{i,j}$.

1. If NO constraints are given among the tasks to be scheduled, then the best schedule is obtained by performing all tasks in parallel, using thereby relations of priority 1 or 2 indifferently (although priority 1 suffices). The resulting schedule will have

$$\text{Sched}(S) = \max\{|T_i|\}.$$

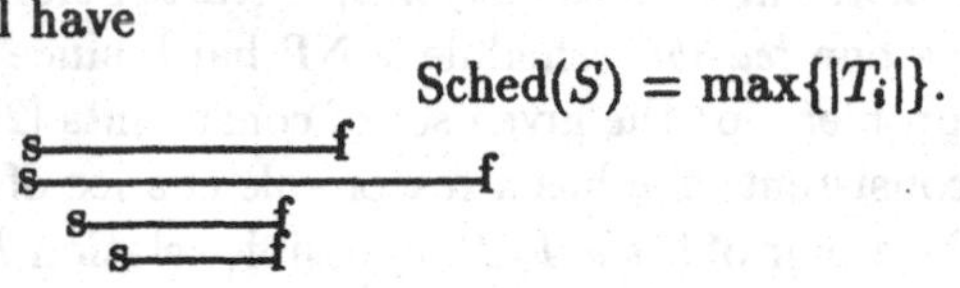

2. If ALL constraints consist of priority 5 disjuncts, then there can be *no* parallelism, and $\text{Sched}(S) = \sum_i |T_i|$.
3. If ALL constraints consist of priority 3 disjuncts *or* ALL of priority 4 disjuncts, then $\text{Sched(S)} = \max\{|T_i|\}$, as all the tasks can be "pushed back" to start (in the priority 3 case) or finish (in the priority 4 case) within any ϵ of the same time.

4. If ALL constraints contain only priority 3 and 4 disjuncts, the optimal schedule length will be somewhere between $= \max\{|T_i|\}$ and $2 \times \max\{|T_i|\}$. Since in any critical chain the first and last tasks must overlap (priority 3 or 4), the chain length cannot exceed the sum of the lengths of two tasks. The latter bound is approached by the case illustrated below with $|T_1| = |T_2| = 2$, $|T_3| = 0.1$, and $T_3\ o^{\smile}\ T_1$, $T_3\ o\ T_2$, and $T_1\ o\ T_2$.

   Notice that, had the constraints all been of the form $o + o^{\smile}$, the heuristic would have consistently chosen o and produced a shorter schedule:

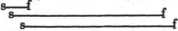

5. An algorithm based solely on ***importance*** and ***priority*** does not always make the best choice of relations, since a good choice for $T_1\ r\ T_2$ may force a worse choice for $T_1\ r\ T_3$. Consider, for example, the collection of tasks A, B, C, and D with $A\ m\ C$, $D\ m\ B$, $A\ (<+d)\ B$, $A\ (>+d)\ D$, but otherwise unconstrained. Assume further that B, C, and D have the same length, but $|A|$ is smaller. In this case the ***importance*** of the completely determined relations $A\ m\ C$ and $D\ m\ B$ is 0 while that of all others is 4 (5 − 1). If the algorithm selects the pair (A, B) first, it will set $A\ d\ B$, (priority 1) which forces $A > D$ and results in

A C
s f
s f
D B

and a schedule length $> |D| + |A| + |C|$. Had the algorithm selected the pair (A, D) first, however, it would have set $A\ d\ D$, resulting in

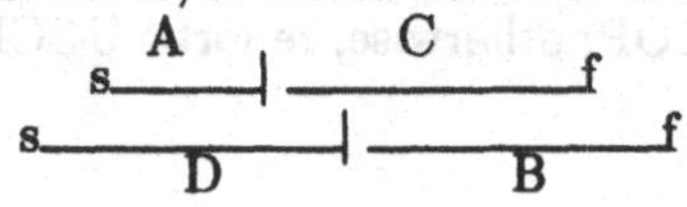

and a schedule length of $|D| + |B| = |D| + |C|$, strictly shorter. This is accomplished by forcing the algorithm to look at the network "from left to right." It first determines which tasks could potentially be scheduled at time 0 and resolves their relationships; ignoring those, it finds the tasks which could be scheduled "next," and so on.

One further notation: to describe the set, M_0, of tasks that could possibly be scheduled at time 0 we say that "*T starts-later* than *T'*," or $T\ sl\ T'$, if $T\ (d + f + o^{\smile} + m^{\smile} + >)\ T'$. The complementary relation, "*T starts no later* than *T'*", or $T\ \neg sl\ T'$, would be the disjunction of the remaining 8 atomic relations. To express the fact that a non-atomic relation $T\ r\ T'$ allows the case $T\ \neg sl\ T'$, we can write "$r\&sl \neq r$."

The algorithm to obtain a near-optimal choice of a completely determined collection of tasks and ***atomic*** relations consistent with the given constraints will utilize the recursive routine Find_Net given below, which finds a compatible choice of atomic relations for a given subset, $\mathcal{M}$, of tasks using the ***importance*** heuristic.

Input Parameters :
A Network, $\mathcal{N}$, of Tasks $T_1, T_2, \ldots, T_n$ and temporal interval relations $T_i\ r_{i,j}\ T_j$ for some or all pairs of tasks.
The lengths, $|T_i|$.
A subset, $\mathcal{M}$, of tasks.

Algorithm Find_Net

1. Run Constraint Propagation on the Network of tasks and relations. If inconsistent, return NO SOLUTION and STOP.

2. Replacing $r_{i,j}$ by $r_{i,j}{}^{\smile}$ if $|T_i| > |T_j|$, calculate the *importance* of each $r_{i,j}$ (as modified by the constraint propagation) for tasks T_i and T_j in $\mathcal{M}$.
3. While some relation $r_{i,j}$ between tasks in $\mathcal{M}$ is non-atomic
   (a) Select a non-atomic relation $r_{i,j}$ of highest *importance* in $\mathcal{M}$.
   (b) Pick one of the atomic disjuncts, r, of $r_{i,j}$ of smallest priority number (highest priority) and Call Find_Net, passing as parameters the network $\mathcal{N}$ with $r_{i,j}$ replaced by r and the same subset $\mathcal{M}$.
   (c) If Find_Net reports success, return network found by Find_Net and STOP.
   (d) Otherwise, replace $r_{i,j}$ by $r_{i,j} - \{r\}$ in the network, recalculating its *importance*.
4. {At this point, all relations in the sub-network generated by $\mathcal{M}$ are atomic.} Run Constraint Propagation on the network of relations. If inconsistent, return NO SOLUTION and STOP; otherwise, report SUCCESS and return the network obtained.

For the main algorithm, Short_Net, we will assume that the data has already been separated into subsets such that there is no path from a task in one subset to a task in another which does not pass through a completely unconstrained pair of tasks. (These subsets are just the *connected components* of the graph on the tasks in which two tasks are called *adjacent* if there is a non-trivial constraint between them.) The algorithm is only applied to each such subnet separately, as the optimal schedule is just the union of optimal schedules for these (unrelated) subproblems. The main algorithm, Short_Net, starts by determining the set $\mathcal{M}$ of all tasks which could possibly be scheduled at time 0:

$$\mathcal{M} = \{T | T \text{a task, and} \,\forall\, T\, r\, T' \; (r \& sl \neq r)\}.$$

Find_Net($\mathcal{N}$, $\mathcal{M}$) is called, and the process is repeated finding the tasks that start later than no more than those of $\mathcal{M}$. The process continues until all relations are atomic or until the network is determined to be inconsistent.

Input Parameters :

A Network, $\mathcal{N}$, of Tasks $T_1, T_2, \ldots, T_n$ and temporal interval relations
$T_i\; r_{i,j}\; T_j$ for some or all pairs of tasks.
The lengths, $|T_i|$.

Algorithm Short_Net

1. Set $\mathcal{M} = \emptyset$
2. While not all relations in $\mathcal{N}$ are atomic
   (a) Determine by testing all network relations the set
   $$\mathcal{M}' = \{T | T \notin \mathcal{M} \;\text{ and}\, \forall\, T\, r\, T' \;\text{ with }\; T' \notin \mathcal{M},\; (r \& sl \neq r)\}.$$

(b) Set $\mathcal{M} = \mathcal{M} \cup \mathcal{M}'$

(c) Call Find_Net($\mathcal{N}$, $\mathcal{M}$)

(d) If Find_Net returns No Solution, Return NO SOLUTION and STOP.

3. Return SUCCESS and the network obtained and STOP.

Although the heuristic is quite effective, as stated the above algorithm is quite simplistic. Many improvements in strategy and data structures can be implemented. Recently Freuder suggested a particularly attractive strategy [10] based on *inferred disjunctive constraints* (IDC). When incorporated into the Find_Net algorithm, the IDC strategy prunes vast portions of the search space while guaranteeing the finding of a solution if one exists.

4 Conclusion

The activities of cooperating robots, the steps of a project, or the actions of a distributed real-time system may be partially circumscribed by a collection of temporal constraints which specify the durations as well as some possible relations between activities. Scheduling such operations efficiently can be challenging, especially when there exists a degree of uncertainty concerning the relationships among them. The results obtained indicate how to calculate the greatest lower bound (*glb*) of the total time for all feasible schedules when a collection of tasks is constrained only by imposed temporal interval relations. Although tempting in the fully determined case to reduce the constraints to relations between start times and finish times of tasks, extension to the case of incomplete information requires that the problem be expressed directly in terms of temporal interval relations. In general, the network of such intervals and temporal relations is not reducible to a single network of temporal point relations. Theorems 1 and 2 combined demonstrate the equivalence of the *glb* of the schedule lengths to the *Size* of a complete critical chain of intervals; such chains are defined in terms of how each task *supports* the next. Critical chains display a role analogous to critical paths in traditional planning and scheduling in that they are minimal subsets of tasks and relations that require as long to complete as the whole set of tasks. The heuristic algorithm Find_Short uncovers, in the case of an underdetermined network (non-atomic interval relations), a "good" choice of completely determined network compatible with the given constraints: one which maximizes the parallelism in the task schedule. An algorithm to encounter a complete critical chain is then able to determine the *glb* of the schedule lengths, consequently enabling—in designing a real-time system cyclic executive—the shortest possible time for each cycle to be determined when the temporal interval relations are given between the required tasks. Similarly, a planning system may now establish the minimum time needed for a job even under uncertainty with respect to the relations between the tasks, thus concentrating resources on the tasks of a complete critical chain.

Acknowledgment: This research was partially supported by the *National Science Foundation* and a *Florida High Technology and Industry Council* research grant.

References

[1] Allen, J. Maintaining Knowledge about Temporal Intervals. *Comm. of ACM 26*, 11 (1983), pp. 832-843.

[2] Anger, F., Hwang, J., and Chow, Y. Scheduling with Sufficient Loosely Coupled Processors. *Jour. of Parallel and Distributed Computing 9*, 1 (May 1990), pp. 87-92.

[3] Anger, F., Rodriguez, R., and Hadlock, F. Temporal Consistency Checking of Natural Language Specifications, *Applications of Artificial Intelligence III: Proceedings of SPIE 635*, Orlando, FL, (Apr 1990), pp. 572-580.

[4] Anger, F., Ladkin, P., and Rodriguez, R. Atomic Temporal Interval Relations in Branching Time: Calculation and Application. *Applications of Artificial Intelligence IX, Proceedings of SPIE*, Orlando, (Apr 1991), pp. 122-136.

[5] Anger, F., Allen, J., and Rodriguez, R. Determined or Under-Determined Temporal Interval Relation Constraints: A Scheduling Strategy. *UWF Technical Report CSD-TR No. 93-009*, Univ. of W. Florida, Pensacola, FL, Oct 93.

[6] Anger, F., Allen, J., and Rodriguez, R. Optimal and Heuristic Task Scheduling under Qualitative Temporal Constraints. *Proceedings IEA/AIE-94*, (Jun 1994), pp. 115-122.

[7] Baker, K. *Introduction to Sequencing and Scheduling.* Wiley & Sons, New York, 1974.

[8] Dean, T. Using Temporal Hierarchies to Efficiently Maintain Large Temporal Databases. *Journal of ACM 36*, (1989), pp. 687-718.

[9] Dechter, R., Meiri, I., and Pearl, J. Temporal Constraint Networks. *Artificial Intelligence Journal 49*, (1991), pp. 61-95.

[10] Freuder, E. and Hubbe, P. Using Inferred Disjunctive Constraints to Decompose Constraint Satisfaction Problems. *Proceedings of the 13th IJCAI*, Chambery, France, Sep 1993, pp. 254-260.

[11] Ladkin, P. Satisfying First-Order Constraints about Time Intervals. *Proceedings of 7th National Conf. on Artificial Intelligence*, St. Paul, MN, (Aug 1988), pp. 512-517.

[12] Ladkin, P. and Reinefeld, A. Effective Solution of Qualitative Interval Constraint Problems. *Artificial Intelligence 57*, (1992), pp. 105-124.

[13] Lageweg, B., Lawler, E., Lenstra, L., and Rinnooy Kan, A. Computer Aided Complexity Classification of Deterministic Scheduling Problems. *Technical Report BW 138-81*, Sichting Mathematisch Centrum, Amsterdam, 1981.

[14] Lawler, E., Lenstra, J., and Rinnooy Kan, A. Recent Developments in Deterministic Sequencing and Scheduling: A Survey. In *Deterministic and Stochastic Scheduling*, M. Dempster, *et al.*, eds., D. Reidel Publ., Dordrecht, Holland, 1982, pp. 367-374.

[15] Ligozat, G. On Generalized Interval Calculi. *Proceedings of the Ninth National Conference on Artificial Intelligence*, Anaheim, CA, Jul 1991, pp. 234-240.

[16] Rodriguez, R. and Anger, F. Intervals in Relativistic Time. *Proceedings on Information Processing and Management of Uncertainty in Knowledge-Based Systems: IPMU*, Mallorca, (Jul 1992), pp. 525-529.

[17] Rodriguez, R. A Relativistic Temporal Algebra for Efficient Design of Distributed Systems. *Journal of Applied Intelligence 3*, (1993), pp. 31-45.

[18] Ullman, J. NP-Complete Scheduling Problems. *Journal of Computer and System Sciences 10*, (1975), pp. 384-393.

[19] van Beek, P. *Exact and Approximate Reasoning about Qualitative Temporal Relations.* PhD Thesis, University of Alberta, 1990.

[20] van Benthem, J. Time, Logic and Computation. *Linear Time, Branching Time and Partial Order in Logics and Models for Concurrency*, G. Goos and J. Hartmanis, eds., Springer-Verlag, New York, 1989, pp. 1-49.

[21] Vilain, M., Kautz, H. Constraint Propagation Algorithms for Temporal Reasoning. *Proceedings of the 5th Natl. Conf. on Artificial Intelligence*, Pittsburg, PA, (Aug. 1986), pp. 377-382.

Introducing Contextual Information in Multisensor Tracking Algorithms

NIMIER V.

ONERA BP 72 CHATILLON CEDEX, France

Abstract. Perception systems used in fields like robotics and aircraft tracking have recently been developing toward the use of combined sensors of different kinds. Nevertheless, the measurement management of such systems has to be adapted to the sensors used, and particularly to all modifications in the utilization conditions. For this task we propose here to use a set of heuristics to modify the operation of a Kalman filter dedicated to air vehicle tracking.

1 INTRODUCTION

Perception systems used in fields like robotics and aircraft tracking have recently been developing toward the use of combined sensors of different kinds. Such systems offer the particular advantages of enhanced performance, greater robutness in the environment, and more complete data concerning the situation to be analyzed [2]. But these multisensor systems, which have received so much attention over recent years [1], [2], [3], are hampered by difficulties stemming from the use of asynchronous, heterogeneous measurements, as well as from the difficulty in choosing which sensor, in a given situation, will be able to generate the pertinent and least disturbed information.
In the tracking field, algorithms have been proposed for multisensor systems [4], [5], and [10]. These algorithms generally use a Kalman filter along with its various variants (P.D.A.F., I.M.M.). However, an approach like this always aims to combine the measurements in order to satisfy a certain criterion bearing upon the precision of the target precision estimation. That is, the optimum solution to the mean quadratic error minimization problem is given by the Kalman filter.
It now seems obvious that this criterion alone, while it is important, is not the only one that should guide the conception of a multisensor tracking algorithm. All the measurements generated by the various sensors must be used, not only to obtain a precise track, but also to achieve greater robutness in the various situations that may occur in a real context.
The measurement management has to be adapted to the sensors used, and particularly to all modifications in the utilization conditions. The weather, and the formation of natural or artificial disturbances, are elements that can have a direct effect on a combination of sensors, and must be taken into account in order to get the expected benefit from a multisensor system.
Contextual analysis is therefore needed to bring out all the characteristics that have any concrete influence on system operation. From this analysis we derive rules, or heuristics, which will assign the most appropriate combination of sensors to a given situation.
However, the rules used are the result of complex and subjective reasoning, and so are approximate and uncertain, even considering only the definition of the affirmations to

which they are applied. The fuzzy set theory proposed by Zadeh [6], [7] does give a formal framework for dealing with this type of uncertainty, and it can also be put to profitable use in combination with probability theory. The classical techniques of filtering, based mainly on a probabilistic approach, may for this reason benefit from a contextual analysis for adapting to the changing measurement generation conditions.
We propose here to use a set of heuristics to modify the operation of a Kalman filter dedicated to air vehicle tracking
This paper includes two parts. The first describes a Kalman filter using measurements generated by different sensors operating asynchronously. The second develops the method using fuzzy sets to adapt the operation of this filter to real conditions.

2 MULTISENSOR KALMAN FILTER

2.1 Hypotheses and General Formulas

This algorithm is a sequential Kalman filter using data input generated asynchronously by the sensors. A synchronous version used for targets maneuvering through clutter can be found in [5]. We will be describing the filter in its nonlinear (EKF) form. The general scheme remains the same for a linear filter.
The state equations for a system with two sensors, a radar sensor and an optronic sensor, take the following general form:

$$X_k = A(T_k)\, X_{k-1} + v_k$$

$$y_k^r = h^r(X_k) + b_k^r \qquad \text{(Radar)}$$

$$y_k^{op} = h^{op}(X_k) + b_k^{op}, \qquad \text{(Optro)}$$

in which X_k is the state vector containing the target coordinates and speed:

$$X_k = [\, x_k, y_k, z_k, v_{xk}, v_{yk}, v_{zk} \,]^T,$$

and T_k is the time difference separating two successive measurements. If t_{k-1} is the date at which the last measurement was acquired, and t_k that of the current measurement, with the two measurements not necessarily coming from the same sensor, we define:

$$T_k = t_k - t_{k-1}.$$

$A(T_k)$ is the state transition matrix between the two times t_{k-1} and t_k. This matrix can be developed in the form:

$$A(T_k) = \begin{bmatrix} 1 & 0 & 0 & T_k & 0 & 0 \\ 0 & 1 & 0 & 0 & T_k & 0 \\ 0 & 0 & 1 & 0 & 0 & T_k \\ 0 & 0 & 0 & 1 & 0 & 0 \\ 0 & 0 & 0 & 0 & 1 & 0 \\ 0 & 0 & 0 & 0 & 0 & 1 \end{bmatrix}$$

Lastly, the functions $h^r(X_k)$ and $h^{op}(X_k)$ appearing in equation (1) can be developed in the form:

Radar $h^r()$

$$\rho^r = \sqrt{x^2+y^2+z^2}$$
$$\theta^r = \tan^{-1}\left(\frac{x}{y}\right)$$
$$\psi^r = \tan^{-1}\left(\frac{z}{\sqrt{x^2+y^2}}\right) \quad (3)$$

Optro $h^{op}()$

$$\theta^{op} = \tan^{-1}\left(\frac{x}{y}\right)$$
$$\psi^{op} = \tan^{-1}\left(\frac{z}{\sqrt{x^2+y^2}}\right) \quad (4)$$

In the above formulas, the subscripts indicating the time dependency of the various coordinates have been omitted.
The observation noises of the various sensors are assumed to be independent, normal distributions with zero mean, and covariance matrix R^r and R^{op}, respectively, for the radar and the optronic sensor.
The noise covariance matrix of state v_k takes the form:

$$Q_k = \sigma_v^2 \begin{bmatrix} \frac{T^3}{3} & 0 & 0 & \frac{T^2}{2} & 0 & 0 \\ 0 & \frac{T^3}{3} & 0 & 0 & \frac{T^2}{2} & 0 \\ 0 & 0 & \frac{T^3}{3} & 0 & 0 & \frac{T^2}{2} \\ \frac{T^2}{2} & 0 & 0 & \frac{T^2}{2} & 0 & 0 \\ 0 & \frac{T^2}{2} & 0 & 0 & \frac{T^2}{2} & 0 \\ 0 & 0 & \frac{T^2}{2} & 0 & 0 & \frac{T^2}{2} \end{bmatrix}$$

We use the following notation in subsequent developments:

$$\mathring{H}_{k-1} = \frac{\partial h^r}{\partial X}\Big|_{X_{k-1}}$$

The prediction phase is performed using the two relations:

$$\hat{X}_{k/k-1} = A(T_k)\,\hat{X}_{k-1/k-1} \quad (6)$$
$$\hat{P}_{k/k-1} = A(T_k)\,\hat{P}_{k/k-1}\,A(T_k)^T + Q_k \quad (7)$$

and the estimation phase with the relations:

$$\hat{X}_{k/k} = \hat{X}_{k/k-1} + K_k^i\,(\,y^i - h_i(\,\hat{X}_{k/k-1}\,)) \qquad (8)$$

$$\hat{P}_{k/k} = (\,I - K_k\,\hat{H}_{k-1}^T\,)\,\hat{P}_{k/k-1} \qquad (9)$$

in which:

$$K_k^i = \hat{P}_{k/k-1}\hat{H}_{k-1}^T\,[\,\hat{H}_{k-1}\hat{P}_{k/k-1}\hat{H}_{k-1}^T + R^i\,]^{-1} \quad (10)$$

The i subscript in the above equations designates the sensor delivering the measurement, which therefore belongs to the set {r, op}.

When one of the components of the observation vector is missing, as is the case for the optronic sensor, the corresponding components of the innovation vector $y^i - h_i(\,\hat{X}_{k/k-1}\,)$ are replaced by zeroes. This procedure is equivalent to replacing the missing measurement with its prediction.

2.2 Sequential Organization of Measurements

We will assume that, in an initial phase referred to by the date t_0, we have already acquired the target's positional and kinematic coordinates as well as the variances and covariances that go along with them. This initial phase consequently provides an initial estimate of the state vector $\hat{X}_{k/k}(t_0)$ and the associated covariance $\hat{P}_{k/k}(t_0)$.

Starting at the moment when a measurement y^i reaches the computer center at time t_1, and regardless of the sensor that generated this measurement, the following two phases are then run in succession:

1) Prediction phase

This consists in computing, at time t_1, the predicted state vector $\hat{X}_{k/k-1}(t_1)$ as well as the associated covariance $\hat{P}_{k/k-1}(t_1)$ from the state vector $\hat{X}_{k/k-1}(t_0)$ and from the associated covariance $\hat{P}_{k/k-1}(t_0)$. This is computed using equations (6) and (7), after computing the variable $T_k = t_1 - t_0$.

2) Estimation phase

This step computes the estimated state vector $\hat{X}_{k/k}(t_1)$ and the covariance $\hat{P}_{k/k}(t_1)$ at time t_1 from equations (8), (9), and (10). The Kalman gain K_k^i computed depends on the sensor delivering the measurement, so, in the general formula (10), the observation noise covariance matrix is replaced by the one corresponding to the sensor designated.

2.3 Remarks

When writing an algorithm for a multisensor system, a new difficulty arises that did not exist before, for an obvious reason, in the algorithms for a single measurement source. As soon as a system includes several sensors, it becomes necessary to consider the relative importance of the sensors when the global estimate is updated. In the previous filter, two parameters determine this importance: first, each sensor's measurement refresh frequency; secondly, the variance associated with the measurement from each sensor.
The frequency of the sensors is a system given that can only be modified by redefining another type of sensor, and consequently at a price that is often high. So the parameter accessible to the algorithm writer is rather the variance parameter, since this enters into the definition of the observation noise covariance matrix and can therefore be modified at will. However, these matrices are generally set once and for all in consideration of the intrinsic performance of each sensor.
For a multisensor system that operates in an environment subject to little or no disturbance, it is often possible to have a relatively good knowledge of the variance parameters that have to be entered in the algorithm. If a disturbance affects one of the sensors, it is absolutely necessary that it be taken into account in the algorithm, in order to avoid too great a degradation of the fused estimate. The algorithm therefore has to be "adaptive" as far as concerns the importance it attributes to each sensor.
This modification of the algorithm can be done in two ways. The first is consistent with statistical reasoning and consists in modifying the measurement noise covariance matrices in the Kalman filter (simply by increasing the variance, if a disturbance arises) in accordance with changing conditions. The second way is to proceed with an expert analysis of the situation and then, comparing with a few typical pre-established situations. This strategy is generally based on a sensor combination that is planned and adapted to the situation.
The former of these solutions turns out to be rather complex to implement, for a number of reasons. First, there can be no optimum solution to the problem, at least in a good many applications requiring real-time processing. Then, when a sensor's utilization conditions degrade very greatly, the number of false alarms increase proportionally, which introduces a confusion between the useful signal and extraneous data, thereby distorting the computation of the signal variance.
Hereafter we will develop the latter solution, using the fuzzy set theory.

3 INTRODUCING CONTEXTUAL INFORMATION

3.1 Introduction

In a real environment, different disturbance may affect the signal a sensor generates. So it is desirable not to use the data this signal is carrying if we also know that this data is greatly distorted. This knowledge can therefore be used in defining the combination of sensors and the relative importance assigned to each of them, as can be seen in the following example.
The signal-to-noise ratio, s, computed as the value of the main correlation peak between the received signal and a copy of the emitted signal divided by the variance of the signal remaining after subtraction of the peak, may be a parameter indicating the presence of disturbances. These disturbances may be due to the existence of obstacles on the ground,

or to the appearance of a multipath phenomenon. It is then possible to distinguish two regions on the axis of the values taken by the parameter s, where two different combinations may be made between the optronic sensor and the radar.
If the signal-to-noise ratio of the radar signal is *high*, the two sensors are likely to generate measurements of good quality, and it is desirable to "average" the set of measurements in order to increase the target position precision. On the other hand, if the signal-to-noise ratio is *low*, the radar signal has a strong chance of being disturbed by spurious echoes (obstacles, vegetation, and so forth), making it unusable. In this situation, the strategy is therefore to use only the signal generated by the optronic system (visible or IR camera, laser rangefinder).
This type of approximate reasoning, or what we agree to call this "rule", may be implanted in the system while conserving the imprecision of the formulation, by the use of fuzzy sets.

3.2 Definition of a Rule

Let us consider the following linguistic variable $\{s^r, S, T_s{}^r\}$, in which s^r is the measure of the radar signal-to-noise ratio, S the interval $[0, \infty]$ and $T_s{}^r$ a set of fuzzy subsets forming a fuzzy partition of S and defined by $T_s{}^r = \{\mathbf{FA}, \mathbf{FO}\}$. The membership functions $\mu_{FA}(s^r)$ and $\mu_{FO}(s^r)$ are represented in Figure 1. Insofar as **FA** and **FO** form a fuzzy partition of S, their membership functions verify the condition:

$$\mu_{FA}(s^r) + \mu_{FO}(s^r) = 1 \qquad \forall s^r$$

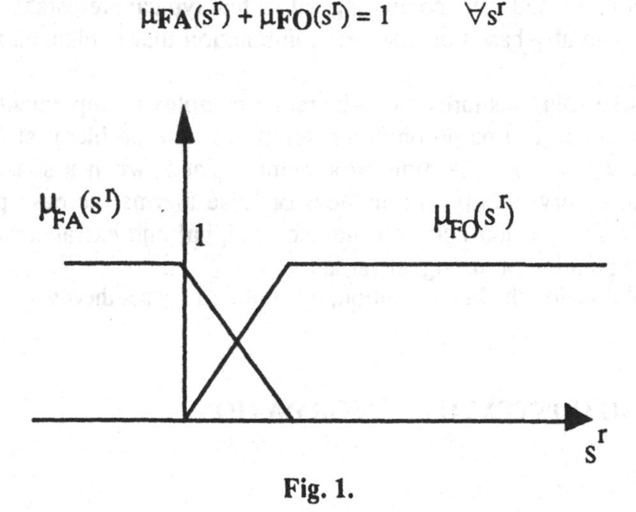

Fig. 1.

Membership function of fuzzy subset FO and FA.

The rule defined in the above example can be formulated as follows:

If s^r is **FO**, then "radar data must be considered" (R1).
If s^r is **FA**, then "radar data must not be considered" (R2).

The two actions, "radar data must be considered" and "radar data must not be considered" have a concrete effect on the Kalman filter. That is, if a measurement arrives at a given date t_n and totally verifies ($\mu_{FO}(s^r) = 1$) the condition of rule R1, a full iteration of the Kalman filter will be run, which will refresh the state in full consideration of this new data. On the other hand, if the measurement totally verifies the condition of rule R2 ($\mu_{FA}(s^r) = 1$), only the predictive phase will be run. The above two rules can then be expressed concretely by:

If ψ^r is **FO** R1
then $\hat{X}_{k/k} = \hat{X}_{k/k-1} + K_k^r\,(y^r - h_r(\hat{X}_{k/k-1}))$

If ψ^r is **FA** R2
then $\hat{X}_{k/k} = \hat{X}_{k/k-1}$

When the measurement of s^r only partially verifies the two conditions, or equivalently if $\mu_{FO}(s^r)\neq 1$, the state is calculated from the formula:

$$\hat{X}_{k/k} = \beta[\hat{X}_{k/k-1} + K_k^r\,(y^r - h_r(\hat{X}_{k/k-1}))] + (1-\beta)\hat{X}_{k/k-1}$$

$$\hat{X}_{k/k} = \hat{X}_{k/k-1} + \beta\, K_k^r\,(y^r - h_r(\hat{X}_{k/k-1})) \qquad (11)$$

where we have used the notation : $\beta = \mu_{FO}(s^r)$.

3.3 Remarks

It can be seen in formula (11) that the coefficient β multiplies the Kalman gain, with the consequence of a decrease in the influence of the measurement on the state estimate update when β is smaller than 1. If β is zero (and the condition of R2 is then totally verified), the radar measurement is therefore purely and simply ignored.
Even if this algorithm may be used on a system having only one radar sensor, it usefulness is of course all the greater for a multisensor system. That is, the weakening of the data generated by the radar sensor has the consequence of increasing the importance of the measurements generated by the optronics in the state refresh. For an air vehicle flying constantly in a disturbed environment, *i.e.* such that the condition of rule R2 is always satisfied, only the data from the optronic camera will be used for estimate the target position. The rule suggested by the operator designated above is therefore entirely realized.
The example presented is given for illustration, but is absolutely not limitative. It is entirely possible to apply this same method using a greater number of rules. Other sensor associations may also be used, for example two radars of different wavelengths and resolutions, or a visible and infrared camera.

3.4 Probabilistic Interpretation

The probability of a fuzzy event was defined by Zadeh [7] and can be used to give a probabilistic interpretation to the "rule system" that has just been described. The estimate of the state such as it is computed by a Kalman filter is the conditional mean of the state, knowing all the observations acquired in the past. We use Y_r and Y_{op} to denote all these observations for the two sensors (radar and optro.), and $P(X_k/Y_r,Y_{op})$ to denote the conditional probability of X_k, knowing all these observations. We can applied the following rule:

$$P(X_k/Y_r,Y_{op}) = \sum_{i=1}^{i=p} P(X_k/Y_r,Y_{op}, A_i)P(A_i) \quad (12)$$

in which the A_i are a collection of fuzzy subsets that form a fuzzy partition of a reference system S. The membership functions verify:

$$\sum_{i=1}^{i=p} \mu_{A_i}(s) = 1,$$

and $P(A_i)$ is the probability of the fuzzy event A_i defined by the formula:

$$P(A_i) = \int \mu_{A_i}(s)P(s/s^r)ds, \quad (13)$$

in which $P(s/s^r)$ is the probability density that models the dispersion of the values measured by the sensor. In this expression, s is the true value of the parameter and s^r is its measure. In the case of an additive gaussian noise of variance σ_s^2, then:

$$P(s/s^r) = \frac{1}{2\pi\sigma_s} e^{-\frac{1}{2}\frac{(s-s^r)^2}{\sigma_s^2}}$$

Applying formula (12) to the above example gives:

$$P(X_k/Y_r,Y_{IR}) = P(X_k/Y_r,Y_{op}, FO)P(FO) + P(X_k/Y_r,Y_{op}, FA)P(FA)$$

When the signal-to-noise ratio of the radar signal is high, the knowledge we have of the state can be modeled by the probability density $P(X_k/Y_r,Y_{op},FO)$ which takes the form of a gaussian probability density having mean $\hat{X}_{k/k} = \hat{X}_{k/k-1} + K_k^r (y^r - h_r(\hat{X}_{k/k-1})$ and covariance matrix $P_{k/k}$. On the other hand, if the signal-to-noise ratio is low, $P(X_k/Y_r,Y_{op},FA)$ can be modeled by a gaussian probability density of mean $\hat{X}_{k/k-1}$ and covariance matrix $P_{k/k-1}$. The estimated state is then the conditional average, which takes the form:

$$\hat{X}_{k/k}=E(X_k/Y_r,Y_{IR})=E(X_k/Y_r,Y_{IR},FO))P(FO) +E(X_k/Y_r,Y_{IR},FA))P(FA)$$

$$\hat{X}_{k/k} =[\hat{X}_{k/k-1} + K_k^r (y^r - h_r(\hat{X}_{k/k-1}))]P(FO) + \hat{X}_{k/k-1}P(FA)$$

$$\hat{X}_{k/k} = \hat{X}_{k/k-1} + P(FO) K_k^r (y^r - h_r(\hat{X}_{k/k-1}) \quad (14)$$

Formula (14) is to be compared with formula (11). It can be seen that P(FO) plays the same role as $\mu_{FO}(s^r)$. Total equivalence is obtained if we consider that the site is marred with no error in its measurement. In this case, $P(s/s^r)$ can be modeled by a Dirac distribution:

$$P(s/s^r) = \delta\,(s - s^r),$$

and, replacing $P(s/s^r)$ with its value in (13), we get:

$$P(FO) = \mu_{FO}(s^r)$$

This probabilistic interpretation is used to compute the *a posteriori* covariance matrix associated with this estimate [9]:

$$\hat{P}_{k/k} = (I - \beta)\,\hat{P}_{k/k-1} + \beta\,\hat{P}^c_{k/k} + \beta\,(1-\beta) K^r_k \nu^r_k \nu^{r^T}_k K^{r^T}_k$$

with

$$\hat{P}^c_{k/k} = (I - K_k \hat{H}^T_{k-1})\,\hat{P}_{k/k-1}$$

and

$$\nu^r_k = y^r - h_r(\hat{X}_{k/k-1})$$

It may be pointed out that the filter constructed with this method is very much like the PDAF filter. The fundamental difference between the two resides in the use of fuzzy events in the filter just described, while in the PDAF it is classical probabilistic hypotheses that are used. This difference is fundamental because it provides a way of introducing contextual data and heuristics in the parameter estimations.

4 CONCLUSION

This paper therefore presents an application of fuzzy sets to the management of the different sources of measurements provided by a multisensor system. The original feature of this approach is that the algorithm is divided into two processing levels. The first is oriented toward processing the data from the sensors with the usual signal processing techniques, generally using a probabilistic approach. The second level is dedicated to the management of contextual information, and its purpose is to introduce heuristics into the operation of the algorithms of the first processing level. This level uses fuzzy set theory, which is entirely suitable to this type of data. The importance taken by each of the sensors on the final estimate therefore draws upon the context and circumstances in which the system is used.

We have illustrated the method with a very simple example, which does not preclude its being used in a more difficult context, and in particular with a larger system of rules. Research is currently under way at ONERA to extend and apply this algorithm to different fields.

REFERENCES

[1] Bar-Shalom Y. "Multitarget - Multisensor Tracking: Application and Advances", Artech House, 1992.
[2] Appriou A. "Perspectives liées à la fusion de données" Science et défense 90, Dunod, Paris, Mai 1990.
[3] Haimovich A.M. et all "Fusion of Sensors with dissimilar Measurement/tracking accuracies", IEEE trans. on AES, Vol 29, Jan 1993.
[4] Roecker J.A., and McGillem C.D. "Comparaison of two-sensors tracking methods based on state vector fusion", IEEE trans. on AES, vol 24, July 1988.
[5] Houles A., Bar-Shalom Y. "Multisensor Tracking of a Maneuvring Target in Clutter", IEEE trans. on AES, Vol 25, March 1989.
[6] Zadeh L.A., "Fuzzy set", Information and control, vol 8, 1965
[7] Zadeh L.A., "Probability Mesures of Fuzzy Event", JMAA, vol 23, 1968
[8] Popoli R., Mendel J., "Estimation Using Subjective Knowledge With Tracking Application", IEEE trans. on AES, vol. 29, N°3, July 1993.
[9] Bar-Shalom, Y., and Fortman, T.E., Tracking and data association, New York: Academic Press, 1988.

FAM on Grain Scheduling Control of Highly Nonlinear and Disturbed Processes

Ramon Ferreiro Garcia

Dept. of Electronics and Systems. E.S. Marina Civil. Univ. La Coruña
Paseo de Ronda 51, 15011. La Coruña. Spain.

Abstract. This paper describes an adaptive computer-based controller by gain scheduling strategy on the basis of associative memories for the control of nonlinear disturbed processes. The proposed algorithm, search for a rule-base during a training phase for which is applied a pattern input/output signal by means of a set of well adjusted regulators distributed along the full operating range (from 0 to 100/% input reference). After such phase the obtained rule-base will be processed by classical fuzzy methods in on line feedback control.This algorithm is highly efficient for each disturbance under different operating conditions. Good results are obtained by computer simulation using the simulation language Simnon as well as in real time computer control on a training board.

1 Introduction

A classical PI controller may be adjusted for an operating point but in a high non linear process, the deviation of such operating point bring immediately poor performance. If the classic controller is applied by means of a computer algorithm it is possible and it has been done [1], the application of a well adjusted set of controller parameters for every operating point along the controlled variable range. The basic idea consist in the use of that set of controllers not as the real time process controller rather the application of such robust control method is being used as a trainer device to achieve a complete set of data condensed into an associative memory which will be used as a real time controller after a training phase. The proposed algorithm, is a deterministic associative memory which contains the complete data of an extensive training period in which a set of entries due to reference and disturbance changes were applied. Learning capabilities of the deterministic associative memory has been implemented with appropriated software developed specially for such algorithm. The trainer algorithm which is based on the polynomial output feedback, is selected so that the closed-loop poles are satisfying rapid input following avoiding conflicts with system sensitivity, that is tracking performance and robustness under different operating conditions avoiding the effect of modelling errors because of the experimental nature of the training phase. The associative memory contains a set of rules achieved from the training phase with reference and error inputs and an output as shown at figure 1.
The two phases of the proposed strategy (knowledge acquisition by training procedure and control application) are applied sequentially. The first stage consist in the training procedure based in the application of a set of input reference signals till steady state are reached with properly adjusted regulators. With the input/output data a fuzzy associative memory is being achieved. The second stage is the real time control application based in the acquired knowledge for every operational condition.

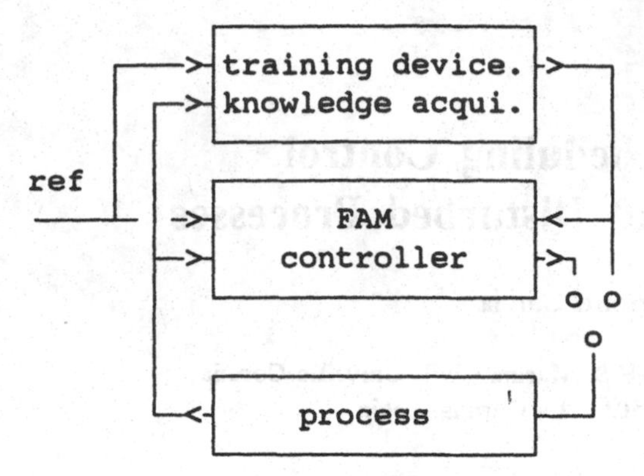

Fig. 1. The controller based on a associative memory.

A schematic block diagram of such control stages is shown at figure 2, where a discrete number of well adjusted controllers are used to training-learning phase, where process control knowledge is acquired and a FAM system obtained as a rule-base to be applied at the second phase which is the real time control phase.

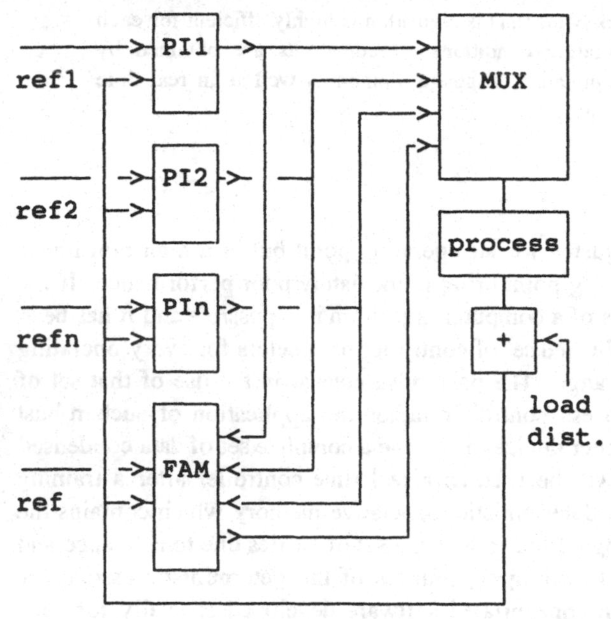

Fig. 2. Detailed arrangement of training computer controllers

The training procedure is completed by applying a set of reference input signals each one with a properly adjusted controller. The set of input references may vary from two to any necessary number in order to ensure the training along the full possible control, variable range. In section 2 a tank level control process is described.In the section 3 it is shown the process of knowledge acquisition through the proposed training-learning algorithm which leads to a rule-base synthesis. The section 4 illustrates the control application to a tank system and finally in section 5 it is shown the graphic results and conclusions.

2 Tank level process

Consider a tank where the cross section varies with height h. The model is [1]

$$q_i = d/dt[A(h)h] + a(2gh)^{1/2}$$

where q_i is the input flow and a is the cross section of outlet pipe valve in a defined load. Let q_i be the input and h the output of the system. The linearized model at an operating point, q_{i0} and h_0, is given by the transfer function

$$G(s) = k_{nl}/(S+k_{dl})$$

where $k_{nl} = 1/[A(h_0)]$
and $k_{dl} = q_{i0}/[2.A(h_0)h_0]$

A good PI control of the tank is given by

$$u(t) = Kp[e(t) + (1/T_i)Se(t)]$$

where Se(t) is the integral of control error with respect to time and the pair K,T_i are the controller parameters for every reference signal under a wide range of tank loads and perturbations as a function of its natural frequency (nf) and relative damping (rd). The values of Kp and T_i that satisfy the desired natural frequency and relative damping are [1]

$$Kp = (2.rd.nf-k_{dl})/k_{nl}$$

$$T_i = (2.rd.nf-k_{dl})/nf^2$$

Unfortunately, due to saturation of control valve and controller output the range of application of such controller parameters is rather restricted to short limits reason by which we define a complete set of PI controllers adjusted experimentally by Ziegler & Nichols techniques for the full range of tank reference levels, covering the complete universe of discourse in that variables.

If such pair of controller parameters are adjusted so that satisfies the requirements of desired natural frequency and relative damping for a discrete number of reference inputs and operating points, lets say the full universe of discourse of the control variable, then the training is possible under the proposed conditions.

3 Rule-base synthesis by a training/learning procedure

In the task of training/learning procedure it is the main objective to synthesise a knowledge base which will be applied to control the system by deffuzification of the generated rule-base. We wish to adjust the level dynamics in two dimensions. The tank level control is a classical control problem and admits a math-model control solution. There are a lot of input variables but we will consider only two state fuzzy variables, and one control fuzzy variable. The first fuzzy variable is the error level. Zero level corresponds to the empty tank. The second state fuzzy variable is the reference level. The fuzzy control variable is the flow to the tank. We can quantize each universe of discourse into any desired number of fuzzy-set values. The level dynamics control FAM bank is a (m x n) matrix with linguistic fuzzy-set entries (Kosko, 1992). For the general case, the FAM bank is an hypercube. We index the columns by the m fuzzy sets that quantize the error angle universe of discourse. We index the rows by the n fuzzy sets that quantize the reference level, that is the level universe of discourse. Since a FAM rule is a mapping or function, there is exactly one output flow value for every pair of error level and reference level values. That means, if the input variable I1 is m and input variable I2 is n then output variable U is U(m,n).

The training/learning algorithm is synthesised under real time control of tank level as follows [3]:

- Loop until time for training phase spires.
- Read I/O values
- Calculate the fuzzy-set values into each
  universe of discourse for I/O variables. That is, I1 belongs to the j value from 1 to m and I2 belongs to the k value from 1 to n.
- Store the result of adding the actual to the old data as,
  $U_T(j, k) = U_T(j-1, k-1) + U(j, k)$
  $N(j, k) = N(j-1, k-1) + 1$
  where j,k are the dimensions to define the degree of freedom of the associative memory.
- End loop
- Compute the arithmetic mean and store the results to make the hypercube rule-base as,
  $U_F(j,k) = U_T(j,k)/N(j,k)$

If the I/O variables satisfy the desired performance criteria, then the knowledge base is assumed as correct and will be used as a fuzzy controller. Then, for the tank level control phase, the following sequence must be applied:

- Loop until an external entry stops control sequence
- Read input variables and associate each value to the proper membership function
- Defuzzification FAM rules.
- Apply the unique output to the level control.
- End loop

4 Application to the tank level control

Application of training/learning algorithm acquires the knowledge and stores it into a rule-base under the discrete model $Up = f(HR, E)$, $K_i = f(HR)$ which is shown at figure 3. After the training phase we have obtained a FAM system.

In gain scheduling procedure, the FAM systems are applied so that the reference and error generates a proportional action, and the reference itself generates an integral action parameter. So that, the FAM controller is a combination of a proportional action and an external integral action added according figure 4,to obtain the controller output which is automatically corrected by the FAM system by the expression Uc=Up+Ui.

	-6	-4	-2	0	2	4	6	Ki
1	-6	-4	-2	0	2	4	6	066
2	-12	-8	-4	0	4	8	12	066
3	-15	-10	-5	0	5	10	15	050
4	-12	-8	-4	0	4	8	12	025
5	-9	-6	-3	0	3	6	9	013

a)

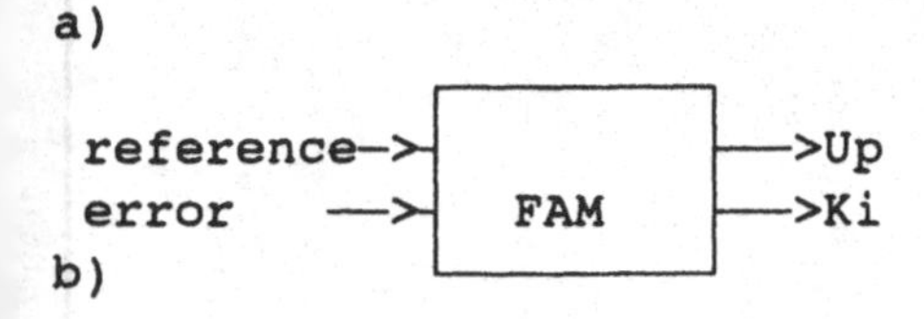

b)

Fig.3. (a) Rule-base achieved by the learning algorithm. (b) FAM system

The deffuzzification task is carried out by means of polynomial regression [3], so that control output is achieved. The effect of external integral action helps a lot in order to avoid persistent error due to parameter variations, load disturbances and environmental disturbances are also compensated by such integral action.

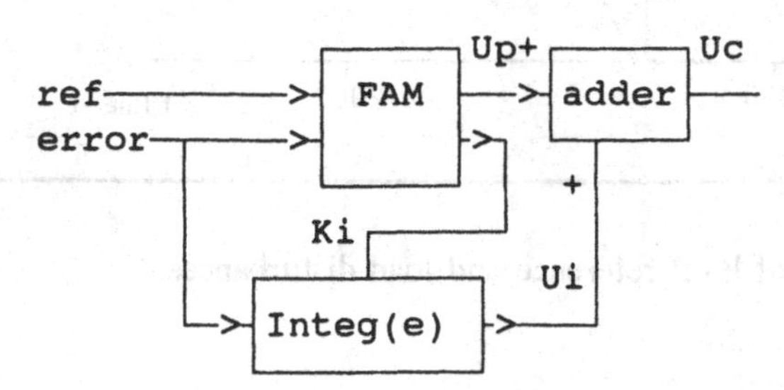

Fig. 4. FAM system with adaptive integral action included.

5 Simulation results and conclusions

Validation of the proposed adaptive algorithm was carried out in two ways. The first one, by simulation based on the mathematical model of the process under simulation language Simnon.

The second way is a training board equipped with computer control. In figure 5 it is shown the time response of level (curve 3) the reference level (curve 1) and controller output variable (curve 2) to a sequence of level reference inputs and a load disturbance sequence.

With a initial load (tank discharge valve) some step inputs in the reference were applied. At 2000 sec. the load was increased till two times the initial value. The tank characteristics are non linear depending on the operating point and are such that tank level surface is a function of the level. In adition, control variable actuator valve are limited. In this situation FAM added to the integral action gives an acceptable response when compared with the well prooved and typical gain scheduling method.

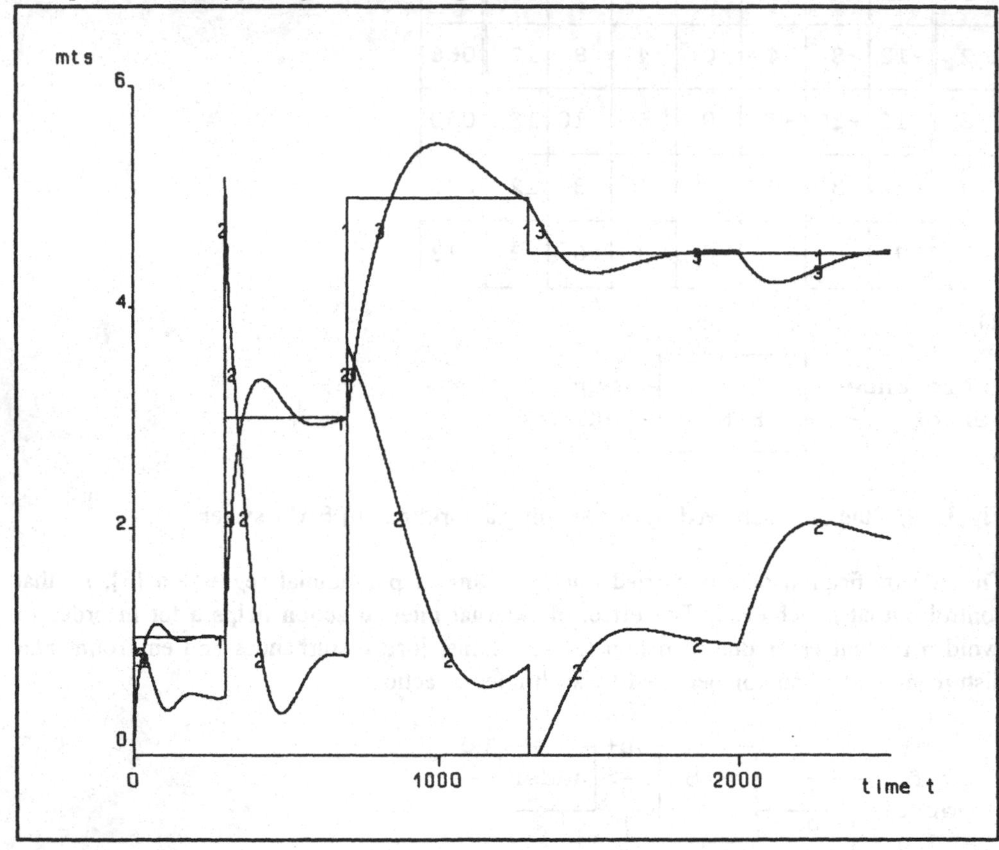

Fig. 5. Time response to a sequence of level reference and load disturbances.

References

1. Astrom, K.J, Bjorn Wittenmark (1989): Adaptive Control. Addison-Wesley Publishing Company, Chap.8, pp 348-349
2. Bart Kosko (1992): Neural networks and fuzzy systems. Prentice Hall International, Inc. pp. 299-335. Chap. 8.
3. Ferreiro Garcia, R. (1994): Associative memories as leaning basis applied to roll control of an aircraft. pp. 23-29. Proceedings of Multivariable system techniques (AMST'94).Ed. by R.Whalley. University of Bradford. U.K.

How to choose according to partial evaluations ?

A. Guénoche

LIM-CNRS, 163 Av. de Luminy
13288 Marseille Cedex 9

Abstract. In a decision problem we have to select one or more candidates among several, according to different criteria. Often some data are missing, but each criterion gives a partial linear order over the candidates. We develop a method based on paired comparisions. We first establish a weighted tournament and then we enumerate total linear orders which are as close as possible to it. Thus we use the transitivity of the majority preference relation to rank all the candidates and to select the winners from these orders.

Many decision problems in economics can be formulated in the following terms : One would like to select some products among several candidates according to marks given them. These marks can be obtained according to criteria or given by one or many judges. We shall admit that the highest marks are given to the best products. Often these evaluations cannot be done systematically and the notes are partially assigned. A missing mark may not be interpreted as a bad one.
To illustrate this text, we use data provided by the INRA (National Institute of Research in Agronomy) which deal with rapeseed. During the year 1990, 14 varieties were tested in 60 trials and the average yields were measured. But for each trial, the number of tested varieties varies between 7 and 14. So we get an array of notes with many missing values, since we have only 601 notes instead of the 840 possible ones. Faced with this experimental data, the question is what are the best varieties and how to select them ?
Experts from the INRA have defined an ordinal model for yield, depending on parameters, and they try to fit the parameter values using the data. But the missing marks make the model unreliable. Another approach, closest to artificial intelligence, would be to estimate these notes from trials having similar yield values in common. We here propose a combinatorial strategy, derived from social choice theory [5] and preferences aggregation [11], which we present in overview.
Each trial establishes a partial order for the different varieties tried. There may be equal or equivalent notes and so this order is a preorder which indicates an individual opinion or a preference. The whole set of preorders constitutes a *profile* Π that allows us to realize paired comparisons. For each pair we decide if collectively variety x_i is better than variety x_j. Thus we build a weighted *tournament*, that is a complete directed graph. Vertices {1, .. n} correspond to varieties $\{x_1, .. x_n\}$, and there is an arc between i and j if a majority of trials ranks x_i before x_j. Let T(i,j) be the weight of this arc which is as great as the number of preferences for x_i. If T(i,j) > 0 we have T(j,i) = 0. If varieties x_i and x_j have never been tested simultaneously, or if there are as many appreciations for x_i as for x_j, we get T(i,j) = T(j,i) = 0.
Then we enumerate all the linear orders at minimum distance of this tournament, that are those having a sum of distances to preorders that is minimum; this is why they are called *median orders*. If the tournament is transitive, there is only one linear order for which all the preferences are supported by a majority; this order is compatible with profile Π. If this

is not the case, for any linear order $O = (o_1, o_2, .. o_n)$, there is at least one arc (o_j,o_i) with $j>i$ that belongs to a cycle; it is the "effet Condorcet". Classically verticesare placed on an axis, from left to right, following this order, and we only draw the arcs of the tournament that disagree with the transitive order relation. Because they go from right to left, they are called *reverse arcs* [8]. Thus the distance between profile Π and a linear order is the sum of weights of reverse arcs, that is

$$d_k(O,\Pi) = \sum_{i=1}^{n-1} \sum_{j=i+1}^{n} T(o_j,o_i).$$

To make a tournament transitive, we must remove all its cycles, reversing some arcs. They correspond to individual opinions, and to best respect the set of preferences is to reverse arcs having a sum of weights as small as possible and giving a transitive tournament. The linear orders at minimum distance of the profile, or equivalently at minimum distance of the tournament, are those that minimize $d_k(O,\Pi)$. They are the most compatible with the set of partial preorders, since they refute a minimum number of individual preferences.
We must emphasize that generally there are several linear orders at a minimum distance in a given tournament. There may be a lot of ties, and even an exponential quantity of the number of candidates [9]. As a collective opinion they are all equivalent, that is why it is necessary to enumerate them. Our answer to the decision problem is to select the first candidates from these median orders. They are called the *winners* according to Kemeny [10].
Having realized an application of this method to varieties of rapeseed, we are going to detail these two steps to highlight specific problems and to develop general solutions, from a practical and a theoretical viewpoint.

1 From notes to tournament

Let S be the set of trials, that correspond to different locations for the experiments, let X be the set of varieties and $R_s(i)$ the yield realized by variety x_i on the location s. If it has not been experimented, $R_s(i)$ is undefined. The roughest way to compare varieties is to consider that x_i is better, for this trial, than x_j if $R_s(i) > R_s(j)$. Consequently, the tournament could be weighted starting from :

$$S(i,j) = | \{s \in S \text{ such that } R_s(i) > R_s(j)\} |$$

But there may be very small differences which do not allows us to conclude that one variety is better than another, and the large differences are treated as the small ones. As a consequence, the weight is the same for hazardous and sure preferences. This practical problem led Frey and Yehia-Alcoutlabi [6] to modify the previous distance between orders and tournament; it is much easier to change the weight function.

1.1 From notes to preorders

To quantify preferences one can estimate, for each location, a least significant difference between yields, but this is from a statistical point of view, which is not in accordance with our combinatorial approach. The argument is the same if we treat each yield as the mean of an interval, and if we decide that one value is greater than another one only if the two intervals are without intersection. Another possibility is to realize, for each trial, a partition of the varieties in ranked clusters and to take into account the gap between them.
If we decide a priori that each trial identifies three classes of varieties, good, average and bad, two thresholds must be determined. They might be uniformly fixed, but in that case the partition is rather arbitrary and does not take into account the distribution of yields.

A better alternative is a partition such that each variety is in the same cluster as those having the closest yield. Therefore, the thresholds are put in the two largest intervals between consecutive yields. This strategy gives the same results as the famous single linkage method in clustering, since in both cases a maximum split partition is built. In fact, the smallest interclass distance is maximized, and the partition is then optimal according to this criterion.

There are other optimal clustering methods for a one-dimensional euclidian space, but the split criterion seems more accurate than the diameter or some normalized standard deviation. When there is one very good and one very bad variety, the resulting partition, two singletons on both sides, and the remaining part, is very natural.

The number of clusters may be adapted according to the difference between the extreme values. However, this would tend to give more importance to trials where large variations are observed.

Let $C_s(i)$ be the rank of the cluster containing variety x_i with the convention that the best varieties are in the class ranked 1, and the worst in the class ranked 3.

1.2 From preoders to tournament

Now, since we have ordered classes of varieties for each trial, the natural process is to weight the tournament taking into account the gaps between cluster ranks. For any couples (x_i, x_j) we count the number of trials in favor of x_i and those which are for x_j. If, in one trial, two varieties are in the same class, it has no contribution to the orientation and to the weight of the arc linking i to j. If these varieties are in two different classes, the trial contributes to the tournament through the difference between cluster ranks. The tournament is weighted

$$T(i,j) = \mathrm{Max}\ \{0, \sum_{s \in S} C_s(j) - C_s(i)\}.$$

So a positive value for T(i,j) shows a collective preference for x_i over x_j and in this case T(j,i) = 0. At most one of the two symmetrical values is positive. If both are null, it means collective indifference. The resulting tournament over the 14 varieties is shown in Table 1.

	1	2	3	4	5	6	7	8	9	10	11	12	13	14
1 :	0	3	19	6	15	1	0	11	1	0	0	13	2	0
2 :	0	0	16	2	11	0	0	12	0	0	0	24	13	0
3 :	0	0	0	0	0	0	0	0	0	0	0	16	11	0
4 :	0	0	14	0	9	0	0	12	0	0	0	22	13	0
5 :	0	0	5	0	0	0	0	1	0	0	0	19	13	0
6 :	0	7	23	9	18	0	0	19	2	0	0	29	17	0
7 :	3	8	17	6	12	2	0	10	2	0	9	22	8	0
8 :	0	0	3	0	0	0	0	0	0	0	0	14	6	0
9 :	0	7	9	8	4	0	0	11	0	0	3	16	8	1
10 :	3	15	17	16	10	8	5	22	3	0	7	26	15	3
11 :	3	2	15	6	6	1	0	14	0	0	0	17	3	0
12 :	0	0	0	0	0	0	0	0	0	0	0	0	0	0
13 :	0	0	0	0	0	0	0	0	0	0	0	4	0	0
14 :	1	6	5	6	5	1	4	10	0	0	3	19	16	0

Table 1 : Fourteen varieties of rapeseed have been compared. Each value T(i,j) quantify the preference for x_i over x_j.

2 Enumeration of median orders

Our algorithm is a *branch and bound* method which follows from an approximation of the minimum distance of a linear order to the tournament, using heuristics. This method has been described in detail in [2] and we recall it briefly here, with some improvements to save computation time and place and also due to the fact that we are only interested in the best varieties that correspond to the first ranked ones.
The first step consists of the estimation of an upper bound Dmax for the distance between a median order and the tournament, thus defining a rejection threshold. The smaller the bound, the greater the efficiency of the algorithm, the narrower the place occupied by calculus and the quicker the solution. We first search for orders close to the tournament adapting heuristics based on the elimination of cycles of length 3 [13] to weighted tournament. For the resulting linear orders, we try local optimizations testing all the transpositions.

Example : From the tournament in Table 1, we get the linear order :

$$(10 > 14 > 7 > 11 > 1 > 6 > 9 > 2 > 4 > 5 > 8 > 3 > 13 > 12)$$

which is at distance 4 with reverse arcs (9,14) et (9,11). The distance of a median order will be ≤ 4.

As for any branch and bound method, we build a tree where each node is a beginning section of a linear order. Each section can be extended to a median order and each node has a value which is a lower bound of the distance of any linear order beginning with this section. At each step we extend the section corresponding to the node having the minimum value, in all the ways possible, if each resulting section has a value lower than or equal to Dmax.
Let $O = (o_1, \dots o_p)$ be a terminal node of the tree, that is, a linear order on p vertices such ranked. Its value $V(O)$ is equal to

$$V(O) = \sum_{x \in X} \sum_{i=1}^{p} T(x, o_i).$$

It corresponds to the sum of weights of arcs having their end in section O wherever their origins are. Its possible successors are sections obtained adding to O one of the n-p vertices that are not placed in O.
One step of the algorithm is :
(i) to find a terminal node of minimum value $O = (o_1 \dots o_p)$; the tree is developed using a *least cost strategy*.
(ii) If $p = n-1$, it is sufficient to add the remaining vertex and the order $(o_1 \dots o_{n-1} y_1)$ is at minimum distance from the tournament according to this strategy. So we have found a solution at distance $\Delta = V(O)$. If $\Delta <$ Dmax, the upper bound is updated.

Else we extend this node, linking to O the nodes corresponding to this section prolonged by some of the n-p vertices $Y = \{y_1, ..y_{n-p}\}$ that are not present in O. These vertices must not dominate o_p since a median order is a hamiltonian path of the tournament [3], and they must lead to sections having values that do not overpass Dmax.
The value of section O extended by y_i is equal to $V(O)$ plus the cost corresponding to y_i, that is the sum of weights of arcs coming from Y and going to y_i.

$$V(o_1 \dots o_p y_i) = V(O) + \sum_{k=1}^{n-p} T(y_k, y_i).$$

If we look for just one solution, we can stop after the first one but if we need all the median orders, we continue to extend the tree until no terminal node can be prolonged without overpassing Δ.

Example : From the tournament in Table 1, there are 3 median orders at distance 4. The second one has been found by the heuristic.

```
1 :   10 > 14 > 7 > 9 > 11 > 1 > 6 > 2 > 4 > 5 > 8 > 3 > 13 > 12
2 :   10 > 14 > 7 > 11 > 1 > 6 > 9 > 2 > 4 > 5 > 8 > 3 > 13 > 12
3 :   10 > 14 > 7 > 6 > 9 > 11 > 1 > 2 > 4 > 5 > 8 > 3 > 13 > 12
```

Sometimes this kind of method encounters difficulties, due to the complexity of computing a consensus function in order relations [1]. Given a tournament and an integer k, to decide if there exists a linear order at distance not greater than k is a NP-hard problem. The amount of computer memory needed is proportional to the size of the tree. Even efficient programs [2, 12] can fail on tournaments as small as this one (14 candidates); one cannot enumerate nor produce a single median order. The following paragraphs are devoted to this difficulty.

2.1 Decreasing the number of nodes

A single set of vertices, ranked in different ways, can create several nodes in the tree. Our aim is to keep, for any subset ordered in a beginning section, the value of the best order. Let the section (i>j>k>l) have value v. If in the further development of the tree, the same subset, differently ordered is obtained, for instance (j>k>l>i) with value v', there are three possibilities :
i) if v' = v, this new section is kept in the tree,
ii) if v' < v, we also update the stored value for subset W({i,j,k,l}) := v'. As the node with value v' will be examined before the one with value v, the latter cannot be extended, because
iii) if v' > v, this section cannot lead to a median order, and will not be added to the tree.
Testing this idea for the enumeration of median orders shows that the resulting tree has roughly two or three times less nodes, depending the number of candidates. However, the data structure must be adapted (a characteristic function of any subset of vertices is used) to store the best value of a section, but the global result is better by far.
With a tournament for which a median order is hard to find, a first calculus may be the search for a single solution. In this case possibilities i) and iii) can be merged, since we already have an optimal order on this subset; it is useless to get another equivalent order. If we get a solution it will provide the real distance Δ, and in a further computation, with this smallest bound, one can enumerate all of them.

2.2 Fitting values for nodes

To give a value to a section, which is a lower bound of the distance between the tournament and any order beginning by this section, we only use reverse arcs having their end in this part. One can also count other arcs having both their origins and ends in the complementary part, as I. Charon et al. [4] do for unweighted tournaments.
Concerning orders at minimum distance from a given tournament, we have the obvious property : Any ending section of a median order is a median order of the sub-tournament corresponding to vertices included in this section. To evaluate a beginning section, one cannot determine the best order for its complementary part Y, but one can minorate the part of distance due to an order on Y using only cycles with length 3. For any 3-cycle (x,y,z), the sub-tournament is written :

	x	y	z
x	-	a	0
y	0	-	b
z	c	0	-

This cycle will have at least the weight v'{x,y,z} = Min {a,b,c}. So before the enumeration process, one can evaluate the cost of all the 3-cycles. This can be done efficiently, since a subset {x,y,z} is a 3-cycle if and only if T(x,y) > 0 and T(x,z) = 0.
For a beginning section O, we want to use the maximum weights of 3-cycles in its complementary part Y. Concerning these cycles, two remarks can be made:

- First, they must contain only vertices belonging to Y; because if a vertex of a 3-cycle belongs to O, the arc of minimum weight of this cycle can have its end in O, and it has been previously counted in V(O).
- Secondly, if the weight of a 3-cycle (x,y,z) has been counted, one cannot use another cycle using one of its arcs; because reversing the common arc, we remove two cycles, making them transitive.

So we need a set of 3-cycles in Y, being arc-separated and having a sum of weights as great as possible. This problem being NP-hard, for the complementary part Y of a beginning section O, we define V'(Y) as the sum of weights obtained using a greedy algorithm : We first sort all the 3-cycles in decreasing weight order, and for a given subset Y, we will run through this list picking up all the separated cycles included in Y. In this way, a begining section value can be improved by :

$$V(O) = \sum_{x \in X} \sum_{i=1}^{p} T(x,o_i) + V'(Y).$$

This lower bound is better than the previous one, and make it possible to build much smaller trees, keeping the same results; Testing it with an optimal heuristic distance value, the tree may have between 5 and 20 times less nodes !

2.3 Restricting ranks for vertex

Suppose we know the distance of a median order, or any approximation D of this distance; we term a D-order as a linear order having a distance to the tournament not greater than D. One can determine a minimum (resp. maximum) rank for any candidate which is the lowest (resp. greatest) position it can have in a D-order. Even if D has the minimum value Δ, for which the D-order set is not empty, for one candidate there may be no median order where these positions are reached, but in any D-order, this candidate cannot be placed before (resp. after) that rank.
Let us examine candidate x_k in a tournament coded in a T matrix as previously. It can be ranked first in a D-order iff $\sum_i T(i,k) \leq D$. If not, it can be the second, after x_j, iff $\sum_{i \neq j} T(i,k) \leq D$. But may be T(j,k) is the greatest preference over x_k, and so x_k can be ranked at the second place iff

$$\sum_i T(i,k) - \mathrm{Max}_i\, T(i,k) \leq D.$$

More generally : Let Ip be the set of index of the p greatest values of {T(i,k), i=1, ..n}. The candidate x_k can be ranked (p+1) in a D-order iff

$$\sum_i T(i,k) - \sum_{i \in I_p} T(i,k) \leq D,$$

that is, if the sum of the n-p+1 smallest values does not overpass D.

Using the same reasoning for the rows of matrix T, if x_k is ranked p+1 in a D-order, then the sum of the p smallest values in {T(k,i), i=1,..n} is lower than or equal to D. So, for every candidate, one can calculate its minimum and maximum ranks in a D-order.
For the minimum rank, which is the only interesting one in a decision problem as we are looking for the best candidates, the values of {T(i,k), i=1, ..n} have only to be sorted by increasing order and the smallest values are added until D is reached. If p values can be added, there are n-p candidates that must be ranked before x_k and so its best rank is n-p+1.

Example : After the heuristic, we get Dmax=4 and we look for 4-orders. The variety x_1 cannot be first in such an order, since it is dominated by x_7, x_{10}, x_{11} and x_{14} and so an order beginning with x_1 will at least be at distance 10 from the tournament. Howerer if x_7 and x_{10} are placed before, to place x_1 in the third position will provide a cost exceeding T(11,1) + T(14,1) = 4. Without estimating the value of a section starting with x_7 and x_{10} one can claim that x_1 cannot be ranked better than third.
In the worst case, the same variety will come eighth, because x_7, x_{10}, x_{11}, et x_{14} can be ranked before x_1 without any cost. One can also place x_6, x_9 and x_{13} before x_1, with a distance of 4, but no other variety, without taking off one of these. The minimum and maximum ranks of the varieties in the 4-orders, that has been shown to be median, are the following :

Varieties	1	2	3	4	5	6	7	8	9	10	11	12	13	14
Minimum rank	3	7	11	8	9	3	2	10	4	1	4	13	12	1
Maximum rank	8	9	12	9	11	7	5	12	7	2	7	14	14	5

We note that only varieties x_{10} and x_{14} can be ranked 1, but only x_{10} is the first one in a median order. We may also note that bothx_{12} and x_{13} can be last, but only x_{12} ends up there. The variety x_1 never reachs its extreme ranks, since it is between 5th and 7th; on the other hand x_9 is fourth in the first median order and seventh in the second one.
The average of minimum and maximum ranks can also be used to order varieties and this constitutes a new heuristic method for obtain a linear order and an upper bound to the optimal distance from the tournament.

2.4 Reducing tournament to the k best candidates

The distance value between an order and the tournament contains a part of distance coming from the last ranked candidates. As these latter are of little interest, we would like to consider the sub-tournament of the candidates that are always or at least once, in the first part or even in the first place of a median order. We would like to eliminate those which will never have a good rank. If we search for winners in a median order, we may consider only the sub-tournament of candidates with minimum rank equal to 1; all the others can be eliminated.
For our example, only varieties x_{10} and x_{14} would remain in competition, and as T(10,14) > 0, the variety x_{10} is certainly the best (according to this type of evaluation).
If we decide to select varieties that are in the first half part, we consider the sub-tournament corresponding to those with minimum rank seventh, that are the height varieties which are always the first height ones in median orders.The sub-tournament of the remaining varieties being transitive, median orders of the height first will be the beginning sections of median orders of the whole tournament.

3 Conclusions

This decision method, based on paired comparisons, has several advantages. First it makes it possible to treat incomplete data, using the transitivity of the collective preference relation. Secondly, it takes into account some uncertainty concerning the notes without fitting the usual statistical framework of parametric models. Different strategy to weight the tournament can be compared and the final decision can be made according to common winners.

At last the risk of getting an untreatable tournament, which is regarded as a major disadvantage for this kind of method, is not one, when you have to select the best candidates. It is sufficient to estimate an upper bound of the distance between a linear order and the tournament. The better this bound, the narrower the interval between extreme ranks for candidates.

The elimination of the always badly ranked ones can then be performed until all the median orders of a sub-tournament can be enumerated. Let us mention that these winners are not necessarily the first ranked candidates of the whole tournament. But this selecting strategy is particularly accurate if you consider as hazardous the individual preferences that place a bad candidate before a good one.

References

1. J.P. Barthélemy, B. Monjardet: The median procedure in Cluster Analysis and Social Choice Theory. *Math. Soc. Sci.*, 1, 235-267 (1981)

2. J.P. Barthélemy, A. Guénoche, O. Hudry: Median linear orders : Heuristics and a branch and bound algorithm. *European Journal of Operational Research*, 42, 313-325 (1989)

3. J.C. Bermond: Ordres à distance minimum d'un tournoi et graphes partiels sans circuits maximaux. *Math. et Sc. Hum.*, 37, 5-25 (1972)

4. I. Charon-Fournier, A. Germa, O. Hudry: Utilisation des scores dans les méthodes exactes déterminant les ordres médians de tournois. *Math. Inf. et Sc. hum*, 119, 53-74 (1992)

5. P.C. Fishburn: *Mathematics of decision theory*, Mouton, Paris, 1972.

6. J.J. Frey, A. Yehia-Alcoutlabi: Comparaisons par paires : Une interprétation et un généralisation de la méthode des scores. *R.A.I.R.O.*, 20, 3, 213-227 (1986)

7. A. Guénoche: Un algorithme pour pallier l'effet Condorcet *RAIRO*, 11, 1, 73-83 (1977)

8. O. Hudry: *Recherche d'ordres medians : complexite algorithmique et problèmes combinatoires*. Thèse ENST Paris, 1989.

9. O. Hudry: New results on tournaments, Conference "Combinatorics in Behavioral Sciences", Irvine, 15-19 August 1994.

10. J. Kemeny: Mathematics without numbers. *Daedelus*, 8, 571-591 (1959)

11. B. Monjardet: Tournois et ordres médians pour une opinion. *Math. et Sc. Hum.*, 43, 55-73 (1976)

12. G. Reinelt: *The linear ordering problem : algorithms and applications*, Helderman Verlag, 1985

13. A.F.M. Smith, C.D. Payne: An algorithm for determining Slater's i and all nearest adjoining orders. *Br. J. math. stat.*, 27, 49-52 (1974)

The Role of Measurement Uncertainty in Numeric Law Induction

Marjorie Moulet

Laboratoire de Recherche en Informatique,
Bâtiment 490, Université Paris-Sud, 91405 ORSAY Cédex
Email: Marjorie.Moulet@lri.fr

Abstract. Empirical induction of numeric laws aims at finding numerical laws from an experimentation set, using symbolic and inductive techniques. Faced with experimental data, these techniques must be adapted in order to deal with noisy data. In this article, we present the solutions for dealing with numerical uncertainty which are implemented in our system ARC [16]. These solutions are inspired by experimental science and join simplicity to efficiency. Moreover, when used in a more realistic way, numerical uncertainty can significantly improve not only the accuracy of the results but also the search algorithm's efficiency.

1 Introduction

Numerical law discovery is a field of Machine Learning which aims at finding one or more numerical laws from a set of measurements of given symbolic and numeric attributes (or variables). ARC [16] is an improvement of the system ABACUS [4] itself a descendant of the precursor system BACON [8, 9].

Numerical law search is performed in a pure Artificial Intelligence way, by heuristically exploring the space of all possible combinations of initial variables that one can create with a given set of mathematical operators. This method is complementary with data analysis or statistical techniques based on statistical regression: regression, the role of which is to approximate a data plot with a law of a given form, requires to previously define a model (the simplest one is the linear model: $y = \beta_1 x + \beta_0$) or a set of models or at least a first approximation of the model. On the opposite, an exploratory approach, like ARC's, allows to discover the model and the law automatically, which is necessary as soon as the expert does not have or gets very few information on the model. A study of different inductive methods for discovering numerical laws is presented in [17].

Heuristics are necessary to guide the inductive system among the huge solution space. ARC combines iteratively initial variables until a combination is found identical on a given minimum percentage of examples. Let us shortly describe the algorithm [18, 14]:

1. For each couple of variables (x, y):
   (a) study its **monotonic dependency**: the values of y increase (positive dependency) or decrease (negative dependency) when the values of x are in increasing order.

(b) create $x+y$ and $x*y$ if there is a negative dependency, $x-y$ and $\frac{x}{y}$ if there is a positive one.

2. Evaluate these new combinations according to an evaluation function and separate them in two sets: the **active nodes** and the **suspended nodes**. The active nodes are the most promising combinations and are first developed, the process bactracking to the suspended nodes when no solution is found with the active nodes.
3. The process (1-2) is repeated iteratively until a combination has a constancy higher than the required threshold **constancy**, or until the maximal number of combinations have been created.

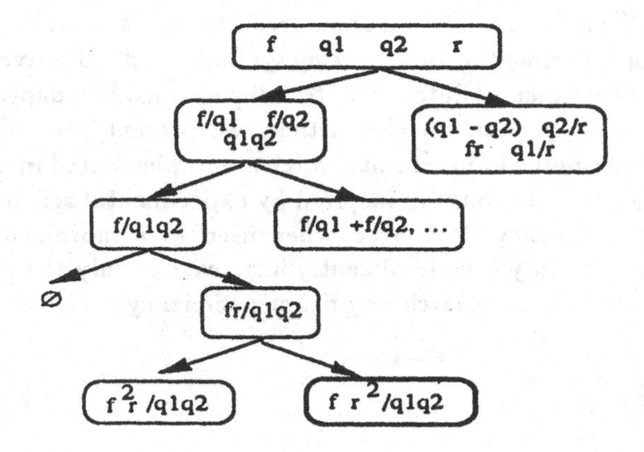

Fig. 1. Example of a search graph

Figure 1 illustrates a search graph that ARC explores when looking for Coulomb's law: $f = k * \frac{q_1 q_2}{r^2}$, where k is a constant. The law is found after the successive creation of the following combinations: $\frac{f}{q_1}, \frac{f}{q_1 q_2}, \frac{fr}{q_1 q_2}$ and lastly $\frac{fr^2}{q_1 q_2}$ which is found constant. ARC improves on ABACUS mainly by introducing a more efficient new evaluation function [15].

2 Uncertainty Representation

ARC includes a complete management of numerical uncertainties[1] which makes it more adapted than its predecessors to real data. Despite the self evident necessity to specify data accuracy during the acquisition step, we must emphasize that this requirement is not always respected by machine learning researchers. Even in data analysis or in statistics [5, 21] the dependent variable can be noisy but the dependent variables are supposed to be either without error or with an additive noise which is supposed with null mean and equal variance, and these conditions are not reasonable hypotheses in experimental data.

[1] We suppose that symbolic values are without error. Since symbolic variables are only used when looking for a possible classification, this choice does not interfere with the numerical law induction step itself.

2.1 Related approaches.

ABACUS deals with uncertainty through a unique parameter **uncertainty** which represents the *maximum* relative error of any numerical value, and by default equals 2%. This choice corresponds to a large approximation where initial values as well as any computed value have the same uncertainty error. On one hand, it assumes that the different initial variables have been measured with the same accuracy. On the other hand, it contradicts the general idea that error increases according to computations.

FAHRENHEIT [12, 6] requires that the same experiment is done many times in order to estimate the uncertainty attached to numerical measured values. However, according to [11], this process only computes the fidelity of the measurement device, while its accuracy is also a function of its correctness and its sensibility. Moreover, this process does not allow to estimate the uncertainty attached to the dependent variables. Statistical techniques like E* [22], or KEPLER [23], assume only errors on the dependent variable and they are not able to take into account the knowledge of this error when it is available.

2.2 ARC's Hypotheses.

In ARC, uncertainties are given for each initial numerical data, either for each variable under the *relative* form and noted $\rho(x)$, or for each value under an *absolute* form, and noted $\delta(x)$, with $\rho(x) = \frac{\delta(x)}{x}$. According to the definition used in physics, we consider that *the absolute uncertainty of a value corresponds to the maximal deviation that it can verify.*

In this way, we assume that the real value x_0', the measure of which is x_0, belongs to the interval $[x_0 - \delta(x_0); x_0 + \delta(x_0)]$, which is called the **value interval** associated with x_0'. In the following, we confuse the real value x_0' and the measured value x_0, and we note $\delta(x_0)$ the absolute error of both x_0 and x_0'.

With respect to physical principles and error computing theory [19, 20], we assume that each value has the same possibility to be the right measure, like in a fuzzy set [1, 2, 7]. No probability law is associated to errors because that would require to know the probability law of each initial variable, which could be quite difficult to acquire as soon as it is not a classical law. Moreover, that would require to compute in turn each new variable's probability law. Beside, all these computations seem to be too expensive with regard to the additional information they could provide concerning the values uncertainty.

2.3 Computation of the Value Interval.

Still remains the following question: how to represent these value intervals in order to optimize the memory space as well as the computation time? In ARC, basic operators are arithmetical ones: +, -, *, / but trigonometric functions (and logarithmic ones in ARC.2) are also allowed. The uncertainty of new variables must be computed automatically according to the previous relative or absolute uncertainties. Table 1 presents the computation formulas that we have chosen for arithmetical operators. These formulas are inspired by classical physics [10, 11, 20].

op	$\rho(x \text{ op } y)$ (relative form)	$\delta(x \text{ op } y)$ (absolute form)
+/-	$\frac{\delta(x)+\delta(y)}{\lvert x+y\rvert}$	$\delta(x)+\delta(y)$
*	$\rho(x)+\rho(y)+\rho(x)*\rho(y)$	$\delta(x)*\lvert y\rvert+\lvert x\rvert*\delta(y)+\delta(x)*\delta(y)$
/	$\rho(x)+\rho(y)+\rho(x)*\rho(y)$	$\frac{\delta(x)}{\lvert y\rvert}+\lvert\frac{\lvert x\rvert*\delta(y)}{y^2}\rvert+\frac{\delta(x)*\delta(y)}{y^2}$

Table 1. Uncertainty formulas for Arithmetical Operators

In opposite to the addition's formula $\delta(x+y)=\delta(x)+\delta(y)$ which is the physical one, the product and ratio 's formulas are more accurate than the usual ones: $\rho(x*y)=\rho(x)+\rho(y)$. We compute the absolute error of the product $x*y$ as the maximum error on $x*y$, given the value intervals of x and y [2]:

$$(x-\delta(x))*(y-\delta(y)) \leq x*y \leq (x+\delta(x))*(y+\delta(y)) \Leftrightarrow$$

$$x*y-x*\delta(y)-y*\delta(x)+\delta(x)*\delta(y) \leq x*y \leq x*y+x*\delta(y)+y*\delta(x)+\delta(x)*\delta(y)$$

The absolute error of $x*y$ is therefore the maximum of the two deviations: $|x*\delta(y)+y*\delta(x)+\delta(x)*\delta(y)|$ and $|-x*\delta(y)-y*\delta(x)+\delta(x)*\delta(y)|$, and is given by the formula given in Table 1. In physics, $\delta(x)*\delta(y)$ is simply neglected with regard to $x*y$, that leads to the formula $\delta(x*y)=x*\delta(y)+y*\delta(x)$. This physical formula is an approximation and is usually used only to evaluate the result's quality. On the opposite, in ARC, these uncertainties are in turn used to compute other uncertainties, and we prefer to keep the right formula.

However, the formulas presented in our table still represent an approximation since by taking the maximum of the two deviations, we "shift forward" in a certain way the value interval of the result. In general, when a non linear function is applied to a symmetric value interval, the resulting value interval is no more symmetric around the resulting value. For instance, if $x = 10 \pm 0.1$, the value interval of x is [9.9; 10.1], and x^2 is in the value interval [98.01; 102.01], which is not symmetric around 100. One solution could be to compute the interval by applying the same function to the two extrema of the initial value intervals, but that would require to store three values: the two extrema in addition to the value itself, and to make at each operation three computations.

[2] The demonstration is given for x and y having identical signs but in the opposite case, we would obtain a similar formula.

A second solution could be to apply the derivative of the function, as soon as the function is derivable. Indeed, the relation $df = f'(x).dx$ corresponds to $\delta((f(x)) = f'(x).\delta(x)$. This formula presents the advantage to only require one computation but it does not allow to retrieve the real interval: for $x = 10 \pm 0.1$, the absolute error would be found equal to $\delta(x^2) = 2 * 10 * 0.1 = 2$, that defines the value interval [98, 102] "shift forward" with regard to the right interval.

Interval calculus and possibilistic theory propose to store the value of the two interval's extrema and then to always compute the operations on these two values. However, since physicists prefer in general to separate the value and its corresponding error, we conclude that they attach a higher importance to the measure than to its error. Therefore and with care of interpretability, we prefer to store the result of the operation computed on the measured value rather than the results attached to the interval value's extrema.

To summarize, we choose to loose some accuracy (that we can expect to be negligible with regard to the computed error itself) and in counterpart to gain in rapidity. In the case of arithmetical operators, the formulas shown in Table 1 allow to compute the interval value with only one computation. Concerning other continuous functions on $\Re$, like logarithm or trigonometric functions, we choose to apply derivative, which also approximates error in only one computation.

3 A new equality function

As soon as one deals with noisy numerical values, a main problem arises which is to decide whenever two values can be considered as equal. In ARC, the quality of a combination is a function of the "**constancy rate**", which is the percentage of examples which have the *same* value for this combination. The comparison function therefore influences directly the evaluation function and in this way the search efficiency itself. We have thus introduced a new equality function which takes into account the associated errors of the two values under study.

3.1 Definition

Let us call δ-equality the function which indicates if two values x_i and x_j have the "possibility" to be equal, whithout specifying the probability they have to have exactly this same value. [2] compute the possibility and the necessity that a given interval is "higher" to another, given the distribution functions of each interval. The possibility that two intervals M= $[\underline{m}, \overline{m}]$ and N= $[\underline{n}, \overline{n}]$ are equal is defined as follows: Pos(M = N) = min (Pos($\overline{m} \geq \underline{n}$), Pos($\overline{n} \geq \underline{m}$)).

If we consider real values, Pos(x,y) =1 if $x \geq y$, or 0. Thereby, if we consider M and N as two value intervals associated to x and y: $M = [x - \delta(x), x + \delta(x)]$; $N = [y - \delta(y), y + \delta(y)]$, M and N can be considered as equal under the following condition:

$$\text{if } x \leq y, \text{Pos}(M = N) = 1 \text{ if } y - \delta(y) \leq x + \delta(x)$$
$$\text{if } x \geq y, \text{Pos}(M = N) = 1 \text{ if } x - \delta(x) \leq y + \delta(y)$$

In both cases, the condition under which the two intervals can be equal is $|x - y| \leq \delta(x) + \delta(y)$. This condition is graphically illustrated by figure 2:

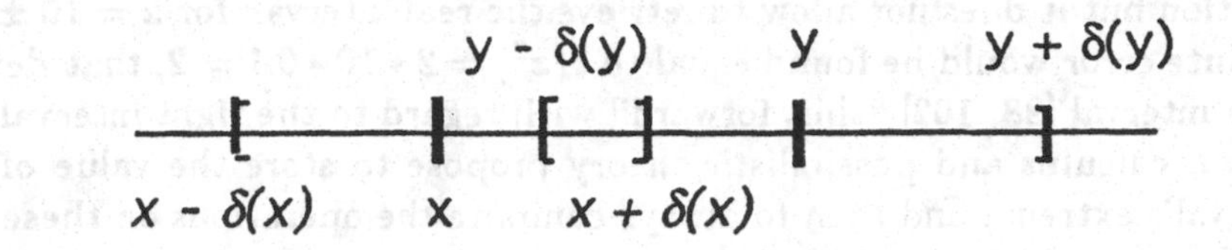

Fig. 2. Representation of two δ-equal values x and y

We propose then to define a new equality, the $\delta - \textit{equality}$, which represents the possibility for two values to be equal:

two values x and y are $\delta - \textit{equal}$ iff $\lvert x - y\rvert \leq \delta(x) + \delta(y)$

Example: $x = 10 \pm 3$ *is* $\delta - \textit{equal}$ *to* $y = 12 \pm 1$ *since* $|10 - 12| \leq 3 + 1$

This function is reflexive and symmetric but not transitive: for instance, let $x = 10 \pm 3$, $y = 14 \pm 2$ and $z = 16 \pm 1$: x is $\delta - \textit{equal}$ to y, y is $\delta - \textit{equal}$ to z but x is not $\delta - \textit{equal}$ to z.

It can be useful to test sometimes if a value x can be considered as equal to zero (for instance in the aim of a division). Since we do not know a priori the error associated to zero, we consider that x is $\delta - \textit{equal}$ to 0 when 0 belongs to the value interval associated to x:

x is $\delta - \textit{equal}$ to 0 iff $x - \delta(x) \leq 0 \leq x + \delta(x)$

In the same way, a set of values can be considered as $\delta - \textit{equal}$ to 0 if and only if 0 belongs to the intersection of all their value intervals.

3.2 Choice of a representative value

Whenever ARC finds a combination which has $\delta - \textit{equal}$ values on a sufficient percentage of examples, it needs to select a value able to represent as most accurately as possible the set of all these values. Let us note M this value that we qualify as *representative*.

If x and y are $\delta - \textit{equal}$, by definition that means that they are possibly equal, in such a way that the possible equal value belongs to their value interval, and therefore belongs to their intersection. It seems therefore natural to choose M in this intersection, that we call the *minimum interval*: $I_{min} = [y - \delta(y), x + \delta(x)]$.

Moreover, it is necessary to compute the uncertainty of this representative value M, that is to say its own value interval. Besides, by definition, its value interval represents all the values to which M can be equal. This set corresponds therefore to what we have called the *minimum interval*. In order to respect the symmetry of the value interval, we must moreover choose the representative value as the middle of this interval:

The representative value M of x and y such that $x \leq y$ with respective absolute uncertainties $\delta(x)$ and $\delta(y)$ is: $M = \frac{\lfloor x + \delta(x) + y - \delta(y) \rfloor}{2}$ and $\delta(M) = \frac{\lfloor x + \delta(x) - y + \delta(y) \rfloor}{2}$

Example: $x = 10 \pm 3$ *is* $\delta -$ *equal to* $y = 13 \pm 2$, $M = 12 \pm 1$.
This definition can be easily generalized to a set of values [13].

4 New Heuristics

4.1 Revising Basic calculi

The introduction of uncertainties associated to initial measures as well as the iterative computation of the uncertainties associated to the computed next values both allow to refine each step of ARC's algorithm:

- In order to study each variable pair (x, y), ARC needs to define the groups of examples called *projection sets* such that all variables except x and y are fixed. Besides, in the case of noisy data, when the uncertainties are not taken into account, the system has a chance to find empty sets. In this case, some good combinations risk not to be created and consequently the solution may not be discovered. In ARC, the $\delta -$ *equality* function allows to compare the values according to their uncertainties and thereby to compute accurately the projection sets even when given noisy data.
- Once computed the projection on two given variables, the discovery system computes their monotonic dependency, by studying each of the projection set. With this aim, ABACUS counts the number of increasing and decreasing segments. ABACUS decides that there is a monotonic dependency if 70% of segments have the same monotony. For instance, in Figure 3, ABACUS counts 7 increasing segments and 5 decreasing segments, and therefore concludes that there is no monotonic dependency, while undoubtedly one exists.

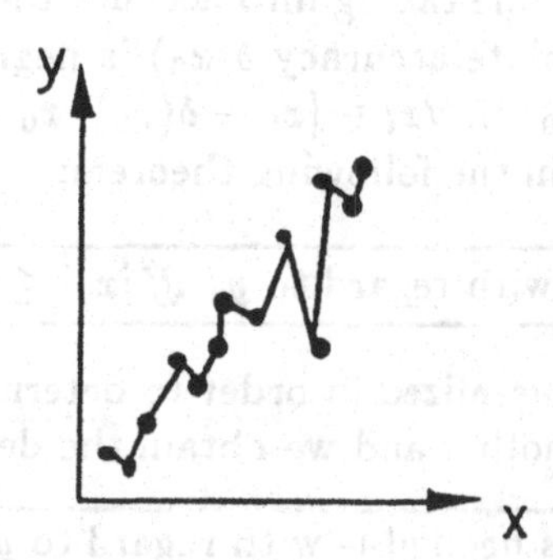

Fig. 3. monotonic dependency discovery

Thanks to the uncertainty knowledge, ARC can distinguish, during the monotonic dependency detection, the changes of slope which are significant from the ones implied by pure noise. ARC applies here the following heuristic: given points (x_i, y_i) ordered according to the increasing values of x,

if x_i is $\delta - equal$ to x_{i+1} and y_i is $\delta - equal$ to y_{i+1}, then *do not take into account the segment joining these two points.*

On the example illustrated by Figure 3, ARC determines in this way that there are 8 changes of slope among which only one is negative, and therefore concludes that there is a positive global dependency.

- In the same way, ARC's evaluation function is defined with the "constancy rate" of a combination and is computed by replacing the classical function of equality by the new $\delta - equality$ function. This leads the system to select more accurately the "best" combinations.
- Lastly, a solution is defined as a combination which is "constant" on a sufficient high number of examples. This "constancy rate" is also computed from now according to the $\delta - equality$ function. This leads ARC to find different solutions than those found by ABACUS and in some cases to find new ones. We show how the uncertainty influences and determines the success of the discovery [13].

4.2 Negligibility

Pruning uninteresting paths in the search graph is a source of another kind of difficulty. In order to perform some pruning, we have introduced the notion of negligibility of a variable with regard to another one. The idea is the following: if a variable x is negligible with regard to another variable y, it is not useful to create their sum or their difference. This notion has been also introduced in fuzzy theory [3], and a simple definition can be found concerning numerical values.

Since the $\delta - equality$ function must replace the standard equality, we must also exchange the standard definition of negligibility (*x is negligible with regard to y iff $x + y = y$*) by a new one taking into account the value intervals. We now consider that x_0, with absolute accuracy $\delta(x_0)$ is negligible with regard to y_0, with absolute accuracy $\delta(y_0)$ if: $\forall x_i \in [x_0 - \delta(x_0), x_0 + \delta(x_0)]$, , $y_0 + x_i \in [y_0 - \delta(y_0), y_0 + \delta(y_0)]$. We obtain the following theorem:

x_0 is negligible with regard to y_0 *iff* $\lvert x_0 \rvert \leq \delta(y_0) - \delta(x_0)$

This theorem can be generalized in order to determine if a given variable is negligible with regard to another and we obtain the definition:

x is negligible with regard to y *iff* for all examples E, the value of x at E is negligible with regard to the value of y at E.

4.3 Elimination of unaccurate variables.

Following this notion of negligibility, we are interested in variables with a very low accuracy. It can happen indeed that, by many successive combinations, different values of a same variable present a relative uncertainty higher than 100%, like 0.013 ± 0.02. Even if such a variable can lead to a solution, since error only increases, it can only lead to a solution with a still higher uncertainty, without any real interest. It should then be very useful to detect such uninteresting variables in order to delete them.

With this aim, ARC applies a new heuristic controlling the elimination of variables the uncertainty of which is very low, determined as variables such that a majority of its values have a relative uncertainty rate higher than 100%. This majority is given by a parameter called *rate-elimination* (by default equal to 50%) indicating the percentage of uncertain values which is considered sufficient for eliminating a variable.

$$\text{If } \frac{card\{x_i/\delta(x_i) \geq |x_i|\}}{card\{x_i\}} \geq \text{*rate-elimination* Then eliminate } x$$

5 Conclusion

This work underlines the numerous different problems that uncertainty implies and which were not solved by previous numerical law induction systems. More than any other learning tool, it seems essential that such a system takes into account numerical uncertainties, in order both to be more efficient and to deal with real data.

Initial information being only data itself, it is essential that the system is able to interpret, to compare and to manipulate it. Moreover, when looking at any experimental scientist, we can note that he always uses data uncertainty. In a numerical law induction system, priority must also be given to the integration of data uncertainty. In order to deal with uncertainty in each of the different reasoning steps, we have reviewed some main notions like *equality* or error computation associated to arithmetical operators. Moreover, we have introduced new notions such as the negligibility.

In conclusion, the simple introduction of data uncertainty and the operationalisation of real physical sense notions allow to improve and to refine the numerical law induction process. The study of the monotonic dependency between a pair of variables is now more robust to noise, the evaluation function becomes more accurate and allows to optimize search. ARC becomes in this way not only able to find more accurate results, but also to solve additional problems.

References

1. Dubois D. and Prade H. *Théorie des Possibilités - Applications à la Représentation des Connaissances en Informatique.* Masson, Paris, 1985.

2. Dubois D. and Prade H. Fuzzy numbers: an overview. In Bezdek J.C., editor, *Analysis of Fuzzy Information*, volume 1: Mathematics and Logic. CRC Press, 1987.
3. Dubois D. and Prade H. Order-of-magnitude reasoning with fuzzy relations. *Revue d'Intelligence Artificielle*, 3(4), 1989.
4. B.C. Falkenhainer and R.S. Michalski. Integrating Quantitative and Qualitative Discovery in the ABACUS System. In Y. Kodratoff and R.S Michalski, editors, *Machine Learning: an Artificial Intelligence Approach*, volume 3. Morgan-Kaufmann, 1990. Also published in Machine Learning Journal, vol. 3, 1986.
5. Saporta G. *Probabilités, analyse de données et statistique*. Editions technip, 1990.
6. Zytkow J.M., Zhu J., and Hussam A. Automated Discovery in a Chemistry Laboratory. In *Proceedings of the 8th National Conference on Artificial Intelligence*, 1990.
7. Zadeh L.A. Fuzzy Sets. In *Information and control*, volume 8, pages 338–353, 1965.
8. P. Langley. Rediscovering Physics with BACON.3. In *Proceedings of the 6th IJCAI*, pages 505–507. Morgan-Kaufmann, 1979.
9. P. Langley, H. Simon, G.L. Bradshaw, and J. Zytkow. *Scientific Discovery. Computational Approaches of the Creative Process*. MIT Press, 1987.
10. Eurin M. and Guimiot H. *Physique*. Classiques Hachette. Hachette, 1953.
11. Joyal M. *Cours de physique*, volume 3: Electricité. Masson & Cie, 1956.
12. Zytkow J. M. Combining many searches in the FAHRENHEIT discovery system. In *Proceedings of the 4th International Workshop on Machine Learning*. Morgan-Kaufmann, 1987.
13. M. Moulet. Using accuracy in numerical law discovery. In Y. Kodratoff, editor, *Proceedings of the 5th European Working Session on Learning*, Lecture Notes in Artificial Intelligence, pages 118–136. Springer-Verlag, 1991.
14. M. Moulet. ARC.2 : Régression Linéaire en Découverte Scientifique. In *Actes des 1res Journées sur l'Apprentissage et l'explication des Connaissances*, 1992.
15. M. Moulet. ARC.2: Linear Regression in ABACUS. In *Proceedings of the ML 92 Workshop on Machine Discovery*, 1992.
16. M. Moulet. *Découverte empirique de lois numériques ou ABACUS Revu et Corrigé*. PhD thesis, Thèse de Doctorat, Université de Paris-Sud, France, 1993.
17. M. Moulet. Current Limits and Potentials of a BACON-like System. In G. Nakhaeizadeh and C. Taylor, editors, *Workshop Notes on Machine Learning and Statistics*. MLnet Familiarization Workshops, 1994.
18. M. Moulet. Iterative Model Construction with Regression. In T. Cohn, editor, *Proceedings of the 11th European Conference on Artificial Intelligence*, pages 448–452. Wiley and Sons, 1994.
19. Moore R. *Interval Analysis*. Prentice Hall Int., 1966.
20. Moore R. *Methods and Applications of Interval Analysis*, volume 2 of *SIAM Studies on Applied Mathematics*. Philadelphia, 1979.
21. Tomassonne R., Lesquoy E., and Millier C. *La régression: nouveaux regards sur une ancienne méthode statistique*. Editions INRIA, 1986.
22. C. Schaffer. A proven domain-Independent Scientific Function Finding Algorithm. In *Proceedings of the 8th National Conference on Artificial Intelligence*, pages 828–833, 1990.
23. Y.H Wu and S. Wang. Discovering Functional Relationships from Observational Data. In G. Piatetsky-Shapiro and W.J. Frawley, editors, *Knowledge Discovery in Databases*. MIT Press, 1991.

Springer-Verlag and the Environment

We at Springer-Verlag firmly believe that an international science publisher has a special obligation to the environment, and our corporate policies consistently reflect this conviction.

We also expect our business partners – paper mills, printers, packaging manufacturers, etc. – to commit themselves to using environmentally friendly materials and production processes.

The paper in this book is made from low- or no-chlorine pulp and is acid free, in conformance with international standards for paper permanency.

Lecture Notes in Computer Science

For information about Vols. 1–871
please contact your bookseller or Springer-Verlag

Vol. 907: T. Ito, A. Yonezawa (Eds.), Theory and Practice of Parallel Programming. Proceedings, 1995. VIII, 485 pages. 1995.

Vol. 908: J. R. Rao Extensions of the UNITY Methodology: Compositionality, Fairness and Probability in Parallelism. XI, 178 pages. 1995.

Vol. 909: H. Comon, J.-P. Jouannaud (Eds.), Term Rewriting. Proceedings, 1993. VIII, 221 pages. 1995.

Vol. 910: A. Podelski (Ed.), Constraint Programming: Basics and Trends. Proceedings, 1995. XI, 315 pages. 1995.

Vol. 911: R. Baeza-Yates, E. Goles, P. V. Poblete (Eds.), LATIN '95: Theoretical Informatics. Proceedings, 1995. IX, 525 pages. 1995.

Vol. 912: N. Lavrac, S. Wrobel (Eds.), Machine Learning: ECML – 95. Proceedings, 1995. XI, 370 pages. 1995. (Subseries LNAI).

Vol. 913: W. Schäfer (Ed.), Software Process Technology. Proceedings, 1995. IX, 261 pages. 1995.

Vol. 914: J. Hsiang (Ed.), Rewriting Techniques and Applications. Proceedings, 1995. XII, 473 pages. 1995.

Vol. 915: P. D. Mosses, M. Nielsen, M. I. Schwartzbach (Eds.), TAPSOFT '95: Theory and Practice of Software Development. Proceedings, 1995. XV, 810 pages. 1995.

Vol. 916: N. R. Adam, B. K. Bhargava, Y. Yesha (Eds.), Digital Libraries. Proceedings, 1994. XIII, 321 pages. 1995.

Vol. 917: J. Pieprzyk, R. Safavi-Naini (Eds.), Advances in Cryptology - ASIACRYPT '94. Proceedings, 1994. XII, 431 pages. 1995.

Vol. 918: P. Baumgartner, R. Hähnle, J. Posegga (Eds.), Theorem Proving with Analytic Tableaux and Related Methods. Proceedings, 1995. X, 352 pages. 1995. (Subseries LNAI).

Vol. 919: B. Hertzberger, G. Serazzi (Eds.), High-Performance Computing and Networking. Proceedings, 1995. XXIV, 957 pages. 1995.

Vol. 920: E. Balas, J. Clausen (Eds.), Integer Programming and Combinatorial Optimization. Proceedings, 1995. IX, 436 pages. 1995.

Vol. 921: L. C. Guillou, J.-J. Quisquater (Eds.), Advances in Cryptology – EUROCRYPT '95. Proceedings, 1995. XIV, 417 pages. 1995.

Vol. 922: H. Dörr, Efficient Graph Rewriting and Its Implementation. IX, 266 pages. 1995.

Vol. 923: M. Meyer (Ed.), Constraint Processing. IV, 289 pages. 1995.

Vol. 924: P. Ciancarini, O. Nierstrasz, A. Yonezawa (Eds.), Object-Based Models and Languages for Concurrent Systems. Proceedings, 1994. VII, 193 pages. 1995.

Vol. 925: J. Jeuring, E. Meijer (Eds.), Advanced Functional Programming. Proceedings, 1995. VII, 331 pages. 1995.

Vol. 926: P. Nesi (Ed.), Objective Software Quality. Proceedings, 1995. VIII, 249 pages. 1995.

Vol. 927: J. Dix, L. Moniz Pereira, T. C. Przymusinski (Eds.), Non-Monotonic Extensions of Logic Programming. Proceedings, 1994. IX, 229 pages. 1995. (Subseries LNAI).

Vol. 928: V.W. Marek, A. Nerode, M. Truszczynski (Eds.), Logic Programming and Nonmonotonic Reasoning. Proceedings, 1995. VIII, 417 pages. 1995. (Subseries LNAI).

Vol. 929: F. Morán, A. Moreno, J.J. Merelo, P. Chacón (Eds.), Advances in Artificial Life. Proceedings, 1995. XIII, 960 pages. 1995 (Subseries LNAI).

Vol. 930: J. Mira, F. Sandoval (Eds.), From Natural to Artificial Neural Computation. Proceedings, 1995. XVIII, 1150 pages. 1995.

Vol. 931: P.J. Braspenning, F. Thuijsman, A.J.M.M. Weijters (Eds.), Artificial Neural Networks. IX, 295 pages. 1995.

Vol. 932: J. Iivari, K. Lyytinen, M. Rossi (Eds.), Advanced Information Systems Engineering. Proceedings, 1995. XI, 388 pages. 1995.

Vol. 933: L. Pacholski, J. Tiuryn (Eds.), Computer Science Logic. Proceedings, 1994. IX, 543 pages. 1995.

Vol. 934: P. Barahona, M. Stefanelli, J. Wyatt (Eds.), Artificial Intelligence in Medicine. Proceedings, 1995. XI, 449 pages. 1995. (Subseries LNAI).

Vol. 935: G. De Michelis, M. Diaz (Eds.), Application and Theory of Petri Nets 1995. Proceedings, 1995. VIII, 511 pages. 1995.

Vol. 936: V.S. Alagar, M. Nivat (Eds.), Algebraic Methodology and Software Technology. Proceedings, 1995. XIV, 591 pages. 1995.

Vol. 937: Z. Galil, E. Ukkonen (Eds.), Combinatorial Pattern Matching. Proceedings, 1995. VIII, 409 pages. 1995.

Vol. 938: K.P. Birman, F. Mattern, A. Schiper (Eds.), Theory and Practice in Distributed Systems. Proceedings,1994. X, 263 pages. 1995.

Vol. 939: P. Wolper (Ed.), Computer Aided Verification. Proceedings, 1995. X, 451 pages. 1995.

Vol. 940: C. Goble, J. Keane (Eds.), Advances in Databases. Proceedings, 1995. X, 277 pages. 1995.

Vol. 941: M. Cadoli, Tractable Reasoning in Artificial Intelligence. XVII, 247 pages. 1995. (Subseries LNAI).

Vol. 942: G. Böckle, Exploitation of Fine-Grain Parallelism. IX, 188 pages. 1995.

Vol. 943: W. Klas, M. Schrefl, Metaclasses and Their Application. IX, 201 pages. 1995.

Vol. 944: Z. Fülöp, F. Gécseg (Eds.), Automata, Languages and Programming. Proceedings, 1995. XIII, 686 pages. 1995.

Vol. 945: B. Bouchon-Meunier, R.R. Yager, L.A. Zadeh (Eds.), Advances in Intelligent Computing - IPMU '94. Proceedings, 1994. XII, 628 pages.1995.

Vol. 946: C. Froidevaux, J. Kohlas (Eds.), Symbolic and Quantitative Approaches to Reasoning and Uncertainty. Proceedings, 1995. X, 420 pages. 1995. (Subseries LNAI).

Vol. 947: B. Möller (Ed.), Mathematics of Program Construction. Proceedings, 1995. VIII, 472 pages. 1995.

Vol. 948: G. Cohen, M. Giusti, T. Mora (Eds.), Applied Algebra, Algebraic Algorithms and Error-Correcting Codes. Proceedings, 1995. XI, 485 pages. 1995.